McGRAW-HILL YEARBOOK OF
Science &
Technology

1998

McGRAW-HILL YEARBOOK OF
Science & Technology

1998

Comprehensive coverage of recent events and research as compiled by the staff of the McGraw-Hill Encyclopedia of Science & Technology

McGraw-Hill

New York   San Francisco   Washington, D.C.   Auckland   Bogotá   Caracas   Lisbon   London   Madrid
Mexico City   Milan   Montreal   New Delhi   San Juan   Singapore   Sydney   Tokyo   Toronto

Library of Congress Cataloging in Publication data

McGraw-Hill yearbook of science and technology.
1962–   . New York, McGraw-Hill Book Co.

  v. illus. 26 cm.
  Vols. for 1962–   compiled by the staff of the
McGraw-Hill encyclopedia of science and technology.
    1. Science—Yearbooks. 2. Technology—
Yearbooks. 1. McGraw-Hill encyclopedia of
science and technology.
Q1.M13   505.8   62-12028

ISBN 0-07-052418-1
ISSN 0076-2016

McGraw-Hill

A Division of The McGraw·Hill Companies

McGRAW-HILL YEARBOOK OF SCIENCE & TECHNOLOGY
Copyright © 1997 by The McGraw-Hill Companies, Inc.

1 2 3 4 5 6 7 8 9 0   DOW/DOW   9 0 2 1 0 9 8 7

This book was printed on acid-free paper.

*It was set in Neue Helvetica Black Condensed and Garamond Book by
The PRD Group, Shippensburg, Pennsylvania. The art was prepared by
North Market Street Graphics, Lancaster, Pennsylvania. The book was
printed and bound by R. R. Donnelley & Sons Company, The Lakeside Press.*

Editing, Design, & Production Staff

Roger Kasunic, Director of Editing, Design, and Production

Joe Faulk, Editing Manager

Ron Lane, Art Director

Thomas G. Kowalczyk, Production Manager

Suzanne W. B. Rapcavage, Senior Production Supervisor

Consulting Editors

Prof. Eugene A. Avallone. *Consulting Engineer; Professor Emeritus of Mechanical Engineering, City College of the City University of New York.* MECHANICAL AND POWER ENGINEERING.

A. E. Bailey. *Formerly, Superintendent of Electrical Science, National Physical Laboratory, London, England.* ELECTRICITY AND ELECTROMAGNETISM.

Prof. William P. Banks. *Chairman, Department of Psychology, Pomona College, Claremont, California.* PHYSIOLOGICAL AND EXPERIMENTAL PSYCHOLOGY.

Dr. Allen J. Bard. *Department of Chemistry and Biochemistry, University of Texas, Austin.* PHYSICAL CHEMISTRY.

Dr. Alexander Baumgarten. *Director, Clinical Immunology Laboratory, Yale-New Haven Hospital, New Haven, Connecticut.* IMMUNOLOGY AND VIROLOGY.

Prof. Richard D. Berger. *Plant Pathology Department, University of Florida, Gainesville.* PLANT PATHOLOGY.

Prof. Gregory C. Beroza. *Department of Geophysics, Stanford University, California.* GEOPHYSICS.

Dr. Robert T. Beyer. *Hazard Professor of Physics, Emeritus, Brown University, Providence, Rhode Island.* ACOUSTICS.

Prof. S. H. Black. *Department of Medical Microbiology and Immunology, Texas A&M University, College Station.* MEDICAL MICROBIOLOGY.

Prof. Anjan Bose. *Director, School of Electrical Engineering and Computer Science, Washington State University, Pullman.* ELECTRICAL POWER ENGINEERING.

Ronald Braff. *Principal Engineer, MITRE Corporation/Center for Advanced Aviation System Development, McLean, Virginia.* NAVIGATION.

Dr. Chaim Braun. *Bechtel Corporation, Gaithersburg, Maryland.* NUCLEAR ENGINEERING.

Robert D. Briskman. *President, CD Radio, Inc., Washington, D.C.* TELECOMMUNICATIONS.

Michael H. Bruno. *Graphic Arts Consultant, Sarasota, Florida.* GRAPHIC ARTS.

Dr. John F. Clark. *Director, Graduate Studies, and Professor, Space Systems, Spaceport Graduate Center, Florida Institute of Technology, Satellite Beach.* SPACE TECHNOLOGY.

Ross A. Clark. *Executive Vice President, Search Energy, Inc., Calgary, Alberta, Canada.* PETROLEUM ENGINEERING.

Prof. David L. Cowan. *Chairman, Department of Physics and Astronomy, University of Missouri, Columbia.* CLASSICAL MECHANICS AND HEAT.

Dr. C. Chapin Cutler. *Retired; formerly, Ginzton Laboratory, Stanford University, California.* RADIO COMMUNICATIONS.

Dr. Jay S. Fein. *Division of Atmospheric Sciences, National Science Foundation, Arlington, Virginia.* METEOROLOGY AND CLIMATOLOGY.

Dr. William K. Ferrell. *Professor Emeritus, College of Forestry, Oregon State University, Corvallis.* FORESTRY.

Prof. Lawrence Grossman. *Department of Geophysical Science, University of Chicago, Illinois.* GEOCHEMISTRY.

Dr. Ralph E. Hoffman. *Associate Professor, Yale Psychiatric Institute, Yale University School of Medicine, New Haven, Connecticut.* PSYCHIATRY.

Prof. Stephen F. Jacobs. *Professor Emeritus, Optical Sciences Center, University of Arizona, Tucson.* ELECTROMAGNETIC RADIATION AND OPTICS.

Dr. S. C. Jong. *Senior Staff Scientist and Program Director, Mycology and Protistology Program, American Type Culture Collection, Rockville, Maryland.* MYCOLOGY.

Prof. C. L. Liu. *Associate Provost and Professor of Computer Science, University of Illinois, Champaign.* COMPUTERS.

Prof. Karl E. Lonngren. *Department of Electrical & Computer Engineering, University of Iowa, Iowa City.* PHYSICAL ELECTRONICS.

Dr. Philip V. Lopresti. *Retired; formerly, Engineering Research Center, AT&T Bell Laboratories, Princeton, New Jersey.* ELECTRONIC CIRCUITS.

Prof. Craig E. Lunte. *Department of Chemistry, University of Kansas, Lawrence.* ANALYTICAL CHEMISTRY.

Dr. Michael L. McKinney. *Department of Geological Sciences, University of Tennessee, Knoxville.* INVERTEBRATE PALEONTOLOGY.

Dr. George L. Marchin. *Associate Professor of Microbiology and Immunology, Division of Biology, Kansas State University, Manhattan.* MICROBIOLOGY.

Prof. Melvin G. Marcus. *Deceased; formerly, Department of Geography, Arizona State University, Tempe.* PHYSICAL GEOGRAPHY.

Dr. Henry F. Mayland. *Soil Scientist, Northwest Irrigation and Soils Research Laboratory, USDA-ARS, Kimberly, Idaho.* SOILS.

Dr. Orlando J. Miller. *Center for Molecular Medicine and Genetics, Wayne State University School of Medicine, Detroit, Michigan.* GENETICS AND EVOLUTION.

Prof. Conrad F. Newberry. *Department of Aerospace and Astronautics, Naval Postgraduate School, Monterey, California.* AERONAUTICAL ENGINEERING AND PROPULSION.

Prof. Jay M. Pasachoff. *Director, Hopkins Observatory, Williams College, Williamstown, Massachusetts.* ASTRONOMY.

Prof. David J. Pegg. *Department of Physics and Astronomy, University of Tennessee, Knoxville.* ATOMIC, MOLECULAR, AND NUCLEAR PHYSICS.

Dr. William C. Peters. *Professor Emeritus, Mining and Geological Engineering, University of Arizona, Tucson.* MINING ENGINEERING.

Dr. Donald R. Prothero. *Associate Professor of Geology, Occidental College, Los Angeles, California.* VERTEBRATE PALEONTOLOGY.

Prof. W. D. Russell-Hunter. *Professor of Zoology, Department of Biology, Syracuse University, New York.* INVERTEBRATE ZOOLOGY.

Dr. Andrew P. Sage. *Founding Dean Emeritus and First American Bank Professor, University Professor, School of Information Technology and Engineering, George Mason University, Fairfax, Virginia.* CONTROL AND INFORMATION SYSTEMS.

Mel Schwartz. *Materials Consultant, United Technologies Corporation, Stratford, Connecticut.* MATERIALS SCIENCE AND ENGINEERING.

Prof. Marlin U. Thomas. *Head, School of Industrial Engineering, Purdue University, West Lafayette, Indiana.* INDUSTRIAL AND PRODUCTION ENGINEERING.

Prof. John F. Timoney. *Department of Veterinary Science, University of Kentucky, Lexington.* VETERINARY MEDICINE.

Dr. Shirley Turner. *U.S. Department of Commerce, National Institute of Standards and Technology, Gaithersburg, Maryland.* GEOLOGY (MINERALOGY AND PETROLOGY).

Prof. Joan S. Valentine. *Department of Chemistry and Biochemistry, University of California, Los Angeles.* INORGANIC CHEMISTRY.

Prof. Frank M. White. *Department of Mechanical Engineering, University of Rhode Island, Kingston.* FLUID MECHANICS.

Prof. Richard G. Wiegert. *Institute of Ecology, University of Georgia, Athens.* ECOLOGY AND CONSERVATION.

Prof. Frank Wilczek. *Institute for Advanced Study, Princeton, New Jersey.* THEORETICAL PHYSICS.

Prof. W. A. Williams. *Department of Agronomy and Range Science, University of California, Davis.* AGRICULTURE.

Dr. Terry L. Yates. *Chairman, Department of Biology, University of New Mexico, Albuquerque.* ANIMAL SYSTEMATICS.

Contributors

A list of contributors, their affiliations, and the titles of the articles they wrote appears in the back of this volume.

The 1998 *McGraw-Hill Yearbook of Science & Technology* provides recent outstanding achievements in science, technology, and engineering. This tradition began in 1962 when, with its first such Yearbook, McGraw-Hill presented a 1961 review and a 1962 preview of "the most significant scientific and technical events." The 1998 Yearbook continues this 36-year tradition. The Yearbook still serves as an annual review of what has occurred and a preview of work that is currently taking place. It also supplements the *McGraw-Hill Encyclopedia of Science & Technology*, updating the fundamental information in the eighth edition (1997).

Topics for the Yearbook are selected by our consulting editors and our editorial staff based on the present significance and the potential applications of the topics. Each Yearbook article is authored by one or more specialists on the subject. Noted researchers continue to lend their support by taking the time to share their knowledge with our readers.

Librarians, students, teachers, the scientific community, and the general public continue to find in the *McGraw-Hill Yearbook of Science & Technology* the information they want and need in order to follow today's rapid accomplishments in science and technology, as advances mushroom and the pace from research to development continues to quicken at the dawn of the twenty-first century.

Sybil P. Parker
EDITOR IN CHIEF

McGRAW-HILL YEARBOOK OF
Science &
Technology

1998

A–Z

Adaptive signal processing

Signal processing is a discipline that deals with the extraction of information from signals. The devices that perform this task can be physical hardware devices, specialized software codes, or combinations of both. In recent years the complexity of these devices and the scope of their applications have increased dramatically with the rapidly falling costs of hardware and software and the advancement of sensor technologies. This trend has made it possible to pursue sophisticated signal-processing designs at relatively low cost. Some notable applications, in areas ranging from biomedical engineering to wireless communications, include the suppression of interference arising from noisy measurement sensors, the elimination of distortions introduced when signals travel through transmission channels, and the recovery of signals embedded in a multitude of echoes created by multipath effects in mobile communications.

Statistically based systems. Regardless of the application, any functional system is expected to meet certain performance specifications. The requirements, as well as the design methodology, vary according to the nature of the end application. One distinctive design methodology, which dominated much of the earlier work in the information sciences, especially in the 1950s and 1960s, is based on statistical considerations. This framework assumes the availability of information in advance about the statistical nature of the signals involved, and then proceeds to design systems that optimize some statistical criterion. The resulting optimal designs are, in general, complex to implement. Only in special, yet important cases have they led to successful breakthroughs culminating with the Wiener and Kalman filters.

Moreover, in many situations a design that is motivated by statistical considerations may not be immediately feasible, because complete knowledge of the necessary statistical information may not be available. It may even happen that the statistical conditions vary with time. It therefore may be expected, in these scenarios, that the performance of any statistically based optimal design will degrade, the more so as the real physical application deviates from the modeling assumptions.

Adaptive systems. Adaptive systems provide an attractive solution to the problem. They are devices that adjust themselves to an ever-changing environment; the structure of an adaptive system changes in such a way that its performance improves through a continuing interaction with its surroundings. Its superior performance in nonstationary environments results from its ability to track slow variations in the statistics of the signals and to continually seek optimal designs.

Adaptive signal processing deals with the design of adaptive systems for signal-processing applications. Related issues arise in control design, where the objective is to alter the behavior of a system, and lead to the study of adaptive control strategies; the main issue is the stability of the system under feedback.

Operation. The operation of an adaptive system can be illustrated with a classical example in system identification. The **illustration** shows a plant (or system) whose input-output behavior is unknown and may even be time variant. The objective is to design an adaptive system that provides a good approximation to the input-output map of the plant. For this purpose, the plant is excited by a known input signal, and the response is taken as a reference signal. Moreover, a structure is chosen for the adaptive system, say a finite-impulse response structure of adequate length, and it is excited by the same input signal as the plant. At each time instant the output of the finite-impulse response system is compared with the reference signal, and the resulting error signal is used to change the coefficients of the

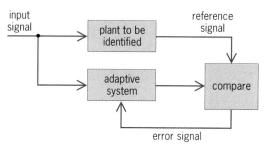

input signal

reference signal

error signal

Adaptive system identification.

finite-impulse response configuration. This learning process is continued over time, and the output of the adaptive system is expected to provide better tracking of the plant output as time progresses, especially when the structure of the plant is time invariant or varies only slowly with time.

Characteristics. Apart from emphasizing one particular application for adaptive systems, the above example also highlights several characteristics that are common to most adaptive designs:

1. An adaptive system adjusts itself automatically to a changing environment. This is achieved by changing the parameters of its internal structure in response to an error measure. Several adaptive structures have been used in practice, but the most frequent ones are linear combiners, finite-impulse response filters, infinite-impulse response filters, and linear combiners that are followed by nonlinearities such as sigmoid functions or nonlinear decision devices.

2. The interaction of an adaptive system with its environment takes place through an input signal and a reference signal. The reference signal is used to evaluate the performance of the adaptation process through the computation of an error signal. The adaptive system responds to the error signal in such a way as to minimize a predetermined cost function that is computed either from the error signal directly or from a filtered version of it.

3. An adaptive system is inherently nonlinear and time variant. For this reason, it is in general considerably more difficult to analyze its performance than that of a linear time-invariant system. Nevertheless, adaptive systems offer more possibilities than conventional nonadaptive designs, and many analysis methods have been developed that offer reasonable approximate methods for performance evaluation.

4. An adaptive system can be trained to perform certain tasks. This usually involves a learning phase where the adaptive system is exposed to typical input and reference data and is left to adjust itself to them. At the end of the learning procedure, the system can be exposed to new data and will be expected to provide a reasonable response. Typical examples of this scenario abound in neural-network applications and in the equalization of communication channels. In the latter application, the outputs of communication channels (at the receiver) are equalized to compensate for the distortion of signals

sent over long distances, and to thereby estimate the input (or transmitted) signals.

Adaptive algorithms. The performance of an adaptive system is critically dependent not only on its internal structure but also on the algorithm used to automatically adjust the parameters that define the structure. In general, the more complex the internal structure of the adaptive system, the more complex the algorithm that is needed for its adaptation. Moreover, several algorithms may exist for the same adaptive structure, and the choice of which algorithm to use is dictated by several factors that include the following:

1. The speed with which the algorithm learns in a stationary environment; that is, how fast the algorithm converges to the optimal statistical solution under stationarity assumptions. This factor determines the convergence rate of the algorithm.

2. The speed with which the algorithm tracks statistical variations. This factor determines the tracking capability of an algorithm and is a major motivation for the use of adaptive schemes, especially in recent mobile communications applications where adaptive equalizers are used to compensate for (that is, adapt to) changes in time-variant transmission channels, and thereby give good estimates of the transmitted signal from the received signal despite these changes.

3. The manner in which the performance of an adaptive scheme, operating in steady-state conditions, deviates from the performance of a statistically optimal design. This factor measures the so-called misadjustment of an adaptive scheme and serves to compare its performance with that of an optimal design in a statistical framework.

4. The amount of computational effort required to carry out each adjustment of the parameters of the adaptive system structure. Applications that require a large number of adaptive parameters tend to dictate a preference for computationally fast algorithms at the expense of other performance factors.

5. The reliability of an algorithm in finite-precision implementations. This factor is concerned with the numerical behavior of an algorithm when implemented in finite-precision arithmetic, and whether numerical effects might lead to erroneous behavior.

6. The robustness of an adaptive system to disturbances and unmodeled dynamics. This factor determines whether small disturbances can result in large errors and therefore compromise the performance of an adaptive scheme.

The above factors are often competing requirements, so that it is usually necessary to seek a compromise solution that best suits a particular application.

Derivation procedures. There have been many procedures for the derivation of adaptive algorithms, but the most frequent, at least in having had the most applications, are procedures that are based either on the method of stochastic approximation or on the least-squares criterion. In both cases, and especially for finite-impulse response adaptive structures, each criterion has led to several different variants that address in one way or another the above performance factors.

In stochastic-approximation algorithms, a recursive procedure is devised for the minimization of a predetermined cost function by iteratively approximating its minimum through a gradient descent procedure (a standard procedure in optimization theory). While this class of algorithms generally suffers from slow convergence rates, it still enjoys widespread use due to its simplicity, low computational requirements, and often observed robustness under different operating conditions. The most prominent member of the stochastic-approximation algorithms is the least-mean-squares algorithm, which is undoubtedly the most widely used adaptive filtering algorithm. Other members include filtered-error variants that are useful in active noise control and infinite-impulse response system identification problems, as well as several frequency-domain adaptive schemes.

In least-squares algorithms, a recursive procedure is devised for the minimization of a quadratic cost function. This family of algorithms is based on the least-squares criterion, which was developed in the late eighteenth century in work on celestial mechanics. Since then, the least-squares criterion has enjoyed widespread popularity in many diverse areas as a result of its attractive computational and statistical properties. Notably, for linear data models, least-squares solutions can be explicitly evaluated in closed form, can be recursively updated as more input data are made available, and are optimal maximum likelihood estimators in the presence of gaussian measurement noise.

Many recursive least-squares algorithms have been developed. Several of them are computationally more demanding than least-mean-squares-type algorithms, but variants exist that are computationally competitive, although more complex. They have better convergence properties but are less robust to disturbances.

Array algorithms. Following a trend initiated in the late 1960s in Kalman filtering methods, least-squares adaptive schemes are currently more often implemented in convenient array algorithms. These algorithms are closely related to the QR method, a numerically stable algorithm for solving systems of linear equations, and have the properties of better conditioning (that is, lower sensitivity to errors in the initial data), reduced dynamical range (which favors better conditioning), and orthogonal transformations, which typically lead to better numerical performance in finite-precision arithmetic. In the array form, an algorithm is described as a sequence of elementary operations on arrays of numbers. Usually, a prearray of numbers has to be triangularized by a rotation, or a sequence of elementary rotations, in order to yield a postarray of numbers. The quantities needed to form the next prearray can then be read off from the entries of the postarray, and the procedure can be repeated. The explicit forms of the rotation matrices are not needed in most cases. Such array descriptions are more truly algorithms in the sense that they operate on sets of numbers and provide other sets of numbers, with no explicit equations involved.

For background information *see* ADAPTIVE CONTROL; ALGORITHM; ELECTRIC FILTER; ESTIMATION THEORY; LEAST-SQUARES METHOD; NEURAL NETWORK; OPTIMIZATION; STOCHASTIC CONTROL THEORY in the McGraw-Hill Encyclopedia of Science & Technology.   Ali H. Sayed; Babak Hassibi; Thomas Kailath

Bibliography. S. Haykin, *Adaptive Filter Theory*, 3d ed. 1996; J. G. Proakis et al., *Advanced Digital Signal Processing*, 1992; B. Widrow and S. D. Stearns, *Adaptive Signal Processing*, 1985.

Agriculture

Deterioration of plant products after harvest is a major limitation on food resources worldwide. Estimates of losses for all food production range from 10 to 30%, but much greater losses are common in developing countries. Enough food is lost annually to feed a population equal to that of the United States. Negative effects are not confined to economic losses and hunger. Since many of the pathogens responsible for deterioration of foodstuffs also produce substances highly toxic to humans and livestock, food safety is an issue as well. The impact of postharvest plant pathogens in food deterioration and food safety has become a top-ranking issue for plant pathologists in the United States.

Postharvest pathogens and susceptible crops. Organisms responsible for deterioration of harvested plant material include rodents, insects, fungi, bacteria, and occasionally viruses. Affected crops include foodstuffs, fiber (including wood products), and ornamental plants. In practice, however, postharvest pathology usually refers to deterioration of stored foodstuffs and florist stocks by microorganisms. Because the stored products are often acidic in nature (such as most fruits) or are stored under dry conditions (as are grains and cereals), fungi are usually the agents responsible for damage. However, bacteria can be important on specific crops.

Fungi. Large numbers of fungi cause postharvest decay. Among those attacking stored fruits and vegetables are species of *Alternaria, Botrytis, Botryodiplodia, Botryosphaeria, Cladosporium, Colletotrichum, Fusarium, Mucor, Penicillium, Phoma, Phytophthora, Pythium,* and *Rhizopus.* Harvested grains or nuts are frequently attacked by species of *Aspergillus, Fusarium,* and *Penicillium.* This deterioration is often accompanied by the production of mycotoxins, fungal substances injurious to humans and animals. The fungus noted for attacking stored nursery and florist stocks is *Botrytis cinerea,* causing the disease gray mold. Similar descriptive names are given to other common postharvest diseases, such as brown rot on peaches, caused by *Monilia* species; blue mold of apples and pears, caused by *Penicil-*

lium expansum; and ear rot of corn, caused by *Fusarium moniliforme*.

Bacteria. Although some species of bacteria are important postharvest pathogens of specific fruits (for example, *Erwinia* on guava), bacteria are more important for causing soft rots of a variety of vegetables. Noteworthy are bacterial soft rots of carrots, cabbage, and potatoes (caused by *E. carotovora* and other *Erwinia* species). Other species of bacteria reported to cause postharvest soft rots of vegetables include *Xanthomonas campestris* and *Pseudomonas syringae*. Unlike several of the bacteria causing spoilage of meat or dairy products, these plant-pathological bacteria are not pathogenic to healthy humans.

Mycotoxins. Most plant-pathological fungi are not human or animal pathogens. However, the toxins produced by several postharvest fungal pathogens can be highly deleterious when ingested by humans or animals. The effects are termed mycotoxicosis, of which there are several types.

Aflatoxins. Aflatoxicosis is caused by ingestion of aflatoxins, produced by the fungi *Aspergillus flavus* and *A. parasiticus*, ubiquitous storage molds of peanuts, corn, and other nuts and grains. As is the case with numerous other postharvest pathogens, the original infections in the crop may occur in the field prior to harvest. Epidemics of aflatoxicosis in humans have occurred in India and Kenya. Aflatoxins attack the liver and can induce gastrointestinal hemorrhage. Long-term consumption of low levels of aflatoxin may also be a serious problem; aflatoxin B_1 is a potent carcinogen.

Ochratoxin. Ochratoxin is a mycotoxin produced by *Aspergillus ochraceus*, *Penicillium viridicatum*, and related fungi. These fungi grow on cereals, grains, nuts, and other foodstuffs. A primary indicator of ochratoxicosis is kidney dysfunction. The effects of ochratoxicosis on poultry and pigs are documented in several countries in Europe and North America. In addition, ochratoxin (and aflatoxin) has been found in dairy products and breast milk.

Other mycotoxins. *Fusarium* species can produce mycotoxins whose effects range from infertility to death. Zearalenone, produced by *F. graminearum* in stored corn and other grains, induces abnormalities in the genitals and reproductive organs of swine. *Fusarium moniliforme* produces the toxic moniliformin and a potent mutagen, fusarin C. Fumonison B_1, produced by *F. moniliforme*, causes leukoencephalomalacia in horses. There are indications that *F. moniliforme* lives asymptomatically inside corn plants and can be transmitted to the next generation of plants via the seed. *Fusarium sporotrichioides* produces T-2 toxin and other trichothecenes in grain that overwinters in the field or is improperly stored. Consumption of overwintered grain was responsible for outbreaks of alimentary toxic aleukia in which thousands of persons perished in the Soviet Union during the final years of World War II.

Stachybotrys chartarum produces trichothecenes on harvested straw and hay; consumption by livestock can be fatal. Inhalation of the spores may cause substantial injury to humans that come in contact with contaminated straw.

Health concerns. The ability to produce mycotoxins is ascribed to a growing number of fungi. Whether or not all of these fungi and their mycotoxins represent an actual threat to health is controversial (for example, *Alternaria* toxins in human or animal foods). Most research concentrates on characterizing the environmental sources, chemical structures, and biological activities of mycotoxins. Biological assays may involve animal testing with rats, mice, rabbits, guinea pigs, or poultry, but human and animal cell cultures, brine shrimp, and microorganisms are increasingly useful. The basis for toxicity to animals includes mutagenicity (aflatoxins), interferences with ribosomal functions (trichothecenes), inhibition of renal function (ochratoxins), affinity for cytosolic estrogen receptors (zearalenone), and alterations in lipid biosynthesis and immune response (fumonisins).

Many mycotoxins are well characterized biochemically. Establishing the production of a given toxin by a fungal isolate can assist in identification of the isolate. Mycotoxin production furnishes a basis for chemotaxonomy of some postharvest pathogens.

Control. Because many postharvest pathologies have origins in preharvest infections, control often begins in the field. Typically, this involves integrated pest management, in which pesticide use is integrated with cultural practices. The objective is economical control with minimal use of pesticides. Integrated pest management may include biological control, that is, the use of one organism to control another. Experimental control strategies include utilization of fungi or bacteria that are nonpathogenic to the crop and incapable of toxin production. Examples include the use of non-toxin-producing *Aspergillus* strains to displace toxin-producing strains in cotton seed, and the displacement of *F. moniliforme* by a bacterial competitor inside growing corn plants. Breeding for resistance to fungal infection of corn kernels is an additional research objective.

Control may also be exerted subsequent to harvest. Fruits and vegetables are often subjected to fumigants, such as methyl bromide, or rinses with sodium hypochlorite (bleach) disinfectants to reduce the number of fungal and bacterial propagules entering storage with the crop. Fungicides such as thiabendazole are also used.

Pathogen resistance and concerns over pesticide safety have directed research toward alternatives, such as sulfur dioxide, sodium carbonate, butylated hydroxyanisole, and sugar analogs. Storage of produce in atmospheres modified in proportions of oxygen, nitrogen, and carbon dioxide levels is common. Some countries allow the consumption of foods preserved by irradiation.

Recently, some fruit rots have been experimentally controlled by use of naturally occurring yeasts and bacteria. When applied in sufficient numbers, these microorganisms, present on fruit surfaces in nature, can act as effective competitors to postharvest pathogens. Experimental systems have employed strains of bacteria belonging to *Erwinia cypripedii, Pseudomonas fluorescens,* and *P. putida* against bacterial soft rots of fruits and vegetables. A variety of bacteria (for example, *Bacillus subtilis* and nonpathogenic strains of *P. syringe*) and yeasts (for example, *Cryptococcus laurentii, Candida oleophila*) have also been utilized against fungal postharvest pathogens. The Environmental Protection Agency has registered some of these microorganisms for use on pome and citrus fruits.

Monitoring and regulation. Because control cannot be assumed to be effective, monitoring and regulation are needed. For crops which represent no toxicological hazard, detection of infection and early marketing is an applicable strategy if the objective is selling for the fresh market. For example, the incidence of *Botrytis* on the flower parts of kiwi fruit has been found to be an accurate predictor of the incidence of mold in stored kiwi fruits. Similarly, the incubation of harvested produce under artificial conditions favorable for accelerated decay is increasingly useful as a predictor for incidence of storage rot. Blemished lots of fruits and vegetables are usually used for processed foods.

Such alternatives are often not suitable for products such as grains, feeds, and nuts because of the potential for contamination by mycotoxins. Developed countries have established legal limits for levels of mycotoxins in foods and animal feeds. For example, there is an upper limit of 10 micrograms per kilogram of aflatoxin in feed for dairy cattle in the United Kingdom; corn with more than 20 parts per billion aflatoxin cannot be sold through interstate commerce in the United States. However, very serious problems with mycotoxin-contaminated feeds and foodstuffs persist in developing countries. The mycotoxins enter the food chain through postharvest deterioration of grains, nuts, and other foods. In some locations in western Africa, concentrations of mycotoxins in human breast milk far exceed those established for animal feeds in developed countries.

For background information *see* AGRICULTURAL SOIL AND CROP PRACTICES; MYCOLOGY; MYCOTOXIN; PLANT PATHOLOGY in the McGraw-Hill Encyclopedia of Science & Technology.   Frank Dugan

Bibliography. F. E. Jonsyn, S. M. Maxwell, and R. G. Hendrickse, Ochratoxin A and aflatoxins in breast milk samples from Sierra Leone, *Mycopathologia,* 131:121–126, 1995; J. E. Smith and R. S. Henderson, *Mycotoxins and Animal Foods,* 1991; A. L. Snowdon, *Color Atlas of Post-Harvest Diseases and Disorders of Fruits and Vegetables,* vol. 1, 1990, vol. 2, 1992.

Air pollution

Air pollution is a major environmental concern worldwide. High regional emissions of various gases and particles contribute to measurable increases in global levels. Rising carbon dioxide (CO_2) levels, which are implicated in global warming, have led to international efforts to curtail CO_2 emissions. Other large-scale regional changes in the atmosphere have been observed, such as the phenomenon of the Arctic Haze, which has been traced to large-scale industrialization processes in North America and Europe. In air-pollution research, it has become apparent that political boundaries often cross regional airsheds. These airshed regions generally correspond to variations in the relief of the Earth's surface, but they generally expand and contract in dimension with the vertical mixing extent of the atmosphere during a typical day and over various seasons. Thus, there are often disparate relationships between airshed regions and politically contiguous areas. These disparities challenge policy makers and officials who are carrying out environmental air-quality policies and regulations.

Air pollution across national borders. Along the United States' northern and southern borders are examples of problems of air pollution that have been recognized and addressed both at the scientific level and in governmental circles. Along sections of the United States–Canada border, for example, there have been several air-pollution incidences and long-term problems related to transboundary transport of gases, smoke, and other particles. Acid rain deposition, polluted lakes, damage to structures, influences on health, and effects on aquatic life have been studied. For example, the acid rain phenomenon was linked to industrial sulfur dioxide emissions from fossil-fuel-burning power plants and nitrogen oxides produced by motor vehicles.

In the United States, owing to local and long-range transport, visibility-impairing particles periodically find their way across state boundaries and into national parks, such as Grand Canyon National Park and the Great Smoky Mountains National Park. Debate in the scientific and public domains centers on sources, sinks, and pathways of emitted particles, and regional climate effects on air-quality levels. Ways to ameliorate resultant pollution problems and to preserve the relative pristine nature of the atmospheric environment are major goals of industry and government.

United States–Mexico border. With the enactment of the North American Free Trade Agreement (NAFTA), there is a rapidly emerging economic and political focus on the United States–Mexico border. On the United States side, national air-quality legislation has been enacted, albeit differentially at local levels. On the Mexican side, however, little or no major environmental legislation and controls on pollution have been developed. Border air-pollution problems are exacerbated by rapid population growth in the region, large population differences on opposite sides of the border, and disparate economic levels of twin towns and cities along the border. It is also difficult to predict when an environmental problem may emerge, especially if it was not regarded as a serious difficulty in the past.

Prior to the North American Free Trade Agreement, three major geographical areas of concern over air pollution were the San Diego–Tijuana region; the copper-smelting region of Cananea, Mexico, and nearby Douglas, Arizona; and the large urban region of El Paso–Ciudad Juarez. Various problems existed in each of these areas in the past, and most persist today. Particulate emissions from unpaved streets, open burning in Tijuana, and large-scale ozone generation (for example, nitrogen oxides and volatile organic compounds which interact with ultraviolet radiation to form ozone) in San Diego remain issues. Sulfur dioxide gas emissions from copper smelters of Cananea and Douglas have been historically significant environmental problems. In the El Paso–Juarez region there are multiple pollutants for which levels exceed health standards, such as particulates, carbon monoxide, and ozone. In this region, there is a long history to the problems, which relate to open burning; unpaved roads; thousands of small industries; automotive, truck, and bus emissions; and, in the recent past, smelter operations. As part of the 1983 La Paz Agreement between Mexico and the United States, the problems of smelter emissions and operations on both sides near Cananea and Douglas were identified and partially resolved, but a large array of problems remain along this portion of the border. The Douglas operations have been closed down, while on the Mexican side, operations continue.

Atmospheric factors. Studies linking air quality to the effects of meteorological conditions and climate have a long history, but are only recently emerging in the scientific literature relative to the United States–Mexico border. The general synoptic weather variability (day-to-day weather patterns and circulation of the air) has been recently linked to fluctuations in historical ozone concentrations from sites near the border—at San Diego, California, at Tucson, Arizona, and at El Paso and San Antonio, Texas. The United States–Mexico border region is a large expanse, and impacts of regional weather and climate must be taken into account in order to understand the spatial variations of air pollution. Geographic differences in the frequencies of high-pressure patterns, high-temperature extremes, and calm wind periods occur across the region, leading to and accentuating the seasonal and annual frequency of high ozone levels from place to place.

In any local area along the border, considerable variations of topography, surface wind regimes, and dispersion potentials of pollutants seem to be the rule. It is at the local scale that dispersion and pollution problems are evident and perhaps best understood. Pollutant emission variations; population distributions; and economic, industrial, and social functions together promote local variations of air quality across the border. An example of a recently recognized air-quality problem is that in Nogales, Arizona.

Arizona and Sonora. Nogales, Arizona, and Nogales, Sonora, are examples of different populations and neighboring cities straddling an international border. These towns are situated in rugged terrain in a desert wash environment (elevation about 3600–4200 ft or 1200–1400 m) aligned perpendicular to the international border and sloping gently downward from Mexico into the United States. The wash drains from south to north and into the Santa Cruz River, which flows by Tucson, Arizona. This location is characterized by complex topography on both sides of the border. Nogales, Arizona, has a population of 20,000 but the Nogales, Sonora, area is ten times as large.

These cities are growing due to the increased economic activity along this border locale. For example, the *maquiladora* (industrial twin plant program) and the presence of an accessible Mexico-to-United States transportation corridor for cars, buses, and trucks have stimulated interchange and economic activity in the region. Major concerns regarding air pollution in the area—concerns that are now being addressed—include burning of garbage, manure burning, household wood burning, dirt road emissions, and vehicular emissions.

The Arizona Department of Environmental Quality, the agency mandated to deal with state air-quality issues, during the winter of 1988–1989 observed high concentrations of fine-particulate matter on the United States side of the border. The 24-h U.S. National Ambient Air Quality Standard (NAAQS) for particulate matter smaller than 10 micrometers is 150 micrograms per cubic meter. It is quite often exceeded at a monitoring station located right at the border crossing. At 0.3 mi (0.5 km) and 2.5 mi (4 km) from the border, there is a distance decay of particulate surface concentration values, and few observations exceed the 24-h National Ambient Air Quality Standard. For example, on average, during October–December 1988, particulate levels were 95.2 μg/m^3 at the border crossing; 70 μg/m^3 at 0.3 mi (0.5 km) away; and only 47.3 μg/m^3 at a county sensor 2.5 mi (4 km) from the border. At the border crossing, 29% of the observations were above the 24-h standard. Half a kilometer away, only 6% exceeded the standard; at 2.5 mi (4 km) away, none exceeded the standard for the measurement period.

The particulate problem at the border can be partly explained by appreciating the nature of the seasonal wind regime of the area and its connection to the broader weather patterns of the region. Long-term hourly wind data, taken in Nogales, Arizona, demonstrate the local problem.

Wind speeds are generally low in winter months (less than 3 mi/h or 1.5 m/s), especially in early-to-mid winter, and during evening and early morning hours. This is characteristic of the southwestern United States and Sonoran desert region, which is dominated by high pressure, clear skies, and little or no wind. A classic upvalley and downvalley terrain-induced wind regime is prevalent in the Nogales wash environment. Nighttime and morning winds are generally from the Mexican side toward the north, or downvalley, into and through Nogales, Arizona. During the day, a wind shift typically occurs, with winds coming from the north (upvalley) and west after 10 a.m. until about sundown. Nighttime atmospheric inversions (often below 330 ft or 100 m in depth) trap emissions and reduce vertical mixing of the lower levels of the atmosphere. Under these conditions, gentle downvalley drainage winds and periodic large-particulate emissions from Mexico often combine to contribute to high evening-to-10 a.m. exceedances of the 24-h (150 $\mu g/m^3$) and annual (mean of 50 $\mu g/m^3$) health standard in Nogales, Arizona.

Pollution dispersion modeling of these conditions, employing numerical diagnostic wind models and airshed models, is in progress by United States and Mexican researchers. This work is supported by many entities in both countries, such as the U.S. Environmental Protection Agency, the Southwest Center for Environmental Research and Policy, and other United States and Mexican federal and state governments and agencies. The purpose is to determine emission fluxes and resolve accountability of the particulate problem and other pollutants in the air. As expected, initial modeling results of the particulate dispersion patterns conform to terrain orientation. Generally, dispersion patterns correlate with the observed particulate distributions throughout Nogales, Arizona, and Nogales, Mexico. Detailed chemical analyses of particles in the air, to pinpoint types of sources, are also in progress.

Scientists from both sides of the border are cooperating in environmental programs designed to develop an awareness and appreciation for mutual air-quality problems and to work toward resolving border air-pollution (and also water-pollution) issues.

For background information *see* AIR; AIR POLLUTION; DUST AND MIST COLLECTION; ENVIRONMENTAL ENGINEERING in the McGraw-Hill Encyclopedia of Science & Technology.   Anthony J. Brazel; Mark J. Fitch

Bibliography. N. S. Berman et al., Classification and physical model experiments to guide field studies in complex terrain, *J. Appl. Meteorol.*, 34:719–730, 1995; A. C. Comrie, An all-season synoptic climatology of air pollution in the U.S.–Mexico border region, *Prof. Geog.*, 48:237–251, 1996; J. Shields (ed.), *Air Emissions, Baselines, and Environmental Auditing*, 1993; P. A. Solomon and T. A. Silver (eds.), *Planning and Managing Regional Air Quality Modeling and Measurement Studies: A Perspective Through the San Joaquin Valley Air Quality Study and AUSPEX*, 1994.

Alzheimer's disease

Alzheimer's disease accounts for approximately 50% of all forms of dementia. Conservative estimates indicate that severe dementia may affect only 1.1–6.2% of those individuals over the age of 65, while mild and moderate dementia may affect another 2.6–15.4%. Severe dementia increases from about 1% of the population over 40 years of age to 7% of the population over 80. One study reported that demen-

tia affects 10.3% of the population aged 65, and increases to 47.2% of the population aged 85 and over. Currently, Alzheimer's disease is the fourth leading cause of death in the United States, accounting for 150,000 deaths per year. By the year 2040, 14 million Americans will be directly affected by this disease.

Clinical signs. Alzheimer's disease, like many dementias, is progressive with subtle indicators (for example, memory lapses and irritability) leading to end-state conditions (such as mutism, rigidity, and incontinence). Survival averages 8.5 years from initial diagnosis, although some individuals live for more than 20 years after diagnosis. Early-onset Alzheimer's disease, afflicting individuals 40–50 years of age, may have a different etiology from late-onset Alzheimer's. Genetic factors as well as environmental hazards have been implicated in the etiology of Alzheimer's disease.

Diagnosis. Alzheimer's disease is diagnosed postmortem by the presence of neurofibrillary tangles, protein filaments within neuronal cells that twist together to form helixes, and clumps of degenerating neurons surrounding a core of amyloid protein (neuritic plaques). These tangles and plaques contribute to neuronal loss, resulting in a 40–50% loss of cortical volume in frontal, temporal, and parietal association areas of the brain, and in the subcortical limbic system including the hippocampus and amygdala. A variety of other degenerative diseases, including Pick's disease and Parkinson's disease, demyelinating diseases such as multiple sclerosis, hydrocephalic conditions, and vascular infarcts, also result in dementia. Not all forms of dementia are progressive; 10–15% may be reversible dementias due to depression, metabolic disorders, toxic factors, and drugs.

Clinical markers of the onset of Alzheimer's disease are difficult to distinguish from nonclinical age-related lapses of attention or memory, that is, benign senescent forgetfulness or nonpathological age-associated memory impairments. Distinguishing normal age-related changes to cognition and language from abnormal or pathological changes may be important for the early diagnosis, hence possible treatment, of Alzheimer's disease and related disorders.

Language and communication. Communication problems are often the first symptoms of a progressive dementia such as Alzheimer's disease. Such problems are frequently noted by spouses and other family members. By studying how language and communication are affected by progressive disorders such as age-related dementias or Parkinson's disease, researchers may be able to determine the boundaries between what is a pattern of normal aging and what is a pathological process. Unlike language impairments such as aphasias, which can be traced to a specific and acute biological trauma, the onset of dementia is slow and subtle.

Dementia and speech. Studies of speech reveal that older adults with dementia are less likely to spontaneously produce complex grammatical forms than their healthy peers; however, their speech does not exhibit a progressive degeneration into baby talk. Syntactic production appears to be buffered from the effects of Alzheimer's disease. Some simplifications of syntax may occur, but the speech of individuals with Alzheimer's disease does not appear to become agrammatic. The retention of basic syntactic structure with the progression of Alzheimer's disease is evidence that syntax is a separate module, although the decline in content and length indicates that cognitive limitations associated with Alzheimer's disease also affect language production.

Semantic disruptions. In contrast, semantic and lexical processes do appear to be disrupted by Alzheimer's disease. Two hypotheses have been put forth to account for these disruptions. Some researchers have concluded that the structure of semantic knowledge is destroyed by Alzheimer's disease. As a result, an individual's performance on verbal fluency tasks, such as generating exemplars of categories or words beginning with a specific letter, is impaired; picture and object naming is hindered; and word associations are destroyed. Others have concluded that the semantic network of adults with Alzheimer's dementia is intact but becomes inaccessible. From this perspective, semantic priming in reaction-time tasks involving word naming and lexical decisions is preserved.

Impairments to discourse. Many of the impairments which have been observed in adults with Alzheimer's dementia may stem from their gross word-finding problems, whereas other problems may originate from attentional deficits and cognitive confusions. The heavy use of words such as ''this'' and ''that,'' the loss of specific reference and cohesion, the prevalence of vague terms and empty speech, a loss of detail, an increase in repetition and redundancy, and confusing shifts in topic and focus have all been noted as characteristic of the speech of adults with Alzheimer's disease.

Other communication problems have also been linked to Alzheimer's disease. For example, the personal narratives told by healthy older adults supply related setting information, complications, the protagonist's actions, and a resolution. However, the spontaneous narratives of adults with Alzheimer's disease characteristically supply only setting information unless they are prompted by their conversational partner to provide additional information. The ability to use or follow a familiar script, or a series of temporally and causally linked events such as eating in a restaurant, going to a movie, or holding a wedding, is also impaired by Alzheimer's disease, as are spontaneous turn taking, topic initiation, topic maintenance, topic shifting, conversational repairs, and speech acts, such as requesting, asserting, clarifying, and questioning. A breakdown of self-awareness and metalinguistic skills may contribute to the discourse problems of adults with Alzheimer's dementia, because additional clinical signs of the progression of the disease are indicated by an erosion of metalinguistic skills marked by declines in re-

quests for clarification, references to memory problems, and self-evaluation of skills and abilities. End-stage Alzheimer's disease is often characterized by mutism, inappropriate nonverbal vocalizations, and frequent failures to respond to others.

Nun Study. The ongoing Nun Study has already provided some clues to the development and onset of Alzheimer's disease. The study involves examining autobiographies written by participants at approximately age 20 and approximately 60 years later, examining their health, physical function, and cognitive ability. The participants, 678 members of a religious congregation, were all born before 1917; since 1988, they have been undergoing periodic assessments of their health, physical function, and cognitive ability, and all have agreed to postmortem brain donation for neuropathological examination. These autobiographies revealed that two measures, one of grammatical complexity and one of idea content, were predictive of cognitive impairment in late life, and the idea content measure was predictive of Alzheimer's neuropathology. In other words, those nuns whose autobiographies exhibited a simplified style at age 20 were at greater risk for the development of cognitive impairment 60 years later and at greater risk for the development of Alzheimer's disease. In contrast, nuns whose early autobiographies were written in a more elaborated, complex style were more likely to have intact cognitive skills in late life and to show little of the neuropathology characteristic of Alzheimer's disease after death.

A number of interpretations of this finding are possible: (1) individuals with reduced linguistic ability may exhibit the signs of Alzheimer's disease at younger age because they lack many of the linguistic skills that are critical to perform well on the tests for the diagnosis of dementia; (2) Alzheimer's disease may be a lifelong process whereby neuropathological damage gradually accumulates, and hence an early sign of the disease may be reduced linguistic ability in young adulthood; or (3) reduced linguistic ability in young adulthood may make individuals more vulnerable to the onset of the disease later in life because they fail to develop resistant, neurologically complex brains as a result of a reduction in cognitive and social stimulation throughout adulthood. Ongoing analysis of language-production data from these nuns may help to distinguish among these interpretations. This information may help refine criteria for the early diagnosis of Alzheimer's and related dementias and may help ensure that individuals receive appropriate interventions and treatments.

For background information *see* ALZHEIMER'S DISEASE; BRAIN; COGNITION; INFARCTION; LINGUISTICS; NERVOUS SYSTEM DISORDERS; SENILE DEMENTIA; SPEECH DISORDERS in the McGraw-Hill Encyclopedia of Science & Technology. Susan Kemper

Bibliography. F. I. M. Craik and T. A. Salthouse (eds.), *Handbook of Cognitive Aging,* 1992; F. A. Huppert, C. Brayne, and D. O'Connor (eds.), *Dementia and Normal Aging,* 1994; S. Kemper et al., On the preservation of syntax in Alzheimer's disease:

Evidence from written sentences, *Arch. Neurol.*, 50:81–86, 1993; D. A. Snowdon et al., Cognitive ability in early life and cognitive function and Alzheimer's disease in later life: Findings from the Nun Study, *JAMA*, 275:528–532, 1996.

Anergy

Lymphocytes (the B and T cells of the immune system) act to identify and eliminate foreign antigens such as bacteria, toxic substances, and cancer cells that endanger the individual. The recognition of these variable threats relies on the random production and expression of a unique antigen-binding site on almost every lymphocyte. Lymphocytes whose receptor recognizes a foreign antigenic substance are subsequently stimulated by that antigen to multiply. These special lymphocytes play a role in the elimination of the antigen or of the cells carrying it. Lymphocytes with a surface antigen receptor that specifically recognizes one of an individual's own self-antigens can also be generated, thus creating the potential for autoimmunity (reaction against one's own cells). Nevertheless, the clinical appearance of autoimmune disease is relatively rare because most individuals possess effective self-tolerance mechanisms to control such autoreactive cells. One major mechanism is clonal deletion: lymphocytes are killed if they recognize a self-antigen during their maturation in the thymus gland or bone marrow. However, not all human self-antigens are expressed in these central lymphoid organs while the lymphocytes are developing. Thus, tolerance to an individual's own antigens must also depend on other means. Clonal anergy has been proposed as one self-tolerance mechanism. Theoretically, autoreactive lymphocytes lose their capacity to multiply following their recognition of a self-antigen in the peripheral immune system. *See* IMMUNE SYSTEM.

Induction. The clonal anergy self-tolerance mechanism has been investigated by using a laboratory culture system of mature T lymphocytes. Normally, these T cells secrete the growth-promoting factor called interleukin 2 and begin to multiply after they detect the presence of their specific antigen localized on the surface of specialized antigen-presenting cells such as tissue macrophages (cells that remove antigens) or lymph-node dendritic cells. However, if these antigen-presenting cells (macrophages or lymph-node dendritic cells) are first treated with a chemical that prevents their expressing important co-stimulatory structures on their surface, these cells fail to stimulate the T cells to multiply in the presence of the antigen. Instead, they induce the T cells into a state of antigen unresponsiveness called clonal anergy. Such inactivated T cells can no longer multiply in response to the presence of antigen because they cannot be stimulated to secrete the interleukin 2 growth factor. Anergy is also induced in T cells when their antigen receptors are bound and physically cross-linked by specialized proteins such as

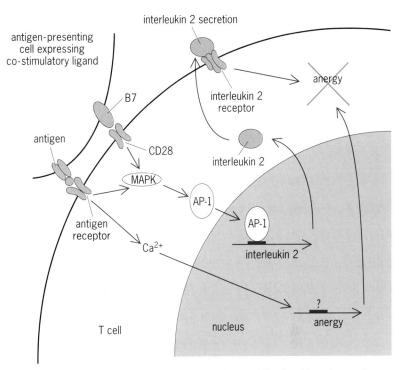

Fig. 1. T-cell activation model illustrating the stimulation of T cells with antigen and normal antigen-presenting cells which result in the production of interleukin 2 (IL-2) and the continuance of normal antigen responsiveness. The induction of an *Anergy* gene is purely hypothetical.

mitogenic lectins or antibodies. These observations have led to a two-signal model of T-cell activation. The interaction of antigen with the T-cell antigen receptor and the occupancy of other co-stimulatory receptors on the T cell by substances (ligands) found on the normal antigen-presenting cell promote interleukin 2 secretion and multiplication. Signaling by the T-cell antigen receptor alone (in the absence of co-stimulation) fails to initiate multiplication and instead induces clonal anergy (**Fig. 1**).

Intracellular signaling. In order to understand how these molecules can regulate the responsiveness of mature T cells, a primary consideration is the intracellular signals that can develop in the lymphocyte following the interaction of co-stimulatory receptors and ligands. The best-characterized co-stimulatory receptor is CD28. The binding of CD28 to its ligand B7 (a protein that is expressed on the surface of the activated antigen-presenting cell) greatly increases the amount of interleukin 2 secreted by the T cell when it detects antigen. The CD28 molecule does this by regulating the activity of certain intracellular signaling enzymes important to the initiation of interleukin 2 production. The result is that CD28 signaling leads to higher interleukin 2 messenger ribonucleic acid (mRNA) levels, both by increasing the levels and function of critical nuclear transcription factors (for example, the activating protein-1, or AP-1) at the interleukin 2 gene promoter site, and by enhancing the survival of the interleukin 2 mRNA transcripts. Higher levels of interleukin 2 mRNA

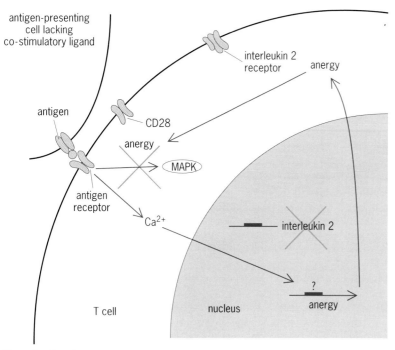

Fig. 2. T-cell activation model illustrating the ligation of the T-cell antigen receptor in the absence of co-stimulatory signals which fail to induce the production of interleukin 2. Clonal anergy is induced without opposition from the IL-2 receptor. The induction of an *Anergy* gene as shown is purely hypothetical.

means that greater amounts of interleukin 2 protein will be secreted. Signals mediated by CD28 also allow for continued responsiveness of the T cell after it has recognized an antigen. CD28 signals increase the secretion of interleukin 2, resulting in subsequent stimulation of the T cell's own interleukin 2 receptors (IL-2R). Signals from the interleukin 2 receptor then act to maintain a high level of antigen responsiveness of the T cell. This is why the treatment of anergic T cells with purified interleukin 2 is sufficient to restore their antigen responsiveness.

Anergy-inducing stimuli. All anergy-inducing stimuli share two important features. They initiate proximal biochemical signaling events such as the mobilization of intracellular calcium ions, and yet they fail to induce the secretion of interleukin 2 and the multiplication of the T cells. T cells also become anergic when they are exposed to a calcium-mobilizing ionophore. It has been observed that the development of clonal anergy in response to these stimuli can be prevented either by pharmacological inhibitors of the calcium-signaling pathways or by protein synthesis inhibitors. Since an increase in calcium-ion concentration can lead to the initiation of gene transcription, the data suggest that clonal anergy can develop as a consequence of new gene expression. However, the gene products responsible for the development of clonal anergy remain entirely unknown.

Defective signal transduction in anergic T cells. Experiments will help to determine how the clonal anergy mechanism keeps T cells from secreting interleukin

2 and multiplying. It is known that anergic T cells cannot produce interleukin 2 upon reexposure to antigen because of an inability to transcribe the interleukin 2 gene. This is because stimulation of the antigen receptor in anergic cells fails to activate enzymes (called mitogen-activated protein kinases, or MAPK) that regulate the AP-1 transcription factor. The nature of the coupling defect between the antigen receptor and the mitogen-activated protein kinases remains uncertain (**Fig. 2**).

The block in interleukin 2 secretion and multiplication that develops in anergic T cells does not indicate a defect in signal transduction by the antigen receptor itself. The receptor remains competent to induce the mobilization of calcium ions in anergic T cells and to regulate the transmission of signals into the nucleus. Consistent with this, the production of interleukins 3, 4, 5, and 6 and interferon-gamma is not as significantly affected as that of interleukin 2. Anergic T cells can assist B cells in the generation of an antibody response, and they remain capable of directly killing antigen-bearing target cells if the T cells are present at high number. The experiments indicate that the production of the interleukin 2 growth factor relies on a signal transduction pathway that is sensitive to anergy and is not utilized by these unaffected T-cell responses. This suggests that the clonal anergy mechanism has evolved to control only the expansion of potentially autoreactive T cells in the peripheral immune system, while leaving some of their functional responses intact.

Clonal anergy in intact immune system. Clonal anergy represents one potential mechanism utilized by the immune system to regulate the behavior of T cells with potentially autoreactive antigen receptors. A number of living animal models have confirmed the existence of clonal anergy in T cells as a result of their exposure to antigen on antigen-presenting cells that express few co-stimulatory ligands. Furthermore, treatment of mice that are prone to autoimmune disease with CD28 or B7 antagonists can halt the progression of disease, consistent with the induction of anergy in the offending autoreactive T-cell population. These observations indicate that a greater understanding of the mechanisms underlying the development of clonal anergy in the peripheral immune system could prove important in the design of both vaccines and future therapies for autoimmune diseases.

For background information *see* ANTIGEN; AUTOIMMUNITY; CELLULAR IMMUNOLOGY; CYTOLYSIS; HISTOCOMPATIBILITY; IMMUNOSUPPRESSION; LIGAND; LYMPHATIC SYSTEM in the McGraw-Hill Encyclopedia of Science & Technology.   Daniel L. Mueller

Bibliography. P. E. Fields, T. F. Gajewski, and F. W. Fitch, Blocked Ras activation in anergic CD4+ T cells, *Science*, 271:1276–1278, 1996; W. Li et al., Blocked signal transduction to the ERK and JNK protein kinases in anergic CD4+ T cells, *Science*, 271:1272–1276, 1996; D. L. Mueller and M. K. Jenkins, Molecular mechanisms underlying functional T-cell unresponsiveness, *Cur. Opin. Immunol.*,

7:375–381, 1995; A. Sundsted et al., In vivo anergized CD4+ T cells express perturbed AP-1 and NF-kB transcription factors, *Proc. Nat. Acad. Sci. USA*, 93:979–984.

Animal evolution

Recent work on animal evolution has focused on the role of development in influencing evolutionary radiations. Evolutionary radiations are significant rises in biological diversity in comparison with previous or later times, and frequently entail a pronounced increase in number of biological groups or taxa (taxonomic diversity) and in morphological variety (disparity). The radiation of mammals after the mass extinction at the end of the Cretaceous, 65 million years ago (Ma), is a classic example; the radiation of flowering plants in the Cretaceous (145–65 Ma) is another. Lately, much research has been directed at the description of Paleozoic (545–250 Ma) animal radiations in the marine realm, of which the Cambrian explosion (initiated 545 Ma) of skeletonized body plans is the most prominent. Improved quantitative analyses have confirmed the reality of the patterns involved, while more refined geological dating of the timing of events has led to the conclusion that the events occurred very rapidly.

Hypotheses. As more detailed patterns of evolutionary radiation throughout the history of life are being revealed, evolutionary biologists and paleobiologists grapple with different hypotheses on the processes involved. Most proposals reduce to one of two alternatives: an ecological (externalist) hypothesis, whereby radiations are driven by the availability of empty ecological niches; and a developmental (internalist) hypothesis, whereby radiations are primarily the result of experimentation on a large number of morphological themes available because of an inherent flexibility of developmental systems. Evolutionary biology for a long time has assumed the primacy of the ecological view. However, studies on the role of development in evolutionary radiations have strengthened the developmental view.

Morphological evidence. Bursts in evolutionary activity are often reflected in a temporal asymmetry (an "arrow of time") in the pattern of morphological disparity, by long-term reduction or maintenance of high levels of morphological distinctness achieved early in the evolution of a clade or group of clades (phylogenetically defined groups). The failure to expand morphological variation over long periods of time, despite continual environmental change, suggests a temporal asymmetry in developmental flexibility. For example, crinoids (sea lilies and allies) barely expanded their levels of variation after an initial radiation, despite 250 million years of evolution in the Paleozoic Era. Middle Cambrian arthropods (trilobites, crustaceans and allies) have at least the same morphological disparity as Recent ones, despite more than 500 million years separating the two samples. In light of extensive reorganizations of ecosystems, as during mass extinctions, these results challenge the extent to which morphological diversification is driven by ecological forces, and are consistent with higher developmental flexibility in the Early Paleozoic.

More definitive tests, however, demand developmental data. One approach is to partition data sets into characteristics that are prone to be developmentally limited and characteristics that are prone to be ecologically limited. With such an approach, the temporal asymmetry in variation of early gastropods (snails) was shown to be the result of a reduction in developmental flexibility. Another approach is to assemble explicitly developmental data in a group (for example, juvenile morphologies, malformations) and compare morphological distributions based on such developmental data with those based on adults. Thus, the diversity of certain adult echinoids (sea urchins) has been described as a simple reshaping of developmental sequences already present in primitive species. Also, the range of normal among-species variation in taxa as different as tetrapods, fruit flies, and sea urchins matches in remarkable ways within-species variation in malformations, even though on functional grounds a match between patterns of variation in fit and highly unfit forms would not be expected. These results lend support to the claim that there are intrinsic bounds to morphological diversification as a result of developmental constraints that channel variation along biased pathways. This poses limits to the power of natural selection in evolution.

Phylogenetic evidence. Developmental characteristics (such as pathways of tissue interaction, rates of development, and patterns of expression of genes involved in development) can be mapped onto a phylogenetic tree and then tested to determine whether clades can also be defined by developmental features, suggesting that the origin of such groups is associated with changes in development. Most major groups of echinoids, for example, have been shown to be uniquely defined by larval morphology. Most animal phyla, in turn, are associated with a distinct pattern of developmental gene architecture and expression. Correlation is not necessarily causation, but such results suggest that developmental flexibility has been a major determinant in the radiation of higher taxa.

Theoretical evidence. The developmental hypothesis to account for evolutionary radiation and accompanying temporal asymmetry generates the theoretical prediction that radiations must be associated with higher developmental flexibility, with later quiescence signaling lower flexibility. Essentially, this hypothesis predicts an intrinsic change in the rate of production of new characteristics through time. An extrinsic change in the rate of success of novelties is predicted by the extrinsic (or empty ecospace) hypothesis. Recently, though, an important role for development has been codified in a third alternative, the rugged fitness landscapes model. This model is

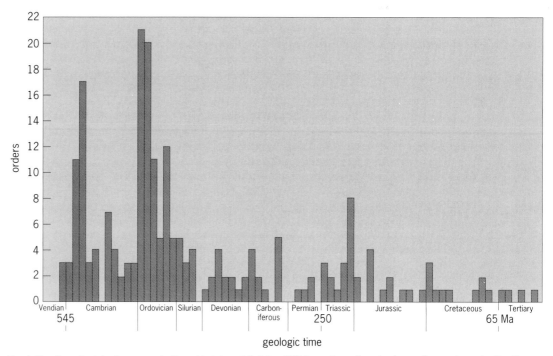

Fig. 1. Number of originations per stratigraphic interval (total = 219) for orders of marine invertebrates from the Vendian (last Precambrian period, 610 Ma) to the Pleistocene. Each histogram bar corresponds to a geologic stage. Numbers on the x axis correspond to the base of the Paleozoic, Mesozoic, and Cenozoic eras. Data were culled to account for taxonomic and sampling biases.

a modern rendition of the metaphor of adaptive landscapes, a surface of adaptive peaks and valleys, where the position of evolving entities is determined by the interplay of different evolutionary forces. The rugged fitness landscapes model adds to the metaphor limitations arising from the topography of the landscape, equivalent to inherent limitations imposed by individual development. The model predicts an intrinsic change in the rate of success of new characteristics through time as a result of evolution happening in a developmentally limited fitness landscape. Thus, evolutionary radiations would be driven by selection favoring occupation of ever higher adaptive peaks, but the process would be regulated by an intrinsic availability of peaks. Microevolution (for example, population differentiation) should occur in smooth regions of a fitness landscape and proceed more slowly; macroevolution (for example, the origin of body plans) should occur more rapidly in highly rugged regions. The rugged fitness landscapes model, backed by computer simulation studies, has been used to explain the observed asymmetry in origination of major groups (phyla, classes, orders) through geological time (**Fig. 1**). However, its main prediction, an exponential decline in the frequency of appearance of major innovations, has only recently been tested rigorously with fossil data and shown not to hold. Although evolutionary biologists are aware of the fact that major evolutionary radiations are bound to be influenced by both developmental and ecological constraints, no clear consensus has yet emerged on their relative importance or

how they should interact. The rugged fitness landscapes model was a first step in that direction.

Constraining the Cambrian explosion. When the Early Paleozoic radiations, and in particular the Cambrian explosion of skeletonized animal body plans, are put in developmental perspective, it is necessary to consider the timing of three components of the overall pattern: the splitting of lineages that ultimately give rise to body plans; the complexification of developmental controls; and the origin of the body plans themselves. Accordingly, different models arise for the sequence of evolutionary events involved (**Fig. 2**). Extensive research is being conducted by paleontologists, evolutionary developmental biologists, and molecular evolutionists in an attempt to better understand the underlying pattern of the Cambrian explosion and to distinguish alternative models.

Regardless of the details, the uniqueness of the Cambrian Period is, in retrospect, an expression of the way that constructional themes became available. Such constructional themes, being very conservative, imply a high degree of developmental entrenchment (stability and resistance to change), and it is possible that at that time characteristics with high potential for entrenchment were as readily produced as any other. The implication would be that an inherently smaller number of such characteristics existed then, and that the Early Paleozoic radiation reflected the rapid exploration of a finite set of opportunities for characteristics with high potential for entrenchment. Concomitantly, other characteristics

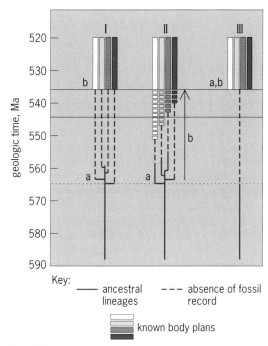

Key:
— ancestral lineages   - - - absence of fossil record

known body plans

Fig. 2. Models of the sequence of evolutionary events surrounding the Cambrian explosion as constrained by the fossil record. The events include (a) the divergence of the major lineages and (b) the origin of body plans. Model I: The lineages that eventually lead to animal phyla branch early, but a burst of body-plan diversification occurs near 530 Ma. Genomes achieve complexity early but undergo possible secondary expansions near 530 Ma. Model II: Animal lineages, genomes, and body plans branch, diversify, and increase in complexity progressively during the Late Precambrian, with a final burst of advanced body-plan diversification near 530 Ma. Model III: Animal lineages, genomes, and body plans all diversify near 530 Ma. The solid horizontal line at 545 Ma represents the boundary between the Precambrian and Cambrian periods. (After J. W. Valentine, D. H. Erwin, and D. Jablonski, Developmental evolution of metazoan body plans: The fossil evidence, Dev. Biol., 173:373–381, 1996)

would accumulate through time and move integrated character complexes farther and farther away from the opportunity of evolving into fundamentally different body plans, therefore reducing the chances of origination of major innovations. More studies of the evolutionary dynamics of characteristics with different degrees of entrenchment are needed in order to support these considerations, especially through the construction of morphological spaces of possible and actual forms and through phylogenies of development itself.

The evolution of animal development was an important event, prerequisite for all the evolutionary activity that ensued. The extent to which its primitive condition could be elaborated and reorganized defined the finiteness of the Cambrian explosion. That this was so is suggested by the remarkable commonality in developmental and genetic control mechanisms across the animal kingdom from nematodes to mice, which was discovered by research beginning in the mid-1980s. It is now generally recognized that the evolution of developmental and genetic regulatory mechanisms must have played a crucial role once an environmental barrier (for example, oxygen concentration in the atmosphere) was crossed. Whether the diversification of animal development was a result of selection pressures associated with the occupation of empty ecological niches, whether it proceeded for developmental reasons by experimentation of many morphological alternatives available because of less regulated developmental controls, or whether some combination of both occurred, is still a matter of active empirical research. But the supporters of the developmental hypothesis seem to have strengthened their case.

For background information *see* ANIMAL EVOLUTION; CAMBRIAN; DEVELOPMENTAL BIOLOGY; MACROEVOLUTION; PALEONTOLOGY; PHYLOGENY in the McGraw-Hill Encyclopedia of Science & Technology.   Gunther J. Eble

Bibliography. D. H. Erwin, The origin of metazoan development: A paleobiological perspective, *Biol. J. Linn. Soc.*, 50:255–274, 1993; M. L. McKinney (ed.), *Biodiversity Dynamics: Turnover of Populations, Taxa, and Communities*, 1998; R. A. Raff, *The Shape of Life: Genes, Development, and the Evolution of Animal Form*, 1996; J. W. Valentine, D. H. Erwin, and D. Jablonski, Developmental evolution of metazoan body plans: The fossil evidence, *Dev. Biol.*, 173:373–381, 1996.

Antimatter

Early in 1996, scientists working at the CERN accelerator laboratory in Geneva, Switzerland, reported seeing 11 electrical signals that seemed to come from antihydrogen atoms, and estimated that two of these were noise. Even though most physicists were quite certain that antihydrogen could be formed, this report generated great interest because for the first time an antimatter atom had been directly observed. The antihydrogen atom is the antimatter counterpart of the simplest atom, the hydrogen atom. Within the hydrogen atom, a single electron orbits a single proton. Within the antihydrogen atom, the electron and proton are replaced by their antiparticle counterparts, the positron and antiproton. A single positron thus orbits a single antiproton.

CPT theorem. The fundamental CPT theorem of physics predicts that particles and antiparticles have nearly identical properties. All positrons and all electrons are predicted to have exactly the same mass and exactly the same amount of charge, but with the opposite sign of charge. All protons and antiprotons are believed to be similarly related. Scientists thus expect that the opposite charges of the positron and antiproton (forming antihydrogen) will attract each other in exactly the same way as do an electron and proton (forming hydrogen). The CPT theorem thus predicts that antihydrogen and hydrogen atoms will have exactly the same mass and the same atomic properties.

Undoubtedly, the best reason to make and study antihydrogen is to test whether antihydrogen and hydrogen atoms actually do have identical properties as expected. For example, the electron in a hydrogen atom can be excited from its inner orbit to a larger orbit when laser light of a particular, well-defined color (that is, frequency) irradiates the atom. Shining the same laser on an antihydrogen atom would test to high accuracy whether laser light of the same frequency excites the positron. The laser techniques are so accurate that this comparison would be one of the two most stringent tests of the fundamental CPT theorem.

Production of cold antihydrogen. Unfortunately, the nine observed antihydrogen atoms lived for less than a few billionths of a second and traveled at nearly the speed of light. Even if many more such atoms could be produced, there is little hope of accurately measuring any of their properties under these conditions. What is needed are antihydrogen atoms so cold that they can be confined for much longer times for precise study.

One difficulty in forming cold antihydrogen is the rarity of the exotic antimatter particles that make up antihydrogen. Most of the world is composed of electrons, protons, and neutrons, so these particles are readily available. Their corresponding antiparticles are not. Positrons exist naturally most often as the occasional by-products of radioactive decays. Antiprotons occur naturally only as products of collisions of cosmic-ray particles with gas molecules as the particles enter the Earth's atmosphere.

Another difficulty is that an antiparticle and particle annihilate when they collide. A colliding positron and electron cease to exist, their masses being converted to light energy by A. Einstein's formula. A colliding antiproton and proton similarly annihilate, except that pion particles are produced instead of light. The collision of an antihydrogen atom and a hydrogen atom therefore causes both to annihilate. The energy release per fuel mass is so tremendous that it makes fusion seem inefficient in comparison.

Synthesis of ingredients. The two ingredients for cold antihydrogen are now available in the extremely good vacuum needed to avoid annihilations. Antiprotons are synthesized artificially at large accelerator facilities. Their energy is then subsequently reduced by a huge factor of 10^{10}, until they have a temperature of only 4 K above absolute zero ($-459.67°F$). With magnets and battery voltages, they are then stored inside a container called a Penning trap, being suspended so that they never encounter any material and annihilate.

More than 10^6 positrons have recently been confined in a similar so-called container without walls at the same low temperature and extremely high vacuum. These positrons originate from the radioactive decay of sodium. They are slowed first by collisions within a cold tungsten crystal (they leave the crystal before they have time to annihilate), and eventually also are stored in a Penning trap.

Binding of ingredients. Even with both cold ingredients now available, cold antihydrogen is very difficult to produce. The difficulty may be understood by considering a positron and an antiproton that are approaching each other. Both particles have energy, and this energy must be removed if the two particles are to become bound to each other. If only these two particles participate in this interaction, there is no particle available to carry off the excess energy. If the excess energy is not carried off, antihydrogen is not formed.

Possible methods for producing cold antihydrogen differ in the way they provide a third particle and in the way they make available a third particle to carry off the excess energy. One promising method involves arranging a collision between two positrons and one antiproton. One positron and the antiproton form an antihydrogen atom while the second positron carries off the excess energy. In another process, the excess energy is carried off by a light particle, the photon. In this case a colliding positron and an antiproton form an excited state of the antihydrogen atom, which then radiates the photon. This process can be enhanced or stimulated by illuminating the collision with a powerful laser. Another proposed antihydrogen production mechanism involves arranging a collision between an antiproton and a positronium atom. Positronium is the artificial atom within which a positron and an electron orbit each other. When the antiproton and the positron form antihydrogen, the excess energy is carried off by the released electron.

LEAR experiment. The first attempt at producing cold antihydrogen was carried out in late 1996. The positrons were obtained from an intense radioactive source, 1 curie (3.7×10^{10} becquerels) of an isotope of cobalt. This intense source was carefully shielded by many lead bricks, since a direct exposure to this source at a distance of 1 m (3 ft) gives the yearly radiation dose allowed at CERN in only 3 s. To accumulate positrons, the source was lowered under remote control out of its lead housing and down into the experimental area. After the accumulation was over, and before any personnel entered the shielded volume, the source was returned to its lead enclosure. As in the initial observation of antihydrogen, the antiprotons again came from the Low-Energy Antiproton Ring (LEAR) at CERN. This unique facility was scheduled to close at the end of 1996, and the attempt to make antihydrogen was scheduled for the last $1\frac{1}{2}$ weeks before LEAR shut down. Enough was unknown about the recombination processes to make it difficult to confidently predict whether clear signatures of cold antihydrogen would be observed. The experiment was sufficiently complicated that it was difficult just to get all of its different components to work reliably and simultaneously in such a short time period. There are many technical challenges to obtaining the two antimatter species, the positrons and the antiprotons, so that most of the experimental time was focused on identifying and solving problems that are not present when the matter counterparts, electrons and protons, are studied in the same apparatus.

Further experiments. With the closing of LEAR, there is no facility in the world making available antiprotons suitable for cold antihydrogen formation. The CERN laboratory, however, now has a plan for providing the needed antiprotons in a way that is far less costly than the current methods. The three large storage rings which are currently used only to deliver low-energy protons for experiments, including LEAR, will all be shut down. Over a 2-year period, one of these rings will be modified so that it will perform the function of the three rings. After the initial capital expenditure, there is a considerable saving in operating costs. To be sure, there is also a performance penalty. Each pulse of antiprotons delivered to experiments will contain at least 10 times fewer antiprotons than current pulses from LEAR. This approach should be workable for antihydrogen experiments, however, because techniques have been developed to accumulate antiprotons in tiny ion traps rather than in a large storage ring. Tiny ion traps are much less expensive to construct and to operate than are large storage rings.

If cold antihydrogen can be formed, research will be undertaken to optimize its formation and accurately measure its properties. The techniques to slow, cool, and accumulate cold antiprotons and positrons in a Penning trap have already been mentioned. Other challenges include confining antihydrogen atoms in a magnetic trap, accomplishing laser spectroscopy of antihydrogen, and using lasers to cool the antihydrogen. Long and difficult experiments are in prospect. If successful, however, this long-term research program should help answer questions about distinguishing the worlds of matter and antimatter.

For background information *see* ANTIMATTER; ANTIPROTON; CPT THEOREM; ELEMENTARY PARTICLE; LASER COOLING; LASER SPECTROSCOPY; PARTICLE ACCELERATOR; PARTICLE TRAP; POSITRON; POSITRONIUM in the McGraw-Hill Encyclopedia of Science & Technology.   Gerald Gabrielse

Bibliography. G. Baur et al., Production of antihydrogen, *Phys. Lett. B*, 368:251–258, 1996; G. Gabrielse, Extremely cold antiprotons, *Sci. Amer.*, 267(6):78–89, December 1992.

Astronomical observatory

The *Solar and Heliospheric Observatory* (*SOHO*), a project of international cooperation between the European Space Agency (ESA) and the National Aeronautics and Space Administration (NASA), was launched on December 2, 1995, and began full scientific operation on February 14, 1996. It obtains a continuous, uninterrupted view of the Sun from its halo orbit around the point 1.5 million kilometers (930,000 mi) sunward from the Earth, where the gravitational forces of the Earth and the Sun balance

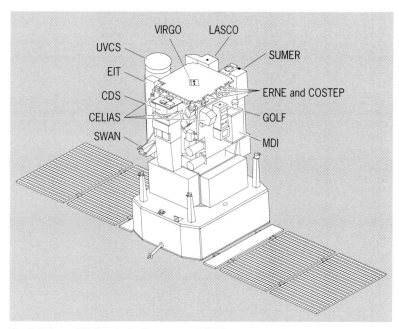

Fig. 1. *Solar and Heliospheric Observatory* (*SOHO*) spacecraft. Acronyms of scientific instruments are identified in the text and table.

each other. The spacecraft orbits this equilibrium point to keep Earth-based tracking telescopes from looking directly at the Sun.

Scientific objectives. *SOHO* (**Fig. 1**) examines the hidden interior of the Sun, the hot transparent solar atmosphere, and the solar wind of charged particles and magnetic fields that continuously flows outward from the Sun. Altogether, 12 instruments are on board *SOHO*, each designed to study one or two of these regions in a different, yet complementary way (see **table**). Their combined data will link events in the Sun's atmosphere and solar wind to changes taking place deep within the Sun. Never before has it been possible to study all three regions of the Sun simultaneously.

The *SOHO* mission, then, has three principal scientific objectives: study of the structure and dynamics of the solar interior; study of the heating mechanisms of the Sun's million-degree atmosphere, or solar corona; and investigation of the solar wind, its origin, and its acceleration processes.

Solar interior. *SOHO* illuminates the unseen depths of the Sun by recording widespread throbbing motions of the Sun's visible surface, or photosphere. These oscillations are caused by sounds that are trapped inside the Sun. On striking the surface and rebounding, the sound waves cause the gases there to move up and down.

By considering a sequence of oscillations with longer and longer period, describing sound waves that penetrate deeper and deeper, *SOHO* determines physical properties inside the Sun's deep interior. Since the technique is similar in scientific principle to using earthquakes, or seismic waves, to decipher the Earth's internal structure, it has become known as helioseismology.

Scientific instruments on *SOHO*		
Investigation	Acronym	Measurement
Helioseismology		
Global Oscillations at Low Frequencies	GOLF	Global Sun velocity oscillations (*l* = 0–3)*
Variability of Solar Irradiance and Gravity Oscillations	VIRGO	Low-degree (*l* = 0–7) irradiance oscillations and solar constant
Solar Oscillations Investigation/Michelson Doppler Imager	SOI/MDI	High-degree (up to *l* = 4500) velocity oscillations
Solar atmosphere remote sensing		
Solar Ultraviolet Measurements of Emitted Radiation	SUMER	Temperatures, densities, and velocities in chromosphere and corona
Coronal Diagnostic Spectrometer	CDS	Temperatures and densities in transition region and corona
Extreme-Ultraviolet Imaging Telescope	EIT	Evolution of chromosphere and corona; full-disk observation
Ultraviolet Coronagraph Spectrometer	UVCS	Coronal electron and ion temperatures, densities, and velocities (1.3–10 solar radii)
Large-Angle Spectroscopic Coronagraph	LASCO	Coronal evolution, mass, momentum, and energy transport (1.1–30 solar radii)
Solar Wind Anisotropies	SWAN	Solar-wind mass-flux anisotropies and temporal variations
Solar wind on site		
Charge, Element, and Isotope Analysis System	CELIAS	Energy distribution, and mass and charge composition (0.1–1000 keV per unit charge)
Comprehensive Suprathermal and Energetic Particle Analzyer	COSTEP	Energy distribution of ions (*p*, He), 0.04–53 MeV per nucleon; and of electrons, 0.04–5 MeV
Energetic and Relativistic Nuclei and Electron Experiment	ERNE	Energy distribution and isotopic composition of ions (*p*-Ni), 1.4–540 MeV per nucleon; and of electrons, 5–60 MeV

* Here, *l* is the spherical harmonic degree, or the number of times that the sound wave reflects off the solar surface in one trip around it.

There are three helioseismology experiments onboard *SOHO*, each acquiring long uninterrupted observations of solar oscillations. Two of them emphasize global oscillations and sound waves that can penetrate the deep solar interior. They are known as GOLF (Global Oscillations at Low Frequencies) and VIRGO (Variability of Solar Irradiance and Gravity Oscillations). The third *SOHO* helioseismology instrument obtains precise oscillation data with high spatial resolution unobtainable from the ground, investigating surface oscillations of relatively small spatial scales (**Fig. 2**). It is called the Solar Oscillations Investigation/Michelson Doppler Imager (SOI/MDI). The three instruments determine the radial distribution of density, pressure, and temperature; establish the depth and latitude variation of rotation and other motions; and determine interior conditions that lead to the development of solar magnetic activity.

GOLF and SOI/MDI employ the Doppler technique for measuring motions of the solar photosphere. When part of the visible surface heaves upward toward the Earth, the wavelength of a spectral line formed in that region is shortened; if the region moves away from the Earth, back toward the solar interior, the wavelength is lengthened.

VIRGO measures variations in the Sun's irradiance, or its total luminous output, with extremely accurate, precise, and stable radiometers. As the Sun fades and brightens, VIRGO obtains a sensitive record of global oscillations, refining knowledge of the physical and dynamical properties of the deep solar interior.

The precise, long-duration measurements from GOLF and VIRGO may also lead to the unambiguous detection of solar gravity waves for the first time. These waves are largely confined to the Sun's energy-generating core, and the force of gravity determines

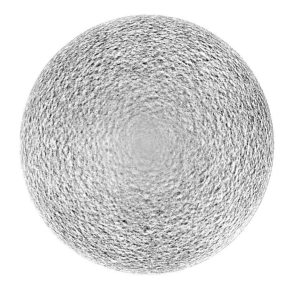

Fig. 2. Average of 30 radial-velocity maps obtained by the SOI/MDI instrument on *SOHO*, showing motions on the Sun's surface. Areas shown bright are moving toward the Earth; areas shown dark are moving away. Large, chaotic areas of horizontal flow, called supergranulation, dominate the view. They are invisible near the center because there the motion is across the view from Earth. (*SOHO MDI/SOI Consortium; P. Scherrer; Bernhard Fleck*)

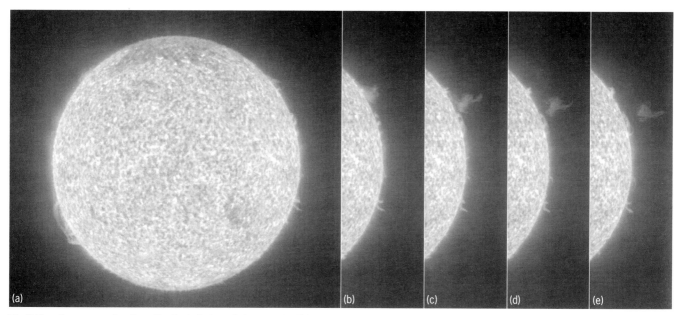

Fig. 3. Prominence erupting from the Sun's limb on February 11, 1996, viewed in extreme-ultraviolet light (wavelength of 30.4 nm) by the EIT instrument. The blob of 60,000 K gas, more than 100,000 km (60,000 mi) long, is constrained by magnetic field lines. (*a*) 16:50 Universal Time. (*b*) 18:05 UT. (*c*) 19:45 UT. (*d*) 20:35 UT. (*e*) 21:50 UT. (*SOHO EIT Consortium; J.-P. Delaboudinière; Bernhard Fleck*)

how quickly they rise and fall. Gravity waves are expected to have long periods, of an hour or more, and to reveal conditions at the very center of the Sun.

Solar atmosphere. Despite more than a half century of investigation, the exact mechanism for heating the corona remains a mystery. Elucidation of this mechanism is a main scientific objective of the *SOHO* mission.

The solar atmosphere is being studied by six *SOHO* instruments. Three of them study the chromosphere and the transition region into the low corona. They are known as SUMER (Solar Ultraviolet Measurements of Emitted Radiation), CDS (Coronal Diagnostic Spectrometer), and EIT (Extreme-Ultraviolet Imaging Telescope). Two *SOHO* instruments examine the middle corona between 1.1 and 10–30 solar radii from the Sun's center. They are known as UVCS (Ultraviolet Coronagraph Spectrometer) and LASCO (Large-Angle Spectroscopic Coronagraph). The remaining SWAN (Study of Solar Wind Anistropies) instrument maps out the corona's large-scale structure.

Five of these instruments detect invisible radiation at ultraviolet or extreme-ultraviolet wavelengths. Since this radiation is partially or totally absorbed in the Earth's atmosphere, it must be observed by using telescopes in space.

SUMER, CDS, and EIT observe spectral lines over a temperature range from 10,000 to a few million degress Kelvin, and determine velocities down to 1 km/s (0.6 mi/s). SUMER and CDS obtain images of the chromosphere and corona with high spatial and temporal resolution (down to 1 second of arc and as brief as 1 s) with a field of view of about 4 minutes of arc; EIT provides full-disk images with coarser resolution (**Fig. 3**). UVCS is an occulted telescope

equipped to measure ultraviolet line intensities and profiles, determining physical parameters of the solar corona from 1.2 to 10 solar radii from the Sun's center with an angular resolution down to 12 seconds of arc. SWAN makes complete sky maps of the hydrogen density in the expanding solar corona.

LASCO uses an occulting disk to mask the Sun's photosphere and view the dim visible sunlight scattered by free coronal electrons. (At 10^6 K, several electrons are set free from each atom, leaving an ion behind.) Since the sky's light confuses such images, the finest detail is obtained from space where the daytime sky is truly black. The LASCO instrument contains three such coronagraphs with nested and overlapping annular fields of view from 1.1 to 30 solar radii from the Sun's center, looking closer to, and further from, the Sun than all previous space-borne coronagraphs.

The coronagraph images provide electron densities, or the number of electrons per unit volume, specifying their global distribution and radial variation. The inner coronagraph also permits high-resolution imaging spectroscopy from 1.1 to 3 solar radii. It measures the intensities and wavelength (Doppler) shifts of visible lines emitted by coronal ions, providing temperature, density, and velocity information that will also be used to understand the currently unknown mechanism for heating the Sun's corona.

Solar wind. The hot solar atmosphere, or corona, is expanding into interplanetary space, filling the solar system with a perpetual flow of electrified matter called the solar wind. Unlike any wind on Earth, the solar wind is a rarefied mixture of protons, electrons, and magnetic fields, streaming radially outward from the Sun. As the corona disperses, it must be replaced by gases welling up from below to feed the solar wind.

Spacecraft have made on-site measurements of the solar wind near the Earth, showing that it has two components, one moving at a relatively slow speed of about 400 km/s (250 mi/s), and a high-speed stream moving at about 800 km/s (500 mi/s). The acceleration and origin of the solar wind are not completely known, and are included as principal scientific objectives of the *SOHO* mission.

Given the high observed temperatures, the ordinary slow-speed wind is a consequence of its expected outward flow. So, the basic mystery for the acceleration of the slow-speed component is the unknown heat source of the corona. No one really knows what gives the high-speed stream its additional push.

The high-speed component of the solar wind apparently squirts out of extended regions of low density and low temperature in the solar corona. These regions, called coronal holes, appear as large dark areas in extreme-ultraviolet or x-ray images, seemingly devoid of radiation. The magnetism in the coronal holes stretches radially outward, providing a fast lane for the high-speed wind. However, the source of the slow-speed wind remains a mystery which, it is hoped, will be solved by future *SOHO* observations.

Coronal remote sensing and on-site experiments onboard *SOHO* will provide a comprehensive data set to study the solar wind from its source at the Sun to the Earth. Two of them, UVCS and LASCO, obtain temperature, density, and velocity information in regions near the Sun where the solar wind is accelerated and has its origin. Three *SOHO* instruments, CELIAS (Charge, Element, and Isotope Analysis System), COSTEP (Comprehensive Suprathermal and Energetic Particle Analyzer), and ERNE (Energetic and Relativistic Nuclei and Electron Experiment), analyze the charged particles in the solar wind passing by the spacecraft.

CELIAS measures the mass, ionic charge, and energy of the low-speed and high-speed solar wind, as well as energetic particles emitted during explosions on the Sun. COSTEP and ERNE form a collaboration to study the energy release and particle acceleration processes in the solar atmosphere as well as particle propagation in the interplanetary medium.

Energizing space near Earth. Fortunately for life on Earth, the terrestrial magnetic field provides a shield from the full effect of the solar wind, deflecting it from the Earth and hollowing a cavity in it. Yet, this magnetic cocoon, called the magnetosphere, is constantly being buffeted, distorted, and reshaped by the variable solar wind, and some of it manages to penetrate the Earth's magnetic shield at its weak points. The Sun thereby feeds a vast and shifting web of energetic particles, electric currents, and magnetic fields that encircle the Earth in space.

The Sun's gusty wind can therefore affect the Earth's environment significantly. It can disturb the Earth's magnetic field, producing geomagnetic storms; create the northern and southern lights (the auroras); disrupt navigation and communication systems; destroy electronics; endanger astronauts; and create electrical power blackouts on Earth. *SOHO's* investigations of the acceleration, evolution, and origin of the solar wind therefore have a direct impact on human activity. Indeed, all of these effects are of such vital importance that national centers employ space weather forecasters and continuously monitor the Sun from ground and space to warn of threatening solar activity. *SOHO* samples part of the solar wind just before it impacts on the magnetosphere, and also looks back at the solar wind's ultimate source, the Sun, in order to identify and analyze the solar features that cause terrestrial effects.

For background information *see* CORONAGRAPH; HELIOSEISMOLOGY; MAGNETOSPHERE; SOLAR WIND; SUN; ULTRAVIOLET ASTRONOMY in the McGraw-Hill Encyclopedia of Science & Technology.   Kenneth R. Lang

Bibliography. K. R. Lang, *Sun, Earth and Sky*, 1995; K. R. Lang, Unsolved mysteries of the Sun—Part I, *Sky Telesc.*, 92(2):38–42, Part II, 92(3):24–28, 1996.

Behavioral genetics

Genetic factors can shape an individual's personality, and can even cause potentially life-threatening psychiatric disorders. This article discusses a receptor gene that influences the personality trait of novelty seeking, and the genetic factors that are responsible for the onset of bipolar disorder.

Determinants of Novelty-Seeking Behavior

Personality or temperament refers to the characteristic style or manner of a person's behavior. Personality is distinguished from two other important determinants of behavior: motivation, or the goals that an individual hopes to accomplish by carrying out certain acts; and the particular skills that an individual uses to accomplish an objective. Personality or temperament can be described as the energy, pace, or persistence that distinguishes an individual's behavior. Personality also includes the emotional aspects of behavior, such as fearfulness, aggressiveness, anger, or self-restraint. Although the role of genetics in determining intellectual skills and motor abilities has been investigated for many decades, the importance of heredity in the determination of personality has only recently been investigated by behavioral scientists and geneticists. Large-scale twin studies carried out since the 1970s demonstrate that 30–60% of the variance between individuals, tested by standard personality instruments, is accounted for by additive genetic factors.

Linkage analysis. Although twin studies support the contention that heredity plays a moderate role in the determination of personality, until recently little was known about the number or nature of the responsible genes. One approach to identifying such genes is linkage analysis, based on the principle that if a gene influences a trait, family members who share a chromosomal region harboring that gene will be more similar with respect to that trait than relatives who have different versions of that region. The advantage of linkage analysis is that it scans large segments of deoxyribonucleic acid (DNA) and does not require prior knowledge of the mode of inheritance or biochemical mechanism of the relevant genes. The disadvantage is that gene polymorphism explains only a small percentage of the differences between individuals.

Allelic association. A second approach for identifying genes that are related to personality or other behavioral characteristics is allelic association, based on the principle that if a gene influences a trait, individuals who share a particular allele variant of the gene should be more similar to one another than are individuals with different alleles. This approach is most powerful when candidate genes that are ''biologically reasonable'' and have known functional polymorphisms are available. The main advantage of allelic association studies is that they are remarkably sensitive to genes with small effect sizes.

Neurotransmitter systems. Which genes are likely candidates for determination of human personality? One of the most widely studied neurotransmitter systems in the brain uses dopamine, and a family of dopamine receptors are known which carry the dopamine signal into the interior of the neuron for further processing. Five known dopamine receptors have been cloned since the late 1980s; the D4 dopamine receptor (*D4DR*) is a good candidate for contributing to behavioral traits since the *D4DR* gene is expressed in limbic brain regions involved in cognition and emotion. Moreover, the third exon of *D4DR* contains a highly variable 48-base-pair repeat sequence coding for 16 amino acids, and the length of the expressed receptor differs in length across individuals and population groups. The size of the *D4DR* receptor polymorphism (different structural forms) in the third exon of the gene ranges from 32 amino acids (2-repeat) to more than 320 amino acids (10-repeat) depending on the particular genetic allele (48–480 base pairs) that the individual possesses. Differences in dopamine binding to its receptor and the accumulation of cyclic adenosine monophosphate (AMP) second-messenger signal have been observed for this receptor repeat. Although the physiological significance of these effects are small, the 7-repeat form of the receptor, when expressed in laboratory ware, mediates a cellular response to dopamine that is blunted relative to the other major forms of *DRD4*, including the 4-repeat form of the receptor. The *D4DR* receptor is therefore a likely component for determining normal personality.

Personality profiles. In one study, personality profile scores were analyzed by using the Tridimensional Personality Questionnaire (TPQ) and *D4DR* exon III alleles in 124 normal volunteers. The TPQ, which was designed to reflect primarily genetic components of temperament (and integrates concepts of neuroanatomical and neurophysiological underpin-nings of behavioral tendencies) measures four dimensions of personality: novelty seeking, harm avoidance, reward-dependent behavior, and persistence. The TPQ is a 100-item self-report (true or false) personality inventory. The data from the TPQ confirmed that novelty seeking is a dopaminergic trait (that involves the use of dopamine as a neurotransmitter). Novelty seeking was significantly associated with the *D4DR* polymorphism, because individuals with the longer 7-repeat allele had higher novelty-seeking scores. The study was quickly confirmed and extended by researchers from the National Institutes of Health (NIH). The convergence of these results is particularly striking considering that the two groups use different personality questionnaires and sampled ethnically disparate populations. In addition, the study by the NIH group included family studies which showed that the association of the long form of the *D4DR* receptor and novelty seeking is the result of genetic transmission rather than that of population stratification.

Individuals with high novelty-seeking scores are described as excitable, impulsive, extravagant, and disorderly. Individuals with low novelty-seeking scores are stoic, rigid, reflective, reserved, and regimented. Novelty-seeking behavior is characterized by exhilaration or excitement in response to novel stimuli.

Dopamine-related behavior. Dopamine was originally hypothesized by behavioral geneticists to underlie novelty seeking because of its pivotal role in the stimulation of euphoria in humans and exploratory approach behavior in animals. Dopamine-deficient individuals with Parkinson's disease were found to be lower in novelty seeking than others but were not abnormal in other aspects of temperament. The imaging technique of positron emission tomography showed that novelty-seeking behavior is correlated with increased brain blood flow in the dopamine-rich corpus striatum (a brain area concerned with sensory information and motor coordination) and with striatal uptake of a dopamine precursor, particularly in the left caudate nucleus (a brain region that is part of the basal ganglia).

Other behaviors that are potentially linked to dopamine and the long allele of the *D4DR* gene are being investigated. A recent study suggests that polymorphism is overabundant in heroin addicts. Many studies show an association between novelty-seeking behavior, as defined by the TPQ, and addictive forms of behavior. Attention deficit hyperactivity disorder (ADHD) affects 3–6% of school-age children, and in most cases its symptoms of inattentiveness, impulsiveness, and hyperactivity can be ameliorated by administration of dopaminergic drugs such as methylphenidate or amphetamine. It was recently shown that the *D4DR* polymorphism is associated with this disorder in a group of American children.

Allelic contributions to novelty seeking. The results from a genetic association study suggests that 5–10% of the genetic variation in the personality dimension of novelty seeking is accounted for by carrying the long form of the *D4DR* exon III allelic polymorphism. Since about 50% of the differences between individual scores on the TPQ are due to genetic effects, the reported findings suggest that several other genes are also contributing to novelty-seeking behavior. *D4DR* polymorphism results in a small gene effect, and the group of individuals carrying this particular long form of the gene score higher (by about one-half of a standard deviation) on novelty seeking than the group not carrying this gene. Moreover, not all individuals carrying the gene score high on novelty seeking, and some people scoring high on novelty seeking do not carry this particular gene.

Richard P. Ebstein

Genetics of Bipolar Disorder

Bipolar disorder (BP), also known as manic-depressive illness, is a severe psychiatric illness characterized by fluctuating mood episodes in the form of major depressions and manic or hypomanic episodes. This common illness has a prevalence of approximately 1%, a mean age of onset in the third decade of life, equal proportions of men and women affected, and 10–15% increased risk of death from suicide. The major depressive episodes consist of 2 weeks of distressing or impairing depressed mood or loss of interest along with four accessory symptoms such as suicidality or irregular appetite, sleep, activity, energy, self-esteem, or concentration. Manic episodes consist of 1 week of abnormally elevated, expansive, or irritable moods sufficiently severe to cause marked impairment, accompanied by three accessory symptoms such as grandiosity, decreased need for sleep, overtalkativeness, racing thoughts, distractibility, increased activity, or excessive risk taking. Hypomanic episodes are less severe but otherwise similar to manic episodes. In order to meet the diagnostic criteria for bipolar disorder, an individual must have had a history of one or more manic episodes (BPI), or a history of one or more hypomanic episodes and of one or more major depressive episodes (BPII).

Family studies. Studies have been carried out for family members in general, for twins, and for adoptees.

Familial clustering has been demonstrated in bipolar disorder by a two- to threefold increased rate of mood disorders (including bipolar disorder and recurrent major depressions) in family members of bipolar disorder probands (the initial family contact) compared to family members of normal control probands. Relatives with recurrent major depressions without any evidence of a clinical history of manic or hypomanic episode, often referred to as recurrent unipolar disorder, are overrepresented in the families of bipolar disorder probands. The more closely genetically related a family member is to the affected proband, the greater is the risk of that family member having the illness. For example, a child of a bipolar proband has a greater risk for developing the illness than a cousin has.

Twin studies can demonstrate that a trait or illness has a genetic component. Identical (monozygotic) twins share all of their genes identically by descent from the same parent, whereas fraternal (dizygotic) twins share half of their genes, just as any other sibling pair would. Studies have shown that there is a 60–70% concordance rate (diagnosis common to both twins) for bipolar disorder in identical twins and only 20% in fraternal twins, indicating that susceptibility to this disorder is largely genetic. Recurrent unipolar disorder was overrepresented in identical co-twins of bipolar probands. Since the concordance rate for identical twins is not 100%, nongenetic (namely, environmental) factors must also play an important role. Examples of environmental factors which might affect concordance rates include exposure to certain stresses, toxins, and developmental insults.

Adoption studies attempt to separate genetic, prenatal, and perinatal events from postnatal environment as contributors to illness by examining the rate of illness in biological relatives of affected adoptees compared to the rate of illness in biological relatives of control adoptees. There have been few studies in this area, but findings are generally consistent with a large genetic influence in the transmission of bipolar disorder.

Complex genetic disease. Attempts to match epidemiologically observed transmission patterns of an illness to known modes of inheritance are referred to as segregation analyses. Segregation analyses have been unable to demonstrate the transmission of a single highly penetrant major gene leading to bipolar disorder, thus putting it in the category of complex genetic diseases which do not exhibit classic Mendelian recessive or dominant inheritance. There are several other complexities which may apply to the genetics of bipolar disorder. Genetic heterogeneity may be present, meaning that mutations in any one of several genes may lead to the same phenotype (observable characteristic). Alternatively, the genetic mechanism could be caused by oligogenic inheritance, in which mutations in multiple genes are required in order for the disease phenotype to be expressed. Another possibility is that the disease-causing alleles may have a high frequency in the population. This can interfere with traditional linkage analysis by introducing families with the same gene being passed down through both maternal and paternal lines. Other genetic mechanisms such as anticipation (increasing genetic predisposition in succeeding generations) and a parent-of-origin effect have been postulated in bipolar disorder.

Positional cloning. Positional cloning represents the most plausible approach to identifying susceptibility genes for bipolar disorder, because there is no strong candidate gene known to underlie the pathophysiology of this illness. In positional cloning, the location of the gene on a chromosome is first determined, and then the susceptibility gene is detected within that narrowed region. A wide variety of polymorphic genetic markers are examined across all of the chromosomes in order to blanket the entire genome in a search for an area of a particular chromosome region where the illness and the markers cosegregate (travel together through the generations) in the families studied. The segment of a positional cloning strategy is referred to as genetic linkage analysis.

After a particular area of a chromosome is found to contain markers cosegregating with illness and the genetic region is narrowed down to a tractable size, the effort is made to isolate genes in this chromosomal area and to search for the one which exhibits mutation in affected individuals compared to unaffected controls.

Meiotic recombination. The key biological phenomenon exploited in linkage analysis is meiotic recombination or crossing over during which both the maternally and paternally derived chromosomes lie in proximity and undergo exchange of genetic material between the homologous chromosomes (for example, between the paternally derived chromosome 1 and the maternally derived chromosome 1). Genes and other genetic markers that are close together are less likely to be separated by this process than are those that are farther apart. Therefore, they are usually inherited together by the progeny cells and are said to be genetically linked. The two general types of linkage analysis are parametric logarithm of the odds ratio (LOD) score analysis and nonparametric allele-sharing methods.

Parametric analysis. Parametric LOD score analysis involves testing the likelihood of the observed markers and illness in a family, under a model of inheritance, when linkage is present versus the likelihood when such linkage is lacking. Assumptions are made because the correct model of inheritance is unknown for bipolar disorder. As an attempt to increase the chance of specifying the correct genetic model, several are often tested on the same data set by varying parameters such as mode of inheritance, gene frequency, and alternative diagnostic categories.

Nonparametric allele-sharing methods. Nonparametric allele-sharing methods are exemplified by the affected sibling pair (ASP) method, where the frequency with which a genetic marker allele (or variant) is inherited from a particular parent (referred to as identical by descent, or IBD) in a pair of siblings both affected with illness is known. The presence of a disease-causing gene is revealed by exhibition of more than the expected 50% identical-by-descent allele sharing between affected siblings. While LOD score analysis is referred to as parametric, allele-sharing methods are referred to as nonparametric (model independent). Hence, affected sibling pair methods are more robust than traditional linkage methods in mapping complex diseases because no single model of inheritance is assumed.

Confirmation of linkage in bipolar disorder. The first replicated report of linkage using allele-sharing methods in bipolar disorder was to markers in the pericentromeric region of chromosome 18. Initial reports from a pedigree series of 22 families indicated that an independent replication using the same methods was achieved in an independent pedigree series of 30 families. It has been shown that much larger sample sizes are needed to consistently replicate specific linkage findings in oligogenic diseases. This is due to the initial reports relying on finding any one of many susceptibility genes, and replication efforts focusing on finding evidence of only that one particular susceptibility gene.

Diagnosis and treatment. Prognosis for individuals afflicted with bipolar disorder is favorable with proper diagnosis and treatment. Diagnosis of bipolar disorder has primarily derived from clinical observations, with family studies providing some assistance in limiting the diagnostic boundaries between bipolar disorder and other illnesses. The main medications used to treat this illness are lithium, carbamazepine, valproic acid, and antidepressants.

Long-term goals. As the evidence solidifies around certain areas of particular chromosomes as containing susceptibility genes for bipolar disorder, there will be a predictable shift of effort toward narrowing down the large linkage regions. The primary methods for achieving this will likely be larger sample size through increased recruitment of suitable families and pooling of data through multicenter collaborations.

For background information *see* AFFECTIVE DISORDERS; BEHAVIOR GENETICS; COMPUTERIZED TOMOGRAPHY; DEVELOPMENTAL BIOLOGY; GENE ACTION; GENETICS; HUMAN GENETICS; LINKAGE (GENETICS); MOTOR SYSTEMS; RECOMBINATION (GENETICS); TWINS (HUMAN) in the McGraw-Hill Encyclopedia of Science & Technology.   Alan R. Sanders; Elliot S. Gershon; Sevilla D. Detera-Wadleigh

Bibliography. American Psychiatric Association, *Diagnostic and Statistical Manual of Mental Disorders: DSM-IV*, 4th ed., 1994; J. Benjamin et al., Mapping personality traits to genes: Population and family association between the D4 dopamine receptor and measures of novelty seeking, *Nat. Genet.*, 12:81–84, 1996; R. P. Ebstein et al., *D4DR* exon III polymorphism associated with the personality trait of novelty seeking in normal human volunteers, *Nat. Genet.*, 12:78–80, 1996; F. K. Goodwin and K. R. Jamison, *Manic-Depressive Illness*, 1990; H. I. Kaplan, B. J. Sadock, and J. A. Grebb, *Kaplan and Sadock's Synopsis of Psychiatry: Behavioral Sciences, Clinical Sciences*, 7th ed., 1994; M. Kotler et al., Excess dopamine D4 receptor (*D4DR*) exon III seven repeat allele in opioid dependent subjects, *Amer. J. Hum. Genet.*, 59(suppl.):A92, 1996; G. J. LaHoste et al., Dopamine D4 receptor gene polymorphism is associated with attention deficit hyperactivity disorder, *Mol. Psychiat.*, 1:121–124, 1996; E. S. Lander and N. J. Schork, Genetic dissection of complex traits, *Science*, 265:2037–2048, 1994.

Biological clocks

Virtually all eukaryotes have the capacity to organize their metabolic lives on a daily basis. The resulting rhythms in metabolism and behavior are known as circadian rhythms, and the mechanism underlying this rhythmicity as the biological clock. In various human systems, circadian clocks dictate the timing of cell division, regulate reproduction, and contribute to the regulation of almost every measurable physiological parameter. Biological clock malfunction is linked to a variety of psychiatric disorders, including seasonal affective disorders and some forms of manic-depressive illness.

Cellular clocks. There are three aspects to all circadian systems: input, the oscillator, and output. The input refers to how signals from the environment are perceived by an organism and how these signals act to reset the internal biological clock. In all organisms the chief resetting agent is light, although temperature changes are also universally effective in synchronizing internal circadian clocks with the external world. The oscillator refers to how the biological timekeeper actually works, that is, the biochemical functions of the component parts (genes, enzymes, ion channels, and membranes), and how these components are assembled within the cell to form an accurate timekeeper. The output refers to how time information that is generated by the oscillator is transduced to change the behavior of the oscillator cell and ultimately (in a multicellular organism), to change the behavior of the organism. The concept of an oscillator allows a theoretical distinction between the molecular gears and cogs which actually describe the mechanism, and the parts of a cell or organism that are specifically devoted to synchronizing the rhythm with the environment (input) or expressing the rhythm (output).

Genetic variants. The use of genetic variants has been central in unraveling the biological clock. Genetics was first applied to *Drosophila* (a fruit fly) and *Neurospora* (a fungus) where approximately 12 genes were identified as encoding potential clock components. Although clock mutants have subsequently been identified in a number of organisms, including algae (*Chlamydomonas*), higher plants (*Arabidopsis*), cyanobacteria (*Synechococcus*), and rodents (mice and hamsters), nearly all that is known about the molecular mechanism of the clock has arisen from work on the *frequency* (*frq*) gene of *Neurospora* and the *period* (*per*) and *timeless* (*tim*) genes of *Drosophila*. In each organism, molecular genetics was used to clone the corresponding genes. Studies of the regulation of these genes suggested that there is a mechanism through which the clocks are assembled. In both cases the evidence is consistent with transcription/translation-based negative-feedback-loop oscillators that involve the rhythmic expression of ribonucleic acids (RNAs) that encode proteins that act to shut off the genes that encode the proteins. Thus, the *frq* gene encodes the corresponding protein (FRQ) which acts to shut off *frq* in *Neurospora*, and the *per* and *tim* genes encode PER and TIM, which act together to shut off *per* and *tim* in the fly (see **illus.**).

Circadian cycle in Neurospora. For example, the *Neurospora* clock cycle may be presumed to begin at

midnight. The *frq* and FRQ levels are low, but *frq* RNA is beginning to rise (a process that will take about 10–12 h to reach peak levels). Following a 2–4 h lag, two forms of FRQ protein (a long and a short form) begin to appear. After a time, FRQ enters the nucleus, where it acts to suppress the expression of its own and probably other genes—a process that begins quickly and can be complete within 3–6 h of when *frq* RNA is first expressed. As soon as either form of FRQ can be observed, it is partially phosphorylated. By midday, *frq* RNA is at peak levels, and the level of partially phosphorylated FRQ is rising. Since FRQ indirectly regulates its own expression, it is likely that it also affects the expression of other genes (*ccg*'s, or clock-controlled genes), thereby providing a mechanism whereby time information might leave the clock to organize the behavior of the cell and the organism as a function of time. After noon, *frq* RNA levels fall; and FRQ, becoming extensively phosphorylated, declines in the late afternoon and early evening hours, consistent with a model in which the phosphorylation triggers its turnover. This describes in broad outline the negative feedback loop of the clock.

Theory predicted that light would modify circadian clocks by causing rapid changes in the level or activity of a central clock component, which was confirmed by analysis of *frq* regulation. Within minutes, light causes a substantial rise in the level of *frq* transcript, and the dose response for this light induction matches the dose response for clock resetting. The machinery for perceiving light works at all times of day in all organisms, and gets the signal to the pertinent clock cells at any time of day. However, the unidirectional molecular light responses are interpreted by the dynamics of the feedback-loop oscillator, and serve to advance the clock into the next day when seen in the late night, and to delay the clock back to the previous day when seen during the early evening: same signal (light), different response. The response of *frq* to light indicates that late-night advances result from a precocious rise in *frq* and FRQ to midday levels, while early-evening delays result from the (light-induced) return to high midday *frq* levels. Two global regulatory genes, *white collar-1* (*wc-1*, a transcription factor) and *wc-2*, govern all known light responses in *Neurospora*, including the induction and maintenance of *frq* RNA levels in light. Surprisingly, at night, *frq* levels are also low in *wc-1* and *wc-2* mutant strains (novel arrhythmic clock mutants) which display no overt rhythms. The working model demonstrates that *frq* is normally activated at night by the WC-1 protein acting perhaps in concert with the WC-2 protein. Since *frq* can always be induced by light, WC-1 is probably always there, but not necessarily always active.

Circadian cycle in Drosophila. During the midday (or subjective day) the *Drosophila* circadian oscillator remains in a state of rest but poised for activation. During the afternoon and early evening, transcription of *per* and *tim* begins so that by dusk there is

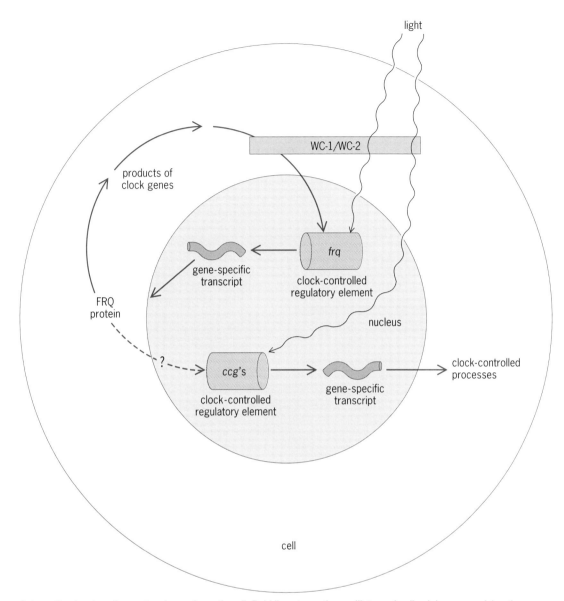

Schematic of a circadian system in a eukaryotic cell. Bold lines trace the oscillator, a feedback loop comprising the products of clock genes. With the exception of the FRQ protein, nothing is known about where in the cell these might act. Light acts through the *wc* genes and resets the clock by inducing the *frq* gene; the *wc-2* gene is also required for rhythmicity in the dark. Genes that produce gene-specific transcripts (wavy lines) eventually direct the synthesis of proteins that are involved in different cellular processes. Clock-controlled genes (*ccg*'s) are output genes, regulated by the clock but not part of the feedback loop. (Adapted from S. C. Crosthwaite, J. J. Loros, and J. C. Dunlap, Light-induced resetting of a circadian clock is mediated by a rapid increase in frequency transcript, Cell, 81:1003–1012, 1995)

already a substantial amount of transcript from these genes in the cytoplasm. The amount of transcript for both genes rises together, but not the proteins. The proteins PER and TIM interact in a 1:1 heterodimeric complex. TIM, although unstable in the light, is required to stabilize PER; thus, the complex can accumulate only when TIM is present. After several hours in the dark the levels of PER:TIM rise, and the complex (perhaps acting through other proteins) enters the nucleus and turns off *per* and *tim* expression. Within a few hours after the peak in PER:TIM, when levels of both are still substantial, transcription of each gene restarts.

Light entrains this feedback loop in a manner that provides an interesting variation as compared to *Neurospora*. Light also acts quickly to change the level of a central clock component, although the primary clock-specific response is not induction (as with *frq*) but light-enhanced turnover of a clock protein. The protein TIM is light sensitive; therefore its levels are kept low in light which prevents PER from accumulating because it requires TIM to stabilize it. When the lights go off, TIM accumulates because its messenger ribonucleic acid (mRNA) is already there. When the clock is cycling in constant darkness, delays occur during the early evening when light effects the degradation of TIM. Once TIM:PER gets into the nucleus and its action is nearly complete, light advances the clock by accelerating TIM turnover. TIM is the initial response element in this clock, with the changes in PER appearing only as secondary responses.

For background information *see* BIOLOGICAL CLOCKS; CELL (BIOLOGY); EUKARYOTAE; GENETICS; MOLECULAR BIOLOGY; PROTEIN in the McGraw-Hill Encyclopedia of Science & Technology. Jay C. Dunlap

Bibliography. B. Aronson et al., Negative feedback defining a circadian clock: Autoregulation in the clock gene *frequency*, *Science*, 263:1578–1584, 1994; S. C. Crosthwaite, J. J. Loros, and J. C. Dunlap, Light-induced resetting of a circadian clock is mediated by a rapid increase in *frequency* transcript, *Cell*, 81:1003–1012, 1995; J. C. Dunlap, Genetic and molecular analysis of circadian rhythms, *Annu. Rev. Genet.*, 30:579–601, 1997; K. Siwicki, Circadian rhythms, *Semin. Neurosci.*, 7:1–70, 1995.

Biotechnology

Two recent developments in biotechnology have the potential for far-reaching applications. One is the formation and organization of genome resource banks for conserving rare bioresources and promoting biodiversity within a specific ecosystem as well as throughout the global community. The other is the deoxyribonucleic acid (DNA) microchip which accelerates DNA hybridization in multiple assays in order to rapidly detect changes in genetic sequences associated with disease.

Genome Resource Banks

Biodiversity, the variety of biota on the Earth, helps to regulate climate, absorb pollutants, safeguard watersheds, and maintain soils. It transforms solar energy into the biochemicals allowing life to proceed for all species. Humans depend on diverse organisms for food, medicine, chemicals, fiber, clothing, structural materials, energy, and recreation.

Biodiversity crisis. At the 1992 Earth Summit in Rio de Janeiro, Brazil, the Convention on Biological Diversity formally recognized the global biodiversity crisis and the importance of maintaining biological resources in nature and in zoos and botanical gardens. More than 100 countries were parties to the Convention, and have begun debating about genetic resources, including intellectual property rights, regulations governing collection and dissemination of biomaterials, and equitable sharing of benefits with indigenous communities. The parties have established that genetic resources lie within the sovereign jurisdiction of individual nations. The Convention's language is broad, and no specific statements are given on accessing, transferring, using, and sharing benefits from genetic resources. Rather, the Convention calls on the parties to agree to mutual terms, practice informed consent, and ensure that national legislation on intellectual property rights are supportive of the Convention objectives. Although some countries are moving toward formulating legal policies to regulate use of bioresources, few models exist, and enforcement of regulations will remain weak for some time.

Protecting biodiversity. The preferred approach for safeguarding biodiversity is preserving entire ecosystems that contain a rich array of species and populations. Unfortunately, this strategy leaves 96% of the Earth's biodiversity unprotected and often deprives local people of user rights over forests, waters, fisheries, and wildlife. There is a need to continue emphasizing habitat conservation while implementing other levels of protection. Some taxa can be conserved in zoos and botanical gardens, but enough individuals must be maintained to ensure sufficient genetic diversity to keep these populations healthy. Because of space limitations, only about 1000 animal species can be conserved in modern zoos.

Genome Resource Banks (GRBs) are one approach to assisting in the conservation of biodiversity and genetic diversity. A GRB is a repository of biomaterials that may include germ plasm (sperm, oocytes), embryos, tissue, blood products, waste products (feces, urine), saliva, milk, and deoxyribonucleic acid. Because GRBs are based on the organized collection, storage, and use of biomaterials, this approach could provide (1) a high degree of security against the loss of species or populations from catastrophes, such as a disease epidemic or flood; (2) a powerful method for exchanging genes among populations, thereby minimizing effects of inbreeding and genetic drift; and (3) biomaterials for research in genetics, disease, reproduction, and other life sciences. A GRB is not a replacement for the natural environment; it simply is another means of assisting in preserving and protecting available resources.

Wildlife biomaterials. In 1994 a canine distemper epidemic killed as many as 300 lions in the Serengeti ecosystem in Tanzania. The availability of cryopreserved serum from this population before, during, and after the epidemic allowed epidemiologists to identify the onset and cessation of the disease and the likely vector. Stored blood products have been useful for refining taxonomic classifications and phylogenetic relationships, measuring genetic variation, and resolving paternity issues in a wide variety of species, including crocodiles, giant pandas, and whales. Assisted reproductive techniques, especially artificial insemination, are helping manage endangered species in zoos. There now are many examples of producing offspring by this means, including the wolf, black-footed ferret, cheetah, ocelot, giant panda, addax, Eld's deer, bison, eland, gaur, chimpanzee, baboon, and marmoset.

Genome banks for plants and animals. The Consultative Group on International Agricultural Research (CGIAR) represents a consortium of donors that supports 17 International Agricultural Research Centers (IARCs) in developing countries. The Centers' gene banks hold more than 500,000 accessions of crop germ plasm from more than 3000 species. CGIAR has signed an agreement with the United Nations Food and Agricultural Organization (FAO), placing these collections into a trust. CGIAR centers do not seek intellectual property rights or claim ownership, but manage these germ plasm accessions for the benefit of the international community. The Food and Agricultural Organization assists in making germ plasm available to users.

Gene banks for livestock largely involve the commercial distribution of sperm and embryos from genetically superior individuals to improve growth efficiency and meat or milk production. There is movement to develop a global program for managing farm animal genetic resources under the guidance of the Food and Agricultural Organization. Gene banking also has become important in biomedical research, because many new animal genotypes have been developed to better study human diseases. The value of cryotechnology has been recognized by the Jackson Laboratory (Bar Harbor, Maine) and the National Institutes of Health (Bethesda, Maryland), which now store embryos and sperm from genetically valuable rodents. Customized collections also are in place for microorganisms used in environmental, food, and biomedical industries to produce valuable products. The American Type Culture Collection (Manassas, Virginia) maintains more than 80,000 cultures of algae, protozoa, bacteria, bacteriophages, cell lines, hybridomas, fungi, yeasts, recombinant DNA materials, viruses, and plant tissue cultures.

Organizing GRBs. Despite the many advantages of genome banks for wild biota, there is no large-scale scheme to ensure that biomaterials collection, storage, and use occur in an orderly fashion that enhances conservation. The Conservation Breeding Specialist Group of the International Union for the Conservation of Nature–World Conservation Union's Species Survival Commission serves as a neutral catalyst and facilitator for conservation planning for animals, plants, and their habitats. The Conservation Breeding Specialist Group has a global network of 700 members, and has coordinated species and habitat recovery workshops in 45 countries. One of its objectives is to develop new tools for conservation, such as wildlife genome banks.

Community-based genome banking. In the future, it may be possible to build megalithic structures that hold a vast array of living plant and animal biomaterials. Rather than a ''top-down'' administrative approach, the Conservation Breeding Specialist Group has advocated a grass-roots strategy for wildlife GRBs. It appears unrealistic to establish centralized regional or national repositories because of expense and legal and safety issues. A ''bottom-up'' approach is more compatible with modern approaches to conservation in which wildlife authorities, managers, curators, and researchers work in partnerships at community-based levels. The result is a multidisciplinary, science-based approach to solving problems. For GRBs, there is support for preserving rare biomaterials at the level of a species or taxon—an attainable objective because many species are monitored, and in some cases managed, by taxonomic specialist groups at the level of the World Conservation Union

or by regional zoo associations. The World Conservation Union's Species Survival Commission comprises about 100 taxon-based groups, dealing with organisms ranging from African elephants to water beetles to fungi. Many regions also have sophisticated species management committees (for example, North America supports more than 70 Species Survival Plans). It also has been concluded that conventional crop seed banks could diversify activities to include rare plant species, and museums could expand collections to include frozen but living specimens.

Creating a GRB. The first step toward developing a GRB is agreement by species stakeholders that a repository can contribute to conservation. Then a GRB Action Plan is advocated; this written document explicitly justifies the repository while providing information on species biology, accessibility to living specimens, type and amount of biomaterials to collect, technical aspects of collection, storage and use, research needs and accessibility, and ownership and funding of the resource.

Sampling. Modern genetic and population biology theory is used to determine which individuals from a given species should be selected for sampling, and how many sperm are needed from a given male to ensure that his genes are available for future generations. Recent calculations have confirmed the usefulness of a GRB for perpetuating genetic variation in natural and zoo-maintained populations. It also has been determined that three banks should be developed for each species: an ''in-perpetuity'' repository (for use only when the species approaches extinction); another repository for routine management of living animals; and another as a research resource for scientists.

Ownership of biomaterials. Five recommendations have emerged from Conservation Breeding Specialist Group workshops on the right and title to rare biomaterials: (1) Nations have sole responsibility for determining the ownership and value of their genetic resources within international and legal limits. A country always has the option of declining to provide samples to any GRB program. (2) Biomaterials ideally should be donated to each respective GRB program, thereby uncoupling the resource to commercial interests. There is reluctance to commercialize genetic materials, fearing that putting a price on such materials could encourage illegal trade in endangered species. At the same time, these biomaterials have potential for helping local communities. It is especially important that developing countries receive some type of compensation, such as technology transfer, educational outreach, or equipment or cash for specific conservation programs. (3) Some proportion of every collection should be placed in a central GRB within the region in which it is collected, and it should be suitably regulated. (4) Ownership of germ plasm and embryos should remain with the owner, unless given to a species coordinator or management committee. Decisions about apportionment of offspring resulting from the use of these biomaterials should be made by the owners of the donor and of the recipient at the time of need. (5) The right and title to all other biomaterials (tissues, blood products, DNA, and other bodily fluids or excretions) should be given by the owner of the donor to a regional coordinator or species specialist group.

Accessibility to biomaterials. In general, collectors of the genetic resources are stewards only, releasing biomaterials only upon the request of coordinators of specialist groups and management committees. For germ plasm or embryos, the coordinator should receive approval from the owner that release is granted and that ownership of resulting offspring has been negotiated. The coordinator supplies the GRB facility with precise information needed to safely ship biomaterials to the proposed user. Given sufficient amounts, biomaterials should be provided for all legitimate purposes, including routine management of genetic diversity, and applied and basic research. Requesters should, however, provide scientific justification for proposed use.

Other concerns. Prominent among other issues is the funding to support the development, long-term maintenance, and use of stored biomaterials. Even conventional crop seed banks have difficulty in securing financial support. Accurate identification and tracking of biomaterials require the development of new computer programs. There also is a need to set standards for safe monitoring and quality control of these specialized accessions. A high priority is developing policies, procedures, and cooperative arrangements that ensure safe acquisition and use of biomaterials without transferring pathogens that could contaminate wild or domestic stocks.

Goals. Creating a protected area is not the only means of preserving and protecting biodiversity. Partnerships among wildlife authorities, managers, scientists, and local communities are key, and novel approaches such as collecting, storing, and using biomaterials can contribute by offering insurance, basic knowledge, and resources to solve practical problems. Organized Genome Resource Banks are needed that have clear delineations about ownership, accessibility, and use for the ultimate goal of conserving biodiversity and genetic diversity.

David E. Wildt

DNA Microchips for Genetic Analysis

DNA microchips may revolutionize the way that many important tests and assays are performed in clinical, diagnostic, and biomedical research laboratories. The evolution of DNA microchips has been brought about by a synergistic combination of diverse disciplines which include microelectronics, microfabrication, organic chemistry, molecular biology, and genetics. DNA microchips and other unique devices (integrated microlaboratories) will have a variety of applications in the clinical laboratory for the diagnosis of cancer, infectious diseases, and genetic diseases. In addition, these new microelectronic chips and devices will be used in human genome work, drug discovery, biomedical research, and many other research areas. Significant efforts by academic and industrial workers are being carried out in this area. Companies are developing biochip arrays with large numbers of DNA test sites by using photolithographic combinatorial synthesis techniques. Electronically active DNA chip devices, which use electric fields to rapidly transport DNA and accelerate the DNA hybridization reactions occurring on the chip surface, are being developed. The Genosensor consortium is involved with various techniques for improving the detection of hybridization on DNA chips. Other groups are developing a technique called "sequencing by hybridization" in microchip formats.

Background DNA diagnostics. The basic DNA chip is usually an array of microscopic test sites on a silicon or glass substrate material. Each microscopic test site has several million specific sequence DNA molecules attached to its surface. The DNA sequences at each site are made complementary to the target DNA sequences in the test sample. The DNA sequences on the test sites can be of natural or synthetic origins. These DNA sequences consist of the complementary base sequences that selectively bind the target DNA sequences in the sample [adenine (A) binds to thymine (T), and guanine (G) binds to cytosine (C)]. The target DNA sequences are the unique genetic sequences in bacteria, virus, oncogenes (cancer genes), or genetic disorders which allow a particular microorganism or disease to be identified and diagnosed. For example, the sequence GTG-GGC-GCC-GGC-GGT-GTG-GGC represents a section of the *H-ras* oncogene found in normal human cells. In cancerous cells, a single-point mutation frequently occurs in codon 12 [the twelfth group of three nucleotide sequences of messenger ribonucleic acid (mRNA)], where the G base is replaced by a T base; the sequence now becomes GTG-GGC-GCC-GTC-GGT-GTG-GGC. DNA diagnostic assays are designed

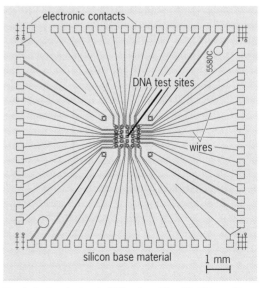

Fig. 1. Active electronic DNA microchip with 25 test sites (80-μm-diameter microlocations) on a 1-cm-square silicon base substrate.

to detect these types of changes in a genetic sequence which relate to a disease. The process by which the complementary DNA probe sequence on the chip test site and the target DNA sequence bind together is called hybridization. The specific complementary DNA strands hybridize and form a DNA double-helix structure via the base-pairing properties. The specific genetic code in each DNA strand and hybridization represent the most important aspects of information storage and processing in all living organisms. In an actual DNA diagnostic assay procedure, a reporter (signal) group, which allows the hybridization reaction to be detected with high sensitivity, is incorporated into the DNA sequences. Reporter groups can be fluorescent molecules, chemiluminescent molecules, enzymes, or radioisotopes. A reporter group is usually chemically attached to the DNA probe sequence (in some assays the target DNA is labeled with the reporter group). When a sample is tested and the DNA probe hybridizes with a target DNA molecule, the reporter group provides a detectable signal. This reporter group signal (fluorescence, chemiluminescence, color, or radiation) both identifies and quantifies (provides a measure of the amount of) that particular target DNA (mutant oncogene, bacteria, virus) which is present in the test sample.

DNA chip design and fabrication. An advanced form of these DNA chips is the active electronic DNA chip device. This device is a multisite, electronically controlled array of independently addressable test areas, each capable of attracting, binding, and repelling DNA under specific conditions of charge polarity, current, and voltage. These microchips take advantage of the principles of electrophoretic transport of charged molecules in an electric field, but on a very miniaturized scale. The chip is designed and constructed on silicon/silicon dioxide (Si/SiO_2) substrate materials with standard photolithography techniques common to the microelectronics industry. Chips with 5×5 arrays of 25 circular microlocations and 8×8 arrays with 64 microlocations have been developed. DNA arrays with fewer test sites will be used for infectious-disease diagnostic panels, while devices with larger arrays (hundreds of test sites) will be used for genetic disease and genomic applications. **Figure 1** is a schematic representation of the 25-addressable-microlocation electronic DNA chip. The 25 microlocations are positioned over independently controlled platinum microelectrodes. The microlocations are 80 micrometers in diameter, and the four auxiliary control electrodes are 100 μm in diameter. The radiating lines are gold wires which connect the microelectrodes to electrical contact pads on the perimeter of the device. The contact wires are covered with insulating layers of silicon dioxide and silicon nitride.

Electronic DNA hybridization analysis. Active DNA microchips are able to control the transport of charged reagent and analyte molecules on the surface of the device. Since DNA molecules in solution carry a net negative charge, they can be moved in an electric field to an area (microlocation) which is biased posi-

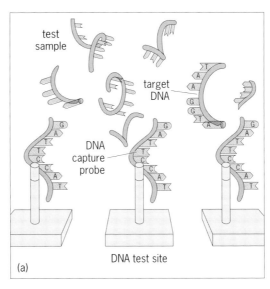

(a)

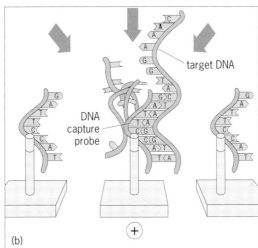

(b)

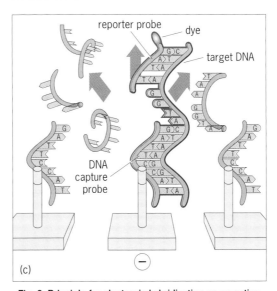

(c)

Fig. 2. Principle for electronic hybridization on an active electronic DNA chip device. (*a*) Step 1: application of target DNA onto a DNA chip which contains test sites for identifying specific DNA sequences. (*b*) Step 2: biasing a specific test site at a positive potential, which causes transport, concentration, and accelerated hybridization to occur at this site. (*c*) Step 3: application of a negative potential which removes the nonspecifically bound DNA, improving the specificity and sensitivity of the assay.

tive. Conversely, DNA is repelled and transported away from negatively biased microlocations. In an assay format, target DNA sequences in a sample solution can be concentrated in a short period of time at a positively biased microlocation. If a complementary DNA capture probe is attached to the microlocation, hybridization of the target DNA can occur. The electric field can be used to produce a concentrating effect which facilitates the hybridization process because of the law of mass action (the rate at which hybridization proceeds is directly proportional to the product of the reaction of the microlocation and the complementary DNA capture probe). The directed transport and addressing process can be carried out simultaneously at test sites which have different DNA capture sequences, permitting rapid multiplex tests on a single sample.

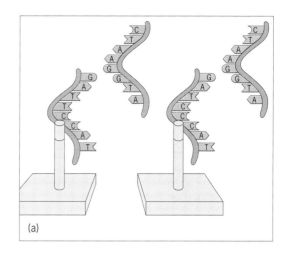

More important to overall hybridization specificity and sensitivity is the reverse process, in which DNA is repelled from the test site by reversing the field polarity. This parameter is called electronic stringency control, and the three-step process allows nonspecific sample DNA and unhybridized probes to be selectively removed from the specific microlocation, leaving the hybridized target DNA (**Fig. 2**). In step 1, the sample DNA (containing the target DNA sequence) is placed onto the chip which contains test sites for identifying specific target DNA sequences (Fig. 2a). The figure shows only one DNA test sequence chemically attached to each test site; however, on an actual device millions of DNA molecules are chemically attached. Step 2 involves biasing a specific test site (or sites) at a positive potential, which activates the transport, concentration, and an accelerated hybridization process (Fig. 2b). After hybridization is complete, step 3 is carried out by applying a negative potential to the test site in order to detach the nonspecifically bound DNA (Fig. 2c). This step significantly helps to improve the specificity and sensitivity of the hybridization assay.

Point mutation analysis. The ability to detect point mutations in a DNA molecule can be important in clinical diagnostics, particularly for many types of cancers and genetic diseases. A single-base mismatch in an oncogene (for example, the *H-ras* oncogene) can be an indicator of a potential cancer. Many other cancer oncogenes (such as *myc* and *P53*) and genetic diseases (such as sickle-cell anemia and thalassemia) are due to single point mutations. A main application of DNA microchips will be in the area of point mutation analysis. With these active electronic DNA microchips, precise control of the electric field is used to aid in the discrimination of the match and the mismatch DNA sequences (**Fig. 3**). Active DNA chips have been used to discriminate both one-base (G → A) and two-base (G → A and G → T) mismatches in the *H-ras* oncogene sequences. The electronic DNA chips can carry out these types of discriminations in less than 5 min, compared with 1–2 h for conventional hybridization diagnostics.

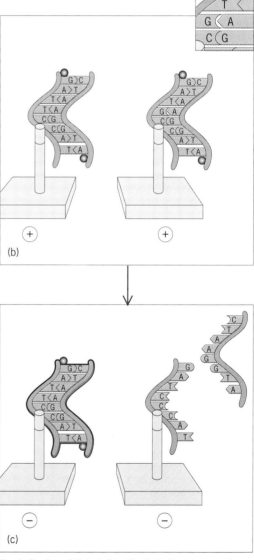

Fig. 3. Single-point-mutation discrimination on an active electronic DNA chip device. (*a*) The DNA to be tested (*b*) is first hybridized to two test sites which contain DNA sequences differing by one base pair. (*c*) The sites are biased at a specific negative electric potential, which removes the DNA from the site with the one-base mismatch, thus identifying the nature of the sequence in the target DNA.

For background information *see* BIOTECHNOLOGY; CANCER (MEDICINE); CONSERVATION OF RESOURCES; DE-OXYRIBONUCLEIC ACID; GENETIC ENGINEERING; MICRO-COMPUTER; MUTATION; PHYLOGENY in the McGraw-Hill Encyclopedia of Science & Technology.

Michael J. Heller

Bibliography. W. Bains and G. G. Smith, A novel method for DNA sequence determination, *J. Theor. Biol.*, 135:303–307, 1988; W. G. Conway, The practical difficulties and financial implications of endangered species breeding programs, *Int. Zoo Yearb.*, 24/25:210–219, 1986; R. Dermanac et al., Sequencing of megabase plus DNA by hybridization, *Genomics*, 4:114–128, 1989; M. Eggers et al., A microchip for quantitative detection of molecules utilizing luminescent and radioisotope reporter groups, *Biotechniques*, 17:516–524, 1994; S. P. A. Fodor et al., Light directed spatially addressable parallel chemical synthesis, *Science*, 251:767–773, 1991; S. P. A. Fodor et al., Multiplex biochemical assays with biological chips, *Nature*, 364:555–556, 1993; M. J. Heller, An active microelectronic device for multiplex DNA analysis, *IEEE Eng. Med. Biol.*, 15(2):100–104, 1996; A. Karow and J. Critser (eds.), *Tissue Banking in Reproductive Biology*, 1997; M. S. Strauss, D. L. Plucknett, and J. Hodges, The convention on biological diversity and domestic animal diversity, in R. H. Miller et al. (eds.), *Biotechnology's Role in the Genetic Improvement of Farm Animals*, 1996.

Fig. 1. Zebra mussels may have typical stripe patterns as well as many other color patterns. Average adult length is 0.4–0.8 in. (1–2 cm). *(Courtesy of J. Ellen Marsden)*

Bivalvia

Bivalvia is one of the two largest classes of aquatic organisms in the phylum Mollusca. This article focuses on several species of zebra mussel which have invaded the fresh-water habitats of North America; and on the migration of donacid clams by riding waves during tidal cycles.

Invasion of North America by Zebra Mussels

The zebra mussel (*Dreissena polymorpha*; **Fig. 1**) has invaded many fresh-water habitats in North America, beginning in the 1980s. By adhering to hard surfaces with threads made in a specialized gland in its foot, this bivalve pest causes fouling of water intakes, boats, navigation buoys, fishing nets, and other shellfish. Zebra mussels cost utilities and other industries several hundred million dollars in additional maintenance and capital expenses annually. They have also killed or displaced native North American fresh-water mussels in several locations. Because the zebra mussel is a filter feeder and occurs in high densities (often more than 10,000 individuals/ft^2 or 100,000 individuals/m^2), it has significantly reduced plankton densities, increased light penetration, and changed other ecological variables that affect growth and reproduction of water plants and fish.

Distribution of zebra mussels. These extensive economic and ecological impacts prompt much interest in where zebra mussels may spread and what may limit the invasion. Their original distribution (prior to 1800) was in Eurasia in the Black Sea and its tributaries. During the era of canal building and industrialization in the 1800s, zebra mussels spread throughout Europe. It is believed that these organisms were carried to North America in ship's ballast water, which may have been dumped into Lake St. Clair, between Lake Huron and Lake Erie. In less than 10 years, zebra mussel populations have spread throughout the Great Lakes, down the St. Lawrence River, through the Erie Canal to the Hudson River basin, and through the Chicago River and connecting waterways to the Mississippi River basin (**Fig. 2**).

Adaptations. The zebra mussel has adaptations that enable it to spread rapidly by both natural and human-mediated mechanisms. First, it has an enormous reproductive capacity. Males and females occur in about equal numbers, with each female being capable of producing as many as a million eggs per year. Second, fertilization is external, and developing immature stages (larvae) remain in plankton (floating aquatic organisms) for several weeks before juvenile mussels settle on a hard substrate. This enables zebra mussel progeny to float downstream and colonize new locations far from the original populations. During the early years of the North American invasion, the range of zebra mussels spread downstream by about 125 mi (200 km) per year, most likely by the passive dispersal of zebra mussel larvae. Third, mussels can spread upstream by at least two mechanisms: the larvae and young juveniles may be taken up and later dumped in ballast water of ships traveling upstream, and adults may adhere to ships or cargo being transported upstream. Finally, mussels may be transported unintentionally overland in bait buckets, attached to boats or trailers. Thus, appearance of mussels in the Susquehanna River, which is not connected to previously infested waters, is probably due to this type of mechanism. The poten-

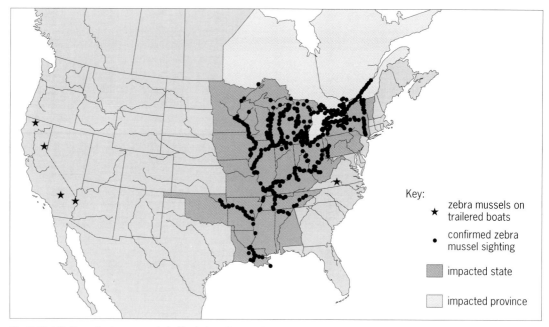

Fig. 2. Distribution of zebra mussels in North America as of November 1996. *(After Southeastern Biological Sciences Center, Zebra mussels here to stay: Learning to live with the invaders, Upwellings, 18(2):1–4, 1996)*

tial for spread across the continential divide is highlighted by the observation of mussels on boats entering California at border agricultural inspection stations.

Although zebra mussels were introduced originally into northern, temperate fresh-water regions of the United States, their spread down the Mississippi River to Louisiana has focused attention on their temperature and salinity tolerance. Recent studies have indicated that North American varieties of zebra mussels have adapted to temperatures higher than is the case for most zebra mussels that have been studied in Europe. Furthermore, zebra mussels can adapt to moderate brackish-water conditions, suggesting that estuaries and other saline waters may be subject to zebra mussel colonization.

Survival and reproduction. The contrast in temperature adaptation of North American zebra mussels with many European studies is apparent in the temperatures at which the mussels begin to spawn during spring or early summer. In several European studies, major spawning activity began at 54°F (12°C). In contrast, several studies in North America do not exhibit major peaks of spawning activity until water temperature exceeds 68°F (20°C). Thermal tolerance limits also appear to be higher in North American populations. Although studies in northern Europe indicated that these bivalves have tolerance limits of 81–82°F (27–28°C), North American mussels could adapt to temperatures above 86°F (30°C). It has been speculated that North American zebra mussels, having originated near the Black Sea, the warmest region of their European distribution, may be genetically more thermally tolerant than mussels from colder, northern European waters on which most European studies have been conducted.

High salinity. Data in Europe and North America also indicate that zebra mussels can adapt to higher saline conditions than the fresh waters in which they predominantly occur. Although zebra mussels cannot survive in ocean water (salinity of approximately 35 parts per thousand), some species are found in the Baltic Sea region in brackish lagoons having salinities as high as 5 ppt. In the Aral Sea, a sea which became highly saline in the former Soviet Union, zebra mussels thrived in salinities as high as 10 ppt; however, they disappeared when the salinity reached 14 ppt. Evidence suggests that abrupt increases in salinity may prove lethal but zebra mussels can acclimate to higher saline levels when salinity increases gradually. For example, in experiments on responses to serotonin, a known activator of zebra mussel spawning, zebra mussels from Lake Erie failed to spawn when placed directly in a salinity of 7 ppt; however, Lake Erie animals which had been gradually acclimated to higher salinities by progressive transfers over several days into intermediate salinities were able to spawn at high rates in 7 ppt.

Genetic selection. Genetic selection for higher salinity tolerance may also occur. The number of zebra mussel larvae flowing into brackish waters of Mississippi River estuaries, where they are subject to salinity tolerance selection, has been estimated at greater than 10^9 living larvae per second during the summer reproductive season. While it is difficult to predict the outcome of this intense selection pressure, the survival of zebra mussel populations with higher thermal and salinity tolerance than is the case with northern European populations seems likely. The successful invasion and reproduction of zebra mussels in the lower Mississippi indicates that waters in the southern United States are more vulnerable to infestation by zebra mussels than was believed during the early stages of the zebra mussel invasion.

Jeffrey L. Ram

Swash Riding by Donax variabilis

Coquina clams (family Donacidae) are similar to many other bivalve mollusks in their general morphology and biology but are spectacularly different in their mode of locomotion. Similar to other bivalves, the fleshy body, foot, and gills of these donacid clams are enclosed between two shells joined dorsally by an elastic hinge. Also similarly, they eat by filter feeding, that is, by pumping seawater through a ventral inhalant siphon, across filtering gills, and out through a dorsal exhalant siphon (**Fig. 3**). As in many other bivalves, they are infaunal (live in sand or mud), and the foot is protracted, anchored, and retracted to pull a clam through sand. However, unlike most other bivalves that spend their lives buried, donacid clams jump out of the sand to ride waves several times during each tidal cycle.

Tidal migrations. One species, *Donax variabilis*, lives on the wave-swept, exposed outer beaches of the outer barrier islands of the southeastern United States. During a rising tide, thousands of these clams can be seen jumping in unison out of the sand just in front of incoming swash from waves. A clam jumps by using rapid downward thrusts of its foot. Swash carries clams toward the shore several times per rising tide, but during falling tides clams jump into backwash and ride seaward. The population thus migrates tidally, remaining in the wetted part of the beach in well-stirred water at the very edge of the sea.

Just as human surfers pick only the best waves to ride, these clams choose only the largest swash to ride. A wave approaching a beach starts breaking as the water becomes gradually shallower. Human surfers ride at the crest of the breaking wave. After breaking, a wave rolls up on shore first as a bore and then as swash. These clams ride the swash and during a rising tide choose only 20% of the waves—those that are the largest. The swash that carries clams the farthest toward the shore requires the clam to make fewer jumps to traverse the whole intertidal zone.

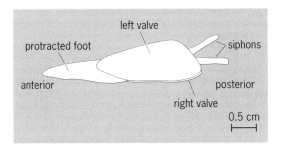

Fig. 3. General morphology of a clam of the species *Donax variabilis*. The retractable muscular foot and siphons are shown in protracted positions. The left and right valves enclose the viscera, with the gape of the shell being ventral and the hinge being dorsal. 0.5 cm = 0.2 in.

Wave sounds. *Donax variabilis* clams choose the largest waves by listening to the sound of approaching waves. Larger waves make louder sounds as they crash onto shore. In laboratory studies, these clams can be stimulated to jump out of sand in the absence of waves when presented with wave sounds. They are most responsive to low-frequency sounds (20–150 Hz) typical of the rumbling sound that an incoming breaking wave produces. Clams are less responsive to higher-frequency sounds that are dissimilar to wave sounds. For example, an 800-Hz sound must have a sound pressure that is 10 times higher (that is, a sound pressure level that is 20 dB higher) than an 80-Hz sound to elicit the same response.

Tidal rhythm. Responsiveness to wave sounds varies on a tidal rhythm. When presented with low-frequency sounds in a laboratory setting, and in the absence of direct tidal cues, more clams jump as the time of high tide approaches. Fewer clams jump as low tide approaches, and none respond at the time of low tide. That biorhythm of responsiveness corresponds to activity levels observed at those times on a beach. Although clams do not migrate within approximately 1 hour of low tide, they migrate at all other times.

Sound detection. How these clams detect sounds is unknown. There are, however, several structures that have mechanosensory potential for detection of vibration or pressure from sound waves. These structures are sensory hairs (cilia) at the tips of the siphons; possible stretch receptors in muscles crossing the gape of the shell, which are stretched as the shell is jiggled by the passing sound wave; and statocysts, ciliated sacs containing calcite balls, located near the base of the foot. In many other invertebrates, statocysts detect gravity and vibrations.

Morphology. Swash riding entails some morphological specializations of the shell associated with fluid dynamics. *Donax variabilis* clams are both denser (heavier relative to their volume) and more wedge-shaped than most closely related bivalves. These two morphological features determine shell orientation and confer stability during swash riding. Similar to a weathervane, hydrodynamic forces acting on the clam cause the pointy anterior end to rotate upstream. A clam's shell is sufficiently heavy that it maintains contact with the sand at a pivotal point. The wedgelike shape exposes more surface area to flow on one side of the pivotal point, thus causing the clam to rotate until it is oriented.

Once oriented, *D. variabilis* experiences downward hydrodynamic forces due to the shell's negative angle of attack. This is the same principle used in the design of racing cars, where rear wings create downward aerodynamic forces that enhance stability. Oriented clams can stabilize their slide, which enables them to dig in easily by sticking out their foot to grab the sand. These morphological features enable clams to control how long they ride the swash.

Habitat. There are many *Donax* species that live on warm temperate and tropical sandy beaches all over the world. They range in maximum size from 1 to 8 cm (0.5 to 3 in.) long. There are 17 eastern Pacific species and 11 western Atlantic species, and perhaps 50 species worldwide. While many *Donax* species swash-ride in a manner similar to that of *D. variabilis*, other species may be specialized for different wave regimes. For example, at Punta Chame, Panama, *D. californicus* is found on protected beaches with small waves, whereas only 1940 ft (500 m) away on the other side of this peninsula the more wedge-shaped *D. ecuadoriensis* and *D. assimilis* are found on exposed beaches with larger waves. Wave exposure is correlated with beach slope: steeper slopes indicate larger waves on reflective beaches, whereas shallow slopes indicate smaller waves on dissipative beaches; bivalves that live on reflective beaches tend to be denser. Species also vary in the extent of their migrations. Instead of migrating tidally, South African *D. serra* migrate to follow the position of the low-tide water table, the location of which changes with phases of the Moon. Migratory habits may also change with time of year. *Donax variabilis* migrates intertidally during summer in North Carolina but lives on offshore sand banks during winter.

Ecological roles. The ecological roles of other sandy beach inhabitants that ride the swash and backwash are quite diverse. Between swash rides, mole crabs *Emerita* filter-feed organic particles by protruding their antennae into the backwash. Swash-riding snails (*Terebra*, *Hastula*, *Olivella*, and *Bullia*), which use an expanded foot as a sail, also occupy a diversity of ecological niches. Some species (some *Olivella* species) are filter feeders that extend pedal tentacles into the backwash. Others, such as *Bullia digitalis*, are predators that ride swash to track down prey. Despite disparate ecological roles, swash riders share several features, such as a wedge shape, the ability to orient and slide stably in flow, and responsiveness to pressure changes such as those produced by wave sounds, that correspond to and enable their swash-riding mode of life.

For background information *see* BIVALVIA; MOLLUSCA; SOUND; TIDE in the McGraw-Hill Encyclopedia of Science & Technology.   Olaf Ellers

Bibliography. O. Ellers, Behavioral control of swash-riding in the clam *Donax variabilis*, *Biol. Bull.*, 189:120–127, 1995; O. Ellers, Discrimination among wave-generated sounds by a swash-riding clam, *Biol. Bull.*, 189:128–137, 1995; O. Ellers, Form and motion of *Donax variabilis* in flow, *Biol. Bull.*, 189:138–147, 1995; A. McLachlan et al., Adaptations of bivalves to different beach types, *J. Exp. Mar. Biol. Ecol.*, 187:147–160, 1995; T. F. Nalepa and D. W. Schloesser, *Zebra Mussels: Biology, Impacts and Control*, 1993; F. J. Odendaal et al., *Bullia digitalis* Gastropoda actively pursues moving prey by swash-riding. *J. Zool. (London)*, 228:103–113, 1992; J. L. Ram et al., The zebra mussel (*Dreissena polymorpha*), a new pest in North America: Reproductive mechanisms as possible targets of control strategies, *Inv. Repro. Devel.*, 22:77–86, 1992; J. L. Ram, P. P. Fong, and D. W. Garton, Physiological aspects of zebra mussel reproduction: Maturation, spawning, and fertilization, *Amer. Zool.*, 36:326–338, 1996.

Black hole

Black holes are the remnants after heavy stars burn out and collapse. The gravitational pull of a black hole is so strong that not even light can escape. Moreover, craters and other distinctive features are impossible because of the crushing forces exerted on matter in the black hole. The contrast between the initial complexity of a star and the final, apparently structureless state of a black hole is dramatic and puzzling. Indeed, if the black hole shows no hint of its complex past, there is an apparent conflict with determinism. According to determinism, the present state of the world developed from some past state that should be deducible from the present one, at least in principle. Theoretical physicists are strong believers in determinism and therefore are inclined to suspect that the complexity of the star has been transformed into some kind of internal structure of the black hole, rather than having disappeared. Recent research may have uncovered the appropriate hidden structure.

Black-hole thermodynamics. An important hint came with S. Hawking's discovery that black holes are not completely black but emit thermal radiation. The temperature is extremely small: that of a black hole with the same mass as the Sun is 6×10^{-8} K (11×10^{-8}°F) above absolute zero. Nevertheless, this thermal behavior leads to the expectation that black holes support microscopic structure because such behavior in ordinary matter is due to the apparently random motion of large numbers of molecules.

Entropy is a quantitative measure of structure, defined as the logarithm of the number of possible microstates (molecular motions) consistent with a given macroscopic state of the matter (total energy, and so forth). Its value can be determined from the temperature by application of standard thermodynamics. The entropy of a black hole is $Ac^3/(4G\hbar)$, where A is the surface area, c is the speed of light, G is Newton's constant, and $\hbar$ is Planck's constant divided by 2π. This is an extremely large number: For a black hole with the mass of the Sun, it is approximately 10^{77}, whereas the Sun itself has entropy only around 10^{46}. This suggests that black holes have much more structure than is ordinarily given by molecules. However, the form is mysterious because, as explained, black holes show no evident structure.

Superstring theory. Black holes are described by Einstein's general theory of relativity. This theory is accurate in the astronomical setting but inadequate at extremely small distances, where the effects of quantum mechanics are important. In the microscopic setting, it is presumably superstring theory that becomes appropriate. In this theory all the basic constituents of matter are small loops of strings. The strings acquire different properties by vibrating in various patterns. For example, strings that vibrate a lot correspond to very heavy states. It is therefore natural to identify rapidly vibrating strings with black holes.

Superstrings are extremely small, approximately 10^{-20} times the size of an atomic nucleus. Ordinary experiments would therefore fail to observe their vibrating nature, and would be able to discern only some gross features, such as the mass. In this description black holes appear structureless, but in reality they have an intricate internal structure. This is just the situation hoped for. It is remarkable that string theory gives rise to hidden features in heavy objects in such a natural way. *See* SUPERSTRING THEORY.

Entropy and area. There is a precise test of these ideas. It is possible to calculate the entropy of the vibrating fundamental string. The result should be compared to the entropy of the corresponding black hole, that is, its surface area (multiplied by some constants). A black hole that corresponds to a string is unfortunately so small that its size is completely obscured by fluctuation effects which dominate at small distances, and preclude a normal space–time interpretation. Nevertheless, its area can be estimated and an entropy inferred. The result is consistent with the evaluation in string theory. This supports the identification between string states and internal structure in black holes. However, the black holes associated with fundamental strings are too small to be worthy of the name. Specifically, their area should be precisely determined, without need of estimates.

That string theory contains objects known as solitons has only recently been appreciated. The solitons exist in different forms, including membranes and particles. They can vibrate, but they are heavy even if they do not. This immediately suggests a relation to black holes. Unfortunately, the simplest black holes of this kind are so small that their size is poorly defined, just as for the ones that correspond to ordinary strings.

Different kinds of solitons can form bound states with each other and with strings. For example, a particle can lie on top of a membrane and, moreover, strings can stretch between the particle and the membrane. Black holes identified with certain bound states have a proper surface area that is not greatly affected by the vibrations of the solitons and strings. Bound states of solitons and strings agree better with intuitive notions of black holes than simple strings or isolated solitons. Specifically, they give a definite prediction for the entropy.

Bound states have an internal structure that is more involved than that of a single string. The important qualitative feature remains that the microscopic structure is extremely small so the macroscopic observer is sensitive only to some bulk features, such as the mass. This is in perfect analogy with ordinary thermodynamic systems, where the molecular structure is invisible to macroscopic observers.

In the case of heavy bound states of solitonic membranes and particles, with strings attached, the membranes can be considered to consist of a large number of coincident fundamental membranes, each endowed with a certain minimal mass. Similarly, the particles are really many coincident particles. The strings that connect membranes with particles can attach to any one of the fundamental membranes and particles. Clearly, this kind of bound state has a much more complex structure than a single string. Indeed, it is related to a large number of strings.

It is a challenge to calculate precisely how much entropy this kind of structure gives rise to. The problem has been solved in some cases, including the bound state described above. The result agrees precisely with the one predicted from the area of the black hole. The agreement does not rely on any controversial estimates or approximations. This gives great confidence that string theory indeed describes the internal structure of black holes.

Quantum effects. An ordinary electron may be considered a tiny black hole. However, the position of an electron is uncertain because of quantum effects, and this uncertainty is much larger than the associated black hole. Thus the identification is purely formal.

The relation between black holes and bound states may have the same problem. However, for so-called supersymmetric black holes there are no quantum fluctuations at all, so in this case there is certainly no reason to distrust the above discussion. Supersymmetry is unfortunately so restrictive that it implies zero temperature for black holes. It would therefore be desirable to understand black holes even in the absence of supersymmetry. Some evidence suggests that quantum effects remain benign in this case. However, this issue is somewhat controversial, and it is still possible that entirely new ideas are necessary to take quantum effects into account.

Ten-dimensional world. String theory is actually a theory in ten space–time dimensions (nine space dimensions and one time dimension). Of course, visible space has three spatial dimensions, so realistic models must assume that six spatial dimensions are very small. In the ten-dimensional world the membranes considered above actually have four spatial dimensions. Both the membranes and the strings wrap around the small dimensions so that, from the three-dimensional point of view, they are pointlike except for a small size created by vibrations. It is intriguing that the internal structure of black holes becomes hidden in the extra dimensions in this very precise sense. This circumstance may eventually lead to a deeper understanding of the ten-dimensional world.

Even though there is confidence in the relevance to black holes of the structure embodied in bound states of solitons and strings, how this structure manifests itself in experiments is not yet clear. For example, a daring scientist that jumps into the black hole could see direct evidence of the ten-dimensional world. Moreover, some effects may be accessible even to the scientists that stay outside. Such questions are currently under intense investigation. Their answers may profoundly change the understanding of space and time.

Significance of results. The theoretical motivations for superstring theory and the thermodynamics of black holes are quite different, so their agreement is by no means self-evident. The precise quantitative agreement of their predictions for the internal structure of black holes can therefore be construed as evidence that the theories do indeed describe real phenomena. These kinds of consistency arguments are crucial because black-hole thermodynamics and especially superstrings will, at best, be accessible to experiments only in a very indirect form. Since the confidence in black-hole thermodynamics is quite strong, the precise accounting for the internal structure of black holes is perhaps best characterized as a significant triumph for superstring theory.

For background information *see* BLACK HOLE; ENTROPY; QUANTUM MECHANICS; SOLITON; SUPERSTRING THEORY; SUPERSYMMETRY; THERMODYNAMIC PRINCIPLES in the McGraw-Hill Encyclopedia of Science & Technology.   Finn Larsen

Bibliography. S. Hawking, *A Brief History of Time: From the Big Bang to Black Holes*, 1988; E. Witten, The holes are defined by the string, *Nature*, 383:215–216, 1996.

Bose-Einstein statistics

All particles have an intrinsic quantum property called spin. Particles with half-integer spin are known as fermions (for example, electrons and neutrons), and those with integer spin are known as bosons (for example, photons). The classification of a particle as a boson or fermion has a profound effect on its behavior in the microscopic quantum realm. In accordance with the Pauli exclusion principle, fermions share the property that two identical particles may not occupy the same quantum state. For bosons, not only is the exclusion principle inapplicable, but the probability of making a transition to an occupied quantum state is enhanced by a factor equal to one plus the number of bosons already there. This characteristic of bosons is important and leads to the somewhat ubiquitous phenomenon of Bose-Einstein condensation.

This phenomenon was first predicted in 1925 by A. Einstein, who realized that if an ideal gas of noninteracting bosons is cooled sufficiently the gas will undergo a phase transition to a distribution with a macroscopic occupation of particles in the lowest available energy state. If the system is infinitely large, the lowest available energy is the state with the particle velocity equal precisely to zero. It is surprising that such a state could be occupied by a finite fraction of all of the particles when the temperature

is above absolute zero. Einstein predicted this would happen as a result of purely quantum-statistical effects associated with the particle spin.

In order to form a condensate, the temperature would need to be reduced below the point where the thermal de Broglie wavelength was equal to the interparticle spacing, and then multiple occupancy of the lowest energy quantum state would be inevitable. The threshold for condensation may be stated as the requirement that the dimensionless phase-space density, equal to the product of the number density of particles and the cube of the de Broglie wavelength, be greater than 2.6. This requirement is difficult to fulfill for a dilute gas with a typical density of 10^{14} atoms/cm^3 since the temperature must be reduced to the nanokelvin region.

Phenomenon related to condensation. A number of related phenomena are linked with the concept of Bose-Einstein condensation. In a laser, the generation of coherent light usually arises from the stimulated emission of photons into a high-quality mode of an optical resonator. This phenomenon is a consequence of the boson nature of photons, but a laser is a steady-state device with constant pumping and output coupling rather than a system at equilibrium. The superfluidity of liquid helium at low temperature was interpreted in 1938 as Bose-Einstein condensation, although in this case strong particle interactions exist. The phenomenon of superconductivity is also associated with condensation, although the elementary boson here is constructed of paired electrons. The strong interactions in the solid or liquid limit the condensate fraction and make a complete theoretical treatment difficult, but give rise to many interesting phenomena. Recently evidence for condensation has been observed in a system of excitons in copper(I) oxide (Cu_2O).

Condensation in dilute gases. Since the early 1980s, a major goal in atomic physics has been to produce a condensate in a dilute gas. Such a condensate would be much closer in spirit to the situation originally considered by Einstein. The motivation for this effort is that such a system can be well understood theoretically with a straightforward treatment of the elastic binary atomic collisions. Work toward generating a condensate in a weakly interacting gas was first pursued in spin-polarized hydrogen experiments. In these experiments, the necessary evaporative cooling techniques were developed to approach the stringent requirements for low temperature and high density. The search for Bose-Einstein condensation in laser-cooled alkali atoms has a short but dramatic history. In 1995, a group using rubidium atoms and another using sodium succeeded in producing a Bose-Einstein condensate in an alkali vapor in which the average interaction between the particles was repulsive. A third group, using lithium, also reported cooling into the quantum degenerate regime, in this case with net attractive interactions between the atoms.

Cooling neutral atoms. The task of cooling room-temperature atoms to temperatures of a few nanokelvins

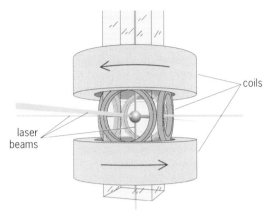

Fig. 1. Layout of the apparatus for achieving Bose-Einstein condensation. Six laser beams intersect in a high-vacuum glass cell. The coils generate the magnetic trap.

is accomplished by a hybrid combination of laser cooling, magnetic trapping, and evaporative cooling. Atoms are loaded from either a background vapor or an atomic beam into a magnetooptical trap in the center of a high-vacuum chamber or glass cell (**Fig. 1**). The atoms are thereby completely isolated from room-temperature surfaces and can be stored for minutes at a time.

Laser cooling. The magnetooptical trap cools the atoms and significantly increases the phase-space density. The mechanism behind this is the ponderomotive force on the atom from the scattering of photons. The trap consists of six laser beams intersecting in the center of the cloud and the magnetic field generated by the two current-carrying coils aligned in the horizontal plane, one below and one above the cloud. A complicated sequence of optical pumping between internal atomic states gives both an efficient cooling cycle and a confining potential for the atomic motion. The phase-space density achievable at this stage is still typically six orders of magnitude below that required for condensation.

Magnetic trapping. Alkali atoms have a magnetic dipole moment and feel a force in an inhomogeneous magnetic field. This phenomenon may be used to confine the cloud in a region of space. A trap is created by a geometry in which the field strength increases with distance from the center. A number of different configurations have been used, including time-averaged harmonic potentials, magnetic traps combined with off-resonant light, Ioffe-Pritchard traps, and fields created by permanent magnets.

Evaporative cooling. Evaporative cooling is achieved by extracting the highest-energy particles from the distribution and allowing thermalization to refill this part of the spectrum. The cooling principle is that when such an atom is removed it takes away significantly more than the average energy per particle of the cloud. Consequently, if the cloud is allowed to rethermalize, it forms an equilibrium distribution with a lower temperature. The high-energy atoms are removed by tuning a radio-frequency field to

selectively spin-flip those atoms at the edge of the cloud. As the sample cools and decreases in size, the frequency is reduced to continue the evaporation. The thermalization rate (the collision rate per atom) may increase, allowing the evaporative cooling to run away. The collision rate rises, even though the cooler atoms have lower velocity because of the rapid increase in density as the atoms are localized in the trap center.

Bose-Einstein condensate. When the phase-space density exceeds the requirement for Bose-Einstein condensation, the cloud separates into two components. The condensate in **Fig. 2** is a narrow density peak in the center of the trap on top of a much broader thermal background. The peak exhibits an anisotropic velocity distribution consistent with the lowest available energy state of the confining potential, which is quite distinct from the isotropic thermal distribution of the noncondensate fraction. This condensate contained approximately 1500 atoms and formed at a temperature of approximately 170 nK. Subsequent experiments have produced much larger condensates with up to 5×10^6 atoms.

Elementary properties. The availability of reproducible condensates in dilute vapors has allowed the study of the properties of this novel system.

Excitations. Elementary excitations of the condensate have been observed by modulating the confining potential. The excitations are measured as shape oscillations in the condensate observed during ballistic expansion. The frequencies of the normal modes of the condensate measured in this way agree well with theoretical prediction.

Self-consistent condensate. The condensate wave function is not simply the ground state of the confining potential, but must be calculated self-consistently by

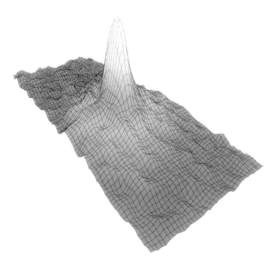

Fig. 2. Bose-Einstein condensate. The chart shows the velocity distribution of a cloud of rubidium atoms. Both the shading and the height of the surface indicate the density of atoms having the velocities specified by the projection of the surface on a horizontal plane (with axes indicated by the grid). The central feature is the condensate, which rises above the plateau of atoms in the thermal component.

taking into account interactions. Collisions between the particles lead to a repulsive or attractive mean field. The theoretical description of this situation is straightforward and has achieved a high level of accuracy in predicting the shape and extent of the condensate, the excitation frequencies, and the average energy.

Coherent matter waves. In many ways, the most interesting analogy of Bose-Einstein condensation is the comparison with the coherence properties of the laser. The presence of a phase for the atomic-matter field is one of the most interesting features of the Bose-Einstein condensate. Experiments have been proposed to probe such phase properties. Interesting possibilities are to build an atom laser or to use the condensate as a source for holographic imaging or atomic interferometry. This new system will undoubtedly hold rich and novel physics, yet to be discovered.

For background information *see* BOSE-EINSTEIN STATISTICS; COHERENCE; EXCLUSION PRINCIPLE; LASER; LASER COOLING; LIQUID HELIUM; PARTICLE TRAP; SUPERCONDUCTIVITY in the McGraw-Hill Encyclopedia of Science & Technology.   Murray J. Holland

Bibliography. M. H. Anderson et al., Observation of Bose-Einstein condensation in a dilute atomic vapor, *Science*, 269:198–201, 1995; C. C. Bradley et al., Evidence of Bose-Einstein condensation in an atomic gas with attractive interactions, *Phys. Rev. Lett.*, 75:1687–1690, 1995; K. B. Davis et al., Bose-Einstein condensation in a gas of sodium atoms, *Phys. Rev. Lett.*, 75:3969–3973, 1995.

Brachiopoda

Brachiopods form one of very few phyla that have been present on Earth for over 550 million years, since early Cambrian times, when the first widespread appearance of animals with mineralized skeletons occurred. They were the most abundant filter-feeding group for much of that time, and dominate fossiliferous deposits in many localities. Brachiopods are still present in all of the oceans, but are common and abundant only in restricted areas, such as the deep sea, fiords, polar seas, some parts of New Zealand, and the west coast of North America. They also are often found in cryptic habitats (areas which provide protection from physical disturbance and include crevices, underwater caves, and deep rocky overhangs). Many of the temperate and tropical shallow sea localities where they were formerly dominant are now occupied by bivalve mollusks. Whether this replacement is due to competition or whether bivalves more successfully colonized space made available after the dramatic decline of the brachiopods, during the mass extinction event at the end of the Permian period 150 million years ago, is strongly debated. In some areas, however, they still dominate local sites, with densities sometimes exceeding 1000 individuals per square meter.

Low-energy life-styles. In recent years, investigations of the physiology and ecology of this group have begun to help explain their current distribution patterns, and their success in areas where food supplies are restricted. Compared to other marine invertebrates, brachiopods have very low organic tissue content, and around 97–99% of their bodies is either water or skeletal calcium carbonate. They also have very slow growth and metabolic rates, contract their muscles slowly, pump water through their filtration apparatus slowly, and have several other adaptations which are interpreted as minimizing energetic losses; so they have been termed low-energy life-style, or marginal, species.

The low tissue contents of brachiopods have important ecological consequences. The body design comprises an animal composed of a small amount of tissue spread around the inside of a large shell. Analyses of costs and benefits to predators feeding on brachiopods show that the total amount of organic material available compared with the effort required to gain access to the tissues via crushing or drilling holes through the shell is similar to the same measure in bivalve mollusks. Thus the brachiopod shells are easier to crush and drill than bivalves, but the reduction in organic matter available to a predator balances the reduced effort needed to gain access to the tissues. A given amount of brachiopod tissue occupies 5–25 times more space than bivalve mollusk tissues. These factors, plus the finding that tissues of some brachiopods possibly contain noxious chemicals, explain why predators usually avoid eating them and why no species which specifically consume brachiopods have been identified.

Another factor in categorizing brachiopods as low-energy organisms is their limited ability to generate metabolic power. Following a meal, an animal's metabolic rate rises. In some snakes, such as the Burmese python, metabolism increases 45 times after consuming a large meal; this is the same as the maximum difference between rest and full-speed metabolism in racehorses. However, in most marine invertebrates the rise following feeding is of the order of 2 to 3 times the prefeeding metabolic rate; and in sedentary or inactive species, the rise is a measure of metabolic scope, or the maximum attainable metabolic rate. A study of brachiopods determined that their postprandial metabolic rise is 1.6 times prefeeding levels—one of the lowest rises recorded for a marine ectotherm. The absolute rise in metabolism in brachiopods was even lower compared with other species.

Antarctic brachiopods. Antarctica may be the area where brachiopods currently are most successful. Over 50 species have been described from this region, where they are widespread and abundant. In caves at Signy Island *Liothyrella uva* (**Fig. 1**) has been found at densities of nearly 1800 per square meter. This species grows up to 2.3 in. (58 mm) in length and often lives in clumps of up to 50 specimens attached to a central individual. At higher densities, rock surfaces are completely covered by a

Fig. 1. Antarctic brachiopods (*Liothyrella uva*) from a cave at 45 ft (15 m) at Signy Island. Growth rings on the shell valves are visible; and the tips of the filter-feeding apparatus, the lophophore, are also visible through the anterior, exhalant opening of the shell.

layer of brachiopods two or three animals deep. This is reminiscent of the dominance they commonly had in many parts of the world in earlier geological periods. Other brachiopod species in the Antarctic are widespread and, although less locally dense, form very large populations over wide areas. For example, *Magellania fragilis* occurs extensively throughout the Weddell Sea at densities of up to 26 individuals per square meter.

Growth rings and growth rates. Brachiopods have growth rings on their shells, and in *M. fragilis* these rings have been used to assess ages and growth rates. The largest specimens found in a study of growth in this species were 1.4–1.6 in. (35–40 mm) in length, and these brachiopods were nearly 50 years old. Maximum growth rates, found in small individuals, were 0.10–0.12 in. (2.5–3 mm) per year, whereas 25-year-old specimens, in the middle of the age

range, were growing at around 0.02 in. (0.5 mm) per year. These growth rates are slower than those of temperate brachiopods, and much slower than bivalve mollusks from similar habitats. Thus, the Antarctic brachiopods grow slowly and live to great age. A common estimate of the performance of a given species is the ratio of the productivity to standing stock, or biomass of a population of that species. Productivity is the sum of the amount of growth of all individuals in a population, and the biomass is the sum of the organic mass of all those individuals.

More recently, growth rates, rings, and age have been studied in *L. uva* at Signy Island. Growth rates were assessed from animals measured, returned to their normal habitats in the sea, and remeasured after varying periods between 6 months and 2 years. The results showed brachiopods were growing faster in terms of shell length during winter periods but were increasing in body mass during the summer. This indicated either that shell growth and soft tissue growth are decoupled or that the brachiopods store resources obtained from feeding to be used in growth of tissues and shell at other times of the year, which would decouple all growth from periods of feeding. In either case, the growth strategy adopted by *Liothyrella* is an unusual one since most animal species grow when they are fed, with skeletal growth and tissue growth occurring in parallel.

The largest *Liothyrella* in the growth assessment was approximately 50 years old (**Fig. 2a**). However, the oldest specimens had only 25 growth rings on their shells, and if the growth rings were laid down annually, the specimens could be no more than 25 years old. To solve the problem, pieces of shell were

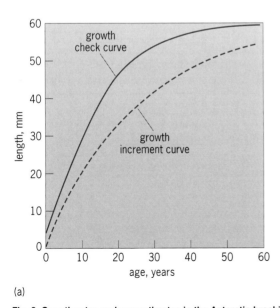

(a)

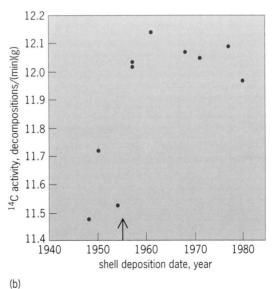

(b)

Fig. 2. Growth rates and age estimates in the Antarctic brachiopod *Liothyrella uva*. (*a*) Length versus age from growth ring analyses. (*b*) Radiocarbon (^{14}C) content of shells of different age. The data points represent the year when shell samples were laid down. The rise in ^{14}C content coincides with the earliest records of ^{14}C associated with aerial nuclear bomb tests found in ice cores from glaciers in the Antarctic peninsula (indicated by the arrow).

taken from the oldest parts of shell valves of specimens from a range of sizes and were analyzed for radiocarbon (^{14}C) content. Small amounts of ^{14}C are present in all environments on Earth, and ^{14}C is incorporated into skeletons of all animals that utilize calcium carbonate for their skeletal material, which is the case for most marine species.

Levels of ^{14}C in the environment increased in the mid-1950s, when large quantities were released from the earliest aerial nuclear bomb tests. Such increases in ^{14}C have been found in ice cores from glaciers in Antarctica, and the first rises occurred in 1954 and 1955. The *Liothyrella* shell samples also showed an increase in ^{14}C content (Fig. 2b), which indicated that ages approaching 50 years for the oldest specimens were correct. This also meant that growth rings were not laid down with an annual periodicity, and a comparison between numbers of growth rings and the growth rate study suggested that growth rings are laid down only every 1.84 years. This is the first time that growth rings with greater than annual periodicities have been found in any animal species. The underlying reason for a near-biennial growth ring periodicity remains unclear, but the two main alternatives are entrainment to a 2-year environmental cycle which occasionally breaks down, or to an internal cycle.

Cycles in the physical environment such as El Niño in the South Pacific are already known, but no such phenomenon with an approximate 2-year cycle has been identified in Antarctica. The alternative endogenous biological cycle appears more likely, as several Antarctic invertebrate species have reproductive cycles of greater than annual periodicity.

Implications. If a near-biennial growth ring periodicity is a characteristic of Antarctic species in general, many of the previous growth assessments based on growth rings are in error, and cold-water species would be even older and more slow growing than previously thought. If it is a representative finding for brachiopods, whether it is true for all brachiopods or only for those living in cold water has to be determined. If it is true only for Antarctic brachiopod species, the earlier estimates of age in *Magellania* should be revised to maximums of around 90 years. There are also powerful implications for the interpretation of fossil brachiopod populations, where shell characteristics are the most common attributes used for the analysis of evolutionary trends and paleoenvironmental reconstructions.

For background information *see* ANTARCTICA; BIVALVIA; BRACHIOPODA; CAMBRIAN; FOSSIL; PALEOECOLOGY; PALEONTOLOGY in the McGraw-Hill Encyclopedia of Science & Technology.   L. S. Peck

Bibliography. T. Brey et al., Population dynamics of *Magellania fragilis*, a brachiopod dominating a mixed-bottom macrobenthic assemblage on the Antarctic shelf, *J. Mar. Biol. Ass. U.K.*, 75:857–869, 1995; M. A. James et al., The biology of living brachiopods, *Adv. Mar. Biol.*, 28:175–387, 1992; L. S. Peck et al., in A. Williams (ed.), *Treatise on Invertebrate Palaeontology, H. Brachiopoda*, 1997; L. S. Peck and T. Brey, Bomb signals in old brachiopods, *Nature*, 380:207–208, 1996.

Brain

Centuries ago, phrenologists studied the conformation of the human skull in an attempt to learn about variation in brain structure and mental function. They viewed the brain as a collection of parts, just as the body can be described as a collection of organs, each with a particular function and a characteristic location. While neuroscientists doubt the details of the phrenological view, the search for the cerebral basis of conceptual knowledge remains a project of functional localization, supported by theories of cognition and behavior. In 1929, it was proposed that brain function depends upon the allocation of fairly homogeneous neural resources to particular tasks. This uniformity of functional capability was called equipotentiality. Extreme localization and equipotentiality lie at two ends of a continuum describing functional organization in the brain.

The selective impairment of mental function that often follows localized brain injury or disease, and the extreme functional specificity of neurons in birds, rodents, and nonhuman primates, even at the very earliest stages of development, indicate that the brain does not function as a uniform pool of resources. However, given the reorganization of human cortex that accompanies recovery of function, and the evidence provided by electrophysiological studies of functional plasticity in the brains of other animals, it is clear that the function of each region of the brain cannot be determined by its anatomic location. In order to investigate conceptual organization, neuroscientists have used evidence from brain lesions and more recently evidence from neuroimaging to identify localized systems in the brain that subserve particular behaviors or cognition. It is not yet known how far these processes can be localized in the brain.

Neuroimaging. Since the late 1980s, researchers have developed methods of measuring changes in the brain that shed light on long-standing questions concerning the cerebral basis of conceptual organization. Using positron emission tomography (PET) and functional magnetic resonance imaging (fMRI), these methods reveal localized changes in cerebral blood supply that accompany task-dependent changes in neural activity.

In order to acquire neuroimaging data, changes in regional cerebral blood flow (rCBF) are measured with a PET scanner by using an injected or inhaled radioactive tag, or local blood oxygenation changes are measured by using a magnetic resonance imaging (MRI) scanner to detect intrinsic blood-oxygenation level-dependent (BOLD) contrast. Changes in BOLD contrast and in regional cerebral blood flow arise from a localized vascular response in the brain that is caused by increased oxygen utilization associated with neural activity. Because of the difficulty inher-

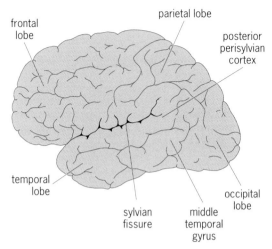

frontal lobe

parietal lobe

posterior perisylvian cortex

temporal lobe

occipital lobe

sylvian fissure

middle temporal gyrus

Sagittal view of the brain. The inferior frontal lobe includes Broca's area, and the posterior perisylvian cortex includes Wernicke's area.

ent in measuring small changes in BOLD contrast and in regional cerebral blood flow, subjects in neuroimaging studies must continuously perform a behavior of interest to the researcher for tens of seconds or even minutes. Brain-imaging studies therefore employ behavioral tasks that engage subjects in prolonged, uniform activities. The measurements made in an active condition are contrasted with regional cerebral blood flow or blood oxygenation in a baseline condition. The baseline condition employs a task that is postulated to lack key functional processing demands involved in the active condition. As a consequence of the inherently contrastive nature of neuroimaging studies, their results are open to multiple interpretations arising from the relationships among properties of the active and baseline tasks. Nonetheless, neuroimaging provides a powerful tool for studying brain function.

Conceptual organization. Before neuroimaging, studies of human brain organization depended upon functional dissociations among individuals with impaired brain function. Such investigations provided a foundation for cognitive neuroscience. Three findings are especially relevant to the study of language and conceptual knowledge in the brain: (1) Language is primarily impaired by damage to the left hemisphere, involving parts of the frontal, temporal, or parietal lobes near the sylvian fissure (see **illus.**). (2) Damage to the frontal lobe disturbs primarily the processing of syntax (the structural properties of utterances in a language), while the temporal lobe appears to have a more important role in lexical processing (that is, mental functions related to the use and understanding of words). (3) Information about syntax, semantics (the relationships among the meanings of words and between words and the objects or events they describe), and lexical form (both phonology and orthography) may be represented in distinct regions of the brain. This conclusion is based on the observation that brain damage

can result in selective impairment of words of only one grammatical class (for example, a deficit in producing nouns but not verbs) or one semantic category (for example, a deficit in naming animals but not other categories of objects).

Brain-imaging research has begun to investigate whether or not these functional dissociations in the brain can be anatomically localized, and consequently whether the anatomic localization based on lesion evidence, or the more general finding of hemispheric asymmetry in the effects of damage, will correspond to elements of the function of uninjured brains.

Grammatical class and regional function. It has long been known that the ability to produce or understand function words (articles, prepositions, auxiliaries, and so on) can be damaged independently of such ability relating to nouns and verbs. More recently, many studies have documented clear dissociations between the ability to produce or understand nouns versus verbs. Damage to the frontal lobe is typically associated with greater impairment in processing verbs and function words than nouns, while damage to the anterior and medial parts of the temporal lobe is associated with greater impairment in processing nouns than verbs and function words. Recent PET and event-related potentials (ERPs) investigations have attempted to confirm this general characterization of the roles of the frontal and temporal lobes in processing words of different grammatical classes. The results are generally encouraging, although the overlap is not perfect. For example, a study by researchers at the National Institutes of Health found that the inferior frontal lobe is activated in verb production relative to object naming, but so is the posterior third of the middle temporal gyrus. These conflicting results show that researchers do not have a clear map of the organization of grammatical knowledge in the brain, but they have been able to confirm the important role of grammatical class in the organization of lexical knowledge in the brain.

Lexical form versus meaning. One major discovery from the study of aphasia (an inability to speak, to write, or to understand language because of a brain lesion) is that knowledge of the sound and spelling of words is represented independently of the meanings of the words. The general consensus that has emerged from these studies is that knowledge concerning lexical form is represented in neural tissue adjacent to the relevant sensory and motor cortices, while semantic knowledge is represented more widely, perhaps involving parts of the frontal, temporal, and parietal lobes. Research with PET, functional magnetic resonance imaging, and event-related potentials has generally confirmed this picture. A recent PET study found that making semantic judgments about the referents of words or of pictures produced blood flow changes in the left hemisphere in perisylvian areas consistent with those changes found by neuropsychologists to be critical for object identification or for decisions about the semantic properties of objects presented verbally or pictorially. By comparing judgments of words and those of pictures to a visual judgment baseline, the researchers found that there are common areas involved in both tasks, presumably devoted to semantic processing in parts of the temporal, frontal, and parietal lobes. Moreover, the researchers were able to identify areas involved in the processing of word forms as opposed to pictures. This suggests that there may be localized brain regions devoted to processing information about word form that communicate with brain areas devoted to the representation of semantic information, whether that information is activated by word form, pictures, or other modes of access.

Category-specific function in brain. Neuroimaging studies also provide support for the hypothesis that category-specific deficits (for example, a deficit in naming inanimate objects but not animals) may imply localized category-specific function in the intact brain. Several research groups have used PET to measure local changes in blood flow for the identification of objects in specific semantic classes (such as animals, tools, or vehicles). All of these studies have found activity in the temporal lobe, predominantly in the left hemisphere in normal subjects, providing additional support for the idea that the temporal lobe plays an important part in the representation of conceptual knowledge. The studies also found divergent patterns of blood flow change for stimuli in different semantic classes. However, the pattern of these divergences differed considerably across the studies. These disparities may result from the use of different baselines in the different studies, or from slight differences in the behavioral demands placed on subjects by the stimuli or decisional tasks. Nonetheless, as in the case of grammatical class distinctions, the neuroimaging studies confirm a basic principle of the organization of lexical and conceptual knowledge in the brain: abstract linguistic and conceptual properties play a crucial role in determining the functional organization of the human brain.

Language and hemispheric asymmetry. Evidence from the pattern of lesions in individuals with disorders related to language and conceptual organization, such as aphasia, anomia (difficulty in naming recognizable objects), and agnosia (an inability to recognize objects), suggests that language processing depends upon the brain's left hemisphere. In spite of the existence of homologous structures in Broca's area, Wernicke's area, and other temporal and inferior frontal areas in the left and right hemispheres, it is primarily damage to left-hemispheric cortex that provokes language impairments. Lateralization of language function to the left hemisphere has been observed in some brain-imaging studies, but a number of studies have shown an unexpected degree of bilateral activation. Many brain areas are highly interconnected in the left and right hemispheres by way of the corpus callosum. It is possible that regions in the left hemisphere may be needed in language processing, and they may provoke increased activity in their right-hemispheric analogs simply because there are many connections between analogous left- and right-hemispheric regions. This would explain bilateral changes in a manner consistent with the primacy of left-hemispheric processing in language. Another possibility is that homologous areas in the two hemispheres are involved in the comprehension and production of language but damage to the left hemisphere is differentially debilitating because of that hemisphere's contribution to language processes.

For background information *see* BRAIN; COMPUTERIZED TOMOGRAPHY; HEMISPHERIC LATERALITY; MEDICAL IMAGING; MEDICAL ULTRASONIC TOMOGRAPHY; NERVOUS SYSTEM (VERTEBRATE); PSYCHOLINGUISTICS in the McGraw-Hill Encyclopedia of Science & Technology.   Benjamin Martin Bly; Alfonso Caramazza

Bibliography. H. Damasio et al., A neural basis for lexical retrieval, *Nature*, 380:499–505, 1996; A. Martin et al., Discrete cortical regions associated with knowledge of color and knowledge of action, *Science*, 270:102–105, 1995; A. Martin et al., Neural correlates of category-specific knowledge, *Nature*, 379:649–652, 1996; R. Vandenberghe et al., Functional anatomy of a common semantic system for words and pictures, *Nature*, 383:254–256, 1996; E. K. Warrington and T. Shallice, Category-specific semantic impairment, *Brain*, 107:829–853, 1984.

Brown dwarf

The core of the Sun is sufficiently hot and dense to host nuclear fusion reactions. The energy released by these reactions eventually emerges from the Sun's core as sunlight. In the early 1960s, S. S. Kumar noted that, if they existed, stars with masses less than one-tenth that of the Sun would not have the high temperature in their cores necessary to sustain nuclear fusion reactions. These objects, called brown dwarfs, would not truly be normal stars, because they would not engage in the most basic stellar behavior: nuclear fusion. Some astronomers suggested that Jupiter might be a brown dwarf, because it is like the Sun in its composition but too small to host nuclear fusion. Other astronomers refuted this claim, stating that Jupiter probably formed differently than did stars. This drew a distinction between planets and brown dwarfs.

Largely because they did not know what these objects would look like, astronomers settled on the name brown dwarf. (Brown is a mixture of many colors and represents the lack of agreement on the dwarfs' appearance.) Other proposed names included lilliputian stars, black dwarfs, infrared dwarfs, failed stars, super-Jupiters, and substellar objects.

Such objects, which are neither planet nor star, had fascinated astronomers even before Kumar's theoretical work. Brown dwarfs have been invoked to explain the dark-matter problem. However, many astronomers believed brown dwarfs did not exist because none had been found. Finally, in 1995 the

Images of the brown dwarf Gliese 229B. (*a*) Discovery image, taken at Palomar Observatory in California (*T. Nakajina and S. Durrance*). (*b*) Hubble Space Telescope image (*S. Kulkarni and D. Golimowski, NASA*).

first definitive detection of a cool brown dwarf was announced; also a young brown dwarf was detected.

Formation and evolution. It is generally assumed that brown dwarfs are formed in a manner similar to that of stars, through the condensation of interstellar gas. As an interstellar gas cloud collapses by the force of gravity, it heats up. As a result, young brown dwarfs are quite hot and luminous. This interior heat is the principal store of energy for the brown dwarf. Over the age of the universe, about 1.5×10^{10} years, the brown dwarf, like a hot rock, will cool and deplete the central store of energy, growing dimmer and dimmer.

Calculations by several groups have established that the minimum mass for an object that shines as a star (through nuclear fusion) is about 80 times the mass of Jupiter. These calculations reveal that these lowest-mass stars will also have a minimum luminosity, the energy lost or emitted per unit time, of about 10^{-4} times the luminosity of the Sun. Young brown dwarfs can have luminosities above this value, but as they age and cool their luminosities plummet. These theoretical calculations show that all brown dwarfs have essentially the same radius, about one-tenth that of the Sun. This is also approximately the radius of Jupiter.

Search for brown dwarfs. Brown dwarfs are expected to shine primarily in the infrared region of the electromagnetic spectrum. Searches for brown dwarfs were formerly limited by the availability of large, sensitive detectors. Infrared detectors of a reasonable size and high sensitivity became available in the 1980s. This development provided the principal technological impetus to the search for brown dwarfs. The *Infrared Astronomical Satellite* (*IRAS*) conducted the first sensitive all-sky survey in the long infrared bands from 12 to 100 micrometers. No single or isolated brown dwarfs were found despite

diligent searches of the entire database. This gave additional impetus to searches for brown dwarfs as companions to nearby stars and for young brown dwarfs in star-forming regions. It was only recently that both these approaches finally led to success. Clearly, searching for companions to nearby stars makes eminent sense. The primary difficulty is that brown dwarfs, especially old brown dwarfs, are dim and cool objects. The brown-dwarf companion could be lost in the glare of the bright star. This difficulty makes it hard for astronomers to image companions, be they brown dwarfs or planets, around nearby stars. Fortunately, two new revolutions in astronomical instrumentation, adaptive optics and interferometry, over the next decade will enable astronomers to image directly large planets around nearby stars. *See* INTERFEROMETRY; SATELLITE (SPACECRAFT).

In 1987, astronomers announced the discovery of a faint companion to a white dwarf, GD 165 (a star which is much less luminous than ordinary stars). The luminosity of this companion, GD 165B, was estimated to be around about 10^{-4} times the solar luminosity, the minimum luminosity of a star. This object thus lies at the border between stars and brown dwarfs.

Hot, young brown dwarf. The Pleiades (also known as the Seven Sisters) comprise one of the nearest clusters of young stars and accordingly have been the focus of a number of searches for isolated young brown dwarfs. Young brown dwarfs are luminous and thus very difficult to distinguish from stars. Fortunately, the fragile element lithium can provide a means to distinguish cool stars from hot, young brown dwarfs. Lithium is transmuted in stars by high-temperature nuclear reactions in the stellar core. Thus any lithium in these stars is expected to be rapidly destroyed by the nuclear reactions in the core. However, the interiors of brown dwarfs with mass less than 65 times that of Jupiter never attain the minimum temperature to transmute lithium. Therefore, although young brown dwarfs might look identical to low-mass stars, they will exhibit the signs of lithium, which the low-mass stars will not.

Two searches for candidate young brown dwarfs in the Pleiades were undertaken in 1995. Sensitive spectroscopic observations with the giant Keck telescope in Hawaii revealed the telltale features of lithium in two objects. This identified two young brown dwarfs, called PPL 15 and Teide 1.

Cool, old brown dwarf. Simultaneously was reported the discovery of a companion of the nearby star, Gliese 229, located a mere 17 light-years (1.0×10^{14} mi or 1.6×10^{14} km) from the Sun (see **illus.**). With a luminosity of 6×10^{-6} times that of the Sun, this was the first unambiguous discovery of an old brown dwarf, one that had absolutely no similarity to any star. The discovery of Gliese 229B (Gl 229B) was the result of a systematic survey of nearby stars that used an instrument tailored for imaging faint objects next to bright stars.

The infrared spectrum of Gliese 229B showed a remarkable resemblance to the spectrum of Jupiter, and no similarity to any star. Subsequent studies have shown that the spectrum of this object exhibits strong features due to methane and water. Finally, observations with the Keck telescope appear to indicate that trace elements, such as titanium, vanadium, and iron, that show strong spectroscopic features in the coolest stars are not visible in Gliese 229B. Astronomers speculate that in the relatively cool atmosphere of Gliese 229B these metals form dust particles and then sink below the atmosphere, becoming impossible to see from the Earth. A good analogy is condensation of water vapor into water (that is, rain). In this case, the rain droplets will be particles such as corundum or estatite, for example.

The mass of Gliese 229B is at the present unknown but generally considered to be 20–50 times that of Jupiter. The illustration shows the discovery image of Gliese 229B, taken at the Palomar Observatory in California, and a picture taken by the Hubble Space Telescope. Turbulence in the atmosphere makes images obtained from ground-based telescopes, such as telescopes at Palomar, fuzzy. The Hubble Space Telescope is located well above the atmosphere and thus can obtain very clear images. The spike seen in the Hubble image is the diffraction spoke of Gliese 229A, an artifact caused in the telescope. *See* SPACE TELESCOPE.

Brown dwarfs and exoplanets. Currently, only one old brown dwarf and two young brown dwarfs are known. However, there is every reason to believe that the Milky Way Galaxy contains a large number of brown dwarfs, though probably not enough to allow brown dwarfs to solve the dark-matter problem. Even more notable is the possible discovery of planets in radial-velocity studies of bright stars. The masses of these (inferred) objects range about 1–10 times that of Jupiter. It appears that the objects with

mass greater than a few times that of Jupiter are found only in eccentric orbits, just like stellar companions but unlike planets. It is possible that these are low-mass brown dwarfs. *See* PLANET.

The National Aeronautics and Space Administration (NASA) has concluded studies defining the search for planetary systems around nearby stars. The final goal of this program, called ExNPS, is to image Earth-like planets around Sun-like stars. On the way, the program envisions the imaging and detailed study of giant planets. The discovery of Gliese 229B, along with its remarkable spectrum, is a glimpse of the sort of detailed studies that will take place in the near future on even lower-mass objects, namely planets.

For background information *see* ADAPTIVE OPTICS; BROWN DWARF; INFRARED ASTRONOMY; PLANET; PLEIADES; STELLAR EVOLUTION in the McGraw-Hill Encyclopedia of Science & Technology.

Shrinivas R. Kulkarni; Ben R. Oppenheimer

Bibliography. T. Nakajima et al., Discovery of a cool brown dwarf, *Nature*, 378:463–465, 1995; B. R. Oppenheimer et al., Infrared spectrum of the cool brown dwarf Gl 229B, *Science*, 270:1478–1479, 1995; R. Rebolo et al., Brown dwarfs in the Pleiades cluster confirmed by the lithium test, *Astrophys. J.*, 469:L53–L56, 1996; C. G. Tinney (ed.), *The Bottom of the Main Sequence—And Beyond*, 1994; B. Zuckerman and E. E. Becklin, Companions to white dwarfs: Very low-mass stars and the brown dwarf candidate GD 165B, *Astrophys. J.*, 386:260–264, 1992.

Calcite

Calcite ($CaCO_3$), the most common simple salt at the Earth's surface, is present in almost all geological environments. Depending on the nature and extent of impurities, calcite is usually white or light in color, and its physical and optical properties make it attractive when polished. It is not a very hard mineral (3 out of maximal 10 on the Mohs hardness scale), so rocks composed of calcite are relatively easy to quarry, carve, and polish. Calcite-bearing rocks have, therefore, been favored as building stones for centuries, and include marble, limestone, and calcite-cemented sandstone.

Characteristic reactions. In the natural environment, calcite-bearing rocks are broken down by chemical reactions involving carbonic acid (H_2CO_3), which is formed when carbon dioxide (CO_2) dissolves in water. The weathering can be written as reactions (1) and (2), where (l) refers to liquid, (g) to gas, (s)

$$CO_{2(g)} + H_2O_{(l)} \rightleftharpoons H_2CO_{3(aq)} \qquad (1)$$

$$CaCO_{3(s)} + H_2CO_{3(aq)} \rightleftharpoons Ca^{2+}_{(aq)} + 2HCO^{-}_{3(aq)} \qquad (2)$$

to solid, and (aq) to aqueous, meaning dissolved. From these reactions it can be seen that the partial pressure of CO_2 in the atmosphere affects the extent of weathering. In the natural environment, rainwater

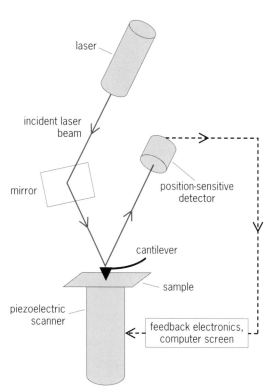

Fig. 1. Schematic diagram of the function of a typical scanning force microscope operating in constant force repulsive mode. *(After S. L. S. Stipp, Understanding interface processes and their role in the mobility of contaminants in the geosphere: The use of surface sensitive techniques, Eclogae geol. Helv., 87:335–355, 1994)*

becomes acidic enough from dissolved CO_2 to slowly dissolve marble and limestone; when the calcite cement is dissolved from a sandstone, the rock simply crumbles.

With increases in the burning of fossil fuels and the smelting of sulfur-bearing ores, atmospheric concentrations of carbonic acid, sulfuric acid (H_2SO_4), and nitric acid (HNO_3) have increased. These acids likewise dissolve in water, increasing the acidity of rain and snow and hastening the weathering of statues and building stones. If a more thorough understanding could be gained of the processes occurring at the interface between calcite and the solution or atmosphere in contact with it, strategies might be developed to improve protection. Some recent observations using very high resolution techniques have shed light on the molecular-level behavior of calcite during exposure to the atmosphere.

Scanning force microscopy. This is one of a family of techniques, the scanning probe methods (SPM), that originated with the Nobel Prize–winning work of B. Binnig and H. Rohrer. Scanning force microscopy (SFM), also known as atomic force microscopy (AFM), records the atomic-scale forces that act between two bodies as they are brought very close together. **Figure 1** shows the function of a typical scanning force microscope when operating in constant force repulsive mode; that is, the image

produced is a map of the height over the surface, where the sum of the atomic-scale repulsive forces is held constant at some value set by the experimenter.

A sample is fixed onto a piezoelectric scanner (Fig. 1). The scanner is made from a semiconducting ceramic so that when a voltage is applied across it, it bends slightly or extends. When the voltage is removed, it regains its original shape. By precisely controlling voltages, the sample can be made to scan in x and y directions over ranges as small as a few nanometers. To measure the local distribution of forces at the sample surface, a sharp tip is brought very close. The tip, often having a radius of curvature of about 50 nm or less, is mounted on a cantilever about 100–200 micrometers long. The cantilever has a very light spring constant, so that the force it exerts is very small, in the nanonewton range. A laser beam is focused onto the shiny back of the cantilever, then reflected into a position-sensitive detector. A force, set by the experimenter, determines the position of the laser reflection on the detector that defines zero deflection for the cantilever.

As the tip approaches the surface, atomic-scale forces cause the cantilever to deflect either up or down, and the location of the reflected laser beam in the detector changes. The detector senses this change in location and sends a signal through feedback electronics to a computer, which calculates the up or down movement required by the piezoelectric element in order to relax the cantilever and return the spot to zero deflection. The appropriate voltage is sent to the scanner so that the z position is adjusted, and the spot is recentered on the detector; then the scanner adds increments in x or y to move the sample to the next probing point. A record of

Fig. 2. Scanning force microscope image of a freshly cleaved calcite surface after only 10 min exposure to air. Light color is highest, or closest to the observer, whereas dark is lowest, or farthest away. The terraces are flat, and the step edges are only 0.3 nm high, which is the thickness of one layer of Ca and CO_3.

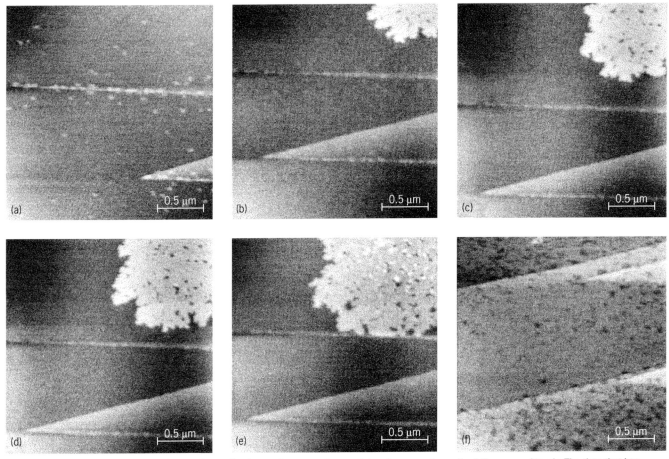

Fig. 3. Series of scanning force microscope images taken from a surface of calcite exposed to air. The time that has elapsed since cleavage is (a) 2 h 15 min; (b) 3 h 30 min; (c) 4 h; (d) 4 h 30 min; (e) 6 h 45 min; (f) 22 h. *(From S. L. S. Stipp, W. Gutmannsbauer, and T. Lehmann, The dynamic nature of calcite surfaces in air, Amer. Mineral., 81:1–8, 1996)*

each adjustment in z that is required to maintain constant force at each of the x,y points results in an image of the apparent topography for the surface. Height differences of fractions of an ångström (0.1 nm; the approximate radius of a calcium atom) can be recorded.

Scanning force microscopy is particularly useful for studying geochemical reactions such as weathering, because it can be used in air or under controlled solution or gas composition. Many other techniques that are used to study structure or composition at very high resolution need analytical chambers that are held under vacuum, or the sample must be prepared by coating with carbon or gold, so the experimental conditions are very different from those where the reactions actually take place in nature. Scanning force microscopy has been used to study many mineral surfaces, at molecular-scale resolution, during the time and for the actual conditions under which the reactions were taking place.

Observations. Figure 2 shows a scanning force microscopy image, taken in air over an area 2 μm wide, on a freshly cleaved calcite surface. The surface is atomically smooth on the terraces, and the steps are 0.3 nm high, which is the height of one layer of

$CaCO_3$ molecules. Such a surface, however, maintains its fresh appearance for only a few minutes. As it is observed, step edges retreat, grow, and change shape; holes appear on the terraces; and sometimes large, dendritic formations appear. All of this happens while the calcite surface is apparently dry, exposed only to air, out of contact with water. This same general behavior has been observed on hundreds of calcite samples.

Figure 3 provides one example of the changes that can occur on a series of images taken over time. The surface after cleavage was as flat as that in Fig. 2. In Fig. 3a two atomically flat terraces are separated by a 0.3-nm-high cleavage step running from left to right. The triangle on the lower right is also 0.3 nm taller than the terrace that it sits on. After 2 h, small spots have accumulated on the terraces and along the top of the cleavage steps—one of the typical forms of accumulation. Ordered crystals grow out from terrace edges or in small islands, with height of 0.3 or 0.6 nm, the thickness of one or two layers of $CaCO_3$ molecules, respectively. From other experiments, some of the material contributing to the accumulations can be seen to come from tiny particles of calcite dust that were created during fracture and held at the surface by static charge; some of the material comes from the edges of other cleavage steps, or from holes that form in the terraces.

In the next and subsequent images (Fig. 3b-f), the analysis area has been moved 1 μm to the right, so more of the cleavage triangle is visible. After more time has passed, the spots seem to disappear again and a large dendritic accumulation begins to form, with edges of branches and holes that run parallel to the crystal's preferred directions of cleavage. The accumulation has a height of about 10 layers of $CaCO_3$ molecules, and the holes extend to the original surface. It spreads until it reaches step edges. Although they are many layers tall, such accumulations seem incapable of moving easily up or down over edges that are only one layer tall. Growth extends along the terrace (Fig. 3e), and separately formed accumulations generally cover the adjoining terraces. After 22 h, a site very close to the earlier images shows that the surface has been completely covered (Fig. 3f). Each of the terraces has accumulated a layer of the same thickness (3.0 nm or 10 layers), so that the height difference from one terrace to the next is still 0.3 nm.

Experiments with many samples, under various conditions of humidity and gas composition, show that the rate of change of surface morphology is affected by the relative moisture content of the atmosphere in contact and, to a lesser extent, by the partial pressure of CO_2. Surfaces that were first exposed to water, then dried, and examined during exposure only to air undergo the same sorts of changes in morphology.

Results from experiments using other high-resolution techniques contribute to the interpretation of these observations. It has been shown that mineral surfaces adsorb a layer of water several molecules thick from the humidity present in air. These new observations suggest that such a thin water film dissolves some ions from the calcite step edges and terraces. When saturation is reached in this invisible capillary layer, the ions precipitate again somewhere else on the surface; thus, the principle known as dynamic equilibrium can be directly observed. Higher humidity means that more water is adsorbed; thus a thicker solution layer is produced, resulting in increased mobility for the calcium and carbonate ions. However, experiments have shown that even in 20–25% humidity, which feels fairly dry at room temperature, significant changes occur within minutes.

Implications. One application of these results is especially important to the study of weathering processes. In the past, it was assumed that in the absence of liquid water, apparently dry surfaces remained static. Calcite-bearing statues and building stones were thought to be damaged only during the time when acidic precipitation (rain, snow, or fog) fell directly on them. These new observations suggest that destruction occurs even when surfaces appear dry. If sulfuric and nitric oxides are present in the atmosphere, they are also dissolved in the invisible

water that adsorbs on seemingly dry calcite. The acidic solution then dissolves and reorders the surface. The next rainfall washes away this saturated solution and continues to dissolve more calcite. Stone with pore spaces (such as limestone) or higher permeability along grain boundaries (such as calcite-cemented sandstone) weathers faster because each event increases the surface area available to the acidic capillary layer. Additional research on calcite surfaces at molecular-scale resolution is in progress. Its aims are to gain a better understanding of this and other processes that occur during exposure to solutions and atmospheres of varying composition.

For background information *see* ACID RAIN; CALCITE; CRYSTAL DEFECTS; CRYSTAL GROWTH; LIMESTONE; MARBLE; PIEZOELECTRICITY; SCANNING TUNNELING MICROSCOPE; WEATHERING PROCESSES in the McGraw-Hill Encyclopedia of Science & Technology.   Susan L. S. Stipp

Bibliography. B. Binnig and H. Rohrer, The scanning tunneling microscope, *Sci. Amer.*, 253:50–56, 1987; C. M. Eggleston, in K. R. Nagy and A. E. Blum (eds.), *Review of Scanning Probe Microscopy, Clay Minerals Series*, 1994; A. J. Gratz, P. E. Hillner, and P. K. Hansma, Step dynamics and spiral growth on calcite, *Geochim. Cosmochim. Acta*, 57:491–495, 1993; R. Reeder, *Carbonates: Mineralogy and Chemistry, Reviews in Mineralogy 11*, Mineralogical Society of America, 1983; S. L. S. Stipp, C. M. Eggleston, and B. S. Nielsen, Calcite surface structure observed at microtopographic and molecular scales with atomic force microscopy (AFM), *Geochim. Cosmochim. Acta*, 58:3023–3033, 1994; S. L. S. Stipp, W. Gutmannsbauer, and T. Lehmann, The dynamic nature of calcite surfaces in air, *Amer. Mineral.*, 81:1–8, 1996.

Cancer (medicine)

The increased frequency of newly discovered types of inherited cancers has influenced researchers to focus on determining the origin, location, and mechanisms that create errors (mutations) within the deoxyribonucleic acid (DNA) replication sequences of certain genes. This article discusses how the mismatching of certain base-pair sequences leads to the induction of several types of colon cancer, and the genetic mutations which are believed to be responsible for the production of breast cancers.

Hereditary Nonpolyposis Colorectal Cancer

Hereditary nonpolyposis colorectal cancer (HNPCC) is an autosomal dominant disease and accounts for as much as 10% of colorectal cancers reported in industrial nations. HNPCC is difficult to distinguish from common forms of sporadic colon cancer; but it is generally characterized by early-onset colon cancer in closely related individuals, and the frequent occurrence of other specific cancers in close relatives, including cancer of the endometrium, stomach, pancreas, and urinary tract. At least four HNPCC genes have been identified; each encodes a protein that is a key part of the cellular machinery for DNA replication and repair.

HNPCC gene. The first HNPCC gene was mapped by A. de la Chapelle, B. Vogelstein, and their colleagues using genetic linkage analyses on related persons with HNPCC. The search highlighted a region of chromosome 2 as the probable location of a gene for HNPCC susceptibility in two families. In nearly simultaneous reports, three different groups observed that the majority of tumors from individuals with HNPCC showed widespread alterations in short repeated DNA sequences, particularly the dinucleotide repeat sequences called microsatellites. These observations of a generalized instability phenotype recalled earlier descriptions by a number of groups of allelic losses and accumulation of specific point mutations in some colorectal cancers. The phenotype was termed replication error positive (RER+) because it suggested a common underlying failure of a gene (or genes) important in fidelity of replication. The RER+ phenotype is now known to occur in almost 90% of HNPCC tumors and about 15% of sporadic colorectal tumors.

Well-defined pathways that facilitate DNA repair in the bacterium *Escherichia coli* and in the yeast *Saccharomyces cerevesiae* suggested a variety of candidate genes for HNPCC. Most notable were studies showing that, when mutated in yeast, any of three genes known to be involved in DNA mismatch repair (*PMS1*, *MLH1*, and *MSH2*) resulted in yeast with essentially the same RER+ phenotype that had been observed in HNPCC tumors. Similar results had been reported years earlier for the *E. coli mutS* and *mutL* genes, which are homologs of the yeast *MSH2* gene and the *PMS1* and *MLH1* genes, respectively.

Mismatched repair genes. In DNA the nucleotide base adenine normally pairs with thymine, and cytosine with guanine. Mismatched base pairs are generated in a variety of ways, including errors in DNA replication or genetic recombination, and damage to the nucleotide precursors incorporated during DNA replication. Damage may arise from a variety of sources, including environmental factors. In *E. coli*, the *mutS* gene product is involved in recognition of the mismatch; the *mutH* gene product is a methyl-directed endonuclease that cleaves at the site of mismatch; and the *mutL* gene product mediates formation of complexes between the *mutS* and *mutH* proteins. Together, the *E. coli mut HLS* repair system is responsible for recognizing errors in a newly synthesized unmethylated strand of DNA, excising the errors, and then resynthesizing the corrected double-stranded DNA. These observations suggested that the human homologs of these mismatch repair genes carry out the same function, and that mutation in these genes prevents mismatch repair and predisposes to HNPCC. *See* DEOXYRIBONUCLEIC ACID (DNA).

This hypothesis was proven correct in subsequent studies. First, the human homolog of the bacterial *mutS* and yeast *MSH2* genes, called *hMSH2*, was identified and mapped to chromosome 2p. Second,

studies showed that the expression of this protein in *E. coli* interfered with the normal bacterial mismatch repair pathway. Finally, germline mutations in the gene were found in affected individuals of several HNPCC families. Inherited mutations in the *hMSH2* gene may be responsible for 40–50% of HNPCC cases.

Human homologs of MutL gene. A second HNPCC locus that was mapped to chromosome 3p proved to be a homolog of the yeast *MLH1* and *E. coli mutL* genes. The chromosome 3p gene was cloned by two research groups using different approaches. One particularly productive set of experiments involved searching for homologs of bacterial and yeast mutator genes in a database of 65,000 human genes identified by their expressed sequence tag, or EST. (ESTs are 200–400 base pairs of sequence that define part of a gene that is expressed in same tissue. They are cataloged and traded by the primer or EST sequence.) Three clones of interest were identified: one with homology to *MLH1*, and two with homology to *PMS1*; they were termed *hMLH1*, *hPMS1*, and *hPSM2*, respectively. The *hMLH1* gene was of particular interest because it mapped to chromosome 3p21. Confirmation that *hMLH1* was a second HNPCC gene came from multiple reports of protein-disrupting mutations of this gene in affected individuals from linked families. Inherited mutations in *hMLH1* appear to account for another 40–50% of classic HNPCC.

Other genes of interest. There is evidence that the first two HNPCC genes account for the majority of inherited HNPCC and about 15% of apparently sporadic cases of colorectal cancer. However, HNPCC in some families appeared unlinked at either locus, which suggested that additional candidates, such as *hPMS1* and *hPMS2*, needed further investigation. After examination, it was determined that both of these genes are associated with HNPCC in at least a few families. Mutations have also been found in other relevant genes, such as DNA polymerase delta (a subunit of DNA polymerase), in colon cancer cell lines that exhibit microsatellite instability. However, to date no mutations in these genes have been reported in HNPCC families.

Mutations, mechanisms, and clinical implications. Although a link between colon cancer and defects in DNA repair seems to occur naturally, the exact mechanism by which germline mutations of DNA mismatch repair genes cause tumors remains unclear. Some tumors show loss of heterozygosity, that is, loss of the unaffected wild-type allele and retention of the mutated allele. This suggests that mismatch repair genes function as classic tumor suppressors, with two mutations being necessary for cancer development. Presumably, loss of both copies of the relevant repair gene results in the inability to correct genomewide mutations, and increases the likelihood that random mutations will occur in other critical oncogenes or tumor suppressors and lead to cancer.

Genetic screening. Genetic screening of individuals at risk for HNPCC is a topic of considerable interest.

The lifetime risk for colorectal cancer for mutation carriers is nearly 80%, and risk for some other associated cancers, such as endometrial cancer, is also elevated. Because the average age of diagnosis for HNPCC-related colorectal cancer is about 45, most centers recommend that screening begin between the ages of 20 and 25. Individuals considered at risk are defined by an agreed-upon set of characteristics called the Amsterdam Criteria, including (1) people with three or more close relatives with colorectal cancer, (2) colorectal carcinoma involving at least two generations, and (3) one or more colorectal carcinoma cases diagnosed at less than 50 years of age.

Genetic penetrance. Few studies have addressed the important issue of phenotypic heterogeneity for different HNPCC alleles. Currently, over 50 distinct mutations have been reported in the *hMSH2* and *hMLH1* genes and include deletions, splicing mutations, and transversions. Population-based epidemiologic studies will be needed to determine whether mutations in different genes or even different mutations in the same gene predispose individuals to the particular characteristics of the disease such as an earlier age at onset, occurrence of other cancers, or sensitivity to specific environmental exposures. It is of particular interest that not every individual with HNPCC reports a family history of cancer.

Mismatch repair genes in nongastric cancers. The role of mismatch repair genes in the etiology of nongastrointestinal cancers has been studied extensively by analysis of tumor collections for the RER+ phenotype. For some cancers, such as brain and oral cancer, there is little evidence of microsatellite instability. For others, such as lung, breast, and prostate cancer, the data are inconsistent. The confusion may simply reflect the difficulty in distinguishing RER+ phenotype from random loss of chromosomes that is characteristic of many tumors. Alternatively, it may reflect real differences relating to the grade or stage of tumors examined. For example, some studies indicate that the level of microsatellite instability in breast tumors is minimal, while others suggest that it may be as high as 21%. In one study, the instability was significantly correlated with lobular histotype (histology of one type of breast cancer), lymph node involvement, and large tumor size.

Inherited Mutations in Breast Cancer Susceptibility Genes

Breast cancer is a disease of immense public health significance that will occur in 1 in 8 women in the western world by age 85. Epidemiological data suggest that 5–10% of all breast cancers are attributable to rare, highly penetrant, inherited mutations in autosomal dominant genes (genes located on chromosomes other than the *X* and *Y*); in early-onset breast cancers, this proportion is estimated to be much greater. Currently, two major breast cancer susceptibility genes demonstrating autosomal dominant modes of inheritance have been identified: *BRCA1* and *BRCA2*.

BRCA1 gene. The first of these genes to be identified, *BRCA1*, was mapped in 1990 to a region of chromosome 17q21 by linkage analysis in those families with large numbers of early-onset cases of breast or ovarian cancer. Cloning and sequencing revealed that *BRCA1* encodes a single 5592-nucleotide message spanning 24 exons and producing an 1863-amino acid protein. *BRCA1* has little homology to other cloned genes, except for a RING finger motif (a specific set of 40 amino acids that appear in a particular order and define a structure common in proteins involved in regulating transcription), which has been previously observed in several transcription factors and suggests that *BRCA1* may play a role in transcription.

Observations. Several observations have suggested that *BRCA1* functions as a tumor suppressor gene. Haplotype analyses of the tumors of individuals with inherited *BRCA1* mutations indicated that the wild-type (normal) allele was typically lost and the inherited deleterious allele was retained. Loss of heterozygosity (LOH) in the *BRCA1* region in sporadic breast tumors is not uncommon, with 40–80% of tumors showing it. This suggests that *BRCA1* may also have a role in sporadic breast cancer, although somatic mutations of *BRCA1* have yet to be detected in sporadic breast tumors.

Experiments are being conducted in an attempt to identify the mechanisms by which *BRCA1* acts. For example, analysis of the effect of antisense oligonucleotides on mammary epithelial cells in culture and studies of expression in early mouse development suggest that *BRCA1* may play a role in the regulation of proliferation. There is some debate whether normally functioning *BRCA1* protein is localized to the nucleus or whether it is a secreted protein that exhibits the properties of a granin. Granins are proteins that localize to secretory vesicles, are secreted by a regulated pathway, are posttranslationally glycosylated, and are typically responsive to hormones.

Patterns of autosomal dominant inheritance. The pattern of breast cancer observed in families with a high incidence of breast or ovarian cancer is most consistent with autosomal dominant inheritance and high levels of penetrance. In these families, *BRCA1* mutation carriers appear to have an 80% or greater lifetime risk of breast cancer and a 40–60% lifetime risk of ovarian cancer by age 85. In addition, carriers of mutant *BRCA1* alleles in these families appear to have an increased risk of prostate and colon cancer, although further study is needed to confirm this point.

Germline mutations. More than 100 distinct, germline mutations (mutations within a lineage of cells that form gametes) in *BRCA1* have been described. Mutations are widely dispersed throughout the coding sequence. All classes of mutations are observed, such as missense, nonsense, deletions, and insertions; over 75% of reported mutations result in production of a truncated protein. These proteins are not full length because of introduction of a premature stop codon during translation. Truncated proteins are seldom functional. Whether the biological consequences—such as cancer occurrence, severity, and type, and age at onset—vary by specific mutation is still under investigation.

Identical mutations. The presence of identical mutations in unrelated families that share ethnic origin suggests the presence of founder mutations that have been propagated through several generations. Identification of such mutations can be highly useful for the development of genetic testing programs, because it reduces the regions of the gene that must be screened within particularly well characterized populations. For example, one of the most common mutations reported in *BRCA1* is a single mutation that occurs with unusually high frequency among women of Ashkenazi (central or east European) Jewish descent. Compelling data suggest that this mutation (a two-base pair deletion at nucleotide 185) occurs in about 0.9% (95% confidence limit, 0.4–1.8%) of all Ashkenazi Jewish individuals, and accounts for a large proportion of inherited breast cancer in this population.

Inherited mutations. Studies of women from the general population (not selected on the basis of family history) who were diagnosed with breast cancer before age 35 suggested that about 7.5% of early-onset breast cancer was due to inherited mutations in the *BRCA1* gene. These studies also found a number of women with germline mutations in *BRCA1* who lacked a significant family history of breast or ovarian cancer. The absence of correlation between *BRCA1* mutation status and the presence of a distinctive family history may reflect variation in family structure, incomplete family history information, varying allele penetrance, modification by other genes, or the influence of environmental factors.

BRCA2 gene. The *BRCA2* gene, mapped to chromosome 13q12-13 in 1994 and cloned in 1995, has a coding region of approximately 10,256 base pairs. *BRCA2* does not show strong homology to known genes, with the exception of a small region of weak homology to the *BRCA1* gene. As with *BRCA1*, analyses of high-risk families revealed mutations associated with disease throughout the *BRCA2* coding region, most of which cause protein-truncation events. Unlike *BRCA1*, *BRCA2* mutations do not appear to confer a significantly increased risk of ovarian cancer but do seem to increase the risk of male breast cancer. In addition, preliminary data suggest that *BRCA2* mutations predispose to prostate, pancreatic, and colon cancers.

There are indications that *BRCA2*, like *BRCA1*, may also be a tumor suppressor gene. Loss of heterozygosity has been seen in many tumors from families with strong evidence of linkage to *BRCA2*. Tumors from individuals with inherited mutations in *BRCA2* show consistent loss of the wild-type allele. *BRCA2* mutations appear to be infrequent in sporadic breast cancers. At this time, however, there is little specific information on *BRCA2* function or its role in sporadic breast cancers.

Female carriers of a *BRCA2* mutation are estimated to have a more than 80% lifetime risk of breast cancer by age 85. This single mutation—in this case a dele-

tion of the pyrimidine base thymine (T) at nucleotide 6174—occurs with high frequency in women of Ashkenazi Jewish descent.

Public health investigations. While inherited mutations in the *BRCA1* and *BRCA2* genes are not especially common in the general population, they occur with sufficient frequency to be a public health concern. There is mounting interest in genetic testing for breast cancer susceptibility. However, the development of comprehensive population-based screening strategies is contingent on the development of efficient mutation detection systems with reasonable sensitivity and specificity, an increased understanding of the risk associated with specific mutations, and the formulation of screening and treatment recommendations. Investigations are ongoing to determine the function and contribution to cancer etiology of the *BRCA1* and *BRCA2* genes. It is hypothesized that other genes and environmental factors may affect the penetrance of both of these genes. Large-scale population-based studies will be needed before the frequency, spectrum, and penetrance of mutations in *BRCA1* and *BRCA2* can be determined.

For background information *see* BREAST DISORDERS; CANCER (MEDICINE); GENE ACTION; GENETIC MAPPING; HUMAN GENETICS; LINKAGE (GENETICS); MUTATION; ONCOLOGY; SEX-LINKED INHERITANCE in the McGraw-Hill Encyclopedia of Science & Technology.

Elaine A. Ostrander; Kathleen E. Malone

Bibliography. M. Aarnio et al., Lifetime risks of different cancers in hereditary non-polyposis colorectal cancer (HNPCC) syndrome, *Int. J. Canc.*, 64:430–433, 1995; R. Fishel et al., The human mutator gene homolog MSH2 and its association with hereditary nonpolyposis colon cancer, *Cell*, 75:1027–1038, 1993; J. Jiricny, Colon cancer and DNA repair: Have mismatches met their match?, *Trends Genet.*, 10:164–167, 1994; S. Neuhausen et al., Recurrent *BRCA2* 6174delT mutations in Ashkenazi Jewish women affected by breast cancer, *Nat. Genet.*, 13:126–128, 1996; J. P. Struewing et al., The carrier frequency of the *BRCA1* 185del AG mutation is approximately 1 percent in Ashkenazi Jewish individuals, *Nat. Genet.*, 11:198–200, 1995; R. Wooster et al., Identification of the breast cancer susceptibility gene *BRCA2*, *Nature*, 378:789–792, 1995.

Carbon dioxide

Carbon dioxide (CO_2) is naturally abundant, relatively nontoxic, and nonflammable. It is one of the least expensive organic solvents currently available. Because of its green characteristics, that is, it is environmentally benign, carbon dioxide has been employed commercially as a solvent in food processing (for example, decaffeination of coffee and tea and extraction of hops), in reduced-emission coating systems, in production of polyurethane and polystyrene foams free of chlorofluorocarbons, and in extraction of organic contaminants from dredged material. These processes can involve high-capacity plants—

over 50×10^6 lb (22.5×10^6 kg) per year in the case of decaffeination processes.

Characteristics. In addition to its environmental advantages, CO_2 possesses several physical attributes that render it a useful extraction solvent. First, liquid CO_2 exhibits a viscosity of only 0.1 centipoise (one-tenth that of water or conventional organic solvents). Because diffusivity (and thus mass-transfer rates) are inversely proportional to viscosity, extractions in CO_2 can be accomplished faster than in conventional solvents, provided that solubilities are sufficient. Second, CO_2 is a gas under normal atmospheric conditions; therefore, dissolved material can be completely recovered from a CO_2 solution through depressurization, generating dry product and allowing straightforward recycling of the CO_2. Finally, CO_2 has a readily accessible supercritical regime, with its critical temperature being 31°C (88°F), allowing supercritical fractionation of complex mixtures containing thermally labile components; that is, they tend to decompose when subjected to elevated temperatures.

Carbon dioxide is particularly appropriate for separations involving compounds of biological origin in that many such products are produced in dilute aqueous solution (culture medium). Use of CO_2 in liquid-liquid extraction from water is not burdened by concerns over contamination of the aqueous phase with trace amounts of the organic solvent (typical liquid-liquid extraction would require remediation of the aqueous phase prior to discharge). Further, CO_2 is useful in situations requiring rapid extraction and ease of concentration, as is the case with biological materials. Unfortunately, liquid CO_2 has a dielectric constant of approximately 1.5, and thus will not solubilize any significant quantities of polar or hydrophilic solutes. In fact, the solubility of even single amino acids in CO_2 is only fractions of a microgram per liter, and solubility of biological compounds such as proteins with higher molecular weights can safely be assumed to be zero.

Extraction of proteins. Several research groups have shown that surfactants (and thus reverse micelle formation) can be used to prompt proteins to migrate from aqueous solution to an organic solvent, in which protein solubility is ordinarily zero. Typically, the pH or ionic strength of the aqueous solution is adjusted so that the protein can be readily solubilized by the emulsion (organic) phase. Despite successful extractions from both aqueous buffer and cell culture medium, use of surfactant-based extractions is not practiced commercially because of (among other things) the difficulties in reuse of the emulsion. In emulsion extractions, protein is recovered from the organic phase via addition of salt (to dewater the micelles) or of micelle-disrupting agents such as ethyl acetate. As these techniques are not easily reversible, recycling of the emulsion phase is problematic. In addition, contact of large volumes of aqueous and organic phases will inevitably lead to contamination of the aqueous phase, requiring subsequent remediation steps.

In the late 1980s, a research group at Battelle's Pacific Northwest Laboratories found that emulsions could be formed (using an anionic surfactant) in near- and super-critical ethane and propane, and thus proteins could be extracted from an aqueous solution. Further, by careful control of the pressure, a series of proteins could be induced to migrate one by one into the organic phase, suggesting that a selective extraction process for biological compounds could be created. It was later demonstrated that a protease named subtilisin Carlsberg could be extracted into a propane-based emulsion (by using a commercial nonionic surfactant, which is an ethoxylated sugar ester) and then recovered by depressurization with only a small loss in activity. Because protein solubilization and precipitation from the emulsion was controlled simply through manipulation of the pressure, use of a compressible fluid essentially solved the problem of reuse of the emulsion phase; yet use of propane on large scale was impractical.

Although it would seem that the next step in this research should have been use of emulsions in CO_2, this proved to be a difficult task. Investigations by research groups at a number of institutions approached the problem from various angles. In 1991, several groups found that commercially available hydrocarbon surfactants exhibit very poor solubility in CO_2; however, during this same period, others found that fluorinated compounds dissolve in CO_2 in much higher quantities than their hydrocarbon analogs at comparable pressures. Subsequently, researchers determined that fluorinated and silicone-functional surfactants dissolve in CO_2 at mild pressures. A recent demonstration showed that such surfactants do indeed form reverse micelles in CO_2 and that bovine serum albumin can be solubilized in these micelles.

Extraction of subtilisin Carlsberg into carbon dioxide. Eventually, in an experiment a series of anionic and nonionic surfactants based on a fluoroether hydrophobe were generated. The hydrophobic portion of each of the surfactants was an oligomer of hexafluoropropylene oxide, while the hydrophilic portion was either a sulfonate group (anionic) or an oligomer of ethylene oxide (nonionic). These surfactants exhibit significant solubility in CO_2 with concentrations above 1 mmol at pressures below 200 atm (20 megapascals) at room temperature. The behavior of these fluoroether surfactants in CO_2/buffer mixtures revealed some interesting features. As expected, as pressure increases, there is an increased tendency to form Winsor II systems; these are emulsions in organic phase, plus an excess water phase. As the aqueous to organic phase ratio (volume/volume) increased, the tendency for formation of Winsor II emulsion increased as well. This is significant for potential applications suggesting that a given volume of water can be extracted with less than a 1:1 ratio of CO_2.

While the anionic surfactants readily formed Winsor II systems, only some of the nonionics (those with the longer polyethylene glycol chains) formed

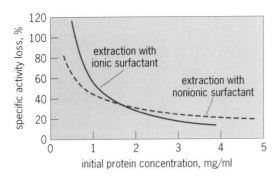

Specific activity loss of subtilisin Carlsberg versus initial buffer concentration following extraction into carbon dioxide at room temperature and 125-bar (12.5-MPa) pressure, followed by isolation of the emulsion phase and recovery via slow depressurization.

Winsor II systems in CO_2/buffer mixtures. It was found that inclusion of protein in the buffer affected the behavior of the emulsion phase; those nonionic surfactants that would form only Winsor I systems (emulsion in the aqueous phase plus excess organic) or Winsor III systems (middle phase emulsion plus excess water and organic phases) in CO_2/buffer mixtures could form Winsor II systems when protein was present. Possibly, some of the protein interacts with the surfactant in such a way as to act as a cosurfactant, helping to stabilize the inverse emulsion. For each type of surfactant, the presence of the protein allows formation of Winsor II systems at lower surfactant concentrations.

In performing extractions, (1) the surfactant was added to liquid CO_2 and stirred to mix; (2) protein was mixed with buffer and pumped into a high-pressure view cell; (3) the CO_2/surfactant mixture was pumped into the view cell while part of the aqueous solution was withdrawn, in order to contact the phases at constant pressure; (4) the system was stirred, and the phases allowed to separate (settle); (5) the aqueous phase was withdrawn from the cell, while sufficient CO_2/surfactant mixture was added to maintain constant pressure; and (6) the CO_2 phase was slowly depressurized, and the amount of protein recovered and its activity were measured.

As shown in the **illustration**, the amount of protein present in the initial buffer solution plays an important role in determining both the amount of protein that can be extracted into CO_2 and the activity that is retained by this protein following recovery. Possibly, a certain amount of the protein is sacrificed as a cosurfactant to the system, helping to stabilize the emulsion. It was observed that the presence of the protein in the system changes the phase behavior relative to the case where only buffer is present. Higher amounts of protein can be extracted by using the anionic surfactant. This is not surprising in that at high pressure the pH of the aqueous phase (despite a nominal buffer pH of 7) is likely to be between 4 and 5, given that CO_2 dissolves in water, forming carbonic acid (H_2CO_3). At low pH's, subtilisin exhibits numerous positive charges on its surface, provid-

ing the potential for favorable interaction with the anionic surfactant.

In the case of the fluoroether surfactants described here, the protein apparently interacts with the surfactant in such a way as to help stabilize the emulsion, which does lead to the loss of a certain amount of the protein during extraction.

For background information *see* BUFFERS (CHEMISTRY); CARBON DIOXIDE; EMULSION; MICELLE; SOLVENT; SOLVENT EXTRACTION; SURFACTANT in the McGraw-Hill Encyclopedia of Science & Technology.

E. G. Ghenciu; E. J. Beckman

Bibliography. J. F. Ely and T. J. Bruno (eds.), *Supercritical Fluid Technology: Reviews in Modern Theory and Applications*, 1991; J. L. Fulton et al., Aggregation of amphiphilic molecules in supercritical carbon dioxide: A small angle light scattering study, *Langmuir*, 11:4241–4250, 1995; K. P. Johnston et al., Water-in-carbon dioxide microemulsions: An environment for hydrophiles including proteins, *Science*, 271:624–626, 1996; B. D. Kelley, D. I. C. Wang, and T. A. Hatton, Affinity-based reverse micellar protein extraction: I. Principles and protein-ligand systems, *Biotech. Bioeng.*, 42:1199–1208, 1993; V. J. Krukonis and M. A. McHugh, *Supercritical Fluid Extraction: Principles and Practice*, 1995; A. V. Yazdi et al., Highly carbon dioxide soluble surfactants, dispersants, and chelating agents, *Fluid Phase Equil.*, 117:297–303, 1996.

Chaos

The branch of science known as chaotic dynamics originated with work done around 1900 by the mathematician H. Poincaré. His work and that of others have led to a basic understanding that even very simple systems can evolve with time in a very complex manner. In viewing such time evolution, words such as wild, random, and turbulent come to mind. Furthermore, this basic type of motion is extremely common and occurs in a vast variety of fields. Examples are the motion of thermally convecting fluids, the fluctuations of light intensity in certain laser systems, the motion of celestial bodies (for example, two planets circling a star), and even the irregular beating of a diseased heart. Recently, researchers have realized that chaotic motions can be controlled by means of small perturbing forces. That is, the natural, free-running chaotic motion can be altered in a predetermined way so as to improve the performance of an otherwise chaotic system. This article describes the principles of chaos control and some recent developments in this area. The key defining attribute of chaos that allows the achievement of control with only small perturbations is exponential sensitivity.

Exponential sensitivity. Mathematically, chaotic motion can be defined as behavior that is exponentially sensitive to small perturbations. This may be illustrated by considering the evolution of two identical systems with time. For example, the systems might be two balls rolling without friction, each on one of two identical surfaces with hills and valley. Up to a certain point in time the motion of the two balls is the same. One of the two balls can be given a very tiny kick, and the subsequent motion of the two balls can then be followed and their positions compared. There is said to be exponential sensitivity if the small kick leads the states of the two systems (the positions of the two balls) to steadily separate with time, and furthermore this separation is exponential. An example of exponential separation may be given by supposing that 1 second after the kick the distance between the two balls is $\frac{1}{16}$ mm and that this distance initially doubles every second thereafter. Thus, after 2 s the separation distance is $\frac{1}{8}$ mm, after 3 s it is $\frac{1}{4}$ mm, and so on. Even though the initial separation was very small ($\frac{1}{16}$ mm), the separation rapidly becomes large. After, say 15 s, if doubling would occur each second, the separation would be roughly 1 m. Thus a very tiny kick eventually has a large consequence. The meteorologist E. N. Lorenz referred to this as the butterfly effect: Whether or not a butterfly flaps its wings in India, say, might eventually determine completely different weather patterns in the United States some long time thereafter (say, a month or two). Lorenz suggested that this effect might limit long-range weather forecasting. Paradoxically, this extreme sensitivity can be an advantage in attempting to control a chaotic system.

Unstable periodic orbits. Another attribute of chaos that has proven to be crucial for its control is that chaotic motion typically has embedded within it unstable periodic motions. Before explaining this, it should be emphasized that this attribute is very much connected with the attribute of exponential sensitivity and should not be viewed as independent. A periodic orbit is a motion that precisely repeats itself in a fixed duration of time, called the period. Thus, if the period is 1 s, the system starting at a state on the periodic orbit will go through a sequence of states as time increases, returning precisely to its initial state after 1 s, and then going through precisely the same sequence of states as before, and so on forever. A periodic orbit is unstable if small perturbations from it cause the system state to move away from the periodic orbit exponentially with increasing time. Chaotic motions typically come very close to unstable periodic orbits in their erratic wandering. However, because of the unstable nature of these periodic orbits, the chaotic orbit typically does not remain in the vicinity of any one of them for long. However, if the system state could be placed exactly on an unstable periodic orbit, it would, in principle, undergo periodic (not chaotic) motion forever. In practice, small system noise (which is inevitably present) perturbs the system from its potentially exact periodic orbit, and chaotic motion thus always prevails.

Controlling chaos. In the past the predominant reaction of researchers to the occurrence of chaos was to study its properties as they existed in any given

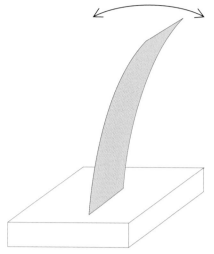

Fig. 1. Magnetoelastic ribbon clamped at its base.

system, while the reaction of engineers and designers was most often to avoid it. Although these attitudes are still prevalent, work in 1990 by E. Ott, C. Grebogi, and J. A. Yorke showed that there is another, more active approach to the occurrence of chaos. In particular, they demonstrated theoretically that when a chaotic orbit wanders close to an unstable periodic orbit embedded in the chaotic motion, the system control can be adjusted so as to give the system a small carefully chosen sequence of perturbations which eventually land the system state on the periodic orbit. If noise in the system is present, the noise tends to jostle the system state off the periodic orbit, and since the periodic orbit is unstable, the system state would move away and resume its chaotic motion. However, with a small control available, this can be counteracted: When the orbit is observed to move slightly off the periodic orbit, the small control can be reapplied to return it to the periodic orbit. Thus the general approach is: (1) Observe the free-running chaotic motion and perform an analysis of it to extract knowledge of the unstable periodic orbits. (2) Assess the system performances that would apply if the system followed the time evolution of each of the unstable periodic orbits found in step 1. (3) Choose the periodic orbit that gives the best performance and use the small control to keep the system on that orbit. This basic approach has proven to be widely applicable, as demonstrated in a variety of laboratory experiments.

Example of chaos control. The first demonstration of controlling chaos in an experiment was done by using a thin ribbon of magnetoelastic material that was clamped at its base but was otherwise free to move (**Fig. 1**). The stiffness of the ribbon material was nonlinearly dependent on an applied magnetic field. When the strength of this field was oscillated periodically, the ribbon alternately buckled under gravity and stiffened in a complicated chaotic manner. The position of the ribbon was measured by an

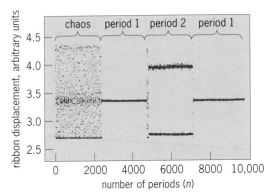

Fig. 2. Experimental control of a chaotic ribbon. *(From W. L. Ditto, S. N. Rauseo, and M. L. Spano, Experimental control of chaos, Phys. Rev. Lett., 65:3211–3214, 1990)*

optical sensor. By making very small perturbative adjustments in the applied magnetic field, the motion could be controlled and made to assume a variety of periodic motions. **Figure 2** shows the ribbon displacement sampled once every oscillation of the applied magnetic field as a function of the number of periods, n. Initially no control was applied, leading to chaotic variations in the ribbon displacement ranging between 2.7 and 4.4 on the vertical axis. Then, at about time $n \cong 2200$, the control was turned on, and a periodic orbit with the same period as the magnetic field oscillation was controlled. This orbit is labeled period 1 in Fig. 2. It shows up on the plot as a cluster of points around the horizontal line extending from $n \cong 2200$ to $n \cong 4800$ at a displacement of approximately 3.4. At time $n \cong 4800$ the control of this orbit was turned off, and a different periodic orbit (labeled period 2 in Fig. 2) was controlled. This orbit has a period that is twice as long as the period of the applied magnetic field oscillation, and shows up in Fig. 2 as points clustered around two horizontal lines located at displacement values of about 2.8 and 3.9. Then at time $n \cong 7100$ the control was switched back to the original orbit with period 1.

Applications. The experiment just described clearly shows that the basic idea of controlling chaos by means of unstable periodic orbits embedded in the chaotic motion works very well for the particular case tested. In addition, many other experiments have been done on other systems. Examples include control of a chaotically behaving laser, control of a fluid undergoing chaotic thermal convection, control of arrhythmically beating heart tissue, and control of chaotic behavior in a chemical reactor. In summary, controlling chaos is a broadly applicable concept, and laboratory experiments illustrate its feasibility. It thus seems that these ideas hold promise of significant technological application.

For background information *see* CHAOS in the McGraw-Hill Encyclopedia of Science & Technology.

Edward Ott

Bibliography. E. Ott, C. Grebogi, and J. A. Yorke, Controlling chaos, *Phys. Rev. Lett.*, 64:1196–1199, 1990; E. Ott, T. Sauer, and J. A. Yorke (eds.), *Coping with Chaos*, 1994; E. Ott and M. Spano, Controlling chaos, *Phys. Today*, 48(5):34–40, May 1995; T. Shinbrot et al., Using small perturbations to control chaos, *Nature*, 363:411–417, 1993.

Chlorofluorocarbon

The use of chlorofluorocarbons results in the release of undesirable volatile substances into the atmosphere, and much research has been focused on how to deal with the problem. Recent research has involved seeking substitutes for chlorofluorocarbons to serve as cleansing agents in industrial processes, and finding a simple reaction that could be used to destroy chlorofluorocarbons before they escape into the atmosphere.

Alternative Cleaning Methods

A diverse group of industrial manufacturing processes require a cleaning step in at least one phase of the operation. Such industries include manufacturing of electronic circuit boards and electronic devices, metals fabrication, and aerospace manufacturing. The types of surface contaminants vary greatly, ranging from simple particulate dust to oils, greases, and rosin fluxes. Until about 1992, the primary cleaning solvents used in many of these applications were the chlorofluorocarbons 1,1,2-trichloro-1,2,2-trifluoroethane (CFC-113) and 1,1,1-trichloroethane (TCA). In the United States, production bans for these and other chlorofluorocarbon (CFC) materials were put into effect on January 1, 1996, due to environmental concerns. Over the past few years the various industries using CFCs for cleaning have been encouraged to find alternative agents.

The general procedures used for cleaning include manual wiping, unheated immersion dip tanks, and vapor degreasing. In vapor degreasing, the substrate to be cleaned is suspended over a tank of heated solvent. Solvent vapors condense on the surface of the part. Liquid solvent then drips back into the main bath, carrying contaminants with it. CFC-113 and TCA were very popular solvents in these applications because of their nonflammability, moderate boiling points, solvency for a broad range of nonpolar organic material (such as oils, waxes, and greases), and low toxicity. Nearly 200×10^6 lb (90×10^6 kg) of CFC-113 and over 1×10^9 lb (4.5×10^8 kg) of TCA were used annually on a worldwide basis in cleaning operations from the 1960s into early 1990.

Decline in use. It has been widely established that CFCs as well as a host of other chlorinated, fluorinated, and brominated hydrocarbons are responsible for large cyclic losses in stratospheric ozone levels. The ozone layer is necessary to protect life on Earth from damaging ultraviolet radiation. The degree to which a compound depletes the stratospheric ozone layer is given by a relative index called the ozone depletion potential value. In addition to ozone loss, CFCs pose unacceptably high global warming poten-

TABLE 1. Initially proposed hydrochlorofluorocarbon replacements for CFC-113 and TCA

Compound	Formula	Ozone depletion value*	Global warming potential
CFC-113	CCl_2FCClF_2	0.80	1.35
TCA	CH_3CCl_3	0.10	0.02
HCFC-141b	CH_3CCl_2F	0.15	0.07
HCFC-123	$CHCl_2CF_3$	0.02	0.02
HCFC-225ca	$CHCl_2CF_2CF_3$	0.01	0.04
HCFC-225cb	$CHClFCF_2CClF_2$	0.04	0.04

* Referenced to CFC-11 (CCl_3F), which is given an ozone depletion value of 1.0.

tials. Global warming is caused by entrapment of heat by organic molecules in the lower atmosphere. Concern for the Earth's environment has led to increased environmental scrutiny in the areas of solvent use and emissions, with thousands of regulations covered in the U.S. Clean Air Act of 1990 and in international agreements of the 1987 Montreal Protocol and subsequent amendments. Manufacturers and users of CFCs must now consider the many regulations dealing with human safety, environmental release, and waste management, in addition to the solvency, surface compatibility, and flammability of any new system being offered.

Transitional solvent replacements. In many applications it has been possible to replace CFC-113 and TCA with the classical chlorinated solvents trichloroethylene (CCl_2CHCl), methylene chloride (CH_2Cl_2), and tetrachloroethylene (C_2Cl_4, also known as perchloroethylene, or PERC). They remain the best replacements for hand-wipe cleaning and vapor degreasing but are highly regulated as potential health hazards and hazardous air pollutants. Serious commercial effort to introduce new solvents did not occur until the early 1990s, with use of hydrochlorofluorocarbons (HCFCs). The chemical strategy here is to replace some of the fluorine and chlorine atoms with hydrogen to reduce atmospheric lifetimes. **Table 1** compares the ozone depletion and the global warming potentials of some HCFCs with that of TCA and CFC-113. The HCFCs provide some environmental improvement over that of CFC-113, but their use has been limited.

Halogenated solvents. A main attraction for using halogenated hydrocarbon materials in cleaning applications is that they tend to be nonflammable. This is important in cleaning operations that require heating or in hand-cleaning operations where evaporation is desirable. One way to minimize ozone damage by volatile materials is to remove the chlorine atoms from the solvent molecule. This was accomplished with the introduction of perfluorocarbons (PFCs), which replace all hydrogen with fluorine. Perfluorocarbons turn out to be very poor solvents but have found use in some limited applications. When hydrogen is substituted for a few of the fluorine atoms in a perfluorocarbon molecule, solvency is greatly improved. These materials are known as hydrofluorocarbons (HFCs). The most promising hydrofluorocarbon molecule is HFC-43-10 (2,3-dihydrodecafluoropentane). Solvency of HFC-43-10 is still more selective than CFC-113 but can be improved by use of appropriate azeotropic blends. Another recently introduced class of solvents is the hydrofluoroethers. The structural strategy with hydrofluoroethers is to have partial fluorination in the molecule, keeping the hydrogen spatially away from the very electronegative fluorine atoms. **Table 2** shows a selected listing of the properties of perfluorocarbons, hydrofluorocarbons, and hydrofluoroethers.

Additional halogenated compounds that have been made available to industry include *para*-chlorobenzotrifluoride (PCBTF) and 3,4-dichlorobenzotrifluoride (DCBTF) and their solutions with ring-substituted monochlorotoluene (MCT) and perchloroethylene (PERC). Neat (pure and undiluted) *para*-chlorobenzotrifluoride has moderate solvent properties, but blends with monochlorotoluene and perchloroethylene are greatly improved. The blends are also less combustible but are highly regulated because of the presence of monochlorotoluene and perchloroethylene.

Nonhalogenated solvents. Many more options are found with this class of solvents, but they are generally classified as flammable or combustible. The most commonly used solvents for hand cleaning at room temperature are the aliphatic mineral spirits (petroleum distillates) and the aromatic mixtures of naphthenes and xylenes. These are widely used in automotive repair and industrial machine shops. Here the industry approach has been to offer higher-boiling fractions (lower vapor pressure) to minimize release of volatile organic compounds and to reduce flammability concerns. Recently, several high-boiling-

TABLE 2. Selected properties of perfluorocarbons (PFC), hydrofluorocarbons (HFC), and hydrofluoroethers (HFE)

Compound	Atmospheric lifetime, years	Ozone depletion potential	Global warming potential	Boiling point, °C (°F)
CFC-113 (CCl_2FCClF_2)	100	0.80	1.35	48 (118)
HCFC (CCl_2FCH_3)	9.4	0.11	0.14	32 (90)
PFC (C_6F_{14})	2900	0	0	45 (113)
HFC-4310 ($C_5H_2F_{10}$)	20.8	0	0.25	55 (131)
HFE ($C_4F_9OCH_3$)	4	0	0.1	60 (140)

TABLE 3. Properties of selected nonhalogenated chlorofluorocarbon alternative solvents

Compound	Boiling point, °C (°F)	Flash point, °C (°F)
Mineral spirits	157–196 (315–385)	105–137 (221–279)
n-Methyl-2-pyrrolidone (NMP)	203 (397)	187 (369)
Gamma-butyrolactone (GBL)	204 (399)	98 (208)
Ethyl lactate	154 (309)	59 (138)
Furfuryl alcohol	170 (338)	75 (167)
Volatile methylsiloxane (OS-20)	152 (306)	34 (93)
d-Limonene	163 (325)	119 (246)

point organic solvents have been advanced for degreasing applications; however, the characteristic low vapor pressures increase drying time. **Table 3** lists some common materials presently being used and tested. n-Methyl pyrrolidone and gamma-butylacetone are miscible with water and alcohols, and can be used neat or as formulated blends with other solvents. Typically, they are not effective on hydrophobic materials such as straight- and branched-chain hydrocarbons. Much attention has been directed to naturally occurring terpenes, unsaturated hydrocarbons that occur in most essential oils. d-Limonene belongs to this class and has been found to be effective on a wide range of aliphatic lubricating oils. Either neat or as a formulated composition, d-limonene has found use in hand-wipe applications; however, if not stabilized with antioxidants, it tends to polymerize in air, forming nonvolatile tacky residues which require removal. High-boiling-point alcohols such as furfuryl alcohol have also been of interest. Like terpenes, furfuryl alcohol is derived from plant matter, such as oat husks and rice hulls. Other high-boiling, low-flash-point solvents include organic esters such as ethyl lactate, methyl decanoate, butyl butyrate, and isopropyl myristate; however, there is a potential concern with the formation of acid by-products via hydrolysis. Additional materials based on the diesters of glutaric, adipic, and succinic acids are also of interest. Traditional low-flash-point solvents, such as isopropyl alcohol, cyclohexane, and acetone, have made a comeback in degreasing operations. These materials are among the most flammable, however.

Siloxanes. A new class of short-chain linear polydimethylsiloxanes has been advanced for cleaning applications. Known commercially as volatile methylsiloxane fluids (VMS), they lend themselves particularly well to hand cleaning applications because of their rapid evaporation rates. They can also be used at ambient temperature in soak baths, as azeotropic blends with various alcohols, or as neat solvent. These materials are based on the repeating unit $-[(CH_3)_2SiO]_n-$, where n is 0–4. Volatility of these materials decreases with increasing chain length, with a corresponding decrease in flammability. The volatile methylsiloxane fluids have excellent toxicological profiles with short atmospheric lifetimes (10–30 days) and are very good solvents for nonpolar soils.

Solvent formulation. The recent trend in solvent use has been to combine two or more of them to prepare solutions that have a desired physical property. Some blends are quite complex. The goal here is to maximize solvency across a broad spectrum of soils while improving safety by reducing the amount of volatile organic compounds and the flash points. For example, terpenes have been blended with higher-boiling-point esters or short-chain aliphatic hydrocarbons to boost cleaning. Furfuryl alcohol can be blended with isopropyl alcohol to reduce volatile organic compounds and flammability. Hydrocarbon solvents such as heptane, cyclohexane, or n-hexane can be microemulsified with water by using propylene glycols and surfactants. Co-solvent cleaning may also employ the use of perfluorocarbons, hydrofluorocarbons, or hydrofluorethers as rinsing or flammability inerting agents.

Aqueous surfactant systems. Aqueous cleaning systems have proven to be the most popular alternative to CFC and TCA because of their safety characteristics. Many aqueous systems are nontoxic and nonflammable, contain no volatile organic compounds, and do not release atmospheric emissions; however, key issues to be considered include foaming, drying, corrosion of metal surfaces, and wastewater handling. Cleaning with heated baths helps to reduce drying time but can also accelerate corrosion rates. Hot forced air and air knives are used in some applications to promote quick drying times. Unlike the relatively simply blends and azeotropes of many solvent-based systems, aqueous products will contain a balance of several ingredients taken from classes of materials such as builders and alkalinity agents (phosphates, silicates, carbonates, hydroxides, and so forth), surfactants (alkyl and aryl ether sulfates and sulfonates, ethoxylated or alkoxylated alcohols, and so forth), corrosion inhibitors (triazoles, polycarboxylic acids, and phosphonates), emulsifiers, and coupling agents. Most aqueous compositions are quite complex, requiring formulation technology to develop robust systems that will provide acceptable cleaning. The development of new surfactant molecules has been an ongoing process since the mid-1960s, and the commercial offering to formulators is quite immense. In addition to the common ethoxylated alcohols, there have been regular advances in propoxylated-ethoxylated alcohols, ethoxylated mercaptans, amines, and monoglycerides. For industrial cleaning, the choice of surfactants must include those that provide low foam with good oil penetration and emulsification, requiring a blend of several surfactant types over a range of hydrophobicity values.   Steven A. Bolkan

Atmospheric Problems

Chlorofluorocarbons contain carbon, fluorine, and chlorine; the most common are $CFCl_3$ and CF_2Cl_2. CF_2Cl_2, a gas at room temperature and pressure, is commonly used as a refrigerant in air conditioners and has been manufactured in large quantities since the 1950s. When refrigeration equipment leaks, the CF_2Cl_2 is released into the atmosphere. The central carbon atom forms unusually strong bonds to two chlorine and two fluorine atoms:

$$\begin{array}{c} F \quad\quad F \\ \diagdown \quad\diagup \\ C \\ \diagup \quad\diagdown \\ Cl \quad\quad Cl \end{array}$$

Reactions in the stratosphere. The strong bonds in CF_2Cl_2 prevent the atmospheric reactions that normally degrade foreign molecules. This molecule survives to reach the stratosphere, where the strong, high-energy solar ultraviolet light [wavelength (λ) < 215 nanometers] breaks it up to release chlorine atoms. The atmospheric lifetime of typical CFCs is of the order of hundreds of years. The chlorofluorocarbons degrade the protective ozone layer that extends from 9 to 18 mi (15 to 30 km) altitude. This layer is important to life on Earth because it prevents excessive amounts of the Sun's ultraviolet light (with wavelengths of 240–320 nm) from reaching the Earth's surface. This ultraviolet radiation is lethal to certain organisms, damages the deoxyribonucleic acid (DNA) of all organisms, and can cause skin cancer and cataracts in humans. As ozone (O_3) absorbs ultraviolet light, it is photodissociated into dioxygen (O_2) and monatomic oxygen (O), but this does not normally lead to net ozone depletion because the odd oxygen atoms recombine with O_2 as in reaction (1).

$$O_2 + O \rightarrow O_3 \tag{1}$$

According to the proposal of F. S. Rowland and M. J. Molina, the reactions involved in degrading ozone into normal dioxygen are believed to be as follows. First, chlorine atoms are released from the CFC [reaction (2)]. The resulting chlorine atoms destroy ozone as in reactions (3) and (4). An important

$$CF_2Cl_2 \rightarrow CF_2Cl + Cl \tag{2}$$

$$Cl + O_3 \rightarrow ClO + O_2 \tag{3}$$

$$ClO + O \rightarrow O_2 + Cl \tag{4}$$

feature is that the chlorine atom used in reaction (3) is regenerated in reaction (4). Thus, one chlorine atom can destroy many ozone molecules, typically thousands, before it is removed irreversibly by some other process. The O atoms that take part in reaction (4) arise from ultraviolet irradiation (< 240 nm) of normal oxygen (O_2) at an altitude of >9 mi (15 km)

as in reaction (5). The ozone layer is vulnerable

$$O_2 \rightarrow O + O \tag{5}$$

because the total amount of ozone in the atmosphere is very small; if the ozone were collected in pure form at ground level, it would be no more than a few millimeters thick.

The polar regions are most affected by ozone depletion as a result of the presence of polar stratospheric clouds, which absorb chlorine during the dark winter season but release it when the spring rays of the Sun strike the clouds. The resulting ozone hole was first observed by the British Antarctic Survey in 1985. This event alerted industry and governments to the reality of the Rowland-Molina hypothesis, which previously had negligible practical effect on the manufacture and sale of CFCs. By 1993, the Antarctic ozone hole reached a record 9.4×10^6 mi^2 (24×10^6 km^2). A decrease in atmospheric ozone has now been observed in both hemispheres and at all latitudes. In addition, circumstantial evidence makes it very likely that CFC release is the ultimate cause of the problem. As a result, government action was taken, and the Montreal agreement (1987 and 1992) banned production of the materials in advanced countries, although third-world countries are still allowed to manufacture CFCs, with a phase-out date of 2010. Replacement hydrochlorofluorocarbons, which are less ozone depleting, were developed during 1985–1990, but these usually need new specially designed equipment. Older automobiles, for example, still need the original CFCs. With the recent price increase as a result of the ban on manufacture, illegal CFC imports into the United States have sharply increased. The resulting buildup of stockpiles of seized CFCs have to be destroyed—by no means easy to do. The scale of the problem is significant: in 1995, 139 tons (126 metric tons) of illegally imported CFCs were seized by the U.S. Department of Justice.

Destruction. A new way to destroy CFCs was recently suggested. The CFC is passed as a gas over hot sodium oxalate, which causes it to be converted into the easily disposable solids carbon, sodium chloride, sodium fluoride, and carbon dioxide, as in reaction (6). It is unusual that such a mild reagent as

$$CF_2Cl_2 + 2Na_2C_2O_4 \xrightarrow{\text{heat}}$$
$$C + 2NaF + 2NaCl + 4CO_2 \tag{6}$$

sodium oxalate is able to react with CFCs, which resist attack by metals at red heat and tend to extinguish the flame in attempts to destroy them in a fuel–air flame. The innocuous character of sodium oxalate is shown by the fact that it is present in rhubarb leaves.

The way that the reaction happens is still unknown, but a plausible speculation has been suggested: The oxalate acts as a source of two electrons, as in reaction (7), on the surface of the salt. The

$$C_2O_4^{2-} \rightarrow 2CO_2 + 2e^- \tag{7}$$

resulting electrons either are located at imperfections on the crystal surface or remain attached to small graphite particles derived from decomposition of the oxalate. The CFC approaches the salt surface and binds to it. The surface electrons are then believed to be transferred to the CFC, which causes a C-Cl bond to lengthen, with the ultimate expulsion of the negatively charged chloride ion (Cl^-). This process is assisted by the presence of the positively charged sodium ion (Na^+), which receives the chloride ion to form sodium chloride. The process releases the neutral CF_2Cl molecular fragment, or radical, from the surface. This radical is known to be highly reactive and must be quickly converted to carbon and sodium salts by further reaction with the salt. This type of chloride expulsion reaction is known to require a polar medium, which is provided by the presence of the ions on the surface of the salt, instead of the more usual polar solvent.

If development work on such procedures proves successful, the next stage will be scaling up the process in a pilot plant. It is hoped that eventually the process will become a practical way of breaking down CFCs before they enter the atmosphere.

It is estimated that if current trends continue, ozone depletion will reach a maximum in the first decades of the twenty-first century, after which the measures taken to restrict CFC production and use should help restore the atmosphere to its usual ozone content. The effect of any CFC ban is necessarily delayed, because CFCs released in the preceding decade would still be making their way to the stratosphere. It is also possible that the legal restrictions on CFCs will not be universally observed, and so the situation will have to be carefully monitored in the next few decades. *See* STRATOSPHERIC OZONE.

For background information *see* ATMOSPHERIC OZONE; AZEOTROPIC MIXTURE; HALOGEN ELEMENTS; HALOGENATED HYDROCARBON; SOLVENT in the McGraw-Hill Encyclopedia of Science & Technology.

Robert H. Crabtree

Bibliography. J. Burdeniuc and R. H. Crabtree, Mineralization of chlorofluorocarbons by a convenient thermal process, *Science*, 271:340–341, 1996; F. R. Cala and A. E. Winston, *Handbook of Aqueous Cleaning Technology for Electronic Assemblies*, 1996; E. M. Kirschner, Environment, health concerns force shift in use of organic solvents, *Chem. Eng. News*, 72(48):13–20, June 20, 1994; A. F. Whitaker (ed.), *Aerospace Environmental Technology Conference*, Huntsville, Alabama, NASA Conf. Publ. 3298, 1994.

Chlorophyll

Chlorophyll is essential for photosynthesis in all higher plants and algae. The core of the chlorophyll molecule has four nitrogen-containing pyrrole rings that are bonded to a central magnesium atom, and a fifth ring containing only carbon atoms. Various side chains are attached to the pyrrole rings. Chlorophylls *a* and *b* are the most common of the several chlorophyll molecules and are present in all higher plants. They absorb visible radiation in the blue (600–700 nanometers) and red (400–500 nm) regions, and transmit and reflect green light, resulting in the characteristic green color of plant leaves.

Chlorosis. Chlorosis, a deficiency of chlorophyll in leaves, results in pale yellowish green or yellow leaves. A number of factors, including senescence, diseases, toxicities, and mineral deficiencies, can cause chlorosis. Iron is a coenzyme in the chlorophyll synthesis pathway, and iron deficiency results in rapid chlorosis of young leaves. Magnesium and nitrogen are part of the chlorophyll molecule, and deficiencies of these elements result in chlorosis. Manganese, potassium, zinc, and copper deficiencies can also cause chlorosis in plants. The pattern and distribution of symptoms in plants are influenced by the cause of the chlorosis. These patterns, first recognized in the nineteenth century, have been used by agriculturists to diagnose mineral deficiencies and toxicities.

Estimating chlorophyll in plants. Conventional methods of estimating chlorophyll in plant material have involved tissue extraction and spectrophotometric measurement. Plant tissue is extracted with an organic solvent, and the absorbance of the solution is measured with a spectrophotometer at wavelengths of 645 and 663 nm for chlorophyll *a* and *b*. However, this procedure can be time consuming and requires destructive sampling. Estimating plant chlorophyll content by measuring the reflectance of the plant part at approximately 550 nm has also been studied, but this method still requires the use of a benchtop instrument.

In the 1970s investigators developed instruments capable of estimating the amount of chlorophyll nondestructively in leaves under laboratory conditions. In 1982, the Japanese Ministry of Fishery and Agriculture initiated the Soil and Plant Analysis Development Project (SPAD) in conjunction with a commercial manufacturer to develop a hand-held meter to nondestructively measure the chlorophyll concentration in rice leaves. The SPAD 502 Chlorophyll Meter is a small, lightweight (0.5 lb or 225 g), hand-held instrument that can store and average up to 30 readings. It measures the transmittance of the radiation from two light-emitting diodes (LEDs) through a 0.08 by 0.1 in. (2 by 3 mm) area of leaf. The wavelength of one light-emitting diode has a peak at 660 nm, which chlorophyll absorbs, and the peak from the other is in the near-infrared (940 nm) region, which chlorophyll transmits. The instrument uses the ratio of the transmittance from the two light-emitting diodes to calculate a numerical value that is proportional to the amount of chlorophyll in the leaf.

Research has shown that SPAD meter values are linearly correlated with extractable chlorophyll measurements from a wide range of crop species, including rice, small grains, corn, soybeans, grapes, apples, cotton, and tropical and subtropical fruit trees. Correlations of SPAD meter values with extractable amounts of chlorophyll tend to be higher within a species or within a cultivar than across species or cultivars. Chlorophyll meter readings seem most useful for comparing plant chlorophyll contents rather than determining absolute concentrations.

Nitrogen deficiency. Worldwide, the most common cause of leaf chlorosis is nitrogen deficiency. Essentially all agricultural crops require some type of nitrogen input for near-maximum yields, except for legumes, which obtain their nitrogen from the reduction of atmospheric N_2 to organic nitrogen through their symbiotic relationship with *Rhizobia* bacteria. Because fairly high nitrogen concentrations are required for near-maximum yields of most crops, more nitrogen fertilizer is used in the world than all other fertilizer nutrients combined.

The nitrogen that crops obtain from soil is derived from the mineralization of organic nitrogen to inorganic nitrogen, and from inorganic ammonium and nitrate present in the soil. In humid areas it is difficult to predict the amount of nitrogen that a given soil will supply to crops, because the nitrogen mineralization rate depends on the nature of the soil's organic nitrogen and on environmental factors such as temperature and moisture content. In addition, soil nitrate can be leached from the root zone or denitrified. Researchers have attempted to develop nitrogen availability tests for crops in humid regions since 1900, but because of the problems listed above, they have been generally unsuccessful. One test developed in the late 1970s has been somewhat successful for corn, the crop which receives the most nitrogen fertilizer in the United States. This presidedress nitrate test measures the nitrate concentration of the surface foot of the soil when the corn plants are approximately a foot high. (Sidedress refers to the practice of applying nitrogen fertilizer to the side of the corn rows when the plants are 1–1.5 ft tall.) However, this test is labor intensive and must be conducted at a time when most farmers are busy harvesting other crops, so it has not been widely used.

Nitrogen deficiency in cereal crops is characterized by yellow-green foliage, with the oldest leaves often being partially to completely yellow or dead. These symptoms begin in the oldest leaf as a yellow V-shaped pattern starting at the tip and progressing toward the base of the leaf. As the severity of the nitrogen deficiency increases, the symptoms progress upward to younger leaves, and the older leaves turn brown and die. Agriculturists use these symptoms to judge the severity of nitrogen deficiency in cereal crops.

Studies have determined that SPAD chlorophyll meter readings of plant leaves are correlated with the amount of leaf nitrogen within a number of species. The correlations between SPAD meter values and leaf nitrogen levels often depend on factors such as plant part, plant growth stage, cultivar or hybrid, and growing season. Generally, the correlation is linear. However, at very low leaf nitrogen levels, when leaves are yellow or brown, SPAD meter values may be meaningless. At high leaf nitrogen levels near and above that needed for maximum dry-matter production, SPAD meter values reach a plateau with increasing leaf nitrogen concentrations.

Nitrogen fertilizer needs. Much of the recent research on the applied use of the chlorophyll meter in the United States has focused on using the meter to predict nitrogen fertilizer needs of corn. Most research has indicated that SPAD meter readings of leaves from young corn plants cannot be used directly to predict adequacy of nitrogen fertilization because hybrid characteristics and local growing conditions can affect chlorophyll content and its relation to plant nitrogen status. However, calculating a relative SPAD meter reading, which is the average SPAD meter reading of corn leaves in the test area divided by the average from an area in the same field receiving a high nitrogen rate, results in a more accurate test value. A field with a relative SPAD meter reading of approximately 0.95 or higher usually has an adequate supply of nitrogen for a near-maximum grain yield. SPAD meter readings must be taken when six to eight leaves have emerged from the stalk to allow time to apply nitrogen fertilizer, if needed, as a sidedress. For irrigated corn, SPAD meter readings can be taken throughout the growing season prior to silking, and additional nitrogen can be applied in the irrigation water.

Generally, the later in the growing season that chlorophyll meter readings are taken, the more accurately they reflect plant nitrogen status and predict fertilizer requirements. Readings taken on leaves as the corn grain is maturing can accurately indicate whether the crop has received adequate nitrogen for near-maximum yields. This information allows farmers to evaluate their nitrogen fertilization practices and to plan adjustments for the next growing season.

Readings for other crops. It has been shown that chlorophyll meter readings could assist in making nitrogen fertilizer recommendations for a wide range of other cereal crops, potatoes, cotton, and some horticultural crops. Both actual SPAD meter readings and relative SPAD meter readings of young plant leaves have been used to accurately predict whether rice, wheat, and other small grains will respond to additional nitrogen fertilizer. In crops such as potatoes and peppermint, relative SPAD meter readings were shown to be able to predict response to nitrogen fertilizer.

Because of their ease of use and overall accuracy, chlorophyll meters are beginning to be used more generally by farmers and crop consultants to assist in making more accurate nitrogen fertilizer applications to crops. In spite of its limitations, the chlorophyll meter is a valuable new instrument that can reduce the cost of growing crops and lower the potential for nitrate pollution of waters worldwide.

For background information *see* AGRICULTURAL SOIL AND CROP PRACTICES; CHLOROPHYLL; NITROGEN;

PHOTOSYNTHESIS; PLANT GROWTH; SOIL; SOIL FERTILITY in the McGraw-Hill Encyclopedia of Science & Technology.   Richard H. Fox; William P. Piekielek

Bibliography. P. E. Bacon (ed.), *Nitrogen Fertilization in the Environment*, 1995; T. M. Blackmer and J. S. Schepers, Use of a chlorophyll meter to monitor nitrogen status and schedule fertigation for corn, *J. Prod. Agric.*, 8:56-60, 1995; O. A. Monje and B. Bugbee, Inherent limitation of nondestructive chlorophyll meters: A comparison of two types of meters, *Hort. Sci.*, 27:69-71, 1992; W. P. Piekielek and R. H. Fox, Use of a chlorophyll meter to predict sidedress nitrogen requirements for maize, *Agron. J.*, 84:59-65, 1992; H. Schneer, *Chlorophylls*, 1991.

Circuit (electronics)

Since the mid-1980s there has been dramatic growth in the communications field and a major shift from fixed point-to-point analog communications to mobile person-to-person digital communications. Wireless personal communication systems (PCS) are the fastest-growing sector of the telecommunications industry. The two primary applications of personal communication systems are cordless telephony and cellular telephony. While the two systems share many similarities, they have some important differences which affect their design (see **table**). Because the telephone handsets must rely on limited battery power, the design of low-power systems and circuits is an active research area. Decreasing the handset power consumption allows for either longer operating and standby times between recharges or reduction in battery size and weight for the same battery lifetime. The minimum size and weight of the telephones are usually set by the batteries.

Digital communications. Until recently, virtually all forms of radio-frequency communication (radio, television, and cellular and cordless telephony) employed analog modulation. To transmit a telephone conversation digitally, the analog voice signals must first be changed to digital data by analog-to-digital converters. For transmission, the digital data modulate the amplitude, phase, or frequency of a radio-frequency carrier.

To counter the increase in signal bandwidth that occurs when converting from an analog to a digital representation, data compression techniques are used. Depending upon the type and amount of compression, there can be either a small amount of degra-dation in perceived voice signal quality or none at all. Adding redundancy to the transmitted signal allows the receiver to detect and correct errors which occur during communication. This redundancy is called forward error correction coding. The penalties incurred by error correction coding are increases in the bandwidth, the computational complexity, and the power required.

Wireless operating environment. The most basic limitation on successful transceiver (transmitter and receiver) operation is set by the combination of the received signal power and the receiver noise power. As the power of the received signal decreases in relation to the noise power, the signal is degraded and eventually communication becomes impossible. Receiver sensitivity is a measure of the minimum received power level that provides acceptable communication. Typically, receivers in personal communication systems can operate with received power levels of less than 10^{-9} watt. For digital communications, the maximum bit error rate is often used to specify a receiver's performance. A bit error rate of 10^{-6}, or one error in every 10^6 bits received, is a common level of acceptable performance.

The upper limit on the received power level is set by the power of the transmitted signal and the distance between the transmitter and the receiver. Assuming direct signal propagation, the received signal power decreases with the square of the separation distance. Since cellular telephones operate over greater distances, they require greater transmit power than cordless telephones.

Scattering elements, such as buildings, cars, walls, and people, can prevent reception of the direct signal from the transmitter. The receiver must then recover one of the scattered, or reflected, signals. Reflection results in reduced power due to attenuation of the signal during reflection and increase in the propagation path length. Scattering elements create another problem called multipath (Rayleigh) fading. The same signal can arrive at the receiver following two or more different paths. If these path lengths differ by a multiple of a half wavelength, the two signals can cancel each other. This cancellation is referred to as a fade. Also, interference from television, radio, and microwaves often overpowers the weak signals of personal communication systems.

The dynamic nature of the operating environment makes dealing with noise, scattering, fading, and interference even more difficult. The mobility of the receiver is the most obvious cause of changes in the

Comparison of cordless and cellular telephone handsets					
Telephone handset	Range, km*	Mobility speeds	Speech quality	Transmit power, W	Battery weight
Cordless	Less than 0.1	Walking	High	Less than 0.025	Low
Cellular	1–20	Vehicular	Low	Greater than 1.0	High

* 1 km = 0.6 mi.

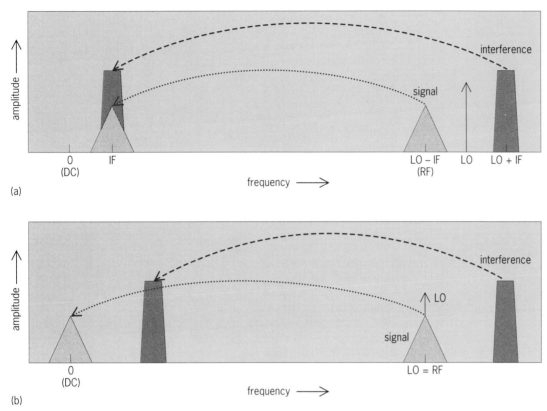

Fig. 1. Comparison of heterodyne and direct-conversion schemes. (*a*) Heterodyne without radio-frequency image filter. (*b*) Direct conversion.

operating environment, but there are many other dynamic effects. The number and power level of interfering signals change, other objects in the environment move, and weather conditions change. The wireless system must be able to respond to these changes.

Low-power design. Intelligent power control is one of the best techniques for increasing battery life. Handsets must be designed to operate far from the fixed base stations. When a handset is closer to a base station, transmitting maximum power is wasteful. To conserve handset power, the base station periodically sends information instructing the handset to increase or decrease its transmit power. These instructions are based upon the strength of the signal received at the base station. This allows the handset to transmit minimum power at all times.

Standby and sleep modes. These are two other common techniques for reducing power consumption. When no call is in progress, turning off unused portions of the circuit reduces power consumption. Similar measures can be taken during the call, such as when the speaker is silent. Although it might seem relatively easy to turn portions of the handset off and on as desired, it generally is not. Many circuits, including power amplifiers and phase-locked loops, take an unacceptably long time to restart operation.

Algorithm design. Traditionally, digital signal processing algorithms were designed with little regard for the fabrication technology or the power con-

sumption of the final circuit. However, an integrated approach to algorithm design and circuit implementation can yield power savings. Algorithm transformation techniques, such as pipelining, parallel processing, and retiming, can create low-power digital signal processing circuit designs. These digital signal processing circuits are used in many functions performed by personal communication systems, including adaptive filtering, error correction coding, and data compression.

Direct conversion. To further conserve power, traditional heterodyne receiver architectures are being replaced by direct conversion designs. Heterodyne receivers use an offset local-oscillator (LO) signal to move the radio-frequency (RF) signal down to a lower intermediate frequency (IF) [**Fig. 1***a*]. This technique requires a radio-frequency filter to prevent undesired interference signals from overwhelming the desired signal at the intermediate frequency. This filter cannot be implemented on an integrated circuit and, thus, significantly increases power consumption. Direct-conversion designs eliminate the need for the filter by moving the radio-frequency signal directly to zero-frequency direct current (DC) [Fig. 1*b*]. A similar technique, called direct modulation, is used in the transmitter to create the radio-frequency signal for transmission.

Integrated circuits. The handsets typically contain several integrated circuits and many discrete components on a printed circuit board. The current trend is to increase the integration level by including more of the functions on a single integrated circuit. This reduces power consumption by eliminating the integrated-circuit packaging and the interconnect capacitance of printed circuit boards. While the low-frequency digital signal processing is done with inexpensive silicon complementary metal oxide semiconductor (CMOS) integrated circuits, the analog radio-frequency functions are usually implemented in either silicon bipolar or gallium arsenide (GaAs) integrated circuits. However, CMOS radio-frequency circuits are beginning to make inroads into the analog radio-frequency portions of the transceiver, particularly for cordless telephones. Additionally, low-power, high-speed analog-to-digital converters allow more of the transceiver functionality to be implemented in CMOS digital signal processing circuits. The transmitter's output power amplifier and the receiver's input low-noise amplifier are the most difficult circuits to implement in CMOS.

Low-noise amplifier. In the design of the receiver, the noise performance and sensitivity are limited by the low-noise amplifier. The low-noise amplifier provides the first amplification of the weak signal from the antenna. Noise created by the low-noise amplifier adds directly to this weak signal. The amount of noise contributed by the low-noise amplifier is quantified by the noise figure. Subsequent stages have less effect on the receiver noise performance since the signal is much stronger after amplification by the low-noise amplifier. Bipolar and gallium arsenide circuits are better suited for low-noise amplifier design. However, the ability to integrate the low-noise amplifier and digital signal processing functions on a single integrated circuit and the lower cost of fabrication are making CMOS more popular.

To provide low-noise performance and a good impedance match to the antenna, a CMOS low-noise amplifier incorporates a source-degenerating inductor, L_s (so named because its effects include a reduction in gain), between the input (or source) transistor, M_1, and ground (**Fig. 2**). The input transistor, M_1, and the inductor, L_s, generate a real input impedance, Z_{in}, to match the resistive antenna impedance. While resistors can be used to generate the impedance match, the source-degenerating inductor provides a lower noise figure. The input and output inductors, L_{in} and L_{out}, tune out the circuit capacitances to allow the low-noise amplifier to achieve high gain at radio frequencies. (The other component of the amplifier is the output, or cascade, transistor, M_2.) Unfortunately, inductors fabricated on silicon integrated circuits suffer from high resistance and large parasitic capacitance. Work continues to improve the operation of these inductors.

Supply voltages. Lowering the power supply voltages is another common technique for reducing total power consumption. For CMOS digital circuits, the power consumption scales with the square of the supply voltage. The integrated-circuit fabrication technologies are optimized for these lower voltages. Additionally, better and faster integrated-circuit tech-

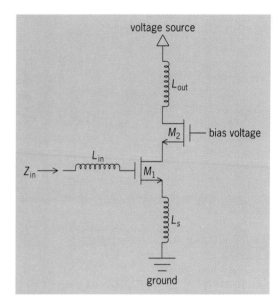

Fig. 2. Low-noise amplifier with inductor source degeneration.

nologies reduce total power consumption. These faster integrated circuits consume less power in performing the same functions.

Applications. Personal communication systems are driving many of the current advances in low-power circuit design. These advances are migrating to other applications, including wire-line communications and portable laptop computers. Decreasing the power consumption reduces operating costs, improves circuit reliability, and reduces the need for expensive cooling fans and heat sinks.

For background information *see* ANALOG-TO-DIGITAL CONVERTER; DATA COMPRESSION; ELECTRICAL COMMUNICATIONS; ELECTRICAL INTERFERENCE; INTEGRATED CIRCUITS; MOBILE RADIO; RADIO RECEIVER; TELEPHONE SERVICE in the McGraw-Hill Encyclopedia of Science & Technology.   Bang-Sup Song

Bibliography. A. A. Abidi, Low-power radio-frequency IC's for portable communications, *Proc. IEEE*, 83:544–569, 1995; D. C. Cox, Wireless personal communications: What is it?, *IEEE Pers. Commun.*, 2(2):20–35, April 1995; J. D. Gibson, *The Mobile Communications Handbook*, 1996.

Client-server systems

A number of forces are bringing about fundamental changes in the way in which computing technology is applied in contemporary organizations. These include the transformation of society from one primarily dependent on the production of goods to one also much concerned with information and knowledge. Information technology has become a major support to humans and organizations in the development of new products and new services. One major change in organizations is that they have become more horizontal in structure. Reengineering, especially busi-

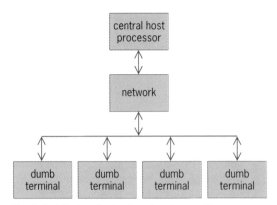

Fig. 1. Architecture for host-centered computing.

ness process reengineering, is often used to accomplish this transformation. Other approaches, such as those based on total quality management (TQM), lead to organizational changes for improved quality and responsiveness. The technological enabler of each of these transformations is advanced information technology. It has enabled the construction of the information-system architecture of an organization to match the reengineered organizational structure and the desire for rapid communications throughout real and virtual organizations. In many cases, these are horizontal structures that require massively parallel communications capabilities between personnel in the organization. Client-server architectures and client-server systems are names generally given to these enablers of these transformations to new structural forms.

Types of computer and architecture. In the traditional host-centered architecture (**Fig. 1**), a user of the system enters transactions at a dumb terminal, or one that has no processing capability. When data entry by the user is completed, the transaction is processed by the central processor, or host processor. The computer results are then returned to the user, completing the transaction cycle. A communications network, generally a wide-area network, is the communications medium that enables the information from any of a large number of users to be transmitted to and from the host processor. With the advent of minicomputers and microcomputers, host processors may be replaced with minicomputers, and dumb terminals replaced with microcomputers. However, each microcomputer in a host-centered computing system acts as if it were a dumb terminal in terms of its interactions with the host processor.

In a client-server architecture, there is no central host. There is a network of computers, acting as intelligent processors, that interact to complete various transactions. For a given application, one or more of these processors act as a server, and one or more of the other processors act as a client. A client is basically a user, or a program, that requests services from a server. In a typical client–server system, one or more clients and one or more servers,

who respond to the requests of clients, reside on various host sites of a computer-communications network. The combination of clients and servers allows for distributed computation and distributed presentation of results. A client, or user, of the system is provided with an interface that allows the client to request services of a server and to display the results that are returned to the client by the server.

Clients and servers. There is a technical distinction between the client and the user. Generally, the former term is used to include the latter in a client–server system. Clients usually do some active interpretation of inputs by a user, such as translating the commands entered by a user into the format that is required by the server. Clients may also check the validity and integrity of user commands. Servers, however, are passive. They will never initiate communications. They wait for the arrival of service requests from clients and then respond to these requests. A server should provide a standardized transparent interface to its clients. Thus, clients are not necessarily aware of all detailed hardware and software actions of the client–server system that provides a service. Sometimes the machines on which client and server processes reside are referred to as clients and servers. However, the server for one process may be the client for another process.

This permits computational intelligence and knowledge resources to be distributed throughout an organization and to be interconnected and aligned. A client–server architecture consists of a client process and a server process which can be distinguished from each other for any given application activity. Generally the client and the server interact in a seamless manner. The client portion and the server portion of the effort are usually conducted on separate platforms for a given application, although this is not necessarily the case.

A formal definition of client-server systems and client-server architectures is useful: A client-server system is based upon an architecture in which a single application is intelligently partitioned between a number of cooperating processors which act in such a way as to complete the required processing in a unified manner, just as a single task would be processed. This client–server architecture is recursive in that the client for one application can become the server for another application.

Features of client–server systems. A number of important features are incorporated in a client–server system (**Fig. 2**). Clients may access any of a variety of servers, and servers may provide many services. Local- and wide-area communication networks may be used to link geographically distributed servers and clients. Bonding software, or middleware, is used to handle the necessary message routing between clients and servers in the client–server system and is necessary to provide a single system image to a user. This is needed in order for a variety of user applications to run in an environment with

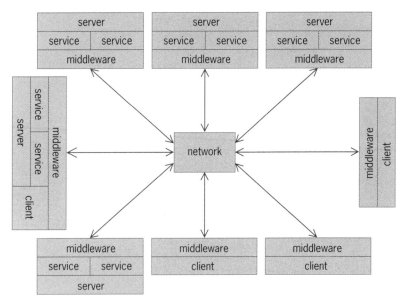

Fig. 2. Architecture for client–server computing.

multiple vendors such that the system user is isolated from the details of the variety of network communication protocols that may be in use, and such that system users do not need to know where a given set of data is physically located. A server may be a general-purpose computer or a special-purpose machine, or it may be a computer workstation. Client–server computing enables work to be divided across cooperating processors such that clients may use shared resources from many servers to complete a task. Action to engage in an application is generally initiated by a client, and a user-friendly graphical interface is incorporated at each client site. Most client–server systems provide a standard query language at each client site to assist clients in making queries of the system.

Architectures and environments. Client–server systems potentially possess a number of advantages, which follow from the architecture of the system and the environment in which it is designed to serve. Most architectures and environments of client–server systems may be characterized as follows. They are oriented toward multiprocessors and distributed processing such that computing and information capabilities may be obtained from the most appropriate source. Interoperability, portability, and reconfigurability are major features such that a wide variety of platforms from various vendors may be connected to the system. The system is open-systems oriented and standards compliant. It is possible to modify, generally to upgrade, a client platform or a server platform without the necessity of modifying the other platform, or without the necessity of the platform even being recognized as having been modified.

These features should be contrasted with the features associated with most host-centered computing environments, which are hardware rather than software oriented. Most features of the system are proprietary, and there is a primary or dominant hardware vendor. As a result, there is little portability, interoperability, or reconfigurability. Much of the mainframe computing community is adopting various attributes and features of client–server systems in order to remain competitive with the newer client–server systems and to stay abreast of contemporary information technology and organizational needs.

In general, a three-tiered client–server architecture is called for in most implementations of client–server systems. These are a presentation tier, which is composed of a graphical user interface and other components that aid visualization; a functionality and connectivity tier, which supports the various processing activities; and a data tier, which supports both new data systems and legacy systems, and which is generally disbursed across the various client–server locations in the most appropriate distributed fashion. Computer security issues are very important in client–server systems to ensure that only authorized users have access to sensitive information. Balancing security with ease of use is an important systems design consideration.

Communication networking technology. Much attention is being devoted to communication networking technology for client–server systems. New forms of connectivity and transmission capacity are required in order to satisfy the demand for distributing the computing environments needed for these systems. Configuration management is especially needed to ensure the primarily horizontal communications and the diverse computing and intelligence attributes that are required of these systems. Very impressive local-area network and broadband integrated services digital network (ISDN) capabilities, as well as associated protocols for their efficient and effective use, have resulted from the continuing effort to satisfy the needs of the customers for client–server systems. Middleware is the term often given to the services made available to application developers through consistent interfaces in order to assist them by reducing the complexity of application development. Open systems interconnections (OSI) is the term generally given to the layers of software that separate the physical-layer network infrastructure from the distributed-layer application services. The utilization of middleware provides major support for rapid and effective application development. This closely ties the development of client–server systems to the development of the Internet and other modern networking efforts. Such groupware applications as electronic mail, project management, group conferencing support, group decision support, and a variety of system development support tools are made possible by client–server systems. There are a plethora of potential application areas for client–server systems, such as the use of groupware in computer-aided design and manufacturing. *See* INTEGRATED SERVICES DIGITAL NETWORK; INTERNET.

Results of implementation. There have been a number of case studies of client–server system implementations; generally these indicate organizational improvement when these systems are developed and deployed in an efficient and effective manner, using requirements that have been defined in accord with organizational needs and associated goals and objectives. It seems a safe projection that the coming years will see a continued evolution of both the technology and applications of this new information technology, which seems ideally matched to contemporary business and organizational drivers for technological progress, quality, and increased responsiveness.

For background information *see* COMPUTER SYSTEMS ARCHITECTURE; DISTRIBUTED SYSTEMS (COMPUTERS); HUMAN-COMPUTER INTERACTION; INFORMATION SYSTEMS ENGINEERING; LOCAL-AREA NETWORKS; WIDE-AREA NETWORKS in the McGraw-Hill Encyclopedia of Science & Technology.   Andrew P. Sage

Bibliography. B. H. Boar, *Cost Effective Strategies for Client/Server Systems*, 1996; B. H. Boar, *Implementing Client/Server Computing*, 1993; B. Bocnenski, *Implementing Production-Quality Client Server Systems*, 1994; J. J. Donovan, *Business Reengineering with Information Technology*, 1994; B. Elbert and B. Martyna, *Client Server Computing*, 1994; K. Watterson, *Client Server Technology for Managers*, 1995.

Climate history

Recent work in the Mojave Desert of southern California increases understanding of sediment transport by wind in response to climatic changes in arid regions. Studies of the geomorphology and sediments of sand dunes have been combined with luminescence dating of these deposits to develop a chronology of periods of eolian deposition. A comparison of this chronology with the proxy record of the region's climatic change that has been derived from lake and alluvial-fan deposits, paleobotanical studies, and tree rings aids in understanding the conditions that favor eolian sediment accumulation there and in developing a process-response model for such eolian systems.

Eolian sediment transport. Accumulation of eolian (wind-transported) sand deposits requires a source of sediment and sufficient wind energy to transport that sediment. Most eolian sand in arid regions occurs in relation to well-defined regional- and local-scale sediment transport systems in which sand is moved by wind from source areas (for example, fine-grained alluvial deposits) via transport pathways to depositional sinks (such as dune fields or sand seas). Changes in climate may impact eolian sediment transport systems via changes in sediment production, availability, and mobility. Variations in flood magnitude and frequency, river sediment load, and lake levels determine the rate of sediment production, and the source of material for the system. Climatic changes impact sediment availability (the susceptibility of a surface to entrainment of material by wind) and mobility (sand transport rates) via changes in the magnitude and frequency of winds capable of transporting sediment, in vegetation cover, and in soil moisture.

Chronology. To understand the response of eolian sediment transport systems to climatic change, the key is the development of a chronology of eolian depositional episodes. Such chronology was rarely possible until the application of luminescence dating to sediments. This technique measures the amount of nuclear radiation (from naturally occurring potassium-40, uranium, and thorium in the surrounding sediment) to which a crystalline material (for example, sand grains) has been exposed. The passage of radiation through the crystal lattice displaces electrons that become trapped at defects in the lattice. When an electron is ejected from a trap, it recombines with a positively charged hole at a luminescence center, and a photon is released. Measurement of the amount of luminescence and the rate of production per unit of radiation yields the amount of past radiation exposure or the equivalent dose, in grays. The natural radiation dose rate, in grays/year, is derived from independent measurements of radioactivity in the field or laboratory. The age of a sample equals the equivalent dose divided by the dose rate. The luminescence signal can be stimulated by heating the sample to 500°C (930°F; thermoluminescence), by exposure of potassium feldspar grains to infrared photons (infrared-stimulated luminescence), or in quartz by exposure to a green laser (optically stimulated luminescence). The luminescence signal is reduced to a residual value by heating to more than 500°C (930°F) or by exposure to sunlight. In most sediments, the event dated is the last exposure to sunlight, equivalent to the time of deposition.

Kelso system. The Kelso eolian sediment transport system (**Fig. 1**) extends for some 60 km (36 mi) eastward from the fan delta of the Mojave River as it exits the Afton Canyon to the Kelso Dunes. The Mojave River rises in the San Bernardino Mountains in southern California (**Fig. 2**), a tectonically active area of mainly granitic rocks, and is the principal source of sediment for this system. Following heavy winter rainfall, the Mojave River may flow through the Afton Canyon to the Mojave River sink and the adjacent east Cronese Basin or as far as Silver Lake. During periods of high rainfall in the Late Pleistocene and Early Holocene epochs, these areas were occupied by a shallow lake (Lake Mojave). Sand is transported east from the source zone through the Devils Playground (between the Old Dad and Bristol mountains), a 10-km-wide (6-mi) area of active crescentic dunes (barchans), climbing and falling dunes, and sand sheets that are stabilized by vegetation.

Kelso Dunes is the major depositional sink for this system. The dune field occupies an area of 100 km² (40 mi²) on the piedmont slopes (bajadas) of the

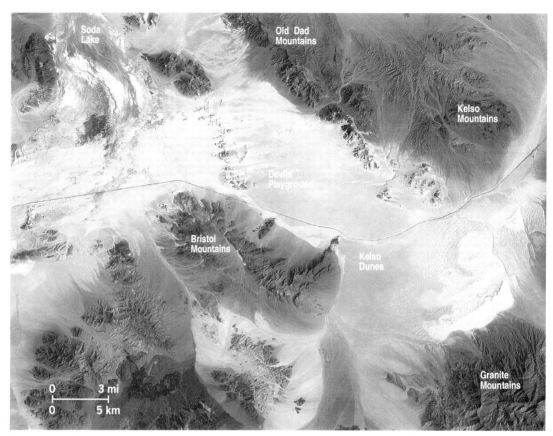

Fig. 1. *Landsat* image of the Kelso Dunes area, Mojave Desert. Sand is transported from the Mojave River sink at the left to Kelso Dunes. Note areas of different dune types at Kelso.

Granite and Providence mountains and represents a sand accumulation of approximately 1 km³ (0.2 mi³). The dune field consists of 40 km² (16 mi²) of active dunes surrounded on the west, north, and east sides by a 60-km² (24-mi²) area of low vegetation-stabilized dunes. A characteristic feature of Kelso Dunes is the juxtaposition of areas of dunes of distinctly different morphological types in terms of size, spacing, and alignment. The dune field has developed by the coalescence of a series of smaller dune fields, each of which represents an episode of sediment input or reworking of preexisting dunes.

Several periods of Late Pleistocene and Holocene eolian deposition and dune remobilization can be identified in the Kelso Dunes eolian sediment transport system. The earliest period of accumulation identified is called Phase I, spanning from 30,000 (or earlier) to 15,000–17,000 years before present (B.P.). This period was terminated by a period of geomorphic stability and soil formation on dunes that lasted for 3000–4000 years. The next major period of eolian accumulation (Phase II) occurred from around 13,000 to 4000 years B.P. The accumulation can be subdivided into Phase IIa, extending 13,000 to 8000 years B.P.; and Phase IIb, continuing to 4000 years B.P. in certain localities. Phase II eolian accumulation was also terminated by a period of geomorphic stability and soil formation. Phase III of

eolian accumulation is found only at West Cronese and Kelso, and spans 2000 to 1500 years B.P. Likewise, Phases IV and V are restricted to these localities and span 350–800 and 150–250 years B.P., respectively. Regional studies involving luminescence dating of periods of eolian accumulation elsewhere in the Mojave also identify two major periods of eolian activity: 20,000 to 30,000, and 15,000 to 7000 years B.P.

Information on past conditions of hydrologic and sediment supply affecting the Kelso eolian sediment transport system is provided by the record of fluctuations in Paleolake (ancient lake) Mojave, which occupied the principal sediment source area at two main intervals during the Late Pleistocene and Holocene epochs. The timing of high stands of Pluvial Lake Mojave is based upon radiocarbon-dated shell and tufa from the beach ridges of Silver and Soda lakes, and upon organic carbon from sediment cores taken from the Silver Lake playa. The radiocarbon-dated evidence suggests that there was an intermittent lake lasting from 22,000 to 8700 calibrated years B.P., with two persistent high stands from 18,000 to 16,000 years B.P. (Lake Mojave I) and 13,700 to 11,400 years B.P. (Lake Mojave II), and final desiccation at 8700 years B.P.

Phase I of eolian accumulation in the Kelso system predates and, in some cases, postdates the period

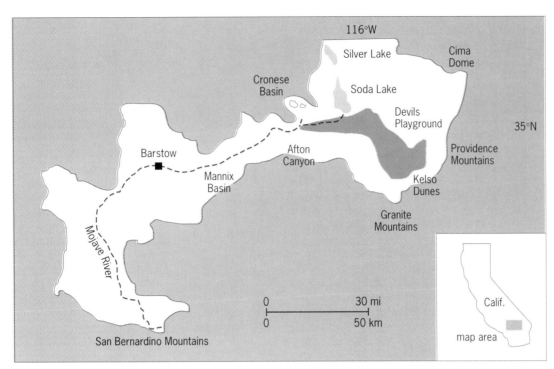

Fig. 2. Map of the Mojave River region.

of Lake Mojave I, whereas Phase II postdates Lake Mojave II. Phase IIa is directly associated with the desiccation phase of Lake Mojave II. The Holocene record of eolian accumulation and dune reactivation and stabilization can be compared to several sources of paleoclimate data. Phase IIb accumulation appears to be associated with a period of dry climates 6800 to 5060 years B.P. This period of eolian accumulation was brought to an end with stabilization of dunes at Kelso and elsewhere in the region after 4000 years B.P. as a result of cooler and wetter regional climates that resulted in increased vegetation cover. The period 4000 to 3000 years B.P. appears to have been cooler and wetter in many parts of the southwestern United States, with lowering of the woodland–desert scrub boundary, increased ground-water recharge in southern Nevada, and shallow lakes in Death Valley, Searles Lake, and Silver Lake Basin. Eolian accumulation in Phases III–V was a response to reworking of sediment emplaced during this wetter period. Paleobotanical and dendroclimatological evidence from southern California suggests that relatively dry conditions persisted at around 1450 years B.P., and from 850 to 450 years B.P. A permanent lake again developed in the Silver Lake Basin around 400 years ago as a result of cooler and wetter conditions and increased rainfall coincident with the duration of the Little Ice Age in Europe, giving rise to renewed stabilization of dunes at Kelso and elsewhere in the region and terminating Phase IV accumulation. Eolian accumulation 150 to 250 years ago in the West Cronese Basin (Phase V) was the result of deflation of sediment deposited by shallow water bodies in these wetter periods.

Studies in the Mojave Desert suggest that accumulation of eolian deposits in this desert region (and possibly many others) is strongly controlled by changes in sediment production. Major past periods of eolian accumulation are associated with reworking of sediment produced during periods of increased flow in the Mojave River. Except in periods of significantly increased precipitation, sediment transport capacity and availability are not limiting factors. Today, for example, the Kelso eolian sediment transport system is starved of sediment, because the flow of the Mojave River is regulated by dams upstream. These studies emphasize the close links between fluvial and eolian processes in arid regions and provide new models that can be tested in other deserts.

For background information *see* CLIMATE HISTORY; DENDROCHRONOLOGY; DUNE; EOLIAN LANDFORMS; LUMINESCENCE ANALYSIS; PALEOCLIMATOLOGY in the McGraw-Hill Encyclopedia of Science & Technology.   Nicholas Lancaster

Bibliography. M. L. Clarke, C. A. Richardson, and H. M. Rendell, Luminescence dating of Mojave Desert sands, *Quaternary Sci. Rev.*, 14(7/8):783–790, 1996; N. Lancaster, Development of Kelso Dunes, Mojave Desert, California, *Nat. Geog. Res. Explor.*, 9(4):444–459, 1993; N. Lancaster, *The Geomorphology of Desert Dunes*, 1995; K. Pye (ed.), *Dynamics and Environmental/Context of Aeolian Sedimentary Systems*, Geological Society of London, 1993; R. E. Reynolds, S. G. Wells, and R. H. I. Brady (eds.), *At the End of the Mojave: Quaternary Studies in the Eastern Mojave Desert*, San Bernardino County Museum Association, 1990.

Combinatorial chemistry

Combinatorial chemistry is a new subfield with the goal of synthesizing very large numbers of chemical entities by condensing a small number of reagents together in all combinations defined by a small set of reactions. This article provides a general overview of the subfield, describing the chemical library, and then examines how the necessary screening and assaying is carried out.

Chemical Libraries

Combinatorial chemistry is sometimes referred to as matrix chemistry. If a chemical synthesis consists of three steps, each employing one class of reagent to accomplish the conversion, then employing one type of each reagent class will yield $1 \times 1 \times 1 = 1$ product as the result of $1 + 1 + 1 = 3$ total reactions. Combining 10 types of each reagent class will yield $10 \times 10 \times 10 = 1000$ products as the result of as few as $10 + 10 + 10 = 30$ total reactions; 100 types of each reagent will yield 1,000,000 products as the result of as few as 300 total reactions. While the concept is simple, considerable strategy is required to identify 1,000,000 products worth making and to carry out their synthesis in a manner that minimizes labor and maximizes the value of the resulting organized collection, called a chemical library.

The earliest work was motivated by a desire to discover novel ligands (that is, compounds that associate without the formation of covalent bonds) for biological macromolecules, such as proteins. Such ligands can be useful tools in understanding the structure and function of proteins; and if the ligand meets certain physiochemical constraints, it may be useful as a drug. For this reason, pharmaceutical applications provided early and strong motivation for the development of combinatorial chemistry.

Peptide libraries. In 1984, H. M. Geysen reported on the synthesis of a library of peptides and the screening of that library to probe how changes in a single amino acid in a peptide would change its association strength with an antibody. Biochemists assume that the detailed chemical structure of any ligand will affect that ligand's binding strength to a macromolecular receptor, but even with powerful modern computer modeling methods it is normally not possible to predict either the magnitude or even the direction of the effect. By synthesizing a large number of peptides, each varying from another by only one amino acid, it was possible to determine empirically which amino acid substitutions made binding stronger and which made it weaker. The synthesis of this library benefited from the fact that chemical methods for linking amino acids together to make peptides has been optimized such that little optimization of the reaction conditions was required. Instead, the challenge was how to conveniently make thousands of peptides in a format that facilitated their use in the subsequent binding studies. The solution was to synthesize the peptides on a rack of plastic pins and to test their binding ability while the peptides remained chemically attached to the pin. While solving technical challenges, perhaps more importantly this approach highlighted a philosophy in studying ligand–receptor binding problems: start with a source of molecular diversity (that is, lots of compounds) organized in a way that makes their empirical testing straightforward; then test them all and analyze the results at the end.

Because there are 20 naturally occurring amino acids, the synthesis of a linear peptide that is n amino acids long can be done in $20n$ different ways. Thus, there are 64,000,000 possible hexapeptides ($n = 6$). It would not be convenient, perhaps not even possible, to synthesize that many peptides on individual pins. In 1985, a synthesis method was reported in which the plastic used for synthesis was encased in an inert mesh resembling a tea bag. Collections of such tea-bag reactors were subjected to the chemical addition of an amino acid at the same time. After the addition, the bags were washed thoroughly and the bags (not the bag contents) mixed. The bags were redistributed to new beakers, and another amino acid was chemically added. In this method, each bag contains only one peptide. By chemically cleaving the peptide from the polymeric support, the peptide itself can be obtained. In a variant of this procedure, a mixture of reagents can be used in any given beaker such that each bag contains a mixture of peptides bound to the polymer. While the production of such mixtures complicates the subsequent assay step, it does provide for a much greater number of assays to be accomplished. Still unresolved is the question of whether the time economy afforded by using mixtures compensates for the pitfalls inherent in their testing for binding activity.

To make very large numbers of individual peptides would require very small bags for purely practical reasons. In 1988, work involving the use of polymer resin beads was reported; approximately one-tenth the width of a pinhead, these beads served as a kind of bag. By utilizing the "split-pool" approach, it was possible to synthesize extremely large peptide libraries in which each bead possessed a single peptide. The amount of peptide on one bead is only around 200 picomoles; however, this is enough for both a simple ligand–receptor binding assay and for the analytical techniques required to establish the exact chemical structure of that peptide.

Organic libraries. Because much of the practical application in discovering tightly binding ligands derives from the pharmaceutical industry, the combinatorial synthesis of druglike compound libraries is of great interest. Two practical considerations make this a greater experimental challenge than the synthesis of peptide libraries.

The synthetic methods required to make druglike molecules (that is, low-molecular-weight, organic molecules) on a polymer support have not been optimized. While solid-supported peptide synthesis

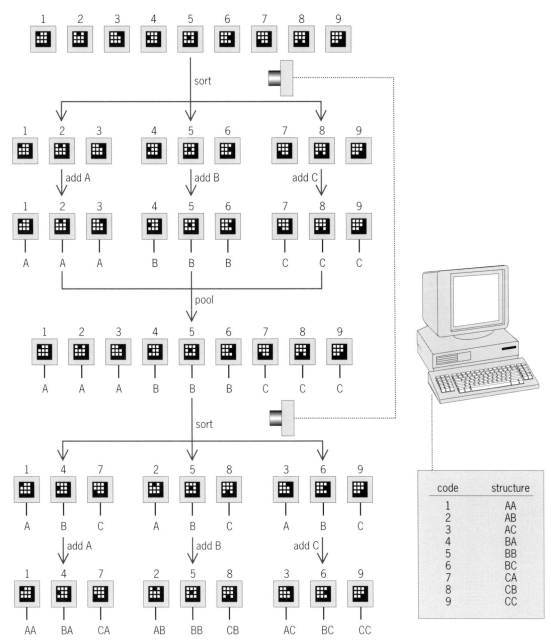

Scheme for using optical-pattern encoding for the synthesis of a combinatorial library. *(Courtesy of X. Xiao)*

originated in the early 1960s and has been extensively developed since, initial experiments with organic solid-phase synthesis in the early 1970s were not followed up widely. In addition, while there are only 20 naturally occurring amino acids and therefore a finite number of reactions required to use them efficiently, an almost infinite number of organic chemical reagents exists and a very large number of reaction types. However, by the early 1990s several groups had reported the synthesis of moderately sized organic libraries by the solid-phase synthesis method. Each approach utilized a strategy like that of the Geysen pin method and thus was amenable to the parallel synthesis of hundreds to thousands of compounds.

The synthesis of much larger-membered libraries using the polymer-resin-bead approach was inhibited by the second practical issue: while 200 picomoles of an organic compound is enough for the ligand–receptor study, it is not enough to identify the structure of the ligand. The reason is that analytical methods for structure determination are both easier and more advanced for use with biological macromolecules such as polypeptides and polynucleotides. Beginning in the mid-1990s, the solution of bead tagging, or encoding, solved this issue. The strategy is simple: if the result of a chemical synthesis step cannot be easily read at low concentration, something should be added to the bead that conveniently encodes the reaction history of that bead for later analysis. The first reported methods of bead encoding involved the use of biological macromolecules themselves. After each step in the organic synthesis, either an amino acid or a nucleotide was added to a growing oligopeptide or oligonucleotide on the same bead so that the specific sequence could be read later, and from that sequence the reaction history could be elucidated. However, neither oligopeptides nor oligonucleotides are chemically inert enough to survive the conditions required for organic synthesis. Later, tags were introduced that were much more inert chemically.

More recent methods of encoding include the use of radio-frequency memory microchips and optical bar-coding strategies. The **illustration** shows a scheme that uses optical-pattern encoding for the synthesis of a combinatorial library. Each of the nine chips has a different pattern written on it with a laser (a 3×3 matrix pattern as shown could encode for 2^9, or 512, unique chips). With the use of pattern-recognition software, the chips are sorted into appropriate reaction vessels so that chemical groups A, B, and C can be covalently added to the core organic molecule on them. After pooling the chips, optical scanning again permits sorting into appropriate reaction vessels so that a new set of three chemical groups can be added. At the end of this combinatorial synthesis involving two steps with three reagents at each, a total of 3^2, or 9, different products is obtained. Larger libraries are obtained by employing additional reaction steps, additional reagents per step, and an appropriately larger matrix pattern to enable the unique encoding of each product in the library.

Major challenges in combinatorial chemistry focus on both the characterization of compounds and the screening of very large compound libraries. In addition, the use of combinatorial chemistry methods is under active study for the discovery of new catalysts, chemosensors, and other chemical substances in which prior binding of the cognate ligand (the molecule that fits into another, usually larger molecule) is necessary.

Anthony W. Czarnik

Assaying Problems

In combinatorial chemistry attention is now being focused on the problem of how to identify the set of molecules that possess the desired combination of properties. In a drug-discovery effort, the library members that strongly bind to a particular biological receptor are of interest. In a search for new materials that behave as superconductors at relatively high temperatures, the special combination of elements yielding the best electrical properties is a goal. In each case, the library might consist of up to a million members, while the subset of target molecules might consist of several thousand contenders or just a single highly selective binder. This subset could then be studied in more detail by conventional means.

To take advantage of the parallel nature of the combinatorial chemistry strategy, rapid screening and assaying protocols are necessary. For example, gas chromatography/mass spectrometry of bead-encoded libraries is currently limited to several analyses per hour, a difficult proposition when the researcher is faced with analyzing several thousand samples. The problem is to find ways of keeping track of each library member so that parallel screening can be coupled with parallel assaying.

Spatial array. Several emerging strategies promise to address this problem. In the first case, a library is constructed in a spatial array such that the chemical composition of each location in the array is noted during the construction. The binding molecules, usually labeled with a fluorescent tag, are exposed to the entire assay. The locations that light up can then be immediately identified from their spatial location. This approach is being actively developed for libraries of proteins and nucleotides. A problem is that the chemistry required to attach various molecules to the solid surface, usually silicon, is quite tricky and difficult to generalize. The assaying strategy is intertwined with the available procedures for synthesizing the libraries themselves.

Polystyrene-bead support. A conceptually straightforward approach is to first synthesize the library by using polystyrene beads as the solid support. The product molecules are then stripped from the support and pooled together into a master solution. This complex mixture consisting of a potentially large selection of ligand molecules could then be exposed to an excess of a target receptor. The next step is to devise a method for identifying the ligand–receptor pairs that point to molecularly specific binding. One approach is to examine a part of the mixture en masse by using affinity capillary electrophoresis. With this technique, the migration times of the ligand–receptor pair are significantly longer than the unreactive ligands, and can be interrogated by electrospray mass spectrometry.

The mass spectrometric method often provides a direct structural identification of the ligand, either by determination of its molecular weight or by collision-induced dissociation experiments. In the latter case, the molecular ion is selected by a primary mass spectrometer and is driven into a region of high-pressure inert gas for fragmentation. The fragment ions are then used to reconstruct the original molecular structure. This direct approach to screening and assaying has the advantage that the screening is carried out in solution rather than on a solid support, and it avoids steric problems associated with resin-bound molecules. At present the approach seems limited to libraries of about 1000 compounds because of interference from unbound ligands and by sensitivity issues. New strategies using mass spectrometry may eliminate this limit.

Mass spectrometry. A different tack involves assaying the polystyrene beads one by one after the resin-bound molecules are exposed to a receptor. With this approach, active beads may be identified by color or by fluorescence associated with the receptor, and are subsequently indexed in standard

96-well titre plates. Identification is then possible by using a variety of spectroscopic techniques; at present, the most popular methods are electrospray mass spectrometry and matrix-assisted laser desorption ionization mass spectrometry.

Successful structure determination using mass spectrometry is often routine, particularly for protein libraries where sequencing strategies have been worked out; however, ambiguities often arise for nonpeptide libraries. Encoding strategies promise to resolve this issue. For example, it is now possible to construct the library such that the first molecule attached to the resin can be identified by its ratio of carbon-13 (^{13}C) to carbon-12 (^{12}C). By using isotope enrichment protocols, a particular isotope ratio can be associated with a specific molecule. Then, for a library synthesized with three steps, the molecule in the first step can be identified from its isotope ratio, the molecule in the third step is known from the synthesis, and the molecular weight is measured by mass spectrometry. Hence, the molecular weight of the molecule used in the second step can be determined by the difference. This approach has been successfully employed to assay a 1000-member library. There are many ways that this approach could be extended to larger systems.

Another advantage of the isotope-encoding strategy is that the code is actually carried with the molecule after it is cleaved from the bead. This property may have important implications when performing bioassays. This scheme appears to be functional in a practical sense. The mass-spectrometry methods are generally sensitive enough (≈ 100 femtomole) to provide a reliable assay, and robotic techniques can be developed to increase the number of assays to perhaps a few dozen per hour.

Another level of sophistication involves the assay of a single polystyrene bead by using a method such as infrared spectroscopy. The advantages of optical techniques are that they are nondestructive and can be employed without removing the target molecules from the linker. There are, however, several obvious drawbacks, including lack of specificity and sensitivity for single bead studies. Recently, however, experiments using infrared microspectroscopy have partially overcome the sensitivity limitations. The selectivity issue has been addressed by using a carbon-deuterium stretching frequency to determine the deuterium content of a given compound. This signature can lead to identification under suitable circumstances.

Improving assaying power. It would be desirable to speed up the whole process. To fully capitalize on the elegant concept implicit in combinatorial chemistry, parallel screening methods capable of identifying large numbers of library members need to be developed. When the libraries are synthesized on beads, it would also be preferable to assay the compound without cleaving it from the polymeric resin support. This process inherently destroys the library and opens the possibility of incomplete cleaving reactions. One approach is to take advantage of a special type of mass spectrometry where it is possible to record spectra from very small areas of a solid surface. With this technique, the sample is bombarded by a focused energetic ion beam having a kinetic energy of several thousand electronvolts. The energetic particle loses some of its momentum in the top layers of the solid, causing desorption of molecular ions near the point of impact. If the energetic ion beam is formed in a short (nanosecond) pulse, the secondary ions may be measured by using time-of-flight detection. The resulting mass spectra can be recorded from an area that is much less than 1 square micrometer. Hence, it is feasible to spatially resolve the chemical components on a single resin particle that is typically in the 20–300-μm size range.

Experiments using this idea have been attempted in the last few years. The first attempts were successful only if the covalent bonds attaching the molecules to the resin were first clipped. This was accomplished by exposing the bead to a vapor of trifluoracetic acid, a standard release agent for acid-sensitive linking moieties. After clipping, the molecules were found to remain on their respective beads, even if prior to their treatment they were essentially touching. For example, in this method, two distinct types of coated beads can be used—one with phenylalanine and one with leucine. It might be possible to extend this method to a collection of thousands of beads arrayed onto a plate.

None of the above approaches provides assaying power that satisfies the need to characterize massive combinatorial libraries. Many schemes appear to have enough sensitivity or selectivity to perform the job, but whether any will be truly practical remains to be seen. It is likely, given the high level of activity in this field, that one of the above methods (or perhaps a completely new one) will become practical.

For background information *see* ANALYTICAL CHEMISTRY; ELECTROPHORESIS; LIGAND; MASS SPECTROMETRY; MOLECULAR RECOGNITION; OPTICAL INFORMATION SYSTEMS; ORGANIC SYNTHESIS in the McGraw-Hill Encyclopedia of Science & Technology.   Nicholas Winograd

Bibliography. C. L. Brummel et al., A mass spectrometric solution to the address problem of combinatorial libraries, *Science*, 264:399–402, 1994; Y. Chu et al., Affinity capillary electrophoresis: Mass spectrometry for screening combinatorial libraries, *J. Amer. Chem. Soc.*, 118:7827–7835, 1996; S. H. DeWitt and A. W. Czarnik (eds.), *A Practical Guide to Combinatorial Chemistry*, 1997; P. A. Fodor et al., Light directed spatially addressed parallel chemical synthesis, *Science*, 251:767–773, 1991; H. M. Geysen et al., Isotopes or mass in coding of combinatorial libraries, *Chem. Biol.*, 3:679–688, 1996; K. Russell et al., Analytical techniques for combinatorial chemistry: Quantitative infrared spectroscopic measurements of deuterium-labeled protecting groups, *J. Amer. Chem. Soc.*, 118:7941–7945, 1996; S. H. Wilson and A. W. Czarnik (eds.), *Combinatorial Chemistry: Synthesis and Application*, 1997.

Communications satellite

Two types of satellite communication systems are being planned that will use the Ka-band of radio frequencies at 20–30 GHz. One type of system will use geosynchronous satellites, and the other will employ large numbers of satellites in low Earth orbits.

High-Bandwidth Geostationary Communication Satellites

Ka-band satellite systems are being planned to provide faster, higher-data-rate communications using small, inexpensive ground terminals (antennas or dishes). Use of the Ka-bandwidth has emerged because of the significant growth in the need for satellite services and the scarcity of remaining C-band and Ku-band frequencies now in use (**Table 1**).

Need for satellite systems. The need for satellite solutions for telecommunications is the result of several key developments.

Ka-band satellite systems are seen as an effective way of providing telecommunications services: they are complementary to fiber-optic systems but have a lower last-mile installation cost, particularly for regions of low population density. Satellites are well suited for providing service to large areas of the Earth. They are also seen as a fast and economical way to expand data communications services without the need to substantially add to the telecommunications infrastructure.

User acceptance. Cellular phone usage has grown rapidly because of its unique ability to provide fast telephone access. Direct-to-home television service via satellite has also raised the acceptance level of wireless communications by providing service of higher quality relative to local cable and local television services. These developments combine to increase the acceptance of wireless communications among a broad segment of the population.

Growth in Internet use. Internet users require fast, or high-data-rate, access to information. Satellite systems can deliver pictures and video 200 times faster than a typical modem and telephone line. *See* INTERNET.

Emergence of teleworker. Less expensive high-performance computers are enabling people to work effectively at home. A telephone modem now provides a maximum transmission rate of 56 kilobits per second (kbps) to and from the computer. To send data to

TABLE 2. Potential Ka-band satellite applications

Applications	Data rate from user, kilobits per second	Data rate to user, kilobits per second
Internet user access	14–128	128–3000
Worldwide collaborative work	1500–10,000	1500–10,000
Private data networks	1500–10,000	1500–10,000
Remote learning/education	384–2000	384–2000
Telemedicine	1500–2000	1500–2000
Video conferencing	384–2000	384–2000
Traffic trunking	>144,000	>144,000
Videophone	128–384	128–384
Content distribution	128	>3000
Video backhauling	384	>8000
Interactive games	2500	2500
Remote site support	1500–100,000	1500–100,000
Temporary site support	1500–100,000	1500–100,000

customers or to exchange information with members of a collaborative work team, teleworkers will need 10 times the capability of their present telephone modem system. To view each other at the same time, using services such as video telephone or video conferencing, home users will need high-speed lines that are full duplex and provide equal data rates simultaneously in both directions. The growing range of applications for Ka-band satellites is listed in **Table 2**.

Use of Ka-band GEO satellites. Geostationary (GEO) satellites remain in the same position above the Earth. They must be located in a circular orbit 22,236 mi (35,786 km) above the Equator, where one orbital period matches the Earth's 24-h rotation time. This position allows a satellite to provide communications links over vast regions. One geostationary satellite can theoretically access millions of customers. In particular, satellites that use the Ka-band can operate with very small spot beams throughout the total coverage area; thus the system can reuse the same radio frequencies many times. For example, a user in spot beam number 11 in **Fig. 1** can transmit over the same frequency as a user in spot beam 2. This is called frequency reuse and is a particular advantage for Ka-band systems because, with very small spot beams and over 1 GHz of available bandwidth, one satellite can approach 8–16 times reuse to provide a total simultaneous throughput of more than 5 gigabits per second of transmission capacity. This equates to 350,000 users receiving Internet data simultaneously at the present average modem speed of 28.8 kbps.

The Ka-band ground terminals or dishes require no tracking mechanisms since the satellite does not move relative to the user. Under clear weather conditions, data can be transmitted and received at 10 times current telephone-line speeds with a dish that is 2 ft (0.6 m) in diameter, as opposed to over 4 ft (1.2 m) for Ku-band and over 10 ft (3 m) for C-band.

TABLE 1. Frequency bands used for satellite commercial communications services

Frequency band	Frequency range, GHz	
	Uplink	Downlink
C	5.9–6.4	3.7–4.2
Ku	12.7–14.8	10.7–12.57
Ka	28.35–28.6	18.55–18.8
	29.0–30.0	19.2–20.2

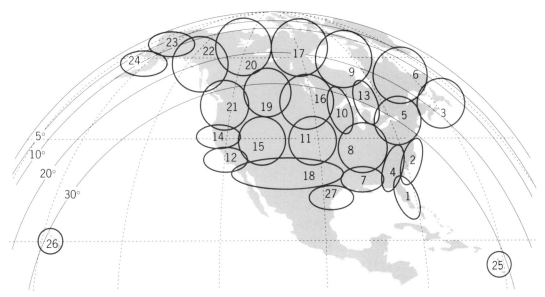

Fig. 1. Cyberstar Ka-band system spot-beam pattern for North America. Beams are numbered 1 through 27. Contours where the altitude of the geosynchronous satellite is 5°, 10°, 20°, and 30° are shown.

Technology challenges. To provide cost-effective and reliable service, a Ka-band satellite system must compensate for the additional space attenuation and rain attenuation that occur at higher radio frequencies. Roughly 60% more satellite power is required to provide the same radio-frequency power density on the ground as in the Ku-band. Other key technologies required include onboard processing and switching to handle many users and data rates, inter-satellite communication links to transfer traffic data from one satellite to another, inexpensive user dishes, and delay-compensation techniques for standard computer interfaces such as the standard transport control protocol and Internet protocol (TCP/IP).

Global Ka-band system. The Cyberstar system (**Fig. 2**) will use recent advances in space processing and the availability of the Ka-band frequency to provide

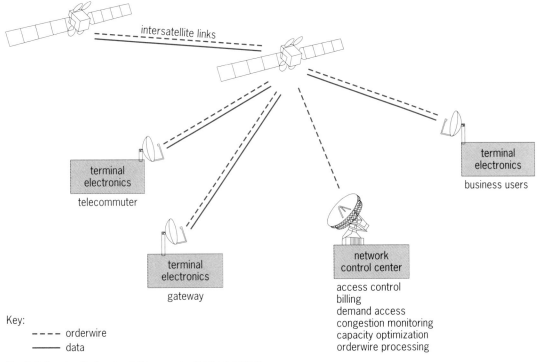

Fig. 2. Cyberstar Ka-band satellite communications system.

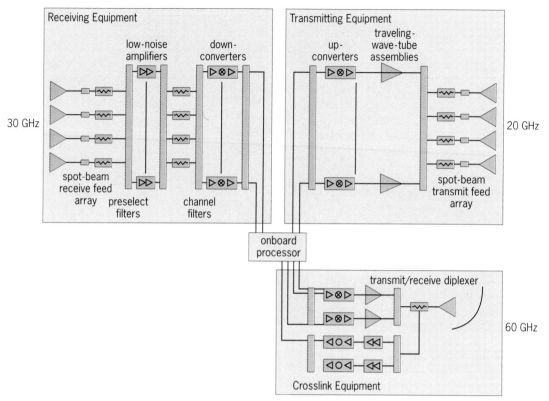

Fig. 3. Communications transmitting and receiving equipment on board a Cyberstar satellite.

			World coverage					
Systems	Services	Number of satellites	North America	South America	Europe	Asia	Africa	Australia
Cyberstar	Interactive media	3	Yes	No	Yes	Yes	No	No
Astrolink	Voice, data, video	9	Yes	Yes	Yes	Yes	Yes	Yes
Galaxy/Spaceway	Two-way interactive, direct-to-home†	20	Yes	Yes	Yes	Yes	Yes	Yes
Voicespan	Telephone, electronic message, software distribution	12	Yes	Yes	Yes	Yes	Yes	Yes
GE Star	Video, audio, video conferencing, video telephone	9	Yes	Yes	Yes	Yes	Yes	Yes
Morning Star	Broadband communications	4	Yes	No	Yes	Yes	No	No
Orion F7/8/9	Transatlantic high-speed digital services†	3	Yes	Yes	No	No	Yes	No
Millennium	Video, audio, fax, data	4	Yes	Yes	No	No	No	No
Panamsat	Voice, data, video	2	Yes	Yes	Yes	No	Yes	No
Echostar	Data, video, video telephone	2	U.S. only	No	No	No	No	No
KaSAT	Data, video, video telephone	2	U.S. only	No	No	No	No	No
Netsat 28	Broadband services	1	U.S. only	No	No	No	No	No
Visionstar	Complement to terrestrial Ka-band system	1	U.S. only	No	No	No	No	No
Teledésic	Multimedia	288‡	Yes	Yes	Yes	Yes	Yes	Yes

TABLE 3. Ka-band satellite systems*

* Systems are planned for operation in the United States.
† Both Ku-band and Ka-band services are provided.
‡ Low-Earth-orbit system requires many small satellites to provide the same world coverage.

high-bandwidth digital communications service to the growing number of multimedia and Internet users. Expected to be in operation by the year 2000, the system will offer a variety of services, including desktop videophone, computer network access, and distance learning (a technique that uses broadcast television to teach classes in remote areas, and enables the teacher and the students to interact).

The system will provide service directly to desktop computer users via a 2-ft (0.6-m) receiving and transmitting antenna terminal located on the roof of a home or business site. The satellite will provide access to the Internet, business networks, or other wired sites through a gateway. A gateway is a 10–15-ft (3.0–4.5-m) terminal located within the view of the satellite that provides the connection to the terrestrial (ground) network. Control of the satellite and the billing and access management will be performed at a central ground-station location. A special operation mode at a low transmission rate, called orderwire, will be used to manage the requests for access to the system.

The satellite communications transmitting and receiving equipment in **Fig. 3** operates by having the satellite capture the 30-GHz radio waves emitted by ground terminals, using a spacecraft radio-frequency receiving antenna. The weak signal is amplified through a low-noise amplifier, filtered to remove noise, and down-converted to a lower frequency. Next, the signal goes to the onboard processor, where thousands of radio-frequency waves from Earth are modified in amplitude and other wave characteristics, and then combined into data channels. From this very compact channel format, the data are reprocessed back into separate signals and up-converted to 20 GHz. The new signals are transmitted back to Earth via traveling-wave-tube amplifiers. If one of the new signals must travel to a location not accessible by the particular satellite, the signal is switched to the satellite crosslink, which up-converts the signal to 60 GHz and sends the signal to another satellite within view of the desired location. Repackaging the data on board the spacecraft through the processor allows the satellite system to manage a very large number of users and many types of ground telecommunications interfaces.

Other Ka-band systems. In 1995, the Federal Communications Commission received 14 requests for Ka-band satellite radio frequencies to access Earth stations in the United States. (No additional filings have been allowed since then.) **Table 3** provides a summary of these systems, which include one low-Earth-orbit system, Teledesic, discussed below.

Other systems are being planned for operation outside the United States. In all, over 60 systems have been proposed for operation in the Ka-band to the International Telecommunications Union.

Arnold Friedman; Kathy A. Shockey

Large-Number, Low-Earth-Orbit Satellite Systems

Satellite systems are unique in their indifference to users' locations and in their ability to provide extremely robust networks of high efficiency. In principle, a communication network made up of a large number of satellites can serve all of the people and businesses on Earth's land surface, on its seas, and in the air, without regard to proximity to cities or other communication hubs. If such a network has many identical satellites of equal rank, in low orbits, scattered more or less evenly around the world, the users' communication paths will be the shortest available, and the loss of one or a few satellites will have a negligible effect on system performance. Time lags in conversation will also be negligible, allowing conversation and data exchanges as though the users were in the same room, even if they are on opposite sides of the Earth. If that system also has a large capacity and is able to provide broadband connections to users, it can bring the full range of communications services enjoyed by urban users in developed nations to people and businesses in every part of the world, regardless of the level of economic and technical development of the nations.

Teledesic is the first large-number, broadband, low-Earth-orbit (LEO) satellite system to be proposed. It was designed under the following constraints:

1. Bandwidth-on-demand architecture, allowing services ranging from basic telephone service through broadband Internet access, high-speed data, and video conferencing.

2. Inexpensive user terminals, with the service defined by the terminal.

3. Network architecture that is geodesic (that is, allows the shortest possible paths between users) and robust, with many parallel paths and nodes for high-traffic conditions.

4. Identical satellites, for low cost (implying a nonhierarchical network).

5. Negligible environmental impact, both to the Earth and to the space environment.

6. Very large capacity, allowing low user charges.

7. Quality of speech and data comparable to the best terrestrial circuits.

8. Low round-trip delay in conversations and data exchanges.

9. Compliance with domestic and international frequency allocations and other rules.

10. Essentially complete coverage of the Earth's surface.

Choice of altitude. The requirement of low round-trip delay demands that the satellites be close to the Earth's surface. A signal must travel from a user to the nearest satellite, then to a satellite near the user's correspondent, and then down to the correspondent. The correspondent's reply must return by the same path. Since radio signals travel at the speed of light, it takes about 67 milliseconds for a signal to go from one side of the world to the other (134 ms, round trip). Conversation becomes difficult and data exchanges slow when the total round-trip delay exceeds 200 ms; this leaves only 16.5 ms for each of the four uplinks and downlinks, including processing delays. On this basis, the satellites can be no higher than about 900 mi (1500 km).

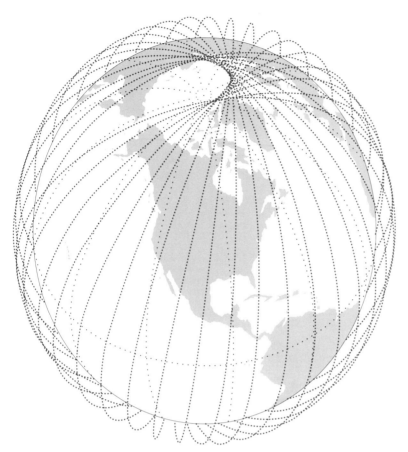

Fig. 4. Near-polar orbits in the Teledesic system. There are equal numbers of satellites in each orbit.

The satellites' altitude affects the cost of the satellites and the users' terminals. Because radio signals, like light, are subject to the inverse-square law, doubling the satellites' altitude requires that transmitted power be increased by a factor of four if nothing else is changed. This increases the cost and weight of the satellites and terminals, and increases the cost to launch the satellite, because of the satellite's greater weight and because of the extra energy the launch vehicle must expend to reach a higher orbit.

Finally, the satellites' altitude determines the number of satellites in the system. If reliability in rainy weather sets the minimum usable angle above the horizon at, say, 40°, and polar or near-polar orbits are used to get full Earth coverage, then the maximum spacing between satellites is set by the need to have at least one satellite in view at all times from any point on Earth. For an altitude of 435 mi (700 km), this requires about 840 satellites; 620-mi (1000-km) orbits allow the number to be reduced to perhaps 500 satellites, but with a concomitant increase in satellite and launch costs.

Network architecture and geometry. The network architecture of the Teledesic system was designed to be extremely robust and to provide many alternate routes to accommodate traffic peaks and equipment failures. **Figure 4** shows the constellation's organization into circular polar orbit planes with equal numbers of satellites in each plane. Each satellite has high-capacity intersatellite links to the two satellites ahead and the two satellites behind it in its own orbital plane, and to one satellite in each of its four nearest orbital planes. This rich interconnection pattern approximates that of the ideal geodesic network as nearly as possible, given the geometric restraints of polar orbits. The polar orbits were selected to provide service to high latitudes, which are not well served by geosynchronous satellites.

Low-Earth-orbit satellite systems have orbital periods of the order of 100 min, so that each satellite is more than 40° above the horizon at any given point on Earth for about 3 min. It is usual in wireless systems to divide the Earth's surface into cells, and to hand off terminals from cell to cell as the terminal or the satellite moves. High capacity can be obtained only if the cells are small, so that the same frequencies can be reused many times; if the cells were to move with the satellites, a user would be in each cell for only a few seconds. The necessary coordination messages for a satellite to perform hand-offs with tens of thousands of terminals every few seconds would entirely consume the satellite's communication capacity. Therefore, large-number, low-Earth-orbit systems must regard communication cells as though they were painted on the Earth's surface, and steer their communication beams to those cells; as a new satellite comes into view, all communications in progress in each cell are transferred in a group to the new satellite.

Choice of radio frequencies. As discussed above, modern telecommunications service must accommodate broadband connections for Internet access and data. To serve a large number of users with broadband service requires at least 500 MHz total bandwidth in each direction. This amount of bandwidth is available only at millimeter-wave frequencies. Teledesic has chosen frequencies in the internationally allocated bands at 27.5–30.0 GHz in the Earth-to-space direction, and 17.8–20.2 GHz in the space-to-Earth direction.

At 30 GHz, the wavelength of the radio signal is 0.4 in. (1 cm). This is of the same order of size as a raindrop; thus much of the signal is reflected or scattered by rain. Since there is rarely rain above the freezing level in the atmosphere, and the freezing level is usually less than 3 mi (5 km) above the surface, the amount of rain in the signal's path can be minimized by using high angles of radiation. The Teledesic system is therefore designed so that terminals do not communicate with satellites that are lower than 40° above the horizon. This assures that the signal will not have to pass through more than a few kilometers of rain.

Service and signal design. The richly interconnected low-Earth-orbit network will provide low-delay digital connections between terminals and, through regional gateway switches, connections to the public

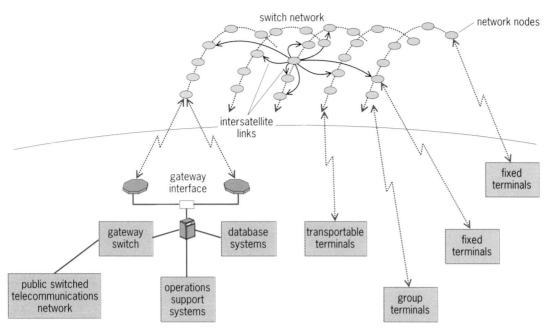

Fig. 5. Connections in a large-number, low-Earth-orbit communications satellite network.

switch telecommunications network (PSTN; **Fig. 5**). A wide range of services will be provided, from 16-kbps voice and data service through services at several megabits per second, and in special cases, high-speed data services of up to 1 gigabit per second. These services will be provided in a bandwidth-on-demand format, in which users or their applications select the data rate or bandwidth they need for a particular connection, and may vary that bandwidth dynamically during the life of the connection.

This kind of service is best provided by a packet system, in which the sequences of bits making up digitized voice or data are broken into groups of a fixed number of bits (called the payload) to which an address preamble (called the header) is added. Each satellite contains a packet-switching machine which reads the header of each packet and routes it appropriately so that it takes the shortest available path to its destination. There is no switching hierarchy: Each satellite is a node in the network, and each is of equal rank. The packet structure and protocols used are similar to, but not identical to, those in the asynchronous transfer mode (ATM) systems now in use for terrestrial packet switching. Teledesic's format was selected so that translation to and from asynchronous transfer mode and other packet systems at Earth gateways and terminals is straightforward.

For background information *see* COMMUNICATIONS SATELLITE; DATA COMMUNICATIONS; DIRECT BROADCASTING SATELLITE SYSTEMS; MOBILE RADIO; PACKET SWITCHING in the McGraw-Hill Encyclopedia of Science & Technology.   Edward F. Tuck

Bibliography. G. Gilder, Telecosm ethersphere, *Forbes ASAP Technology Supplement*, October 10, 1995; T. Logsdon, *Mobile Communication Satel-* *lites: Theory and Applications*, 1995; *The Red Herring Guide to the Digital Universe*, 1996.

Computer-integrated manufacturing

Recent developments in computer-integrated manufacturing have involved improvements in computerized methods and approaches for planning, scheduling, and monitoring processes specific to batch and continuous processing. Other research has focused on applying the concept of artificial intelligence within the computer-integrated manufacturing environment.

Process Industries

Manufacturing can be classified as discrete-parts manufacturing, batch processing, and continuous processing. In discrete-parts manufacturing, products or parts are manufactured one or more at a time. Each product can be monitored individually for quality control, and the production planning is based on an economical order quantity. Discrete-parts manufacturing is used for automobiles, computers, copiers, appliances, and similar products. Batch processing is usual when products require many reaction steps or when isolation and sterility problems occur, such as in the production of dyes and pigments, pharmaceuticals, and food products. In continuous processing, the process usually lends itself to steady-state conditions, and both the stability and the market life of the product are expected to be long. Products that are manufactured by continuous processing include chemicals such as ammonia, sulfuric and nitric acids, gasolines, fuel oil, and aromatics (fragrances and herbs).

In discrete-parts manufacturing, a large quantity of raw materials is required to manufacture a single product; however, in batch or continuous processing, small amounts of material are transformed into many types of products.

Complete computer-integrated manufacturing systems apply computer technology to all operational and information-processing functions, including demand forecast, order receipt, design and production, quality control, product shipment, and field support of the product after sales. Although computer-integrated manufacturing systems are applicable to most, if not all, industries, the actual implementations and procedures vary according to the classification of manufacturing. In batch and continuous systems, emphasis is on production planning and scheduling, process monitoring and control, and process automation and sensors.

Production planning and scheduling. Multipurpose batch-processing plants are used for the economical production of chemical, pharmaceutical, and food products in small quantities. Different products can be processed concurrently within the same facility. Flexible batch production originated in specialized branches of the chemical industry, such as pharmaceuticals and paints. This technique has been extended to other products such as polymers and petrochemicals. It is also rapidly replacing small or medium continuous production plants.

A multipurpose batch plant consists of different vessels (tanks and reactors), auxiliary equipment (mixers, fillers, and pumps), and pipes to transport products. Each product is defined by a recipe that includes the quantities of ingredients, conditions of mix, operator and process control instructions, process parameters to be monitored, and stopping conditions.

Unlike continuous processes, the batch processes are characterized by the lack of a steady state or equilibrium. This is attributed to two causes: the operations in the single process units are not stationary but characterized by a trajectory, for example, the progress of a polymerization reaction; and the dynamics and the couplings of the plant are different and, in most cases, the sequence of batches in the plant is variable. Different products are produced in different quantities and under specific conditions. Unlike discrete-parts manufacturing, both batch and continuous processes require nonpreemptive processing, in which processing of a batch cannot be interrupted by a higher-priority batch. Given these conditions, it becomes crucial to provide efficient scheduling algorithms in order to increase the utilization of the processing facility and to reduce delays in product delivery.

In the scheduling of a batch-processing facility, products are differentiated from each other by due date, importance, and size (volume). In addition, each part may have a set of characteristics, or a profile, including temperature, processing time, cooling rate, and so forth. When two products have identical profiles, they may be processed together in the same facility and are classified as belonging to the same class or batch. There are other constraints that make scheduling in batch processing a complex task, and the role of computer-integrated manufacturing becomes more evident.

The scheduling of batch production is performed in two stages. The first stage consists of forming batches; the second involves sequencing the batches so that some specific objectives are realized. For example, batches must be sequenced so that the deviations from meeting the due date are minimized. Batch formation and sequencing are usually coupled with each other and should be treated as such.

Assigning product to batches for processing is beneficial when the number of products to be produced is large, because it minimizes nonproductive time used for each setup (time to change over from one product type to another). Optimizing the batch-processing sequence on a processing facility will further decrease setup time.

Production planning and scheduling are relatively simple for the continuous processing industry; in a typical continuous process the number of product types is small while the number produced from every type is usually high. The main factors that affect production scheduling are product demand, setup cost, and inventory costs. For example, large-scale products such as most petrochemicals use raw materials that have proved to be in fairly steady supply, and both continuous and batch production could be equally suitable. For seasonal products such as agrochemicals, the operating plant should be designed to work at different production levels during the year, and shutdowns are anticipated. Batch processing is less severely affected by plant stoppage than is continuous processing in which efficient production scheduling procedures are needed to reduce the likelihood of plant shutdown or of production of excessive inventory.

Process monitoring and control. Process industries utilize physical and chemical changes in materials to produce one or more final products. To ensure that the characteristics of the products and the process parameters are within prespecified target values, the process parameters and product characteristics should be continuously measured, monitored, and adjusted. Measuring the parameters of the manufacturing process and the characteristics of the product is achieved by using sensors and transducers. Monitoring of the process and products is achieved by providing the sensors' data to a control system such as statistical process control, which includes the control charts, control limits, and rules and conditions that signal when the process is out of control. Adjusting the process parameter is an action that is performed based on the recommendations of the monitoring process.

In addition to the conventional quality problems, batch and continuous process industries face specific problems related to quality control: (1) Since the measurement methods used in the process industries are themselves miniature chemical processes requiring control, the reproducibility of the test methods should be controlled. (2) The quality-control samples should be protected from contamination with air, freezing, container contamination, and such. (3) In-process samples for quality-control purposes may differ considerably in composition and characteristics from the finished product; therefore, it is important to investigate the relationships between in-process samples and finished products. (4) Testing time of the sample may be relatively long compared with the batch reaction time, requiring anticipation of control decisions.

The traditional methods of statistical process control commonly used in discrete-parts manufacturing are also applicable to batch and continuous processing. Some modifications of these methods may be needed to ensure their applicabilities. For example, modifications of the standard Shewhart Control Chart are quite common in chemical and pharmaceutical quality control. These include cumulative sum charts which are sensitive to small shifts in the process mean (or level), narrower control chart limits, individual charts, and moving-range charts.

In batch processes, individual measurement and moving-range control charts are particularly suitable for process quality control. The individual measurement chart may be used with average and range control charts or alone. Characteristically, in batch processes the data accumulate slowly, and it is advantageous to use the moving-range method for calculating the control-chart limits for the individual measurements. This consists of averaging the differences between consecutive pairs of measurements and substituting constants in an appropriate formula.

In continuous processes, the traditional quality-control methods may be applied without modifications. This is possible because continuous processes are relatively simple and because two or three factors, all under operator control or computer control, are decisive in controlling product quality.

Automation and sensors. Most continuous processes operate under computer control. Hardware is usually quite simple to design, and software can be easily developed to deal with steady-state conditions. In continuous processes, many of the operations utilize highly stable and well-developed equipment such as furnaces and large extraction and distillation columns. Batch processes are more difficult to automate than continuous processes, because the conditions that characterize them are not steady state. Recent advances in computers and sensors have significantly impacted process industries by allowing a wider range of variables to be measured in or close to real time. In Europe, the Community Bureau of Reference has recently focused on the need to develop implementation for direct measurements in the process industries and in medical and environmental monitoring.

The sensors needed for process industries are significantly different from those required for discrete-parts manufacturing. On-line instruments differ from off-line instruments, since the samples "seen" by both types of instruments are different. First, the sampled volume represents just a small proportion of the process stream; it is often difficult to assess whether samples submitted to the quality-control laboratory are indeed representative. Second, the samples may change during and after removal from the process line; during sampling, transfer, and off-line measurement, it is rarely possible to mimic the conditions within the process vessel.

The on-line determination of compositional properties offers the widest scope for instrument development. Instrumentation for the on-line measurement of pressure, temperature, pH, mass flow, volume, viscosity, and density will become more accurate and more widely applicable; development of sensors for odor, color, and moisture content will continue to be a challenge.

Both the food processing and pharmaceutical industries are in dire need of on-line measurement of microbial contamination and growth. The speed and accuracy of the measurements are two key requirements of the sensors to be developed.

Advancements in sensor technologies will enable a wider range of measurement tasks closer to ideal on-line measurement to give real-time output. This, in turn, will reduce the amount of reworking of batches of out-of-specification products and will improve the quality of many products.

Elsayed A. Elsayed

Intelligent Manufacturing

There are two major goals in current research dealing with computer-integrated manufacturing processes. One is to develop integrated, self-adjusting systems that are capable of manufacturing various products with minimal supervision and assistance from operators. The other is to improve product quality while maintaining the production cost at low level. A large number of new technologies have been implemented in the manufacturing industry in the last few decades to achieve the above goals. They include advances in microelectronics and computer systems, new machine tools, robots, chemical vapor deposition, physical vapor deposition, and plasma-aided manufacturing. Computer-integrated manufacturing is a very wide field encapsulating the applications of artificial intelligence and related technologies, including expert systems, fuzzy logic, neural networks, generic algorithms, and pattern recognition. Hybrid systems have been developed that combine one or more techniques of real-time material planning, process control and manufacturing, sensors, signal processing, process modeling, and machine tools. Considerable attention has been paid by many investigators to the development and implementation of artificial intelligence in computer-integrated manufacturing.

On-line process monitoring. On-line process monitoring is one of the most important requirements for achieving the primary goals of computer-integrated manufacturing. Artificial intelligence can be successfully applied in the area of the automated on-line monitoring of the processes. Monitoring tasks have three aspects: sensing, signal processing and monitoring, and decision making. In general, there are two types of monitoring methods: model-based methods and feature-based methods. For many manufacturing processes, sensor signals can be considered as the outputs of a dynamic system in the form of time series; consequently, process monitoring can be conducted based on system modeling and model evaluation. A linear time-invariant dynamic system can be described by a number of models such as state space model, input-output transfer function model, autoregressive model, and autoregressive and moving average model. Among various models, the methods of dynamic data systems are particularly effective for monitoring manufacturing processes. In comparison to other techniques, the methodology of dynamic data systems often provides a better description for the process; in addition, a more accurate spectrum estimation can be obtained. When a model is found, monitoring can be performed by detecting changes of the model parameters or changes of expected system responses (for example, prediction error). Model-based monitoring methods are also referred to as failure detection methods. Model-based methods have two significant limitations. First, many manufacturing processes are nonlinear time-variant systems. Second, sensor signals are dependent on process working conditions. It is often difficult to identify whether a change in sensor signals is due to the change of process working condition or to the deterioration of the process.

Feature-based monitoring methods use less suitable features of the sensor signals to identify the process conditions. These features could be time- or frequency-domain features of the sensor signals, such as mean, variance, skewness, kurtosis, crest factor, or power in a specified frequency band. Often normalized indices (the indices that are independent of unit) are recommended. Monitoring indices may be continuous numbers (such as the mean of sensor signal) or discrete events (such as a local logical signal, "on" and "off"). The process working conditions can also be used as monitoring indices.

Choosing appropriate monitoring indices is crucial. In practice, monitoring indices should be selected based on analytical study and computer simulation of the process, as well as systematic experiments. However, in many applications, selection of the monitoring index remains an art. The selection of appropriate monitoring indices often involves various signal-processing techniques. When a monitoring index is obtained, monitoring can be done by simple comparison with a threshold value. In many applications the determination of the threshold is important; in general, the threshold may be determined by using hypothesis testing or empirical methods. The more effective way is to use multiple sensor signals and multiple monitoring indices. This is also referred to as sensor fusion, classification, or decision making.

The feature-based methods consist of two phases: learning and classification. There are two types of learning methods: learning from samples and learning from instructions. For monitoring of manufacturing processes, learning from samples is usually more effective, since precise instructions typically are unavailable or are rather limited. The various feature-based monitoring methods are pattern recognition, fuzzy systems, decision tree, expert systems, and neural networks.

Applications of artificial intelligence. The concepts of artificial intelligence have been applied to control numerous manufacturing processes. In metal removal processes, the level and spectral distribution of acoustic energy emitted during machining is related to both failure phenomena occurring in the workpiece material and processing conditions. The utilization of pattern-recognition capabilities of neural networks in signal interpretation has made it possible to avoid obstacles associated with traditional techniques. Neural network systems provided with acoustic emission signals as input are employed in machining to estimate missing process descriptors, such as cutting-tool position and tool condition. A feature-based design system has been developed that integrates product design and process planning; it uses a generic algorithm for machining intricate components. In the area of forging, a diagnostic expert system is available that identifies the cause of various manufacturing defects in hot forging and suggests remedies. In the field of welding, the pattern recognition capabilities of neural networks are utilized to obtain models relating indirect welding parameters (such as applied welding voltage and torch travel speed) to direct welding parameters (such as weld pool penetration and bead width). Marked advantages in flexibility, speed, and simplicity are realized with the artificial neural system. Provided that sufficient traning data are available, a neural network model could capture all of the true complexities of arc welding processes, including nonlinearities and parameter cross-coupling. A method using pattern recognition has been developed for on-line monitoring of martensite transformation and porosity during welding of steels. Neural-network technology has also been applied for monitoring an injection molding process.

Intelligent machine tools. Intelligent machine tool systems, also known as cells or centers with advanced software and controllers, have recently become operational. They include a multiaxis machining cell that can shape plastics, a high-speed machining center for machining automotive parts, and a high-precision computer control cell for intelligent die-molding operations. These cells have enabled end users to reduce cycle times of part machining without compromising part precision. Laser systems and electronic indicators to measure the spindle alignment are used to achieve the design goal of transferring parts from one spindle to the other within 2.5 micrometers. The x, y, and z axes of the cells are controlled by using electrical direct drives with linear motors. These motors link the machine base and its driven components with a magnetic field rather than a ball screw, providing tool traverse rates of 80 m (260 ft) per minute, compared to earlier rapid traverse rates of 15 m (50 ft). Intelligent, digital alternating-current servo drives supply the linear motors. These drives are connected by optical fiber to a serial real-time communications-system digital interface that links them to the computer numerical control system. The combination of intelligent servo drives and swift, modular computer numerical control enables the cells to look ahead, prepare blocks, and conduct geometrical and mathematical interpolation in 2 milliseconds in order to perform position compensation for contouring and acceleration errors and to provide high-precision resolution.

Industrial rolling mills equipped with networks of sensors, personal computers, and mainframes are being developed to automate their process lines. Augmenting the computer hardware in automating process lines are process models. Computers compare actual process conditions with the model and quickly make changes to maintain a high degree of quality. Rigorous planning and scheduling coupled with automation have enabled the process to proceed from scrap to finished product in 4-5 h, whereas the best conventional steel mills require a minimum of 72 h. This improved process control has also reduced the mill's energy requirements to 25% of that usually needed and has significantly reduced the time requirements of personnel.

For background information *see* ARTIFICIAL INTELLIGENCE; COMPUTER-INTEGRATED MANUFACTURING; CONTROL CHART; MODEL THEORY; NEURAL NETWORK; PROCESS CONTROL in the McGraw-Hill Encyclopedia of Science & Technology.   Jay Gunasekera; S. Venugopal

Bibliography. J. P. Coulter, L. I. Burke, and H. H. Demirci, Neural networks in material processing and manufacturing: A review, *J. Mat. Proc. Manufact. Sci.*, 1:431-443, April 1993; R. Du, M. A. Elbestawi, and S. M. Wu, Automated monitoring of manufacturing processes, Part 1: Monitoring methods, *J. Eng. Ind.*, 117:121-131, May 1995; J. M. Juran and F. Gryna, *Quality Control Handbook*, 1988; E. Kress-Rogers (ed.), *Instrumentation and Sensors for the Food Industry*, 1993; M. Valenti, Machine tools get smarter, *Mech. Eng.*, 117(11):70-75, November 1995; G. Werniment, Statistical quality control in the chemical laboratory, *Qual. Eng.*, 2:59-72, 1990.

Computer system

Modern computer-based information systems have become increasingly complex because of networking, distributed computing, distributed and heterogeneous databases, and the need to store large quantities of data. People are relying increasingly on computer systems to support daily activities. When these systems fail, significant breakdowns may ensue. A computer system can fail in two major ways.

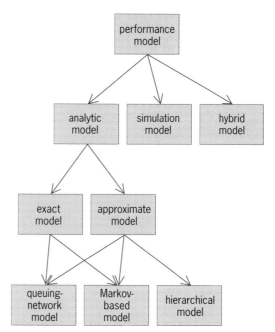

Taxonomy of performance models.

First, functional failure occurs when the system fails to generate the correct results for a set of inputs. For example, if an information system fails to retrieve records that match a set of keywords, or if an air-missile tracking system fails to distinguish between a friendly and enemy missile, a functional failure has occurred. Second, performance failure occurs when the system operates correctly but fails to deliver the results in a timely fashion. For example, if an information system takes a longer time than users are willing to wait for the records they requested, the system is said to fail performancewise even though it may eventually retrieve the correct set of records. Also, if the air-missile tracking system fails to detect an enemy missile in sufficient time to launch a counterattack, the system manifests performance failure. Therefore, in designing a computer system it is necessary to guarantee that the end product will display neither functional nor performance failure. It is then necessary to predict the performance of computer systems when they are under design and development, as well as to predict the impact of changes in configurations of existing systems. This requires the use of predictive performance models.

Performance models. The input parameters of performance models include workload intensity parameters, hardware and system parameters, and resource demand parameters. The outputs generated by performance models include response times, throughputs, utilization of devices, and queue lengths. There are analytic, simulation, and hybrid performance models (see **illus.**). Analytic models are composed of a set of equations, or computational algorithms, used to compute the outputs from the input parameters. Simulation models are based on computer programs that emulate the behavior of a system by generating arrivals of so-called customers through a probabilistic process and by simulating their flow through the system. As these simulated entities visit the various system elements, they accumulate individual and system statistics. Hybrid models combine both analytic and simulation approaches by, for example, replacing an entire subsystem in an analytic model by an equivalent device whose input–output behavior is obtained by simulating the subsystem. Analytic models can be exact or approximate. Approximations are needed either when there is no known mathematically tractable exact solution or when the computation of an exact solution is very complex. Modern computer systems are very complex because of ubiquitous networking, distributed processing using client–server architectures, multiprocessing, and sophisticated input–output subsystems using network-attached storage devices. For this reason, most computer system performance models are approximate models.

Exact performance models. Exact models may be further classified into queuing-network and Markov-chain-based models. Queuing-network models represent a computer system by a network of devices (also called queues) through which customers (for example, transactions, processes, and server requests) flow. Queues may form at each device because of the finite service rate of the device. Solving a queuing network amounts to finding the probability that the network is found in each of the possible states, or at least finding the performance metrics (for example, response times, throughputs, and queue lengths) associated with the queuing network. The solution to a class of queuing networks, called product-form queuing networks, can be obtained through very efficient computational algorithms such as convolution and mean-value analysis. In Markov-based models, the system is modeled by a Markov chain, which represents the set of all possible states of the system with the possible transitions between these states. Each state gives rise to a linear flow-equilibrium equation. Solving the Markov chain amounts to solving the resulting system of linear equations, called the global balance equations. The unknowns in the global equations are the probabilities of finding the system at each of the possible states. In some cases, the Markov chain is not explicitly generated from an analysis of the system, but is derived from a higher-level model used to represent the dynamic and temporal behavior of the system. Examples of these higher-level models include generalized stochastic Petri nets (GSPNs), which are timed stochastic extensions to Petri nets. Algorithms exist that generate the Markov chain automatically from the generalized stochastic Petri net.

Approximate performance models. Product-form queuing networks (PFQNs) cannot model many features of modern computer systems because the conditions for a queuing network to have a product-form (efficient) solution are violated in most cases of interest. For instance, priority-based scheduling disciplines

for the devices in the computer system, such as those implemented by many operating systems for scheduling of central processing units, cannot be modeled directly by product-form queuing networks.

Product-form conditions are also violated in a queuing network that exhibits simultaneous resource possession, that is, a situation where a customer is allowed to hold more than one resource at the same time. This occurs when a file is being transferred from a network-attached tape drive to a disk through a HIPPI (High Performance Parallel Interface) switch. In this case, the tape drive is held by the customer simultaneously with the HIPPI switch or the disk during the transfer. Another example of simultaneous resource possession occurs in shared-memory multiprocessors. In this case, a request to access shared memory needs to hold the processor, the bus, and the shared memory bank during the memory access at the same time.

Queuing networks also do not satisfy product-form conditions when fork-and-join synchronization is present. This situation is characterized by a customer being split into more than one customer at the fork point. These customers follow independent paths within the system and are reassembled into a single customer at the join point when all of them arrive. An example of fork-and-join synchronization occurs in the modeling of a RAID-5 disk (a Redundant Array of Inexpensive Disks, with five disks in the array). Here, a read or write request to a RAID-5 disk is split into five requests to the physical disks that compose the RAID-5 disk.

Another case of violation of the product-form conditions occurs when the routing of customers in a queuing network is a function of the number of customers at other parts of the network. This may occur in the modeling of computer network routing algorithms which decide on the next link to which a packet should be sent based on the level of congestion at the set of possible links in the path to the destination.

Finally, another case where product-form conditions do not hold is when there is a constraint on the maximum number of customers in a part of the queuing network. An example occurs in the modeling of multiprogrammed computer systems where there is a maximum degree of multiprogramming.

Many approximate models have been developed to deal with queuing networks with non-product-form solutions. Some of these approximations are based on modifying the equations of the mean value analysis algorithm. Other approximations are based on formulating the problem as a fixed-point equation and solving it iteratively.

Even queuing networks with product-form solutions may be more efficiently solved with the use of approximate algorithms. The reason is that the computational complexity of solving a queuing network is proportional to the product of the number of devices in the computer system times the number of different classes of customers multiplied by the number of customers in each class. This may render models of moderately large computer systems unmanageable. Several approximations have been developed that drastically cut down the number of operations and memory requirements of the algorithms to solve product-form queuing networks. The most widely used approximation technique in this category is the Schweitzer approximation that can be used to derive an efficient iterative algorithm to solve product-form queuing networks.

Approximation techniques are also used when solving large Markov chains with an extremely large number of states. These techniques are aimed at cutting down the number of states without significant loss of accuracy. Finally, an important category of approximate models comprises hierarchical models. In this case, the complete model is decomposed into a hierarchy of models, $M_1, \ldots, M_n$. The output of the model at level i is used as input to the model at level $i + 1$. In some cases, a dependency between models M_n and M_1 is established requiring an iterative fixed-point procedure to solve the series of models. The modeling techniques used at each level may vary. It is possible to use queuing networks at one level and Markov-based models at another level of the hierarchy. Hierarchical models are very useful when dealing with client–server systems.

Performance of software systems. The design and development of complex software systems is a time-consuming and expensive task. Performance modeling techniques must be integrated into the software development methodology. This integrated approach is called software performance engineering. One goal is to estimate the resource consumption of software under development so that performance models can be used to influence the architecture of the software under development. Better estimates on the resource consumption are obtained as the software development process evolves.

For background information *see* COMPUTER SYSTEMS ARCHITECTURE; DISTRIBUTED SYSTEMS (COMPUTERS); INFORMATION SYSTEMS ENGINEERING; LOCAL-AREA NETWORKS; SOFTWARE ENGINEERING in the McGraw-Hill Encyclopedia of Science & Technology.

Daniel A. Menascé

Bibliography. D. A. Menascé et al., *Capacity Planning and Performance Modeling: From Mainframes to Client-Server Systems*, 1994; R. D. Nelson, The mathematics of product-form queuing networks, *ACM Comput. Surv.*, 25:339–369, September 1993; *Proceedings of the ACM 1996 Sigmetrics Conference*, Philadelphia, May 23–26, 1996; *Proceedings of the 1st IEEE International Conference on Engineering of Complex Computer Systems*, Fort Lauderdale, November 6–10, 1995.

Coral reef

To marine biologists, a reef is a distinct ecosystem epitomizing fragility and the interaction of an incredible variety of corals, algae, and other tropical marine organisms. Over 15% of the total shallow sea floor is occupied by living coral. Reefs provide habitat for diverse fishes and other economically important marine life. However, like the tropical rainforest, coral reefs are experiencing rapid degradation.

To geologists, a reef is not only a living phenomenon, but an ancient one as well. Organic reef deposits occur throughout most of the fossil record (the past 600 million years), testifying to the remarkable longevity of the reef ecosystem. Like the dynamic Earth itself, reefs have undergone remarkable changes which are attributed to mass extinctions, resulting in the collapse of reef ecosystems. Dramatic recoveries and diversifications which followed resulted in distinct patterns. Further investigation of these patterns may answer questions about the future of current reef crises.

Ancient reefs. Animals with hard shells and skeletons evolved approximately 590 million years ago. The first reefs were small mounds bound into coherent masses by calcifying microbes called cyanophytes. These masses were supplemented during the Cambrian Period by archeocyathids, small calcified sponges resembling perforated cones, vases, tubes, and sheets (see **illus.**). Archeocyathids diversified to produce impressive-sized reefs. Toward the end of the Early Cambrian Period, they experienced a mass extinction and declined worldwide, marking the first reef eclipse. In their place came simple microbial and algal communities which dominated through Early Ordovician time.

Ordovician to Devonian time was an interval dominated by reefs of tabulate and rugosan corals and stony sponges called stromatoporoids. At least eight global-scale reef-building episodes are recognized in Silurian reefs and corals, and stromatoporoids continued to dominate Devonian reefs. Compared to Cambrian examples, these reefs contained much larger reef builders and thus more closely resembled modern reefs. The Late Devonian extinction disrupted reef-building ecosystems to such an extent that they never recovered. Following the Late Devonian extinction, there was a protracted nonreef interval lasting 27 million years, and the succeeding Carboniferous and Permian were dominated by a different sort of reef composed of microbial communities, algae, sponges, sea lilies, brachiopods, lacy bryozoans, small noncolonial shelly invertebrates, and many kinds of problematical organisms such as *Tubiphytes*. Conspicuously lacking were large colonial constructors lost during the Devonian extinction. The greatest mass extinction, a disaster to reef communities, occurred by the end of the Permian Period. Nearly every major marine group was affected. The Early Triassic reef eclipse lasted 5–10 million years. It seems clear that a long-duration phenomenon, rather than a sudden jolt, shocked Earth's biosphere.

The recovery of sponges and algae during the Middle Triassic Period reestablished the reef ecosystem and led to the formation of the first coral reefs by the Late Triassic. Scleractinian corals first arose long and conical, while the other was reduced to a small cap or perforated sievelike structure. Diverse forms evolved. After coexisting for 30 million years with corals, these intruders began to replace them within reef communities. Later in the Cretaceous, rudists eventually claimed most of the shallow-water reef habitats. In this unprecedented takeover, possibly rudists were able to outcompete corals (which they came to mimic) or they were just better adapted to the extreme conditions (high temperature and salinity) which were eliminating the corals.

Rudists dominated the reef environment for over 50 million years. Then, near the end of the Cretaceous Period, they declined and succumbed to another mass extinction. As the Cretaceous ended, the surviving rudists became extinct; there was also a near extinction of corals, algae, and many other reef and nonreef organisms. This extinction also eliminated all dinosaurs and many other land animals.

The end-Cretaceous extinction which eliminated the last rudists was followed by another reef eclipse lasting 8–10 million years. The demise of rudists gave surviving corals a clear opportunity at the reef, but why corals waited so long to make a recovery is an intriguing question.

Later, during the Tertiary Period, scleractinians resumed constructional reef roles with subsequent rapid radiation. An association with coralline algae became important, providing ecologic similarity to modern reefs. During the Tertiary, the dominance of coral continued to be strong despite small extinction pulses and major episodes of cooling leading to the ice ages. Corals evolved into very porous, rapidly growing colonies adapted to the reef crest. They even survived rapid sea-level changes during the ice ages. Modern corals remain the dominant constructional organism of the reef ecosystem.

Reef ecosystem evolution. Over a period of 600 million years, globally significant events delivered shocks to the biosphere that destroyed most reefs. Mechanisms proposed to explain perturbations include glaciation and climate cooling, changes in the sea including sluggish circulation and anoxia, meteorite or comet impacts, and sea-level changes. Extinctions in the Ordovician, Devonian, and Permian periods correlate with lowered temperatures. For example, an extended cooling of the ocean surface of only 5–6°F (2.8–3.3°C) would devastate reefs. Many changeovers among reef intervals seem correlated with changes from icehouse to greenhouse supercycles related to outgassing of carbon dioxide (CO_2), global warming, and plate tectonics. The end-Cretaceous extinction has been correlated with a meteorite crater 65 million years old, but rudist reefs disappeared 1.5–2.0 million years prior to that event. Anoxic killing water has been suggested as a cause for the end-Triassic and Permo-Triassic extinctions.

Mass extinctions. Although mass extinctions reduce the number of species, they create new ecological space, permitting rapid speciation as new organisms adapt and radiate to fill that space. Rather than seeking causes of mass extinctions, many researchers

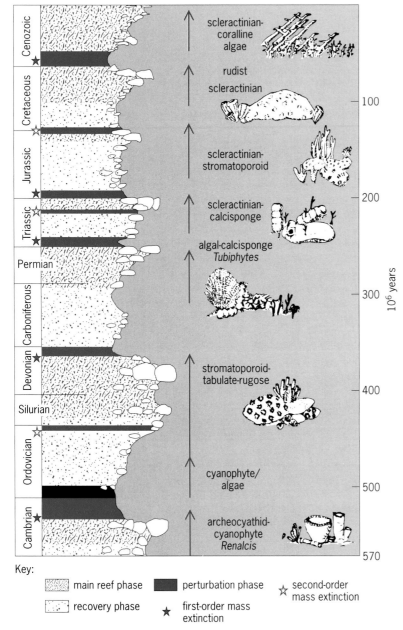

Cambrian to Cenozoic history of reefs, showing the principal organisms inhabiting and building normal marine reefs. Reefs are shown (to the right) as irregular white areas. Arrows show major reef development phases.

during the Middle Triassic but did not become important as reef builders until much later. An extinction at the end of the Triassic was followed by a 4–10-million-year reef gap during Early Jurassic time. A Middle Jurassic recovery led to a Late Jurassic-to-Cretaceous reef ecosystem dominated by corals, calcareous sponges, and algae. Coral reefs continued their evolutionary momentum into the Early Cretaceous Period and were hardly affected by a smaller extinction event and cooling at the end of the Jurassic Period.

Corals of the Early Cretaceous Period were joined by a peculiarly adapted group of bivalve mollusks known as rudists. One shell of these clams became investigate the dramatic recovery intervals. By investigating these critical intervals, they may find clues that help explain the dynamic nature of evolution. At least seven reef intervals are considered long-term, ecologic-evolutionary units. Each interval begins with the establishment of a reef ecosystem, followed by an environmental crisis and a major mass extinction. Global destruction is succeeded by a protracted nonreef interval eventually leading to recovery and a new reef ecosystem. The longest interval following the demise of archeocyathids occurred during the Cambrian–Early Ordovician.

Reef eclipses. Reef eclipses are attributed to extended environmental perturbations precipitated by global ecologic shocks. It is likely that reef ecosystems did not disappear because of a single shock; rather, something in Earth's biosphere was disturbed for an extended period of time. When environmental perturbations ended, conditions favorable to reefs returned, and surviving organisms quickly reestablished reef ecosystems. This type of event occurred after the Late Devonian, Early Triassic, Middle Jurassic, and Early Tertiary. However, the long Cambrian-to-Ordovician eclipse is difficult to explain in terms of continuous environmental perturbation. The rudist reef transition is unique since it was preceded by neither a mass extinction nor a perturbation interval.

Future of Earth's reef ecosystem. Modern reef ecosystems epitomize fragility and delicate balance, although they have survived destructive physical forces, mass mortality, and recent sea-level rises of over 330 ft (100 m). Ironically, reef ecosystems are being undermined by humans who stand to benefit most from their resources. Reef destruction is caused by mismanagement of marine resources, overnutrification by sewage, pollution by agrichemicals and pesticides, and sediment runoff due to poor land-use practices, especially clearcutting. Long-term climate change and the predicted greenhouse warming are of world concern. The strong consensus is that species are currently entering a new mass extinction.

Extinctions occurred repeatedly during the ancient past. A study of the 600-million-year history of reefs will shed light on the disturbing problems that affect living reefs and other ecosystems. While the fossil record does reveal remarkable recoveries after each mass extinction, postextinction perturbation intervals lasted 5–27 million years. Thus from a short-term human perspective, the implications for *Homo sapiens* and most other short-lived mammal species are certainly unfavorable.

For background information *see* ALGAE; BRACHIOPODA; BRYOZOA; EXTINCTION (BIOLOGY); GREENHOUSE EFFECT; REEF; SCLERACTINIA in the McGraw-Hill Encyclopedia of Science & Technology.

George D. Stanley, Jr.

Bibliography. P. Copper, Ancient reef ecosystem expansion and collapse, *Coral Reefs*, 13:3–11, 1994; J. A. Fagerstrom, *The Evolution of Reef Communities*, 1987; G. D. Stanley, Jr., Tropical reef ecosystems, *Encyclopedia of Earth System Science*, 1992; S. M. Stanley, *Extinction*, 1987.

Cosmology

The expanding universe is described by the Friedman-Lemaître cosmological model. This elegant and simple theory, derived independently by A. Friedman in 1924 and G. E. Lemaître in 1927, predicted the expansion of the universe that was observed by E. P. Hubble in 1929. According to this model, usually referred to as the big bang theory, the rate of deceleration of a sufficiently large shell of matter in the universe is determined by the total mass contained within the shell.

In the simplest case of a homogeneous and isotropic universe, it is necessary to specify the mean density of matter defined by the density parameter Ω_M and the present expansion rate or Hubble constant H_0. The parameter Ω_M is the ratio of the present mean density to the critical value, $3H_0^2/(8\pi G)$. Here G is the gravitational constant, for a universe that has zero total energy, or just succeeds in expanding forever. H_0 is the ratio of recession speed to distance, a constant according to Hubble's law. A third parameter must also be specified, since there may be contributions to the density that are not in the form of ordinary matter. In particular, the vacuum may have an intrinsic energy density that is important over cosmological scales.

A. Einstein introduced such a concept in the form of the cosmological constant Λ as a repulsion force to explain why a static universe would avoid collapse. After the expansion of the universe was discovered, the cosmological constant remained as a parameter of cosmology that is equivalent to a contribution $\Omega_\Lambda (\equiv \Lambda/3H_0^2)$ to the density parameter that tends to accelerate the universe. In the absence of a cosmological constant, the deceleration of the universe is specified by $\Omega_M/2$. More generally it is the combination $(\Omega_M - \Omega_\Lambda)/2$ that must be determined in order to decide whether or not the universe is decelerating.

Armed with the parameters H_0, Ω_M, and Ω_Λ, the spatial curvature can be determined. In the Friedman-Lemaître model as described by general relativity, specifying the total matter content is equivalent to specifying the global geometry, and in particular the curvature of space. The curvature parameter is defined by $\Omega_\kappa \equiv 1 - \Omega_M - \Omega_\Lambda$, so that in a flat or Friedman universe, without any spatial curvature, since Ω_κ equals zero, the total density parameter $\Omega_M + \Omega_\Lambda$ is equal to unity. Recent progress has succeeded in reducing much of the uncertainty in these fundamental parameters that has plagued the field of cosmology since the 1980s.

Hubble constant. Determination of the Hubble constant relies on using type Ia supernovae as standard candles that can be seen at great distances. These are used in conjunction with Cepheid variable stars as the more local distance measure. Only with supernovae is it then possible to go to more distant galaxies beyond the Cepheid range and use galaxies at great enough distances that there is minimal uncertainty from the effect of local streaming motions on

H_0. The calibration of the supernova luminosity at maximum light has been established for supernovae in six nearby galaxies for which Cepheid variables have been monitored with the Hubble Space Telescope. A major source of distance uncertainty is removed once Cepheid variables are studied in host galaxies where supernovae have been found. The Cepheid luminosity varies with period, and the known period–luminosity relation enables the distance to be inferred.

A difficulty is that the historic supernovae used for which adequate light-curve data are available occurred in 1875, 1937, 1960, 1972, 1981, and 1990. The quality of the data has improved historically, and opinions differ as to possible systematic uncertainties. Two rival groups of astronomers assign different weights to the calibration procedure, which is compounded by subtle intrinsic variations between the supernovae themselves. The influx of new Cepheid data has brought the two determinations of H_0 into reasonable agreement. A. Sandage and G. Tammann find that $H_0 = 60$ km s^{-1} Mpc^{-1} to within 10% (1 megaparsec equals 3.26×10^6 light-years, 1.9×10^{19} mi, or 3.1×10^{19} km), whereas W. Freedman and collaborators argue for $H_0 = 70$ km s^{-1} Mpc^{-1}, again to within 10%.

Age of universe. From the oldest stars, an age of 14 $\pm$ 2 Gyr can be inferred (1 gigayear equals 10^9 years). With a gestation period for galaxies of 1 Gyr, the true age of 15 $\pm$ 2 Gyr can be compared with the Hubble time H_0^{-1}, equal to 15 Gyr for $H_0 = 65$ km s^{-1} Mpc^{-1}. The age of the universe is approximately equal to H_0^{-1} if the universe is curved or open ($\Omega_\kappa > 0$), a model which is thereby favored. In a flat universe ($\Omega_\kappa = 0$), the age is $(2/3)H_0$ and not easily reconcilable with stellar ages even when the very lowest feasible stellar age scale is adopted.

Density of universe. The luminous matter in the universe amounts to a density in the form of stars of $\Omega_* \approx 0.005$. The contribution in the form of intergalactic gas is determined by measuring absorbing clouds toward distant quasars, and provides a contribution that amounts to $\Omega_{gas} \approx 0.03$. An indirect method appeals to primordial nucleosynthesis of light nuclides, such as deuterium and the isotopes of helium and lithium, in the first minutes of the big bang. Comparison with observed abundances yields the total baryon density, which must include both stars and intergalactic gas, to be $\Omega_B \approx 0.05$, implying that most of the baryons in the universe have been accounted for. The remaining baryons may be in the form of dark compact baryonic objects in the galactic halo, for which there is some evidence from gravitational microlensing studies. Such objects could contribute to the present baryon density at a level several times that of the stellar contributions.

Dynamical measurements provide a robust approach to measuring dark matter on larger scales. These methods arise from the so-called weighing of large regions of the universe, notably galaxy clusters and superclusters. Determination of galaxy peculiar motions makes it possible to estimate the local mass overdensity that is responsible for driving these motions. Without such an overdensity, only a uniform Hubble flow would be observed. Studies of galaxy clusters measure the mass on scales of about 1 Mpc, but galaxy peculiar motions can be used to measure the mass density on scales of up to 30 Mpc. It is inferred that Ω_M is at least 0.2. Most of the matter in the universe is therefore not baryonic. Since only the component of mass density that is nonuniform over the sampled region is measured, an even higher density could be present in the form of a uniform dark-matter component.

Global measurements of density. Supernovae of type Ia (the most luminous variety) have been detected out to a redshift of 0.8. Utilization of these supernovae as standard candles provides a measure of Ω_M and Ω_Λ in combination. Approximately 20 type Ia supernovae have been studied out to this distance. If the universe is assumed to be flat (that is, the curvature parameter, Ω_κ, equals zero), the currently available data yield $\Omega_M = 0.9 \pm 0.3$. Because the luminosity distance to a distant supernova varies with a changing dependence on Ω_M and Ω_κ as a function of redshift, both Ω_M and Ω_κ can, in principle, be measured independently. This goal should be achievable once about 100 type Ia supernovae have been studied out to a redshift of 1.

Curvature of universe. Measuring the curvature of the universe requires the deepest possible probe of cosmology. Temperature fluctuations in the cosmic microwave background radiation, the relic photons from the big bang, have been used to measure the curvature parameter, Ω_κ. In the standard cosmological model, the microwave background radiation is seen back to a redshift of about 1000, at which the last scattering of photons by ionized gas occurred. Subsequently, the diffuse matter in the universe was too sparse to induce any further scattering. Study of the variations between different regions on the sky of the cosmic background temperature in effect provides a mapping of the primordial density fluctuations in the early universe.

Density fluctuations are inferred to be present, since all large-scale structure in the universe has formed by gravitational instability of primordial density fluctuations. Indeed, this leads to the prediction that associated temperature fluctuations of the order of 10^{-5} of the temperature itself must be induced at the last scattering of the radiation. These were measured in 1992 by the *Cosmic Background Explorer* (*COBE*) satellite, at within a factor of 2 of the predicted level. *COBE* measured fluctuations over angular scales of 10–90° and found approximately equal power spread over the different scales. This is precisely the prediction of inflationary cosmology, the standard model of the very early universe.

On a scale of about 1°, however, the inflationary model predicts that there should be a distinct peak in the temperature fluctuations. The growth of density fluctuations can be depicted as being controlled by the interplay between gravity, which provided the driving force, and radiation pressure, which in the early universe tended to disperse the fluctuations when scattering occurred. After the last scattering epoch, density fluctuations grew freely by gravitational instability and eventually formed galaxies and large-scale structures. A natural scale was imprinted on the distribution of primordial temperature fluctuations that corresponds to the horizon scale of the universe at last scattering. This horizon scale is the distance that light had traveled since the big bang. Only on this scale, and on smaller scales, at last scattering did density fluctuations have time to respond at the sound (or acoustic) speed to the release of pressure as photons first traveled freely. There is a resulting enhancement in temperature fluctuations, by about a factor of 3, that occurs on an angular scale of roughly 1°, which is the angle subtended by the horizon of last scattering. The finite duration of the period over which scattering ceases means that the fluctuations in temperature are erased on smaller angular scales.

There have now been more than 20 experiments that have mapped the cosmic microwave background at angular resolutions as high as a few arcminutes. The acoustic peak predicted in the temperature fluctuations has been seen at an angular scale that is near 1°. This angular scale directly measures the curvature of the universe. The angular projection of any physical scale imprinted in the early universe depends only on geometry, and the angular size of the associated feature scales as the square root of the curvature parameter, $\Omega_\kappa^{1/2}$. The last scattering occurred at an epoch of about 300,000 years, and the horizon size was 300,000 light-years. The angular scale of the acoustic peak is simply this scale at a redshift of 1000 seen projected on the universe as it is now observed.

The curvature of the universe can therefore be read off from a map of the cosmic microwave background radiation. Current observations above an angular scale of 1° restrict the curvature parameter, Ω_κ, to be less than 1, and tentative evidence of a decline in temperature fluctuations at 0.25° requires that Ω_κ exceed 0.3. Greatly refined data relative to presently available maps are needed for a definitive measurement. Such data should be forthcoming in experiments that are planned by the National Aeronautics and Space Administration (NASA) in 2000 and the European Space Agency (ESA) in 2005.

For background information *see* CEPHEIDS; COSMIC BACKGROUND RADIATION; COSMOLOGY; HUBBLE CONSTANT; RELATIVITY; SUPERNOVA; UNIVERSE in the McGraw-Hill Encyclopedia of Science & Technology.   Joseph Silk

Bibliography. R. Canal, P. Ruiz-Lapuente, and J. Isern (eds.), *Thermonuclear Supernovae*, 1996; N. Sanchez and A. Zichichi (eds.), *Current Topics in Astrofundamental Physics*, 1995.

Crayfish

Intersex organisms, possessing both male and female characteristics, have been documented mainly in hermaphroditic species. Intersex characteristics may be limited to the external morphology or may extend to gonadal differentiation. Hermaphroditism, the condition of having both male and female functional reproductive organs in one individual, occurs among a variety of crustaceans. Some cases of sequential hermaphroditism, in which intersex individuals serve as a transitional form during the process of sex change, have also been reported. Although sex changes occur in only a small percentage of crustacean species, protandry (a change from male to female) represents approximately 82% of crustacean sequential hermaphroditic species. Protogyny (a change from female to male) is very rare in crustaceans, and not a single case has been reported among higher crustaceans such as decapods.

Crayfish are generally regarded as gonochoristic, that is, they produce distinct male and female offspring; however, intersexuality has been reported in several crayfish species. Most reports, however, were based on the description of a single individual or a small number of specimens; and they provided only a morphological and a partial anatomical description that did not clearly explain the functional sexuality of the specimen or, indeed, of the species. Two crayfish species in which intersexuality has been adequately reported represent either partial protandric hermaphroditism or nonfunctional hermaphroditism, including functional males with immature ovaries.

Intersexuality in burrowing crayfish. The South American crayfish *Parastacus nicoleti* is a burrowing species (family Parastacidae), inhabiting the swampy grounds of southern Chile. It builds burrows that converge into a living chamber located at the level of underground water and, like other burrowing crayfish, lives in families. A recent report of intersexuality in *P. nicoleti* suggests that the crayfish is an example of a species that can experience partial protandric hermaphroditism. The cohabitation of different generations in the same burrow could be the cause of the sex change in this species. Morphological and histological studies of the reproductive system of *P. nicoleti* described two basic sexual types: primary females and protandric hermaphrodites. Thus, the species has been categorized as a partial hermaphrodite. Primary females lack male gonadic tissue and sperm ducts, and may be divided into two types: juvenile (or prepubertal) and adult, in which the abdomen acquires reproductive organs that allow eggs to be incubated and the gonopores to open to the exterior. During the male phase, protandric hermaphrodite individuals have an ootestis and an androgenic gland and various degrees of residues of testicular tissue and of sperm ducts. Transitional protandric hermaphrodites and hermaphrodites in the female phase represent a gradual passage from male to female that occurs dynamically in the population within a burrow, while some of the females in the overall population are primarily females.

Intersexuality in red claw crayfish. The Australian red claw crayfish (*Cherax quadricarinatus*) is a large, tropical fresh-water decapod crustacean (family Parastacidae) native to the rivers and streams of Queensland, Northern Territory (Australia), and Papua New Guinea. It has recently been introduced as an aquaculture species to most continents. *Cherax quadricarinatus* is a gonochoristic species, attaining sexual maturity within 7–9 months and developing a bilaterally symmetrical reproductive system. In males, this consists of pairs of testes, sperm ducts, androgenic glands, and genital openings at the bases of the fifth walking legs. Females have pairs of ovaries, oviducts, and genital openings at the bases of the third walking legs.

During the process of maturation in *C. quadricarinatus* females, morphological changes of the pleopods occur, allowing them to hold the newly deposited eggs. The endopod of mature (vitellogenic) females is longer and wider than the exopod, and a mixture of plumose setae and long, thin, simple setae are present on the endopod. Intersex individuals with both male and female genital openings have been recorded in cultured *C. quadricarinatus* populations in a frequency of 1.2–17%.

Seven types of intersex individuals (five are shown in **illus**. *a–e*), based on the observation of both male and female openings in the same specimen, have been described in *C. quadricarinatus* populations. Intersex individuals exhibit male secondary characteristics, such as the typical red patch on the claw, but they do not have developed female secondary sexual characteristics, such as typical maternal setation. The most frequent type of intersex individuals, comprising 67% of the intersex population, has two female openings at the bases of the third walking legs and one male opening at the base of one of the fifth walking legs.

Reproductive system of intersex red claw crayfish. All *C. quadricarinatus* intersex individuals are functioning males because they possess a male opening, a sperm duct, and a testis (containing viable sperm) on at least one side. An androgenic gland is attached to the subterminal region of each sperm duct. When placed with receptive females, intersex individuals are able to mate and fertilize the eggs, producing viable progeny. However, not all visible female openings indicate the presence of a female reproductive system. Individuals that have both male and female openings on one side do not have an ovary on that side. An ovary with an oviduct is found in only those cases in which a female opening is present in the absence of a male opening on the same side.

In intersex individuals that have both an ovary and a testis, the ovary exhibits a low gonadosomatic index (the weight of the ovary relative to the body weight). In this type of intersex individual, the diameter of the oocytes is small, and the ovarian cytosolic polypeptide profile shows either a low concentration or no accumulation of yolk protein (vitellin). In all of the above-described features, the intersex ovary resembles that of the immature previtellogenic females.

Unlike true hermaphroditism, in which both male and female reproductive systems are functional, either simultaneously or sequentially, the intersex phe-

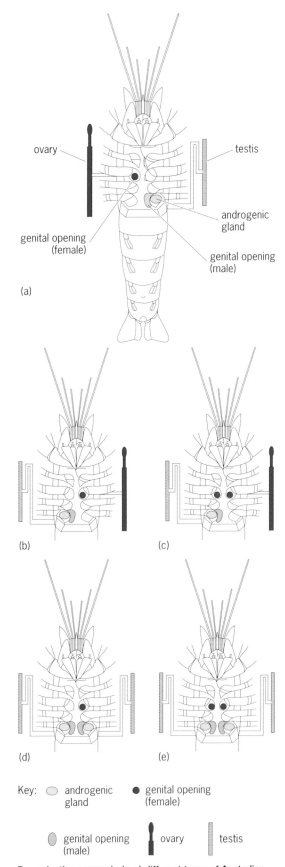

Reproductive organs in (*a–e*) different types of Australian red claw crayfish (*Cherax quadricarinatus*) intersex individuals. *(After A. Sagi et al., Intersex red claw crayfish, Cherax quadricarinatus (von Martens): Functional males with pre-vitellogenic ovaries, Biol. Bull., 190:16–23, 1996)*

nomenon in the red claw crayfish represents a stable, rather than a transient, state since no changes were reported in intersex individuals monitored from 7 to 19 months of age.

Endocrine regulation. The intersex model casts light on hormonal interactions with respect to the sexual differentiation process, which in higher crustaceans is generally accepted to be controlled by an epithelial endocrine gland, the androgenic gland. In male crustaceans, unlike vertebrates, the endocrine and gametogenic functions are separated into distinct organs: the androgenic gland and the testis. The androgenic gland is thought to be the exclusive source of the hormone responsible for sex differentiation in crustaceans. The discovery of the androgenic gland and the experimental demonstration of the control exerted by the androgenic hormone over male morphogenesis in crustaceans may suggest that sequential hermaphroditism, such as the partial protandric hermaphroditism found in *P. nicoletti*, results from the degeneration and disappearance of the androgenic gland, permitting expression of the female phase. The expression of intersexuality in *C. quadricarinatus*—in which one-half of a crayfish has male primary characteristics (including male gonads) and the contralateral half has female primary characteristics, while the secondary external characteristics are masculine on both sides—supports the hypothesis that early male differentiation in decapods is mediated by a secretion from the primordia (embryonic cells) of the androgenic glands. A threshold level of androgenic hormone, which diffuses along the genital tract, causes the expression of male differentiation in the developing gonad. This could explain the presence of the male reproductive system and the absence of the female system on one side, that is, the side on which the androgenic gland exerts its local effect through diffusion. On the other side, in the absence of an androgenic gland, differentiation of the ovary is permitted. However, the ovary of the intersex red claw crayfish never matures, because the developed androgenic gland present in the adult intersex individual exerts an effect on remote target organs (such as the ovary) via the circulation (in addition to the androgenic gland's effect, by local diffusion, on the early sexual differentiation process). A similar circulatory effect is exerted on the secondary sexual characteristics, resulting in their masculine expression.

For background information *see* CRUSTACEA; HERMAPHRODITISM; REPRODUCTION (ANIMAL) in the McGraw-Hill Encyclopedia of Science & Technology.   Amir Sagi; Isam Khalaila

Bibliography. H. Laufer and R. G. H. Downer (eds.), *Endocrinology of Selected Invertebrate Types*, 1988; E. H. Rudolph, Partial protandric hermaphroditism in the burrowing crayfish *Parastacus nicoleti* (Philippi, 1882) (Decapoda: Parastacidae), *J. Crust. Biol.*, 15:720–732, 1995; A. Sagi et al., Intersex red claw crayfish, *Cherax quadricarinatus* (von Martens): Functional males with pre-vitellogenic ovaries, *Biol. Bull.*, 190:16–23, 1996.

Dendritic macromolecules

During the past decade a fourth new class of polymer architecture, exhibiting unprecedented new properties, has attracted the attention of researchers in disciplines as diverse as chemistry, biology, physics, and engineering. The new polymers mimic the dendritic branching of trees and are referred to as dendritic macromolecules. They comprise three architectural subclasses: dendrons or dendrimers, dendrigrafts, and random hyperbranched polymers (**Fig. 1**).

The most esthetic members of this polymer class are the three-dimensional dendrons or dendrimers. Dendrimers possess three unique architectural components: a defined core, an interior, and peripheral surface components constructed in a concentric, ordered fashion around the core. These materials represent the first examples of structure-controlled polymer architecture that rival the control observed for biopolymers such as proteins, deoxyribonucleic acid (DNA), and ribonucleic acid (RNA). Unprecedented control of critical macromolecular design parameters (CMDPs), such as size, shape, surface chemistry, flexibility, and topology, has been demonstrated in over 50 different compositional types (families) of dendrimers. As a result of this unique architectural control, precise nanoscale sizes, specific shapes, and multifunctional surfaces are formed that produce many unusual physical and chemical properties. These properties clearly differentiate these macromolecules from classical linear, crosslinked, or branched polymers. For example, the first precise example of dendritic polymers, polyamidoamine (PAMAM) dendrimers, promises to be valuable in such diverse applications as biotechnology, electronics, catalysis, and nanotechnology. Their precisely controlled molecular weight, size, and water solubility, and their low toxicity validate them as excellent candidates in the biomedical area for drug delivery, genetic material transport/expression, magnetic resonance contrast agents, as well as medical diagnostic assays. In nonbiological areas, dendrimers have found applications as coatings, electronic/photon conduction/transduction, computer printing, as well as chemical and biological sensor materials.

Synthesis. Two major synthetic strategies have been used to make dendrimers: the divergent method and the convergent method.

Divergent synthesis. The divergent method was pioneered in the early to mid 1980s. In this approach, well-defined concentrically branched layers are covalently connected to a core molecule. As more branches are subsequently built out from a core, the dendrimer is forced into a globular shape. Each iteration, referred to as a generation, essentially doubles the number of peripheral groups as well as the molecular weight. By this reiterative method, various dendrimer families possessing over 50 different interior compositions have been synthesized. These dendrimer families have been advanced to molecular

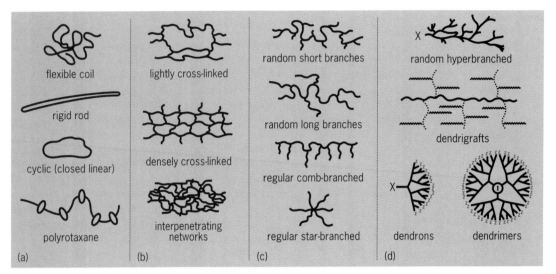

Fig. 1. Major macromolecular architectures of polymer materials. (*a*) Linear. (*b*) Cross-linked. (*c*) Branched. (*d*) Dendritic. I and X represent different core functional groups for dendrimers; Z, on "dendrons," represents a surface functional group.

weights of over 1,000,000 and modified with over 150 different types of surface groups.

Convergent synthesis. Convergent synthesis involves making the peripheral surface groups first, then moving inward to the core. Through a protection-deprotection strategy, monofunctionalized subdendrons are reacted with branching reagents to form monofunctional dendrons. The final, perfect-structured dendrimers are assembled by connecting various monofunctionalized dendrons to a common core. It has been demonstrated that the molecular weight grows at a pace equal to or faster than a divergent synthesis. The major drawback of this approach is that it is limited to laboratory preparation of only lower-generation materials. It is difficult to obtain high-molecular-weight dendrimers with perfect structure because of nanoscopic steric effects associated with final attachment of the dendrons to the core.

Preparation scale. Many classes of dendrimers are still limited to laboratory-scale preparation. Recently the commercial production of dendritic materials was realized for three intensively studied dendrimers: PAMAM dendrimers, polypropylene imine (POPAM) dendrimers, and hydroxylated polyester dendrimers. In all cases they are manufactured by divergent synthetic methods; in some cases they can be produced in hundreds of pounds.

Properties. Dendritic archiecture brings many new properties to the field of polymeric materials. For example, dendrimers offer precise, systematic molecular sizes in the nanoscopic region (1–100 nanometers), and shapes ranging from spheroids to ellipsoids to rods. With these structural features, dendrimers exhibit high solubilities in various solvents, possess low intrinsic viscosities, and present high localized densities of reactive functional groups. Surface functional groups are readily converted in high yield to well over 150 different moie-

ties. This property is uniquely different from classical linear polymers of similar composition, since dendrimers exhibit far less chain entanglement and present their numerous chain ends in an exo (outward) fashion for complete accessibility. Dendrimer terminal (chain ends) reactions may be broadly categorized into two general types: those involving subnanoscopic (small-molecule) reagents, and those involving nanoscaled reactants (**Fig. 2**).

Nanoscopic reagents (for example, DNA, antibodies, and other proteins) have been combined with dendrimers to produce new gene delivery vectors, immunodiagnostics products, or cell-specific targeting devices for biological systems. Known chemistry associated with these nanoscopic constructions clearly differentiates dendrimers from subnanoscopic modules. For example, the valency of a fourth-generation PAMAM dendrimer (NH_3 core) is 48. The number of primary amine surface groups available to react with subnanoscopic reagents (for example, methyl acrylate or acetyl chloride, whose diameters are less than 1 nm) clearly demonstrates this stoichiometry. Reactions of this same dendrimer surface with nanoscopically sized entities such as IgG (immunoglobin G) antibodies (whose diameter is 10 nm) show that the stoichiometry is reduced to 2–4 under the most ideal reaction conditions. These new combining rules have been referred to as sterically induced stoichiometry. They simply evolve from the packing/parking dilemma created by covalently combining two nanoscopically sized reactants.

Another unique feature manifested by dendrimers is their containerlike topology. This topology may be used either to store guest molecules or to serve as a nanoreactor for performing reactions in a micellelike environment. For example, modification of the surface of a POPAM dendrimer to give a dense shell of bulky amino acids allows the entrapment of

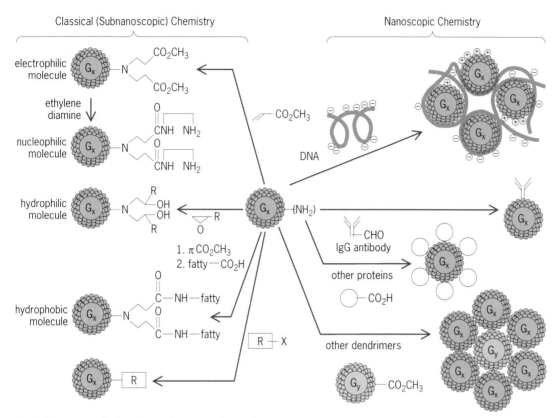

Fig. 2. Subnanoscopic chemistry and nanoscopic chemistry on dendrimer surfaces. Reactants shown along arrows combine with dendrimer at center to form macromolecules at the ends of arrows. R inside a box indicates a surface that is catalytic, photon absorbing, electron conducting, chelating, paramagnetic, or fluorescent. G_x and G_y represent various generations of dendrimers.

many smaller guest molecules inside the dendrimer cavities. As shown in **Fig. 3,** this modified dendrimer behaves as a dendritic box or container molecule. If this normally hydrophilic POPAM dendrimer is modified with hydrophobic alkyl chains, an inverted unimolecular dendritic micelle is obtained. The resulting unimolecular micelle demonstrates varied solution behavior that is dependent upon the surface properties of the dendrimer host, whereas the guest is isolated from the external environment. This phenomenon is referred to as unimolecular encapsulation. These unique features have been used for nanoscopic phase separation in a variety of environments, thus allowing compatibilization of materials with vastly different polarities.

Another important application for dendrimers is the use of these materials as fundamental building blocks for making larger megamolecules. Owing to their controlled nanoscopic size, shape, and reactive surface functionality, dendrimers have been identified as fundamental modules for constructing higher synthetic complexity and new architectures. These more complex molecules are referred to as supermolecular or supramolecular structures. They possess dimensions and properties that are characteristics of a new emerging science referred to as nanotechnology (that is, dimensions that reside between 1 and 100 nm). These new combining rules, unusual

scaffolding effects, unique guest–host relationships, and novel supramolecular assemblies are directly related to the nanoscopic dimensions and unique architecture of dendrimers.

Dendrimer-based conducting polymers. Dendrimers have recently found potential applications in one of the most intensively studied polymer fields: conducting polymers. The first electronically conducting dendritic polymers were prepared by modifying a PAMAM dendrimer periphery with naphthalene diimide anion radicals. The conductivity was realized through a mechanism in which diimide anion radicals aggregate to form an extensive three-dimensional network. In this highly extended network the partially charged aromatic groups of diimide moieties orchestrate each other through strong π-π interactions (π-stacking). As a component of a three-dimensional network, the nanoscale modified PAMAM molecules exhibit isotropic (three-dimensional) bulk conductivity in thin films. These dendrimer-based conductive materials are dramatically different from their classical polymeric counterparts. For example, they exhibit conductivity in three dimensions in contrast to the one dimension for classical systems, such as polypyrrole or polythiophene. They are stable in both moist and dry air. In fact, chemical or electrochemical reductions have been successfully performed in both solution and the solid

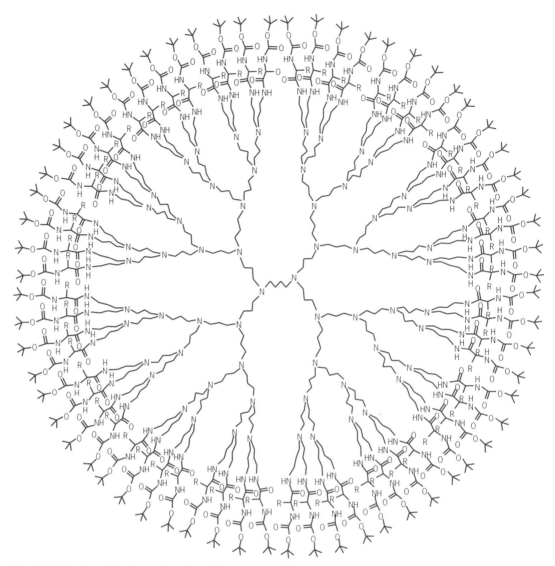

Fig. 3. Modification of dendrimers by amino acids to form a dendritic box.

state (cast thin films). In either case, high loadings of diimide anion radicals can be achieved, and the resulting π-stacking is well distributed in three dimensions within the thin films. The bulk conductivity of thin films depends upon the size and the casting (fabrication) conditions of the dendrimers. The relative size of the dendrimer and diimide groups is important. For that reason, the best conductivity occurs for medium-sized PAMAM dendrimers, at the third-generation stage. Presumably this dendrimer surface combination provides the optimum overlap of π-stacked diimide anion radicals.

It is also found that the dendrimer conductivity of the thin films depends upon the relative amount of anion radical units. Generally, partially reduced dendrimer systems deliver higher conductivity than those with the full amount of anion radicals. The presence of dendrimer groups in this system transforms the conductivity from anisotropic (one-dimensional) to isotropic (three-dimensional). This obser-

vation clearly demonstrates that dendritic architectures manifest completely new properties compared to known linear macromolecular systems. Furthermore, both temperature and moisture levels influence the conductivity of these dendrimer thin films. The mixed valence π-stacking of anion radicals reaches its highest levels at low temperature, while the conductivity of dry thin films increases more than two orders of magnitude when exposed to 90% humidity. These humidity-induced conductivities are very rapid and reversible; however, the character of the conductivity remains unchanged. In all cases, the conductivity is electronic rather than ionic. The highest conductivity of these dendrimer conducting systems is around 10 siemens per centimeter. This is an order of magnitude lower than traditional conducting polymers, such as polypyrrole or polythiophene. However, the dendrimer systems do offer new approaches to molecular-level electronic conducting materials that deserve further examination.

Perhaps these systems may be thought of as artificial metals, albeit with nanoscopic dimensions.

Dendrimer thin films and chemical sensors. The need to identify durable polymer-based chemical and biological sensors for both environmental monitoring and industrial process control, as well as for antiterrorist applications, has focused attention on the use of dendritic materials for the development of agent-specific, real-time sensors. The precise structure and high density of peripheral functional groups in dendrimers have led to the development of new detection devices based on these architectural attributes. Dendrimer molecules can be modified with a wide variety of reactive peripheral functional groups within a confined space.

For example, PAMAM dendrimers were used as surface-modifiable building blocks for sensor applications in the form of organic thin films. In this work, dendrimers were viewed as cooperatively organized chemical surfaces that can probe the targeted organic compounds as a function of their polarity and chemical properties. PAMAM dendrimer molecules were covalently bonded to a gold substrate surface through a self-assembled organic monolayer. The globular-shaped dendrimer was forced to take a flattened shape upon attachment to the surface. The peripheral group reactivity of these immobilized dendrimer molecules (surfaces) was found to be generationally (size) dependent.

A study has shown that these dendritic thin films could be coupled to a surface-acoustic-wave (SAW) mass balance and used to probe the adsorption events of many different volatile organic compounds. This SAW mass balance is a highly sensitive device capable of measuring tiny changes in mass due to molecular adsorption onto the dendritic thin films. Adsorption on the dendrimer films was found to be reversible and fairly rapid, thus exhibiting altered responses for different organic compounds.

Future research. In the unending quest for new properties and applications, it is inevitable that dendritic materials will offer many new architecturally driven features of great scientific and commercial interest. Understanding the role of dendrimers as fundamental nanoscopic building blocks not only will generate new concepts for structure-controlled megamolecules but also will provide unprecedented new materials for society. Just as the first three traditional classes of polymer architecture have so successfully fulfilled the critical material and functional needs of society during the past half century, dendritic macromolecules are expected to play a key role in many new emerging areas such as computer miniaturization, nanotribology, and catalysis, as well as in biomedical applications.

For background information *see* DENDRITIC MACRO-MOLECULES; NANOCHEMISTRY; NANOTECHNOLOGY; ORGANIC CONDUCTOR; ORGANIC SYNTHESIS; POLYMER; SUPRAMOLECULAR CHEMISTRY; SURFACE-ACOUSTIC-WAVE DEVICES in the McGraw-Hill Encyclopedia of Science & Technology.   Donald A. Tomalia; Yong Hsu

Bibliography. J. K. Barton, N. J. Turro, and D. A. Tomalia, Molecular recognition and chemistry in restricted reaction spaces: Photophysics and photoinduced electron transfer on the surfaces of micelles, dendrimers, and DNA, *Acc. Chem. Res.*, 2:332–340, 1991; R. G. Duan, L. L. Miller, and D. A. Tomalia, An electrically conducting dendrimer, *J. Amer. Chem. Soc.*, 117:10783–10784, 1995; D. A. Tomalia, Dendrimer molecules, *Sci. Amer.*, 272(5):62–66, May 1995; N. J. Turro, Supramolecular organic and inorganic photochemistry: Radical pair recombination in micelles, electron transfer on starburst dendrimers, and the use of DNA as a molecular wire, *Pure Appl. Chem.*, 67:199–208, 1995.

Deoxyribonucleic acid (DNA)

Recent research has produced a wealth of information on deoxyribonucleic acid sequences, their role in human disease, and their antimicrobial properties. This article discusses three aspects: (1) A new electrochemical technique has been developed for detecting DNA hybridization, determining DNA sequences, and identifying genes. (2) DNA helicase enzymes can mutate and lead to human disease. (3) A modified form of DNA that mimics transfer ribonucleic acid (tRNA) and can block tRNA binding has potential as a new type of antibiotic.

Electrochemical Detection of DNA Sequences

An understanding of the flow of biological information from DNA to messenger ribonucleic acid (mRNA) to protein provides a broad base for the diagnosis and treatment of disease. Watson-Crick base pairing, where adenine (A) binds to thymine (T) and guanine (G) binds to cytosine (C), provides a basis for storage and replication of information in the DNA double helix; transfer of information to single-stranded mRNA [where now adenine binds to uracil (U) and G binds to C]; and the subsequent readout of information through protein synthesis according to the genetic code (**Fig. 1**). Many illnesses and abnormalities can be linked to changes (mutations) in the DNA sequence of an individual. For example, a predisposition to breast cancer is signaled by mutations in the gene *BRCA1*. In diagnosis, the early detection of disease-causing genes can provide a basis for timely therapeutic intervention. In designing new therapeutics, an understanding of how mutations change the structure or function of the coded protein may reveal the mechanism(s) that causes disease. Detection of mRNA sequences provides information on which genes are being expressed at a given time, and can provide a very informative profile of disease progress. Infectious organisms, such as viruses and bacteria, have unique sequences that can be detected and quantitated in clinical specimens against the background DNA from the human host. For all of these applications, a method of detecting specific DNA sequences is vital.

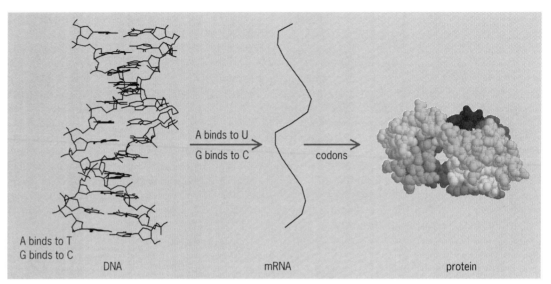

Fig. 1. Flow of biological information from DNA to mRNA to protein. In the process of hybridization, the double helix of DNA (at left) is formed when Watson-Crick base pairs (A binds to T, G binds to C) form between two single strands.

A common method for identifying genes is hybridization, the process by which one strand of DNA forms a double helix with its complementary strand, where A binds to T and G binds to C. To identify a particular sequence in a given sample, a probe strand that is complementary to a gene of interest is immobilized on a solid surface. When this surface is exposed to the sample, the specific target strand hybridizes to the immobilized strand. At this point, an analytical signal is needed to indicate that hybridization has occurred. This indicator is made by attaching a fluorescent or radioactive label to the target strand, but this process is usually expensive, time consuming, and technically challenging.

Mediated electrochemistry. An alternative approach to detecting hybridization is to use a chemical component of DNA itself as a signal. The guanine base of DNA loses an electron at a potential of about 1.25 V versus normal hydrogen electrode (NHE) in aqueous solution. This reaction can lead to DNA damage and provides a partial basis for radiation therapy. The guanine oxidation reaction can also provide a basis for electrochemical detection of DNA, since electron flow from guanine to an electrode poised at a potential positive of 1.25 V should lead to a measurable current, the magnitude of which reflects the amount and structure of DNA in the vicinity of the electrode. However, the abstraction of electrons from double-helical DNA is slow because the guanine base is protected by a sugar-phosphate backbone. To overcome this kinetic problem, a small ion [Ru(bpy)$_3$$^{2+}$] with a similar reduction-oxidation (redox) potential to guanine is used as a mediator. The mediator binds intimately to the DNA, abstracts electrons quickly from guanines in the DNA macromolecule, diffuses freely to the electrode, and then delivers the electrons from

the guanines to the electrode to produce a measurable current. If DNA is attached to an electrode, no current is obtained in the absence of a mediator; however, when the mediator is present, the current obtained reflects the quantity of guanine bases bound to the electrode.

Detecting mismatches. A complication that arises when attempts are made to detect sequences by hybridization is that stable hybrids can form between two complementary strands where there is a single error in the Watson-Crick base pairing (that is, where either A is not bound to T or G is not bound to C). Such an error is called a mismatch, and a single one of these errors in a DNA duplex is called a single-base mismatch. Duplexes that contain one single-base mismatch are only slightly less stable than duplexes that match perfectly. Therefore, in a hybridization assay it is not always possible to determine whether the hybridized DNA is a perfect match or contains a mismatch. This problem is particularly frustrating because most genetic abnormalities of interest involve the mutation of a single base.

The electrochemical detection scheme provides a means for distinguishing perfectly matched duplexes from strands that contain a mismatch at guanine. The current obtained from guanine oxidation by the mediator is a function of the rate at which the electron is abstracted from DNA. It is believed that the rate of guanine in the GC pair is the slowest because the helix packs together best when there are no mismatches and the solvent accessibility of guanine is therefore the lowest. In the mismatches, the packing is not as good, which distorts the DNA and makes guanine more solvent accessible. Subtle changes in the driving force are not believed to be responsible for the change in rate, because the change in driving force required to effect the changes in rate seen are too large to be reasonable. Electron-transfer reactions that occur over relatively long distances proceed by tunneling. (Tunneling is the process whereby an electron can start in one place and end up somewhere else without ever being in between. Because of the uncertainty principle, there is a finite probability of the electron being either on guanine or on the mediator. Thus, the electron could be on guanine at one instant and on the mediator the next.) The rates of electron transfer reactions that occur over relatively long distances vary exponentially with the distance between the two redox partners, in this case, $Ru(bpy)_3^{3+}$ bound to the outside of DNA and guanine situated on the inside of the double helix. The $Ru(bpy)_3^{3+}$ binds to DNA in the minor groove, which precludes intimate contact with the guanine base. As a result, the electron must tunnel across a finite distance. When the guanine is perfectly matched with cytosine, it is most effectively protected from oxidation, so the distance is maximal. However, if guanine is paired with the wrong base, the double helix does not pack together as well. Therefore, the guanine is more exposed to the solvent and consequently is more accessible to the $Ru(bpy)_3^{3+}$ oxidant. As a result, the distance is shorter in the mismatch than in the GC pair, leading to a faster rate of electron transfer and a higher current. Studies of all of the possible mismatches show that the rates of oxidation increase in the order GC < GT < GG < GA < single-stranded G, which correlates with the size of the base opposite G.

Advantages. Mediated electrochemical detection of DNA provides some important advantages over existing fluorescent and radiochemical methods. The first advantage is that labeling of the target strand is not necessary. This may greatly simplify the analysis procedure. Second, the electrochemical method provides a means for distinguishing a single-base mismatch in duplex DNA that is maintained in its native structure. Numerous breakthroughs in technology are required to integrate genetic analysis into the everyday activities of physicians and patients, and mediated electrochemical detection may be one such breakthrough.   H. Holden Thorp

DNA Replication and Repair Helicases

Deoxyribonucleic acid is a complex molecule composed of four bases covalently attached to a phosphorylated deoxyribose sugar. Phosphodiester linkages between sugars form the backbone of the DNA molecule, while the unique order of repeating bases specifies genes which code for proteins. The most stable form of DNA is a double-stranded molecule (dsDNA), composed of two complementary strands held together by interstrand hydrogen bonds between paired nucleotides. However, many aspects of DNA metabolism require the transient generation of localized regions of single-stranded DNA (ssDNA). The individual strands have a chemical and structural polarity defined by the orientation of the sugar; in double-stranded DNA the strands are antiparallel

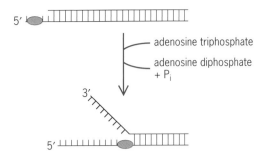

Fig. 2. Mechanism of action for DNA helicases. This helicase requires a short ssDNA region for binding, has 5′ to 3′ polarity, and uses the energy of ATP (adenosine 5′-triphosphate) hydrolysis to translocate along the DNA molecule.

with one having a 5′ to 3′ polarity and the other having a 3′ to 5′ polarity.

Genomic stability. Each somatic cell in a human contains about 6 billion base pairs organized on 46 chromosomes; individual genes specify an estimated 50,000 proteins by first transcribing an RNA copy with the corresponding sequence of bases. The sequence of bases in the RNA is translated into a specific sequence of amino acids in the protein. Only a subset of genes is expressed at a particular time in any given cell type. Since each of these proteins is vital to the normal development and function of the nearly 60 trillion cells which constitute an adult human, it is important that chromosomes are stably maintained throughout an individual's life-span and faithfully transmitted to offspring. Genomic stability requires both the maintenance of intact chromosomes and the correctness of each gene in the DNA.

DNA replication, repair, and recombination are three processes essential for genomic stability. Prior to cell division, DNA polymerases use ssDNA as a template for the synthesis of new DNA molecules (replication), thus ensuring that each daughter cell contains the same genetic information as the parental cell. Cells continuously monitor DNA for the presence of inappropriate or damaged bases and, when found, these errors are corrected by DNA repair enzymes. Failure to repair such errors results in mutations, which are changes in the DNA base sequence that may alter the amino acid sequence and function of encoded proteins. Recombination is a mechanism for the exchange of genetic information and is a prerequisite for the normal segregation of chromosomes during meiosis. Errors in chromosome segregation can result in a condition such as Down syndrome. Some types of DNA repair also include recombination events.

Helicases. Each of these metabolic pathways requires transient ssDNA templates or intermediates. The generation of ssDNA is mediated by DNA helicases, enzymes which bind both DNA and nucleoside 5′-triphosphates (NTP) and couple the binding and hydrolysis of NTP to the unwinding (melting) of duplex DNA (**Fig. 2**). Most helicases require a short region of ssDNA for binding, although some can bind to nicked DNA (dsDNA in which a phospho-diester bond is cleaved in the deoxyribose-phosphate backbone of one strand of duplex DNA). The chemical energy from NTP hydrolysis is transduced into mechanical energy which fuels both the disruption of hydrogen bonds between base pairs and movement (translocation) of the helicase along the DNA molecule. This translocation occurs in one direction in nature, with the direction defined by the polarity of the DNA strand with which the helicase interacts.

Helicases are ubiquitous enzymes. Most organisms have multiple helicases which catalyze essentially the same biochemical reaction. At least 10 helicases have been identified in human cells. This apparent redundancy likely reflects specialization in terms of DNA structure and polarity, as well as protein-protein interactions, and serves to restrict a specific helicase activity to a particular subpathway of DNA metabolism.

Helicases were discovered in the 1970s, and subsequent analyses grouped these enzymes into five related families with characteristic conserved amino acid sequence motifs separated by nonconserved spacer regions of variable amino acid lengths. Nucleotide binding and helix unwinding activities have been demonstrated for many of the family members, and it is presumed that other proteins with these seven motifs (I, IA, and II–VI) are either DNA or ribonucleic acid (RNA) helicases. The presence of these conserved regions is not an absolute indication of helicase activity, as there are some helicases which have only domains I and II, identified as the NTP and magnesium ion (Mg^{2+}) cofactor binding domains, and other proteins which have all seven domains but lack true helicase activity. Nonetheless, these domains have been useful predictors of helicase activity when identified in newly cloned genes of unknown function.

DNA damage, repair, and cancer. Malignancies arise through an accumulation of mutations in genes which encode proteins, such as oncogenes or tumor suppressors. This mutation-causing damage may arise from normal, endogenous events (for example, oxygen metabolites generated by the cell), or may be caused by exogenous, environmental factors (for example, solar ultraviolet irradiation or exposure to certain chemicals). Under normal circumstances (such as low levels of genotoxic agents and functional responses to cellular damage), the agent is either eliminated or neutralized by detoxification pathways before the DNA is altered. In the event that the DNA is modified, there are mechanisms for the removal of DNA damage.

The most prevalent mechanism for damage removal is nucleotide excision repair, a complex pathway which removes damage in the form of short oligonucleotides and then uses the complementary, undamaged strand as a template for resynthesis of the excised oligomer (**Fig. 3**). Recent research has contributed much to understanding the mechanistic details of excision repair in humans, a process requir-

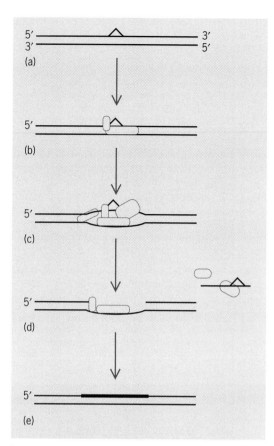

Fig. 3. Model for DNA excision repair. (*a*) DNA damaged by ultraviolet irradiation or chemicals is a substrate for nucleotide excision repair. (*b*) DNA damage is recognized and bound by specific proteins. (*c*) Damage recognition proteins recruit other subunits, including those with helicase activity, to the damaged region, which is locally unwound. (*d*) Dual incisions are made both 3′ and 5′ to the damage, a 27–30-nucleotide oligomer is excised, and this short piece of DNA and some proteins of the excision complex are released from the DNA; release may be mediated by helicases. (*e*) Replication proteins are recruited to the resulting gap, which is filled by using the undamaged complementary strand as a template, and remaining repair proteins are dissociated.

ing as many as 20–25 polypeptides for events that are accomplished by *Escherichia coli* with just six proteins: UVR(A)BC excinuclease, helicase II, DNA polymerase, and ligase.

One of the more intriguing findings of recent years has been the discovery of transcription-coupled repair, a phenomenon in which actively transcribed genes are preferentially repaired. Although the process was first identified in mammalian cells, the molecular details have been reported only for bacterial cells. A transcription-repair coupling factor (TRCF) recognizes and interacts with RNA polymerase stalled at a lesion, releases the polymerase, and recruits the *E. coli* repair enzymes to the damage site. Interestingly, this factor has the seven helicase domains, but the protein lacks true helicase activity.

Associated human disorders. Several rare human diseases are associated with an inherited condition of chromosomal instability and, in some cases, with a predisposition to cancer. These include xeroderma pigmentosum, Cockayne's syndrome, Bloom's syndrome, hereditary nonpolyposis colorectal cancer, ataxia telangiectasia, Fanconi's anemia, and Werner's syndrome. It was once hypothesized that each condition resulted from a defect in processing DNA damage. However, it is now believed that only the xeroderma pigmentosum syndrome is directly due to a defect in basal nucleotide excision repair, while Cockayne's syndrome results when transcription-coupled repair is impaired; nonpolyposis colorectal cancer is correlated with a defect in mismatch repair.

Xeroderma pigmentosum. The underlying cause of xeroderma pigmentosum is a defect in one of seven genes involved in nucleotide excision repair. Clinical manifestation of xeroderma pigmentosum is due to a combination of a genetic defect and environmental factors (for example, exposure to solar irradiation), and includes hypersensitivity to sunlight, an increased frequency of skin cancers or internal cancers and, in some cases, neurologic abnormalities. Two of the proteins associated with xeroderma pigmentosum have the seven helicase motifs and NTP-dependent helix-unwinding activity. Xeroderma pigmentosum group B (XPB), xeroderma pigmentosum group D (XPD), and five other polypeptides make a multiprotein complex called transcription factor IIH (TFIIH) which is a general transcription factor and a repair factor, but only the helicase activity of XPB is essential for transcription. It is not known why TFIIH contains two helicases or even if both of them function as helicases during repair. The precise roles of XPB and XPD in repair are unknown, but two suggestions are localized unwinding of the duplex to permit access of other repair proteins, and dissociation of other proteins following removal of damage.

Cockayne's syndrome. Cockayne's syndrome is characterized by sun sensitivity and progressive neurologic abnormalities, but without the increased frequency of skin cancers. The underlying defect in Cockayne's syndrome is due to a mutation in one of two genes, Cockayne's syndrome group A and group B (*CSA* and *CSB*), both of which have been implicated in transcription-coupled repair. Sequence analysis of *CSB* revealed the seven helicase domains, although DNA unwinding activity has not been reported for the purified protein.

Bloom's and Werner's syndromes. Individuals with Bloom's syndrome exhibit extreme genomic instability resulting in a spectrum of clinical features including sunlight hypersensitivity, immunodeficiency, fertility problems, and a predisposition to cancer. Some cultured cells show an abnormal response to DNA-damaging agents, although no defect in a specific repair pathway has been found. Rather, it is believed that cells associated with Bloom's syndrome are unable to efficiently complete certain, as yet unidentified DNA metabolic events. In 1995 the cloning and sequence analysis of *BLM,* the gene responsible for Bloom's syndrome, indicated that the gene is a putative helicase with homology to the *E. coli* RecQ protein, which is known to be involved in recombination.

In 1996 the gene (*WRN*) that causes Werner's syndrome was cloned. Although sequence analysis of *WRN* revealed seven conserved helicase domains, an enzymatic activity has not been demonstrated. Werner's syndrome is a condition diagnosed in postadolescent individuals with premature signs of aging including hair graying, balding, arteriosclerosis, cataracts, wasting of musculature, and tumors. A specific DNA repair defect has not been identified for individuals with Werner's syndrome, but it has been hypothesized that a consequence of aberrant DNA metabolism may be the early accumulation of mutations in proteins associated with age-related diseases or cancer. *See* CANCER (MEDICINE).   Joyce T. Reardon

DNA Modified to Mimic RNA Function

Many microorganisms produce biochemicals that are natural antibiotics, toxic to competitors within the environmental niche occupied by the organism. Some antibiotics have become part of the physician's conventional repertoire of drugs to combat infections. However, with the increasing use of antibiotics since the 1950s, some clinically important pathogenic microorganisms have evolved mechanisms to evade the effects of antibiotics. Many antibiotics target the protein-synthesizing machinery, the ribosome, of bacteria and fungi. Microorganisms that are resistant to an antibiotic have ribosomes that are mutated at specific locations and thus are no longer affected by the drug. The ribosome decoding site is a location on the ribosome where the genetic information for a protein is translated into that protein's sequence of amino acids. A nucleic acid has been designed to inhibit protein synthesis at the ribosomal decoding site, and it has antibacterial activity.

Ribosome. Living organisms contain two types of nucleic acid molecules: deoxyribonucleic acid and ribonucleic acid (RNA). Ribosomes are composed of RNA and proteins. There is a clear line of demarcation between the biological functions of DNA and those of the different RNAs. DNA, the genetic material of almost all organisms, including viruses, is the inherited blueprint for the required biochemistry, development, and maturation of the organism. For all organisms, RNA is the irreplaceable translator of genetic information into proteins. It is the genetic material for a few mammalian and bacterial viruses. However, the two nucleic acids are very similar. They are polymers composed of monomers called nucleosides. The majority of nucleosides have a single base: adenine (A), guanine (G), cytosine (C), or thymine (T) for DNA, and uracil (U) for RNA. These nucleosides are attached to a five-carbon sugar: deoxyribose for DNA and ribose for RNA. In the polymer, the sugar of each major nucleoside is attached to an adjacent nucleoside by a phosphate.

Three types of RNA are directly involved in protein synthesis: messenger RNA (mRNA), ribosomal RNA (rRNA), and transfer RNA (tRNA). They have

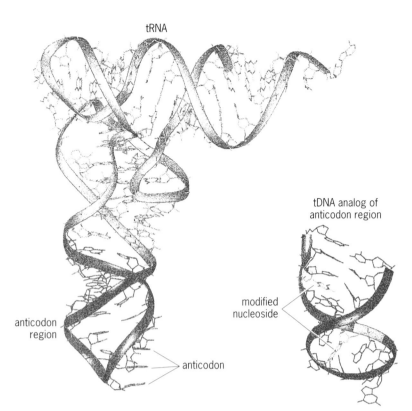

tRNA

tDNA analog of
anticodon region

modified
nucleoside

anticodon
region

anticodon

Fig. 4. Transfer DNA (tDNA), a mimic of the codon binding structure of rRNA. The structure of the small DNA analog of the tRNA anticodon (right) is compared to the entire tRNA structure (left).

many modified nucleosides in addition to the four major nucleosides. Currently, there are over 90 known naturally occurring modified nucleosides in mRNA, rRNA, and tRNA. Additional modified nucleosides have been devised by researchers, including the acquired immune deficiency syndrome (AIDS) drug azidothymidine (AZT). Naturally occurring modified nucleosides alter the chemistry and structure of RNAs, and so enhance and modify the biological function of RNAs. Much of the chemistry contributed to RNA by modified nucleosides has its exact counterpart in the contribution of amino acids to proteins. Some modified nucleosides are hydrophobic and thereby contribute the lipidlike characteristic of being unable to associate with water at a site at which this nucleoside occurs in the RNA. Others are hydrophilic and polar, or hydrophilic and negatively or positively charged. Modified nucleosides have been used to design and produce a mimic to tRNA function.

Transfer RNA. Transfer RNA is responsible for bringing individual amino acids to the ribosome in response to the sequence of three-base codes (codons) in mRNA. Messenger RNA is a copy of the genetic information for a specific protein. Individual amino acids are added to the nascent protein at the ribosome by formation of peptide bonds in the order prescribed by the sequence of codons in the mRNA. At the ribosome's decoding site, tRNAs read the codons correctly at the ribosome by having a complementary set of three bases, an anticodon, that is specific for each amino acid. The anticodon recognizes the codon by virtue of Watson-Crick base pairing (as in DNA). The anticodon GAA for the amino acid phenylalanine binds the codon UUC, with phosphate backbones GAA and CUU running in opposite directions. The reading of the codons is such a fundamental part of protein synthesis that it is considered a virtually immutable function. Thus, this function of tRNA could be a target for the production of a mimic that inhibits protein synthesis.

DNA analogs. A strategy for the development of a novel nucleic acid therapeutic to mimic tRNA function involves the design and production of DNAs that imitate tRNA in order to inhibit the binding of native tRNA to the ribosome (**Fig. 4**). Unlike tRNAs, tDNA is incapable of carrying an amino acid. In contrast to amino acid conjugated tRNA, tDNA enters the ribosome in response to the appropriate codon on the mRNA. It is believed that tDNA displaces tRNA in the ribosome. The DNA analogs are designed to copy the tRNA anticodon specific for phenylalanine (tDNA$_{AC}^{Phe}$). These analogs are made in high yield with standard, modified, and uridine deoxynucleosides by using conventional automated chemical synthesis. All analogs will have uridine deoxynucleoside instead of the thymidine deoxynucleoside normally found in DNA. Transfer DNA$_{AC}^{Phe}$ molecules are relatively small (5500 daltons) and not cleaved by enzymes that cleave RNAs (ribonucleases). The tDNA$_{AC}^{Phe}$ stops protein synthesis by binding bacterial ribosomes and inhibits native tRNA from binding. However, the tRNA binding sites on the ribosome have been shown to be impervious to DNA analogs of tRNA made of conventional deoxyribose nucleosides, A, G, C, and T.

Modified nucleosides. Design of biologically active DNA analogs of the tRNA^{Phe} anticodon required the use of modified nucleosides such as those that occur in natural tRNAs. A deoxynucleoside with a methylated cytosine was required for a magnesium-induced change in the tDNA$_{AC}^{Phe}$ structure that also occurs in the tRNA. A deoxynucleoside with a methylated guanosine was required for disruption of a Watson-Crick-type base pair between this G and a C. The presence of the base pair distorted the anticodon so that the tDNA$_{AC}^{Phe}$ was unable to bind the ribosome. The modifications were introduced at specific nucleosides in the tDNA$_{AC}^{Phe}$ corresponding to nucleosides in tRNA^{Phe} where modifications occur. The structure and dimensions of the resulting molecule was similar to the tRNA^{Phe} anticodon. Modified tDNA$_{AC}^{Phe}$ bound the ribosome in response to the codon for phenylalanine and at the ribosomal decoding site normally occupied by tRNA^{Phe}. In laboratory experiments, modified tDNA$_{AC}^{Phe}$ inhibited translation by 50% at a tDNA$_{AC}^{Phe}$-to-ribosome ratio of 8:1 and completely inhibited translation at a ratio of 23 tDNA$_{AC}^{Phe}$ for each ribosome.

In living organisms, micromolar concentrations of tDNA$_{AC}^{Phe}$ were able to inhibit growth of transformed, antibiotic-resistant bacteria. Transfer DNA$_{AC}^{Phe}$ and its structure could guide the design of translation inhibitors as potential therapeutic agents, but more research is required.

For background information *see* ANTIBIOTIC; BACTERIAL GENETICS; BIOTECHNOLOGY; CELL (BIOLOGY); DEOXYRIBONUCLEIC ACID (DNA); DEOXYRIBOSE; GENETICS; MOLECULAR BIOLOGY; NUCLEIC ACID; RECOMBINATION (GENETICS); RIBONUCLEIC ACID (RNA); SOMATIC CELL GENETICS in the McGraw-Hill Encyclopedia of Science & Technology.   Paul F. Agris; Jacinda Swallow; Richard Guenther

Bibliography. P. F. Agris, The importance of being modified: Roles of modified nucleosides and Mg^{2+} in RNA structure and function, *Prog. Nucleic Acid Res. Mol. Biol.*, 53:79–129, 1996; M. M. Basti et al., Design, biological activity and NMR solution structure of a DNA analog of yeast tRNA[Phe] anticodon domain, *Nat. Struc. Biol.*, 3:38–44, 1996; V. Dao et al., Ribosome binding of DNA analogs to tRNA requires base modifications and supports the "extended anticodon," *Proc. Nat. Acad. Sci. USA*, 91:2125–2129, 1994; S. P. A. Fodor et al., Multiplexed biochemical assays with biological chips, *Nature*, 364:555–556, 1993; E. C. Friedberg, G. C. Walker, and W. Siede, *DNA Repair and Mutagenesis*, 1995; D. H. Johnston, K. C. Glasgow, and H. H. Thorp, Electrochemical measurement of the solvent accessibility of nucleobases using electron transfer between DNA and metal complexes, *J. Amer. Chem. Soc.*, 117:8933–8938, 1995; E. Marshall, Hot property: Biologists who compute, *Science*, 272:1730–1732, 1996; S. W. Matson, D. W. Bean, and J. W. George, DNA helicases: Enzymes with essential roles in all aspects of DNA metabolism, *BioEssays*, 16:13–22, 1994; R. Rawls, Modified DNA molecule mimics transfer RNA, *Chem. Eng. News*, 74:5, 1996; A. Sancar, DNA excision repair, *Annu. Rev. Biochem.*, 65:43–81, 1996; M. Urdea, Branched DNA signal amplification, *Bio/Technology*, 12:926–928, 1994.

Desalination

Desalination is the processing of seawater to obtain pure water through the separation of dissolved saline components. In general, large-scale commercially available desalination processes can be classified into processes that require mainly heat and some electricity for ancillary equipment (distillation processes), and processes that require only electricity (membrane processes).

Any desalination process requires energy, either heat and electrical energy (mainly for pumping), or electrical energy only. (The use of mechanical energy instead of electrical energy is also possible.) For standard seawater (at 25°C or 77°F, with 34,500 parts per million total dissolved solids), the theoretical minimum separation work required to produce 1 m^3 of pure water is about 0.73 kW · h (so that 20.7 W · h is required to produce 1 ft^3). However, the energy consumption of currently available commercial processes is much higher because of thermal losses and inefficiencies that occur during the separation process. The lowest energy consumption, including that for seawater pumping and water pretreatment, is currently obtained with reverse osmosis plants. It amounts to 4–7 kW · h/m^3 (110–200 W · h/ft^3) of electrical energy, depending on fresh-water quality, seawater salinity, and plant configuration. After more than 40 years of intensive research and development in seawater desalination technology, only distillation processes and the reverse osmosis process have achieved commercial large-scale application.

Distillation processes. In distillation processes, seawater is heated to evaporate pure water that is subsequently condensed. With the exception of the mechanical process of vapor compression, distillation processes are driven entirely by low-temperature steam as the heat source. This steam is usually obtained as exhaust or extraction steam from the adjacent turbines of a power plant.

As a result of the high specific heat required to evaporate water, commercial distillation processes are implemented in heat recovery stages placed in series. The larger the number of stages assembled, the better the performance of the distillation processes. However, the overall temperature difference between the heat source and the cooling-water sink limits the number of stages. Typical temperature differences for commercial distillation plants are 2–5°C (4–9°F) per heat recovery stage.

Usually, the thermodynamic efficiency of distillation plants is expressed in kilograms of water produced per kilogram of steam used. This ratio, referred to as the gain output ratio, is in the range of 6–10 for current commercial multistage flash distillation plants and up to 20 for multiple-effect distillation plants.

Multistage flash distillation. In this process (**Fig. 1**), seawater feed passes through tubes in each evaporation stage, where it is progressively heated. Final seawater heating occurs in the brine heater by the external heat source (steam). Subsequently, the heated brine flows through nozzles into the first stage, which is maintained at a pressure slightly lower than the saturation pressure of the incoming stream. As a result, a small fraction of the brine flashes (vaporizes), forming pure-water steam. The heat to flash the vapor comes from cooling of the remaining brine flow, which lowers the brine temperature. Subsequently, the produced vapor passes through a mesh demister, which removes seawater droplets entrapped in the vapor, and into the upper chamber of the evaporation stage. There it condenses on the outside of the seawater feed tubes and is collected in a distillate tray. The heat transferred by pure-water condensation warms the incoming seawater feed as it passes through that stage. The remaining brine passes successively through all the stages at progressively lower pressures, where the process is repeated. To prevent scaling (the formation of chemical deposits) on the brine tubes, the seawater must be pretreated by adding acid or advanced scale-inhib-

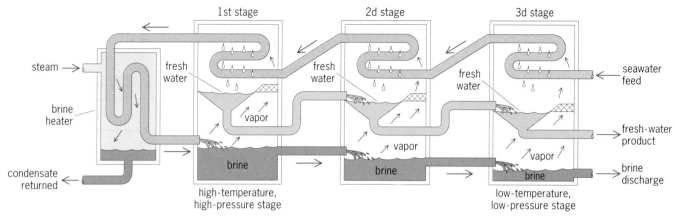

Fig. 1. Flow diagram of a multistage flash distillation system. *(After International Atomic Energy Agency, Use of Nuclear Reactors for Seawater Desalination, IAEA-TEDOC-574, 1990)*

iting chemicals. Some multistage flash plants require separate deaerators (not included in stages) to prepare the seawater. The vent gases from the water deaeration together with any noncondensable gases released during the flashing process are removed by steam-jet ejectors and discharged to the atmosphere.

Multistage flash plants have reached a mature and reliable stage of development. Industrial-scale multistage flash plants typically have up to ten parallel units, each unit producing up to 60,000 m³ (2.1 × 10⁶ ft³) potable water per day. The heat and electricity consumption is in the range of 45–120 kW(th) · h/m³ [1.3–3.4 kW(th) · h/ft³] and 3.0–6.0 kW · h/m³ [85–170 W · h/ft³] respectively. By using polymeric antiscaling additives, the maximum brine temperature is limited to 120°C (248°F) for brine-recycle-mode systems and 135°C (275°F) for once-through-mode systems.

Multiple-effect distillation. **Figure 2** shows the flow diagram of the multiple-effect distillation process. In each stage (effect), heat is transferred from the condensing water vapor on one side of the tube bundles to the evaporating brine on the other side of the tubes. This process is repeated successively in each of the effects at progressively lower pressure and lower temperature, driven by the water vapor from the preceding effect. In the last effect at the lowest pressure and lowest temperature, the water vapor condenses in the heat-rejection heat exchanger, which is cooled by incoming seawater. The condensed distillate is collected from each effect. Some of the heat in the distillate may be recovered by flash evaporation to a lower pressure (not shown in Fig. 2).

Currently, the most dominant multiple-effect distillation processes with the highest technical and economic potential are the low-temperature horizontal-tube multieffect (LT-HTME) process and the vertical-tube evaporation (VTE) process. The main differences between these two processes are in the arrangement of the evaporation tubes, the side of the tube where the evaporation takes place, and the evaporation-tube materials used. In LT-HTME plants, evaporation tubes are arranged horizontally, and evaporation occurs by spraying the brine over the outside of the horizontal tubes, creating a thin film from which steam evaporates. In VTE plants, evaporation takes place inside vertical tubes. In LT-HTME

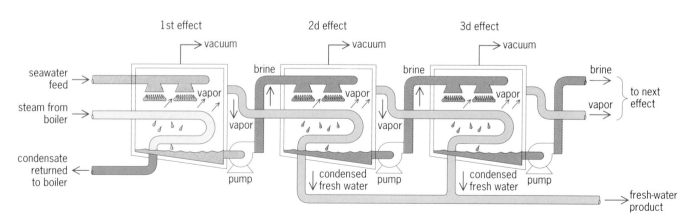

Fig. 2. Flow diagram of a multiple-effect distillation system. Successive effects have progressively lower pressures and temperatures. *(After Electric Power Research Institute, Desalination Technology Evaluation, EPRI TR-101019, 1992)*

plants, the maximum brine temperature is limited to 70°C (168°F), since inexpensive materials such as aluminium for the heat exchanger and carbon steel for the shell are used.

Multiple-effect distillation plants have a much more efficient evaporation heat-transfer process than multistage flash plants. Because of the thin-film evaporation of brine on one side of the tubes and the condensation of vapor on the other side, high heat-transfer coefficients are achieved, thus decreasing the specific heat consumption in comparison to multistage flash plants with identical heat-transfer area and the same temperature difference between heat source and cooling-water sink. In spite of this superiority, the multiple-effect distillation process could not compete with the multistage flash process in the past, chiefly because of the components and materials used in multiple-effect distillation plants, as well as the lack of experience in the operation of large-scale plants. The pretreatment of seawater for multiple-effect distillation plants is similar to that in multistage flash plants.

In some multiple-effect distillation designs, a part of the vapor produced in the last effect is compressed, heating it to a higher temperature so that the energy efficiency of the plant is improved. To compress the vapor, mechanical compressors (with isentropic efficiencies of about 80%) or steam-jet ejectors (with isentropic efficiencies of less than 20%) are employed. These designs, however, are usually not applied in integrated plants for electricity and potable water production.

The heat and electricity consumption of commercial multiple-effect distillation plants is in the range of 30–120 kW(th)·h/m³ [0.85–3.4 kW(th)·h/ft³] and 1.5 to 2.5 kW·h/m³ (42–70 W·h/ft³) respectively, depending on the design and the seawater temperature difference.

Reverse osmosis. Reverse osmosis can be thought of as a filtration process at the molecular-ionic level (**Fig. 3**). It uses a semipermeable membrane (which is generally composed of a polymer), which permits the passage of solvent in either direction but retards the passage of solute. Normal osmosis is a natural process by which water flows through such a membrane from pure water or from a dilute solution into a more concentrated solution. Every solution has a specific osmotic pressure, which is determined by the identity and concentration of the dissolved materials. This flow continues until the resulting osmotic head equals the osmotic pressure of the solution.

If the solution compartment is now enclosed and a pressure higher than the natural osmotic pressure of the solution is applied to it, the direction of water flow is reversed. The solution becomes more concentrated and purified water (called the permeate) is obtained on the other side of the membrane; hence the term reverse osmosis.

The rate of flow of purified water depends on various factors such as the chemical properties of the polymer membrane itself, membrane thickness, membrane area, pressure, concentration, pH, and

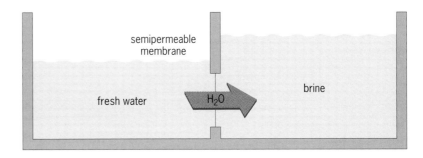

(a)

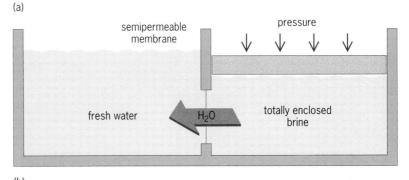

(b)

Fig. 3. Principle of the reverse osmosis process. (*a*) Normal osmosis. (*b*) Reverse osmosis. *(After International Atomic Energy Agency, Use of Nuclear Reactors for Seawater Desalination, IAEA-TEDOC-574, 1990)*

temperature. Under any set of fixed conditions, product flow is proportional to the difference between applied pressure minus the osmotic pressure of the solution, and the permeate pressure.

The saline feed is pumped into a closed vessel where it is pressurized against the membrane. As a portion of the water passes through the membrane, the salt content in the remaining feedwater increases. At the same time, a portion of this feedwater is discharged without passing through the membrane. In practice, the feedwater has to be compressed up to 7–8 megapascals (70–80 bars) since the osmotic pressure of the saline solution is about 6 MPa (60 bars), whereas the osmotic pressure of the permeate is negligible. Membrane modules are of tubular type, plate-and-frame type, spiral-wound type, and hollow-fiber type. The most widely used modules are the spiral-wound and hollow-fiber ones. Reverse osmosis processes need significant preconditioning of the feedwater to protect the membranes.

In recent years, seawater reverse osmosis has become a reliable and commercial process applicable on a large scale. Typical electricity consumption of reverse osmosis plants is in the range of 4–7 kW·h/m³ (110–200 W·h/ft³), depending on the seawater salinity, recovery ratio, required permeate quality, plant configuration, and implementation of energy recovery in the brine blow-down. (The brine blow-down is the remaining brine concentrate which is discharged back to the sea without passing through the membrane. Methods for energy recovery from the brine blow-down include hydraulic turbines that can power electric generators or may be directly coupled to high-pressure reverse-osmosis feed pumps. Integrated turbine-driven centrifugal pumps, where the turbine is in series with the high-pressure pump, are also used.)

Nuclear energy sources. The energy required for seawater desalination can be supplied by either conventional or nuclear energy sources; there are no technical impediments to the use of electricity and heat produced by a nuclear reactor. Fossil energy resources are limited, however, and their increasingly intensive use raises environmental concerns, including the threat of a gradual climate change with far-reaching consequences. At the same time, the worldwide demands for energy and potable water are steadily growing, and appropriate solutions are needed.

A nuclear generating plant could be used for heat production only, for cogeneration of electricity and heat (coupled with a multiple-effect distillation or multistage flash process), or for electricity production only (coupled with the reverse osmosis process). In Kazakhstan a liquid-metal-cooled fast reactor has provided both electricity and heat for seawater desalination since 1973, and in Japan several small-scale seawater desalination plants have been installed in large nuclear power stations. Studies recently performed on the economic assessment of nuclear seawater desalination have shown that nuclear energy would be competitive with fossil energy in a range of situations.

For background information *see* DISTILLATION; OSMOSIS; SALINE WATER RECLAMATION in the McGraw-Hill Encyclopedia of Science & Technology.

Juergen Kupitz

Bibliography. Electric Power Research Institute, *Desalination Technology Evaluation*, EPRI TR-101019, 1992; International Atomic Energy Agency (IAEA), *Options Identification Programme for Demonstration of Nuclear Desalination*, IAEA-TEDOC-898, 1996; IAEA, *Technical and Economic Evaluation of Potable Water Production through Desalination of Seawater by Using Nuclear Energy and Other Means*, IAEA-TEDOC-666, 1992; IAEA, *Use of Nuclear Reactors for Seawater Desalination*, IAEA-TEDOC-574, 1990.

Dinosaur

Although dinosaurs are clearly reptiles, there has been a tendency in recent years to recast their biology and physiology as birdlike or mammallike. The scientific case for warm-blooded dinosaurs was presented in a series of publications beginning in 1971. In 1974, there were proposals for the scientific community to recognize the Dinosauria (including dinosaur descendants, the birds) as a class separate from the reptiles from which they were derived. This has led to confusion in popular literature as to whether or not dinosaurs were actually reptiles.

Endothermic links. In terms of modern cladistic classification, anything which is derived from a reptile is itself a reptile (including birds). Holders of the new viewpoint included endothermy (metabolic thermoregulation) among the defining characteristics of dinosaurs, even though this is an inference, not an observable character. The case for the genealogical linkage of birds to small meat-eating dinosaurs (maniraptoran theropods) was presented in the 1970s and quickly gained acceptance.

Birds are fully endothermic, and the endothermy had to come from somewhere. Therefore, paleontologists had to determine whether the endothermy of birds was carried over from that of dinosaurs, or at least of meat-eating dinosaurs.

Maiasaura. In 1979, the discovery of the duck-billed dinosaur (*Maiasaura peeblesorum*) in Montana was announced. Preserved along with the skull of this female adult hadrosaur was an apparent nest containing the remains of 15 babies, each about 3.3 ft (1 m) long. Additional clusters of young were found, some with babies only 18 in. (45 cm) in length. Abundant eggshell fragments but not eggs were associated with the find, and both eggs and eggshell are common in the Two Medicine Formation of north-central Montana. Because dinosaurs of different size cohorts remained associated in nests, and because the teeth of the young showed wear facets indicating feeding for a length of time, investigating paleontologists concluded that there was some form of extended parental care. They speculated that parental care extended until the hatchlings reached half of adult size: 23 ft (7 m); and that either the parent (or parents) spent considerable time with the young or the young grew very rapidly. Subsequently, 8-in.-long (20-cm) eggs reconstructed from fragments were attributed to *Maiasaura*, and intact 6-in.-long (15-cm) eggs were attributed to a hypsilophodont.

In 1988, the hypsilophodont was described as *Orodromeus*, a dinosaur known principally from embryos within intact eggs. The case was made that the long bones of *Orodromeus* showed a high degree of ossification, and consequently that hatchlings were, like all modern reptiles and some birds, precocial, that is, capable of leaving the nest and fending for themselves. By contrast, the long bones of *Maiasaura* were characterized as cartilaginous and poorly finished, presumably indicating a period of postembryonic helplessness, necessitating extended parental care, as in altricial birds such as all modern song birds.

Bone histology. Since 1990, new findings have emerged that call into question both dinosaurian endothermy and parental care of altricial babies. Bone histology, the study of the microscopic structure of bone, seemed to support the idea of endothermic dinosaurs. It had been recognized for many years that dinosaur bone had a rich blood supply, like that of typical mammals and birds, but unlike that of typical reptiles and other cold-blooded vertebrates. Earlier studies of dinosaur bone histology, however, were principally done on isolated, anatomically random samples of bone. It came as a surprise, therefore, when studies of growth series of femurs (thighbones) of the same species of dinosaur showed growth rings like those of living reptiles.

Growth rings were demonstrated both in the Early Jurassic South African prosauropod *Massospondylus* and in the small theropod *Syntarsus*. Subsequently, growth rings were also demonstrated in the Triassic small theropod *Coelophysis*, in the Late Cretaceous great meat-eater *Tyrannosaurus rex*, and in the Late Cretaceous theropod *Troodon*. Although growth rings appear to be lacking in the Late Jurassic African ornithopod *Dryosaurus*, they are widespread among the Dinosauria, and seasonal interruption in growth seems to have prevailed.

Growth rings. By contrast, no living bird shows growth rings in its skeleton, and fossil birds, including the Late Cretaceous flightless diving bird *Hesperornis*, resemble modern birds in this regard. In 1994, specimens of very primitive birds from the Late Cretaceous of South America were obtained. Both the flightless *Patagopteryx* and the flying bird *Enantiornithes* showed growth rings, suggesting that the reptilian growth pattern, and by inference metabolic rate, that characterized all theropod dinosaurs studied, was carried over into very primitive birds. The elevated growth rates, and by inference metabolic rates, of modern birds evidently evolved somewhat after the appearance of the earliest birds.

Fossil embryos. Birdlike parenting behavior in duck-billed dinosaurs has been called into question as well. In 1994, a new species of the Late Cretaceous duckbill *Hypacrosaurus* was described from a large fossil deposit that straddles the border between Alberta (Canada) and Montana. Included are numerous eggs, containing hatchlings and embryos. The nearly spherical eggs approach the size of a soccer ball and have a volume of 1 gal (3.9 liters) or more. Embryos inside the eggs show evidence of tooth wear. The embryos evidently ground their jaws together as part of the exercises by which the nervous system matured. Thus, tooth wear is not evidence of animals feeding on plant material.

There is a relationship between the size of a mother reptile and the size of her babies. If a 120-in. (3-m) alligator hatches out young 10 in. (25 cm) long on average, the claimed hatchling size of *Maiasaura* of 12 in. (30 cm) or less is suspiciously small. A hatchling size of 3.3 ft (1 m) is expected in a duckbill the size of *Maiasaura*, as is found in *Hypacrosaurus*. It is possible that the presumed hatchlings of *Maiasaura* are actually unhatched embryos. If so, the association of young in the nest, the unfinished textures of the ends of the leg bones, and the wear on the teeth are explained without resorting to parental care. The embryos perhaps perished when the nesting ground was inundated in a flood, and the brittle eggs were fragmented from some other natural mishap before they were buried in protective sediment.

In 1996, it was established that in living precocial and altricial birds there is no appreciable difference in the structure of the limb bones of prehatching embryos. An important difference between the two groups is the degree of ossification of the pelvis, from which the muscles that move the hindlimbs originate. In the comparatively helpless, nest-bound altricial birds, considerable quantities of cartilage are found at hatching, but in precocial birds the pelvis is already well ossified before hatching. In *Maiasaura, Hypacrosaurus*, and *Orodromeus*—in fact, in all dinosaurs in which this characteristic is known—the pelvis is always well ossified before hatching. No dinosaur shows histological evidence of posthatching helplessness that necessitates parental care.

Comparative anatomy. Most recently, several physiologists, in cooperation with some paleontologists, have examined the relationship between the anatomy of the nasal cavity and metabolic rate in warm-blooded vertebrates, cold-blooded vertebrates, and dinosaurs. At first, they focused on evidence for delicate turbinal bones in the nasal cavity. The respiratory turbinals are lacy scrolls of thin bone covered with moist epithelium. The turbinals of birds and mammals make use of countercurrent circulation to warm and humidify incoming air, and to cool and dehumidify exhaled air. By contrast, turbinals are absent in cold-blooded vertebrates, which have much lower respiratory rates. Given the low preservation potential of turbinals, the researchers switched instead to examination of the volume of the nasal cavity, reasoning that since turbinals have a high surface area in order to modify the inhaled air, the nasal cavity of endotherms must be capacious in order to accommodate them. Furthermore, an increase in width of the nasal cavity is very effective in promoting energetically efficient air flow. Birds and mammals have, on average, four times the cross-sectional area of the nasal cavity compared to reptiles of the same size. When the nasal cavities of several dinosaurs are examined by means of computerized tomography scans, the theropods *Nanotyrannus* and *Dromaeosaurus* and the hadrosaur *Hypacrosaurus* plot absolutely on the reptile line, and the ornithomimid *Ornithomimus* plots significantly below the reptile line. Thus, the dinosaurs examined lack the respiratory capacity to have been birdlike or mammallike endotherms.

For background information *see* ANIMAL EVOLUTION; AVES; BONE; COMPUTERIZED TOMOGRAPHY; DINOSAUR; FOSSIL; JURASSIC; METABOLISM; PALEONTOLOGY in the McGraw-Hill Encyclopedia of Science & Technology.   Peter Dodson

Bibliography. K. Carpenter, K. F. Hirsch, and J. R. Horner, *Dinosaur Eggs and Babies*, 1994; A. Chinsamy and P. Dodson, Inside a dinosaur bone, *Amer. Sci.*, 83:174–180, 1995; J. O. Farlow, P. Dodson, and A. Chinsamy, Dinosaur biology, *Annu. Rev. Ecol. Systemat.*, 26:445–471, 1995; N. R. Geist and T. D. Jones, Juvenile skeletal structure and the reproductive habits of dinosaurs, *Science*, 272:712–714, 1996; D. F. Glut, *Dinosaurs: The Encyclopedia*, 1996; J. A. Ruben et al., The metabolic status of some Late Cretaceous dinosaurs, *Science*, 273:1204–1207, 1996.

Earth, age of

Recent research to determine the age of the Earth's core used hafnium–tungsten chronometry. The iron-rich metallic core is the most inaccessible part of the Earth, so determining how it has developed is very difficult. Furthermore, the first 500 million years of Earth history are the most obscure; no indigenous rocks from that period have survived the intense meteorite bombardment of the Earth. The development of a new way of determining the timetable for the growth of the core during the very earliest evolution of the Earth therefore represents a major breakthrough.

Hafnium–tungsten chronometry. The technique utilizes the radioactive decay of a now extinct nuclide of hafnium, ^{182}Hf, which decayed to an isotope of tungsten, ^{182}W. The half-life of ^{182}Hf is 9 million years, which is a short period on geological time scales, since the Earth is approximately 4.5 billion years old. Therefore, no ^{182}Hf remains on the Earth. Nevertheless, variations in the isotopic abundance of tungsten in ancient samples provide clues about the timing of processes that fractionated (separated) hafnium from tungsten. Only one process could have produced this major fractionation—the formation of a metallic core. Therefore, this event can be dated by measuring the isotopic composition of tungsten.

Refractory elements. Hafnium and tungsten are both present at trace levels (parts per billion or million) in rocks and metals from the inner solar system. Theoretical calculations predict that hafnium and tungsten would have condensed into stable solid phases at very high temperatures when the solar nebula first collapsed to form the solar system. Such elements are described as refractory; because of this characteristic, the proportion of hafnium relative to tungsten in the inner solar system is well established. Any loss of volatile elements (such as the noble gases) during the strong heating of the inner solar system that accompanied the early history of the Sun would have had no effect on the concentrations of refractory elements. As the dust and debris of the inner solar system accreted under the influence of gravity to form planetesimals and planets, the kinetic energy would have been converted to heat. Any such heating from accretional energy that was released as the planets and planetesimals of the early solar system collided with decreasing frequency but increasing impact (as they got bigger) would have caused vaporization of volatile elements, but it would not have affected refractory elements such as hafnium and tungsten.

It is essentially clear exactly how much hafnium and tungsten are in the Earth. Although the absolute concentrations may have increased somewhat because of loss of volatile elements, the Hf/W ratio should be about the same in the Earth as it is in primitive, unprocessed solar-system material. Some very primitive meteorites have been found with textures suggesting that they were never part of a major molten planet, and with chemical compositions re-markably similar to that deduced for the Sun on the basis of spectral measurements (if allowance is made for loss of very light volatile elements, such as hydrogen and helium, from meteorites). These particular primitive meteorites are called chondrites, and they provide a reference sample for the average Hf/W ratio and tungsten isotopic composition of the solar system and the total Earth. Unfortunately, because of analytical difficulties, it had not been possible to measure the tungsten isotopic composition of chondrites in order to discover what the tungsten isotopic composition of the total Earth must be. While the isotopic composition of the tungsten found at the Earth's surface was already known, that value is not necessarily representative of the total Earth, because the core could be different depending on when it formed. Without a knowledge of the tungsten isotopic composition of chondrites—the reference for the total Earth—it was impossible to know whether the tungsten isotopic composition found at the surface of the Earth was the same as that for the total Earth or whether it was affected by core formation.

Core formation. Although heating and loss of volatiles would not have affected the ratio of hafnium to tungsten in the Earth as a whole, another important segregation that took place within the planetesimals and planets of the early solar system would have greatly affected this ratio internally. The separation of metal (or metallic liquid) from silicate rock (or magma) appears to have been a very common early process. The two most abundant elements in the Earth are oxygen and iron. From studies of nonprimitive meteorites (silicate achondrites and irons) it is known that segregation of large amounts of iron-rich metallic liquids took place on both a small and large scale. In the latter case the liquid iron, being very dense, would settle toward the center of the planet or protoplanet and form a metallic core. The Earth's core is at a depth of 2900 km (1800 mi), and is still partly molten because of radioactive heating, even 4.5 billion years after the formation of the Earth.

Tungsten is a metal that tends to substitute into iron or iron-rich phases or their liquids; thus, when metal segregated to form the core of the Earth, it incorporated more than 90% of the Earth's tungsten. Hafnium, however, does not fit into iron, iron compounds, or their liquids. Rather, it tends to substitute into silicate rocks and magmas, such as those making up the mantle and crust of the Earth (the outer 2900 km or 1800 mi). For this reason, the ratio of hafnium to tungsten in the silicate Earth is about 20 times higher than its value for the Earth as a whole, whereas the ratio in the core is almost zero. Because the Hf/W ratio of the silicate Earth is so high relative to the starting material from which the Earth was made, it should have generated highly radiogenic tungsten (rich in ^{182}W) if the core formed within the lifetime of ^{182}Hf (effectively about 50 million years), relative to that of the total Earth or that found in primitive solar-system material such as chondrites.

Determining age of the core. A comparison between the tungsten isotopic compositions of the silicate Earth (common tungsten metal) and that of chondrites would indicate when the Earth's core formed relative to the start of the solar system. The later the core formed and the Hf/W ratio of the silicate Earth increased, the less would be the tungsten isotopic effect of radioactive decay of hafnium on the tungsten isotopic composition of the residual silicate Earth. This follows because more of the ^{182}Hf would have decayed into tungsten. The amount of ^{182}Hf relative to stable ^{180}Hf and the amount of ^{182}W relative to nonradiogenic ^{184}W are related by the equation below, where CF refers to the time of core forma-

$$\left(\frac{^{182}\text{Hf}}{^{180}\text{Hf}}\right)_{CF} = \frac{\left(\frac{^{182}\text{W}}{^{184}\text{W}}\right)_{BSE} - \left(\frac{^{182}\text{W}}{^{184}\text{W}}\right)_{CHOND}}{\left(\frac{^{180}\text{Hf}}{^{184}\text{W}}\right)_{BSE} - \left(\frac{^{180}\text{Hf}}{^{184}\text{W}}\right)_{CHOND}}$$

tion, BSE to the bulk silicate Earth, and CHOND to chondrites. Because of the phenomenon of radioactive decay, an equation containing ΔT, the amount of time that has elapsed since the start of the solar system, can be obtained, so that ΔT can be determined in terms of known and measurable isotopic ratios.

New technique. The potential of hafnium–tungsten chronometry has been recognized for many years. However, until recently there was no way of making the tungsten isotopic measurements. Tungsten has a high work function and a very high first ionization potential (7.98 V). Therefore, it is extremely difficult to generate a sufficient proportion of ions relative to neutral atoms to measure the abundances on a modern mass spectrometer at high precision. However, a new technique for successfully ionizing tungsten has been developed. It uses an inductively coupled plasma source connected to a high-precision multiple-collector mass spectrometer. This technique, known as multiple-collector–inductively coupled plasma mass spectrometry (MC-ICPMS), was immediately applied to the high-precision determination of tungsten isotopic compositions in geological materials. The tungsten isotopic composition of the silicate Earth and chondrites is identical. This is now known to very high precision. Therefore, core formation and the associated increase in the Hf/W ratio have had no effect on the tungsten isotopic composition of the silicate Earth. This being the case, these chemical fractionation effects must have occurred late in Earth history, after all the ^{182}Hf had already decayed. For the first time it is known for certain that the Earth's core formed more than 50 million years after the solar nebula collapsed to form the solar system.

Why the core formed so late is unclear. However, some recent models hypothesize that the core formed as a consequence of a magma ocean in the upper mantle. This magma ocean would have developed as a consequence of bombardment by very large impactors during the late stages of accretion.

Perhaps it took such late-stage impacts to produce sufficient melting to result in large-scale metal segregation.

For background information *see* DATING METHODS; EARTH; EARTH, AGE OF; MASS SPECTROMETRY; METEORITE; SOLAR SYSTEM in the McGraw-Hill Encyclopedia of Science & Technology.   Alex N. Halliday

Bibliography. A. N. Halliday et al., Early evolution of the Earth and Moon: New constraints from Hf-W isotope geochemistry, *Earth Planet. Sci. Lett.*, 142:75–90, 1996; D.-C. Lee and A. N. Halliday, Hafnium–tungsten chronometry and the timing of terrestrial core formation, *Nature*, 378:771–774, 1995.

Ebola virus

Ebola viruses are a group of exotic viral agents that cause a severe hemorrhagic fever disease in humans and other primates. The four known subtypes or species of Ebola viruses are Zaire, Sudan, Reston, and Côte d'Ivoire (Ivory Coast), named for the geographic locations where these viruses were first determined to cause outbreaks of disease. Ebola viruses are very closely related to, but distinct from, Marburg viruses. Collectively, these pathogenic agents make up a family of viruses known as the Filoviridae.

Infectious agent. Filoviruses have an unusual morphology, with the virus particle, or virion, appearing as long thin rods. Virions have a diameter of 0.08 micrometer and a minimal length of 0.7–0.9 μm; this length is comparable to that of a small bacterium. However, when these viruses are grown in cell cultures, they can form long filamentous and branched forms that reach 14 μm or more. This is especially true for Ebola viruses.

A filovirus virion is composed of a single species of ribonucleic acid (RNA) molecule that is bound together with special viral proteins, and this RNA–protein complex is surrounded by a membrane derived from the outer membrane of infected cells. Infectious virions are formed when the virus buds from the surface of infected cells and is released. Spiked structures on the surface of virions are formed by three molecules of a single glycoprotein and are firmly embedded in the virion membrane. These structures project from the virion and serve to recognize and attach to specific receptor molecules on the surface of susceptible cells. This recognition and binding allows the virion to penetrate into the cell. Then, with the virion free to operate within the cytoplasm, the genetic information contained in the RNA molecule directs production of new virus particles by using the cellular machinery to drive synthesis of new viral proteins and RNA.

Pathogenesis. Although much is known about the agents of Ebola hemorrhagic fever disease, the ecology of Ebola viruses remains a mystery. The natural hosts of filoviruses remain unknown, and there has been little progress at unraveling the events leading to outbreaks or identifying sources of filoviruses in the wild. Fortunately, the incidence of human disease is relatively rare and has been limited to persons living in equatorial Africa or working with the infectious viruses. However, in this time of rapid transportation over large distances, the threat of agents such as Ebola viruses spreading to remote areas is taken very seriously by public health professionals. People infected with the virus are not as contagious as, say, persons suffering from a cold or measles, and the virus is spread primarily through close contact with the body of an infected individual, his or her body fluids, or some other source of infectious material.

Diagnosis. Ebola virus hemorrhagic fever disease in humans begins with an incubation period of 4–10 days, which is followed by abrupt onset of illness. Fever, headache, weakness, and other flulike symptoms lead to a rapid deterioration in the condition of the individual. In severe cases, bleeding and the appearance of small red spots or rashes over the body indicate that the disease has affected the integrity of the circulatory system. Contrary to popular belief, individuals with Ebola virus infections do not melt or have their organs dissolve, but die as a result of a shock syndrome that usually occurs 6–9 days after the onset of symptoms. This shock is due to the inability to control vascular functions and the massive injury to body tissues.

Ebola viruses can be found in high concentrations in tissues throughout the body and are especially evident in the liver, spleen, and skin. From studies of human cases and experimentally infected animals, it appears that the immune response is impaired and that a strong cellular immune response is key to surviving infections. This immunosuppression may also be a factor in death, especially if secondary infections by normal bacterial flora ensue. No severe human disease has been associated with Reston virus infections, and it appears that this virus may be much less pathogenic for humans than it is for monkeys.

Epidemiology. The first described cases of Ebola virus disease occurred in 1976, when simultaneous outbreaks of two distinct subtypes occurred in northern Zaire and southern Sudan. Many of the human infections occurred in local hospitals, where close contact with fatal cases, reuse of contaminated needles, and a low standard of medical care led to most fatalities. An Asian form of Ebola virus was discovered in a type of macaque, the cynomolgus, exported from the Philippines to the United States in late 1989. It was determined that a single monkey breeding facility in the Philippines was the source of the virus. Named Reston virus after the Virginia city in which infected monkeys were first identified, it represented a new form of Ebola virus. A repeat of this incident took place in early 1992, when monkeys from the same breeding facility were shipped to Siena, Italy.

In 1994, another new species of Ebola virus was identified, associated with chimpanzee deaths in the west African country of Côte d'Ivoire. This episode signaled the reemergence of Ebola virus in Africa, which had not been seen since 1979 when an outbreak of the Sudan subtype occurred. A large outbreak of disease was identified in early May 1995 in and around the city of Kikwit, Zaire. This outbreak of a Zaire subtype was a repeat of the 1976 episode, but this time 600 mi (1000 km) to the south. Another Ebola virus outbreak took place in 1995 in Gabon and resulted in human fatalities, but the details were slow to be revealed. This virus has been isolated, and preliminary reports indicate that it may be a variant of the Zaire subtype. This virus may have also been responsible for human fatalities in Gabon reported to have occurred in early 1996. Soon after this outbreak, yet another Reston virus outbreak occurred when infected monkeys were sent to a United States holding facility in Alice, Texas. That introduction was quickly detected and stopped by quarantine and testing measures implemented after the first Reston virus episode in 1989.

Control and prevention. Outbreaks of Ebola virus disease in humans are controlled by the identification and isolation of infected individuals, implementation of barrier nursing techniques, and rapid disinfection of contaminated material. Diagnosis of Ebola virus cases is made by detecting virus proteins or RNA in blood or tissue specimens, or by detecting antibodies to the virus in the blood. Such testing is important in tracking and controlling the movement of the virus during an outbreak.

Dilute hypochlorite solutions (bleach), 3% phenolic solutions, or simple detergents (laundry or dish soap) can be used to destroy infectious virions. No known drugs have been shown to be effective in treating Ebola virus (or Marburg virus) infections, and protective vaccines against filoviruses have not been developed. However, research is being directed at developing such treatment and should provide insights into the disease mechanisms used by filoviruses.

Evolution. Filoviruses have been shown to be genetically related to two other virus families, the Paramyxoviridae (including measles virus and mumps virus) and the Rhabdoviridae (including rabies virus and vesicular stomatitis virus). These viruses evolved from a common progenitor in the very distant past. These viruses evolved into very distinct lineages, yet have maintained a similar approach to reproducing.

The evolutionary profile, or phylogeny, for Ebola viruses and Marburg viruses has been determined. Ebola viruses have evolved into a separate and very distinct lineage within the filovirus family, and the individual species of Ebola viruses have also evolved into their own sublineages. Filoviruses show a great deal of diversity, indicating that they have likely coevolved with their natural hosts over a long period of time. Ebola viruses within a given subtype did not appear to show a great deal of change over periods of many years. This genetic stability or stasis in the wild suggests that these viruses have evolved to occupy very specific ecological niches.

For background information *see* IMMUNOSUPPRESSION; RIBONUCLEIC ACID; VIRUS in the McGraw-Hill Encyclopedia of Science & Technology.

Anthony Sanchez

Bibliography. Centers for Disease Control, Ebola-Reston virus infection among quarantine nonhuman primates—Texas, 1996, *MMWR*, 45:314–316, 1996; B. N. Fields et al. (eds.), *Fields Virology*, 3d ed., 1996; L. G. Horowitz, *Emerging Viruses: AIDS and Ebola—Nature, Accident or Intentional*, 1996; A. Sanchez et al., Reemergence of Ebola virus in Africa, *Emerg. Infect. Dis.*, 1:96–97, 1995; A. Sanchez et al., The virion glycoproteins of Ebola viruses are encoded in two reading frames and are expressed through transcriptional editing, *Proc. Nat. Acad. Sci. USA*, 93:3602–3607.

Ecological interactions

Cephalopoda is a class of exclusively marine, present-day mollusks that evolved during the Cambrian Period from a benthonic ancestor, probably a herbivore monoplacophoran mollusk. After 500 million years of evolution the result is a group of organisms with a variety of life-styles (there are benthonic forms, such as octopuses; nectobenthonics, such as cuttlefishes; and nectonics, such as squids), high and fast mobility (performed by a characteristic jet propulsion mechanism), a highly developed nervous system, a relatively short life-span, and a very fast growth for invertebrate standards, all supported by their carnivorous habits. Their biomass, estimated in more than 49 billion long tons (500 million metric tons) per year, together with their voracity, makes them an important component of the marine food webs. Moreover, different cephalopod species constitute an important fishery resource, representing more than 2% of world marine captures.

Two cephalopods of commercial interest are *Todaropsis eblanae* and *Illex coindetii*. Both species are short-finned squids belonging to the Ommastrephidae, a family of neritic and oceanic squids. One area of their world distribution is the coast of Galicia (northwest Spain), where they both inhabit the same area and are a by-catch of a multispecies trawling fishery.

Aging and growth. Size of recruitment to this fishery (the size at which a species starts to be fished) is about 1.3 in. (34 mm) of mantle length for *T. eblanae* and 1.9 in. (48 mm) for *I. coindetii*, their maximum sizes being around 8.7 and 14.9 in. (222 and 380 mm) of mantle length, respectively; in both cases the maximum size of females is bigger than that of males. As the life-cycle studies of these species have been based mainly on commercial samples, their prerecruit life is practically unknown. Assuming for *T. eblanae* and *I. coindetii* the common biological features known for other ommastrephids, they probably hatch from small eggs grouped in gelatinous masses floating in midwaters. The planktonic hatching of ommastrephids is called Rynchoteuthis because of its characteristic proboscis, resulting from the fusion of two tentacles, which split with the ontogenic development of the animal.

Oceanographic cruises have failed until recently to find the Rynchoteuthis stage of *T. eblanae*. In the case of *I. coindetii*, the Rynchoteuthis hatches with 0.04 in. (1 mm) of mantle length. Growth of rynchoteuthion and juvenile stages is considered to be exponential, attaining the size of recruitment 3–5 months after hatching. The broad range in recruitment ages is a consequence of the high growth-rate variation between individuals. As a result of this and of an extended period of spawning, ordinary methods of estimating fish growth, such as length–frequency analysis (that is, the analysis of the changes in time of the length–frequency distribution of a population), usually do not give good results with cephalopods. An alternative method, statolith increment analysis, yields better results, especially with ommastrephids. This method consists of counting the number of periodic increments deposited in the statoliths (a pair of structures homologous to the fish otoliths). This increment deposition is normally assumed to be daily; however, it has been validated for few cephalopod species. These do not include either *T. eblanae* or *I. coindetii*; therefore, all of the age estimates considered here (such as those of recruitment age) are made assuming that one increment equals one day of life. By this method, the lifespan of these species in northwest Spanish waters appears to be close to 12 months for *T. eblanae* and 12–15 months for *I. coindetii*.

Reproduction. Full maturity is reached between the seventh and ninth month of life for *T. eblanae* and between the fourth and tenth month for *I. coindetii*. Males of both species mature faster than females. Investment in reproduction, assessed as weight of reproductive tissues with respect to body weight of mature specimens, is higher in females than in males.

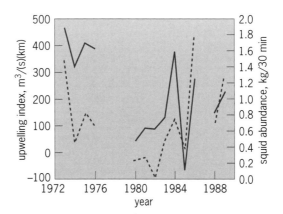

Fig. 1. Annual changes in the intensity of spring and summer upwelling and autumnal abundance of the short-finned squid *Todaropsis eblanae* hatched during spring and summer off the northwest Spanish coast. The upwelling index (solid line) is based on the direction and intensity of coastal surface winds and increases with the rise of upwelling intensity. Squid abundance (broken line) represents the average biomass (1 kg = 2.2 lb) per 30 min of trawling at a speed of 2.5–3.0 knots (4.6–5.6 km/h). *(After M. Rasero, Relationship between cephalopod abundance and upwelling: The case of Todaropsis eblanae (Cephalopoda: Ommastrephidae) in Galician waters (NW Spain), ICES C. M./K:20 (mimeo), 1994)*

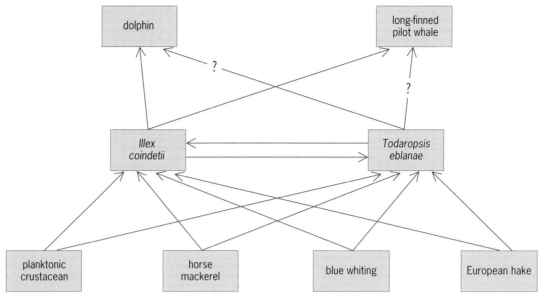

Fig. 2. Some of the prey–predator relationships involving the short-finned squids *Todaropsis eblanae* and *Illex coindetii* off northeast Atlantic waters.

Reproductive tissues of both species represent on average 18–20% of the mature female body weight and 4–6% of the mature male body weight. Ovulation of both squids seems to be partial; that is, there is a significant fraction of oocytes still in vitellogenesis when spawning starts. However, spawning is intermittent in both species.

Despite the similarities in maturation, spawning pattern, and reproductive investment, fecundity of *T. eblanae* and *I. coindetii* off Galician waters seems to be significantly different. Maximum potential fecundity, estimated as total immature and mature oocyte number present in gonad and oviducts of a female, was 143,792 for *T. eblanae* and 287,000 (counting only mature oocytes) for *I. coindetii*. The size of mature eggs is consistently bigger for *T. eblanae* (0.06–0.7 in. or 1.5–1.7 mm) than for *I. coindetii* (0.03–0.5 in. or 0.8–1.2 mm). Therefore, the reproductive strategy of the first species seems to be directed to producing smaller number of hatchlings but of a bigger size than those of the second species.

Spawning of these squids in northwest Spain occurs year-round; however, both species experience a seasonal spawning peak, with the period of highest primary production in the area occurring during spring and summer. This increased primary production is associated with seasonal intensification of the coastal upwelling and radiation. Following this primary production, growth of zooplankton and organisms of higher trophic levels, including paralarvae and prerecruits of these squids, should be favored. This trophic dependence is probably based on the high correlation ($r = 0.731$) found between autumnal abundance of *T. eblanae* and mean spring and summer values of Bakun's upwelling index (based on the direction and intensity of coastal sea-surface winds) from 1973 to 1989 (**Fig. 1**).

Feeding and trophic relationships. The feeding of prerecruits is unknown, although they probably follow the general feeding pattern of young cephalopods that feed mainly on small-sized crustaceans. The diet of commercial sizes is known from identification of gut contents taken from animals caught in the wild. This identification is a difficult task since these squids (like other cephalopods) mince their food before swallowing it. The result is a collection of little pieces of prey tissues inside the stomach of the squid that is further complicated by fast digestion, which is probably a consequence of mincing the food during ingestion. This behavior may imply an advantage since it can permit the process of more food by time unit, obtaining more energy and nutrients for fast growth.

Type of prey. The identification of prey is mainly based on characteristic tiny hard remains (such as vertebrae, scales, otoliths, and pieces of carapace), which can also give information about the size of the prey ingested. The appearance of such debris is more frequent in the stomachs of *T. eblanae* than in those of *I. coindetii*. The diet of *T. eblanae* and *I. coindetii* in Galician waters is based on, in decreasing order of importance, teleosteans, crustaceans, and cephalopods, up to a total of 21 and 23 different prey species, respectively. Twelve of these prey were common for both squids. Within teleosteans, there were some species of commercial interest: blue whiting (*Micromesistus poutassou*), hake (*Merluccius merluccius*), and horse mackerel (*Trachurus trachurus*). *Micromesistus poutassou* is the most important prey in both squids, representing as much as 43% of the diet of *T. eblanae*.

Within cephalopod preys, there were different cases of cannibalism among both squids, being as high as 22% of the diet of immature *T. eblanae* off the west coast of Galicia. However, there was no difference in diet between sexes. Overall, *I. coindetii* seems to forage in a wider variety of habitats than *T. eblanae*, because its diet also includes more coastal and epipelagic organisms. Possible competition for food between both squids is probably relaxed because of the higher mobility of *I. coindetii*. The diet of both squids was dependent on the zone considered, size, and stage of maturity.

Prey size. The relative importance of predator size and maturation are difficult to separate because both factors are strongly linked, and mature specimens are also the biggest. However, size is probably the most influential variable. Estimation of prey size from hard remains in the stomachs shows that small specimens of both species (less than 0.22 lb or 100 g of body weight) feed frequently on prey as big as themselves, a proof of the predatory capacity of these squids. However, very often the prey appears not to be ingested totally because the maximum weight of the stomach contents is several times less than the estimated weight of the prey ingested. The maximum weight of food in the stomach can be more than 14% of the squid body weight for small specimens and 7% for large specimens. These values correspond to a single daily meal. If more meals are taken per day, considering the fast digestion, the daily rations could be higher. Therefore, based on the voracity of these squids and their relative abundance in the area, their impact on prey populations must be significant.

However, like many cephalopods, these squids probably fall prey to fishes, sea mammals, and sea birds. Predators of *T. eblanae* are scarcely known, but *I. coindetii* remains are frequently found in the stomach contents of different species of cetaceans stranded off the Galician coast (**Fig. 2**).

Further studies are necessary to clarify the ecological role of these cephalopods in the marine food web of the northeast Atlantic waters, but from the present data these squids appear as an important channel for fast transference of nutrients and energy from the lowest levels of consumers to top predators.

For background information *see* CEPHALOPODA; ECOLOGY; FOOD WEB in the McGraw-Hill Encyclopedia of Science & Technology.   B. G. Castro

Bibliography. A. F. González, B. G. Castro, and A. Guerra, Age and growth of the short-finned squid *Illex coindetii* in Galician waters (NW Spain) based on statolith analysis, *ICES J. Mar. Sci.*, 53(5):802–810, 1996; R. Hanlon and J. B. Messenger, *Cephalopod Behaviour*, 1995; M. Rasero, Relationship between cephalopod abundance and upwelling: The case of *Todaropsis eblanae* (Cephalopoda: Ommastrephidae) in Galician waters (NW Spain), *ICES C. M./K:20* (mimeo), 1994; M. Rasero et al., Predatory relationships of two sympatric squid, *Todaropsis eblanae* and *Illex coindetii* (Cephalopoda: Ommastrephidae) in Galician waters (NW Spain), *J. Mar. Biol. Ass. U.K.*, 76:73–87, 1996.

Ehrlichiosis

Ehrlichiosis is a tick-borne infection that often is asymptomatic but also can produce an illness ranging from a few mild symptoms to an overwhelming multisystem disease. Ehrlichiosis is included with those infections that are said to be emerging, either because they have been recognized only recently or because they were previously well known but now are occurring more frequently.

History. Prior to 1986, the bacteria of the genus *Ehrlichia* were of interest primarily to veterinarians. These microorganisms have a global distribution and are capable of producing a hemorrhagic disease in dogs and febrile illnesses of horses, cattle, sheep, and bison. A constant feature in all these circumstances is the predilection of the *Ehrlichia* to enter the cellular elements of the blood. The specific *Ehrlichia* species determined whether mononuclear cells, granulocytes, or platelets were invaded. Human illness associated with these organisms was first recognized in southern Japan in the 1950s. Because these illnesses seemed to be a local phenomenon, they received little attention elsewhere.

However, in 1986 a 51-year-old man, after planting some trees while vacationing in Arkansas, discovered some ticks on his neck and removed them uneventfully. Ten days later, he developed fever accompanied by a serious systemic illness that eluded specific diagnosis and required a prolonged hospitalization before his recovery. Careful microscopic examination of a slide preparation of his peripheral blood revealed distinctive inclusion bodies in his lymphocytes and monocytes. When reviewed by microbiologists and pathologists, these inclusions were identified as ehrlichiae. This illness was the first case of ehrlichiosis diagnosed in the United States. To date, over 400 cases of ehrlichiosis have been diagnosed in 30 states, and ehrlichiosis now is considered to be the most common tick-borne infection in this country.

Epidemiology. Human ehrlichiosis is caused by two distinct species: *E. chaffeensis* and an unnamed ehrlichial species. *Ehrlichia chaffeensis* infects primarily mononuclear blood cells; the disease produced by this species is referred to as human monocytic ehrlichiosis. The other ehrlichial species invades granulocytic blood cells, causing human granulocytic ehrlichiosis. The latter organism closely resembles *E. equi*, a species that infects horses.

Both of these ehrlichia species are transmitted to humans by the bite of infected ticks; however, the two species appear to have somewhat different geographic distributions. *Ehrlichia chaffeensis* occurs most commonly in the south-central and southeastern states, where it is associated primarily with the Lone Star tick (*Amblyomma americanum*); it is also transmitted by the common dog tick (*Dermacentor variabilis*). The agent of human granulocytic ehrlichiosis is found in the upper midwestern states of Wisconsin and Minnesota, as well as in several northeastern states. This agent seems to be transmitted principally by the deer tick (*Ixodes scapularis*). Although ticks (the vector) are the mode of transmission of ehrlichial infections to humans, the ticks must acquire the ehrlichial organisms from animal sources (the reservoir hosts). The animal reservoirs for these organisms have not yet been confirmed, but white-tailed deer are a likely reservoir host of *E. chaffeensis* in the southeastern United States.

Clinical signs and pathogenesis. The forms of the disease caused by the two ehrlichial species are indistinguishable. Illness occurs most often during April–September, corresponding to the period when ticks are most active and humans are pursuing outdoor activities. Most individuals relate a history of tick exposure and have been engaged in work or recreation in wooded areas where they encountered ticks.

While obtaining its blood meal by biting an individual, the infected tick inoculates ehrlichiae into the skin. The microorganisms then spread throughout the body via the bloodstream. *Ehrlichia* must achieve an intracellular residence in order to survive and multiply. *Ehrlichia chaffeensis* seeks out tissue macrophages in the spleen, liver, and lymph nodes. It is suspected that the agent of human granulocytic ehrlichiosis has a tropism for granulocytic precursor cells in the bone marrow. Once the organism has entered the cell, it multiplies there for a period of time without producing symptoms (the incubation period). In ehrlichiosis, the incubation period can last from 1 to 3 weeks after exposure to the infected tick. Thereafter, individuals develop fever, chills, headache, and muscle pains. Gastrointestinal symptoms such as nausea, vomiting, and loss of appetite also are common. Laboratory abnormalities regularly include anemia, low white blood cell and platelet counts, and abnormal liver function. More severely ill individuals also may manifest abnormalities of the central nervous system, lungs, and kidneys. Elderly patients are more likely to become seriously ill than children and young adults. Fatality rates up to 4% have been documented. Because the clinical presentation is nonspecific, the diagnosis of ehrlichiosis may not be immediately apparent. Prolonged intervals between the onset of illness and the administration of appropriate therapy can lead to more severe disease symptoms and a greater risk of fatality.

Diagnosis and treatment. An important clue to the diagnosis of human granulocytic ehrlichiosis is the recognition of cytoplasmic vacuoles filled with ehrlichiae (morulae) in circulating neutrophils. Careful examination of stained smears of peripheral blood often yields such findings in human granulocytic ehrlichiosis. In contrast, similar detection of infection in circulating monocytes is a distinctly rare event in disease caused by *E. chaffeensis*. Laboratory diagnosis usually is made by detecting an increase in species-specific antibodies in serum specimens obtained during the acute and convalescent phases of the illness. However, such serologic testing is of no use in establishing the diagnosis before treatment is initiated. In an attempt to shorten this diagnostic lag, some research laboratories are offering to test blood specimens by using polymerase chain reaction (PCR) technology. Unfortunately, this method is not yet widely available. Therefore, therapy must be initiated on clinical suspicion.

Ehrlichiosis closely resembles another tick-borne illness, Rocky Mountain spotted fever, except that the rash, characteristic of spotted fever, is usually absent or modest. Hence, ehrlichiosis has been referred to as spotless fever. Fortunately, both diseases can be treated with tetracycline antibiotics. Most individuals respond to tetracycline therapy within 48–72 h.

Prevention. The avoidance of tick bites is fundamental to preventing ehrlichiosis. Sometimes this can best be accomplished by staying out of tick-infested areas. It has been documented that the regular use of insect repellant by individuals who spend time in grassy or wooded areas offers substantial protection against tick bites. It also is advisable to wear long-sleeved shirts and trousers that can be tucked into boots. Individuals should examine themselves for ticks periodically. Ticks attached to the skin can be removed with tweezers by grasping the tick as close to the point of attachment as possible and pulling gently and steadily.

For background information *see* CLINICAL MICROBIOLOGY; HEMORRHAGE; INFECTION; IXODIDES; RICKETTSIOSES in the McGraw-Hill Encyclopedia of Science & Technology.　　　Steven M. Standaert; William Schaffner

Bibliography. J. E. Dawson et al., The interface between research and the diagnosis of an emerging tick-borne disease, human ehrlichiosis due to *Ehrlichia chaffeensis*, *Arch. Int. Med.*, 156:137–142, 1996; J. S. Dumler and J. S. Bakken, Ehrlichial diseases of humans: Emerging tick-borne infections, *Clin. Infect. Dis.*, 20:1102–1110, 1995; S. M. Standaert et al., Ehrlichiosis in a golf-oriented retirement community, *N. Engl. J. Med.*, 333:420–425, 1995; D. H. Walker and J. S. Dumler, Emergence of the ehrlichioses as human health problems, *Emerging Infect. Dis.*, 2:18–29, 1996.

Electric power generation

Control and stability of the electric supply network can be achieved by rapid deployment of hydrogenerated electricity. However, the ease and flexibility with which a hydrogenerating unit can be controlled compounds the equipment deterioration associated with normal energy conversion and the attendant forces acting on the large structure. Frequent changes in operating conditions produce cyclic stresses on the active components of the generator. These conditions often result in the onset of undetectable failures, emphasizing the limitations of conventional monitoring technology for immense generators. Because most hydrogenerator deterioration is associated with the stationary structure, a new monitoring scheme was implemented by mounting a complement of sensors on the rotor to scan the stator (see **illus.**).

The rotating part of a hydrogenerator is the salient-pole rotor, which ranges 5–15 m (15–50 ft) in diameter and weighs several hundred tons. The stationary part is the stator, which typically is 1–5 m (3–15 ft) tall (see illus.). The clearance between the rotor and the stator ranges 8–45 mm (0.3–1.75 in.).

The stator structure contains the electrical windings that are connected to the utility network. The stator core is a tightly compressed stack of silicon steel laminations that form a cylinder with evenly spaced slots within the bore circumference; these slots extend the full height of the core. The stator coils are tightly mounted in these slots (normally two to a slot). Magnetically, the laminated stator core serves to strengthen the coupling of the magnetic fields produced by the electromagnets mounted on the rim of the rotor and the magnetic fields produced by the flow of current in the stator winding. Thermally, the laminated structure is designed to remove heat that is produced in the stator coils and in the stator core.

The hydrogenerator rotor-mounted scanner uses its complement of sensors to measure the critical operating parameters on the stator. The sensors sweep by the active stator elements and measure radiant thermal energy, the clearance between the rotor and the stator (air gap), radio-frequency radiation, and magnetic fields. This combined information, measured over the entire generator, permits real-time evaluation for generator operation and long-term analysis for scheduled maintenance planning.

Corroborative diagnoses are achieved from the analyses of different phenomena that have their origins rooted in a common cause. For example, current-carrying coils of the stator winding that have become loose in the stator core slots (the result of a large number of load cycles) will become hot because of poor heat transfer to the stator core iron. The coils will be exposed to effects of partial-discharge sparking across the same void that is acting as a thermal barrier between the coil and the side of the stator slot. Relative coil motion also usually leads to electrical distress inside the stator coil.

Requirements of rotor-mounted sensors. Successfully measuring various conditions on the stator from the spinning rotor, transferring the information to a central processor, and presenting useful information to the user require the orchestration of many system elements: a platform for the sensors; a power supply for the on-rotor electronics; telemetry for bidirectional communications; a central process controller; and software for control, data conversion, analysis, and user interface. Additionally, much of the monitoring system must successfully operate in a harsh environment that includes high centrifugal forces, large temperature variations, mechanical vibration, strong electric and magnetic fields, and high relative speeds between the sensors and their targets.

One important requirement of the rotor-mounted scanner is knowing the exact location of each sensor when a measurement is made. Every time the rotor makes one revolution, a start-of-rotation marker identifies the exact location of each sensor with respect to the stator. Precise counting techniques, timing error correction, and a synchronous rotor speed that is magnetically coupled to the stable utility-system frequency ensure accurate spatial reference of all rotor-borne measurements.

Temperature measurements. Infrared techniques are used to measure the stator core surface temperatures. For a typical generator (cited in these examples), the rotor speed is 120 revolutions per minute. There are 504 coils in the stator winding, which means there are 504 stator teeth; these are protrusions of the stator core separating the stator coils. The thermal target is the edge of the tooth, the temperature of which is influenced by heat input from the stator core and stator winding. Accurately detecting the infrared radiant energy at these high relative surface speeds requires two infrared thermal radiant energy measurements and a reference temperature measurement.

The thermopile has a relatively long thermal time constant compared to the relatively short time the thermal target is in front of the infrared sensor. The relative velocity of the sensor and its target is of the order of 70 m/s (230 ft/s). The thermopile is used to measure an average temperature for the circumference of the generator within a band of 125 mm (5 in.), axially. The average infrared temperature measurement is made in reference to the thermopile's junction temperature. A thermal diode is isothermally connected to the thermopile, providing the requisite temperature reference.

The high-speed infrared measurement is accomplished by using a pyroelectric device. This sensor accurately reports the temperature variations of the scanned surface with respect to the surface's average temperature. The device has a high frequency re-

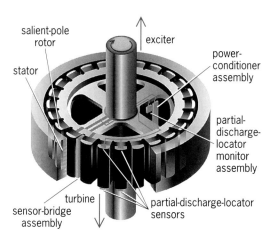

salient-pole rotor
exciter
stator
power-conditioner assembly
partial-discharge-locator monitor assembly
turbine
sensor-bridge assembly
partial-discharge-locator sensors

Rotor-mounted scanner placement on a hydroelectric generator.

sponse, greater than the stator-tooth-passing frequency of the thermal targets. These high-speed measurements are algebraically added to the average base temperature for the circumference of the stator core in the same 125-mm (5-in.) axial band. The number of thermal sensors required to monitor the temperature of the entire stator bore is the length of the stator core divided by 125 mm (5 in.).

True air gap. The air gap is measured by using sensors each of which consists of two coplaner coils that mutually couple to the bore surface at the stator teeth. The sensors travel over the stator teeth along two trajectories located about 150 mm (6 in.) from each end of the core. One of the sensor's coils is operated at a relatively high, fixed frequency. The magnetic flux produced by this coil is directed across the air gap to impinge on the surface of each stator tooth in turn. Because of the sensor's high operating frequency, eddy currents induced at the surface of the stator tooth produce an equal and opposite flux. This flux exclusion phenomenon affects the coupling between the two sensor coils. Therefore, the mutual coupling between the two coils is predictably altered by the interaction of the source flux, the stator tooth, and the air-gap distance between the sensor and the stator bore. The closer the stator iron is to the set of air-gap sensor coils, the stronger is the mutual coupling. This is an accurate and reproducible measure of the distance between the rotating and nonrotating components of the generator.

To understand and locate the minimum air gap, the stator profile and the rotor profile must be known. The same mutual coupling technique between two sensor coils, as discussed above, is used for the air-gap measurement from sensor coils mounted on the stator to the surface of each salient rotor field pole. The combination of the two profiles produces a true, dynamic air-gap measurement.

Partial discharge. The electrical and mechanical barrier between the current-carrying conductors and ground potential is the insulation around the outside of each stator coil. Typically, the voltage between the conductors of the stator winding and the ground is up to 8000 V. In-service thermal cycling, large oscillating magnetomechanical forces acting on the stator conductors, and very small imperfections in the insulation combine to reduce the quality of the insulation. As aging of the insulation takes place, minor sparking begins. Energetic sparking produces ozone, which becomes a fourth element in the process of stator coil deterioration, accelerating insulation deterioration.

This discharge, depending on the nature of the sparking phenomena, is known as partial discharge, corona, or arcing. This discharge activity produces pulses with very steep wavefronts that can be measured as radio emissions in the megahertz range. It is possible to locate and measure this radio-frequency activity by capacitively coupling the radio-frequency energy from the stator coil to the sensor. The rotor-mounted sensor forms one plate of the capacitor and a stator coil forms the other when the probe and coil are in proximity.

There is one set of sensors for a magnetic north pole and another for a south pole. The sensors are placed at the center point, at both the top and the bottom, of the two selected field poles. The sensor remains in synchronism with the near-crest voltage (positive and negative) on each stator coil, near the point of maximum voltage stress on the ground insulation. By knowing the exact location of the sensor, it is possible to map the level of partial discharge (sparking activity) for the entire stator winding and locate individual coils that are producing the discharge. Much information about the trend in partial discharge, the location of sparking activity, the operating and maintenance history of the generator, and knowledge of stator core temperatures from the thermal map are combined to give guidance for future generator operation and maintenance.

Magnetic flux. The combined magnetic flux in a generator represents the coupling phenomenon between the mechanical torque applied to the shaft and the production of electricity in the stator winding. By monitoring discrete portions of the magnetic flux, it is possible to monitor the integrity of the stator winding and the magnetic circuit of the stator core steel. One set of Hall-effect probes (the direct-axis probes) is located near the axis of the main magnetic flux (resultant flux) that produces the voltage on the stator winding. A second set of Hall-effect probes is located near the axis between the magnetic north and magnetic south poles that are rotating in synchronism with the rotor. Three Hall-effect probes orthogonally oriented to one another allow flux measurements in the radial, tangential, and axial directions. If the internal conductors of any of the current-carrying elements of the stator winding begin to break down, the result is a distortion of the magnetic field that surrounds the current flowing in the conductor. This distorted magnetic field is detected with the direct-axis Hall-effect probes. This reading is corroborated by a line on the thermal map of higher temperature at the location of the magnetic anomaly.

Machine condition monitoring. Rapid progress is being made, using advancing computer techniques, to automate and diagnose myriad measurements and corresponding data from advanced monitoring systems, such as the rotor-mounted scanner system. The assembled knowledge and experience will be combined into an expert-system rule base to provide a diagnosis that will be in close agreement with the assessment that would have been made by a generator expert. The corroborative nature of the measured information provided by the rotor-mounted scanner will make the task of machine condition monitoring much more cost effective. Thus, this monitoring technology makes it possible for generator maintenance to be scheduled according to the internal conditions of the machine rather than at fixed time periods. The main advantage is avoidance of repair costs that could not be anticipated without such monitoring of the deterioration of internal conditions.

For background information *see* ALTERNATING-CURRENT GENERATOR; ELECTRIC ROTATING MACHINERY; EXPERT SYSTEMS; HYDROELECTRIC GENERATOR; PYROELECTRICITY; THERMOELECTRICITY; WINDINGS IN ELECTRIC MACHINERY in the McGraw-Hill Encyclopedia of Science & Technology.   James S. Edmonds

Bibliography. T. L. Churchill, *Development of a Rotor-Mounted Scanner for Hydrogenerators*, Final Report of Research Project 2591-5, Electric Power Research Institute, 1993; J. S. Edmonds, On-line condition monitoring for generators, *Int. Water Power Dam Construc.*, pp. 80–82, October 1994; J. S. Edmonds et al., Failure mode testing of a hydrogenerator equipped with a rotor-mounted scanner, *Int. Water Power Dam Construc.*, pp. 45–51, January 1993; J. R. Rasmussen and D. Thompson, On-line monitoring system averts potential generator failure, *Int. J. Hydropower Dams*, pp. 45–50, January 1994.

Electronics

Recently, the power-supply voltage for many commercial very large scale integration (VLSI) chips has decreased to 3 V, and it will continue to decrease to even lower levels. This trend is driven by three factors simultaneously: feature-size scaledown in VLSI technologies, power management in large VLSI chips, and increased demand for mobile or portable battery-operated products. This low-power trend presents great challenges to analog and mixed analog-digital chip designers, particularly since integrated circuits are generally designed in VLSI technologies that are developed primarily to meet the needs of digital applications, such as computation and storage.

Feature-size scaledown. The growth of microelectronics since 1970, from the first successful large-scale integration (LSI) of microprocessor and memory chips to present-day very large scale and ultralarge scale integration (VLSI and ULSI), has been spectacular. The key to this growth has been the drive to much smaller dimensions using the principles of scaling introduced in the early 1970s. Present and future microelectronics products require all of the features of deep submicrometer (that is, with feature sizes much less than 1 μm) complementary metal-oxide-semiconductor (CMOS) technology, namely high integration and more functionality at low power and low cost. Device minimum feature sizes are scaled down to smaller values in all three dimensions. Present VLSI technologies are at around half-micrometer and are scaled down further to around 0.25 μm. Compared with the present 0.6-μm CMOS technology at 5-V power supply voltage, it is expected that scaled CMOS in the year 2004 will be at 0.1 μm, operate from a 1-V supply, and achieve a performance gain of about 7 times, a density improvement of 20 times, and a power-per-function reduction of 12 times. *See* INTEGRATED CIRCUITS.

Power management. This high packing density in large VLSI circuits results in an enormous power consumption per unit area, so that the heat generated is not a negligible factor. Even in the low-power scenario in the year 2004, the active power density will be higher than that of present 5-V technology. Power consumption in commercial chips has increased 4 times every 3 years since 1980, which is the same pace as the increase in bit density of advanced dynamic random access memories (DRAMs). The average power dissipation of digital circuits, which in most mixed analog-digital commercial chips accounts for the majority of the chip area, is proportional to the square of the supply voltage. To ensure device reliability of scaled submicrometer devices, the reduction of supply voltage is mandatory. However, a fundamental issue here is the fact that propagation delay in CMOS digital logic circuits increases rapidly when the supply voltage is reduced to a level close to the threshold voltage of CMOS devices, which is currently around 0.7 V in standard bulk CMOS VLSI technologies.

In the late 1970s and early 1980s, the wide conversion from bipolar to CMOS technology brought the supply voltage down and reduced static power consumption significantly. In order to meet the demands of high-speed digital computing, the supply voltage was kept constant for a long time, at the bipolar transistor-transistor-logic (TTL) compatible level of 5 V, while transistor feature size was being scaled down in all three dimensions. This constant-voltage scaling has created a crisis in power consumption and a quick move toward reducing supply voltage while scaling feature size, which leads to constant electric field scaling. This practice will lead to 0.1-μm features and a 1-V supply by 2004. To maintain high speed under low supply voltage, a considerable effort is under way to develop new logic design techniques, such as pass transistor logic (PTL), with the main goal of reducing parasitic and wiring capacitances through the use of fewer transistors (lean integration) compared to conventional CMOS logic. This will reduce both dynamic power consumption and propagation delay time.

Mobile products. Historically, power consumption has not been a major issue in the semiconductor industry, or a high priority unless some cooling limits were being exceeded. However, low-power design has been used in niche applications, such as in wristwatches, military applications, space hardware, and hand-held calculators. The current emphasis on low power is more than an extension of this early application base; it involves a new domain of applications, particularly in portable, mobile, battery-operated products which are coming into widespread use in the fields of telecommunications, computers, multimedia, medicine, and information processing. The minimal number of batteries that is possible with low supply voltages makes the product compact and light. When low power dissipation is incorporated together with low voltage, the product will be long lived as well. *See* CIRCUIT (ELECTRONICS).

Low-voltage analog circuits. In digital circuits, a low supply voltage usually results in low power consumption while good circuit performance is retained. This, however, is not the case for low-voltage analog circuits, where certain design changes are often required to maintain or restore analog performance criteria, such as gain, bandwidth, and linearity, at acceptable levels under low-supply-voltage conditions. These design changes usually encompass increased supply-current consumption levels and rail-to-rail signal swing (that is, a signal that varies from the higher supply rail or drain supply voltage to the lower supply rail or source supply voltage) at both input and output in order to maintain signal integrity in the presence of noise and keep the dynamic range at an acceptable level. All this may lead to larger power consumption than was originally anticipated. This imposes a tremendous challenge for analog designers, particularly since analog integrated circuits are included in mixed analog-digital chips and designed in VLSI technologies that are optimized and scaled only for digital VLSI. Innovative low-voltage and low-power analog VLSI design solutions are needed, particularly in CMOS technologies.

Micropower analog circuits. Because of the drive toward very low power operation from a low supply voltage, CMOS analog circuits which operate in the weak inversion region, also called the subthreshold region, are becoming increasingly popular. In this region, the current level in a MOS transistor is very small, of the order of nanoamperes, while the transistor's gate-source voltage is kept below the device's threshold voltage. This type of operation results in a significant reduction in power consumption. Further, since the cutoff frequency of a CMOS analog filter, or the frequency of a CMOS oscillator, is proportional to the ratio of current to capacitance, the operation of the transistors in the subthreshold region will also allow the size (area) of the capacitance to be reduced significantly so that integrated-circuit design is possible on a single chip, including the capacitor. Such single-chip design solutions are extremely difficult to achieve at very low frequencies with bipolar or strong-inversion (above-threshold) CMOS technologies.

A CMOS integrated oscillator for noncardiac pacing applications has been designed to operate in the subthreshold region in order to minimize both power consumption and chip area. The oscillator is fabricated in a 2-μm N-well CMOS process and occupies 0.281 mm^2, including a 100-pF capacitor which occupies 77.8% (0.219 mm^2) of the total area. Experimental results show a frequency of oscillation as low as 0.3 Hz and a power consumption of about 0.24 μW at 0.3 Hz and about 0.3 μW at 100 Hz, with a supply voltage as low as 2 V.

Other important applications of such micropower analog technology are in the design of monolithic hearing aids, artificial biologically inspired neural networks such as silicon retinas, and micropower filter banks for mobile microsystems used in the extraction of information from sensory signals, both auditory and visual. While digital technology is still very much in its time of greatest strength, there is tremendous potential in low-power analog VLSI for addressing practical problems. Analog VLSI is well suited for construction of artificial neurons, the functional units of the brain. This means that with analog VLSI a chip can follow the brain's lead with new information-processing applications in a variety of areas such as image processing, speech recognition, associative recall, and handwriting recognition.

One such application has been the fabrication of a family of silicon retinas. Each silicon retina is an analog VLSI chip, built in a standard digital technology, that is a square centimeter in area, contains over 500,000 transistors, weighs about a gram, and consumes about a milliwatt of power. Between arrays of phototransistors etched in silicon, dedicated circuits execute smoothing, and motion processing. The transistors within the chip operate in the subthreshold region.

Compared with a typical charge-coupled device (CCD) camera and standard digital image processor, the silicon retina is a paragon of efficiency in performance, power consumption, and compactness. A special-purpose digital equivalent would be about the size of a standard washing machine. Unlike cameras that must time-sample, typically at 60 frames per second, the analog retina works continuously without needing to sample until the information leaves the chip already preprocessed.

Operations performed with the silicon retina capture some of the functions that real retinas perform; however, real retinas contain many more circuits than the silicon retina's synthetic one. While it makes sense to build chips to maximize efficiency in the three critical elements of power, cost, and density, analog VLSI low-power techniques will still need to be advanced a long way to approximate neural efficiency.

For background information *see* INTEGRATED CIRCUITS; LOGIC CIRCUITS; NEURAL NETWORK; SEMICONDUCTOR MEMORIES; TRANSISTOR in the McGraw-Hill Encyclopedia of Science & Technology.

Mohammed Ismail; Chung-Chih Hung; Changku Hwang

Bibliography. C. Hwang et al., A very low frequency, micropower, low voltage CMOS oscillator for noncardiac pacemakers, *IEEE Trans. CAS I*, 42:962–966, 1995; M. Ismail and T. Fiez, *Analog VLSI Signal and Information Processing*, 1994; T. Kuroda and T. Sakurai, Overview of low-power VLSI circuit techniques, *IEICE Trans. Electron.*, E78-C(4):334–344, 1995; Special issue on low-power electronics, *Proc. IEEE*, vol. 83, April 1995; Special issue on low-voltage low-power analog integrated circuits, *J. Analog Integr. Circ. Sign. Proc.*, vol. 8, no. 1, July 1995.

Elephant

Fossil remains of extinct proboscideans from all continents except Australia and Antarctica have been essential for the phylogeny and classification of the

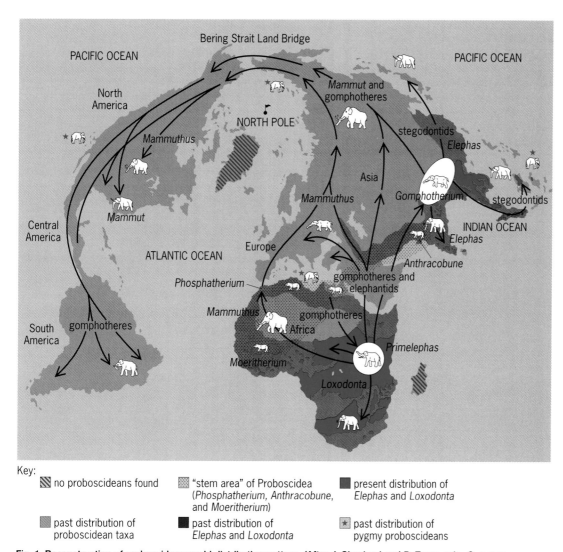

Key:

⊠ no proboscideans found

▨ "stem area" of Proboscidea
(*Phosphatherium*, *Anthracobune*,
and *Moeritherium*)

■ present distribution of
Elephas and *Loxodonta*

▨ past distribution of
proboscidean taxa

■ past distribution of
Elephas and *Loxodonta*

★ past distribution of
pygmy proboscideans

Fig. 1. Reconstruction of proboscidean world distribution pattern. *(After J. Shoshani and P. Tassy, eds., Summary, conclusion, and a glimpse into the future, The Proboscidea: Evolution and Palaeoecology of Elephants and Their Relatives, pp. 335–348, 1996)*

Proboscidea (**Fig. 1**). Currently, there are about 162 species and subspecies classified in at least eight families; the vast majority belong to a group called gomphotheres (**Fig. 2**). Earliest members of Proboscidea are believed to have inhabited the shores of the ancient sea of Tethys in the Eocene Epoch, some 45 million years ago (Ma). The closest relatives of Proboscidea are Sirenia and Desmostylia (classified in Tethytheria). This finding is based on two osteological characteristics: a lateral expansion of bone along the posterior, lateral, and ventral parts of the cranium; and the positioning of the anterior border of orbit over or in front of the first upper molar. The next kin to Tethytheria is Hyracoidea, hyraxes or the biblical conies (classified in Paenungulata). Among the features that enabled living and extinct elephants to cope with daily encounters is the size of the brain, which is relatively large compared to body size, especially the relatively large, convoluted temporal lobe. Other attributes include the ele-

phant's ability to make and use tools and to produce and perceive infrasonic calls not audible to humans.

Field observations on living elephants reveal that they play a pivotal role in their Asian and African ecosystems. They help to maintain biodiversity by continuously shaping the environment in which they live. For example, elephants will overbrowse the trees in their habitats and eventually kill them. Trees will be replaced by grasses, then by seedlings of trees, and successively these will be replaced by trees to complete the cycle. Different stages of the cycle of ecological succession support different animals, and thus elephants contribute to the biodiversity of their habitat. These attributes are also applied to extinct forms. All in all, elephants have been designated as a keystone or superkeystone species (a species responsible for the survival of other species living in the same ecosystem).

Etymology. The trunk (from the French word *trumpe*, meaning to trumpet) is the most distinguish-

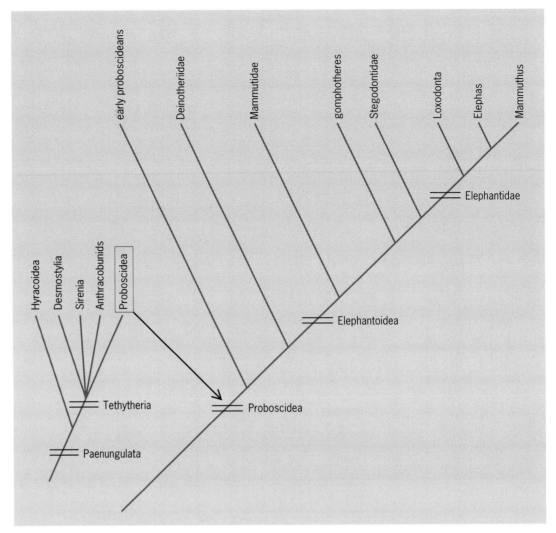

Fig. 2. Simplified phylogram of proboscidean taxa.

ing organ of an elephant, and was used to name the order Proboscidea in the early 1800s. It is now known that some extinct proboscideans did not have a proboscis. Therefore, the proboscis, or the trunk, is not a distinguishing feature of this order; other characteristics that apply to all living and extinct members of Proboscidea are required.

Origin and diversification. The evolutionary history of Proboscidea, which encompasses about 45 million years, has been interwoven with complex adaptation processes, including parallel and convergent paths. One center of radiation may have been the northern shores of modern north Africa, near the Nile River in Egypt. From there, based on fossil evidence, migratory routes into Africa and toward Eurasia may be traced. Central Asia may have been another radiation area from which lineages (groups of related genera and species) dispersed toward other parts of the globe. Anatomical features of one of the earliest proboscideans (for example, *Moeritherium*) reveal that it did not have a trunk, but perhaps a mobile upper lip.

A fully developed proboscis was probably present in a group called gomphotheres, some 25 Ma. These proboscideans were extremely cosmopolitan, and diverse in size, shape, and dental chewing surfaces. These features indicate their habitats and probable diets; mountains, deserts, lake shores, bogs, and swamps are among the interpreted environments. The latest gomphotheres (for example, *Haplomastodon waringi*) were contemporaneous with the earliest humans in South America and possibly on other continents. Other contemporaneous taxa include the American mastodon (*Mammut americanum*) and the woolly mammoth (*Mammuthus primigenius*). Contemporaneity does not necessarily imply close relationships. The genera *Mammut*, *Haplomastodon*, and *Mammuthus* were relatively distantly related on the family tree. Early human species apparently ate the meat of these proboscideans, either by hunting or by exploiting carcasses. Cave paintings in Europe depict images of mammoths. When early explorers discovered skulls with one large hole in the middle of the cranium (the large opening between the eyes where the nose begins), they believed that this opening was for the eye.

Proboscidean characteristics. Of the 162 known species and subspecies of proboscideans, only two species belonging to two genera are living today: the African elephant (*Loxodonta africana*) and the Asian elephant (*Elephas maximus*). There are two subspecies in Africa: the bush or savanna elephant, *L. a. africana*, and the forest elephant, *L. a. cyclotis*; and three subspecies in Asia: *E. m. maximus* (the island of Sri Lanka), *E. m. indicus* (mainland Asia), and *E. m. sumatranus* (the island of Sumatra). The physical characteristics of *Loxodonta* and *Elephas* are more similar than dissimilar. Dissimilarities include concave (in *Loxodonta*) versus convex or straight (in *Elephas*) profile of the back, ears which exceed the width of the neck (in *Loxodonta*) versus ears which do not exceed the width of the neck (in *Elephas*), head without two knobs in the front (in *Loxodonta*) versus head with knobs (in *Elephas*), and two fingers (in *Loxodonta*) versus one finger (in *Elephas*) at the tip of the trunk. The shared-derived characters (or synapomorphies) for all proboscideans include large, hooked scapular coracoid process, astragalus (talus, one of the ankle bones) with a medial process, and lower third molars with two definite cuspids at the posterior end. Enlarged second incisors to form tusks as seen in living elephants may also apply.

Evolutionary trends. Evolutionary changes which resulted in diversification of proboscideans were marked by certain dominant trends: increase in size (many proboscideans were over 13 ft or 4 m at the shoulders; dwarfism, however, is observed in certain lineages); longer limb bones and shorter and broader feet; growth of the skull to extraordinarily large size; and shortening of the neck.

Mandible. An elongated lower jaw (mandible) and a secondary shortening of the skull was an early primary trait among proboscideans. Secondary shortening of the lower jaw (especially the area of the mandibular symphysis) and shift in the center of gravity of the head posteriorly was a trend associated with parallel evolution in advanced proboscideans.

Proboscis. The development of a proboscis is based on the elevated position of the external naris, enlargement of the infraorbital foramen, the connection between frontal and premaxilla bones, and the shape of the premaxilla. It is believed that the combination and elongation of the upper lip and nose have evolved to accommodate the distancing of the head from the ground due to the increase in size of the animal. Subsequently, the proboscis is further elongated to form a very mobile trunk, possibly independently in different lineages.

Teeth. The forward or horizontal displacement of the cheek teeth (premolars and molars of which the earlier teeth are smaller than later ones), positioned as though they are moving on a conveyor belt, is present in all known Neogene (Miocene through Pliocene epochs) proboscideans, from mammutid through elephantid species. The vast majority of other mammals (humans included) have vertical rather than horizontal tooth replacement.

However, throughout the history of the Proboscidea, there has been a decrease in the numbers of premolars, canines, and incisors.

Hypertrophy (excess growth) of the middle incisors formed tusks. Some of these were straight, curved downward, or upward and spiraled; they functioned in food gathering, defense, offense, and display. The enamel covering of tusks decreased to a longitudinal band and then disappeared. Tusks greatly increased in length and diameter; those of proboscideans are the largest known teeth of animals, living or extinct.

Enlargement and specializations of the cheek teeth in proboscideans were achieved by increasing the number of cusps; large teeth of living elephants may weigh over 11 lb (5 kg). This trend was accompanied by molarizing the deciduous premolars and thinning of enamel; it began at the early stages of proboscidean evolution. Parallel evolution, which accounts for the increasing number of lamellae, is found among the three genera of Elephantidae (*Loxodonta*, *Elephas*, and *Mammuthus*), and in *Stegodon.*

Brain development and survival. Relative to the size of the brain, the temporal lobe in living and extinct proboscideans is larger, more convoluted, and denser than that of a human. In humans, functions attributed to the temporal lobe include recognition, storing, and retrieving of information related to senses of sight, touch, smell, and hearing. These are aspects associated with short- and long-term memory, which can allow elephants to find the shortest route to lead a herd to food sources in time of drought, or to use and make tools in order to survive.

Functions of proboscis. The proboscis is one of the most versatile organs among mammals. It is extremely dexterous, mobile, sensitive, and independent of the rest of the elephant. The trunk has two major sets of muscle, longitudinal (external) and transverse (internal), totaling about 150,000 fascicles (muscle units). These fascicles provide the fine tuning which allows the trunk to perform amazingly delicate functions, such as picking up a peanut, cracking it open on the back of the end of the trunk, blowing the shell away, and eating the kernel. The trunk has six or possibly seven functions: (1) breathing, (2) smell, (3) hormone detection in conjunction with the Jacobson's organ, (4) touch, (5) fifth hand, (6) sound production, and (7) possible seismic detection. Sound reception, especially the ability of proboscidean species to produce and perceive infrasonic calls not audible to human ears, is another pivotal change in proboscidean evolution.

Closest relatives. Morphological (soft and hard tissues) and molecular (amino acid and deoxyribonucleic acid sequences) characteristics provide evidence for close affinity between Proboscidea and Sirenia (manatees and dugongs). On superficial examination, there seems to be no reason to ally them, yet their carpal (wrist) bones are arranged one directly on top of another unlike those which are found in most mammals where the bones are staggered, one on top of two. This is only one of many unique (synapomorphic) morphological features uniting Proboscidea, Sirenia, Desmostylia (extinct marine herbivore mammal), and Hyracoidea (hyraxes). Another characteristic used in this analysis is the position and structure of the external auditory meatus, the opening where the soft part of the ear begins. One recent discovery that lends credence to Proboscidea–Desmostylia taxa relationships is a specimen of *Behemotops* from the Miocene Epoch of Washington in the northwest United States.

For background information *see* ANIMAL EVOLUTION; ELEPHANT; MAMMALIA; PROBOSCIDEA; SKELETAL SYSTEM in the McGraw-Hill Encyclopedia of Science & Technology.   Jeheskel Shoshani

Bibliography. R. Delort, *The Life and Lore of the Elephant*, 1992; H. F. Osborn, *Proboscidea*, vols. 1 and 2, 1936, 1942; A. S. Romer, *Vertebrate Paleontology*, 3d ed., 1966; J. Shoshani (ed.), *Elephants: Majestic Creatures of the Wild*, 1992; J. Shoshani and P. Tassy (eds.), *The Proboscidea: Evolution and Palaeoecology of Elephants and Their Relatives*, 1996.

Encephalitis

The arboviral encephalitides comprise several different families of arthropod-borne viruses (arboviruses) that cause encephalitis, an inflammation of the brain tissue in humans and vertebrate animals. They are transmitted from one vertebrate host to another by insects, ticks, or other arthropods. The biology and behavior of the arthropod and vertebrate hosts greatly influence the ecology and epidemiology of these viruses. Surveillance for arboviral encephalitis involves the organized monitoring of levels of virus activity, vector populations, infections in vertebrate hosts, human cases, weather, and other factors to detect or predict changes in the transmission dynamics of the infection.

Infectious agents. There are approximately 500 species of arboviruses throughout the world, the majority being from South America and Africa. More than 100 species have been isolated from humans, although not all cause serious disease. At least 44 arboviruses are known to infect livestock or other domestic animals. There are four common encephalitis viruses in the United States: La Crosse encephalitis, St. Louis encephalitis, and western and eastern equine encephalomyelitis. Venezuelan equine encephalomyelitis (VEE) occasionally moves northward into the southwestern United States. Several other viruses (Jamestown Canyon, Cache Valley, and Powassan) sometimes cause disease in humans or domestic animals in the United States.

Transmission cycles. Arbovirus transmission cycles are usually complex (see **illus**.). The normal cycle may involve several bird or mammal species and several mosquito species (or in a few cases, tick

species, not described in this article). Weather, food resources, predators and parasites, and space resources are a few of the external factors that influence the arbovirus transmission cycle. Enzootic virus transmission (transmission within the natural habitat) may occur at a low intensity among the normal vertebrate hosts and mosquito species within specific habitats in rural or suburban environments. Thus, transmission may remain undetected by most monitoring programs. However, when favorable weather or other environmental conditions lead to an abundance of both vertebrate hosts and mosquitoes and when few vertebrate hosts are immune to the virus, transmission may increase in intensity and expand in distribution, producing an epizootic (transmission to animals outside the natural habitat and host range). If epizootics begin early in the transmission season and if localized centers of epizootic activity (referred to as foci) expand into urban centers with adequate host and vector populations, the risk of human involvement increases. (When humans are involved, the term epidemic is used.)

Zoonotic diseases. Zoonotic diseases (diseases that normally occur in vertebrate hosts but can be transmitted to humans), and particularly the vector-borne zoonoses, are usually separated in time and space from the human population. This means the disease has usually already spread through the wild vertebrate population by the time the disease makes its way into human or domestic animal hosts. To be effective, however, prevention activities usually must target the zoonotic portion of the cycle, often at the level of the vectors. Thus, surveillance programs must monitor the zoonotic portion of the cycle. By the time that cases are diagnosed in humans, the epidemic will be well under way and no longer preventable. Even the smallest delays can diminish the impact of prevention or control measures on the course of a potential epidemic. Virus isolation and identification techniques are rapid, and new sampling methods can quickly define the vector situation. Still, these procedures require considerable time and effort.

Epidemics of encephalitis tend to be cyclic, recurring every 10–20 years. The long time lag makes it very difficult to prepare for the next occurrence. Without an adequate surveillance program, health agencies can be lulled into the assumption that these diseases have disappeared and are no longer a problem.

Surveillance. The process of surveillance can be divided into three phases: data collection, data entry and analysis, and information dissemination.

Data collection. In order to design a surveillance program, the planners must decide what kinds of data to collect, what kinds of methods and equipment to use, and how often and in how many locations to collect the data. They must also decide how they will integrate the surveillance data into the larger mission of health protection. It is preferable to have multiple measures (for example, weather, sentinel flocks, and mosquito surveillance) to get a more

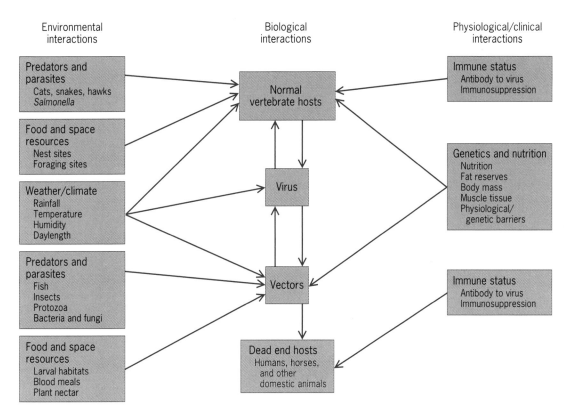

Generalized arbovirus cycle, showing biological and environmental interactions. *(After C. G. Moore et al., Guidelines for Arbovirus Surveillance in the United States, 1993)*

accurate estimate of potential for epidemics. A single measure may not give the full picture of virus activity. Two common surveillance methods are mosquito trapping and sentinel chicken flocks.

Specimen collection must be rapid, economical, and within the capabilities of the agencies that must do the work. Some methods, such as trapping and collecting blood samples from wild birds or other vertebrate hosts, require special training and state and federal permits.

Sampling bias refers to the selective nature of many sampling methods. Thus, mist nets capture only birds that fly into the nets; light traps collect only mosquitoes that are attracted to the light or other bait. Often, biased sampling methods may be used to collect specimens in a particular physiological state. For example, some mosquito traps primarily collect gravid (egg-bearing) females. Gravid females have taken at least one blood meal (since they need the blood to produce eggs), so they are more likely to have become infected with a virus. This can reduce the effort needed to detect virus activity in the community.

What specimens to collect, how often, and over how large an area are issues that depend as much on economics as on biology. Local surveillance programs must fit within the budgetary constraints of the local agency. The kinds of information useful in predicting arbovirus activity include environmental factors (rainfall, snow, temperature, humidity); vec-

tor populations (abundance, age, infection rates); wild (normal) vertebrate hosts (abundance, immune status, infection rates, age); incidental hosts (infection rates); and viral strain differences.

Collected specimens must be tested for evidence of encephalitis viruses. Mosquitoes are usually tested by isolating the virus. Collections of mosquitoes are separated by species and sex into pools of about 50–100 individuals. They are then ground up in a solution containing antibiotics and other ingredients. After centrifugation, the supernatant is inoculated into cell cultures or suckling mice. If a virus is present, cells or mice begin to die within about 2–7 days or more, depending on the virus. By adding a layer of agar that contains a selective dye to the cell cultures, the patches of dead cells can be seen as small, clear circles. When a virus has been isolated, it is identified by fluorescent antibody testing, neutralization testing, or several other methods.

Newer methods, such as antigen capture enzyme-linked immunosorbent assay (ELISA) and polymerase chain reaction (PCR), are growing in popularity as tools to detect virus in mosquito pools. These procedures often take less time than virus isolation, so disease control activities can be started more quickly if necessary. Because ELISA and PCR methods can detect viral particles even if they are not infectious, it is important to run the new tests in parallel with virus isolation for a sufficient period to establish a relationship between the two measures.

Arboviruses tend to be eliminated from the blood of vertebrates within a few days by the immune system. Therefore, virus isolation is not usually attempted from vertebrate hosts. Instead, changes in the immune system (production of antibodies) can be measured to determine if the animal has recently (or at some time in the past) been infected with a particular virus. Standard laboratory methods commonly used to detect antibodies in blood or serum include hemagglutination inhibition (HI), complement fixation (CF), plaque reduction neutralization (PRNT), and ELISA tests. ELISA tests can be used to detect both immunoglobins M and G (IgM and IgG). IgM antibodies usually disappear after a few months, while IgG antibodies often persist for many years. Thus, the presence of IgM can be a good indicator of recent infection, while IgG gives a measure of long-term virus activity in the host population.

Data entry and analysis. Data entry and analysis form a core part of the surveillance system, because they determine what information will eventually reach the end user. Data entry issues, such as staffing, timeliness, and uniformity, make the use of computers and computerized data formats economically attractive.

One of the most important (and most often omitted) data items is the denominator measure. It is crucial to know how many total birds were bled and the number that tested positive for the virus. Similarly, mosquito infection rates are usually expressed as the number positive per 1000 tested.

Information dissemination. The analyzed surveillance information should be quickly transmitted to those who can use it. When epidemics or epizootics occur, they usually cover several states. Thus, it is important to share data throughout a region so a rapid and effective response can be mounted at the earliest sign of major activity. Electronic data sharing is an important component of this rapid data exchange. The U.S. Centers for Disease Control and Prevention maintain the National Electronic Arbovirus Reporting System (NEARS), which provides rapid data exchange at the national level.

Surveillance information is useless unless it is used appropriately. Usually surveillance information is used to decide when to begin specific control measures. It can also be used to justify withholding or delaying control. Ideally, surveillance data will tell health planners if they are dealing with isolated hot spots or with widespread activity. It will also indicate what percentages of hosts and vectors are infected and which species of mosquitoes are most important. All of this information is needed to guide the prevention and control response.

At the first sign of encephalitis virus activity, the typical control response should begin with public service messages designed to reduce human-mosquito contact by encouraging people to use repellents, wear long-sleeved shirts and long pants, repair door and window screens, and avoid certain localities. Veterinarians should be contacted and re-

minded to vaccinate horses if equine encephalomyelitis viruses are active. As indicators become more severe, local agencies should begin mosquito control or other prevention activities.

For background information *see* ARBOVIRAL ENCEPHALITIDES; EXOTIC VIRAL DISEASES; IMMUNOASSAY; IMMUNOGLOBULIN; MOSQUITO; ZOONOSES in the McGraw-Hill Encyclopedia of Science & Technology.   Chester G. Moore

Bibliography. B. J. Beaty and W. C. Marquardt (eds.), *Biology of Disease Vectors*, 1996; C. G. Moore et al., *Guidelines for Arbovirus Surveillance in the United States*, 1993; M. W. Service, *Mosquito Ecology: Field Sampling Methods*, 2d ed., 1993; S. M. Teutsch and R. E. Churchill, *Principles and Practice of Public Health Surveillance*, 1994.

Endangered species

Biological diversity is being depleted worldwide through habitat loss, introduced species, overexploitation, and pollution. These conditions reduce population sizes and increase susceptibility to chance fluctuations (environmental, catastrophic, demographic, and genetic), and so increase the risk of extinction. Human intervention is required to ensure the survival of many species: 11.7% of mammals and 10.6% of birds are categorized as threatened species by the World Conservation Union (IUCN). A recent assessment from the Conservation Breeding Specialist Group of the IUCN classified 38% of vertebrates as threatened.

There are seven major genetic issues in conservation biology: inbreeding, accumulation of deleterious mutations, loss of genetic variation in small populations, genetic adaptation to captivity and its effect on reintroduction success, effects of outbreeding, fragmentation of populations, and taxonomic uncertainties. Since endangered species are unsuitable for experimentation (they are slow breeders, expensive to keep, and present in low numbers), genetic issues in conservation usually have to be resolved by experimentation with laboratory animals, or by combined analyses of data from many wildlife populations (meta-analyses). Laboratory animals, such as fruit flies or mice, have proven to be reliable genetic models for outbreeding wildlife species.

Inbreeding, genetic variation, and extinction. Small population size, and its consequences—inbreeding (matings between relatives) and loss of genetic variation (multiple alleles of many different genes)—are the most important genetic concerns in conservation. These factors increase the risk of extinctions. Over time, all individuals in a small population become related, so that matings between relatives are inevitable. Last century, Charles Darwin noted that inbreeding reduced reproduction and survival (called inbreeding depression) in naturally outbreeding plants. This process has subsequently been found to occur in domestic plants and animals. There was skepticism that wildlife in zoos suffered from inbreeding depression, but compelling evidence has been provided that captive wildlife does indeed experience such depression.

Skepticism about inbreeding depression for wildlife in nature has also been expressed. However, inbreeding depression in wild environments has been reported in fish, snails, lions, shrew, white-footed mice, golden lion tamarins, birds, and outbreeding plants. It is generally more severe in the wild than in captivity.

Inbreeding has been shown to cause extinction in laboratory and domestic animals and plants deliberately inbred by repeated brother–sister mating or by self-fertilization. Extinctions have also been observed with slow inbreeding due to small population size; 25% of 60 fruit fly populations with 67 parents per generation became extinct over 210 generations. Inbreeding in wildlife generally occurs as a consequence of the cumulative effects of modest population sizes over many generations.

Genetic variation is required for populations to evolve, through natural selection, in response to environmental change (for example, altered disease-producing organisms, or climate). Naturally outbreeding species normally have extensive genetic variation. This variation ultimately arises by mutation, and its level within a population is influenced by sampling (through population size), natural selection, and migration among populations.

Simple genetic theory predicts that more genetic variation is lost in smaller populations, and such loss is cumulative over generations. This theory has been verified in fruit flies, and a recent meta-analysis showed that genetic variation is related to population size in wildlife.

The genetic consequences of finite population size depend on the effective population size, rather than the number of individuals present (the census size). Effective size depends not only on the number of sexually mature adults but on the variation in family size, inequalities in sex ratio, and the fluctuations in numbers over generations. Estimates of the ratio of effective population size to actual size that include the effects of all relevant factors have recently been found to average only 11%, much lower than was previously assumed. This means that wildlife populations are genetically only about one-ninth as large as they appear to be.

In the wild, demographic and environmental fluctuations and catastrophes contribute to extinctions. There is considerable controversy as to the importance of these factors compared to genetic factors in wildlife extinctions. However, there is growing circumstantial evidence that genetic factors are involved. The most important evidence comes from island populations, which are particularly prone to extinction; most recorded extinctions are island populations, even though these are a minority of all populations. Many island populations are now known to be inbred, often to levels where laboratory and domestic animals show elevated risks of extinction. Furthermore, island populations typically have lower levels of genetic variation than mainland populations. It is probable that the impacts of inbreeding and loss of genetic variation often interact with environmental extremes and demographic fluctuations (in birth and death rates) to cause extinctions. Loss of genetic variation reduces the ability of wild populations to survive climatic extremes, pollutants, diseases, pests, and parasites. While the susceptibility of island populations to extinction has been interpreted as being due to nongenetic causes, it is most probably due to the interactions of genetic and nongenetic factors.

Endangered species frequently have less genetic variation than related nonendangered species. Inbreeding and loss of genetic variation have been implicated in the decline or extinction of wild populations of Florida panthers, Isle Royale wolves, deer mice, bighorn sheep, Puerto Rican parrots, heath hens, middle spotted woodpeckers, a topminnow fish, and inbreeding colonial spiders.

Minimizing the major issue. Inbreeding depression and loss of genetic variation in small populations can be overcome by immigration, that is, by introducing unrelated individuals from another population (if one exists). However, immigration occasionally reduces reproductive fitness (called outbreeding depression). Outbreeding depression is most common following crossing between isolated populations adapted to different local environments. It may be relatively common in plants, but it is less common in animals as they are more mobile and show less local adaptation. Outbreeding depression generally has a much smaller effect in animals than inbreeding depression.

Deleterious mutations. Deleterious genes (alleles) arise by mutation and are removed by natural selection. In small populations, some alleles become common by chance and reduce reproductive fitness. The accumulation of deleterious mutations has been documented in small, asexually reproducing populations, but its importance in sexually reproducing species is controversial.

Rare recessive deleterious mutations are exposed by inbreeding and so can be more effectively removed from inbred populations through natural selection. Such purging has been documented in many species. However, it lessens, rather than eliminates, the deleterious effects of inbreeding.

Captive breeding. Many endangered species are being bred in captivity in order to save them from extinction, with reintroduction into the wild usually being the desired goal. Three adverse genetic changes occur in captivity: inbreeding depression, loss of genetic variation, and genetic adaptation to the captive environment. Genetic adaptation to captivity (for example, tameness and poor predator avoidance) has been documented in fish, plants, bacteria, and several fruit fly species; it is generally disadvantageous upon return of the population to the natural environment.

Genetic management. Genetic principles have been applied to the management of endangered species, especially in captivity. The objectives are to minimize inbreeding, loss of genetic variation, and genetic adaptation to captivity. This management can be done by using an adequate number of founders (20–30), ensuring that each founder contributes equally to the following generations, equalizing the number of offspring per family used for breeding, and minimizing the number of generations in captivity. Unfortunately, many captive populations of wildlife have been founded when only small numbers are left, thus compromising their genetic future. For example, in the United States the captive population of Speke's gazelle was founded from one male and three females.

Minimizing kinship is being widely applied to endangered species management, as it is considered to be the best procedure for managing small pedigreed populations with unequal founder contributions. Parents are chosen so that the expected inbreeding in their offspring is minimized. It combines the benefits of equalizing family sizes and adjusting founder representation. Experiments in fruit flies have verified the predicted benefits of minimizing kinship in retaining genetic variation.

Cryopreservation. Cryopreservation has the potential to prevent deleterious genetic changes in captivity. Embryo and semen freezing technology is being used in domestic and laboratory mammals, but is not yet available for most wildlife.

For background information *see* ALLELE; EXTINCTION (BIOLOGY); POPULATION ECOLOGY; POPULATION GENETICS in the McGraw-Hill Encyclopedia of Science & Technology.   R. Frankham

Bibliography. R. Frankham, Conservation genetics, *Annu. Rev. Genet.*, 29:305–327, 1995; World Conservation Monitoring Centre, *Global Biodiversity: Status of the Earth's Living Resources*, 1992.

Environmental noise

Some individuals in and around Taos, New Mexico, perceive an irritating low-frequency sound that has become known as the Taos hum. There have been persistent complaints of annoying low-frequency sounds in the United States and elsewhere, but the group in Taos has been particularly outspoken about the perception. Based on interviews with these hearers, it has been possible to develop a behavioral profile associated with perception of the Taos hum: (1) The hum is selective; only a small percentage of Taos residents perceive it. (2) The hum is persistent since most hearers perceive it on a weekly basis. (3) The source of the hum must be widespread since hearers perceive it throughout the Taos area, and individuals from disparate regions in the United States have reported a similar phenomenon. (4) The hum is invariably described as low frequency and reminiscent of a distant pump, an idling diesel truck, or a high-powered bass speaker when receiving no audio signal. Musicians identify the hum as a modu-

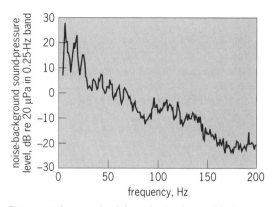

Representative sample of the noise background in the frequency range 1–200 Hz recorded in a rural home in Taos, New Mexico, 7000 ft (2100 m) above sea level, by using a custom microphone for detecting low-frequency sound.

lated tone near E_b (41 Hz), and those with some technical background identify it as an amplitude-modulated sound with a carrier frequency ranging from 30 to 80 Hz.

Matching tests of Taos hum. Psychophysical matching tests were made with 10 hearers to quantify the attributes of the hum. To keep the tests simple, a modulated sine-wave source and a low-frequency speaker were used to generate matching sounds. A custom microphone made from an 18-in. (45-cm) woofer served as an ultrasensitive low-frequency sound detector (able to detect sound-pressure levels of −60 to −80 decibels with respect to a reference pressure of 20 micropascals, from 8 to 80 Hz). Eight of the hearers were able to generate a sound that matched the hum they were perceiving. Individual hearers matched the same sound to the perceived hum quite reproducibly. Surprisingly, many hearers reported beats between the speaker-generated sound and the hum, implying the physical existence of a second tone.

The results of these matching tests were both intriguing and puzzling. The carrier frequencies selected by the hearers to match the hum ranged from 32 to 80 Hz. The modulating frequencies they selected ranged from 0.5 to 2 Hz. All hearers selected amplitude levels near or below the known sensitivity of the ear in this frequency range. The most striking observation was that hearers could still perceive the hum in the presence of a relatively flat noise spectrum when the matching signal was turned off. The **illustration** shows a representative sample of the noise background in the frequency range from 1 to 200 Hz recorded in a rural Taos home at 7000 ft (2100 m) above sea level by using a custom microphone for detecting low-frequency sound. The noise amplitudes, even when converted to one-third octave bands, are still 15–20 dB below accepted hearing thresholds in this frequency range. Therefore, the source of the hum does not appear to be noise in the environment.

External nonacoustic sources. Additional acoustic, geodynamic, and electromagnetic measurements were made at other locations in and around Taos to track down the origin of a strong harmonic component in the 60-Hz power grid around Taos. At first glance this provided a possible source for the hum, but hearers agreed almost universally that the perceived sound was unusually intense at a very remote site near Tres Piedras, where little trace of the 60-Hz magnetic component could be found.

Measurements made with magnetometers in an electromagnetically quiet area showed the Schumann resonances created by lightning-induced electromagnetic radiation trapped in the spherical waveguide formed between the ionosphere and the Earth. The observed resonant frequencies of this waveguide cover the frequency range implicated in the Taos hum, but the Schumann resonances appeared to be normal in amplitude and nothing unique could be found about them that might explain the hum.

Taos is a relatively quiet seismic area, so geophone data showed nothing but background noise. An electrostatic field detector recorded many events, but all could be attributed to lightning strikes or movements of people in the vicinity. Electromagnetic data were collected from 20 Hz to 18 GHz by using an assortment of antennas and receivers. Everything that was found could be attributed to 60-Hz power lines, electromagnetic noise, or known radio and television transmitters.

The extremely low frequency (ELF) submarine tracking system operated by the U.S. Navy is considered by some hearers to be the source of the Taos hum. Consequently, the electromagnetic spectrum recorded in Taos was examined carefully for evidence of peaks in amplitude in the 65–75-Hz region where extremely low frequency signals would be prominent. None were found above the noise in this frequency region. Considering the spread spectrum of the extremely low frequency broadcast, the limited antenna length, and the fact that the broadcast originates from transmitting sites in the Michigan peninsula and in Wisconsin, it is unlikely that the signal would be above the noise level in northern New Mexico.

Possible causes. The study leaves a mystery. There are no acoustic signals that might account for the Taos hum, nor are there any seismic events that might explain it. There are no unusual lines at suspect frequencies in the electromagnetic spectrum recorded near Taos. In fact, other than the signals generated by the power grid or, in the case of the golf course, a power generator located at the course headquarters, no clear lines at all were found in this spectrum. At a more fundamental level, there is no known mechanism whereby weak electromagnetic signals might be transduced selectively by the human ear and perceived as sound. The next step, therefore, was a shift in emphasis from the physical environment to the hearers themselves.

It has been proposed that the hum might represent some kind of low-frequency tinnitus arising from a disturbance in the cochlea of the inner ear. One problem with such explanations is that tinnitus is not well understood. Most individuals with tinnitus match the tone they perceive to frequencies between 3 and 6 kHz, and rarely if ever does a tinnitus sufferer match to a tone below 1 kHz. It is not clear why such a phenomenon should skip from regions of the cochlea where 3–6-kHz tones are represented over intermediate zones to the extreme apical end of the cochlea where the lowest frequencies of sound are represented. Furthermore, many hearers insist that they can hear beats between the perceived tone and the objective tone generated in matching experiments. Beats have not been demonstrated conclusively in matching experiments using subjects with tinnitus.

Survey of hearers. To establish what fraction of the people in the Taos region are affected by the hum and to probe the characteristics of their perception, survey forms were sent to 8000 residents in Taos and Ranchos de Taos and the neighboring communities of Tres Piedras, Questa, Eagle Nest, and Peñasco. These communities represent points to the west, north, east, and south of Taos on a circular perimeter roughly 40 mi (65 km) in diameter with Taos at its center. A total of 1440 residents in these areas responded to the survey, and 161 respondents were classified as hearers. Seventy-two percent of the hearers are 30–59 years of age, 52% are male, and 47% are female. The hum is perceived by 62% of hearers between 8 p.m. and 9 a.m. About 80% of the hearers perceive the hum at least once week, and a hearer's first experience with the hum usually occurs in the immediate vicinity of Taos. Nineteen percent of the hearers perceive the hum at sites more than 50 mi (80 km) from Taos.

If it is assumed that all hearers responded to the survey, the lower limit of the fractional number of hearers can be estimated to be about 161/8000, or approximately 2% of the population. If this ratio holds for other populations in remote or quiet rural areas, the number of individuals who perceive an annoying low-frequency noise would be expected to be very substantial in the United States and, by implication, in other parts of the world.

Annoyance sounds in Great Britain. The subject of annoyance caused by low-frequency sound has attracted much attention in Great Britain since the mid-1960s. Although potential acoustic sources for low-frequency hums were identified in some studies, many unexplained hums were found, and the sources remain as elusive as those in Taos.

For background information *see* ACOUSTIC NOISE; HEARING (HUMAN); HEARING IMPAIRMENT; NOISE MEASUREMENT in the McGraw-Hill Encyclopedia of Science & Technology.   Joe H. Mullins; James P. Kelly

Bibliography. S. D. Arlinger, Normal hearing threshold levels in the low-frequency range determined by an insert microphone, *J. Acoust. Soc. Amer.*,

90:2411–2414, 1991; B. Berglund and P. Hassmén, Sources and effects of low-frequency noise, *J. Acoust. Soc. Amer.*, 99:2985–3002, 1996; L. J .Sivian and S. D. White, On minimum audible sound fields, *J. Acoust. Soc. Amer.*, 4:288–321, 1933; R. E. Walford, A classification of environmental "hums" and low frequency tinnitus, *J. Low-Freq. Noise Vibrat.*, 2:60–84, 1983.

Enzyme

Harnessing the precision and efficiency of enzymes for catalysis of chemical reactions holds great promise for many industries. Enzymes are proteins, linear chains of amino acids that fold into complex three-dimensional structures. They catalyze the myriad chemical reactions in all living organisms. Because enzymes have a strong preference for acting on one particular substrate, they require less starting material and produce less waste than many conventional catalytic processes. In addition, they can catalyze intricate reactions which are inaccessible by any other method.

Enzymes already have a wide range of applications, including chemical synthesis, biodegradation of harmful chemicals, environmental and medical diagnostics, medical therapeutics, food manufacturing, detergents, and agriculture. However, a few obstacles keep the full industrial potential of enzymes from being realized. The current understanding of how proteins fold is too limited to allow enzymes to be designed from scratch in the laboratory. They do not exhibit certain features that are desirable from an industrial standpoint, such as long-term stability or the ability to function in artificial environments. It is sometimes possible to modify natural enzymes to make them more useful by changing their amino acid sequences. However, it has so far proven extremely difficult to predict what structural change a given modification will produce, and how that change will affect the protein's function.

Directed molecular evolution. Directed molecular evolution is a method of generating custom-made enzymes from natural enzymes without knowledge of the latters' structure. This technique begins with the collection of an assortment of randomly mutated genes for an enzyme. Then, bacteria or bacteriophages expressing the genes are selected or screened for desired characteristics. Improved genes are subjected to further mutation and selection or screening until a satisfactory enzyme is created. This technique allows the engineer, rather than nature, to specify the criteria for the enzyme's characteristics. New, more useful enzymes—such as ones that act on modified substrates, or that are active under a new range of conditions such as organic solvents, high temperatures, or lengthened processing or storage times—have been developed with this technique.

Current applications. One example of the many applications which directed molecular evolution has facilitated is synthesis of specialty chemicals such as pharmaceuticals or materials with novel properties.

Subtilisin. The evolution of the enzyme subtilisin, a serine protease, which was optimized for polymerization of natural and synthetic amino acids, has resulted in an enzyme that is 10 times more active in the organic solvent dimethylformamide (DMF) than the naturally occurring enzyme is in water, the solvent in which naturally occurring enzymes are most active. The new enzyme is useful for performing reactions which require a polar aprotic environment to dissolve one of the reactants or to assist the reaction.

Antibiotic synthesis. An enzyme that was evolved for a deprotecting step in a cephalosporin-based antibiotic synthesis may eliminate the traditional deprotection method. That method involves catalytic zinc in organic solvents and generates large amounts of solvent and zinc-containing waste material. The enzyme that was chosen for this evolution had been isolated from a microorganism because it displayed a small amount of activity on the industrial substrate (the substrate which needs to be modified in the industrial process; in this case, an intermediate in the synthesis of an antibiotic). The natural substrate for the enzyme was obviously different from the industrial substrate for which it was optimized by directed evolution. (The natural substrate is transformed by the enzyme in the organism.) This indicates that a subtle change in substrate specificity was generated along with the increase in activity.

This example illustrates a unique strength of directed molecular evolution. An examination of the beneficial amino acid substitutions accumulated during the evolution shows that these changes were two or three layers of amino acids away from the active site, the part of the binding region of the enzyme that performs the catalysis. This suggests that changes immediately adjacent to the active site are too strong, and attests to the difficulty of finding the necessary subtle adjustments by any rational design method.

Mutagenesis methods. The directed evolution technique requires efficient methods of creating large numbers of randomly mutated proteins. Each protein is uniquely encoded by a gene, or sequence of deoxyribonucleic acid (DNA) bases. Several methods exist for introducing random mutations into DNA sequences. These methods derive from the explosion of recombinant DNA technology, which has accelerated the pace of advances in all areas of protein engineering. Some methods involve exposing DNA to various chemicals which react with nucleotides and change the latters' structures. Other methods generate changes in DNA sequences by using the error-prone polymerase chain reaction technique.

Error-prone polymerase chain reaction. The polymerase chain reaction (PCR) technique has revolutionized molecular biology by providing a convenient method of creating millions of copies of a single gene. First, a DNA polymerase, which synthesizes a complementary second strand to a single strand of DNA, is activated at a moderate temperature. Next, the temperature is raised so that all of the new double-stranded DNA strands melt apart into two single strands, and then is lowered again to allow polymerase to synthesize new complementary strands. Finally, a series of heating and cooling cycles is performed until a very large number of copies of the original DNA has been made. Error-prone polymerase chain reaction uses conditions that cause mistakes to be made during the DNA replication process at a controllable rate, generating random mutations in the gene being amplified.

DNA recombination. In 1994, a new technique for directed evolution, known as DNA recombination or sexual polymerase chain reaction, was announced. This method allows the shuffling together of fragments of genes from any number of parents, for example, a set of improved variants. The idea behind DNA shuffling is to fragment genes containing known or potential beneficial mutations and to put them back together randomly so that all possible combinations of the original mutations can be generated and screened efficiently. To do this, DNA is first digested by the enzyme DNAse I, which cuts DNA bluntly at random locations, creating fragments about 200 base pairs long. Then, the fragments are subjected to a polymerase chain reaction in which the overlap of each fragment with a fragment from another gene causes both fragments to be elongated by a DNA polymerase as if the overlapping fragment were a primer. When the elongated fragments separate during the heating cycle, they can again anneal to overlapping fragments from other genes and become further elongated. This reaction continues until most of the DNA has been reassembled to the length of the original gene, with each complete gene containing fragments from different parent genes. This mixture can then be amplified by using a conventional PCR to yield a pool of DNA in which different combinations of the mutations contained in the parent sequences are represented.

Natural mechanisms. DNA recombination approximates natural mechanisms of sexual evolution. Sexual evolution overcomes many of the drawbacks of asexual evolution. In asexual evolution, the best variant in each generation is chosen to begin the next generation; other improved variants are set aside, and their mutations must be rediscovered in another generation in order to become incorporated. Moreover, because linear evolution provides no mechanism for deleting mutations other than random back mutation, deleterious mutations can accumulate in the evolving molecules. Finally, since each beneficial mutation confers a smaller improvement if it must be detected in a mutant already containing a number of beneficial mutations than if it is the first beneficial mutation, recombination of several mutations discovered individually is more efficient than cumulative discovery of mutations.

Screening techniques. The utility of directed evolution relies on the availability of a method for rapidly detecting beneficial changes incurred during the evolution. Two paradigms for the detection of these beneficial changes exist: natural selection and screening. Natural selection requires that the survival of the host organism be linked to the desired enzymatic trait. Many cases of natural selection exploit the breakdown of a compound which the bacteria can be forced to use as a carbon source; then it is possible to choose conditions under which only those organisms using the substrate for food most efficiently will survive.

However, in most cases it is impossible to devise a natural selection. In general, each desired change in activity requires the development of a creative new rapid assay, called a screen. For example, a screen can involve a subtle change in the substrate which allows a chromophore, or colored chemical group, to be generated when the desired enzymatic cleavage occurs. This type of assay has been used in deprotecting enzyme experiments; the structure of the substrate used for the screening has to be modified slightly from the structure of the goal substrate in order to facilitate the assay. The industrial substrate required the cleavage of a *para*-nitrophenyl ester from the rest of the antibiotic, but this reaction could be observed only by high-performance liquid chromatography (HPLC), a method too laborious for a large-scale screen. To overcome this difficulty, a single carbon atom was added to the substrate, so that the enzyme instead cleaved off a *para*-nitrobenzyl ester, colored bright yellow. Each colony to be screened was transferred into one well of a 96-well plate. The intensity of the yellow color, which correlated to the level of activity of the cleaving enzyme, was read by a plate reader. The modified substrate was close enough to the desired substrate that the screen yielded mutants which had increased activity on the real substrate.

Many other screening methods have been used, including visual screening of fluorescent protein mutants, comparison of the sizes of visible halos of product formed by a reaction on a plate containing growth medium, or use of antibodies specific for the desired product and a labeled antigen.

For background information *see* BIODEGRADATION; BIOTECHNOLOGY; CATALYSIS; DEOXYRIBONUCLEIC ACID (DNA); ENZYME; GENE; MUTATION; ORGANIC SYNTHESIS; PROTEIN in the McGraw-Hill Encyclopedia of Science & Technology.　　　　Olga Kuchner

Bibliography. K. Chen and F. H. Arnold, Tuning the activity of an enzyme for unusual environments: Sequential random mutagenesis of subtilisin E for catalysis in dimethylformamide, *Proc. Nat. Acad. Sci. USA*, 90:5618–5622, 1993; J. Maynard Smith, The century since Darwin, *Nature*, 296:599–601, 1982; J. C. Moore and F. H. Arnold, Directed evolution of a *para*-nitrobenzyl esterase for aqueous-organic solvents, *Nat. Biotech.*, 14:458–467, 1996; W. P. C. Stemmer, DNA shuffling by random mutagenesis and reassembly: *In vitro* recombination for molecular evolution, *Proc. Nat. Acad. Sci. USA*, 91:10747–10751, 1994.

Epidemiology

The development of agriculture during the last 2000 years, combined with technological advances in the past 500 years, have enabled humans to permanently convert large areas from natural ecosystems into mosaics of monoculture cropland interspersed with remnant natural habitats. In both North and South America, the written notes of the first explorers and settlers provide evidence that the native New World inhabitants maintained extensive landscapes by using fire and agriculture.

Although disease is a fact of human evolution, large epidemics or pandemics of diseases such as plague, influenza, and smallpox did not occur until technological advances in travel (such as shipping) enabled humans to move great distances relatively quickly. Today, increasingly dense populations and international migrations of humans facilitate rapid spread of emerging infectious disease among people on separate continents.

Emerging infections. An emerging disease or emerging infection is one in which its incidence and prevalence in the human population has recently increased or threatens to increase in the foreseeable future. Three main categories of emerging diseases have been identified: (1) The virus, bacteria, or protist may jump species boundaries and invade humans after evolving in other organisms or other locations. For example, the bacterium *Legionella*, which causes Legionnaire's disease, may infect humans in air-conditioned environments, and there is strong evidence that the human immunodeficiency viruses HIV-1 and HIV-2 originated in populations of monkeys in Africa with subsequent dispersal in humans. (2) A disease may be recognized as having an infectious origin even after being established in the human population for some time. Lyme disease is an example; cases reported now may have gone unreported in the past. (3) A disease may be caused by reappearance of a known agent after a period of decreasing prevalence in the population. Tuberculosis is an example.

Only about 10% of all existing species of organisms have been described. The diversity of viruses, bacteria, and protists (the most important agents of disease in humans) is almost unknown. From 1973 to 1995, more than 30 agents of human disease have been reported, with more appearing each year. This paucity of information on number of potential pathogens and on global biological diversity, and the speed at which new diseases are appearing make it clear that previously unknown diseases and unrecognized infections will continue to emerge in populations of both humans and domestic animals. *See* INFECTION.

Biodiversity. Biological diversity, or biodiversity, is the diversity of life from the level of the molecule to the ecosystem. Studies of biological diversity include investigations into the genetics, taxonomy, and ecological variety in living or fossil organisms. Biodiversity can be considered as the complete assemblage of species in a defined geographic area, with the area being local, regional, or global. Biodiversity includes life from the level of the genetic sequence of the smallest virus to the species composition of the largest plants or the most freely moving mammals.

Detailed knowledge of the diversity of life is limited to geographic areas that have been studied for relatively long periods of time; however, even for the areas that are most well known, knowledge of the smaller animals is far from complete. The microbes, arthropods, and nematodes of most areas are still awaiting taxonomic description, identification, and inventory. There have been recent proposals to perform complete biological inventories of specific areas of the Earth. Some inventories are already under way at one or more of the Long Term Ecological Research (LTER) sites located in various areas of the United States. In these intensively studied research sites, correlations in prevalence of potential or actual disease, climate change, and diversity of life are being recognized and investigated.

Protists and bacteria. The ability of systematic biologists to characterize agents of potential or actual disease in humans has been shown to be extremely important in the identification of three diseases transmitted by ticks that have emerged in the United States since 1975: human babesiosis (first recognized in 1969), caused by *Babesia microti*; human ehrlichiosis (first recognized in 1989), caused by *Ehrlichia* spp.; and Lyme disease (first recognized in 1982), caused by *Borrelia burgdorferi*. *See* EHRLICHIOSIS.

Increased numbers of individuals with Lyme disease and human babesiosis in the United States have been correlated with the greater numerical density of deer. Deer populations in the eastern United States are at record-high levels, as are their ectoparasitic ticks (*Ixodes dammni* in the northeast and *I. scapularis* in the southeast). In two-host ticks, the adults feed on deer and humans, and the nymphs feed on field mice. The ticks are the vectors for the three diseases mentioned above. The incursion of humans into areas frequented by deer and mice predisposes them to tick bites and subsequent infection by protists and bacteria vectored by the ticks.

Although systematists were able to assign names to the ticks responsible for transmission of the disease, response to these emerging protist and bacterial infections was hampered by insufficient knowledge of the distribution and biological diversity of the mammals, and of their arthropod and protistan parasites. These organisms are not studied sufficiently, and this problem is exacerbated by a decrease in numbers of systematic biologists with expertise in this area. Complete knowledge of the biodiversity of the eastern United States would enable health officials to more easily identify disease agents and implement more successful control measures.

Viruses. The outbreak of Sin nombre virus in the southwestern region of the United States is being studied intensively by researchers at the University of New Mexico and the Centers for Disease Control and Prevention in Atlanta, Georgia. Strong correlations among increased rainfall amounts, increased seed production, and increased numerical density of deermice (*Peromyscus maniculatus*, a terrestrial mouse species that is the host of Sin nombre virus) have been identified directly because of the biodiversity studies taking place on the Sevilleta Long Term Ecological Research site in New Mexico. Yearly monitoring of climate and rodent populations on the Sevilleta site has shown that in the year prior to the outbreak of Sin nombre virus in the southwest, there was a corresponding tenfold increase in numerical density of *P. maniculatus* throughout the region. Although Sin nombre virus was unknown in humans in the United States before 1993, material collected from *P. maniculatus* during the course of studies of the mammalian diversity of the southwest is presently being studied. Preliminary results indicate that Sin nombre virus has existed in *P. maniculatus* in the New Mexico region since 1974. Mice and their tissues, stored in ultracold freezers in the Museum of Southwestern Biology, University of New Mexico, were collected as a result of studies of mammalian biodiversity. The fact that the material was collected and stored in a well-curated museum and was later accessed as part of a historical study provides strong evidence of the importance of biological collections in studies of biodiversity.

Human immunodeficiency viruses 1 and 2, the causative agents of acquired immune deficiency syndrome (AIDS), are thought to have originated in central Africa. Present evidence indicates that a species jump by HIV-1 and HIV-2 from nonhuman primates to humans occurred. International transportation of monkeys for use in medical research and the very fast movement of infected humans from extremely remote areas to centers of dense population are also thought to have played a role in the rapid spread of AIDS. The construction of a road from Mombasa, Kenya, to Kinshasa, Zaire, probably accelerated the spread of the viruses into centers of dense human population.

Little information is available regarding the biological diversity of mammals and their symbiotic viruses and parasites in central Africa. Development of vaccines or control measures to be used against newly emerging viruses or other diseases in human populations would be accelerated if researchers had prior knowledge of the existence of these actual or potential agents of disease in animal populations. *See* EBOLA VIRUS.

Biotechnology and bioprospecting. The advent of molecular methods in biotechnology, especially in the area of studies of the genetics and phylogenetic relationships among organisms, has empowered biologists with completely new tools. Very small amounts of material collected during field-oriented studies of biological diversity (such as single insects, other minute arthropods, or small amounts of tissues from an organism with a viral or protistan infection) can now be studied. Advances are rapidly occurring in

molecular biological methods, especially in technologies utilizing deoxyribonucleic acid (DNA) for identification of species or strains. Systematic biologists are now using techniques such as the polymerase chain reaction (PCR) for DNA sequencing and DNA fingerprinting to identify, characterize, and describe new and emerging diseases. Collaborations among scientific institutions are occurring, with important data being quickly shared over national and international computer networks.

Significant effort is also being focused on identification of useful chemical compounds that occur naturally in plants, fungi, and animals. Discoveries from bioprospecting surveys include identification of the anticancer properties of Taxol from the Pacific yew tree (*Taxus brevifolia*) in the Pacific Northwest, and Vinblastin and Vincristine from the Madagascar periwinkle (*Catharantus roseus*) in Madagascar.

Global environmental change resulting from human activity (agriculture, industry, and urbanization) is accelerating the loss of biological diversity worldwide. This loss will profoundly affect the ability of biotechnologists to utilize the vast natural resources which are storehouses of potential biomolecules that are now being used and have the potential to be used to improve food crops, develop new drugs, or combat both urban and agricultural pests.

For background information *see* ACQUIRED IMMUNE DEFICIENCY SYNDROME (AIDS); BACTERIA; BIOTECHNOLOGY; DISEASE; EPIDEMIOLOGY; INFECTION; LEGIONNAIRE'S DISEASE; LYME DISEASE; PROTISTA; PUBLIC HEALTH; VIRUS in the McGraw-Hill Encyclopedia of Science & Technology.   Scott L. Gardner

Bibliography. S. S. Morse (ed.), *Emerging Viruses*, 1993; L. S. Roberts et al., *Foundations of Parasitology*, 5th ed., 1995; M. E. Wilson, R. Levins, and A. Spielman (eds.), Disease in evolution: Global changes and emergence of infectious diseases, *Ann. N.Y. Acad. Sci.*, 740:1–503, 1994.

Equine protozoal myeloencephalitis

Equine protozoal myeloencephalitis (EPM) is a treatable but frequently debilitating central nervous system disease of the horse. It is probably the most commonly diagnosed equine neurologic disease in the Western Hemisphere and results in considerable economic and personal loss. Although nervous disorders of the horse have been known for centuries, the unique pathology associated with EPM was not recognized until the early 1960s. The disease was initially known as segmental myelitis and was found in horses from the northeastern United States. Subsequent cases have been reported among native horses throughout North America, as well as in Central and South America.

Infectious agent. During the early 1970s, a protozoal parasite was found in damaged areas of the central nervous system, and the disease became known as equine protozoal myeloencephalitis. By the mid-

1980s, the organism was isolated from the spinal cord of an affected horse, cultured in the laboratory, and named *Sarcocystis neurona*. The availability of cultured parasites permitted the development of an antemortem diagnostic test using immunoblot analysis in 1991. Sequence analysis of the small subunit ribosomal ribonucleic acid (rRNA) gene of *S. neurona* was used to develop a parasite-specific deoxyribonucleic acid (DNA) assay which utilizes the polymerase chain reaction (PCR) in 1993. The PCR assay was used to tentatively solve the life cycle of the parasite. The small subunit ribosomal RNA gene sequence of *S. neurona* matched that of both *S. falcatula* sporocysts isolated from opossum feces and *S. falcatula* sarcocysts from bird muscle.

Epidemiology. In 1988 affected North American horses ranged in age from 2 months to 19 years, but over 60% were 4 years old or less. Thoroughbreds, Standardbreds, and Quarter Horses were most often affected, although many other breeds and ponies were represented. No geographic or seasonal predilection was determined. Clinical testing indicates that EPM occurs most frequently in the eastern half of the United States and along the west coast. Interestingly, this pattern mirrors the distribution of the opossum population in North America.

The *S. neurona* exposure rate among horses appears to be much higher than the incidence of disease. In 1978, reports indicated that 25% of equine neurologic disease submissions were due to EPM. The number of horses diagnosed with EPM in Lexington, Kentucky, has averaged 8-9% of all neurologic cases over the last several years. The number of horses diagnosed at postmortem represents a small portion of the total number of horses affected, and thousands of cases probably occur in the United States annually. However, this represents a small percentage of the 6 million horses in North America.

Life cycle. Members of *Sarcocystis*, like many other genera of coccidia, have an obligatory, predator-prey or scavenger-carrion life cycle. The parasite must cycle between two hosts to survive. The host range of most *Sarcocystis* species is narrow. Oocysts are produced by sexual reproduction or gametogony in the intestinal lining of the appropriate predator (definitive host). The oocyst wall usually ruptures prior to passage in the feces, thus releasing two infective sporocysts into food and water of the prey animal (intermediate host) by fecal contamination. Birds and insects may serve as passive transport hosts by ingesting sporocysts and disseminating them in the environment.

Once ingested by the appropriate intermediate host, sporocysts excyst, releasing four sporozoites which penetrate the gut and enter the cells lining blood vessels (vascular endothelial cells) in many organs. A small group of parasites (called a meront) forms rapidly by asexual division (merogony) from each sporozoite. The host cell eventually ruptures, releasing new merozoites into the bloodstream. This is followed by a second round of merogony in vascular endothelial cells throughout the body. Second-generation merozoites are released in the bloodstream and may undergo a third merogonous cycle in vascular endothelial cells or may directly enter skeletal muscle cells, where they develop into specialized meronts known as sarcocysts. Mature sarcocysts contain bradyzoites which complete the life cycle when ingested by the appropriate definitive host. *Sarcocystis fayeri* uses this method to cycle between horses and dogs and usually produces little pathology in either host. Estimates of the prevalence of *S. fayeri* in North America are 30% (personal communication). Sarcocysts of *S. neurona* apparently are unable to develop in affected horses, thus blocking transmission of the parasite to the definitive host. The horse is considered an aberrant, dead-end host for *S. neurona*.

Sarcocystis falcatula cycles normally between the opossum and many species of birds. The apparent identification of the parasite life cycle permits the development of an infectivity model to study many of the important questions regarding EPM. Following ingestion of *S. neurona (S. falcatula)* sporocysts, excysted sporozoites penetrate the equine intestinal tract and ultimately reach the central nervous system. It is unknown if merogonous generations occur in vascular endothelial cells outside the central nervous system of the horse. However, it is clear that merozoites must penetrate the vascular endothelium which forms the blood–brain barrier to reach the central nervous system, although the precise mode of entry has not been determined.

The minimum incubation period from sporocyst ingestion to the development of clinical signs is not known. The youngest horse confirmed with EPM was a 2-month-old foal. Transplacental infection has not been reported, but it cannot be completely ruled out. If transplacental transmission does not occur, the minimum incubation period may be less than 8 weeks.

Clinical signs. The clinical signs of EPM are highly variable and directly referable to the site of parasite infection within the central nervous system. Clinical onset may be rapid or prolonged with mild to severe signs that may progress without treatment. Although EPM may mimic virtually any equine neurologic disease, the most common clinical signs have been asymmetric posterior ataxia (incoordination) and weakness. Impaired conscious proprioception (the ability to sense limb position), muscle atrophy, and various signs associated with cranial nerve deficits (facial paralysis, local muscle atrophy, difficulty in chewing or swallowing) frequently occur. Seizures have been reported in horses with histopathologic confirmation of infection.

Diagnosis. Immunoblot testing of equine serum and cerebrospinal fluid samples provides veterinarians with valuable information about exposure to *S. neurona*. The test utilizes cultured merozoites to detect antibodies produced in response to proteins believed to be unique to *S. neurona*. Antibodies directed against proteins shared by *S. fayeri* or other organisms are differentiated. A positive serum test (specific antibody present) indicates exposure only. However, positive cerebrospinal fluid indicates that parasites have penetrated the blood-brain barrier and stimulated a local immune response. A false positive cerebrospinal fluid test could occur if a detectable immune response within the central nervous system successfully eliminated the parasite prior to the development of clinical signs. However, if this occurred routinely, a higher rate of false positive cerebrospinal fluid test results would be expected. Nonetheless, this possibility makes it difficult to interpret positive EPM test results for cerebrospinal fluid samples collected from clinically normal horses. The test is not recommended for use in clinically normal horses until more is known about host–parasite interaction.

Horses with acute neurologic disease that initially test negative should be retested. Some horses simply fail to respond adequately to the specific proteins evaluated. Similarly, chronically affected horses may test negative. However, it is expected that horses which initially test positive will test negative following appropriate treatment.

Parasite-specific PCR testing of cerebrospinal fluid also provides information regarding the presence of *S. neurona* in the central nervous system. Although the sensitivity of PCR testing is relatively low, the demonstrated ability to detect false negative western blot test results makes it a useful adjunct for clinical diagnosis of EPM.

Treatment. Treatment of EPM has been somewhat empirical. Several treatment protocols have been used with variable success. However, prompt, aggressive therapy with appropriate drugs has been reported to result in good clinical response in up to 70% of the horses treated. Treatment recommendations include oral administration of sulfadiazine (20 mg/kg) once or twice daily and pyrimethamine (1.0 mg/kg) once daily. It is advisable to withhold feed (especially grass or hay) for a few hours before and 1 hour after drug administration to avoid competitive inhibition of absorption from the intestinal tract.

Anti-inflammatory drugs are also recommended during the initial phase of the disease. Inflammation within the central nervous system is probably responsible for many of the early clinical manifestations of EPM. Dimethyl sulfoxide, flunixin meglumine, and phenylbutazone have been helpful during the first few days of treatment. Corticosteroid use should be avoided, if possible, because of immunosuppression. Vitamin E supplementation (8000–9000 IU/day) may also have an anti-inflammatory effect and may promote healing of damaged nervous tissue. Folic acid supplementation is also highly recommended for pregnant mares on treatment.

The average length of time required for successful treatment is approximately 12–16 weeks. However, some individuals may require a considerably longer period of time. If treatment is discontinued prematurely, horses may relapse days, weeks, or even months later. This probably represents the emergence of a small, persistent focus of infection that was not eliminated by initial treatment. When relapse occurs, a full treatment regimen should be implemented. A higher dose of the appropriate medications for a longer period may be required to eliminate the infection.

For background information *see* HORSE PRODUCTION; INFECTION; NERVOUS SYSTEM DISORDERS; NEUROBIOLOGY; PARASITOLOGY; VITAMIN E in the McGraw-Hill Encyclopedia of Science & Technology.

David E. Granstrom

Bibliography. J. P. Dubey et al., *Sarcocystis neurona* N. sp. (Protozoa: Apicomplexa), the etiologic agent of equine protozoal myeloencephalitis, *J. Parasitol.*, 77:212–218, 1991; J. P. Dubey, C. A. Speer, and R. Fayer, *Sarcocystis of Animals and Man*, 1989; C. K. Fenger et al., Identification of opossums (*Didelphis virginiana*) as the putative definitive host of *Sarcocystis neurona, J. Parasitol.*, 81:916–919, 1995; D. E. Granstrom et al., Equine protozoal myeloencephalitis: Antigen analysis of cultured *Sarcocystis neurona* merozoites, *J. Vet. Diagn. Invest.*, 5:88–90, 1993.

Erosion

Wind erosion affects 1,066 million acres (430 million hectares) worldwide. Particularly serious wind erosion problems exist in the Sahel, the Maghreb, the United States' Great Plains, parts of the Commonwealth of Independent States (former Soviet Union), China, Mongolia, Iraq, India, Paraguay, and Australia. In the Great Plains, about 5 million acres (2 million hectares) are damaged moderately to severely by wind erosion each year. Blowing soil obscures visibility, pollutes the air, causes automobile accidents, fouls machinery, imperils animal and human health, fills road and irrigation ditches, buries farm fences, deteriorates water quality, reduces seedling survival and growth, lowers marketability of vegetable crops, increases susceptibility of plants to diseases, and contributes to transmission of some plant pathogens.

Causes of wind erosion. Erosion is caused when strong winds blow over a soil surface that is bare (sparse or no vegetation), dry, loose, and finely divided. Soils of coarse (sandy) textures are more susceptible to wind erosion than soils of fine texture (containing more silt and clay), but for a given texture, erosion is related closely to the soil surface condition. During a drought, as in the Great Plains in the 1930s, 1950s, 1970s, and the fall and winter of 1995–1996, the soil surface is dry, vegetation is sparse or nonexistent, and the surface is loose from tillage or freezing–thawing. Under these conditions when the wind blows at greater than threshold velocity, soil begins to move.

Particle transport modes. Wind erosion occurs in three distinct modes based on particle size: suspension, saltation, and surface creep. Very small soil particles, less than 0.004 in. (100 micrometers) in diameter, are lifted from the surface and can be carried in suspension, sometimes for hundreds of miles, before being deposited. These particles constitute the easily observable part of wind erosion, seen as dust storms. Generally, less than 10% by weight of an eroding soil is carried in suspension.

In saltation, wind-blown sand grains or soil aggregates 0.004–0.02 in. (100–500 μm) in diameter are lifted from the surface but are too large to be suspended, so they return to the surface to strike plants or other soil aggregates. These particles, moving close to the ground in short jumps, make up roughly 50–80% of an eroding soil. They usually end up in a fence row, ditch, or vegetated area.

Sand-sized soil particles or soil aggregates 0.02–0.04 in. (500–1000 μm) in diameter, too large to leave the surface in ordinary erosive winds, are pushed, rolled, and driven by the saltating particles. Surface creep makes up 7–25% of an eroding soil. Creep particles seldom move far from their points of origin.

Plant damage. When plants intercept the saltating particles returning to the surface, they can be damaged physically. This damage is referred to as sandblast injury, abrasive injury, or wind erosion damage. In controlled wind tunnels, damage caused by wind-blown soil particles has been studied in field crops (alfalfa, cotton, grain sorghum, millet, soybean, tobacco, and winter wheat); vegetable crops (cabbage, carrot, cowpea, cucumber, green bean, onion, pepper, and tomato); and range grasses (Indian grass, sand love grass, sideoats grama, and switchgrass).

In sandblast injury, leaves and stems are abraded, growth and development are slowed, the quality and quantity of yield are reduced, moisture stress is increased, and some plants are killed. Factors influencing the extent of damage are wind speed, amount of abrasive material, duration of exposure, plant species, age of plant, and environmental conditions after damage (water and nutrient availability as well as air and soil temperatures). Abrasive damage does not account for all the crop losses from wind erosion. Plants also are damaged by the whipping action of the wind, which removes tissue from the end of the leaf, much like a flag is tattered.

Plants often are buried where eroding soil particles accumulate. Buried leaf tissue does not receive sunlight and soon dies. Wind erosion can expose roots to the environment by removing the soil around the base of the plant. If the roots are shallow, as in young seedlings, the plants can be blown from the soil.

Wind speed, amount, and duration. When wind velocity near an erodible soil surface exceeds 14.5–16.8 mi/h (6.5–7.5 m/s), saltation-size particles begin to move. As wind speed increases above this threshold velocity, more particles are entrained into the moving air mass until the wind carries its maximum load. The velocity of the wind also determines the speed and the size of the eroding particles. Material in the air stream moves at a velocity near the wind speed.

The amount of material in the wind stream is referred to as soil flux and has units of mass per unit

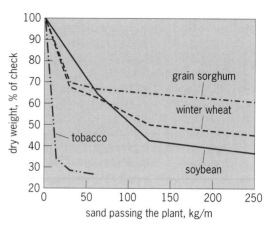

Fig. 1. Effect of the amount of sand passing the plant, in kg/m, on the dry weight of grain sorghum, soybean, tobacco, and winter wheat. The percent of check refers to the dry weight of a damaged plant divided by the dry weight of an undamaged plant. 1 kg/m = 0.67 lb/ft.

width per time, that is, pound per rod width per hour (gram per centimeter width per second) [1 rod equals 16.5 ft or 5.03 m]. The higher the wind velocity, the higher the flux. The quantity of soil moved by the wind is proportional to the cube of the wind velocity. The longer a plant is exposed (duration) to eroding soil particles, the greater the damage. The terms velocity, flux, and duration can be combined into an expression that describes the amount of soil particles passing a plant in a plane. In severe cases, this value can be as high as 25 tons/rod (4500 kg/m); values less than 5 lb/rod (50 g/m) are usually considered negligible.

Figure 1 shows for several crops the reduction in dry-weight production, relative to unexposed plants (check plants), 2 weeks after exposure to various combinations of wind speed, flux rate, and duration. Relatively small amounts of soil passing the plants (332 lb/rod or 30 kg/m) reduced the dry-weight production by 19–71%, depending on the crop species. As the amount of soil passing the plant increases, the reduction in dry weight levels off until, if the amount of wind erosion is very high, the plant dies.

Not only does wind erosion damage reduce the dry-matter production and sometimes cause plant death, but surviving plants show reductions in both quantity and quality of yield and delayed maturity. Tomato seedlings exposed to 277 lb/rod (25 kg/m) produced only 39% of the yield of undamaged plants and only 58% of the number of marketable fruits. Fruits were unmarketable because of defects or small size. Date of first bloom was delayed by 7–10 days.

Species tolerance. Plant species differ in their tolerance to abrasive damage. Although the exact reasons for the greater wind erosion tolerance of some species have not been studied, they may include tougher cell walls, waxy coatings on the leaves and stem, a greater ability to repair damage to cell walls, and a greater ability to withstand moisture stress. What-

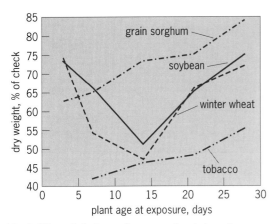

Fig. 2. Effect of plant age at exposure to wind erosion on the dry weight of grain sorghum, soybean, tobacco, and winter wheat.

ever the reason for resistance to sandblast damage, grain sorghum produces more growth than winter wheat, soybeans, or tobacco at similar levels of erosion (Fig. 1). Most vegetable plants are damaged more easily than field crops, and the quality reduction and delayed maturity are of greater importance to vegetable producers.

Plant age. In general, the younger a plant is when exposed to wind erosion, the greater the damage **(Fig. 2).** Some species are most susceptible at the time when food reserves in the seed are exhausted (about 2 weeks after emergence) and the plant is dependent on photosynthate (the product of photosynthesis) produced by the leaves. Loss of any area from the seedling's small leaves greatly reduces food production and, therefore, growth of the plant.

Environment after damage. The environment in which the plant grows after being damaged by wind-blown particles affects the rate at which the plant can repair and recover from damage. Tomatoes exposed to blowing sand survived better, grew faster, and produced more flowers when soil moisture was high at time of exposure or if the soil moisture was increased by watering after exposure. When a plant cell is struck by a soil particle, it is ruptured and the water in the cell is lost to the atmosphere. If the plant can rapidly replace this lost water, abrasive damage can be repaired more quickly and normal growth can be regained.

Physiological effects. Sandblast damage causes physiological changes within plant cells. Decreases in dry-matter production of damaged grain sorghum and winter wheat were due to reduced photosynthesis and increased respiration. Activity of the enzyme responsible for photosynthesis, ribulose-1,5-diphosphate carboxylase, was reduced, so less carbon dioxide was converted to carbohydrates and, therefore, less energy was available for plant growth. Activity of the enzyme nitrate reductase in soybean was decreased by abrasive injury, which results in less nitrogen available for protein synthesis. The activities of enzymes decrease when the plant experiences a shortage of water (moisture stress), such as when plants are damaged by wind-blown soil particles.

Disease transmission. When soil-borne disease organisms are present on blowing soil particles, the plants, if susceptible, can be inoculated with that disease. Incidence of common blight of pinto bean was increased by 25–55% when plants were exposed to blowing soil with the organism present. Also, incidence of bacterial leaf spot of alfalfa increased 6–26%.

Control methods. Effective control methods for wind erosion are known and are continually being improved. Some recommended practices include maintaining the crop residues on the soil surface in a standing position (no till or minimum tillage), planting barriers perpendicular to the prevailing wind direction and at frequent intervals (shelterbelts or rows of grasses), using vegetative mulches (wheat straw or other plant residue anchored to the soil) or nonvegetative materials (asphalt, polyvinyl acetate, and latex films), and roughing the soil surface with ridges and large stable aggregates.

For background information *see* DESERT EROSION FEATURES; DUST STORM; EROSION; PLAINS; SOIL; SOIL CONSERVATION in the McGraw-Hill Encyclopedia of Science & Technology.   Dean V. Armbrust

Bibliography. M. Agassi, *Soil Erosion, Conservation, and Rehabilitation*, 1996; D. V. Armbrust, Wind and sandblast injury to field crops: Effect of plant age, *Agron. J.*, 76:991–993, 1984; D. V. Armbrust, G. M. Paulsen, and R. Ellis, Jr., Physiological responses to wind- and sandblast-damaged winter wheat plants, *Agron. J.*, 66:421–423, 1974; R. Lal (ed.), *Soil Erosion Research Methods*, 2d ed., 1994; D. Pimentel, *World Soil Erosion and Conservation*, 1993.

Extinction (biology)

Of all the species that ever lived, some 99% are now extinct. Extinction is the rule rather than the exception. The vast majority of these extinctions are normal or background extinctions: they constantly occur and have done so since life originated on Earth some 3.5 billion years ago. There are also episodes in Earth history when extinctions are far above the background levels, that is, mass extinctions. Five mass extinctions are recognized during the Phanerozoic Eon, comprising about the last 540 million years. Although constituting only a small percentage of total extinctions during Earth history, these five represent major reorganizations of the Earth's biota. The terminal Permian mass extinction some 250 million years ago (Ma) was the most severe of the five mass extinctions, possibly with species extinction of over 95%, but the terminal Cretaceous mass extinction 65 Ma is best known because it includes the extinction of dinosaurs.

Ultimate and proximate causes. Over 80 dinosaur extinction theories have been proposed, far more than for any other group of animals. Of these, only three ultimate causes of extinction are well enough articulated to be testable with the fossil record. Ultimate causes, although usually not suggested as producing extinction directly, eventually lead to various proximate causes that result in extinction. These three ultimate causes are asteroid impact, volcanism, and marine regression.

Asteroid impact theory. The basic premise of the asteroid impact theory is that a 6-mi-diameter (10-km) asteroid struck the Earth 65 Ma, producing ejecta and a plume far enough into the atmosphere to be spread around the globe, blocking out radiation from the Sun. This resulted in the cessation of photosynthesis, leading to death and extinction of many plants, the herbivores that fed upon them, and the carnivores that fed upon the herbivores. The probable impact crater, named Chicxulub, is located near the tip of the Yucatan Peninsula; it is the second largest such Earth structure thought to be a crater. Other physical evidence supporting such an impact are an increase in the element iridium (Ir) at the Cretaceous/Tertiary (K/T) boundary, and minerals, especially quartz grains, that show shocked lamellae in two directions. A high level of iridium at the Earth's surface and double lamellae are both more indicative of an impact than of volcanism. Some of the more proximate effects of an asteroid impact suggested are acid rain, global wildfire, sudden temperature increases and decreases, tsunamis, and strong hurricanes.

Volcanic theory. The volcanic theory relies on the occurrence of massive eruptions of flood basalts called the Deccan Traps on the Indian subcontinent during the time interval of the K/T boundary. The material estimated to have erupted over a 4-million-year interval would have a volume sufficient to cover both Alaska and Texas to a depth of 2000 ft (610 m). The suggested proximate causes resulting from extensive volcanism have not been as well articulated as those for asteroid impact and thus cannot be as easily tested. Some of the same proximate causes are argued to have occurred under either impact or volcanism, although climatic changes caused by massive eruptions are thought to be longer term.

Marine regression theory. The marine regression theory is based on the loss of epicontinental seas that occurred during the K/T transition. Although not fully understood, major transgressions and regressions of epicontinental seas appear to be driven by changes in the motion of the Earth's geological plates. Sea levels rise and transgression occurs as motions of the plates increase. As margins of the plates collide and subduction occurs, the plates (and continental margins) are pulled down. Low-lying areas become even lower and are inundated. The reverse (regression and a drop in sea level) occurs as the rate of plate collision decreases, allowing the plate margins to rebound. The extent of regression is based upon known and extrapolated marine and terrestrial deposits of the Late Cretaceous and early Tertiary found around the world. Estimates suggest that 11.2 million square miles (29 million square kilometers) of land

Vertebrate species survival across the K/T boundary

Species from the Upper Cretaceous Hell Creek Formation, eastern Montana	Number and percent survivorship	
Elasmobranchii (sharks and relatives)	0/5	(0%)
Actinopterygii (bony ray-finned fishes)	9/15	(60%)
Lissamphibia (frogs and salamanders)	8/8	(100%)
Mammalia		
Multituberculata*	5/10	(50%)
Eutheria (placentals)	6/6	(100%)
Metatheria (marsupials)	1/11	(9%)
Reptilia		
Testudines (turtles)	15/17	(88%)
Squamata (lizards and snakes)	3/10	(30%)
Choristodera†	1/1	(100%)
Crocodilia (crocodiles and alligators)	4/5	(80%)
Ornithischia (bird-hipped dinosaurs)	0/10	(0%)
Saurischia (reptile-hipped dinosaurs)	0/9	(0%)
Total	52/107	(49%)

* A group of completely extinct rodentlike mammals.
† A group of completely extinct gaviallike reptiles.

(approximately the size of Africa) were exposed during this interval, more than twice the next largest such addition of land during the past 250 million years. The more long-term effects include a major loss of low coastal plain habitats, fragmentation of the remaining coastal plains, establishment of land bridges, extension of fresh-water systems, and climatic change with a trend toward cooling on the newly emerged landmasses.

Vertebrate record. The global record of dinosaurs during the Late Cretaceous is probably the most complete record available for this group of vertebrates, with hundreds of dinosaur sites around the globe, including a few in Antarctica and New Zealand. For example, the record is complete enough to show that during the last 10 million years of their reign in western North America, the number of genera of dinosaurs dropped from 32 to 19, a 40% loss. The vertebrate fossil record at the K/T transition, however, is far poorer than is usually realized. The fossil record for this transition is extremely limited, being restricted to localities in the Western Interior of North America. More recently discovered sites in China, South America, and India offer hope for a better K/T record of vertebrates in the near future. For now, however, little can be said about the rate of dinosaur extinction at the K/T boundary. The record is inadequate to address the issue of rates of extinction. Fortunately, the record is good enough to enable analysis of the magnitude and selectivity of the extinction of dinosaurs and their contemporary vertebrates.

The present record from the best-known sites in eastern Montana indicates 107 well-documented species of vertebrates belonging to 12 major monophyletic lineages (see **table**). Of these species, only 19 are dinosaurs. Because of a poor fossil record, pterosaurs and birds (actually a clade of saurischian dinosaurs) could not be included in this sample. It is known, however, that a number of major clades of modern birds had appeared before the K/T boundary extinctions and that pterosaurs had dwindled to one family. Of the 107 species, 52 (49%) survived. This level of survival is only about 10% lower than for similar intervals before and after the K/T boundary. The degree of survival at the K/T boundary for vertebrates in the Western Interior of North America appears to have been higher than for a number of other major groups of plants and animals. Nevertheless, the effects on vertebrates were profound. Dinosaurs, the large land vertebrates, which dominated most of the Mesozoic Era, were replaced by mammals in a very short time—probably only a few thousands or tens of thousands of years. Mammals, including Primates, rapidly diversified, filling niches left by dinosaurs and creating new ones.

The most obvious pattern of extinction among these 12 major vertebrate clades is that extinctions were concentrated in five clades: sharks and relatives, lizards, marsupials, ornithischians, and saurischians. Species in these five clades account for 75% of the extinctions. This demonstrates that the K/T extinctions were highly selective. Therefore any theory of extinction must account for this selectivity.

Testing extinction theories. It is impossible to use the vertebrate fossil record to test directly whether an asteroid struck Earth, whether there were major volcanic eruptions, or whether the seas underwent a major regression. It is possible, however, to test the various proximate causes that have been proposed as the result of these ultimate causes. For example, starting with the asteroid impact, it is possible to examine the suggestions of acid rain, sharp temperature decrease, and global wildfire. Work on extant vertebrate species and habitats has confirmed that acid rain hurts aquatic organisms most; yet except for sharks and relatives, aquatic species did very well through the K/T transition. If a sharp, short temperature spike occurred, it is known from modern vertebrates that wholly or partially terrestrial ectotherms should have been most affected. Yet once again, most of these, except lizards, did well through the K/T boundary. Whether dinosaurs should be considered as endotherms, ectotherms, or as vertebrates having another kind of physiology remains controversial. Finally, global wildfire, which is argued to have consumed 25% of all aboveground biomass, should have been a nearly equal-opportunity killer, burning many creatures, and suffocating others in aquatic systems with a huge influx of detritus. The fossil record, as discussed earlier, is highly selective; therefore, global wildfire is a very unlikely scenario. All of these proximate causes of an asteroid impact fall short when tested with the vertebrate fossil record. What remains of this event as a killing mechanism is the blocking of sunlight, causing a cessation of photosynthesis, enough to kill and cause the extinction of up to 80% of plants species, at least in some areas of North America. This would have had a devastating effect on large herbivores, especially if already stressed by other events such as marine regression and habitat fragmentation.

As global marine regression began in the last few million years of the Cretaceous Period, tremendous new tracts of dry land were added. All known dinosaur-bearing vertebrate localities near the K/T boundary come from coastal-plain habitats that were being drastically reduced during marine regression. This reduced dramatically the size of these habitats, stranding dinosaurs in ever smaller areas. This development stressed dinosaur populations, setting them up for the final biotic catastrophe caused by asteroid impact and massive volcanism. At the same time, the coastlines were retreating from the Western Interior of North America, taking the sharks and relatives with them. These life-forms could traverse fresh-water courses up to a few hundreds of miles or kilometers, but not thousands of miles or kilometers, thus severing the marine connection. Although low coastal streams disappeared, the total fresh-water systems held their own and, in many cases, even increased in length as the coastline retreated. Thus, fresh-water species did very well, with descendants such as paddlefish, sturgeon, and gar still plying the Missouri-Mississippi river systems. The lowering of sea level also reestablished connections between separated landmasses such as eastern Asia and western North America. The earliest ungulate relatives known from North America appear at this time, possibly from ancestors known in Asia 20 million years earlier. These new ungulates have a dental morphology that resembles that of the opossumlike marsupials, which arose in North America some 100 Ma and had been very common for at least the 20 million years leading up to the K/T boundary 65 Ma. It seems likely that the appearance of these ungulates in North America resulted in competitive doom for the marsupials. Interestingly, when they both appeared in South America a few million years after the K/T boundary, the ungulates became herbivores while the marsupials became omnivores and carnivores. The one group whose evolution cannot be explained by this globally testable hypothesis is the lizards in the Western Interior, which underwent a drastic reduction. A more local explanation is that the suggested climatic shift to a wetter, rainier climate in areas such as eastern Montana following the K/T boundary may have driven out the more dry-adapted lizards.

The proximate causes resulting from marine regression explain the highly selective pattern of vertebrate extinctions and survivals through the K/T transition in the Western Interior of North America better than do those attributed to asteroid impact or possibly massive volcanism. These patterns do not argue whether these latter two ultimate cause of extinctions occurred, but that they do not explain what is known of the vertebrate record. In combination with evidence from plants and marine species, however, its appears that all three ultimate causes probably played roles to varying degrees in what is the best-known mass extinction in Earth history.

For background information *see* ANIMAL EVOLUTION; ASTEROID; DINOSAUR; EXTINCTION (BIOLOGY);

MACROEVOLUTION; ORGANIC EVOLUTION; PALEOCEAN-OGRAPHY; VOLCANOLOGY in the McGraw-Hill Encyclopedia of Science & Technology.　　J. David Archibald

Bibliography. J. D. Archibald, *Dinosaur Extinction and the End of an Era: What the Fossils Say*, 1996; D. E. Fastovsky and D. B. Weishampel, *The Evolution and Extinction of the Dinosaurs*, 1996; N. MacLeod and G. Keller (eds.), *The Cretaceous-Tertiary Mass Extinction: Biotic and Environmental Changes*, 1996.

Fluid flow

Recent advances in the study of fluid flow include the observation of novel features in fluid flow in soap films, the accurate modeling of free-surface nonlinear flows, and magnetic resonance imaging of fluid flow.

Fluid Flow in Soap Films

Soap films, in the form of soap bubbles or sheets, have long been of scientific and practical interest. Now new features are being observed that appear only when the film is flowing, either smoothly (laminar flow) or chaotically or turbulently. These films are so thin that the turbulence that can be produced in them is, for practical purposes, two dimensional. Two-dimensional turbulence is expected to be very different from its three-dimensional counterpart. It is of interest because large-scale turbulent eddies in the ocean or in the atmosphere are believed to behave as would an idealized purely two-dimensional fluid. The thickness of the atmosphere around the Earth is only 1% of the Earth's diameter. For soap films the ratio of thickness to lateral size is only 0.001%.

Structure of soap films. A typical soap film is 99% water, with an atomic layer of soap molecules packed closely on the two surfaces. Soap molecules that cannot be accommodated on the surfaces distribute themselves through the water layer in between. The overall thickness of a soap film is roughly 2 micrometers or 10,000 atomic layers.

Two-dimensional flows. Figure 1a shows the reflection pattern produced by a soap film spanning the annulus between a fixed outer ring, 4 in. (10 cm) in radius, and an inner disk, 2 in. (5 cm) in radius, which rotates at 10 revolutions per second. The shades of gray indicate variations in film thickness. The darkest and brightest regions are the result of light waves being reflected from the front and back surfaces of the film out of phase and in phase, respectively. When the film is viewed in room light, it has a multicolored appearance, since the thickness variations produce strong reflection at various wavelengths. Because the film molecules near the ring and the disk are presumably stuck to them by an attractive molecular force, the film is being strongly sheared. In this concentric geometry, the thickness pattern does not change with time. A centrifugal force on the film causes the film thickness to increase with radial distance from the inner rotating disk, producing the observed ring pattern.

The same spinning-disk apparatus can produce flow that might be called chaotic, that is, irregular and time varying but not really turbulent. The inner disk is merely shifted off center, producing the reflection pattern seen in Fig. 1b. Observation of the motion of dust particles on the film reveals that the kidney-shaped pattern is globally and rather

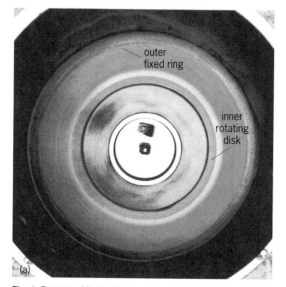
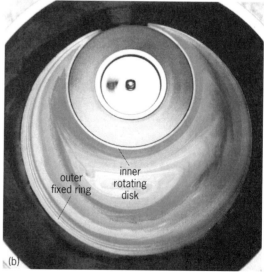

Fig. 1. Pattern of light reflected from a soap film that lies between the inner rotating disk and the outer fixed ring. (*a*) The axis of the rotating disk is at the center of the ring. (*b*) The ring is off center and the pattern moves about. The photographs were made with only a 1-ms exposure time to assure that the image was not blurred by the fluid motion of the film. (*From X.-I. Wu et al., Hydrodynamic convection in a two-dimensional Couette cell, Phys. Rev. Lett., 75:236–239, 1995*)

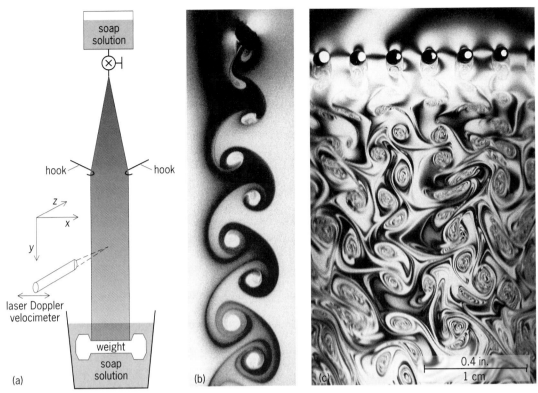

Fig. 2. Experiment for producing a vertically flowing soap film. (*a*) Experimental arrangement. (*b*) Thickness variations in the film produced by a 0.12-in.-diameter (3-mm) rod that penetrates it near the top of the figure. The entire pattern is flowing downward at an average speed of 6.5 ft/s (2 m/s). (*c*) Vortexlike structures produced by the comb at the top of the figure. The separation of the comb teeth is 0.12 in. (3 mm). The mean vertical flow speed is the same as in *b*.

smoothly rotating. Near the center of the pattern, however, the film thickness (and the velocity as well) is fluctuating rather randomly in time. Technically stated, the thickness h varies from point to point in the plane (x, y) of the film; that is, h is a function of x, y, and time t. The observed fluctuation pattern can be explained by a theory arguing that the thickness variations behave like smoke in turbulent air or a dye dispersed in a turbulent liquid.

The setup shown in **Fig. 2*a*** is designed to produce strongly turbulent flow. Soap solution in the square container at the top flows through a valve to the point where two thin wires or fishing lines meet so as to form an inverted V. The wires are under tension because of weights at the bottom and are held apart by hooks. It is easy to form a soap film between the wires and to study the velocity of the flow in the straight vertical section. The reservoir at the bottom receives the spent film. Typically, the parallel section might be 6 ft (2 m) high, with the wire separation being 2 in. (5 cm). However, there is no difficulty in making films 65 ft (20 m) high and several meters wide. Under the action of gravity in the downward direction, the soap film flows at mean speeds of 3–10 ft/s (100–300 cm/s).

Drag forces on flowing film. In one important respect the soap film is not two dimensional: the surrounding air exerts a force on the film that slows it down. This effect can be minimized rather easily by placing the entire apparatus of Fig. 2*a* in a vacuum

chamber where the air pressure is only 3% of atmospheric pressure (although the experiments illustrated in Fig. 2 were all carried out at atmospheric pressure). Further reduction of the air pressure would cause the film to evaporate.

Turbulent flow. If a glass rod of, say, 0.12 in. (3 mm) diameter is wetted and then thrust through the flowing film, it generates vortices downstream that give rise to the thickness variation, $h(x,y,t)$, photographed in Fig. 2*b*. A measurement of the velocity components $v_x(x,y,t)$ and $v_y(x,y,t)$ indeed establishes that this is correct. If the single rod is replaced with a comb, each tooth generates a trail of these von Kármán vortices, and they interact with each other. This interaction causes the flow to become more turbulent, giving rise to randomly appearing thickness variations $h(x,y,t)$ seen in Fig. 2*c*. *See* TURBULENT FLOW.

Two- and three-dimensional turbulence. Everyday fluids such as water are governed by the laws of classical physics, and this statement is true even for complicated turbulent flows in one, two, or three dimensions. For two-dimensional fluids, the Navier-Stokes equation is used for predicting the time variation of $v_x(x,y,t)$ and $v_y(x,y,t)$. This equation is Newton's second law, $\mathbf{F} = m\mathbf{a}$ (force equals mass times acceleration), applied to all the fluid molecules composing the film. It follows from the mathematical structure of the Navier-Stokes equation that two-dimensional turbulence and three-dimensional turbulence are

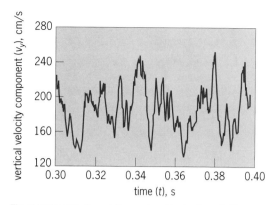

Fig. 3. Time variation of the random fluctuations in the vertical component of velocity in the flowing soap film of Fig. 2c, measured at a point roughly 4 in. (10 cm) below the comb.

very different. In three dimensions the average intensity of vortex motion in the fluid can increase with time, but in two dimensions this turns out to be impossible.

Figure 3 shows the vertical component of the velocity, $v_y(t)$, measured as a function of time at a point roughly 4 in. (10 cm) below the comb of Fig. 2c. The measurements were made half way between the two supporting vertical wires, with a laser Doppler velocimeter, and the time interval spanned by the measurements was only 0.1 s. The measurements were made well after the flow had commenced, so that a "steady state" had been reached. The velocity is seen to vary erratically with time. If v_x and v_y were to be measured at any instant of time, these velocity components would also be seen to be varying erratically from one point to another. It would be a hopeless and useless task to try to calculate the time and position dependence of the velocity of a turbulent fluid. However, it is sensible to extract statistical information about turbulent flow, such as the mean-square velocity difference measured at two instants of time separated by a specified short time interval, or the time-averaged mean-square velocity difference at a single instant of time at two points separated by a specified distance. Such quantities cannot yet be derived from the Navier-Stokes equation, but increasingly trustworthy physical arguments can be made about the expected mathematical form that these statistical averages should have. It is hoped that the study of two-dimensional flow in soap films will contribute to a better understanding of turbulence in three dimensions as well as two.

Walter I. Goldburg

Free-Surface Nonlinear Flows

Before they reach breaking, highly nonlinear free-surface waves (whose amplitude is comparable to water depth) can be quite closely represented by inviscid irrotational flow theory (potential flow theory), provided full nonlinearity is kept in both the kinematic and dynamic free-surface boundary conditions (that is, provided that these conditions, which respectively express the facts that water particles

on the free surface stay on the free surface and that pressures on opposite sides of the free surface are equal in magnitude and opposite in direction, are not simplified by assuming small-amplitude waves). The solution of such unsteady nonlinear problems over domains of complex geometry, however, can be achieved only through numerical modeling. Recent laboratory experiments and computations for deep-water and shallow-water breaking waves in fact show that fully nonlinear potential-flow models can predict characteristics of overturning waves, with surprising accuracy, prior to the time that touchdown of a breaker jet occurs on the free surface. It is thus only after such touchdown that significant vorticity is created, together with intense energy dissipation—the conditions which violate the hypotheses of fully nonlinear potential flow. For shoaling waves in shallow water, bottom friction, which is neglected in fully nonlinear potential-flow computations, can also be shown not to affect wave shape significantly in the region where nonlinearity is large, even for long waves such as solitary waves; the reason is the short distance of propagation over the slope in this region.

Fully nonlinear potential-flow computations. **Figure** 4a and b, for instance, show chronological stages of fully nonlinear potential-flow computations of surface elevation η, as a function of horizontal coordinate x, for the shoaling and overturning of a solitary wave (that is, an exact soliton solution of fully nonlinear potential-flow equations) of incident height $H_0 = 0.2h_0$ on a 1:35 slope (where h_0 denotes a reference depth for an initial constant-depth region in front of the slope). From curve A to D, as time t increases (the dimensionless time is $t' = t\sqrt{g/h_0}$, where g denotes the gravitational acceleration), the wave becomes increasingly high and skewed. At curve D, the wave-front face has a vertical tangent, which defines the breaking point. The comparison of computed surface elevations at various gages with well-controlled laboratory experiments shows discrepancies of less than 2% up to quite close to the breaking point. This confirms the validity of fully nonlinear potential-flow equations and hypotheses in their description of strongly nonlinear wave propagation in shallow water. From curves A to D in Fig. 4a, the horizontal particle velocity under the wave crest, which is originally fairly uniform from bottom to surface for the incident wave, becomes increasingly nonuniform as a function of depth. (This behavior is unlike predictions of nondispersive, nonlinear shallow-water equations, often used to model such wave-propagation problems.) At the breaking point, the wave particle velocity at the crest is equal to the phase velocity (the velocity of the wave crest); for later times, the particle velocity is larger, leading to wave overturning (curves E–G). Experiments indicate that fully nonlinear potential-flow equations are valid up to the stage of curve G, that is, just before touchdown of the jet. From the breaking point to this stage, the wave height slightly reduces without any energy dissipation, because of a decrease in

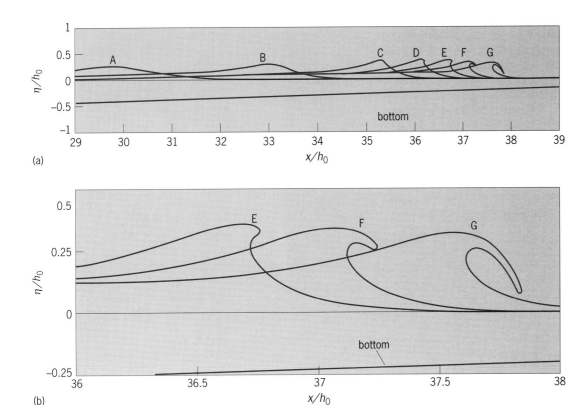

Fig. 4. Fully nonlinear potential-flow computations for the shoaling of a solitary wave over a 1:35 slope. Symbols are explained in text. (*a*) Chronological stages of shoaling, A–G. Dimensionless times, t', of curves A–G, measured from the start of generation by a piston wave maker located at $x/h_0 = 0$, are 37.17, 40.73, 43.48, 44.53, 44.94, 45.40, and 46.00. The slope starts at $x/h_0 = 10$, and the depth is constant in front of the slope. (*b*) Enlargement of a portion of *a*. (*After S. T. Grilli et al., Shoaling of solitary waves on plane beaches, J. Waterway Port Coastal Ocean Eng., 120(6):609–628, 1994*)

wave potential energy and an identical increase in kinetic energy; this is confirmed by many independent laboratory experiments. Water particle accelerations in the breaker jet can reach 2–3 g. Hence, impact pressures from breaking waves on solid structures can be one to two orders of magnitude larger than those from nonbreaking waves of identical height.

Recent two-dimensional fully nonlinear potential-flow models can thus predict characteristics (such as shape, kinematics, and integral properties) of long-crested surface waves, up to the breaking point, with an accuracy close to that of the best-quality laboratory experiments. When equipped with both realistic wave generation at one extremity (such as numerical wavemakers) and active energy absorption at the other one (for example, an "absorbing beach"), such models are referred to as numerical wave tanks and are increasingly used as tools for investigating complex free-surface nonlinear flows (as in Fig. 4). Very accurate numerical computations have now been made for the generation, shoaling up to a very high fraction of the breaking height (crest-to-trough height of the wave at the breaking point), and absorption of fully nonlinear periodic waves. By using fully nonlinear numerical wave tanks, many physical properties of waves, both local (such as height, phase velocity, and particle veloc-

ity) and integral (such as momentum flux, energy flux, and mass flux), have recently been calculated at the breaking point. Finally, the standard mathematical formulation of fully nonlinear potential-flow models, which uses Green's identity in the form of boundary integral equations, leads to particularly efficient numerical algorithms based on the boundary element method. Boundary element method algorithms have recently been extended to three dimensions, and their applicability to real problems is mostly a matter of having sufficient supercomputer power.

Models of postbreaking waves. Despite their recent success, fully nonlinear potential-flow models are still limited to nondissipative, prebreaking waves. Flow separation, for example, over submerged obstacles, also cannot be modeled, as well as discontinuities such as separated jets. For situations where these and detailed postbreaking wave characteristics are needed—for instance, to calculate wave overtopping (in which incident waves form a jet that propagates over the crest of a structure) and impact on coastal or hydraulic structures or to study effects of a dam failure—it is necessary to use models based on the primitive equations of motion, which include both vorticity and viscous-turbulent dissipation (that is, the Navier-Stokes equations). The efficient solution of such equations, over sufficiently large and

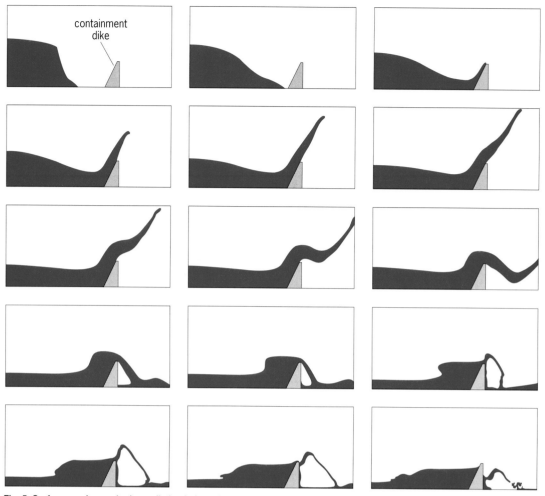

Fig. 5. Surface-marker and microcell simulation of viscous flow from a broken water dam over a containment dike inclined at 60°. (The dam, to the left, is not shown.) The initial height of the water is twice the height of the dike. *(P. Raad)*

resolved complex geometrical domains, possibly three-dimensional, can be achieved only through using either mark-and-cell (MAC) or volume-of-fluid (VOF) type methods, in which both mass and momentum conservation are only globally satisfied within a cell of finite area or volume. For complex highly nonlinear free-surface flows, the use of microcells and surface markers has recently made it possible to achieve both sufficient resolution and accuracy (**Fig. 5**). For highly nonlinear prebreaking waves, both mark-and-cell and volume-of-fluid methods provide results closely similar to fully nonlinear potential-flow results, confirming the validity of the latter simpler model for such waves. Fully nonlinear potential-flow models, however, when applicable, because of the efficiency of their numerical algorithm, can usually achieve better resolution and accuracy than the former models. For postbreaking waves, recent mark-and-cell and volume-of-fluid results are in good agreement with experiments, even when, because of violent motion, breaker jets penetrate the main body of water and expel sizable parcels of water above the free surface. This confirms the relevance of primitive fluid dynamics equations for modeling the physics of such seemingly chaotic postbreaking flows.

Even with the latest mark-and-cell methods, the modeling of postbreaking waves is still quite a formidable numerical task, and computations are often limited to two dimensions and to fairly restricted spatial and temporal extensions. Since the mid-1980s, a more efficient but somewhat restrictive approach has been pursued with high degree of success in the study of surf-zone hydrodynamics. It consists in developing and applying period-averaged governing equations for two- and three-dimensional postbreaking wave propagation, typically from the breaking point to the shoreline onward. Over the years, increasingly detailed levels of analysis and modeling of oscillatory bottom boundary layers, dissipation in wave-surface rollers and advected turbulence patches, and nonlinear interactions between both cross-shore and long-shore mean circulation flows—to name but a few of the physical problems being researched—have been achieved.

Stéphan T. Grilli

Magnetic Resonance Imaging of Fluid Flow

Fluid flow is an essential element of many industrial processes, particularly within the food, chemical, and petroleum industries. The ability to simulate and monitor fluid flow is important in order to design and control associated processes. It has always been a challenge to observe noninvasively fluid states and velocities corresponding to relatively small, local regions. Tomographic imaging provides a means. Data collected by imaging can provide useful insights into the physical laws and can be used to determine empirical descriptions and properties for describing fluid flow, as well as for direct monitoring of industrial processes.

Magnetic resonance imaging (MRI) provides spatial resolution of some properties associated with nuclei within samples. Since the initial report on magnetic resonance imaging in 1973, methods and instruments for this technique have rapidly progressed, largely because of its utility for medical diagnosis. The use of magnetic resonance imaging as a nondestructive probe of structures and flow in scientific and industrial research has attracted considerable attention, and very rapid developments are being made.

Principles of magnetic resonance. Nuclei with nonzero spin angular momentum have a nuclear magnetic moment associated with them. When an ensemble of such nuclei is placed in a magnetic field, the magnetic moments tend to align with the field, ultimately at equilibrium with a Boltzmann distribution of energies, to produce a net magnetization in the ensemble. The most commonly observed nucleus is probably that of hydrogen, ^1H (or the proton), and other observed nuclei include carbon-13 (^{13}C), fluorine-19 (^{19}F), and sodium-23 (^{23}Na). Nuclear magnetic resonance (NMR) experiments employ a radio-frequency antenna placed near the sample to perturb the nuclear magnetization from the equilibrium state and to monitor the behavior of the magnetization after such perturbations. The magnetization precesses about the static magnetic field with a characteristic frequency, the Larmor frequency, which is proportional to, and uniquely determined by, the gyromagnetic ratio of the nuclei and the strength of the field. The precessing magnetization induces an oscillating voltage in the antenna, which may be analyzed to determine component frequencies and phases.

The acquired signal is very sensitive to the dynamics of the nuclear spin system under study. Following a radio-frequency perturbation, the system has to redistribute its excess energy to return to equilibrium, a process known as relaxation. The two relaxation regimes commonly studied are the relaxation of longitudinal and transverse magnetization components, characterized by relaxation times T_1 and T_2, respectively. The relaxation of nuclei, and in turn the acquired signals, is sensitive to the local environments and motions of the associated molecules. Observation of these effects makes it possible to probe the microscopic structures and fluid dynamics within the sample.

Magnetic resonance imaging provides for the spatial mapping of properties of the nuclei by using a static magnetic field whose strength varies with position according to applied field gradients. The nuclei in different regions thereby possess different Larmor frequencies, and the signal intensity at a given frequency is proportional to the nuclear spin density at the corresponding spatial position.

The mapping of the spin density provides for local quantification of the amount of observed nuclei. Various contrast schemes can also be used to probe flow and structures. The contrast schemes are carried out through selection of perturbations in applied radio-frequency pulses and gradient sequences to emphasize differences in relaxation or other properties. Chemical-shift contrasts can be used to selectively observe nuclei associated with different chemical species. Structural heterogeneities can be mapped by observing relaxation contrasts arising from differences in molecular mobilities in different material environments.

The sensitivity of nuclear magnetic resonance signals to molecular translational motion is useful as a contrast scheme to measure fluid velocities. Several different approaches have been used. One is based on the detection of phase shifts that arise from the fact that the nuclei spend different lengths of time in different magnetic fields, depending on their velocity. Another method is based on the observation of time-of-flight effects on image intensity due to, for example, washout of excited spins from the imaging region. All the methods involve observing evolution of the nuclear spin system over a well-defined period of time.

Applications to fluid flow. Applications of magnetic resonance imaging to fluid flow have focused on two areas demanding different experimental approaches: the imaging of fluid concentrations, and the imaging of fluid molecular displacements. Current trends in applications of these techniques are seen in the areas of porous media, filtration processes and transport within reactors and packed columns, and fluid rheology.

Porous media. Many important processes involve the flow of fluids through porous media, including catalytic chemical reactions, hazardous waste treatment, ground-water flow, and petroleum production. Nuclear magnetic resonance provides the means for characterizing a number of properties important for describing flow in porous media, including porosity, permeability, wettability, and pore-size distributions. A very interesting application of nuclear magnetic resonance is downhole well logging, whereby properties of subsurface earth formations are measured in place.

Magnetic resonance images showing spatial resolutions of a fluid phase within a porous rock were first reported in 1984. Since then various studies

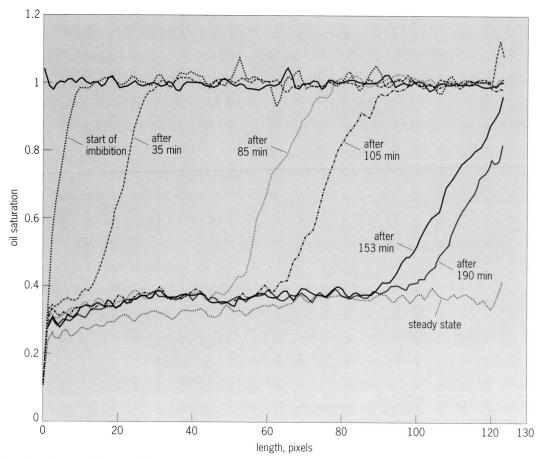

Fig. 6. Proton profile images of oil saturation during primary inhibition in limestone. Deuterated water was used as the aqueous phase. Each pixel is 0.022 in. (0.56 mm) long. Images (the curves) correspond to different times during the experiment. *(After C. T. Chang et al., The use of agarose gels for quantitative determination of fluid saturations in porous media, Magn. Reson. Imag., 11:717–725, 1993)*

have addressed problems related to the flow of one or more fluid phases in porous media. In particular, many enhanced oil recovery methods and subsurface waste remediation processes involve the displacement of one fluid phase by a second, immiscible fluid phase. Magnetic resonance imaging techniques for studying multiphase flow require the ability to resolve the liquid phases. Discrimination between oil and water phases in a rock matrix has been achieved by several approaches. One straightforward approach is to remove the signal of the water phase by substituting water with deuterated water or by doping the water with paramagnetic ions, thereby reducing T_2. **Figure 6** shows oil saturation distributions determined from proton images measured as an aqueous phase (deuterium oxide) is displaced from a limestone core sample by an oleic phase (hexadecane). Another approach is to resolve fluids based on differences in their relaxation times. Chemical-shift imaging techniques use the difference of the chemical environment of protons in the water and oil to distinguish the two phases. Spatially and temporally resolved fluid distributions have been visualized with magnetic resonance imaging during displacement processes.

Nuclear magnetic resonance provides the means for measuring both the velocity field and velocity distributions within porous media. The velocity field is represented by a map of fluid velocities, while the velocity distribution is represented by a histogram of different velocity values present in the flow. Fluid velocity maps and velocity distributions have been measured by using model porous media as well as sandstone samples. **Figure 7a** shows a nuclear magnetic resonance velocity image of a cross section of water flow through a packing of 0.25-in. (6-mm) glass beads in a cylinder with an inner diameter of 1.6 in. (4 cm), at a flow rate of 0.076 in.³/s (1.25 cm³/s). Areas with higher signal intensities (lighter tone) correspond to higher flow velocities. Velocity data from two pores marked with squares are shown in Fig. 7b and c.

Packed columns and reactors. Magnetic resonance imaging has also been used to study chromatographic separations. The signal intensity depends on many parameters such as spin density, T_1, T_2, and diffusion. Variations in the packing density along the length or diameter of the column, channel formation, and differences in the degree of wetting can be clearly recognized and used, for example, to develop an

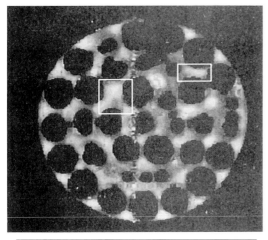

-0.5   0.0   0.5   1.0

velocity, cm/s

(a)

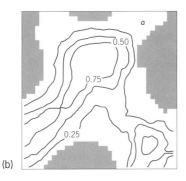

(b)

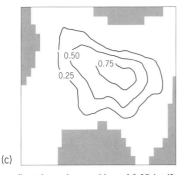

(c)

Fig. 7. Water flow through a packing of 0.25-in. (6-mm) glass beads in a cylinder with an inner diameter of 1.6 in. (4 cm). (*a*) Velocity image of flow. (*b*) Contour plot of velocity in the left pore marked in *a*. (*c*) Plot of velocity in the right pore. (*After Y. E. Kutsovsky et al., NMR imaging of velocity profiles and velocity distributions in bead packs, Phys. Fluids, 8(4):863–871, 1996*)

optimal method for packing columns or to validate analysis assumptions. Recent studies suggest that magnetic resonance imaging may become a valuable tool in the design of membrane bioreactors and filtration devices. Magnetic resonance imaging is capable of imaging both the packing distribution and the flow process. For example, magnetic resonance imaging has been used to study fluid flow in a model bioreactor as a function of biofilm formation. Images of the biofilm density were obtained by relaxation-time-weighted data, and nuclear magnetic resonance velocity imaging revealed how the fluid velocity changed around a biomass growth. With these techniques, the complex interaction between the thickness and roughness of the biofilm, and its influence on the flow pattern through the reactor, may be studied.

Fluid rheology. The complex rheology of heterogeneous fluids such as foams, emulsions, and concentrated particulate dispersions are suitable for study with nuclear magnetic resonance velocity imaging techniques. Of particular interest are the phenomena of apparent wall slip, shear polarization, and extensional effects.

Nuclear magnetic resonance velocity imaging of capillary flows has been demonstrated for newtonian sucrose solutions, shear-thinning polymer solutions, and particulated gel systems. Such experiments are useful for rheometry of opaque fluids and for studying slip phenomena, and can provide data on the shear-rate dependence of fluid viscosity over the complete range of shear rates present in the capillary flow.

Suspensions in flow. Flow measurements of solid–fluid suspensions are of intrinsic importance to materials and food processing. Recent research results have shown that magnetic resonance imaging has a role in understanding the basic rheology of multiphase flows, where it allows the noninvasive quantification of the constituent kinematics and flow-induced structure of newtonian and non-newtonian systems for a wide range of concentrations in laminar flows. For example, nuclear magnetic resonance images showing the spatial variation of particle concentration and fluid velocity have been reported for small, negatively buoyant particles suspended in a gear lubricant oil. In addition, magnetic resonance imaging has been used to gain insight into the behavior of shear-induced particle migration and flow-induced microstructural changes in suspension flows.

Turbulence. Imaging speed can be a critical concern for studying turbulent flows. Echo-planar imaging is a high-speed nuclear magnetic resonance imaging modality with a typical acquisition time of 100 ms for a 128 × 128 image. This technique has been widely used in clinical studies and is especially valuable in parts of the anatomy where there is considerable involuntary motion. The echo-planar imaging flow measurement technique is at a stage of development where its application to dynamic processes of fluid flows can be addressed. An imaging technique which combines echo-planar imaging and a spatial tagging sequence has been applied to turbulent flows in a circular pipe. The nuclear magnetic resonance images have successfully visualized the instantaneous complex fluid motions in a cross-sectional plane perpendicular to the pipe axis.

For background information *see* BIOCHEMICAL ENGINEERING; BIOFILM; CHROMATOGRAPHY; COMPUTATIONAL FLUID DYNAMICS; FLUID FLOW; FLUID-FLOW PRIN-

CIPLES; KÁRMÁN VORTEX STREET; MEDICAL IMAGING; NUCLEAR MAGNETIC RESONANCE (NMR); POTENTIAL FLOW; RHEOLOGY; SOLITON; TURBULENT FLOW; WELL LOGGING in the McGraw-Hill Encyclopedia of Science & Technology. C. T. Philip Chang; A. Ted Watson

Bibliography. Y. Couder, J. M. Chomaz, and M. Rabaud, On the hydrodynamics of soap films, *Phys. D*, 37:384–405, 1989; S. T. Grilli, I. A. Svendsen, and R. Subramanya, Breaking criterion and characteristics for solitary waves on slopes, *J. Waterway Port Coastal Ocean Eng.*, 123(3):102–112, 1997; H. Kellay, X.-l. Wu, and W. I. Goldburg, Experiments with turbulent soap films, *Phys. Rev. Lett.*, 74:3975–3978, 1995; K. Kose, Visualization of turbulent motion using echo-planar imaging with a spatial tagging sequence, *J. Magn. Reson.*, 98:599–603, 1992; Z. Lewandowski et al., NMR imaging of hydrodynamics near microbially colonized surfaces, *Water Sci. Technol.*, 26:577–584, 1992; P. L.-F. Liu (ed.), *Advances in Coastal and Ocean Engineering*, vol. 2, 1996; Proceedings of the Second International Meeting on Recent Advances in MR Applications to Porous Media, *Magn. Reson. Imag.*, vol. 12, no. 2, 1994; P. E. Raad, S. Chen, and D. B. Johnson, The introduction of micro-cells to treat pressure in free surface fluid flow problems, *ASME J. Fluid Eng.*, 117(4):683–690, 1995; D. M. Scott and R. A. Williams (eds.), *Frontiers in Industrial Process Tomography*, 1995; A. Torum and O. T. Gudmestad (eds.), *Water Wave Kinematics*, 1990; X.-l. Wu et al., Hydrodynamic convection in a two-dimensional Couette cell, *Phys. Rev. Lett.*, 75:236–239, 1995.

Food-borne disease

Since the initial association of *Escherichia coli* O157:H7 with an outbreak of bloody diarrhea in 1983, the organisms have emerged as a major cause of food-borne gastrointestinal illness which may lead to life-threatening systemic complications such as renal failure and seizures. In June 1996, the largest documented outbreak of bloody diarrhea caused by *E. coli* O157:H7 occurred in Japan. Over 9000 cases were treated, and 7 people died. The outbreak was centered in the western city Sakai, where 6000 cases (216 hospitalized cases) were reported. Most cases involved school children who had consumed contaminated lunches. The precise food source implicated in this outbreak was not clear, but undercooked beef products were suspected.

The largest outbreak in the United States occurred in 1993 and was associated with the ingestion of undercooked hamburger patties served by a West Coast fast-food restaurant. Over 700 people developed bloody diarrhea, approximately 25% were hospitalized, and 4 children died. Children and the elderly are at increased risk for the development of symptoms.

Escherichia coli O157:H7 is the second or third most common bacterial pathogen isolated from stool, and infections with the organisms are recognized as the most common cause of pediatric acute renal failure. In animals models, *E. coli* O157:H7 causes lesions in the intestinal epithelium. The organisms harbor plasmids that may encode intestinal attachment factors and are lysogenized with bacteriophages encoding potent cytotoxins. Despite an increased understanding of the virulence determinants of *E. coli* O157:H7, the precise mechanisms of pathogenesis leading to diarrhea, renal failure, and central nervous system damage are not fully understood. Effective clinical therapies are not available to intervene in the progression of the disease from the gastrointestinal tract to the renal and nervous systems.

Infectious agent. *Escherichia coli* O157:H7 is serologically categorized based on somatic (O157) and flagellar (H7) antigens. The initial binding of the bacteria to intestinal epithelial cells may be mediated by fimbriae. Following the initial binding event which occurs at microvillar tips, *E. coli* O157:H7 closely adheres to intestinal epithelial cells and induces histopathologic alterations referred to as attaching and effacing lesions. Such lesions are characterized by the dissolution of normal microvilli and the formation of pointed cytoplasmic protrusions that contact adherent bacteria in such a manner that the organisms appear perched on a pedestal. The pedestal is formed by a collection of polymerized host cytoskeletal proteins, and bacteria-mediated transmembrane signaling is necessary to polymerize the cytoskeletal elements.

The ability to induce attaching and effacing lesions maps to a 35-kilobase (kb) chromosomal region of *E. coli* O157:H7 that is conserved among all bacterial pathogens mediating such lesion formation. One gene within the 35-kb region encodes a 97-kilodalton outer membrane protein called intimin. Mutations in the intimin gene eliminate attaching and effacing lesion formation. Other genes within the locus may encode proteins necessary for transmembrane signaling and pedestal formation, and the secretory apparatus necessary for the assembly and transport of the bacterial proteins involved in attaching and effacing lesion formation.

Escherichia coli O157:H7 isolates produce one or more protein toxins which are members of the Shiga-like toxin family of cytotoxins. The Shiga-like toxins are genetically and functionally similar to Shiga toxin produced by the enteropathogen *Shigella dysenteriae*. Shiga-like toxins are alternatively called verotoxins, since their cytotoxic activity was first characterized in laboratory experiments using Vero cells.

Pathogenesis. The attaching and effacing lesions probably alter the normal absorptive function of gut epithelium and disrupt tight junctions. Shiga-like toxins may participate in the development of bloody diarrhea by damaging intestinal capillaries. Histopathologic examination of tissues from individuals who developed hemolytic uremic syndrome after *E. coli* O157:H7 infection demonstrated profound damage to small blood vessels in the kidneys and the central nervous system. This finding led researchers to speculate that Shiga-like toxins may act as vasculotoxins, specifically damaging the vascular endothelial cell lining. Recently, it was shown that endothelial cells derived from small blood vessels in human kidneys express high levels of the membrane glycolipid called globotriaosylceramide that serves as the toxin receptor. Shiga-like toxins may pass through damaged gut epithelium to circulate in the blood, selectively bind to cells lining small blood vessels, and initiate vascular damage in the kidneys and the central nervous system, resulting in hemolytic uremic syndrome.

Individuals with *E. coli* O157:H7–associated bloody diarrhea and hemolytic uremic syndrome often have elevated titers of antibodies directed against the O157 lipopolysaccharides in the bacterial outer membrane. The formation of attaching and effacing lesions and damage to blood vessels in the gut mucosa mediated by Shiga-like toxins may allow other bacterial products, such as the O157 lipopolysaccharides, to gain access to the circulation. The presence of O157 lipopolysaccharides in the circulation is of interest since the molecules are known to induce excessive production of proinflammatory cytokines from human peripheral blood monocytes. Proinflammatory cytokines cause a number of changes in vascular endothelial cells, including an increase in membrane globotriaosylceramide expression. Thus, O157 lipopolysaccharides may act to sensitize vascular endothelial cells to cytotoxic action of Shiga-like toxins by upregulating toxin receptor expression.

Transmission. The most frequently implicated vehicle of *E. coli* O157:H7 transmission is undercooked ground beef, although cases of bloody diarrhea have been associated with the ingestion of dairy products, unpasteurized apple juice, salami, foods contaminated with utensils used in the preparation of beef products, and water supplies contaminated with human or bovine feces. Transmission of *E. coli* O157:H7 between individuals has also been documented and may be extensive in institutional settings. Approximately 1% of cattle harbor *E. coli* O157:H7 in their gastrointestinal tract. Meats used for the preparation of ground beef may be contaminated during processing when they come in contact with feces or viscera. Because the bacteria are mixed into ground beef during processing, it is essential to adequately cook the interior of the meat to kill all the bacteria. Because of the extent and severity of the outbreak which occurred in the western United States in 1993, the U.S. Food and Drug Administration recommended that hamburgers be cooked to an internal temperature of 155°F (68.3°C) rather than 140°F (60°C). The Centers for Disease Control acquired hamburger patties implicated in the West Coast outbreak and found less than 100 *E. coli* O157:H7 bacteria per pattie. The low infectious dose of *E. coli* O157:H7 highlights the need for extreme care in cooking beef products.

Diagnosis. Infection with *E. coli* O157:H7 results in a spectrum of clinical presentations ranging from transient asymptomatic carriage to severe extraintestinal complications and death. However, the majority of cases manifest with a predictable series of signs and symptoms. After a 24–48-h period of nonbloody diarrhea, abdominal cramping, and mild fever, individuals begin to experience bouts of bloody diarrhea. Bloody diarrhea may last for 2–10 days and be accompanied by severe abdominal pain. For most individuals, rehydration therapy results in a complete recovery. Unfortunately, approximately 10% of children under the age of 10 develop life-threatening extraintestinal complications 7–10 days after the resolution of the diarrheal disease. These individuals develop the hemolytic uremic syndrome, which is characterized by the abrupt onset of decreased urine production, loss of kidney function, and anemia, and may be accompanied by edema, hypertension, blood-clotting disorders, and seizures. About 5% of individuals with hemolytic uremic syndrome die, and 5% develop chronic end-stage renal disease or brain damage.

Treatment. Treatment for bloody diarrhea and hemolytic uremic syndrome caused by *E. coli* O157:H7 is largely supportive. The use of antibiotics may shorten the duration of diarrhea, but some antibiotics have been demonstrated to upregulate Shiga-like toxin production in laboratory cultures. The use of antimotility agents and narcotics is contraindicated in that they may delay the clearance of the bacteria from the gastrointestinal tract. Individuals with hemolytic uremic syndrome require aggressive fluid and nutritional support, and appropriate treatments to control anemia, hypertension, and seizures. Peritoneal dialysis or hemodialysis is required in cases with prolonged renal failure.

Future treatments are being designed to block the progression from diarrhea to systemic disease. One approach is to minimize the transport of Shiga-like toxins across gut epithelium. Researchers are developing globotriaosylceramide– or globotriaosylceramide analog–containing compounds that could be orally administered to individuals during the diarrheal illness. A related strategy is to neutralize the activity of circulating Shiga-like toxins. Optimal intervention in the progression to hemolytic uremic syndrome may require the intravenous administration of antibodies directed against Shiga-like toxins and orally administered toxin receptor analogs. Agents which inhibit O157 lipopolysaccharide-induced cytokine production may also interrupt the pathogenic process leading to hemolytic uremic syndrome.

For background information *see* BACTERIA; ESCHERICHIA; FOOD MICROBIOLOGY; FOOD POISONING; FOOD SCIENCE; GASTROINTESTINAL TRACT DISORDERS; KIDNEY DISORDERS in the McGraw-Hill Encyclopedia of Science & Technology.   Vernon L. Tesh

Bibliography. M. A. Karmali and A. G. Goglio (eds.), *Recent Advances in Vero-cytotoxin-producing Escherichia coli Infections*, 1994; R. L. Siegler, The hemolytic uremic syndrome, *Pediatr. Clin. N. Amer.*, 42:1505–1529, 1995; P. I. Tarr, *Escherichia coli*

O157:H7: Clinical, diagnostic, and epidemiological aspects of human infection, *Clin. Infect. Dis.*, 20:1–10, 1995; V. L. Tesh and A. D. O'Brien, The pathogenic mechanisms of Shiga toxin and the Shiga-like toxins, *Mol. Microbiol.*, 5:1817–1822, 1991.

Food manufacturing

Meat has been a part of the human diet since the beginning of recorded history, and in some instances it has been important for survival. It is an excellent source of nutrition because it is high in protein and satisfies specific requirements such as the minerals iron and zinc, essential fatty acids, and some of the B vitamins.

Consumer attitudes. Since the 1960s, the consumers' views about meat have changed. There is public concern that meat consumption is associated with degenerative diseases such as coronary heart disease and cancer, and that meat is a carrier of food-borne illnesses, either minor or fatal.

Since the early 1960s, the consumption of red meat, primarily beef and pork, has generally been declining, while there has been a substantial increase in the consumption of poultry (chicken and turkey). In the United States the annual per capita consumption of red meat fell from about 130 lb to 116 lb (60 kg to 50 kg) while poultry increased from about 26 to 60 lb (10 to 30 kg).

Meat processing has been altered in response to consumer attitudes and purchase patterns. Low-fat meat is being produced and improved processing technologies are providing more consistent quality, and health and safety considerations are continually advancing.

Fresh meat. The process for fresh-meat production includes rearing the animal, transportation, slaughter, chilling, and fabrication of the carcass. Advancements in genetics, nutrition, and management have resulted in carcasses with lower fat content; and improvements in transportation, slaughter technology, and chilling have improved the quality of meat. Fresh meat is identified as "not ready to eat," and must be heated properly prior to consumption. Adequate cooking is the main defense against food-borne illness in fresh meat, and the culinary preparation determines nutritional impact.

Processed meat. Among the many possibilities for processing meat, general practices include nitrite curing and heat processing. The result is a product (such as a wiener) which has a typical pink color and a characteristic flavor. Because of the additives used, heat processing and vacuum packaging extend the shelf life to 6 weeks or more for these products, compared to 3–5 days for a retail fresh-meat package. Such processed products are identified as "ready to eat," and can be consumed directly from the package, without further preparation. The major defense against food-borne illness is to prevent recontamination of the products following manufacture and prior to packaging.

Distribution and preparation. In the United States, distribution channels are complicated, and transport distances are long. This has hindered innovations for fresh meat. Modified-atmosphere packaging and retail-ready packages have been slow to be adopted. Fresh meat is normally distributed as larger pieces, often in vacuum bags, and then converted to retail cuts at the point of sale.

The handling and preparation of fresh meat in institutions and individual households represents a potential weak link in the system of protecting against food-borne illnesses. In order to improve this situation, safe-handling labels are now applied to fresh-meat packages. The label provides the wholesaler, retailer, and consumer with simplified reminders or instructions for storage temperature, proper sanitation, and cookery and handling of leftovers.

Meat inspection. Meat inspection regulations were enacted in the United States around 1900, and they were amended and revised in the 1950s and 1960s. They were designed to prevent human consumption of diseased and damaged meat, and to ensure that the operations necessary to convert animals that were raised for food into meat are conducted in a sanitary and appropriate manner. The system depends on on-site, direct, visual inspection.

In recent years a call has been made to update the system to be more scientifically based and to make greater use of preventive measures. In July 1966, the "USDA Pathogen Reduction; Hazard Analysis and Critical Control Point (HACCP) Systems" was announced. This rule has four major elements.

1. Every plant must adopt and carry out its own HACCP plan. The HACCP plan is a preventive approach to food safety. The effectiveness plan must be demonstrated by the plant and will be continually verified by inspectors from the Department of Agriculture's Food Safety and Inspection Service.

2. Every plant must regularly test carcasses for generic *Escherichia coli* to verify the effectiveness of the plant's procedures for reducing and preventing fecal contamination, which is the major source of contamination with harmful bacteria such as the enterohemorrhagic strain of *E. coli* (O157:H7) and *Salmonella*. Generic *E. coli* is the best microbial indicator of a plant's ability to keep the levels of fecal contamination as low as possible.

3. All slaughterhouses and plants producing raw ground products must ensure that their *Salmonella* contamination rate is below the current national baseline incidence. This is the first regulatory performance standard for a pathogen on raw meat and poultry.

4. As the foundation for the HACCP plan, every plant must adopt and carry out a written plan for meeting its sanitation responsibilities (known as Sanitation Standard Operating Procedures). Effective sanitation in slaughterhouses and processing plants is essential to preventing direct adulteration of meat and poultry products.

Nitrite curing. During the 1970s, the safety of meats cured with nitrate and nitrite was strongly debated.

There was concern as to whether preformed nitrosamines were present at levels of concern and whether the levels of residual nitrite represented a risk to human health. After much study, the use of nitrate was essentially eliminated, the levels of nitrite used were lowered, and much tighter control of manufacturing processes was instituted. Ascorbate and erythorbate were used at maximum levels to inhibit formation of nitrosamines, and a nitrosamine monitoring program for bacon was started by the Department of Agriculture.

Even though it appeared that the problem was solved, concern resurfaced from epidemiological studies suggesting a link between the consumption of cured meat and some forms of childhood cancer. However, recent evidence has suggested that nitric oxide is important to biological function in humans, and that the intake of residual nitrite from cured meat is much lower than previously thought.

Work in the 1990s has shown the residual nitrite level in cured meat is about one-fifth of what it was in the 1970s. Therefore, the contribution of cured meat to total body burden of nitrite has been greatly reduced. In addition, it is known that about 5% of ingested nitrate (from green leafy and root vegetables and drinking water, for example) is reduced to nitrite by the microorganisms in the saliva and ingested as nitrite. It has been established that in humans the enzyme nitric oxide synthase catalyzes the stepwise oxidation of the amino acid 1-arginine to nitric oxide and 1-citrulline. Nitric oxide is a biological messenger that influences the physiological functions of neurotransmission, blood clotting, blood pressure control, and the immune system's ability to kill tumor cells and intracellular parasites. There is also growing evidence that conversion of dietary nitrate into nitrite in the saliva may provide significant protection against pathogens in the human digestive tract.

Bovine spongiform encephalopathy. Bovine spongiform encephalopathy, also known as mad cow disease, is a neurodegenerative and fatal disease in cattle. It has recently had enormous impact on the meat industry, especially in Great Britain, and has been the subject of an intense research effort. Although there have been no documented cases of Creutzfeldt-Jakob disease (a spongiform encephalopathy in humans) linked directly to consumption of beef, it has been suggested that a variant form of the human disease associated with bovine spongiform encephalopathy exists.

For background information *see* CANCER; HEART DISORDERS; FOOD MICROBIOLOGY; FOOD POISONING; FOOD PRESERVATION; FOOD SCIENCE; NITROGEN OXIDES; NUTRITION; OXIDATION PROCESS in the McGraw-Hill Encyclopedia of Science & Technology.

Robert G. Cassens

Bibliography. American Meat Institute, *Meat and Poultry Facts*, 1994; R. G. Cassens, *Meat Preservation: Preventing Losses and Assuring Safety*, 1994; R. G. Cassens, Use of sodium nitrite in cured meats today, *Food Technol.*, 49(7):72, 1995.

Forest ecosystem

Allelopathy is any direct or indirect effect, positive or negative, of one plant (including microorganisms) on another through the production of chemical compounds that pass into the environment. This type of effect apparently has been found in many ecosystems, including natural forests and plantations. More than 100 forest species have been found to produce and release allelopathic compounds (allelochemicals), although there are still few field confirmations of the phenomenon.

Allelopathy usually occurs between plants that have not coevolved. Many species seem to show one allelopathic behavior in their natural habitats but a rather different behavior in habitats where they are foreign. This may be due to the lack of previous coadaptation of microorganisms and plants in the underwood. This variation may provide a partial explanation for the poverty of species and the lack of cover in the understory of some exotic species in many parts of the world in comparison with other indigenous forest species. Also, some environmental conditions can affect microorganisms producing phytotoxins in the soil. Anaerobiosis is such a condition that stimulates microorganisms to produce an accumulation of allelochemicals. Low and extremely high moisture can have the same effect. Some allelochemicals could be degraded by microbial species that grow similarly or even better under those conditions.

It is still not clear if allelopathic effects are simply a secondary effect of processes by which plants evade autotoxicity or if allelochemicals are produced mainly for competition. However, it is clear that nearly every plant produces substances that can affect the growth or development of other plants. Many of these substances are toxic even for the cells of the emitter species. Thus, if the conditions are adequate, those phytotoxic substances that are released to the environment can represent a tool for competition or for cooperation. Most of the identified allelochemicals are produced in secondary metabolism. Phenolics are the most common forms of allelochemicals in humid forest environments; terpenoids are most common in dry-climate forests.

Although not every allelochemical has been identified, a wide variety of molecules have been classified as inhibitors of stimulants of the same species or of different species of forest plants. Examples are water-soluble organic acids, straight-chain alcohols, aliphatic aldehydes, ketones, simple unsaturated lactones, long-chain fatty acids, naphtoquinones, anthraquinones, complex quinones, terpenoids, steroids, simple phenols, benzoic acid and its derivatives, cinnamic acid and its derivatives, coumarins, flavonoids, tannins, amino acids and polypeptides, alkaloids and cyanohydrins, sulfides and mustard oil glycosides, purines, and nucleosides.

The rates of production and release of allelochemicals are affected by many abiotic and biotic factors,

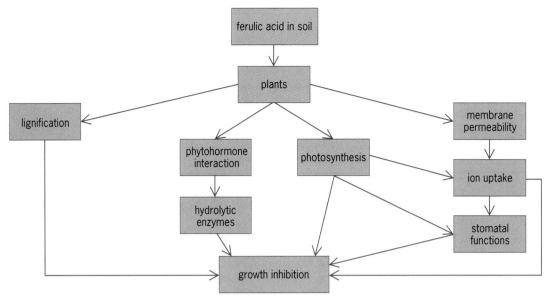

Fig. 1. Physiological effects of ferulic acid on plant growth. *(Courtesy of Rama Devi)*

because stressed plants have higher production and release rates of allelopathic compounds.

Physiological effects of allelochemicals. Physiological effects of allelochemicals are difficult to summarize because of the various modes of action and molecules involved, and the differences in physiology among receptor species.

By comparing the physiological effects of allelo-chemicals released by forest species to synthetic pesticides, several important differences become apparent. The effects of allelochemicals are less specific in mode of action, more diffuse, less concentrated, and less durable in the environment. That is, natural phytotoxins are most biodegradable and affect weakly many physiological processes, while synthetic phytotoxins are less biodegradable and affect strongly one or a few individual physiological processes. In some cases, the effects of natural phytotoxins can be very selective, affecting one or a few receptor species, or can affect all the understory species.

The physiological effects of allelochemicals are very diverse. The effects can be primary (on the basic metabolism) or secondary (on the general growth), reflecting one or several main effects. These effects vary, however, depending on the concentration of the allelochemical. There are many reports of effects of allelochemicals on germination and seedling growth, but also reports about metabolic effects on hormone interactions, cell division and elongation, enzymatic activities, protein synthesis, respiration, photosynthesis, ion and water uptake, transpiration and stomatal conductance, adenosine triphosphatases, and membrane permeability. The physiological effects of the allelochemical ferulic acid on plant growth are shown in **Fig. 1**.

Methods of releasing allelochemicals. Allelopathic substances can negatively affect the species producing them, so that secondary autotoxic molecules must be released into the environment, thus becoming allelopathic. These molecules can be released by raindrops or condensing fog from the aerial parts of the emitter (this being very effective if the molecules are water soluble), or can be released into a hot dry atmosphere if metabolites are volatile (as with many terpenoids) [**Fig. 2**]. In some plants, these molecules

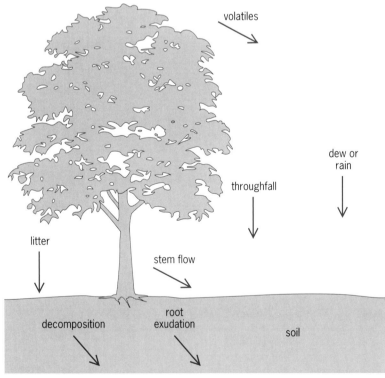

Fig. 2. Release pathways of allelochemicals.

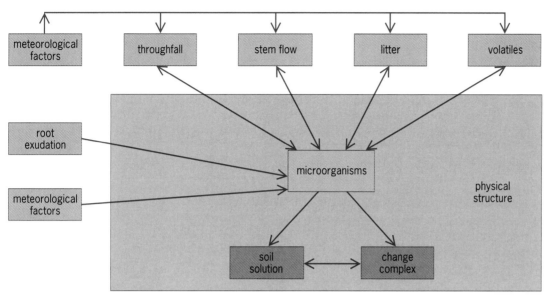

Fig. 3. Processes that affect allelopathy in the forest soil. *(After M. J. Reigosa, C. Souto, and L. Gonzalez, Allelopathic research: Methodological, ecological and evolutionary aspects, in S. S. Narwal and P. Tauro, eds., Allelopathy: Field Observations and Methodology, Scientific Publishers, 1996)*

are released directly through excretion from the roots. Allelochemicals can also reach the soil from decomposing plant material. In forestry, this technique is generally considered the most common method of releasing allelochemicals.

Soil processes. With the exception of volatiles that move directly from the emitter to the receptor (and in some cases, volatiles can also be condensed and washed by the rain to the soil), all other pathways deposit allelochemicals into the soil. The effectiveness of the allelochemicals in the environment can be influenced in the soil by adsorption, desorption, polymerization, and decomposition processes, and also by leaching and microbial degradation. Microorganisms can detoxify soil, using the allelochemical as a source of carbon, energy, or nitrogen, or they can contribute to the phytotoxicity by degrading a nontoxic molecule into a phytotoxic one. Therefore, the same forest species may or may not be allelopathic depending on the soil conditions (pH, organic matter content, texture) and especially on the microbial growth conditions. Moisture and aeration can have a strong effect on the accumulation or degradation of allelochemicals in the soil. Other important abiotic factors include temperature, light, and nutrient content of the soil.

Some allelopathic processes indirectly affect the plants in the forest ecosystems. For example, mycorrhizae, the fungal mycelium that invades the roots of a seed plant, can be affected by the allelochemicals; or the nitrification can be slowed by allelopathy (**Fig. 3**).

Allelopathic processes. Some allelopathic processes in forests affect tree growth on understory species and prevent some forests from regenerating under natural conditions. Other processes in forests are influenced by organisms such as nematodes and my-

corrhizae or biological competition among trees and plants.

Understory species. The effect of different allelochemicals released on the understory species is well known. Some works show how the scarcity of tree species and the lack of cover in the understory of some forest plantations can be attributed to allelopathy. For example, in Santiago de Compostela, Spain, the underwood of oak (*Quercus robur*, a climax species) is different from that of radiata pine (*Pinus radiata*) and very different and richer than that under blackwood acacia (*Acacia melanoxylon*), silver wattle (*A. dealbata*), and eucalyptus (*Eucalyptus globulus*), even when all the species are growing in the same location, on the same soil, and with the same abiotic conditions. The release of allelochemicals when plant residues of the three latter species are decomposing and the dates of release are two of the main explanations for the poverty of the understory. In Korea, the floristic composition of the underwood can be well correlated with the emission of allelochemicals (mainly phenolics) of several pine species. Several species of *Eucalyptus* have shown an understory inhibition in different parts of the world (United States, Australia, Tasmania, Spain, Iraq). However, the results may differ with the same *Eucalyptus* species—for example, showing no allelopathic activity in Australia but strong effects in other countries. In *Eucalyptus*, several potential allelochemicals have been identified, including α-pinene, camphene, cineole, and phenolic acids. Many leguminous trees also show an inhibition area under their canopies, partially due to mimosine or other allelochemicals.

When forest species are not dominant but scattered, a halo effect can be seen around some species, with bare soil under the canopy of individual trees.

Also, a patterning effect can be seen in many multi-species forests, each tree having an associated vegetation. The effect of juglone release by black walnut (*Juglans nigra*) is one of the best-known allelopathic phenomena, affecting understory herbaceous plants. Many other tree species show a similar effect on pattern.

Replanting and regeneration problem. Some forest species fail to regenerate under natural conditions. Some reasons may be the effect of nematodes or biological competition, but in some cases allelopathy is an important factor. This allelopathic phenomenon is being researched actively in Canadian and French forests of spruce (*Picea abies*), black spruce (*Picea mariana*), and Jack pine (*Pinus banksiana*). Sometimes the effect can be due to shrubs (for example, *Kalmia angustifolia* in Canadian forests), and in other cases there can be an autotoxic effect, linked to a high rate of allelochemical production and release coupled with a low rate of soil degradation. Another cause of this failure may be the inhibition of mycorrhizae by the leachates of adult plants and the decomposition of plant residues. More research must be done in order to better understand the regeneration problem and the allelopathy hypothesis.

Allelopathic research is also important in agroforestry, where chemical compatibility can help the sustainability of the system. Beneficial interactions should also be expected. Knowledge of chemical relationships in the forest can also help the traditional forestry plantations.

For background information *see* ALLELOPATHY; BIODEGRADATION; CHEMICAL ECOLOGY; FOREST AND FORESTRY; FOREST ECOSYSTEM; FOREST GENETICS; FOREST SOIL; SOIL in the McGraw-Hill Encyclopedia of Science & Technology.   Manuel J. Reigosa

Bibliography. S. S. Narwal and P. Tauro (eds.), *Allelopathy in Agriculture and Forestry*, 1994; A. R. Putnam and C. S. Tang (eds.), *The Science of Allelopathy*, 1986; E. L. Rice, *Allelopathy*, 1984; S. J. H. Rizvi and V. Rizvi (eds.), *Allelopathy: Basic and Applied Aspects*, 1992.

Forest pests

A forest pest is any organism that adversely affects the functioning of forest trees. Exotic forest pests, which have been introduced into the United States from abroad, have been primarily insects and infectious microorganisms (for example, the gypsy moth and microorganisms that cause chestnut blight). Of greatest concern are exotic forest pests from temperate forests similar to those of the United States, such as in Europe, eastern Russia, China, Australia, New Zealand, and Chile. Much of the recent importation of logs, wood chips, and lumber from temperate forests has been into the Pacific Northwest, including Oregon, Washington, and northern California.

Most of the forestland in the Pacific Northwest is federally owned. Historical harvests were relatively constant until recently, when allowable harvest levels from federal forests were drastically reduced. The wood industry, dependent on federal timber, faced a new challenge: a lack of raw material. Alternative sources of wood were sought. Traditionally, such sources were found just outside the normal industrial working circles, perhaps in an adjacent state. But as competition for raw materials increased domestically and wood prices escalated, the search for raw material expanded abroad. Russia, Chile, New Zealand, and several Pacific Rim countries have increased their exports of raw wood products to the United States.

Previously introduced exotic pests. Chestnut blight, the most destructive forest disease ever recorded in North America, is often given as an example in describing the risks of pest introductions. The disease reduced the American chestnut from the most economically important hardwood species in the eastern United States to mere stump sprouts. Chestnut blight is native to Asia and was probably imported on infected nursery plants around 1900.

Perhaps the most significant insect pest introduced was the European gypsy moth. It was brought into the eastern United States in the 1870s by a French entomologist in an effort to find a silk moth that could survive in North America. This insect now infests more than 123 million acres (50 million hectares) of northeastern forest, with some infestations in the central and western states.

Current threats. In 1991 a delegation of U.S. Department of Agriculture scientists and representatives visited the Russian Far East and Siberia to assess the potential threat of pests being introduced into the United States. The risk assessment team identified 36 groups of potentially damaging pests from Russia, and the following six were addressed in a risk assessment report: Asian gypsy moth, nun moth, pine wood nematodes, larch canker, Eurasian spruce bark beetle, and annosus root disease.

The Asian gypsy moth is capable of faster spread than its European cousin because female moths can fly up to 12 mi (20 km) and the caterpillars eat a wider variety of plant species, including many conifers (more than 600 species in Russia). (European gypsy moth females have no wings so they cannot fly at all.) The nun moth is similar in habits, hosts, and potential damage. Pine wood nematodes and their insect carriers affect weakened trees of several species by attacking and killing these trees. The larch canker is the most destructive disease of European larch. This disease occurs in eastern larch in eastern North America, but has not been found in western North America. The Eurasian spruce bark beetle is considered to be the world's most dangerous exotic bark beetle. Its associated stain fungus is among the world's most pathogenic fungi to conifers. Annosus root disease is caused by a fungus that decays and kills many species of conifers and deciduous trees throughout the world. Although annosus root disease is already present in the United States, the introduction of new strains from Russia could cause serious damage to coniferous forests. Unlike most pests, the causative species (*Heterobasidion annosus*) resides deep within infested logs, making it difficult to eradicate.

Exotic pest introductions from New Zealand or Chile may have less serious consequences than pests from Russia. Tree species that may be imported from New Zealand or Chile are not native to those countries (except for two Chilean hardwoods) and have fewer associated pests than the conifers in Russia. Nevertheless, the Department of Agriculture risk assessment team identified several pests of particular concern that could be imported from New Zealand or Chile: a blue-stain fungus is carried by root-feeding bark beetles that attack stressed trees (trees that are weakened by some agent or condition such as drought, soil compaction, or other pests), and a wood wasp and its fungal associate which causes sapwood rot that eventually results in tree mortality of several species of pine in New Zealand.

Ecological impacts. The ecological impacts from introduced pests are substantial but difficult to quantify. Exotic pests pose a serious threat because they are free of many ecological constraints present in their natural environments. Pests that exist in close harmony with their enemies in their natural environment may increase exponentially in the absence of such controls. Insects and diseases that normally cause only minor damage in native hosts may cause widespread mortality in new host species. Pests of other non-tree species of plants may also be introduced on the logs. These might include fungi, insects, and weed species. These could also be deleterious in a new environment. Such widespread tree mortality, in turn, could benefit some associated species and threaten others, alter water quality and regional hydrology, increase wildfire frequency, and reduce carbon storage capacity.

Nonecological impacts. Exotic pests from imported wood products can affect industries related to forestry, such as the Christmas tree, ornamental, and nursery industries, directly through plant health declines and indirectly through quarantines of infested material shipped out of state. Nearly 85% of Christmas trees and nursery plants in Oregon are exported to other states or countries. The American Association of Nurserymen has strongly advised extreme caution when considering wood product imports, because the nursery industry commonly bears the brunt of new pest introductions. By law, nursery, ornamental, and Christmas tree growers are forbidden to ship plant material infested by arthropods or pathogens. Consequently, each new pest creates problems for regulatory agencies and growers and increases production costs.

Methods to reduce pest introductions. It is nearly impossible to completely eliminate the risks of pest introductions associated with international trade. Instead, regulators use various methods to reduce the probability of introducing nonindigenous pests on logs, chips, and lumber. Methods such as visual inspections, bark removal, topical treatments, heat, fumigation, diffusible fungicides, and irradiation have some effect. Visual inspections can detect the highly visible pest stages, such as insect egg masses or wood decay, but may miss the small organisms and pests under the bark. Debarking removes organisms that reside on or just inside the bark, but debarking is rarely complete, especially around branch stubs or other trunk deformities. Topical treatments applied to wood surfaces provide a barrier against invasion by insects and fungi, and may also keep these organisms from escaping from infested wood. However, these treatments do not eradicate organisms deep within large wood pieces.

Because many potentially harmful organisms reside in the wood, treatments such as heat, fumigation, diffusible fungicides, and irradiation have been suggested for ''sterilizing'' larger pieces of wood. Heating has long been used for eliminating pests from a variety of wood products. Fumigation with methyl bromide is effective against pests deep within wood, but because of concerns about the effects on the ozone layer, use of methyl bromide is scheduled to be phased out. Diffusible materials such as borates are toxic to insects and decay fungi, do not degrade the wood as does heat, and can be verified by a simple color test. Gamma irradiation has been used to eliminate pests from a variety of materials, but its use on wood has been limited to sterilization for laboratory purposes.

For background information *see* FOREST AND FORESTRY; FOREST ECOSYSTEM; FOREST PEST CONTROL; FUNGI; PLANT PATHOLOGY in the McGraw-Hill Encyclopedia of Science & Technology.   Gregory M. Filip

Bibliography. G. M. Filip and J. J. Morrell, Importing Pacific-Rim wood: Pest risks to domestic resources, *J. Forest.*, 94(10):6–10, 1996; D. J. Goheen and B. M. Tkacz, Pest risks associated with importing unprocessed Russian larch logs to the United States, *West. J. Appl. Forest.*, 8(3):77–80, 1993; J. J. Morrell, Importation of unprocessed logs into North America: A review of pest mitigation procedures and their efficacy, *Forest Prod. J.*, 45(9):41–50, 1995.

Forestry

Recent developments in forestry have included the establishment and investigation of mixed- and single-species plantations of trees, and the evolution of sustainable harvesting in tropical forests in order to meet the needs of future generations, as well as the present.

Mixed- and Single-Species Forest Plantations

Many large- and small-scale forestry plantations are being established around the world to meet people's needs for the wide range of goods and services which trees can provide. For example, from 1980 to 1989 a reported annual area of 5.4 million acres (2.6 million hectares) of forest plantations was established in tropical countries, reflecting the need to meet such demands from alternative sources to the natural forest.

The relative merits of mixed- versus single-species plantations have been considered by both foresters and the environmental community. However, the sustainability of the flow of goods and services from any forest plantation, whether composed of mixed or single species, depends on the correct matching of the species to the site and to the objectives of the plantation. Establishment, maintenance, and harvesting must be carried out in accordance with good forestry practice, especially the maintenance of soil properties, and with the requirements of the next crop in mind.

Mixed-species plantations. The choice of species for mixed plantations can be based on natural associations: plant associations that occur naturally, such as pioneers, which rapidly become established and provide soil cover; or climax species, which can develop in the shade of others; or species of intermediate stages of the succession, which may need neighbors to improve their growth or form. If the species have been matched to the site and the other criteria are met, there may be several reasons for meeting plantation objectives through the establishment of several species in intimate mixtures, lines, or small groups. These reasons include ecological stability, synergy, protection, conservation of biological diversity, insurance against failure, production of a wide range of goods and services, fire protection, and appearance.

Ecological stability. Research continues to focus on the use of mixtures of species to increase both the resistance of the stand (its ability to resist change) and its resilience (its ability to return to its former dynamic state after responding to an external influence, whether natural or artificial). The establishment of mixtures of species, or the encouragement of natural regeneration of mixtures of trees or shrubs in plantations, may improve both resistance and resilience through improved nutrient recycling, resistance to pests or diseases, or resilience to climate change.

Synergy. Synergy, in the present context, means that one species enhances the biological performance of another. This phenomenon has been studied most in temperate countries, but there are few documented instances of species mixtures enhancing either total yield or the yield of specific components of the crop. Scandinavian research suggests that the mixed-species effect may be small, and it may be difficult to demonstrate significant practical gains; increases in yield even from nitrogen-fixing species may be restricted to those sites that were initially deficient in nitrogen.

Protection. One species may act as a nurse to another in order to provide protection against frost or other harmful climatic factors, or to improve the form of another species. Nurse species have been used in West Africa, Central America, and Asia to protect species of mahogany (Meliaceae) against attack by the insect *Hypsipyla* spp. by providing shade.

Conservation of biological diversity. Plantations may contribute to biological diversity at the species level where management practices such as thinning encourage a range of tree and shrub species to regenerate. There is growing interest in using plantations of native or exotic species to reestablish forest cover and thus create a forest environment for the gradual restoration of natural species diversity on degraded lands.

Insurance. Although there is very little risk to survival if a species has been correctly matched to a particular site, a mixture of species can act as an insurance policy against market failure (drop in demand) or loss of a single species due to an unexpected insect or disease attack. However, a mixture of species alone is no safeguard against insect or disease attack, and at least for large-scale plantations the presence of a good infrastructure can provide insurance against natural disasters such as fire or insect attack.

Goods and services. Mixtures of species will provide a wide range of goods and services from the same area of land. This may be especially important in community forestry, where a range of different products are required, which may also vary by the type of user. The range of goods and services produced may include wildlife, where mixtures of species can provide the habitat requirements (such as shelter of the source of food) for wildlife species raised for commercial or subsistence purposes or for tourism.

Fire protection. A belt of a different species can create a discontinuity that can retard the spread of a crown fire. Another example is the planting of a broadleaved species, which produces a less flammable litter when mixed with conifers.

Appearance. The irregular appearance of a mixture of species, with a range of height, crown shape, and foliage color, is usually thought to be more esthetically satisfying than a single species.

Disadvantages. There may be disadvantages associated with mixed-species plantation, such as the problems associated with the management of a number of species with differential growth rates. Close monitoring and intervention may be necessary to maintain the desired outcome. The best way of obtaining at least some of the benefits of mixed-species plantations may be to enhance the processes of natural regeneration.

Single-species plantations. The reasons for establishing single-species plantations include simplicity of management, uniformity of product, and high volume per unit area. Most plantations are established with a single species; since many of the fast-growing species which produce industrial roundwood (for example, the pines) are pioneers typical of the early stages of succession that are often species poor in nature, they may not be suited to growth in mixture.

Mixtures in various countries. Although there is no record of the global area of plantations established in species mixture, or even country records, a great many mixtures of tree species have been tested in many countries. For example, the following species mixtures have been used in general forestry practice:

Robinia pseudacacia and poplar in line mixture [Hungary]; *Eucalyptus tereticornis* over *Acacia auriculiformis* [Vietnam]; *Cupressus lusitanica* and *Grevillea robusta* [Kenya]; *E. torelliana* and *E. tereticornis* [Togo]; *Swietenia macrophylla* (mahogany), *Tectona grandis* (teak), and *Artocarpus integrifolius* (breadfruit) [Sri Lanka]; *Nauclea diderichii* as a nurse for Meliaceae [Nigeria]; *Leucaena leucocephalla* as a nurse for *Tectona grandis* (teak)—the nurse is now seriously affected by Leucaena psyllid (*Heterospylla cubana*) [Indonesia].   J. B. Ball

Sustainable Forest Harvesting in Tropical Forests

The concept of sustainability in forestry has evolved from concern with the sustained yield of wood to concern with a much broader range of wood and nonwood forest products. European governments for centuries have been interested in the sustained production of some forest products. At one time, it was important to maintain a reliable supply of wood for warships. While European foresters were developing practices to balance timber utilization with forest growth, large new forests were being discovered. These vast forests were treated as nearly limitless. Populations grew, and the tradition of managing forests for the indefinite future became a guiding principle of forest management.

In 1987, sustainable development was defined as development that meets the needs of the present without compromising the ability of future generations to meet their own needs. Many experts recognized that there would be a continued pressure on the tropical forests caused by land clearing for agriculture and development purposes, and forest degradation from unsustainable forest practices. Several authorities, including the Food and Agriculture Organization of the United Nations (FAO), suggested that policies must be developed to ensure comprehensive land-use planning, improve timber concession agreements, provide incentives for sustainable practices, and provide other fiscal controls.

Between 1980 and 1990 the annual estimated loss of natural forests in the tropical regions amounted to 38 million acres (15.4 million hectares). In 1990 the estimated worldwide area of tropical forests was 4340 million acres (1756 million hectares), or 14% of the Earth's land area.

Recent estimates suggest that nearly two-thirds of the worldwide tropical deforestation is the result of clearing land for agriculture. Forest degradation and forest destruction in the tropics are often caused by careless, unplanned, and uncontrolled harvesting of forest products. This is particularly true of timber harvesting of the rainforests and excessive fuelwood collection in the dry forests.

Objectives. The success of sustainable forest development is closely linked with the importance given to the involvement of the local population in rural development projects. It is essential to create conditions for the dissemination of information that permit the introduction of forest harvesting practices that are environmentally sound and socially accept-

able. Forest utilization activities must, for present and future generations, maintain the regenerative capacity of the forest, maintain the forest biodiversity, continue to provide forest products, provide a multitude of environmental services, and offer a variety of social benefits. Because of the complexity and diversity of the flora and fauna of the forests, it is not easy to introduce the best management practices.

Reduced-impact harvesting. Reduced-impact harvesting techniques have been introduced in a number of tropical forests. Four ingredients are essential to forest harvesting operations if forests are to be managed on a sustainable basis: comprehensive harvest planning; effective implementation and control of harvesting operations; thorough postharvest assessment and communication of results to the planning team and to the harvesting personnel; and development of a competent and properly motivated workforce.

Directional tree felling. The underbrush around the base of the tree to be felled must be cleared away with hand tools. Vines or climbers must be cut about 1–2 years before the tree is to be felled. If the climbers are not cut in advance so they decay and lose their strength, then felling becomes much more dangerous and damaging. The vines connect trees through the crowns so that when one tree is felled many more are pulled down or broken. Experience has shown that on the order of 20 large trees per acre (50 per hectare) will be saved from death or damage by cutting the climbers in advance.

The direction the tree is to be felled is brightly marked on the tree. This facilitates the coordination of felling direction and the location of the planned skid trails. The optimal direction to fell the tree to avoid damage and to provide efficient extraction of the logs does not often coincide with the natural lean of the tree. Overcoming this discrepancy involves more hard work. Wedges and a refined manipulation of the felling cuts can usually do the job. Sometimes more elaborate methods using engines and pulling cables are necessary.

Damage to the residual trees has been found to be reduced to one-tenth that of conventional felling by using these reduced-impact techniques.

Log extraction. Although only a few valuable trees per acre are harvested in tropical forests, there can be an inordinate amount of damage caused by the extraction process. Vegetation is damaged and destroyed, animal habitat is damaged, soil is compacted and eroded, water is polluted, and many additional resources (including money) can be wasted.

Planned skid trails. Ground skidding systems normally have the most impact on the forest environment. They are also the most commonly used extraction system. Both tracked and wheeled skidding machines are used on a network of skid trails connecting the site of the trees being harvested with the truck roads.

Mapping and planning the location of skid trails in relation to the trees to be felled and the road system has produced substantial benefits to the tropical forest environment. For example, a controlled study in Malaysia showed that mapping and planning can reduce the length of skid trails constructed by two-thirds, and reduce the area of land occupied by skid trails by three-fourths. These roads and skid trails were designed to avoid drainage ways and to use the high ground.

Cable systems. The use of cable yarding systems by skilled workers is reappearing in tropical forests. These cable systems can suspend the logs above the ground so soil and water impacts can be greatly reduced. In the 1940s, large cable yarding systems were used in tropical forests. These systems gained a reputation for being very destructive because they did not lift the log completely off the ground but depended upon force to drag logs to roads.

Aerial systems. More sophisticated systems, such as helicopters that lift logs completely above the forest, have been used in the tropics and will be used with greater frequency in the future. Because such systems are very expensive to operate, their use tends to be limited. Research into improving the various flying technologies will continue.

For background information *see* ECOLOGY, APPLIED; FOREST AND FORESTRY; FOREST FIRE; FOREST GENETICS; FOREST HARVEST AND ENGINEERING; FOREST MANAGEMENT; FOREST TIMBER RESOURCES in the McGraw-Hill Encyclopedia of Science & Technology.

Rudolf Heinrich

Bibliography. M. G. R. Cannell, D. C. Malcolm, and P. A. Robertson, *The Ecology of Mixed Species Stands of Trees*, Brit. Ecol. Soc. Spec. Pub. 11, 1992; D. P. Dykstra and R. Heinrich, *FAO Model Code of Forest Harvesting Practice*, 1996; Food and Agriculture Organization, *Forest Resources Assessment 1990*, For. Pap. 124, 1995; Food and Agriculture Organization, *Plantations in Tropical and Subtropical Regions: Mixed or Pure*, 1995; M. J. Kelty, B. C. Larson, and C. D. Oliver (eds.), *The Ecology and Silviculture of Mixed-Species Forest*, 1992; J. B. Larsen, Ecological stability of forests and sustainable silviculture, *For. Ecol. Manag.*, 731:85–96, 1995; N. P. Sharma (ed.), *Managing the World's Forests: Looking for Balance Between Conservation and Development*, 1992; World Commission on Environment and Development, *Our Common Future*, 1987.

Fullerene

The award of the Nobel Prize in Chemistry for 1996 to H. W. Kroto, R. F. Curl, and R. E. Smalley stems from their 1985 discovery that laser vaporization of carbon atoms from a graphite surface can lead to the formation of unusual ball-like clusters of carbon, of which C_{60} is by far the most abundant. The resemblance of the unique soccer-ball-like structure of C_{60} to the geodesic domes designed by R. Buckminster Fuller led to the name buckminsterfullerene for this beautiful, highly symmetric molecule. Exploration of the chemistry of fullerenes, including C_{70}, C_{76}, C_{84}, and larger closed carbon structures, became practical through the 1990 discovery that the soot deposited by an electric arc passing between two graphite rods is rich in fullerenes, particularly C_{60} (popularly known as buckyball). *See* NOBEL PRIZES.

This new field has developed extremely rapidly since 1990. The chemistry of C_{60} and higher fullerenes, which possess alternating single and double bonds in different arrays of five- and six-membered rings, encompasses a broad range of thermal and photochemical addition reactions, which are fundamentally different from the substitution reactions characteristic of benzene and related two-dimensional aromatic systems.

Cycloadditions to C_{60}. The structure C_{60} is a reactive dienophile toward 1,3-dienes in Diels-Alder $[4 + 2]$ cycloaddition reactions, providing access to a large variety of compounds in which a six-membered ring is fused to the C_{60} core across a 6,6-junction (the junction of two six-membered rings on the fullerene surface). This 6,6-bond has high double-bond character and is the usual site for additions to C_{60}. Representative examples are adducts (**1**)–(**4**) in **Fig. 1**. This reaction has been used recently to connect porphyrin derivatives to C_{60} [structure (**5**)]. The fullerene C_{60} is also a reactive dipolarophile in $[3 + 2]$ additions to 1,3-dipoles, such as diazo compounds, azides, and azomethine ylides. (A dipolarophile is a reagent that has affinity for 1,2- or 1,3-dipolar reagents.) These result in methanofullerenes, fulleroazirines, and fulleropyrrolidines, respectively, also from addition across 6,6-bonds [structures (**6**)–(**8**)]. The pyrazolines that result from addition of diazo compounds to C_{60} give a mixture of 6,6-closed methanofullerenes and 6,5-open methanofulleroids [structures (**9**) and (**10**)] upon photo- or heat-induced loss of nitrogen (N_2); the latter rearrange to the former upon irradiation. Although these reactions can frequently be controlled to give principally monoadducts, higher adducts are formed by using excess addend and extended reaction times.

Photochemical reactions can be used to prepare $[2 + 2]$ adducts from α,β-unsaturated ketones (enones) and electron-rich alkenes and alkynes [structures (**11**) and (**12**)]. The enone photoadditions occur through interaction of enone triplet excited states with ground state C_{60}, while the other photoadditions involve C_{60} triplet excited states. It has been proposed that photopolymerization of C_{60} in films or the solid state involves formation of $[2 + 2]$ dimeric structures.

Formation of $[2 + 1]$ adducts to C_{60} from carbenes, silylenes, and nitrenes is also well established.

Formation, differentiation of higher adducts to C_{60}. Since the reactivity of C_{60} monoadducts toward the reagents mentioned above is not appreciably different from that of the unsubstituted fullerene, formation of bis and higher adducts is a serious problem; but it also provides an opportunity to build interesting three-dimensional scaffolds. Since as many as eight isomeric bis-adducts of C_{60} can be formed from

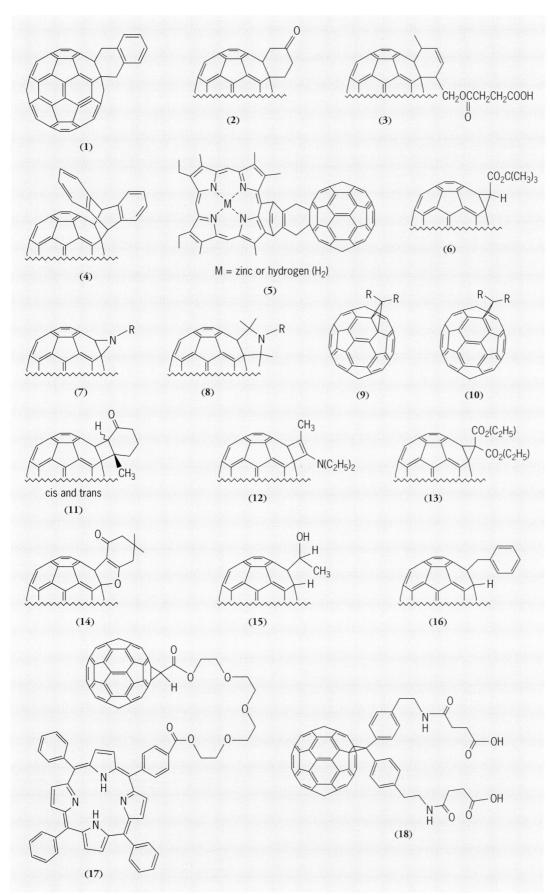

Fig. 1. Representative monoadducts of C$_{60}$.

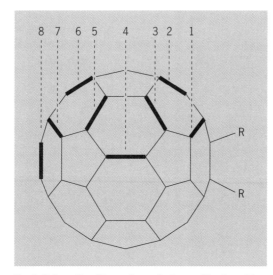

Fig. 2. Schematic of isomeric products resulting from bis-addition to C_{60} across 6,6-bonds.

addition of a symmetrical ligand **(Fig. 2)**, the problem of separating and distinguishing them has been a matter of concern. Separation can usually be achieved by high-performance liquid chromatography (HPLC) on special "buckyclutcher" columns or, sometimes, by classical column chromatography. Structural assignments have been accomplished in a few cases by establishing connectivity patterns derived from two-dimensional carbon-13 nuclear magnetic resonance (^{13}C NMR) spectra. Consistent patterns in ultraviolet (UV) and proton (^{1}H) NMR spectra of bis-adducts have been observed. A new and potentially very useful approach involves determination of helium-3 (^{3}He) NMR spectra of adducts to C_{60} containing an endohedral ^{3}He atom (^{3}He@C_{60}; @ signifies that ^{3}He is inside the C_{60} cage). This technique has the great virtue of giving well-separated sharp peaks spread over a wide range of chemical shifts for each adduct in the mixture. Since ^{3}He NMR spectra can be run on crude reaction mixtures (extraneous materials that do not contain an endohedral ^{3}He atom give no signals), information is immediately attainable as to the number and types of adduct structures generated in the reaction prior to workup. Methods for controlling selectivity in adduct formation have been explored (for example, solid-phase methodology and use of blocking groups).

Other types of addition reactions. Addition of carbon nucleophiles across 6,6-bonds of C_{60} is facile. The best known of these is the Bingel-Hirsch reaction, involving addition of bromomalonate anion to C_{60} followed by intramolecular displacement of bromide to give 6,6-fused dicarbethoxycyclopropanes **(13)**. Similar types of products are produced from sulfonium ylides, involving nucleophilic attack by anionic carbon followed by displacement of dimethyl sulfide. Enolates of 1,3-diketones add to C_{60} to give unusual dihydrofuranylfullerenes **(14)**. Yields of monoadducts in these reactions are generally quite good.

Free-radical and nucleophilic additions to C_{60} are more difficult to control, and generally yield complex product mixtures. The same is true of hydrogenation, which can be carried out via hydroboration, via hydrozirconation, or with diimide. Conditions that optimize formation of $C_{60}H_2$ have been developed. Further reaction gives a number of isomers of $C_{60}H_4$, of which the major product is the 1,2,3,4-isomer, with hydrogens on contiguous carbons. Catalytic hydrogenation leads to products with up to 50 hydrogens on the C_{60} core.

A variety of transition-metal complexes of C_{60} have been prepared, with metal centers including ruthenium (Ru), platinum (Pt), nickel (Ni), palladium (Pd), rhodium (Rh), and iridium (Ir), among others. X-ray crystal structures of such iridium complexes have given insight into the bond lengths, bond angles, and extent of pyramidalization on the C_{60} sphere in a variety of systems. It should be noted that few x-ray structures of fullerenes have been published, principally because of difficulties in obtaining suitable crystals.

Reduction and oxidation. Electrochemical reduction of C_{60} and derivatives leads to reversible incorporation of up to six electrons into three low-lying degenerate unoccupied molecular orbitals. Chemical reduction with alkali metals results in formation of fullerides (M_nC_{60}; $n = 1-6$), many of which behave as superconductors. Interaction of organic donor molecules with C_{60} gives charge-transfer complexes, some of which have ferromagnetic properties at very low temperature.

Oxidation of C_{60} to give $C_{60}^{+}\cdot$ is more difficult, but it can be achieved by photoinduced electron transfer to photoexcited electron acceptors, such as acridinium salts and 9,10-dicyanoanthracene. The C_{60} radical cations, which have been detected spectroscopically, can be trapped by alcohols, hydrocarbons, and ethers to give adducts resulting from initial hydrogen abstraction processes [for example, structures **(15)** and **(16)**].

Reactions of C_{70} and higher fullerenes. To date, much less chemical research has been carried out with C_{70} and the higher fullerenes than with C_{60} because of their limited availability and much higher cost. Since there are four different 6,6-bonds on the oblong C_{70} surface, addition reactions of the types discussed above can, in principle, give up to four isomeric monoadducts. However, chemoselectivity is generally observed, since the 6,6-bonds near the poles of C_{70}, corresponding to regions of higher curvature, are more reactive toward almost all standard reagents than the flatter 6,6-bonds near the equator [**Fig. 3**; structure **(19)**]. For example, sulfonium ylides add only to the 6,6-bond nearest to the pole of C_{70} to give chiral 1,9-methano-C_{70} monoadducts [for example, **(20)**]. Because of polarization of electron density along 6,6-bonds, which are more or less parallel to the long axis of C_{70} [that is, bonds 1 and 3 in **(19)**], regioselectivity (formation of one of several possible isomeric products) upon addition of unsymmetrical addends such as dipolarophiles is

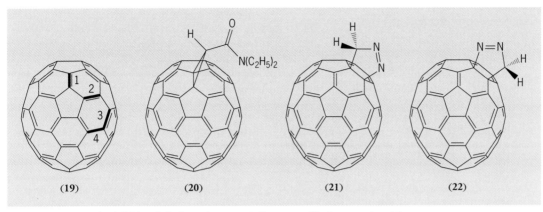

Fig. 3. Four different (1–4) 6,6-bonds in C_{70} and representative monoadducts across bond 1.

observed [such as (**21**) and (**22**) in a 12:1 ratio]. Little has been reported as yet about photochemical [2 + 2] cycloadditions to C_{70} and higher fullerenes. While structural features associated with multiaddition to C_{70} have been analyzed, only limited experimental work in this area has been reported.

The incorporation of a large variety of transition metals within higher fullerenes has been successfully accomplished. Practical applications of these interesting new materials are anticipated; for example, they may find application in electronic and optical devices, as superconductors, and as catalysts.

Biological applications. Fullerenes may have a role in chemotherapy, for they are believed to have antitumor and antiviral properties.

Fullerenes as singlet oxygen sensitizers. Electronic excitation of C_{60} and C_{70} and derivatives results in formation of triplet excited states, which in turn are quenched by triplet dioxygen (O_2) to give singlet oxygen, a chemically reactive excited state of oxygen, with close to 100% efficiency. Singlet oxygen is implicated in a number of biologically important processes, such as cleavage of DNA and destruction of tumor cells. This has stimulated interest in preparing water-soluble fullerene derivatives for site-specific DNA cleavage and photodynamic therapy, a noninvasive approach to cancer treatment. In the latter connection, attention has been directed toward porphyrin-C_{60} dyads [for example, (**5**) and (**17**) in Fig. 1] in which the porphyrin acts as an antenna to absorb light at the red end of the visible spectrum, followed by a series of intramolecular energy-transfer processes and quenching steps, resulting in generation of singlet oxygen. Some advantages of these materials in photodynamic therapy are anticipated with respect to simple porphyrin derivatives.

Antiviral activity. There may be a role for fullerenes in treatment of human immunodeficiency virus (HIV) infection. Molecular modeling indicates that C_{60} fits nicely into the hydrophobic cavity of HIV protease, and indeed a water-soluble derivative of C_{60} (**18**) is an inhibitor of cloned HIV protease and shows virucidal activity in cells infected with HIV. Similar behavior has since been demonstrated with a number of other C_{60} derivatives, with activity in the micromolar range. These compounds also show no cytotoxicity up to 100 μM against various cell lines, and upon intraperitoneal administration to mice at doses of 50 mg/(kg)(day) for 6 days. These encouraging observations could lead to the development of an entirely new class of materials for treatment of HIV and other viral infections.

Outlook. There is interest in the chemistry of fullerenes because of the possibility of constructing topologically novel three-dimensional molecular structures, in contrast to the essentially two-dimensional structures derived from planar aromatic hydrocarbons such as benzene, naphthalene, and anthracene. A huge variety of fascinating molecular frameworks can be envisaged using fullerenes as basic building blocks. Fullerene-derived materials have very interesting electrochemical properties alone and in complexes with metal atoms, both on the surface and encapsulated in the interior of the cage. They can potentially act as optical limiting devices and superconductors, and because of their strong electron-accepting ability they can serve as photosynthetic mimics when joined to strong electron donors such as porphyrins and phthalocyanines. Success has recently been achieved in incorporating fullerene units into polymers, including pure carbon polyfullerenes, ''charm bracelet'' polymers with fullerenes appended to a polymeric backbone, and water-soluble polymeric materials.

For background information *see* CARBON; FULLERENE; MOLECULAR ORBITAL THEORY; ORGANIC CHEMISTRY; ORGANIC REACTION MECHANISM; TRIPLET STATE in the McGraw-Hill Encyclopedia of Science & Technology.   David I. Schuster

Bibliography. W. E. Billups and M. A. Ciufolini (eds.), *Buckminsterfullerenes*, 1993; R. J. Cross et al., Differentiation of isomers resulting from bisaddition to C_{60} using ^{3}He NMR spectrometry, *J. Amer. Chem. Soc.*, 118:11454–11459, 1996; A. Hirsch, *The Chemistry of Fullerenes*, 1994; A. W. Jensen, S. R. Wilson, and D. I. Schuster, Biological applications of fullerenes, *Bioorg. Med. Chem.*, 4:767–779, 1996.

Fuzzy-structure acoustics

Large structures such as ships and airplanes can undergo a variety of complicated vibrations. Such structures typically consist of an outer body made of metal plating (for example, the hull of a ship) or perhaps a massive metallic frame (for example, the chassis of a truck), and a large variety of internal objects that are connected to either the plating or the frame. In designing such structures, it is highly desirable to have some method for predicting how they will vibrate under various conditions. The radiation of sound caused by these vibrations, either into the environment or into the empty portions of the structure, is also of interest because this sound is often either unwanted noise or a means of inferring information about the details of the structure or the excitation. Fuzzy-structure acoustics refers to a class of conceptual viewpoints in which precise computationally intensive models of the overall structure are replaced by nonprecise analytical models, for which the initial information is said to be fuzzy.

External and internal structures. Fuzzy-structure theories divide the overall structure (**Fig. 1**) into a master structure and one or more attached structures, the latter being referred to as the fuzzy substructures, the internal structures, or the internals. (An example of a master structure is the hull and major framework of a ship.) The master structure is presumed to be sufficiently well known at the outset that its vibrations or dynamical response could be predicted if the forces that were exerted on it were known. Some of the forces are exerted on it by the substructures at the points at which they are attached. Such forces can be very complicated; nevertheless, there is some hope that a satisfactory approximate prediction of the vibrations of the master structure itself can be achieved with a highly simplified model.

The fuzzy substructures can be regarded as structures that are not known precisely. The most information-demanding type of fuzzy-structure model has significant drawbacks. Perhaps the most serious disadvantage is that the exploration of such models requires Monte Carlo simulations that are often infeasible or prohibitively time consuming. Consequently, there is some impetus to consider fuzzier models of the fuzzy substructures.

Forces and impedances. The fuzzier models that have been developed in the past few years rest on derivable consequences of the science of mechanics. The substructures are passive, so the displacements of attachment points on a substructure can be related to the forces exerted on them. The same relations, because of Newton's third law, have to hold between the forces exerted on the master structure and the corresponding displacements, except for a simple change in sign. The influences of the substructures on the master structure are analogous to those of springs connecting the master structure to rigid walls, but here the springs are more complicated than those described by Hooke's law (in which force is directly proportional to extension) in that displacements at distant attachment points may affect a given force, the forces are not necessarily in phase with the displacements, and the proportionality and the phase shift vary with the frequency with which the master structure is oscillating.

Nevertheless, there is still a surprisingly simple form for the general relations between the set of forces and the set of attachment-point displacements. Typically, workers in acoustics and vibrations express such relations as a rectangular array, any one element providing a ratio of a force experienced at one attachment point to the velocity of another attachment point, given that the displacements of all the other attachment points are held to zero, that the oscillations are sinusoidal (fixed frequency), and that complex numbers are used for the forces and velocities (so that differences in phases can be accounted for). Ratios of complex force amplitudes to complex velocity amplitudes are termed impedances, and an array of ratios of different forces to different velocities is termed an impedance matrix.

Modes and natural frequencies. The branch of mechanics known as linear vibration theory leads to the prediction that each of the impedance matrix elements can be written as a sum of terms, each term being similar to what would be expected from shaking a simple oscillator (**Fig. 2**), consisting of a mass, a spring, and a dashpot at an attachment point. Such a term, representing the ratio of a force to a velocity, has a frequency-dependent denominator that would go to zero in the absence of damping at an angular frequency (referred to as the resonance frequency or the natural frequency) equal to the square root of the spring constant divided by the mass. For very weak damping, the shaking force and the displacement are in the same direction when the shaking frequency is less than the natural frequency, but oppositely directed when the shaking frequency is higher than the natural frequency.

Typical substructures, when regarded in terms of the impedance matrix that they present to the

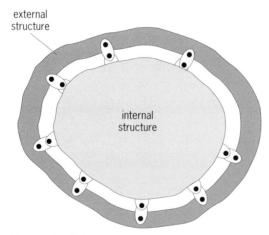

external structure

internal structure

Fig. 1. Mechanical structure conceptually divided into an internal structure (fuzzy substructures) and an external structure (master structure).

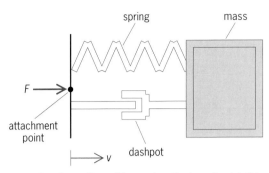

Fig. 2. Simple oscillator driven at its attachment point. *F* is the driving force, and *v* is the velocity at the attachment point. The ratio of the complex amplitude of *F* to the complex amplitude of *v* gives the driving-point impedance.

external environment, behave as a set of independent simple oscillators, each having its own natural frequency. Corresponding terms in the series for the individual impedance matrix elements have the same denominator; consequently they go to zero, in the limit of zero damping, at the same natural frequency. These natural frequencies represent the frequencies at which the substructure would freely vibrate indefinitely in the absence of internal damping, when all of the attachment points are held motionless but the rest of the substructure is free. The overall manner in which the substructure is oscillating when undergoing vibrations under such circumstances is called a natural mode of vibration.

Modal masses. The numerators of the corresponding modal terms in the separate elements of an impedance matrix lead to the identification of a modal mass matrix for each natural mode. Each element of the modal mass matrix has the units of mass. For the special case where there is only one attachment point and the points in the substructure all move back and forth in the same direction, the matrix for each mode is just a one-by-one matrix, so there is a single mass associated with each mode. If there are a number of different attachment points but the points in the substructure still are constrained to move in the same direction, then the elements of the modal mass matrix for a given mode sum to a quantity identified as the modal mass for that mode. Finally, if substructures are considered where the points can move in all three directions and the definitions of the forces and velocity are generalized so that each cartesian component of each force is regarded as a separate force, then the sum of all the modal mass matrix elements for any given mode has a value that is three times as large as what would be regarded as the modal mass associated with that mode. In each such case, the modal masses when summed over all natural modes yield the total mass of the substructure.

Modal mass per unit frequency. Recently developed theories of fuzzy structures lead, after various plausible idealizations, to a formulation that requires only a single function, this being the modal mass per unit frequency bandwidth. Such a function could be

regarded as a quasi-limit. A band of frequencies is chosen that is of narrow bandwidth but nevertheless contains a large number of natural frequencies. A sum is taken over all the modal masses for all of the modes that have frequencies in that band, this sum is then divided by the bandwidth, and the resulting value is associated with the center frequency of the band. This concept is similar to that of modal density, which is the number of resonance frequencies per unit frequency bandwidth—only here there is a weighting by modal mass.

This function can be estimated without the necessity of calculating each resonance frequency and each modal mass separately, with very little detailed knowledge of the design, configuration, or material properties of the substructure itself. Examples considered so far suggest that, for any general category of substructure, a simple functional form can be identified for the dependence on frequency with the necessity to specify only a small number of parameters. A judicious sampling of blueprints or parts lists, in conjunction with a few simple field measurements, might be sufficient to estimate these parameters.

Predictions. In the formulation alluded to above, the influence of fuzzy substructures attached to the master structure tends to resemble that of an added frequency-dependent mass attached to the master structure in parallel with a frequency-dependent dashpot connecting the master structure to a hypothetical rigid wall. The added mass is a frequency-weighted integral over the modal mass per unit natural frequency, the weighting being such that the natural modes whose natural frequencies are less than the driving frequency have a positive contribution, while those for which the natural frequencies are greater than the driving frequency have a negative contribution. The master structure can seem to be less massive than it actually is (**Fig. 3**) when

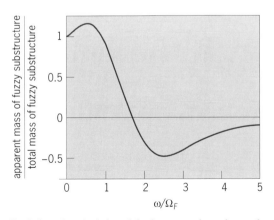

Fig. 3. Sample calculation of the frequency dependence of the apparent mass per unit area added to a master structure (rectangular plate) by a distribution of fuzzy structures when the master structure is oscillating at angular frequency ω. Ω_F is a characteristic frequency for the set of fuzzy substructures. *(After A. D. Pierce, V. W. Sparrow, and D. A. Russell, Fundamental structural-acoustic idealizations for structures with fuzzy internals, J. Vibrat. Acoust., 117:339–348, 1995)*

the bulk of the substructure mass is associated with resonant frequencies less than the excitation frequency.

The apparent dashpot constant, equal to the ratio of the apparent reacting force to the velocity at the attachment point, has a frequency-dependent value that is directly proportional to the mass per unit natural frequency bandwidth evaluated at the driving frequency, with an added multiplicative factor of the square of the driving frequency. Because the passive nature of the attached substructures precludes their acting as an energy source, the dashpot constant is always positive.

The prediction that the actual value of the dashpot constant is independent of the values of any parameters associated with energy dissipation (or damping) within the substructures is only approximate, but it does hold in the limit of sufficiently weak damping. There may be some relevant mechanism by which energy is absorbed within the substructure, but the actual amount of energy absorbed is asymptotically independent of the mechanism's strength. Similar phenomena, in which the amount of energy absorbed seems independent of the strength of the absorbing mechanism, occur in other branches of physics. For example, the energy loss at a shock front is dependent only on the pressure jump at the front in the limit when the shock front is thin and the absorption mechanism is weak. For the fuzzy-structure case, the evidence so far suggests that most internal structures of interest are sufficiently resilient that the asymptotic limit independent of damping mechanisms in the internals is of practical interest. This limit can be realized when (1) the number of resonances is very large and the resonance bandwidth of any given mode, although narrow compared with the resonance frequency itself, is sufficiently wide that there is considerable overlap between the resonance peaks of different modes; or (2) when the excitation of the master structure, while having a narrow bandwidth compared to the central frequency, is sufficiently wide that it includes a large number of natural frequencies of the substructure; or (3) when the force excitation is a short-duration pulse of nearly constant frequency, but the duration of the pulse is short compared to the reverberation time of the substructure, the latter being roughly equal to the reciprocal of the bandwidth of a resonant frequency of the substructure.

One implication of the newly emerging fuzzy-structure theories is that, insofar as there is concern with the vibrations of only the master structure, it is possible to drastically curtail the estimation or measurement of any parameters within the substructures that are associated with internal damping.

For background information *see* ACOUSTIC IMPEDANCE; FORCED OSCILLATION; MATRIX THEORY; MECHANICAL IMPEDANCE; MECHANICAL VIBRATION; RESONANCE (ACOUSTICS AND MECHANICS); VIBRATION in the McGraw-Hill Encyclopedia of Science & Technology.   Allan D. Pierce

Bibliography. A. D. Pierce, Resonant-frequency-distribution of internal mass inferred from mechanical impedance matrices, with application to fuzzy structure theory, *J. Vibrat. Acoust.*, 119:324–333, 1997; A. D. Pierce, V. W. Sparrow, and D. A. Russell, Fundamental structural-acoustic idealizations for structures with fuzzy internals, *J. Vibrat. Acoust.*, 117:339–348, 1995; C. Soize, A model and numerical method in the medium frequency range for vibroacoustic predictions using the theory of structural fuzzy, *J. Acoust. Soc. Amer.*, 94:849–865, 1993; M. Strasberg and D. Feit, Vibration damping of large structures induced by attached small resonant structures, *J. Acoust. Soc. Amer.*, 99:335–344, 1996.

Gamma-ray detectors

A new generation of gamma-ray detectors has been developed, primarily to study the structure of atomic nuclei. Much of the knowledge about the structure of nuclei, as well as atoms and molecules, has come from studying their energy levels through the detection of electromagnetic radiation emitted when the system makes a transition from one state to another. The electromagnetic radiation from atoms and molecules has an energy around 1 electronvolt, while for nuclei it is of order 10^6 eV (1 MeV). Thus, for the study of nuclei it is necessary to design and build detectors aimed at an energy range which is quite different from that of the typical photomultipliers used for detecting electronvolt photons. The recent detectors incorporate new technologies and large arrays to provide high resolution and high efficiency, resulting in an improvement of about a factor of 100 in the ability to resolve rare features of nuclear structure.

The nucleus is a unique quantal system, rather different from atoms and molecules. Atoms exhibit primarily single-particle features of electron excitations in a fixed spherical Coulomb potential. While molecules add collective rotational and vibrational excitations to this behavior, these three modes are so different in energy that they essentially do not interact with each other. In nuclei, a nucleon moves in an average potential created by all the other nucleons and also exhibits single-particle behavior in this potential. However, in this case the potential is not fixed but can assume a variety of shapes and undergo collective rotations and vibrations, and there is a complicated interplay between these and the single-particle motion. Such a system is more difficult to describe theoretically, both because the interaction between nucleons is not well known and because the single-particle and collective motions are not well separated. Correspondingly, it can give different insights into the behavior of quantal systems.

Gamma-ray detector arrays. Generally the most important properties of a gamma-ray detector are high detection efficiency; high energy resolution (narrow energy peaks); high ratio of full-energy to partial-energy events (useless events, in which the gamma ray loses only part of its energy in the detector); and high granularity, to localize individual gamma rays. For gamma rays in the 1-MeV range, by far the best combination of these properties is given by semiconductors made of high-purity germanium single crystals. In these detectors the incident gamma rays lose their energy by Compton scattering (off electrons) or photoelectric absorption, and the resultant electrons and holes created in the germanium detector are collected. The integrated charge is proportional to the gamma-ray energy and gives an excellent energy resolution: full-width-at-half-maximum of about 2 keV for a 1-MeV gamma ray. Many such crystals are assembled into arrays in order to cover large solid angles with the necessary granularity.

This array configuration permits the use of Compton suppressors to reject the partial-energy events. The germanium detectors are surrounded by a dense scintillator (a very efficient low-resolution gamma-ray detector, usually bismuth germanate, or BGO), which detects gamma rays scattered out of the germanium crystal and then suppresses the partial-energy pulse left in the germanium detector. For events where six gamma rays are caught (sixfold events), the fraction with six full-energy gamma rays is about 200 times larger with suppression. This is an extremely important gain, without which such high-fold coincidence measurements would not be practical.

The two leading arrays of this type are Euroball in Legnaro, Italy, and Gammasphere in Berkeley, California. Gammasphere is a spherical shell of 110 Compton-suppressed germanium detectors (**Fig. 1**). These detectors are 25 cm (10 in.) from the center of the shell and cover 47% of the full sphere—the rest is occupied by the Compton suppressors—resulting in a full-energy efficiency for a 1-MeV gamma ray of 12% (compared with about 2% for the previous generation of arrays which had about 20 detectors). Tapered hexagonal bismuth germanate elements fit together to form a nearly complete spherical shell.

For many experiments the energy resolution is dominated by the Doppler broadening due to the recoil motion of the emitting nucleus. A main contribution to the Doppler broadening comes from uncertainty in the gamma-ray direction due to the finite opening angle of the germanium detector, so that more granularity is needed to define this angle better. To accomplish this, 70 of the Gammasphere detectors have been electrically separated into two D-shaped segments, halving the opening angle. In typical experiments this improves the resolving power by a factor of 2.

The new detector arrays, with their increased resolving power, provide the opportunity to explore new aspects of nuclear structure. Two recent advances are the study of superdeformed nuclei and the characterization of drip-line nuclei. However, these detectors have a broader range of applicability. Some experiments use nuclei as a laboratory for

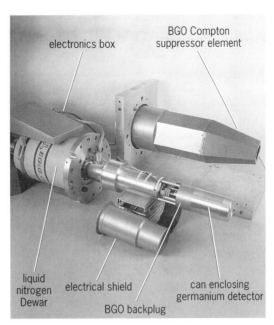

Fig. 1. Gammasphere detector element. The germanium crystal detector is encased in an aluminum can, and slides in to about 3 cm (1 in.) from the front face of the bismuth germanate (BGO) Compton suppressor element. Thus, the germanium is completely surrounded by BGO except at the front, toward the gamma-ray source. A liquid-nitrogen Dewar is required to keep the germanium detector cold. (*Ernest O. Lawrence Berkeley National Laboratory*)

studying other kinds of physics. An example of such a study in progress on Gammasphere is a test of the standard model of fundamental particles. This is accomplished by measuring accurately the beta decay (weak interaction) strength in nuclei and comparing it with that in leptons.

Superdeformed nuclei. Rapidly spinning superdeformed nuclei are the most elongated (football-shaped) nuclei known and are a subject of much current interest. They are produced in reactions when an accelerated ion beam hits a target nucleus off-center, fusing into a compound nucleus with an angular momentum (spin) as high as $60\hbar$, where $\hbar$ is Planck's constant divided by 2π. This angular momentum is removed by 20–30 gamma rays, each of which can carry off $2\hbar$ with an energy proportional to the rotational frequency. This produces energy spectra with long sequences of nearly equally spaced gamma-ray peaks. The yield of superdeformed nuclei produced in this reaction is only about 0.3% of the total yield, and this rotational sequence can be isolated so well from the high background of other gamma rays only because Gammasphere can frequently detect six gamma rays from the 20 or so that are emitted by the superdeformed nucleus.

The first superdeformed rotational band at high angular momentum was found with an earlier generation of gamma-ray detector arrays in the mid-1980s. Since then, more than 200 such bands have been found in several regions of the nuclear chart, provid-

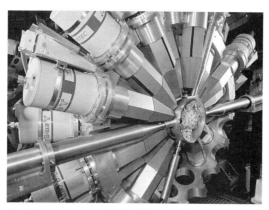

Fig. 2. Gammasphere with the Microball. One-half of the Gammasphere is shown surrounding the target chamber with the Microball inside. The beam line runs from left to right, and the target is at the center of the Microball. (*Ernest O. Lawrence Berkeley National Laboratory*)

ing information about the properties of these elongated nuclei, as well as the effects that stabilize such shapes. Their stability is now understood to result from a subtle interplay of the nonuniform single-particle level spacings (quantal) and the effect of rotation on the nuclear potential (classical). Until recently, superdeformed bands could not be linked to the rest of the energy levels of the nucleus. With Gammasphere, the first of these links has been found with an intensity about 100 times lower than that of the band itself. Such studies will provide some insight on how this system of about 200 nucleons achieves a dramatic change of shape, sometimes in one gamma-ray transition.

Nuclei at the drip lines. Nuclear structure can be studied in a wide range of stable and unstable nuclei out to the point where a nucleon is no longer bound to the nucleus—the so-called drip line. At one extreme, nuclei along the proton drip line exhibit ground-state proton radioactivity, in which unbound protons penetrate the Coulomb barrier around a nucleus and decay to particular states in the daughter nucleus. The decay patterns carry specific information about the nuclear structure, unavailable by other means, and gamma-ray detectors are beginning to be used to extract this information. At the other extreme, along the neutron drip line, very weakly bound neutrons tend to leak out beyond the rest of the nucleons and surround the nucleus with a neutron halo. Both of these drip lines are under intensive study.

Coupling the large gamma-ray detectors to other auxiliary detectors is an excellent way to study these exotic drip-line nuclei. **Figure 2** shows Gammasphere with an additional detector called the Microball inside the target chamber. The Microball is a shell of 96 cesium iodide crystals that surrounds the target and detects alpha particles and protons emitted following the fusion reaction which generates a product nucleus near the proton drip line. In these reactions, so many different product nuclei

are formed that it would be very difficult to separate out and identify a single product without determining what particles are emitted. The Microball is made of light materials so that the gamma rays can pass through it and reach Gammasphere. With this system, properties of nuclei near the proton drip line in the mass 60 and 90 regions are being investigated. Devices that analyze the recoiling nuclei, such as recoil mass separators, can also isolate a single product and will soon be used in conjunction with Gammasphere.

Advanced detector arrays. It may be possible to develop a gamma-ray detector array with a resolving power about 1000 times that of Gammasphere or Euroball. Such a detector would consist of a spherical shell of highly segmented, tapered, hexagonal germanium detectors with no Compton suppressors. The segmentation (about 4000 segments) would be sufficient to resolve the four or five individual interactions of a typical 1-MeV gamma ray, and the scattering sequence could then be reconstructed (a process called tracking). Based on the tracking result, the energy of a gamma ray could be obtained and full-energy events could be distinguished from partial-energy ones. Gamma rays could be readily tracked out of one detector into the next and the appropriate energies added together to recover the full energy, rather than throwing the gamma ray away as Compton suppression does.

Such a Gamma-Ray Energy Tracking Array (GRETA) might achieve an efficiency for a full-energy 1-MeV gamma ray as high as 67%, which would produce the factor-of-1000 improvement in resolving power over current detectors. This might be enough, for example, to trace the change from regular motion in nuclei to the region of chaotic motion at higher excitation energy where the level density is high. Some other important capabilities of such a system would be (1) a more accurate Doppler-shift correction, since the position of the first interaction point gives a very precise direction for the gamma ray; (2) linear polarization determination from the position of the first two interaction points; and (3) an efficiency for full-energy 15-MeV gamma rays of 30% with 15-keV resolution, unprecedented for such high-energy gamma rays.

For background information *see* EXOTIC NUCLEI; GAMMA-RAY DETECTORS; GAMMA RAYS; NUCLEAR STRUCTURE; RADIOACTIVITY in the McGraw-Hill Encyclopedia of Science & Technology.   M. A. Deleplanque;
I. Y. Lee; F. S. Stephens

Bibliography. C. W. Beausang and J. Simpson, Large arrays of escape suppressed spectrometers for nuclear structure experiments, *J. Phys. G: Nucl. Part. Phys.*, 22:527–558, 1996; M. A. Deleplanque, I. Y. Lee, and A. O. Macchiavelli (eds.), *Proceedings of the Workshop on Gammasphere Physics*, 1996; M. de Saint Simon and O. Sorlin (eds.), *ENAM 95: International Conference on Exotic Nuclei and Atomic Masses*, 1995; R. V. F. Janssens and T. L. Khoo, Superdeformed nuclei, *Annu. Rev. Nucl. Part. Sci.*, 41:321–355, 1991.

Gas turbine

Most development activity for gas turbines focuses on air-breathing, internal-combustion configurations. Lower cost, lower emissions, and higher efficiency motivate technology advances. Makers of large gas turbines seek to develop machines optimized for use in combination with steam turbines in combined-cycle power plants. Technologies useful for increasing the volume of airflow, increasing the temperature of the rotor inlet, and lowering emissions account for most of the applicable research. A wider range of applications is pursued with medium and small machines, which employ the same underlying materials and airfoil-cooling technologies that advance large-engine performance, plus some unique ones.

Large gas turbines and combined-cycle power plants. Gas turbines in the output class of 150 to over 250 megawatts offer the lowest specific cost (in terms of dollars per kilowatt) and are found in the world's most efficient fossil power plants. Key goals for the design of large machines are low specific cost, low emissions, and high combined-cycle efficiency. In gas-turbine design, efficiency can be optimized either for simple-cycle or combined-cycle operation. High-pressure-ratio, smaller machines based on aircraft engines have high simple-cycle efficiency, while high-temperature, moderate-pressure-ratio, large machines are optimized for combined-cycle service.

In most gas turbines, the temperature of the gas leaving the combustor is so high that the blades downstream of the combustor, which convert the heat energy into useful work, must be cooled. Normally this is done with air extracted from the compressor. Large air-cooled machines with rotor-inlet temperatures in the 1400°C (2550°F) class were introduced in 1994, and are well along their development path. In combined cycle, these machines produce approximately 350 MW and 450 MW, and should generate electricity at 58% efficiency. Steam-cooled machines with higher temperatures are also in production. These steam-cooled machines, in combined cycle, produce 400 MW and 480 MW at an efficiency of 60%. By replacing cooling air with steam produced with exhaust heat, designers preserve more of the benefits of the high rotor-inlet temperature.

A reheat gas turbine has been operated at nearly 165 MW. In conventional gas turbines, air flows from a compressor, through a combustor, and on to a turbine, from which the flow exits either to the atmosphere or to a heat exchanger or boiler. In reheat gas turbines, air also flows through a compressor, a combustor, and a turbine; but after the first turbine, air enters a second combustor and then a second turbine. For the same exhaust temperature and rotor-inlet temperature—reached twice—a reheat cycle can achieve higher combined-cycle efficiency. The reheat gas turbine may produce combined-cycle efficiency of over 58%.

Medium and small industrial gas turbines. Medium and small gas turbines not only drive electrical generators but also power gas compressors, pumps, and other devices. Medium-output turbines benefit from technologies developed for the largest machines. Designs of these medium machines are similar to those of the large machines, but output sizes are selected to meet the power needs of specific sets of applications, such as industrial cogeneration and district heating.

Small machines with simple-cycle outputs below 25 MW follow a somewhat independent trajectory. Output goals are application driven. Since small gas turbines are often applied independent of an electrical grid, or otherwise where power is consumed locally, the economics often result in a different trade-off between specific cost and efficiency. Power-only combined-cycle plants are rarer in this size. Other exhaust heat-recovery schemes and simple-cycle applications are commonplace.

Current structural technology for ceramics has produced small-scale ceramic turbine blades that can operate at temperatures higher than those for metal blades. Manufacturers are pursuing advanced cycles based on the reduced need for cooling air that is made possible with ceramic blades. Combining a high rotor-inlet temperature with low-magnitude cooling flows increases power generation efficiency approximately 5%. A 5-MW-class machine has been designed to operate with ceramic rotating and stationary blades, and a ceramic combustion liner.

A manufacturer of small gas turbines has recently combined advanced cooling systems and single-crystal casting technology in a design for blades suitable for operating at very high temperatures, and high-efficiency cycles around these components have been developed. By using designs that were developed for components of aircraft engines, it is anticipated that a 15% increase in efficiency can be achieved over 1991 levels by the end of the decade by increasing the pressure ratio and the rotor-inlet temperature.

Aircraft engine derivatives. Industrial engines derived from aircraft engines continue to be improved in the same size range as small and medium heavy-duty designs. The attributes of these engines are similar to those of the flight versions—compact size, low weight, and high simple-cycle efficiency. Advances in compressor and turbine aerodynamics, materials, and cooling technology can be transferred from the flight engines directly into the industrial versions. Larger engines are the first to exhibit the most advanced technology; and for the same reasons, large industrial machines are the most advanced. Aircraft engine manufacturers have introduced new or uprated engines in the 40-MW range with simple-cycle efficiencies of over 40%.

Technologies. Important technologies are being applied, advanced, or investigated for design of gas turbines.

Directionally solidified materials. Turbine rotating and stationary blades are made of nickel- and cobalt-based alloys. Blades of these alloys are made by the investment casting process. By controlling the cooling of the molten metal, it is possible to produce a part with grains oriented in one direction. Components made in this way have superior creep and fatigue life and require less cooling. Directionally solidified blades have been used in aircraft engines for decades. The technique is now being applied to large parts. By substituting directionally solidified materials for conventional castings, manufacturers have produced turbines that run at higher turbine-inlet temperatures; hence output and combined cycle efficiency are increased.

Single-crystal materials. Turbine blades can be made so they contain only one crystal, using a process similar to that used to produce directionally solidified components. The benefit is even longer creep and fatigue life. Some of the newest machines depend on single-crystal blades as well as innovations in the thermodynamic cycle to substantially increase efficiency.

Corrosion- and oxidation-resistant coatings. Coatings composed of cobalt, chromium, aluminum, yttrium, and other elements are applied to turbine blades to prevent surface attack that would result in loss of aerodynamic performance and deterioration of strength-related material properties.

Thermal barrier coatings. Another family of coatings, made from yttrium-stabilized zirconia, has been applied to combustor components and cooled turbine blades to increase the effectiveness of the cooling and to reduce stress-producing thermal gradients that arise from the cooling. The application of this technology to industrial gas-turbine blades is relatively new. Steam-cooled machines rely on these coatings in combination with single-crystal materials.

Structural ceramics. Research in the application of ceramics is accelerating. Small rotor, blade, and shroud elements have been produced; and small gas turbines with such components have operated. Ceramics can operate at higher temperatures than metals. Both monolithic, homogeneous ceramics and composite ceramics involving ceramic fibers in ceramic matrices find application in components for gas turbines. Recent advances have produced ceramics more tolerant of flaws and defects—a feature needed for practical industrial application of these materials. The use of ceramics has progressed to the point that statistically significant bodies of behavior data have been compiled, and credible engine-reliability calculations have become possible, along with reliable inspection processes. The greatest limitation, at least for smaller engines, is the fact that direct substitution of materials (that is, using the same component shape) is not possible because of the difficulties in attaching ceramic to metal components.

Computational fluid dynamics codes. The development of larger and larger computers encouraged the writing of extremely sophisticated programs, or codes, that can deal with complex fluid-flow phenomena. Better understanding of the behavior of the flow of the air and hot gases—the working fluid—has allowed engineers to modify airfoil shapes in compressor as well as turbine blades, resulting in higher compressor and turbine efficiency.

High-pressure-ratio compressors. The appearance of compressors with pressure ratios upward of 30:1, which usually require the two coaxial rotor configuration common in aircraft engines, is the key to high simple-cycle efficiency in aircraft-derivative engines and some small industrial gas turbines. It is also a required technology for high-temperature reheat cycles.

Steam-cooled turbines. Although the concept of using steam rather than air to reduce turbine-blade temperatures dates almost from the birth of the industry, it is only now being applied to standard-model gas turbines under development. When gas turbines and steam turbines are integrated into combined-cycle power plants, the use of steam from the steam turbine or heat-recovery steam generator (that is, a boiler heated by the gas turbine's exhaust) in place of air extracted from the compressor improves efficiency. Steam cooling of combustor components reduces the level of air bypassing the flame zone of the combustor, lowering the rate of production of nitrogen oxides (NO_x) for a given firing temperature. Steam cooling of the first set of stationary blades in the turbine reduces dilution of the working fluid as it passes through these blades, thus permitting higher rotor-inlet temperature for a given level of NO_x production. Steam cooling of the rotating blades and the remaining stationary stages greatly increases efficiency.

Intercooled cycle. Compressors increase the temperature as well as the pressure of the air. By removing some of this heat part way through the compressor, it is possible to reduce the work required to create high pressure. Intercooling also reduces the discharge temperature of the compressor, meaning that more fuel is required to develop the required turbine inlet temperature. The net effect is to increase the output of a gas relative to the size of the turbine. Since the temperature of the compressor discharge is reduced, less air needs to be extracted from the compressor to cool the blades.

Regenerative (recuperative) cycles. In low-pressure-ratio and intercooled gas turbines, the exhaust heat can be used to warm the air exiting the compressor, reducing the amount of fuel needed to deliver the required rotor-inlet temperature. The cost and efficiency of plants based on this cycle are between those of simple-cycle gas turbines and combined-cycle plants. Compared to combined cycle, their lower cost offsets lower efficiency in many situations.

Reheat gas turbine. Standard gas turbine-based cycles feature one heat addition, between the compressor exit and the turbine inlet. By adding a second combustor and a second turbine, a cycle can be developed with high combined-cycle efficiency but with conventional rotor-inlet temperatures. A reheat machine that began operation in 1996 features a 30:1 pressure ratio compressor, an annular combustor, a single-stage high-pressure turbine, a second annular combustor, and a four-stage low-pressure turbine. The compressor and turbines share a common rotor.

Low NO_x combustors. Control of the NO_x emission is now possible by limiting the generation of NO_x within the gas turbine, without consuming water or steam. Such combustors mix fuel with air slightly upstream of the flame zone in the combustor; a cooler flame results, significantly lowering the rate of NO_x formation.

Integrated gasification combined cycle. The processes for efficiently converting solid fuel to useful work by first generating a clean, benign fuel gas have progressed from the technology demonstration phase to the commercial arena. Extremely low emissions, high efficiency, and competitive cost of electricity have been demonstrated on a commercial scale.

For background information *see* CERAMICS; CERMET; COMPOSITE MATERIAL; COMPUTATIONAL FLUID DYNAMICS; GAS TURBINE; METAL CASTING in the McGraw-Hill Encyclopedia of Science & Technology.

Harold Miller

Bibliography. R. Aboulafia, Engines face slow times, *Aviat. Week Space Technol.*, 144(2):95, January 8, 1996; M. S. Briesch et al., A combined cycle designated to achieve greater than 60 percent efficiency, *Trans. ASME*, 117:734–741, October 1995; T. W. Fowler (ed.), *Jet Engines and Propulsion Systems for Engineers*, 1989; M. Kutz (ed.), *Mechanical Engineers' Handbook*, 1986; C. T. Sims, N. S. Stoloff, and W. C. Hagel (eds.), *Superalloys II*, 1987.

Gene

Since the mid-1970s, advances in deoxyribonucleic acid (DNA) technology have led to a less labor-intensive, more automated DNA sequencing process. Instead of relying on a researcher interpreting the sequence of adenine (A), cytosine (C), guanine (G), and thymine (T) nucleotides from bands on an x-ray film image, state-of-the-art DNA sequencers utilize a combination of lasers, high-precision optics with a charge-coupled device, and computer software to determine the sequence of fluorescently tagged DNA molecules. The increased resolution of DNA sequencing, coupled with the highly automated nature of high-throughput sequencing facilities, allows large stretches of DNA to be rapidly sequenced. This has ushered in a new field in molecular biology called genomics, the study of the genome (all the DNA) of an organism. In 1994 the first genome to be completely sequenced with the automated technology was that of the smallpox virus. Sequencing this 186,000-base-pair double-stranded DNA molecule was an important technical and biological achievement, allowing every gene in this dangerous human pathogen to be identified. Because viruses are not considered to be "living" (they require host cells to grow and reproduce themselves), researchers began

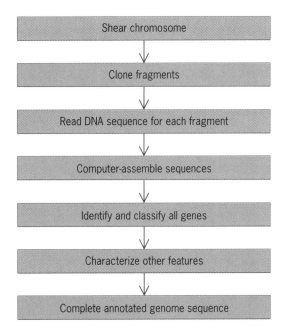

Fig. 1. Whole-genome shotgun sequencing strategy. *(After R. D. Fleischmann et al., Whole-genome random sequencing and assembly of Haemophilus influenzaeRd, Science, 269:496–512, 1995)*

to determine the genome sequence of a free-living organism such as a bacterium.

Haemophilus influenzae. A team of researchers from The Institute for Genomic Research (TIGR), Rockville, Maryland, and Johns Hopkins University, Baltimore, Maryland, attempted to determine the sequence of the 1,830,000-base-pair genome *Haemophilus influenzae*, a bacterium that normally resides in the human upper respiratory tract. The method used by the researchers had never been applied to a segment of DNA of this size. First, the genomic DNA was sheared into small (about 2000 base pairs) fragments. Next, the entire molecule was rapidly sequenced, relying on advanced computer hardware and software. This step is known as shotgun sequencing. Finally, the results were carefully analyzed in order to assemble the random bits of sequence into a large reconstruction of the original DNA molecule (**Fig. 1**). This "bottom-up" strategy contrasted with the accepted "top-down" model for sequencing large stretches of DNA (such as the model used for the *Escherichia coli* and human genome projects). The top-down approach consisted of a lengthy mapping phase, cloning of large overlapping segments of the genome, and shotgun-sequencing each segment until the project was completed. The bottom-up shotgun method allows genomic DNA to be sequenced without prior knowledge of the genome structure, eliminating the time-consuming mapping phase and the inherent inefficiency of shotgun-sequencing overlapping segments shared by the large fragments.

Role category. Approximately 24,000 random sequence reads were assembled for the *H. influenzae* project; and following sequence editing and annota-

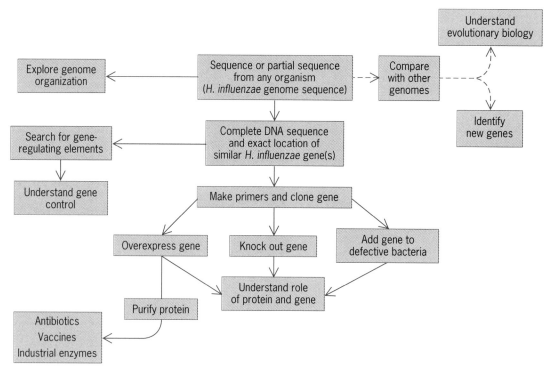

Fig. 2. Downstream applications of whole-genome sequence data. *(After R. Nowak, Bacterial genome sequence bagged, Science, 269:468, 1995)*

tion, the complete double-stranded circular DNA sequence was published in 1995. Publication of the very large amount of data generated by this project would have been difficult in the form of a standard scientific paper. (The 1,830,137 bases of the genome sequence alone would have required over 500 standard pages of text.) Order was imposed by assigning a category to each of the 1743 genes found in *H. influenzae*, based on the role that the gene plays in the cell's life cycle (for example, DNA replication, cell division, and the transport of nutrients). The proposed role for each gene either was known through previous experimentation with *H. influenzae* or, in most cases, was hypothesized based on the strong similarity of the gene to a sequence in a "gene bank" database derived from all known individual gene sequences of other organisms.

Hypothetical genes. Besides the 1007 *H. influenzae* genes for which a role assignment could be made, 347 matched hypothetical genes (having no known function) already in the gene bank database, and 389 genes did not match the previous database. This indicates that about 40% of the *H. influenzae* genes have matches to genes of unknown function, and an additional one-quarter of the genes have never been encountered in previous sequencing studies.

Data management. To manage all the data associated with this study, the investigators set up a *H. influenzae*–specific database on the World Wide Web (http://www.tigr.org) so that researchers could access, analyze, and retrieve the genomic data. This database allows all or portions of the genome sequence to be analyzed and retrieved based on a number of user-specified parameters (gene number, gene name, genome position, or a sequence-similarity search). The hyperlinked format allows users to advance directly from DNA sequence, to sequence similarity alignments, to downloading of the desired sequence. These databases facilitate research by allowing users to receive answers to genome-specific questions. Furthermore, knowledge gained from the entire set of genes is more than the sum of its parts, allowing the posing of broader questions, such as what set of genes makes a bacterium pathogenic or what biochemical pathways are present in this cell. *See* INTERNET.

Mycoplasma genitalium. As a result of the experience gained from the *H. influenzae* project, a second, smaller bacterial genome (*Mycoplasma genitalium*) was completed in the course of a few months. With the publication of the *M. genitalium* genome sequence in 1995, the first whole-genome comparative analysis became a reality. Previously, the ancestor of *M. genitalium* underwent a genome downsizing to its current 570,000-base-pair size (the smallest known genome of any free-living organism); however, it retained enough genes to persist in its human host and grow in a test tube in the absence of human cells. Analysis of the entire DNA sequence indicates the same relative number of genes per 1000 base pairs and the same percentage of unknown and hypothetical hits as *H. influenzae*. However, the determination of a relatively lower percentage of genes dedicated to cellular biosynthetic processes reinforces the notion that the genome reduction made *M. genitalium* more dependent on its host for nutrients. This genome project has therefore established a baseline for determining the minimal set of genes required to sustain life.

Methanococcus jannaschii. Recently, another whole-genome sequence was determined by researchers at The Institute for Genomic Research in collaboration with the laboratory of Carl Woese of the University of Illinois at Urbana-Champaign. In 1977, Woese challenged accepted dogma by advancing the idea that three domains of life exist: prokaryotes (bacteria), eukaryotes (organisms with their genomic DNA in a discrete intracellular structure), and the archaea (archaebacteria). The archaea were originally classified as bacteria and thought to be biological oddities since many of these unicellular organisms can survive at the extremes of life such as hot springs, salt water, or in the case of *Methanococcus jannaschii*, at the bottom of the sea near thermal sea vents. By sequencing a single gene—the conserved ribosomal ribonucleic acid (RNA) gene—from numerous organisms and analyzing the evolutionary distance among the DNA sequences, Woese determined that the archaea constitute an ancient third domain of life: organisms well suited to surviving the inhospitable conditions that were present on the planet 3–4 billion years ago. Evidence has accumulated to substantiate the three-domain theory.

Analysis of the 1,730,000-base-pair *M. jannaschii* genome indicated that half of the 1682 genes did not match the database and approximately 12% of the genes were similar to sequenced genes of unknown function. These data confirm that archaea occupy an evolutionarily distinct position in the tree of life and add a wealth of data to the relatively short list of archaebacterial genes. When the first eukaryotic genome sequence (yeast) becomes available, the comparison of the genes from completely sequenced organisms representing all three branches of life will allow for meaningful, in-depth comparative genomics.

Applications. Many downstream applications of the knowledge gained from whole-genome sequencing are beginning to bear fruit. In addition to the gene and genome comparisons, gene control, genome structure, and gene overexpression and mutation will lead to important discoveries (**Fig. 2**). One application used by the pharmaceutical industry for control of bacterial infections is based on mass-producing the protein product of selected genes and then using that product as a basis for new vaccines. Another approach for killing pathogenic bacteria is based on discovering molecules which inhibit an overexpressed protein's function, either by traditional mass screening of compounds or by a rational drug design approach (such as making compounds which block the working end of the three-dimensional protein molecule). Gene products from thermostable organisms are finding an increasing number of industrial uses. The gene lists generated from high-throughput-sequencing projects will exponen-

tially increase the catalog of novel enzymes for industry or targets for the control of bacterial infections. Finally, by mutating (knocking out) a gene in a living organism, researchers will learn how the gene product functions normally in the cell, with important implications for both the pharmaceutical and biotechnology industries. For all of these applications, the existence of a complete DNA sequence for each gene rapidly accelerates the pace of research. Whole-genome sequencing is revolutionizing biological research, generating a molecular parts list for cellular life, and giving researchers important insight into how the parts function separately and as a whole. *See* ENZYME.

For background information *see* ARCHAEBACTERIA; DATABASE MANAGEMENT SYSTEMS; DEOXYRIBONUCLEIC ACID (DNA); ESCHERICHIA; GENE; GENETIC ENGINEERING; INFLUENZA; RIBONUCLEIC ACID (RNA) in the McGraw-Hill Encyclopedia of Science & Technology.

Brian A. Dougherty; Robert D. Fleischmann; Claire M. Fraser

Bibliography. C. J. Bult et al., Complete genome sequence of the methanogenic archeon, *Methanococcus jannaschii, Science*, 273:1058–1073, 1996; R. D. Fleischmann et al., Whole-genome random shotgun sequencing and assembly of *Haemophilus influenzae*Rd, *Science*, 269:496–512, 1995; C. M. Fraser et al., The minimal gene complement of *Mycoplasma genitalium, Science*, 270:397–403, 1995.

Geochemical prospecting

Detection of gases, associated with all mineral deposits, can lead to the discovery of deposits beneath the ground surface. Carbon dioxide, hydrogen sulfide, and other gases are common constituents of the hydrothermal solutions that transport ore minerals. Other gases are produced when those minerals react with ground water or with microorganisms. Hydrocarbon liquids and gases such as methane and ethane are produced when petroleum and natural gas deposits are formed. Hydrocarbon gases are also found around metallic ore deposits. All gases have a natural tendency to migrate from areas of high pressure to areas of low pressure, that is, from depth to the surface. They migrate slowly by diffusion through rock, or rapidly through permeable zones such as faults, fractures, or breccia zones. The gases are sampled at or near the surface, and are analyzed to measure gas species and their concentration. When the results are plotted on maps, anomalous areas of high or low concentrations are readily seen and may indicate buried mineral deposits or faults.

Production of gases. In addition to gases found in ore-depositing hydrothermal solutions, gases are produced by chemical reactions of ore minerals with the hydrosphere, biosphere, or atmosphere. Sulfide minerals react with oxygenated ground water, producing sulfur gases and sulfuric acid. These acid waters react with carbonate minerals, such as calcite, so that carbon dioxide is produced and oxygen is consumed. Thus, when a soil-gas sample contains increased concentrations of carbon dioxide and decreased concentrations of oxygen, oxidizing sulfide minerals may occur at depth. Hydrocarbon gases are components of petroleum and natural gas deposits but are also found associated with metallic ore deposits. The generally accepted theory that ore metals are transported in hot aqueous (hydrothermal) solutions provides a possible explanation for the association of hydrocarbon gases with metallic ore deposits. Hydrocarbon gases may be produced when the hydrothermal solutions thermally mature organic matter in the host rocks. This process is analogous to the natural process of petroleum formation, where kerogen or sediments rich in organic materials are subjected to increased temperature resulting from deep burial. The transformation of kerogen to petroleum hydrocarbons usually takes place over a long period of time through a process known as catagenesis. However, hydrocarbons can also be produced whenever organic matter is subjected to high temperatures from any source. For example, along the East Pacific Rise in the Gulf of California, gasoline-range hydrocarbons are produced in sediments by thermal waters expelled from the rift zone.

Sampling and analysis. Soil gases are sampled at regular intervals that are determined by the size of the expected target. To explore for small deposits, sample intervals of 30 m (100 ft) may be necessary; if large deposits are expected, sample intervals of 300 m (1000 ft) may be adequate. Gases in interstitial soil spaces are sampled by augering or driving a probe to shallow depth (commonly less than 1 m or 3 ft) and extracting the gas from that depth. The gases are stored in airtight containers for later analysis or are analyzed at the site with portable equipment. Gases are also sampled by collecting soils and heating them to drive off adsorbed gases. Clays and organic matter in soils readily adsorb gases, making soils good natural collectors of gases. Artificial collectors, such as activated charcoal, are also used to

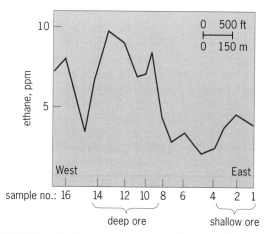

Fig. 2. Plot of ethane along a traverse at the Betze/Post gold deposit, Eureka County, Nevada. Shallow ore is at a depth of less than 500 ft (150 m); deep ore, at a depth of 500 ft (150 m) or more.

collect gases by adsorption. These collectors are buried in the soil and allowed to remain for days or weeks to adsorb gases. The adsorbed gases are driven off the collectors by heating, as with soils.

Analysis of gases is commonly performed by gas chromatography, mass spectrometry, or a combination of both. In gas chromatography, the molecular gas species are separated according to size. The gas sample is injected into a long, thin tube and carried through it by an inert gas to a detector at the end. Separation of gas species is achieved because the smaller molecules pass through the tube more rapidly than the larger molecules. The peak height for each gas is recorded and compared with standards to determine the concentration. The advantage of gas chromatography is that limits of detection are very low; a disadvantage is that only a few gases can be measured at one time.

In projects of the U.S. Geological Survey (USGS), on-site gas analysis is provided by a truck-mounted gas analyzer (**Fig. 1**). The analyzer is a quadrupole mass spectrometer. Analyzer operation is controlled by a program that scans sequentially from 1 to 100 atomic mass units in about 45 s. The analytical data are stored on a disk and printed on paper. The mass spectrometer separates ionized atoms or molecules according to mass, and the concentration of each gas species is determined by electronically counting the number of ions at individual mass peaks. An advantage of mass spectrometric analysis is that a large number of gases can be analyzed at one time.

Latitude and longitude, or other coordinates, are recorded for each sample site to allow plotting individual gas species or combinations of gases on maps. Because soil-gas surveys may involve tens or hundreds of samples and many gas species, the use of computers for storing and plotting of data can save much time. Computer programs can be used to plot and contour the gas data. In addition to saving time, these plots provide excellent visual images of gas anomalies.

Fig. 1. Truck-mounted gas analyzer.

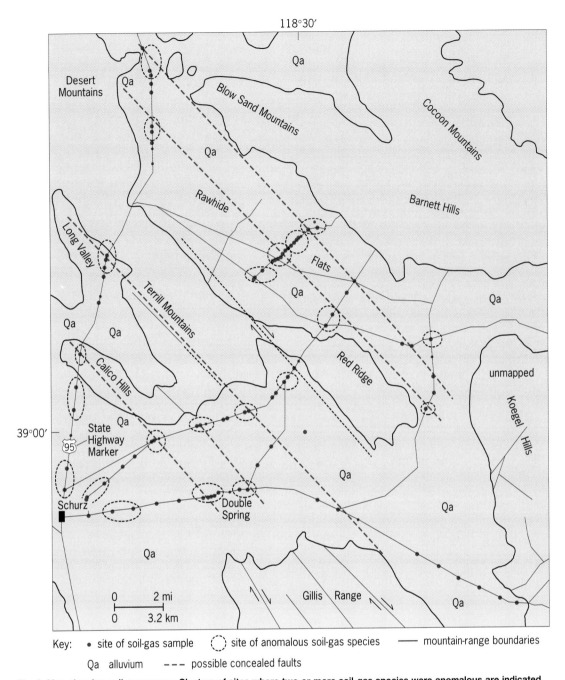

Fig. 3. Map showing soil-gas survey. Clusters of sites where two or more soil-gas species were anomalous are indicated. Lines drawn through these anomalous clusters indicate the locations of possible concealed faults. The area is located in Churchill and Mineral counties, Nevada.

Applications. Soil-gas surveys conducted over Carlin-type gold deposits in Nevada have revealed hydrocarbon gas anomalies that correlate with gold mineralization. In this type of low-grade deposit, micrometer-size gold particles occur in large volumes of sedimentary rock. An example of a hydrocarbon gas anomaly in one of these deposits (Betze/Post) is shown in **Fig. 2**, where the concentration of ethane gas is plotted along a sample traverse that crosses over deep, unoxidized ore and shallow, oxidized ore. The deep ore is overlain by about 300 m

(1000 ft) of alluvium and other sediments that are derived by erosion of rocks in the adjacent mountain ranges. The most prominent anomaly overlies the deep, reduced ore, where hydrocarbons are stable. However, these gases tend to dissociate under oxidizing conditions, resulting in a less prominent anomaly over the shallow, oxidized ore.

Gases migrate from depth most readily along faults because the faults are commonly more permeable to the passage of fluids than the surrounding rock. Thus, it is possible to map faults in bedrock overlain by alluvium or other overburden by soil-gas surveys (**Fig. 3**). Geologic mapping in mountain ranges in this area (north-central Nevada) suggested that faults must exist in bedrock beneath the alluvium-filled basins. Soil gases were measured along several traverses crossing Rawhide Flats and other nearby basins.

Factors affecting soil gases. Many factors affect the application of gases in mineral exploration. Changes in barometric pressure can affect the migration of gases to the surface. Positive pressure changes tend to suppress the outflow of gases while decreasing pressure enhances it, so that a pumping action results. Precipitation tends to fill pore spaces in the soil and block the flow of gases to the atmosphere. Some gases are quite soluble in water; for example, sulfur dioxide readily dissolves in water and may not reach the surface in areas of abundant water. Methane gas, the principal component of deposits of natural gas, is produced by thermal maturation of organic matter; but it is also produced by bacterial action. Approximately 80% of methane in the atmosphere is bacteriogenic; therefore, it is important to be aware of this possible source of methane when interpreting soil-gas data. Additional research is needed to evaluate the factors that influence the application of soil gases to exploration. The association of hydrocarbon gases with metallic ore deposits needs further investigation.

In the Basin and Range Province of the western United States, about two-thirds of the land surface is covered by alluvium-filled basins. In the north-central United States and Canada, large areas are covered by glacial deposits. These and similar areas are a challenge to mineral explorationists. Soil-gas measurements provide one technique that may allow them to "see through" these surficial deposits.

For background information *see* CHROMATOGRAPHY; GEOCHEMICAL PROSPECTING; MASS SPECTROMETRY; MINING; ORE AND MINERAL DEPOSITS in the McGraw-Hill Encyclopedia of Science & Technology.

J. Howard McCarthy

Bibliography. S. E. Kesler (ed.), Soil and Rock Gas Geochemistry (special issue), *J. Geochem. Explor.*, 38(1/2), 1990; R. W. Klusman, *Soil Gas and Related Methods for Natural Resource Exploration*, 1994; A. W. Rose, H. E. Hawkes, and J. S. Webb, *Geochemistry in Mineral Exploration*, 1979.

Geomagnetism

The magnetic field observed at the Earth's surface is produced by internal and external sources. The internal component originates in the liquid outer core, where large convection cells in the iron-nickel alloy work much like a dynamo to produce an approximately dipolar field (**Fig. 1**). The north and south magnetic poles are aligned roughly with the spin axis as a result of the mechanical interaction between the Earth's rotation and core convection. The magnetic field produced outside the core is

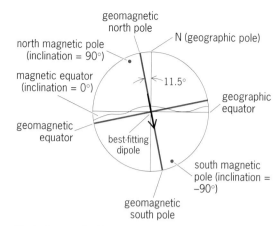

Fig. 1. Geocentric dipole model of the Earth's magnetic field. The best-fitting dipole is the simple dipole magnet which best fits the geomagnetic field at present (in a statistical sense). *(After R. F. Butler, Paleomagnetism, Blackwell, 1992)*

electromagnetically filtered by the intervening mantle. While the mantle comprises mainly silicate minerals such as olivine and pyroxene, it possesses appreciable electrical conductivity.

Magnetic field variations. The external field is produced by strong currents in both the magnetosphere and ionosphere. The magnetosphere is a large region of space surrounding Earth where electric currents produced by charged particles emanating from the Sun (the solar wind) interact with the internal field (**Fig. 2**). The magnetosphere is irregularly shaped, compressed to about 100,000 km (60,000 mi) on the Earth's illuminated side and extending past 200,000 km (120,000 mi) on the Earth's dark side. Much closer to Earth, currents in the ionosphere result from the interaction of solar ultraviolet radiation with atmospheric constituents. As the Earth rotates underneath these irregular currents, observ-

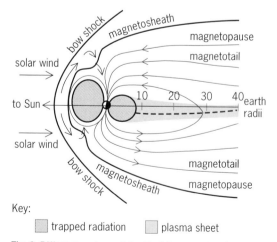

Fig. 2. Different regions of the Earth's magnetosphere. *(After R. T. Merrill and M. W. McElhinny, The Earth's Magnetic Field, Academic Press, 1983)*

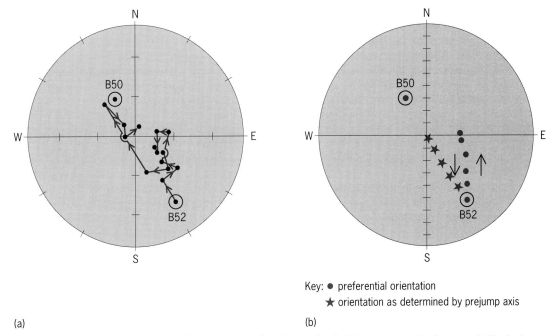

Key: ● preferential orientation
★ orientation as determined by prejump axis

(a)   (b)

Fig. 3. Measurements and predictions of magnetization directions for the first jump in magnetization recorded in the lava flow at Steens Mountain, Oregon. (*a*) Measured directions move from prejump stable direction (B52) to postjump stable direction (B50). (*b*) Calculated directions (arrows) caused by a magnetic storm. (*After A. Jackson, Storm in a lava flow, Nature, 377:685, 1995*)

ers measure time-varying magnetic fields with differing characteristic periods.

External field changes occur with characteristic periods ranging from a few seconds to the 11- to 22-year cycles in solar output and are marked by sunspots. Internal field variations are characterized by much longer periods. The longest internally derived variations occur on million-year time scales and are produced by the motion of the tectonic plates as they move relative to the underlying core. The shortest-period internal field variations are generally taken to be the geomagnetic secular variations produced by changes in flow patterns of the outer core; they have characteristic periods of 0.01–1.0° per year. Polarity transitions lie between these two extremes on the geomagnetic time scale; during a polarity transition the north and south geomagnetic poles reverse hemispheres over an interval of approximately 5000–10,000 years.

The longest-period external variations overlap with the shortest-period internal variations; thus a precise cutoff period is difficult to establish. By using spectral analysis of magnetograms (records of magnetic field variations made at magnetic observatories around the world), variations with characteristic periods shorter than about 4 years are generally considered to be of external origin. All longer-period variations are thought to originate in the core after being electromagnetically screened by the mantle.

Polarity transitions. Changes in the direction and strength of the surface geomagnetic field during polarity transitions have been studied extensively, but the morphology of the transitional field is still widely debated. The dipole field makes up about 90% of the present total field strength, with the remaining 10% constituting the nondipole field, a term that integrates quadrupoles, octupoles, and higher-order multipole components. Detailed studies of polarity transitions show that during a reversal the overall strength of the surface geomagnetic field decreases rapidly and then increases again.

In one model for polarity transitions, the dipole field decays to zero and then grows in the opposite direction, eventually attaining its previous strength after thousands of years. While the dipole decays and grows, the nondipole field remains approximately constant in overall strength. As a consequence, when the dipole field strength is less than or equal to the nondipole field strength, the surface field is highly variable in direction and intensity. An alternative model suggests that the weakened transitional field is not so confused but in fact retains its characteristic dipolar nature.

Rapid field variations. Beginning in 1985, a sequence of papers reported that apparently very rapid field changes were recorded in a thick sequence of Miocene (16.2 million years old) volcanic rocks near Steens Mountain, Oregon. Here, a reverse-to-normal polarity transition occurred while several of the relatively thin lava flows in the 1-km-thick (0.6-mi) pile were extruded. Within hours, the top and bottom margins of each lava flow cooled below the magnetic blocking temperature (the temperature at which the remanent magnetization is frozen in the rock). Theoretical cooling models and actual observation on modern flows suggest that the center of each flow might acquire its magnetization about 10 days later. Observed rates of directional change were as high as 6° per day, orders of magnitude faster than had been suggested previously (**Fig. 3**). These observations were at variance with traditional models for the origin of geomagnetic secular variations and with the inferred electrical conductivity of the mantle. Conventional models hold that even if the core field were instantaneously switched off (which is no mean feat in itself, and probably impossible to accomplish), the maximum directional variation seen at the surface would be on the order of 1–2° per day.

Such an unusually high rate of directional change has provoked considerable interest and scrutiny. Critics have suggested a variety of potential problems, such as the reheating of one flow by a later flow or crosscutting intrusion, uncertainties in the cooling models, and a variety of possible inaccuracies in the magnetic recording process. The research group has addressed most, if not all, of these criticisms in their reports, and their observations seem to have stood the test.

Magnetic storm. Like other polarity transitions that have been studied elsewhere, the lavas at Steens Mountain show that the geomagnetic field strength decreased from a nominal value of about 50 microtesla before the reversal to about 6 microtesla during the reversal. As the reversal continued, the field rebounded to approximately its original strength. Recently, it has been proposed that the rapid field variations seen at Steens Mountain do not owe their origin to an internal source but result from a commonly observed and widely studied external disturbance called a geomagnetic storm.

Geomagnetic storms are large, externally driven field variations, the surface manifestation of much larger storms in the magnetosphere farther from Earth. Storms can last for days, and magnetic field variations as large as 1 have been observed but typical values are more like 0.2–0.3 microtesla. In comparison, the Earth's present magnetic field strength ranges from a high of about 67 microtesla near the south magnetic pole in the Antarctic Ocean to a low of about 24 microtesla in southern Brazil and Argentina. When the magnetic field is strong (that is, the field is not reversing), these storm-driven variations can produce only small directional changes. But during a geomagnetic reversal, if the internally derived field at the surface is as weak as 6 microtesla, an externally derived effect as large as 1 microtesla can have a quite significant effect. What makes the effect even more important is the fact that when the internal field is weak the normal ring currents in the magnetosphere (as well as their storm-driven variations) could be expected to move much closer to Earth, perhaps doubling the magnitude of the storm signal to as much as 2 microtesla, or about one-third the size of the weakened 6-microtesla transitional field. Large storms can take days to grow and decay, and thus it is plausible that the field changes observed at Steens Mountain could have an external origin.

A storm-driven explanation for the rapid directional changes would require stretching a few parameters past their presently accepted limits. One problem is that the large directional shift observed requires a very long storm, 6–10 days, as opposed to the more commonly observed duration of 0.5–3 days. A second problem is that the magnetospheric ring currents must be arranged in just the right orientation to produce the desired effect. Finally, to fully explain the observations, two storms must have been recorded by two independent lava flows. Whether all these things happened in just the right way remains to be seen; however, the possibility of large externally driven magnetic variations has clearly been demonstrated, and it may be easier to accept than the alternative.

For background information *see* EARTH INTERIOR; GEOMAGNETIC VARIATIONS; GEOMAGNETISM; MAGNETOSPHERE; PALEOMAGNETISM; PLATE TECTONICS; SOLAR WIND in the McGraw-Hill Encyclopedia of Science & Technology. Michael O. McWilliams

Bibliography. R. S. Coe and M. Prévot, Evidence suggesting extremely rapid field variation during a geomagnetic reversal, *Earth Planet. Sci. Lett.*, 92:292–298, 1989; R. S. Coe, M. Prévot, and P. Camps, New evidence for extraordinarily rapid change of the geomagnetic field during a reversal, *Nature*, 374:687–692, 1995; M. Prévot et al., How the geomagnetic field vector reverses polarity, *Nature*, 316:230–234, 1985; P. Ultré-Guerard and J. Achache, Core flow instabilities and geomagnetic storms during reversals: The Steens Mountain impulsive field variations revisited, *Earth Planet. Sci. Lett.*, 135:91–99, 1995.

Glaciology

Glaciers form, advance, recede, and disappear in response to changing climate patterns. Alpine glaciers show the most direct response to climatic change of any surface phenomena in the mountains. This is apparent in the Sierra Nevada of California, where several small glaciers of less than 1 km² (0.386 mi²) exist under marginal conditions. Unlike larger ice masses, such as those in Alaska, which may typically display lag times of years or even decades in response to changes in climate, these glaciers respond rapidly, often showing measurable differences on a yearly basis.

Glacial history. The Sierras were extensively glaciated during the Late Pliocene and throughout most of the Pleistocene epochs (about 3 million years to 13,000 years before present). During this time, the glaciers were most active and accomplished much of the rock sculpturing seen today, including carving the walls of Yosemite Valley. Subsequent warming in the Late Pleistocene and Early Holocene (around 13,000 to 8000 years before present) melted these glaciers.

Today's small Sierra glaciers are not remnants of these massive ice bodies but are new glaciers formed as a result of a global cooling referred to as the

Fig. 1. Conness Glacier (*a*) in 1939 *(after Yosemite National Park, Report of Glacier Studies in Yosemite National Park, 1939)* and (*b*) in 1994.

Little Ice Age. This cooling, generally accepted as occurring between 5000 years ago and A.D. 1450–1890, is the primary reason that glacierization, albeit small in scope and scale, returned to the Sierras.

Current glaciation. Glacier size and behavior is a result of processes that add mass (primarily through snowfall) or subtract mass (primarily through snowmelt). A glacier's mass balance is a measurement of the effectiveness of these processes. Positive mass balances indicate more mass (water equivalent of snow and ice) was added than lost during a year, with the opposite being true of negative balances. Similarly, glaciers experiencing a succession of years with positive balances thicken and advance, while those with negative balances thin and retreat.

The linkages between climate and glacier behavior, however, are complex. Any single glacier's behavior is due to a cascade of interactions between the glacier, its topographical setting (affecting microclimate), and the variations in local climate, regional climate, and, finally, global climate. Two neighboring glaciers can exhibit vastly different responses to a changing climate—one advancing, the other retreating, a condition that has been documented in Alaska. Thus, one glacier's response to variations in climate may not be representative of the behavior of all glaciers in a region. Ideally, several glaciers should be investigated before positing any generalizations about changing climates based solely upon one glacier's behavior.

Up to 247 glaciers and peripheral ice masses have been identified in the Sierra Nevada. These small glaciers typically occupy high elevations (greater than 3650 m or 12,000 ft) on the northeastern sides of mountains. These locations are best suited for snowfall and snow preservation; they are high enough and cool enough that most of the annual precipitation falls as snow. They also have a directional aspect that minimizes the amount of radiative energy received from the Sun.

Snowfall in the Sierra Nevada predominantly results from the southward migration of the polar jet stream, which brings Pacific frontal systems into the region during the fall and winter months. This pattern is sometimes augmented by subtropical moisture that greatly enhances snowfall and thus snowpack during El Niño years. Summers at the elevation of Sierran glaciers are relatively short yet warm, with the high mountain passes often not being snowfree until late July or August.

Glaciation studies. Studies investigating these small glaciers have been scant, with most conducted in or near the vicinity of Yosemite National Park. The earliest investigations were conducted by I. C. Russell of the United States Geological Survey (USGS) in the early 1880s. He reported on these glaciers while they were at or near their greatest extent and thickness, close to the end of the Little Ice Age.

The most extensive study was initiated in 1932 by C. A. Harwell, then naturalist of the Yosemite National Park. His primary mission was to photograph and document terminus position (lowest elevational limit of a glacier) and movement, if any, of all glaciers inside and near the park's boundaries. This research initially involved the investigation of six glaciers (Conness, Dana, Koip, Kuna, Lyell, and Maclure); it was conducted on a nearly annual basis by Yosemite Park naturalists until 1961. Thereafter, reconnaissance visits were made to the glaciers in 1969, 1970, and 1975, when the program was ended. The greatest legacy of these surveys was the annual reports and photographic records. These reports, while not containing detailed measurements of the mass balances, provide information as to snow depth and coverage over the ice, as well as terminus location and relative changes since previous visits. Photographs, taken from marked photo stations, document these changes and allow for visual comparisons of glacial extent and variation for over 40 years. **Figure 1** shows the changes in Conness Glacier from 1939 to 1994.

As indicated in these written reports, most glaciers observed during the period of record, 1932–1975, diminished in extent and thickness. In the final 1975 report, several glaciers were reported to be down to as much as one-third to one-half the size noted in 1885.

Fig. 2. Conness Glacier ice loss (*a*) in 1991 and (*b*) in 1992. The rocks in the background indicate the similar views.

The USGS commenced a study on Maclure Glacier in 1965 as part of a contribution to the International Hydrological Decade. Detailed mass-balance measurements were taken from 1965 to 1972. This represents the first detailed effort involving mass balances in the Sierra Nevada glaciers. Results indicate that for most of the period of record Maclure Glacier experienced negative mass balances and thinned slightly.

More recent detailed mass-balance studies on Conness Glacier continue to show a trend of negative to severely negative balances. The California drought of 1987–1992 played a major role in the wastage of this glacier. Conness Glacier lost over 2×10^6 m³ (70.6×10^6 ft³) of water-equivalent ice during this time; and it experienced a 15% decrease in total surface area, with over 5% occurring in 1992 (**Fig. 2**). Commensurate with this areal loss has been an overall thinning of the glacier. In 1987 the ice thicknesses were estimated at 12–20 m (40–65 ft). In 1996 the estimate for ice thickness was 4–10 m (13–33 ft), with piles of rocks emerging through the remaining ice visible on some areas of the glacier (Fig. 1*b*).

Conditions fluctuated in 1993 and 1994. In 1993 the glacier experienced its first positive balance year since 1987, with a net gain of 256,500 m³ (9,058,000 ft³) of water-equivalent snow. However, 1994 saw a net loss of over 340,000 m³ (12×10^6 ft³), effectively melting all of the snow from

the previous winter, all that had remained from the previous year, and again melting down to and into the glacial ice.

Conditions improved somewhat (that is, there was a positive mass balance) in years 1995 and 1996. The winter of 1994–1995, an El Niño year, deposited more snow (about 200% of normal snowfall) on the glacier than at any time since detailed measurements began in the mid-1980s. By the end of summer 1995, an average depth of 2–3 m (6.5–10 ft) of snow remained on the glacier, representing a total positive mass balance of over 424,000 m³ (15×10^6 ft³) of water equivalent. The winter of 1995–1996 deposited nearly normal amounts of snow over the glacier surface; toward the end of 1996 it appeared that the glacier's net annual balance would approach equilibrium, that is, neither gain nor lose mass. This is the first time since recent measurements began that the glacier has gone more than one calendar year without having glacial ice melt occur.

Climatic implications. Most glaciers in the Sierra Nevada have shown nearly continuous thinning and shrinkage since first investigated in 1885. Rates of ice loss have fluctuated through the years, but overall this trend has remained steady or even increased. Perhaps this is indicative of climate change for the western United States. If so, it is necessary to investigate whether the thinning and retreat of these glaciers is due to changes in winter precipitation patterns, an increase in summer temperatures or length of the summer season, or a combination of these conditions. While it is probable that the effect is due to a combination of changes, further investigations are necessary to determine the scope and magnitude of such changes.

There is some indication that these small glaciers tend to respond more rapidly to changes in winter precipitation. Conness Glacier has shown such susceptibility as evidenced by its rapid wastage during the drought years. This idea is supported elsewhere in the West, where mass balances on South Cascade Glacier (in the Cascades of Washington) have been shown to be more dependent on changes in the amounts of winter snowfall.

This may be significant in terms of climate in the western United States. Populations living in the Rocky Mountain states and farther west to the Pacific Ocean depend on snowmelt from mountain snowpacks as their chief source of water. As demands on this water resource increase, so will the need for a stable and adequate water supply. Glaciers may be providing early warning signs that the amount and distribution of snow in the West is changing. Clearly, those conditions that led to glacier formation in the Sierras have changed. Whether these changes are permanent remains to be seen. Increased research is needed to investigate the causes and effects of these fluctuations.

For background information *see* CLIMATE HISTORY; CLIMATIC PREDICTION; EL NIÑO; GLACIAL EPOCH; GLACIOLOGY; HYDROLOGY in the McGraw-Hill Encyclopedia of Science & Technology.   Frederick B. Chambers

Bibliography. R. G. Barry, *Mountain Weather and Climate*, 2d ed., 1992; G. J. McCabe, Jr., and A. F. Fountain, Relationship between Atmospheric circulation and mass balance of South Cascade Glacier, Washington, U.S.A., *Arctic Alpine Res.*, 27(3):226–233, 1995; I. C. Russell, Existing glaciers of the Sierra Nevada, *USGS Fifth Annual Report*, 1885.

Greenhouse effect

The carbon dioxide (CO_2) concentration of the Earth's atmosphere is increasing because of the burning of fossil fuels and the clearing of forests. Prior to the start of the industrial revolution near the beginning of the nineteenth century, the global atmospheric CO_2 concentration was about 280 parts per million by volume (ppmv). It increased to 350 ppmv by 1988 and is currently increasing at about 1.5 ppmv per year. In 1995, the Intergovernmental Panel on Climate Change (IPCC) projected that concentrations would reach about 500 ppmv by the year 2100 if CO_2 emissions were maintained at 1994 rates. However, considering the increasing industrialization of China and other developing countries, as well as the increasing population of the world, it seems more likely that CO_2 emission rates will continue to rise. Consequently, the atmospheric CO_2 concentration may reach considerably higher levels, perhaps 700–1000 ppmv by 2100.

General circulation models. The primary reason for concern about the increasing atmospheric CO_2 concentration is that general circulation models have predicted that it will cause global warming. General circulation models are computer models that strive to simulate the global circulation, energy exchange, cloud formation, precipitation, and other processes of the Earth's atmosphere. Carbon dioxide and some other trace gases, such as methane, nitrous oxide, chlorofluorocarbons (for example, Freon), and ozone, are radiatively active. They are mostly transparent to incoming solar radiation, but they are opaque and do not transmit heat rays emitted by the Earth's surface, thereby acting like an insulating blanket or a greenhouse. Hence, this phenomenon is called the greenhouse effect. The Intergovernmental Panel on Climate Change also reports that the general circulation models project that the surface temperature of the Earth will increase by 1.6–6.3°F (0.9–3.5°C) by 2100 and that precipitation patterns will change. Some areas will become wetter and others drier, an effect that is certainly cause for concern, especially for agriculture. Although it is prudent to prepare for such global climatic changes, large uncertainties (such as effects of CO_2 on cloud formation) associated with the global circulation model predictions remain.

Effects on agriculture. The most favorable production areas for various crops during a period of global warming can be expected to move toward the poles. Countries such as Canada and Russia, whose cool climates currently restrict plant growth, can expect their agricultural production to rise. However, more equatorial regions such as the southern United States and Mexico can expect more difficulty in growing certain crops, such as wheat. Some developing countries that may be most severely affected may be the least able to cope. However, the changing precipitation patterns are probably of even greater concern. If certain areas become drier, nearly all crops within those boundaries are likely to suffer, and at the same time available water resources are also likely to decrease, making it difficult to meet increased agricultural demands for water. Fortunately, precipitation levels are expected to increase worldwide on average, so some areas are likely to become more productive. If these shifts in optimal production areas do occur because of the warming and changing precipitation patterns, there is likely to be considerable disruption of rural society; and if food becomes scarce, other sectors of society will be negatively affected as well.

Although the predictions of global warming are somewhat uncertain, there is general agreement that the global atmospheric CO_2 concentration is increasing and the increase is likely to affect plant growth and crop production directly. In photosynthesis, CO_2 and water are combined in plant leaves utilizing sunlight to produce carbohydrates and oxygen. Thus, CO_2 is a feedstock for life itself, as plants use

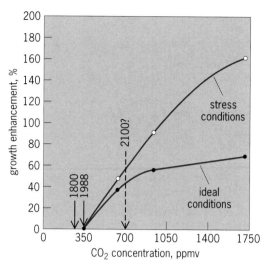

Percentage of enhancement of plant growth versus the CO_2 concentration in the air around the plant leaves. The CO_2 concentrations corresponding to preindustrial times (280 ppmv in 1800), present day (350 ppmv in 1988), and a doubling (700 ppmv projected for 2100) are indicated by the arrows. *(Adapted from K. E. Idso and S. B. Idso, Plant responses to atmospheric CO_2 enrichment in the face of environmental constraints: A review of the past 10 years' research, Agr. For. Meteorol., 69:153–203, 1994)*

the carbohydrates to grow, herbivores eat the plants, carnivores eat the herbivores, and so on up the food chain.

Early experiments. Starting about 1900 but especially in the 1960s and 1970s, many experiments using elevated CO_2 concentrations were conducted in plant growth chambers and greenhouses. These experiments demonstrated that an increase in the CO_2 concentration in the air causes plant leaves to increase their rates of photosynthesis, with consequent increases in plant growth and crop yield. Data from these experiments also showed that a CO_2 concentration of about 700 ppmv (probable for the year 2100) would increase crop yields on the average by about 35%. As a consequence of such experiments, it has become a standard horticultural practice in greenhouses, where growers can easily change the composition of the air around the plants, to enrich the atmosphere inside to about 1000 ppmv CO_2 in order to increase yields.

Typical of greenhouses, most of the experiments were conducted at ideal temperatures, water supply, and fertilizer. However, in recent years much research on global change has focused on whether similar increases in plant growth and crop yield can be obtained under stressful conditions that are more representative of actual fields, where temperatures often are not ideal and where water and soil nutrients may be limited. The results generally have indicated that elevated CO_2 tends to enable plants to withstand environmental stresses (see **illus**.). For a CO_2 concentration of 700 ppmv, the growth stimulation is about 55% for the stressful conditions, which is con-

siderably more than the 40% stimulation observed for the more ideal nonstress conditions.

CO_2 enrichment. Recently a few CO_2-enrichment experiments have been conducted on cotton and wheat under actual field conditions with ample supplies and with limited supplies of soil water. With a new technique called free-air CO_2 enrichment (FACE), CO_2 was released in circular plots in an open field at a controlled concentration of 550 ppmv. Cotton growth was stimulated about 30%, and the yield of cotton lint even more, with both ample and limited water supplies. Wheat growth increased about 20%. Grain yields increased only about 10% at ample water, but with limited water and elevated concentrations of CO_2, yields increased about 20%.

Although the average results of many CO_2-enrichment experiments suggest that a doubling of CO_2 concentration from about 350 ppmv to 700 ppmv will increase plant growth by 40–55% depending on conditions, there have been a wide range of responses among plant species or crops. Faster-growing species have responded more to CO_2 than slower-growing species. Crops adapted to warm climates and growing under high temperatures have responded more than crops adapted to and growing under cooler conditions. In the free-air CO_2 enrichment experiments, summer-grown cotton with an indeterminate growth habit (that is, it continually grows and flowers) responded more than winter-grown wheat with determinate growth habit (it flowers and stops growing after a more or less set time).

C3 and C4 plants. The largest distinction among plant species in their response to elevated CO_2 concentrations, however, is between C3 and C4 plants, so called because of the number of carbon atoms in the first compounds in their photosynthetic biochemical reactions. C4 crops include tropical grasses such as corn, sorghum, and sugarcane, while most other crops are C3 plants. C4 plants are already very efficient in their use of CO_2, and exposure to higher concentrations increases their photosynthetic rates very little. In contrast, C3 plants undergo a process called photorespiration that is wasteful of CO_2 but is suppressed by elevated CO_2 concentrations. Consequently, they respond to atmospheric CO_2 enrichment relatively more than C4 plants. Some of the most noxious weeds are C4 plants, so C3 crops can be expected to compete more effectively against them in the future. However, C4 plants, such as corn, sorghum, and sugarcane, likely will face greater competition from C3 weeds.

Low soil nutrients. Although elevated CO_2 tends to enable plants to endure stress better, it may not help as much for the stress of low soil nutrients. As plants grow, they require nitrogen, phosphorus, potassium, and other nutrients from the soil. Consequently, for plants to grow bigger at elevated CO_2 concentrations, they will need more nutrients; therefore, fertilizer recommendations will have to be increased in order to take advantage of the higher atmospheric CO_2 concentrations. Where fertilizer cannot be applied and nutrients are often limiting, such as is the case for most natural ecosystems and for agriculture in developing countries, there may be little response to higher CO_2 concentrations. Indeed, many laboratory experiments with nutrient solution culture have shown reduced response to CO_2. Yet, in the field a more robust plant growing at high CO_2 concentrations might be able to extract more nutrients from the soil. This is an unresolved question.

Water reduction. Another effect of an elevated level of CO_2 on plants is that it causes partial closing of the tiny leaf pores or stomates through which CO_2 enters and water vapor exits the leaves, thereby reducing water loss per unit of leaf area. However, CO_2-induced increases in plant size and leaf number and other factors compensate, so that measurements of crop water use per unit of land area change only slightly. Although water use may change only slightly with increasing concentrations of CO_2, water use efficiency increases because of the increased growth. Nevertheless, while the direct effect of CO_2 on total water use appears minimal, if global warming occurs or if the climate becomes drier, water requirements could still increase in certain areas.

When stomatal apertures are reduced by elevated CO_2, the slower rate of water loss per unit of leaf area also reduces evaporative cooling. Therefore, foliage temperatures tend to rise, sometimes more than $2°F$ ($1°C$) for a doubling of CO_2. These increases in crop temperature are caused by CO_2, and they are likely to alter the optimum crop production areas regardless of whether there is any global warming.

Photosynthesis. The optimum temperature for photosynthesis and plant growth is also affected by elevated levels of atmospheric CO_2. This optimum temperature increase is as much as or more than the Earth surface temperature increases predicted by the general circulation models. Consequently, plants may have a built-in safety mechanism for weathering the effects of global warming.

Global changes. Impending global environmental changes related to the increasing atmospheric CO_2 concentration are likely to affect future agricultural productivity in a number of ways. Global warming and changes in precipitation patterns, as well as the direct effects of CO_2 on plant foliage temperatures, may shift the geographic production areas of many crops. Crop yields in some areas are likely to increase, and in others decrease. On average, however, the direct effects of elevated levels of CO_2 on photosynthesis and other plant processes are likely to increase yields of C3 crops about 30% or more for a doubling of the current CO_2 concentration, as based on the results from hundreds of experiments, although there is a wide range of responses among crops. In addition, elevated concentrations of CO_2 generally ameliorate the effects of environmental stresses on the plants. Elevated levels of CO_2 result in increased water use efficiency but have minimal direct effects on water use per unit of land area. A warming or drying of climate, however, could increase water requirements. Overall, the ongoing increase in atmospheric CO_2 concentration should benefit agriculture. It should compensate somewhat for the effects of adverse changes in climate, and if such changes are small, significant increases in productivity are likely to occur.

For background information *see* AGRICULTURAL SOIL AND CROP PRACTICES; ATMOSPHERIC CHEMISTRY; CLIMATE HISTORY; CLIMATE MODELING; CLIMATE MODIFICATION; CLIMATE PREDICTION; CLIMATOLOGY; GREENHOUSE EFFECT; PHOTOSYNTHESIS; PLANT METABOLISM; WEEDS in the McGraw-Hill Encyclopedia of Science & Technology.   Bruce A. Kimball

Bibliography. K. E. Idso and S. B. Idso, Plant responses to atmospheric CO_2 enrichment in the face of environmental constraints. A review of the past 10 years' research, *Agr. For. Meteorol.*, 69:153–203, 1994; Intergovernmental Panel on Climate Change, *Climate Change 1995: The Science of Climate Change—Summary for Policymakers*, 1995; B. A. Kimball, Carbon dioxide and agricultural yield: An assemblage and analysis of 430 prior observations, *Agron. J.*, 75:779–788, 1983.

Hall effect

Different states of matter are distinguished by their internal structures or orders. For many years, all the orders were thought to be associated with symmetries (or rather, the breaking of symmetries). For example, when a gas changes into a crystal, the continuous translation symmetry of the gas is reduced to a discrete translation symmetry of the crystal. Based on the relation between orders and symmetries, a general theory of ordering was developed by L. Landau. The success of Landau's theory gave rise to a widespread belief that, at least in principle, all the kinds of order that matter can display were understood.

However, in the early 1980s physicists discovered a new kind of liquid state, quantum Hall state, by placing a two-dimensional electron gas (confined on the interface of two different semiconductors) in a strong magnetic field at low temperature. Many remarkable properties of quantum Hall liquids were soon found. A quantum Hall liquid may be said to be more rigid than a solid (a crystal) in the sense that a quantum Hall liquid cannot be compressed. Thus a quantum Hall liquid has a fixed and well-defined density. More surprises emerge when the electron density is considered in terms of the filling factor, ν, which is the ratio of the electron density to the density of magnetic flux quanta. It has been found that all discovered quantum Hall states have densities such that the filling factors are given exactly by some rational numbers, such as $\nu = 1$, 1/3, 2/3, 2/5, ...

Topological orders. Those filling factors strongly suggest the presence of internal orders or patterns in quantum Hall states despite their liquid property. Indeed, theoretical studies since 1989 reveal highly

Fig. 1. A particle and its quantum wave on a circle.

nontrivial internal orders (or internal patterns) in quantum Hall liquids. However, these internal orders are different from any other known orders and cannot be observed in conventional ways. What is really new (and strange) about the orders in quantum Hall liquids is that they are not associated with any symmetries (or the breaking of symmetries) and cannot be described by Landau's theory. This new kind of order is called topological order, and an entirely new theory is needed to describe it.

Some intuitive understanding of topological order can be gained by trying to visualize the quantum motion of electrons in a quantum Hall state. A single electron in a magnetic field always travels along circles (in what is called cyclotron motion). Because of the wave property of the electron, the cyclotron motions are quantized such that the circular orbit contains an integral (whole) number of wavelengths (**Fig. 1**). The wavelength may be regarded as a step length, and the electron may be said to always take an integral number of steps to go around the circle. If the electron takes n steps to traverse the circle, it is said to be in the nth Landau level. The electron in the first Landau level has lowest energy, and it will stay in that level at low temperatures.

When many electrons combine to form a two-dimensional electron gas, they not only perform their own cyclotron motion in the first Landau level but also go around each other and exchange places. Those additional motions are also subject to the quantization condition. For example, an electron must take an integral number of steps to go around another electron. Furthermore, the electrons also try to stay away from each other as much as possible, because of the strong Coulomb repulsion and the Fermi statistics between the electrons.

Dancing rules. Thus, the quantum motions of electrons in a quantum Hall state are highly organized. All the electrons in a quantum Hall state dance collectively following strict dancing rules:

1. All electrons carry out their own cyclotron motion in the first Landau level.

2. Electrons always stay away from each other.

3. An electron always takes an integral number of steps to go around another electron.

If every electron follows these strict dancing rules, only one unique global dancing pattern is allowed. Such a dancing pattern describes the internal quantum motion in the quantum Hall state. It is this global dancing pattern that corresponds to the topological order in a quantum Hall state. Different quantum Hall states are distinguished by their different dancing patterns (or equivalently, by their different topological orders).

Characterization. A simplest quantum Hall state is a $\nu = 1/m$ Laughlin state, in which an electron always takes exactly m steps around another electron. A Laughlin state contains only one component of incompressible fluid. More general quantum Hall states (called abelian quantum Hall states) with filling factors such as $\nu = 2/5, 3/7, \ldots$ contain several components of incompressible fluid. The dancing pattern (or the topological order) in an abelian quantum Hall state can also be described in a similar way by the dancing steps. The dancing pattern can be characterized by a symmetric matrix, K, with integer entries, and a charge vector, Q, also with integer entries. An entry of Q, Q_i, is the charge (in units of the electron charge, e) carried by the particles in the ith component of the incompressible fluid. An entry of K, K_{ij}, is the number of steps taken by a particle in the ith component to go around a particle in the jth component. All physical properties associated with the topological orders can be determined in terms of K and Q. For example, the filling factor is simply given by $\nu = Q^{\mathrm{T}}K^{-1}Q$, where Q^{T} is the transpose of Q and K^{-1} is the inverse of K. In the (K,Q) characterization of quantum Hall states, the $\nu = 1/m$ Laughlin state is described by $K = m$ and $Q = 1$, while the $\nu = 2/5$ state, which has two components, is described by the equations below.

$$K = \begin{pmatrix} 3 & 2 \\ 2 & 3 \end{pmatrix} \qquad Q = \begin{pmatrix} 1 \\ 1 \end{pmatrix}$$

Edge excitations. It is instructive to compare quantum Hall liquids with crystals. Quantum Hall liquids are similar to crystals in that they both contain rich internal patterns (or internal orders). The main difference is that the patterns in the crystals are static, being related to the positions of atoms, while the patterns in quantum Hall liquids are dynamic, associated with the ways in which electrons may be said to dance around each other. Crystal orders can be measured by x-ray diffraction, and an important problem is that of measuring topological orders of quantum Hall liquids in experiments. The theoretical studies since 1990 suggest that edge excitations in quantum Hall states provide a practical way to do this.

All bulk excitations in quantum Hall liquids have a finite energy gap due to the incompressibility of the quantum Hall liquid. However, quantum Hall liquids of finite size always contain one-dimensional gapless edge excitations, which is another unique property of quantum Hall liquids. The structures of edge excitations are extremely rich, reflecting the

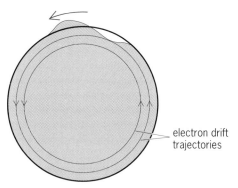

Fig. 2. An edge wave on a quantum Hall droplet, which propagates in only one direction.

rich structure of bulk topological orders. Different bulk topological orders lead to different structures of edge excitations. Thus, it is possible to study and measure the bulk topological orders by studying structures of edge excitations.

To understand edge excitations, the simple $\nu = 1/m$ Laughlin state will first be considered. Although the quantum Hall liquid cannot be compressed, a finite quantum Hall droplet can always change its shape without acquiring much energy. Thus the edge excitations are nothing but surface waves propagating on the edge of the droplet.

Electrons near the edge always feel a force normal to the edge, which holds the droplet together. In the presence of the magnetic field, the force causes the electrons to drift along the edge in one direction (normal to the direction of the force). Therefore, any edge disturbances will also drift in the same direction (**Fig. 2**). Thus, unlike ordinary waves on the surface of water, the edge waves in quantum Hall states can propagate in only one direction.

A more general quantum Hall liquid contains several components of incompressible fluid, and each component can deform independently. Thus a quantum Hall liquid with k incompressible components will have k branches of edge excitations. This correspondence is an example of how a property of bulk topological order (the number of incompressible components) is reflected in a property of edge excitations (the number of edge branches).

Tunneling conductance. In addition to the number of edge branches, the dynamical properties of edge electrons depend on the bulk topological order in a sensitive way. It was found, because of the nontrivial bulk topological order, that electrons at the edges of fractional quantum Hall liquids (with fractional filling factor) form a new kind of correlated state, called a chiral Luttinger liquid. The dynamical properties of chiral Luttinger liquids are very different from those of familiar free-electron systems. For example, the tunneling conductance between two edges (**Fig. 3***a*) is proportional to T^{2g-2} with $g \neq 1$, where T is the absolute temperature. (For free electrons, g is always equal to 1.) The exponent g is quantized and depends only on the topological

order. Thus, measuring g will reveal information about the bulk topological order.

To reach an intuitive understanding of the anomalous exponent g, it is helpful to consider adding an electron to the edge of a $\nu = 1/m$ Laughlin state. The other edge electrons have to take m steps to go around the added electron. Thus for larger m, the added electron causes larger disturbance and it is harder to add the electron to the edge at low temperatures. This behavior will result in a larger anomalous exponent g. In fact, $g = m$ for the $\nu = 1/m$ Laughlin state.

Several experimental groups have successfully measured g through tunneling conductance, and the results agree well with the theoretical prediction. These experiments demonstrate the existence of new chiral Luttinger liquids and open the door for experimental study of the rich internal and edge structures of quantum Hall liquids.

Quasiparticle tunneling. When two edges are separated by a fractional quantum Hall liquid (Fig. 3*b*), quasiparticle tunnelings can also occur. (A quasiparticle is a special kind of excitation in a quantum Hall state which can carry a fractional charge if the filling factor is fractional.) Quasiparticle tunneling has the remarkable property that the tunneling resistance decreases as a power of temperature as the temperature is lowered. This is reminiscent of tunneling between superconductors, which have zero resistance. The analogy goes beyond dc transport. The noise spectrum of quasiparticle tunneling contains a singular peak at a Josephson frequency $f_J = e^*V/h$ associated with the fractional charge e^* of the quasiparticle, where V is the voltage across the gap and h is Planck's constant. Such a singular peak in the noise spectrum is very similar to the ac Josephson effect in tunneling between superconductors.

Absence of general theory. Despite the understanding of the topological orders and edge excitations in abelian quantum Hall states, a theory for topological orders is still lacking for most general quantum Hall states. For example, it is known that some quantum Hall states (called nonabelian quantum Hall states) are beyond the dancing-step description and cannot be characterized by the K-matrix and the charge vector. In any case, quantum Hall liquids open up an entirely new territory for physical exploration, both in experiments and in theory. The studies of

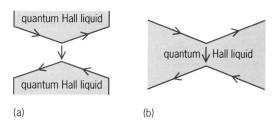

(a)   (b)

Fig. 3. Tunneling phenomena. Arrows indicate particle motion. (*a*) Electron tunneling between edges of two quantum Hall liquids. (*b*) Quasiparticle tunneling between two edges separated by a quantum Hall liquid.

topological orders in quantum Hall liquids will deepen the understanding of orders, and may lead to discoveries of new states of matter beyond quantum Hall systems.

For background information *see* HALL EFFECT; JOSEPHSON EFFECT; MATRIX THEORY; QUANTUM MECHANICS; SYMMETRY BREAKING; SYMMETRY LAWS (PHYSICS); TUNNELING IN SOLIDS in the McGraw-Hill Encyclopedia of Science & Technology.   Xiao-Gang Wen

Bibliography. K. v. Klitzing, G. Dorda, and M. Pepper, New method for high-accuracy determination of the fine-structure constant based on quantized Hall resistance, *Phys. Rev. Lett.*, 45:494–497, 1980; R. Laughlin, Quantized Hall conductivity in two dimensions, *Phys. Rev.*, B23:5632–5633, 1981; F. P. Milliken, C. P. Umbach, and R. A. Webb, Indications of a Luttinger liquid in the fractional quantum Hall regime, *Solid State Comm.*, 97:309–313, 1995; D. C. Tsui, H. L. Stormer, and A. C. Gossard, Two-dimensional magneto transport in the extreme quantum limit, *Phys. Rev. Lett.*, 48:1559–1562, 1982; X. G. Wen, Topological orders and edge excitations in fractional quantum Hall states, *Adv. Phys.*, 44:405–473, 1995.

Hazardous waste

Chemical warfare agents are often called poison gases, although modern ones are liquids. Of the enormous amounts manufactured but little used during World War II and the Cold War, much remains in controlled storage. Still to be destroyed are United States and Russian declared stockpiles of 32,000 tons and 40,000 tons, respectively. Lesser stockpiles of other nations have been destroyed, in part by incineration, in part by neutralization and similar reactions.

Prior to 1970, ocean dumping of old chemical weapons was common. Over 300,000 tons of German chemical weapons were discarded in the relatively shallow Baltic Sea. Danish fishing crews now and then bring up chemical bombs in their nets.

The principal modern technologies for destroying chemical weapons are incineration and a group of chemical reactions often called neutralization (some are not strictly neutralizations as viewed by chemists). Incineration is favored by many chemical munitions engineers but is strongly opposed by some lay groups, especially people who live near proposed chemical munition incineration plants.

The Chemical Weapons Convention, a multinational treaty that should go into effect in 1997, requires all participating nations to destroy their chemical weapons by 2007. Besides controlled stockpiles, vast numbers of chemical munitions lie underground in uncertain locations. Some fired in battle did not burst but burrowed into the earth. Others were buried, often with insufficient documentation, on old military bases, including some in the United States. Estimates are that over 10 million World War I chemi-

cal munitions, most still lethal, are undiscovered in Europe.

Stockpile destruction, known as chemical demilitarization, will take years to accomplish and cost billions of dollars. Adequate technologies have long been available; delays in destruction of chemical weapons are due chiefly to disagreement on how and where to do it. Finding and destroying lost munitions, for which many European nations have active programs, is expected to be even more expensive.

The major modern chemical warfare agents are categorized as vesicants [blister agents; including yperite or mustard gas (**1**), lewisite (**2**)] and nerve agents [including sarin (**3**), soman (**4**), United States VX (**5**), Russian VX (**6**)].

Whether from a stockpile or recovered from the earth, a chemical munition has a metal shell, decontamination of which is necessary after the chemical warfare agent has been drained from it. Some munitions in storage have developed leaks, which occasionally contaminate wood pallets and other packing materials. Sometimes the leaks have wet energetics, that is, rocket propellants and burster explosives.

An advantage of incineration is that it can be adapted to destroy all of these hazardous materials—the chemical warfare agents proper as well as contaminated energetics, packing materials, and drained containers.

Since chemical elements cannot be destroyed, the elements present in an agent appear in the products of destruction. For example, sarin contains phosphorus and fluorine, besides carbon, hydrogen, and oxygen; all of these atoms appear among the wastes from any destruction process.

All chemicals, even water, sugar, and table salt, are toxic to organisms under some circumstances. Rather than classifying substances as toxic versus safe, toxicologists evaluate the degree of toxicity. It is necessary to define what products result from any method used for destruction of a chemical warfare agent and to develop procedures to detoxify or contain, for controlled storage, any products of high toxicity.

Whereas incineration of chemical warfare agents is regarded as a one-stage process, chemical destruction methods generally involve two stages: an initial detoxification reaction to form products that are greatly reduced in toxicity but still unsuitable for release into the environment, and a final polishing treatment.

First-stage neutralization methods. These methods involve reactions with aqueous alkali, alkoxides, ethanolamine, and water.

Reactions with aqueous alkali. Agents (**1**)–(**6**) react with sodium hydroxide (NaOH) in water at temperatures below 100°C (212°F); examples are shown in reactions (1) and (2).

The reaction (1) of sarin with NaOH in water is fast. A low concentration of sarin is half destroyed by 5% NaOH in water at room temperature in less than a second; in 20 s, less than one-millionth of the original sarin remains. In 1973–1982, the U.S. Army explored large-scale use of this reaction. Analyses were performed on what remained at expected completion of reaction. They regularly indicated that small amounts of sarin survive, even under forcing conditions. For this reason the reaction was judged unsatisfactory for destruction of sarin. That was one factor contributing to the Army's choice of incineration as its baseline technology for chemical demilitarization. In 1992, however, research in U.S. Army laboratories showed that the reaction did go to completion; the earlier result was due to a fault of the analytical method used.

Soman resembles sarin in chemical reactivity, but is less soluble in water. It also is destroyed on reaction with aqueous NaOH.

The room-temperature reaction of NaOH with VX forms two phosphorus-containing products, as shown in reaction (2). The one formed in a lesser amount is nearly as toxic as VX itself; therefore the reaction was judged unsatisfactory for chemical demilitarization. Recently, it was found that the extremely toxic by-product is further degraded by aqueous NaOH at temperatures near the boiling

point, yielding materials of low toxicity. Russian VX resembles American VX in both chemical behavior and physiological activity.

Yperite, or mustard gas, reacts with NaOH in water at around 90°C (190°F) to form thiodiethanol [reaction (3)] and other products. The reaction of lewisite with NaOH in water [reaction (4)] breaks both its chlorine–arsenic (Cl-As) and carbon–arsenic (C-As) chemical bonds, yielding exceedingly poisonous sodium arsenite. One method for immobilization of the arsenic is to oxidize it to arsenate; precipitation yields iron(III) arsenate, which is collected and stored in drums in deep, old mines.

Reactions with alkoxides. A generalized symbol for alcohols is R—OH, where R is a moiety (part of a mole-

Reaction (1): Sodium isopropyl methylphosphonate + Sodium fluoride

Reaction (2): Sodium ethyl methylphosphonate (87%) + Sodium S-[2-(diisopropylamino) ethyl methylphosphonothiolate] (13%)

Reaction (3): Thiodiethanol + 2HCl

Reaction (4): $H—C\equiv C—H + Na_3AsO_3 + 3NaCl + 3H_2O$
Sodium arsenite

cule) usually composed of carbon and hydrogen atoms. Alcohols are very weak acids. Their "salts," alkoxides such as RO^-Na^+ and RO^-K^+, are aggressive alkaline reagents. Reactions with alkoxides have been used in Canada and in Russia to destroy sarin and VX. In Canada the products, of low toxicity but nevertheless unsuitable for release, were burned. In Russia it is planned to blend them with lime (calcium hydroxide) into molten bitumen. The cooled, solid bitumen will contain agent-destruction products fixed as calcium salts, which can be buried in landfill or perhaps used for sealing foundations.

Reactions with ethanolamine. Ethanolamine, an industrial chemical, reacts with all the principal chemical warfare agents to form detoxified products. In Russia, ethanolamine is used to detoxify agents in recovered old munitions, with ensuing incineration. The Russians plan to use reactions with ethanolamine to destroy stockpiled sarin and soman, with second-stage bituminization.

Reaction with water. When dissolved in water, sarin decomposes slowly. The reaction is autocatalytic; hydrofluoric acid (HF), a product, accelerates further hydrolysis. Recent research has shown that VX and yperite may also be destroyed by reaction with water; the yperite hydrolysis is shown in reaction (3). In the case of VX, water is added in about one-tenth the volume of the VX, the container is closed, and 2 months elapse. The reaction breaks the P-S bond of VX, forming products of greatly reduced toxicity. (A minor but lethal impurity in U.S. VX, however, resists hydrolysis.)

Low-temperature oxidation methods. Some oxidizing agents detoxify yperite and VX under mild conditions, forming mixtures that require further treatment (polishing). Effective oxidizing agents include sodium hypochlorite (common bleach) and peroxy acids, such as peroxysulfuric acid (H_2SO_5) and peracetic acid (CH_3COOOH).

More vigorous controlled oxidation, to yield the oxides of constituent elements, may be effected by peroxysulfuric acid at temperatures near 100°C (212°F) and by the silver(II) ion [Ag^{2+}] in aqueous nitric acid. Electrochemical systems for regeneration of these oxidants have been devised.

Polishing methodologies. The immediate products of detoxification are not released because they are still toxic, although far less so than the original agents, and release may violate government regulations or terms of the Chemical Weapons Convention.

Biodegradation. Some microbiological populations (bacteria, fungi, and so forth) can sustain life by metabolizing chemicals usually regarded as highly toxic. Such specialized microorganisms have been found in places such as the ground beneath discarded, leaking drums of toxic chemicals. As they consume their special diets, the microorganisms render the toxic chemicals harmless. One application is destruction of the thiodiethanol that is a product of the hydrolysis of yperite; an effective microbiological culture was obtained from a sewage treatment plant. Biodegradation shows potential for destroying products of neutralization or hydrolysis of other chemical warfare agents. Although biodegradation has not been used much in chemical demilitarization, extensive use is probable, in part because biodegradation is popular with people who oppose incineration.

Wet-air oxidation. Low concentrations of carbon compounds dissolved or suspended in water at temperatures of 200–300°C (392–572°F) and under high pressure react readily with the oxygen of air that is pumped in to form simple compounds, among which acetic acid (the active ingredient of vinegar) is prominent. Over 200 plants worldwide use wet-air oxidation to clean up contaminated waters; however, as yet the technology has not been used for the polishing stage of chemical demilitarization.

High-temperature technologies. Technologies shown to be effective for destruction of chemical warfare agents are high-temperature hydrogenation [using dihydrogen (H_2)] and exposure to molten nickel or steel. Experiments indicate promise for pyrolysis of chemical warfare agents in place, that is, by heating strongly the unopened munition. Several suggested high-energy physical technologies are beyond the scope of this article.

Supercritical water oxidation. When a strong vessel containing water is evacuated by a vacuum pump, ultimately only liquid water and water vapor remain. The surface of the water is a boundary between the liquid and vapor phases, but as the temperature is raised, a point is reached at which the boundary suddenly disappears. For water, that temperature is 374°C (705°F). The pressure within the vessel at that critical point is high: 218 times atmospheric pressure, or 3204 lb/in.^{2} The supercritical fluid (fluid heated above its critical point) within the vessel dissolves oils rather well, but common salt (NaCl) is not very soluble in it. Many compounds of carbon, when dissolved in supercritical water, react very rapidly with oxygen to form products of complete oxidation. There is much interest in supercritical water oxidation for destruction of wastes of many sorts. Several pilot plants are testing the technology. However, skeptics believe that problems of corrosion and the blocking of pipes by crystallized salt may make supercritical water oxidation impractical.

For background information *see* BIODEGRADATION; CRITICAL PHENOMENA; EXTRACTION; HAZARDOUS WASTE; OXIDATION-REDUCTION in the McGraw-Hill Encyclopedia of Science & Technology.

Joseph F. Bunnett

Bibliography. J. F. Bunnett et al., Some problems in the destruction of chemical munitions, and recommendations toward their amelioration, *Pure Appl. Chem.*, 67:841–858, 1995; Stockholm International Peace Research Institute, *SIPRI Yearbook*, 1995 (as well as other yearbooks); V. A. Utgoff, *The Challenge of Chemical Weapons*, 1990; Y-c. Yang, J. A. Baker, and J. R. Ward, Decontamination of chemical warfare agents, *Chem. Rev.*, 92:1729–1743, 1992.

High-pressure mineral synthesis

Rocks are made up of individual minerals that, by their own nature or by their association with other minerals, are characteristic of particular depth ranges within the Earth. Pressure rises with increasing depth, so that the specific volumes of minerals must shrink. All the minerals are in the crystalline state; thus the most efficient shrinking mechanism is for the loosely packed crystal structures of typical near-surface environments to be replaced at greater depths stepwise by increasingly denser crystal structures. High-pressure laboratory syntheses and determinations of the pressure–temperature stability fields of individual minerals are indispensable for mineralogists and all geoscientists to allow them to gain insight into the genesis of rocks, the physical state of the Earth's interior, and the dynamic processes leading to earthquakes and magma eruptions.

Surface outcrops of rocks, even those of more deep-seated origin, exhibit mainly the minerals that are stable within the upper 30–40 km (18–24 mi), the thickness of the continental crust (**Fig. 1**). In connection with the tectonics of mountain building, minerals and rocks of the underlying upper mantle of the Earth may become exposed at the surface as well. However, the most deep-seated, yet well-preserved, earth materials that may be encountered at the surface were transported from a depth of about 150–200 km (90–120 mi) by explosive igneous melts, such as kimberlites; these and other related rocks are the common hosts of the high-pressure mineral diamond.

Pressure–temperature stability of minerals. Figure 1 is a pressure–temperature plot of possible environmental conditions prevailing within upper portions of the Earth. The two ordinate axes show a simplified correlation of depth within the Earth and the prevailing pressure (P), assuming a constant density of 2.8 g/cm^3 (1.6 oz/in.^{3}) of the rock overburden. Along the abscissa, temperature (T), which also rises with increasing depth, is plotted, but it may rise in very different proportions. Except for the region on the left generally not verified in the Earth, the pressure–temperature (P-T) conditions may vary from very low geotherms such as 5°C/km (15°F/mi) to very high ones near surface lava flows. In recent laboratory studies using chemical compositions relevant to earth materials, the individual P-T-stability fields of all the common rock-forming minerals were determined. A few examples are given in Fig. 1. They provide important constraints concerning the growth conditions of the minerals in natural rocks. Thus, depending on temperature, minimum pressures for diamond formation may range from 3.0 to 5.0 gigapascals, corresponding to depths of about 110–180 km (66–110 mi; Fig. 1). At lower pressures, elemental carbon must be present as graphite. Along a similar reaction curve, the most common silicon dioxide (SiO$_2$) mineral of crustal rocks, quartz, transforms into the denser polymorph, coesite. The garnet mineral pyrope (Mg$_3$Al$_2$Si$_3$O$_{12}$) requires minimum

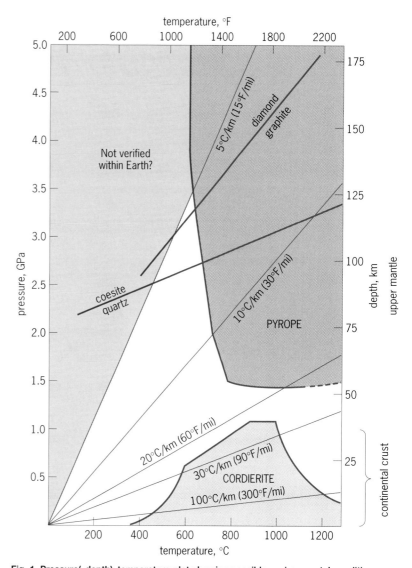

Fig. 1. Pressure(–depth)–temperature plot showing possible environmental conditions prevailing within the upper portions of the Earth, with geotherms indicated. Colored areas delineate the stability fields of two selected minerals, pyrope and cordierite. Heavy lines indicate the transition conditions for the silicon dioxide polymorphs quartz and coesite, and the carbon polymorphs graphite and diamond. 1 km = 0.6 mi.

pressures near 1.5 GPa; but in the presence of water it remains stable above temperatures of about 600–800°C (1100–1500°F) throughout the pressure range of Fig. 1. In contrast, the loosely packed framework silicate cordierite (Mg$_2$Al$_4$Si$_5$O$_{18}$ · xH$_2$O) is confined to low pressures, and can be present only in rocks formed within the continental crust. With such experimental constraints, minerals can be used for geobarometry.

Ultrahigh-pressure metamorphism. Changes of the mineral contents of rocks that are caused by changing pressure-temperature conditions acting on the rocks are summarized as metamorphism. Burial of surface rocks such as sediments to greater depths results in prograde metamorphism. If deeply buried rocks are exhumed to near-surface conditions, they are prone to undergo retrograde metamorphism. If

Fig. 2. Microphotograph of a portion of a pyrope crystal, with radial fractures, showing an inclusion of pure SiO_2, consisting of a core of the high-pressure mineral coesite (high relief) surrounded by the low-pressure mineral quartz (translucent).

chemical equilibrium were always attained during retrograde metamorphism, all rocks exposed at the surface would have to consist of only the lowest-pressure, lowest-temperature minerals relevant for these surface conditions, such as clays. Because of sluggish kinetics of reactions, especially of silicate minerals during cooling and uplift, even metamorphic rocks of high grades may reach the surface practically unchanged.

Since about 1984, mineralogical evidence has been obtained from several regions that metamorphic rocks of the continental crust may contain minerals that are indicative of environmental conditions within the coesite, or even the diamond, stability field (Fig. 1). Those minerals that endured conditions within the coesite field are known as ultrahigh-pressure metamorphic rocks.

The reason why ultrahigh-pressure metamorphic rocks were discovered so late in the evolution of science, here of metamorphic petrology, is that retrograde metamorphism has wiped out the ultrahigh-pressure relics in the rocks to a very large extent. Many normal metamorphic rocks of the continents may once have experienced a history of ultrahigh-pressure metamorphism, but all evidence has since been destroyed. Discovering indications for ultrahigh-pressure metamorphism is thus a detective job for mineralogists and petrologists that requires extremely careful microscopic observations. **Figure 2** is a microphotograph taken from a thin section of the first ultrahigh-pressure metamorphic rock ever discovered, in the Dora Maira Massif of the Western Alps, Italy. The relevant features of this specimen are particularly clear. The garnet pyrope (Fig. 1) hosts a single crystal of coesite that was partly transformed along its periphery into quartz. Because of the volume expansion during this transformation, the garnet host fractured and developed radial cracks around the inclusion. Probably because the rapid exhumation halted the transformation, some coesite survived. Since the mineral coesite is preserved only as inclusions within garnet and since all the external

SiO_2 is represented by quartz, clearly it was the high strength of the garnet crystal that prevented complete coesite-to-quartz transition within the inclusion. Garnet played the role of a pressure vessel during the pressure release caused by the uprising rock unit.

Ultrahigh-pressure metamorphic rocks form a small tectonic unit of about $10 \times 5 \times 1$ km ($6 \times 3 \times 0.6$ mi) in the Western Alps. Other occurrences are recorded from the Central Alps, western Norway, Mali in Africa, northern Kazakhstan, and east-central China. In the last two regions, in addition to coesite, diamond was found as an ultrahigh-pressure mineral; this represents the metamorphic product of organic carbon that was originally present in the sediment precursor. In China, ultrahigh-pressure metamorphic rocks are most frequent as they crop out locally over distances of some 800 km (480 mi).

Tectonic and geodynamic implications. The maximum P–T conditions derived by geothermobarometry from rocks of the various areas of ultrahigh-pressure metamorphism range from about 2.8 GPa, 600°C (1100°F; for the Central Alps) to 4.3 GPa, 1000°C (1800°F; in Kazakhstan), implying depths of metamorphism in the approximate range 100–160 km (60–100 mi; Fig. 1). This is far beyond the thickness of the continental crust, which averages 30–40 km (18–24 mi) but may attain 60–70 km (36–42 mi) in geologically young mountain belts such as the Alps or the Himalayas. Most of the ultrahigh-pressure metamorphic rocks represent metamorphosed sediments or former volcanics formed at the surface of the solid Earth; or they were, at least, initial residents within the continental crust, such as granites. Therefore it is clear that the geological processes of very deep burial of crustal rocks must have taken place intermittently over much of geologic history.

In the modern geoscientific views of plate tectonics, very deep burial of crustal rocks can occur only along the boundaries of colliding plates. Thus an oceanic plate is subducted beneath a continental one, such as is presently occurring along the west coast of the Americas; or two plates carrying continents collide, such as has happened with Europe and Africa for the past 100 million years or so. The ultrahigh-pressure metamorphic rocks of the Western Alps represent parts of the European continental crust that were subducted below the overriding African plate some 35 million years ago. Recent results of absolute age dating indicate that after subduction to 100 km (60 mi) or more, the ultrahigh-pressure metamorphic unit returned to shallow levels (~10 km or 6 mi) of the Alpine mountain belt within a period of only a few million years; this resulted in a rate of exhumation on the order of a few centimeters per year. It seems that such relatively high velocities of uplift are an important prerequisite for the possibility of preserving the mineral records of ultradeep subduction. This order of magnitude of the rate of exhumation coincides with that measured geodetically for the horizontal movements of the present-day plates coating the outer portions of the Earth.

Many questions as to the tectonic mechanisms, geothermal regimes, and ultimate driving forces of the internal dynamics of the Earth are still open. Concerted efforts of all branches of the earth sciences are necessary to further understanding of these processes. High-pressure mineral syntheses provided the fundamental data that were instrumental for a complete reconsideration of the mechanisms of mountain building and the genesis of the continental crust, and they will have to be reconciled with future visions and revisions.

Outlook. The progress of recent years resulted mainly from close interaction between laboratory mineral syntheses and observations from natural rocks. In one set of cases, laboratory results have told petrologists what to look for in rocks. In the other set, these scientists discovered in nature new minerals. For example, the complex magnesium titanium silicate/phosphate, ellenbergerite, after laboratory synthesis, was found to be a high-pressure solid as well, thus providing further support and constraints for ultrahigh-pressure metamorphism. At present, several new synthetic high-pressure silicate phases are known that have not been discovered in natural rocks. If they are found one day, new aspects of ultrahigh-pressure metamorphism will be revealed. If not, their absence will define limits as well.

For background information *see* DIAMOND; EARTH INTERIOR; HIGH-PRESSURE MINERAL SYNTHESIS; METAMORPHISM; PLATE TECTONICS in the McGraw-Hill Encyclopedia of Science & Technology.   W. Schreyer

Bibliography. C. Chopin, Coesite and pure pyrope in high-grade blueschists of the Western Alps: A first record and some consequences, *Contrib. Mineral. Petrol.*, 92:107–118, 1984; R. G. Coleman, and X. Wang, *Ultrahigh Pressure Metamorphism*, 1995; S. L. Harley and D. A. Carswell, Ultradeep crustal metamorphism: A prospective view, *J. Geophys. Res.*, 100 (B5):8367–8380, 1995; W. Schreyer, Experimental studies on metamorphism of crustal rocks under mantle pressures, *Mineral. Mag.*, 52:1–26, 1988; W. Schreyer, Ultradeep metamorphic rocks: The retrospective viewpoint, *J. Geophys. Res.*, 100(B5):8353–8366, 1995.

Hydrogen bond

A hydrogen bond is the interaction of a hydrogen atom bound to one atom with an unshared electron pair of another atom. The hydrogen atom involved is held to one atom (the hydrogen-bond donor) by a strong covalent bond, and it usually has a much weaker interaction with the other atom (the hydrogen-bond acceptor). There is an energy barrier for transfer of the hydrogen ion (a proton) from the donor to the acceptor. In a low-barrier hydrogen bond, the hydrogen is symmetrically spaced between the two atoms, having equal interactions with

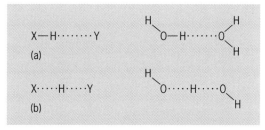

Fig. 1. Hydrogen bonds. (*a*) Normal hydrogen bond. The hydrogen atom forms a covalent bond (solid line) with the hydrogen-bond donor atom (X) and a much weaker noncovalent interaction (dotted line) with the hydrogen-bond acceptor atom (Y). The example shows the hydrogen bond formed between two water molecules. (*b*) Low-barrier hydrogen bond. The hydrogen ion interacts equally with the two hydrogen bond donor/acceptor atoms. The example shows the hydrogen bond formed between a water molecule and hydroxide ion.

each (**Fig. 1**). Hence, the distinction between hydrogen-bond donor and acceptor disappears. Such hydrogen bonds are formed when the hydrogen-bond donor and acceptor have equal affinity for a hydrogen ion. These hydrogen bonds are called low-barrier hydrogen bonds, because the energy barrier for transfer of the hydrogen between atoms is greatly decreased and effectively disappears. Low-barrier hydrogen bonds have bond lengths, defined as the distance between the two donor/acceptor atoms, of 0.23 nanometer or less compared to 0.26–0.28 nm for normal hydrogen bonds. It has also been proposed that these hydrogen bonds are stronger than normal hydrogen bonds; thus the term "short strong hydrogen bond" has been used somewhat synonymously with low-barrier hydrogen bond. The term "symmetric hydrogen bond" has also been used, because the hydrogen atom is centrally placed between the donor/acceptor atoms in a symmetric structure. Recently it has been proposed that low-barrier hydrogen bonds play a major role in catalysis by enzymes.

Evidence for low-barrier hydrogen bonds. Low-barrier hydrogen bonds are observable by several means. The short distance between the hydrogen-bond donor/acceptor atoms has been observed in the solid state by x-ray crystallography. For example, the oxygen–oxygen distance in the hydrogen bond between a water molecule and a hydroxide ion has been shown to be less than 0.23 nm. The hydrogen (deuterium) atom in low-barrier hydrogen-bonded species has also been observed by means of neutron diffraction. These studies show the symmetry of the structure, with the hydrogen atom being equally spaced between the two donor/acceptor atoms as opposed to normal hydrogen bonds, in which the hydrogen atom is nearer the hydrogen-bond donor.

Low-barrier hydrogen bonds in solution have usually been observed in intramolecular cases, in which the donor and acceptor atoms are part of the same molecule in appropriate proximity to accommodate a hydrogen bond between them. Examples of species that form intramolecular low-barrier hydrogen

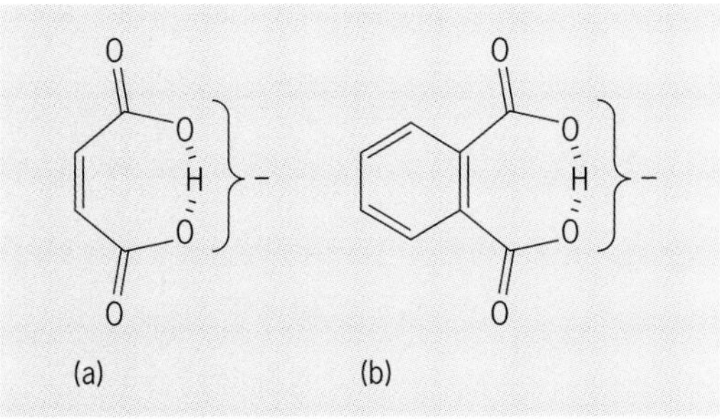

Fig. 2. Examples of intramolecular low-barrier hydrogen bonds (indicated by braces). (*a*) Maleic acid monoanion. (*b*) Phthalic acid monoanion.

bonds are shown in **Fig. 2**. Nuclear magnetic resonance (NMR) experiments have been developed to observe the symmetry of hydrogen bonds in solution. Other notable characteristics of low-barrier hydrogen bonds include the low isotopic fractionation factor (the tendency for deuterium to enrich in a position relative to its content in water) and the low-electron-density environment of the shared hydrogen, as observed by NMR.

Enzyme catalysis. The means by which enzymes achieve their catalytic power has been a longstanding area of investigation. For example, certain enzymes deprotonate very nonacidic compounds, and others hydrolyze amide bonds at neutral pH; both reactions require the formation of high-energy intermediates. Thus, these reactions require highly basic or acidic conditions or high temperatures in order to proceed in the absence of enzyme catalysis. To explain the ability of enzymes to catalyze such reactions under very mild conditions, a major role has been proposed for low-barrier hydrogen bonds. If a normal hydrogen bond in the complex of an enzyme with substrate (ground state) becomes a low-barrier hydrogen bond in the complex of the enzyme with the high-energy intermediate, the low-barrier hydrogen bond could provide 20 kilocalories per mole or more of stabilization energy, based on gas-phase data. The change in the nature of the hydrogen bond could arise from the expected change in the basicity of a hydrogen-bond acceptor group in the substrate upon its conversion to the high-energy intermediate. However, it is unclear how relevant are measurements in the gas phase to species in solution or in the internal environment of an enzyme. The strength of a low-barrier hydrogen bond in solution or in an enzyme complex has been more difficult to address.

Experimental models. Many model compounds have recently been studied to address the issue of the strength of low-barrier hydrogen bonds in solution and the possible importance of these bonds in enzyme catalysis. One study involved the equilibrium between the mesaconic and citraconic acid monoanions in dimethyl sulfoxide solvent (**Fig. 3**). The diacid forms of citraconic and mesaconic acids have the potential to release two hydrogen ions. Release of a single hydrogen ion from the diacid produces

a monoanion, a species having a single negative charge. The citraconic acid anion forms an intramolecular low-barrier hydrogen bond, which should have a stabilizing effect on this species. This equilibrium was compared to the equilibrium for the corresponding diacid, which forms a normal intramolecular hydrogen bond in the citraconic species. The monoanion favors the citraconic species by a ratio of about 160:1, while the diacid favors the mesaconic species by a ratio of 10:1. This corresponds to a difference in relative energy of the citraconic and mesaconic species of about 4.4 kcal/mol. This difference is attributed to the difference in the strengths of the normal and low-barrier hydrogen bonds of the citraconic diacid and monoanion.

Further information regarding the relative strengths of normal and low-barrier hydrogen bonds comes from the first and second acid-dissociation constants for phthalic acid. Phthalic acid is unusually strong for a carboxylic acid, while the deprotonated species (the monoanion) is an unusually weak acid. The presence of the low-barrier hydrogen bond in the phthalic acid monoanion (Fig. 2) provides substantial stability to the monoanion, while this special stability is not present in the diacid or the dianion. This results in an unusual propensity for the diacid to give up an acidic proton to form the stable monoanion, and an unusually low propensity for this stable anion to give up a second proton to form the dianion. As with the studies described above, this work has provided an estimate of about 4–5 kcal/mol for the difference in strength of a normal versus a low-barrier hydrogen bond.

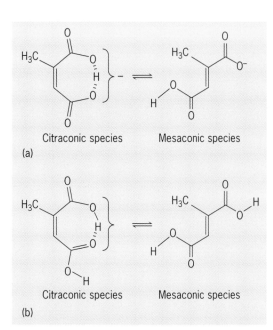

Fig. 3. Interconversion of the (*a*) monoanion and (*b*) diacid forms of citraconic and mesaconic acids. At equilibrium, the monoanion exists as a 160:1 ratio of the citraconic species to the mesaconic species, while the diacid exists as a 1:10 ratio of the citraconic species to the mesaconic species.

Theoretical models. In theoretical computational analysis of the strengths of hydrogen bonds, the energetics of the hydrogen bond is divided into three terms. The covalent and electrostatic terms account for favorable interactions of the hydrogen with the hydrogen-bond donor and acceptor atoms. The third, repulsive term describes unfavorable steric interactions between the hydrogen-bond donor and acceptor atoms as they are brought close together. Low-barrier hydrogen bonds have better covalent and electrostatic interactions, and their shortness is simply a consequence of this. The net effect is that, while hydrogen bonds become stronger with increased sharing of the hydrogen between donor and acceptor atoms, they are not generally made stronger by virtue of being shorter. This argues against a special stability of short hydrogen bonds.

Enzyme complexes. Evidence for low-barrier hydrogen bonds in enzymes and in enzyme:substrate or enzyme:inhibitor complexes has been observed. Nuclear magnetic resonance studies of the digestive enzyme chymotrypsin have demonstrated a hydrogen in a region of very low electron density, which has been attributed to a low-barrier hydrogen bond. Similar NMR evidence for low-barrier hydrogen bonds has been observed in other enzymes. A very short hydrogen bond (less than 0.24 nm) has also been observed in complexes of inhibitors with the enzyme citrate synthase. This short distance is attributed to the formation of a low-barrier hydrogen bond, which is also supported by NMR evidence. Comparison of the affinity of this inhibitor to citrate synthase with the affinity of similar analogs that form only normal hydrogen bonds has provided estimates of about 2 kcal for the difference in energies for the normal versus low-barrier hydrogen bonds in these complexes.

Implications. Experimental model studies have shown that the difference in energies between low-barrier hydrogen bonds and normal hydrogen bonds is about 4–5 kcal/mol in solution, compared to more than 20 kcal/mol measured in the gas phase. Even smaller energy differences between low-barrier versus normal hydrogen bonds have been predicted, based on studies of enzyme:inhibitor complexes and on theoretical studies. These results support the concept that enzymes may indeed utilize low-barrier hydrogen bonds for stabilization of high-energy intermediates or transition states. These low-barrier hydrogen bonds may be stronger by 4–5 kcal than normal hydrogen bonds formed in the ground-state enzyme:substrate complex. These results also suggest that the role of low-barrier hydrogen bonds in enzyme catalysis is likely to be significant, though less dramatic than original proposals based on gas-phase data.

For background information *see* BOND ANGLE AND DISTANCE; CATALYSIS; CHEMICAL BONDING; COMPUTATIONAL CHEMISTRY; ENZYME; HYDROGEN BOND in the McGraw-Hill Encyclopedia of Science & Technology. Dale G. Drueckhammer

Bibliography. J. A. Gerlt and P. G. Gassman, An explanation for rapid enzyme-catalyzed proton abstraction from carbon acids: Importance of late transition states in concerted mechanisms, *J. Amer. Chem. Soc.*, 115:11552–11568, 1993; B. Schwartz and D. G. Drueckhammer, A simple method for determining the relative strengths of normal and low-barrier hydrogen bonds in solution: Implications to enzyme catalysis, *J. Amer. Chem. Soc.*, 117:11902–11905, 1995; S. Shan, S. Loh, and D. Herschlag, The energetics of hydrogen bonds in model systems: Implications for enzymatic catalysis, *Science*, 272:97–101, 1996; A. Warshel et al., On low-barrier hydrogen bonds and enzyme catalysis, *Science*, 269:102–106, 1995.

Hydrothermal vent

Two topics related to the biological communities that live near deep-sea hydrothermal vents are discussed here. The behavior of the chemosynthetic symbionts is described, as well as the use of carbon isotopes to trace the food chains in the hydrothermal vent community. Then, the evolution of the eyes of rimicarid shrimp in order to utilize the dim light emitted from active hydrothermal vents is examined.

Chemosynthetic Symbionts in Hydrothermal Vent Communities

Life in the deep sea is typically scarce and food limited, since these life-forms must rely solely upon the trickling of organic carbon from the productive surface layers thousands of feet above them. The discovery of biological communities around deep-sea hydrothermal vents therefore came as a surprise. These ecosystems support an abundance of life as rich as that found in tropical rainforests. The dense aggregates of animals around deep-sea hydrothermal vents are largely supported by chemosynthesis rather than photosynthesis. Chemosynthesis, a process limited to bacteria, is the ability to utilize reduced inorganic compounds (such as sulfide) as an energy source, and to use this energy to fix carbon dioxide via the Calvin-Benson cycle, the same carbon fixation pathway used in photosynthesis. Chemosynthetic bacteria not only are found as free-living organisms at the vents, but also live symbiotically within the host tissues of a variety of vent animals, such as vestimentiferan tubeworms, snails, clams, and mussels. They are also found living on the surface of some invertebrate species such as shrimp. The symbiotic bacteria are functionally analogous to chloroplasts found within the photosynthetically active cells of green plants and algae, and can provide the host animal with the major portion of its nutritional requirements. Within vent environments, these animals and their symbionts act as primary producers, and along with the free-living bacteria they provide essentially all the food for the community, with all organic carbon being of localized origin. The primary evidence for the origin of this local source of carbon has come in the form of stable carbon isotope data, which have also been of fundamental importance in tracing food chains throughout these and other environments.

Stable carbon isotopes. In nature there are two non-radioactive stable isotopes of carbon that differ only in the number of neutrons found within the atom's nucleus. The predominant isotope, carbon-12 (^{12}C), accounts for approximately 98.9% of nonradioactive carbon, while the heavier isotope, carbon-13 (^{13}C), makes up the other 1.1% in natural systems. The natural abundance of these isotopes in biological tissues can be measured by determining the ratio of ^{13}C to ^{12}C (or converted to a standard value of delta ^{13}C) in a high-precision mass spectrometer. The international standard used is limestone from the Pee Dee formation in South Carolina. Calculated delta ^{13}C values are expressed as parts per thousand with the symbol ‰. For biological samples, the ‰ values are always negative because samples contain less ^{13}C than the standard, with more negative values corresponding to a higher percentage of the ^{12}C isotope.

Chemically the two isotopes are identical, and they can substitute for each other both structurally and in their involvement in chemical reactions of biological importance. However, slightly stronger chemical bonds are formed by the heavier of the two isotopes, and the heavier isotope diffuses more slowly. Therefore, the ratio of ^{13}C to ^{12}C in biological tissues is important, because certain physical and chemical processes discriminate against the heavier of the isotopes, leading to a characteristic signature for some key biochemical reactions. This discrimination is referred to as fractionation, which represents the difference in the isotopic composition of the substrate utilized and the product formed. This fractionation can be quite pronounced for biological reactions, and ecologists use this to trace carbon through food chains in natural environments.

The initial carbon fixation step of the Calvin-Benson cycle, a metabolic pathway shared by photosynthetic and chemosynthetic organisms in which carbon dioxide is converted into 3-carbon sugars used in biosynthesis, is catalyzed by the enzyme ribulose-1,5-bisphosphate carboxylase/oxygenase (Rubisco). This reaction is of particular utility, as Rubisco can discriminate against the isotope ^{13}C by as much as 29‰ if the substrate is not limiting (when a substrate is limited, the enzyme will utilize all of the carbon dioxide molecules available regardless of the isotope). Other processes, such as diffusion of the substrate molecules to the enzyme, also influence this final discrimination and can have a profound effect on the final observed delta ^{13}C value of the autotroph in question. This dramatic difference gives autotrophic organisms a characteristic isotopic signature which can then be followed up the food chain, since the delta ^{13}C of the consumer reflects the delta ^{13}C of its diet.

^{13}C signature groups. In marine environments, photosynthetic organisms typically have delta ^{13}C values of −15‰ to −22‰ (**Fig. 1**). This stable isotope value is reflected in the delta ^{13}C values reported for nonvent deep-sea organisms, as they typically have values between −17‰ and −21‰, which indicates that they are deriving the majority of their food from the surface layers which have percolated down to them. However, within the vent environments, symbiotic organisms generally fall into two isotope signature groups, being either depleted or enriched in ^{13}C relative to the nonvent deep-sea species. This indicated that these deep-sea invertebrates were using locally derived carbon and not carbon generated from surface primary production. Clams and mussels harboring chemosynthetic symbionts show values which range from −27‰ to −36‰, while vestimentiferan tubeworms which also contain symbionts typically have values ranging from −9‰ to −15‰. Free-living bacterial mats also have this bimodal distribution of delta ^{13}C values, presumably due to different species of bacteria within them. Consumers and grazers within these environments, such as crabs, anemones, limpets, and alvinellid polychaetes, have a wide range of stable isotope values (varying by up to 19‰ at a site in the North Pacific). In some cases these consumers have values which fall in either the enriched or depleted groups, and in other cases they fall between the two groups, presumably because of the specific food sources that are available. If both groups are used as food sources, the consumer's delta ^{13}C value will reflect the combined isotopic composition, and the resulting value will reside somewhere in the middle of the distribution, dependent upon the contribution of carbon from each source.

Chemosynthetic alterations. It is apparent that chemosynthetic bacteria fractionate carbon isotopes to degrees different than their photosynthetic counterparts, although both use the same biochemical pathway to fix inorganic carbon. Indeed, there is evidence that laboratory-grown free-living chemosynthetic bacteria fractionate carbon more than photosynthetic organisms do. However, this evidence does little to explain the significance of ^{13}C-enriched values. Several theories have been generated over the years to account for these differences, including different sources of carbon dioxide, different carbon fixation enzymes, multistep biochemical pathways involving the fixation of the carbon into a different form and then refixing it via Rubisco, and substrate limitation encountered by Rubisco. Each of these could explain the observed values, and studies are under way to determine the stable carbon isotope fractionation which may be unique to the Rubisco used by these deep-sea organisms.

Stable carbon isotopes have proved to be invaluable in developing an understanding of carbon sources used by vent chemosynthetic bacteria and animal symbioses, and also in helping to decipher the complexities of feeding interactions within this community. Naturally occurring variations in stable isotope abundance for not only carbon but also nitrogen, oxygen, hydrogen, and sulfur provide the biolo-

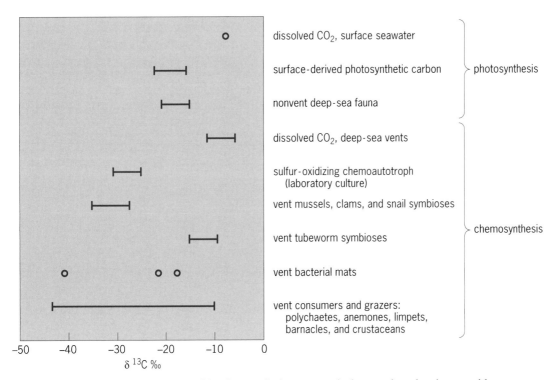

Fig. 1. Stable carbon isotope ratios reported for photosynthetic processes in the ocean's surface layers and for chemosynthesis occurring in hydrothermal vent environments. The delta ^{13}C values of CO_2 in both environments are also provided to show the relative fractionation of both types of autotrophic carbon fixation.

gist with a critical tool in problems encountered in chemistry, biochemistry, ecology, and medical sciences. In the hydrothermal vent ecosystems, this tool often gives clues to processes occurring in environments that are barely accessible to the modern researcher.   Jonathan J. Robinson; Colleen M. Cavanaugh

Decapodan Dim-Light Vision

For a decade following the discovery of hydrothermal vents at the Galápagos rift (formed by the spreading apart of two continental plates) in the Pacific Ocean, it was assumed that animals inhabiting the vent environment lived in total darkness. Most of the vent sites are far too deep for any daylight to penetrate from the surface, and human observers in submersibles saw no local sources of light at the hydrothermal vents. The discovery that processes within black smoker chimneys (tubes of sulfides built up where escaping vent fluids precipitate tiny particles of black smoke) produce very dim light was made after hydrothermal vents were found along the Mid-Atlantic Ridge where newly discovered species of shrimp (order Decapoda) have unusually shaped eyes.

Hydrothermal vent shrimp. Bresiliid shrimp (family Bresiliidae) are members of the fauna that inhabit the hydrothermal vent sites along the Mid-Atlantic Ridge. At the TAG and Snake Pit vent fields, *Rimicaris exoculata* and *R. aurantia* form immense swarms along the sides of active black smoker chim-

neys (**Fig. 2***a, b*). Both animals have large, white, upward-facing dorsal eyes which appear as bilateral wings or lobes that meet at the midline anteriorly near where the stalked spherical compound eyes of surface shrimp would be found. *Chorocaris chacei* (family Bresiliidae) at the TAG site (28°8.3′N; 44°49.5′W; 12,140 ft or 3700 m below sea level) and Snake Pit site (23°20.3′N; 45°0.5′W; 11,480 ft or 3100 m below sea level), *C. fortunata* at the Lucky Strike site (37°17.6′N; 32°16.5′W; 5580 ft or 1700 m below sea level), and a small, possibly new species of chorocarid shrimp from the Broken Spur hydrothermal vent site (29°10.0′N; 43°10.4′W; 10,170 ft or 3100 m below sea level) live in sparse groups and have large, white, forward-facing anterior eyes that meet at the midline (Fig. 2*c, d, e*). The predator species *Alvinocaris markensis* has large white structures in the expected position for anterior eyes, but the eyes are degenerate and the animal is blind. The eyes of other vent shrimp from the Mid-Atlantic Ridge, shrimp from the hydrothermal vents of the back-arc basins in the western Pacific Ocean where one continental plate is disappearing under another, and shrimp from various warm and cold oceanic seeps have not yet been investigated.

Retinal structure. The microscopic structure of the retina of vent shrimp follows a common plan whether the eye is dorsal or anterior. The retina has four layers with a smooth cornea and corneal epidermis on the outer surface. Just inside is a thick

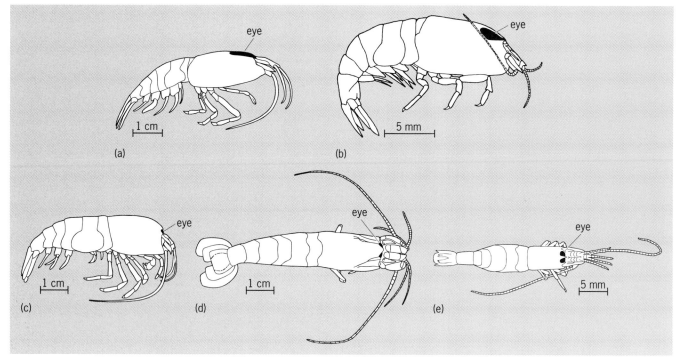

Fig. 2. Bresiliid shrimp found at hydrothermal vents on the Mid-Atlantic Ridge. (a) *Rimicaris exoculata.* (b) *Rimicaris aurantia.* (c) *Chorocaris chacei.* (d) Dorsal view of *C. fortunata.* (e) Dorsal view of unnamed chorocarid shrimp.

layer of photosensitive membrane arranged in photoreceptors as irregular arrays of rhabdomeral microvilli. The rhabdom occupies about 80% of the total volume of the second layer. The third layer is composed of white diffusing cells that give the eye its white color. Thin cylindrical photoreceptor processes (the arhabdomeral segments) penetrate the layer of white diffusing cells to connect the rhabdomeral segments with the photoreceptor nuclei and the axons of the optic nerve. The fourth layer lines the inner surface of the white diffusing cells and consists of the small spherical nuclei of the photoreceptors interspersed with dark ellipsoidal cells which are vestigial screening pigment cells. The screening pigment granules of the photoreceptors are trapped in the axons of the optic nerve below the photoreceptor nuclei. Underneath the fourth layer, the axons of the photoreceptors come together to form the optic nerve that connects the eye to the brain.

Photoreceptors. The photoreceptors are arranged in groups of five to seven cells. In a living animal, the cornea and corneal epidermis are colorless and transparent. However, a dissecting microscope can reveal a honeycombed structure with light pinkish-purple photoreceptor rhabdomeral segments separated by thin walls of white diffusing cells. At the bottom of each ommatidial compartment, a circle of tiny holes indicates where the photoreceptor arhabdomeral segments penetrate the layer of white diffusing cells.

Comparison between the retinal structure of hydrothermal vent shrimp and that of a surface shrimp such as *Palaemonetes* emphasizes the extent to which the retina of the former is adapted to its dim-light environment. The same set of cell types are found in both retinas; however, the arrangements are markedly different. The spherical compound eye of *Palaemonetes* has a clear ommatidial structure with the geometrical packing of the lenses formed by cone cells and underlying photoreceptors necessary for pattern vision in daylight. The bresiliid shrimp retina, however, has no image-forming lenses and must function as a light-intensity detector. Moreover, the size of the eye in bresiliid shrimp is much larger, the light-sensitive rhabdomeral portion of the photoreceptors is enlarged, and the volume density of rhabdom within the rhabdomeral segment is typically eight times greater than in the compound eye of surface shrimp.

Pigment cells. In the imaging eye of *Palaemonetes* the pigment cells serve to isolate the ommatidia optically to preserve the resolution of the image, whereas in the nonimaging eye of bresiliid shrimp the reflecting pigment cells have moved and expanded into a scattering layer of white diffusing cells behind the rhabdomeral segments, doubling the sensitivity of the eye in the same way as the reflective tapetum does in the eyes of nocturnal mammals. The opaque, black screening pigment cells in bresiliid shrimp are scattered globular shapes located below the retina where they cannot interfere with the eye's opportunity to capture incoming light. Thus, the imaging compound eye of surface shrimp, intended for use during the day, has been adapted for vision in a very low level of light in ways that remove the eye's imaging properties but make it a very sensitive photodetector.

Source of light. Although the light emitted from hydrothermal vents is too dim for human observers to see, sensitive electronic imaging techniques have revealed that the light arises at the throats of high-temperature black smoker chimneys where hot vent water emerges into cold seawater. Numerous light-producing mechanisms have been proposed, including blackbody radiation, sonoluminescence, and triboluminescence; however, further measurements will be necessary before the nature of the light source at hydrothermal vents is understood. Photochemical studies of the visual pigment in vent shrimp and preliminary measurements of the spectral sensitivity of the eye using electroretinographic recordings suggest that the eye is maximally sensitive to green light—apparently having the same visual pigment that other marine decapod crustaceans use for viewing bioluminescence and natural daylight. Whatever the mechanism that produces the dim light at hydrothermal vents, the bresiliid shrimp living there have eyes specifically adapted to utilize it.

Implications. The discovery that bresiliid shrimp at hydrothermal vents have eyes adapted to see dim light given off by black smoker chimneys raises interesting questions: (1) Does the existence of a source of light at hydrothermal vents mean that photosynthesis might have originated in methane- or sulfide-utilizing vent organisms which then drifted upward and colonized the surface? (2) Each new hydrothermal vent site on the Mid-Atlantic Ridge seems to yield new species of bresiliid shrimp, each with enlarged eyes that are different in position and shape but have the same basic retinal plan. What is the family tree for bresiliid shrimp, and is the adaptation of the eyes for dim-light vision happening over and over? (3) The photoreceptors of bresiliid shrimp lack any evidence of the daily shedding and renewal of photosensitive membrane once thought to be a ubiquitous property of photoreceptors. Is this absence of rhabdom turnover related to the dimness of the light or to the absence of day–night and summer–winter cycles in light intensity at hydrothermal vents? Answers will emerge as the study of the mid-oceanic ridges continues.

For background information *see* CRUSTACEA; EYE (INVERTEBRATE); GEOLOGIC THERMOMETRY; HYDROTHERMAL VENT; MARINE MICROBIOLOGY; MID-OCEANIC RIDGE; PALEOCEANOGRAPHY; PLATE TECTONICS in the McGraw-Hill Encyclopedia of Science & Technology.   Steven C. Chamberlain

Bibliography. S. E. Humphris et al. (eds.), *Seafloor Hydrothermal Systems: Physical, Chemical, Biological, and Geological Interactions*, 1995; K. Lajtha and R. H. Michener (eds.), *Stable Isotopes in Ecology and Environmental Science*, 1994; P. J. O'Neill et al., The morphology of the dorsal eye of the hydrothermal vent shrimp, *Rimicaris exoculata*, *Vis. Neurosci.*, 12:861–875, 1995; D. J. Nuckley et al., Retinal anatomy of a new species of bresiliid shrimp from a hydrothermal vent field on the Mid-Atlantic Ridge, *Biol. Bull.*, 190:98–110, 1996; L. M. Parson, C. L. Walker, and D. R. Dixon (eds.), *Hydrothermal Vents and Processes*, Geol. Soc. Spec. Pub. 87, 1995; J. J. Robinson and C. M. Cavanaugh, Expression of Form I and Form II ribulose-1,5-bisphosphate carboxylase/oxygenase (Rubisco) in chemoautotrophic symbioses: Implications for the interpretation of stable carbon isotope values, *Limnol. Oceanog.*, 40:1496–1502, 1995; C. L. Van Dover et al., A novel eye in "eyeless" shrimp from hydrothermal vents of the Mid-Atlantic Ridge, *Nature,* 337:458–460, 1989.

Immune system

The immune system attempts to protect the body from harmful infections. This involves neutralizing viruses, killing bacteria and parasites, and dealing with potentially toxic proteins produced by some bacteria during the course of an infection. Immune protection results from the concerted effort of the white blood cells. There are two major groups of immune cells: the myeloid lineage (originating from the bone marrow) comprising macrophages, neutrophils, basophils, eosinophils, and mast cells, and the lymphocyte lineage comprising B lymphocytes (originating from the bone marrow), T lymphocytes (originating from the thymus), and natural killer cells. Lymphocytes are the cells responsible for specific or acquired immunity, while myeloid cells provide nonspecific or innate immunity. The terms "acquired" and "innate" distinguish the two types of immunity. Acquired or adaptive immunity develops in response to an infection and is specific for the infectious agent (that is, it is acquired in response to different infections during the life of an individual). Innate immunity, however, is always present and operates in the same manner for different infections (that is, all humans would have the same system irrespective of the infections to which they have been exposed). These two systems cooperate for the efficient removal of pathogenic microorganisms.

T and B cells. Cells of the lymphocyte lineage can be further divided into B lymphocytes or B cells, which secrete antibodies during an immune response, and T lymphocytes or T cells. There are two main types of T cells: cytotoxic or killer T cells, which kill virus-infected cells; and helper T cells, which help activate other cells such as B cells and macrophages. Lymphocytes are produced throughout life, developing from specialized precursor cells or stem cells. B cells develop in the bone marrow, while T cells of both types develop and mature in the thymus. Millions of B cells and T cells enter the circulatory and lymphatic systems every day, passing through the blood, lymphoid organs, and lymphatics as they patrol the body. T cells and B cells express (on their surface) specialized proteins with which they recognize their predetermined antigen (any structure which can be the object of an immune response). The antigen receptor of B cells is a membrane-bound form of antibody (also called immuno-

globulin or Ig), while for T cells it is the T-cell receptor. Each lymphocyte expresses an antigen receptor which is unique to that cell and specific for a particular antigen. The extraordinary diversity of antigen receptors and the large number of lymphocytes present at any time mean that the acquired immune system is capable of recognizing and responding to a very large number of foreign antigens.

The diversity of antigen receptors expressed by lymphocytes results from the genetic rearrangement process which generates the functional immunoglobulin and T-cell receptor genes. This process, which occurs early during lymphocyte development, involves the joining together of gene segments at random to form the final product. The random nature of the process means a lymphocyte capable of recognizing every possible antigen, including self-antigens, will be generated at some time during the life of an individual. If such self-reactive lymphocytes were to respond to those self-antigens, a state of autoimmunity would develop. Although autoimmune diseases exist, they are uncommon, which means that mechanisms must exist to deactivate self-reactive lymphocytes and thus maintain immunological tolerance.

Tolerance. Tolerance is a functional state in which the immune system does not respond to self. With the complexity of the immune system and the large number of cell and tissue types involved, a large number of different mechanisms appear to be involved in maintaining this state of self-unresponsiveness. The analysis of tolerance has been greatly facilitated by the development of transgenic mouse models. In these model systems, every B or T cell expresses an identical antigen receptor, meaning that all the lymphocytes are specific for the same antigen. If that antigen is a self-protein, then every lymphocyte will be self-reactive. In this way, the fate of self-reactive cells can be determined by following the lymphocyte population. By varying the location and nature of the self-antigen in these transgenic models, a number of tolerance mechanisms have been defined, including clonal deletion, clonal anergy, and clonal ignorance.

As will be described below, immunological tolerance is the result of the deletion of lymphocytes which strongly recognize ubiquitous self-antigens; the silencing (anergy) of lymphocytes which strongly recognize self-antigens restricted to particular tissues; and the ignorance of lymphocytes which recognize self-antigens that are uncommon, expressed in areas of the body where lymphocytes do not normally go, or that are not presented in a way which the immune system recognizes.

Clonal deletion. When a lymphocyte first expresses its antigen receptor in either the bone marrow or the thymus, its potential for self-reactivity is revealed. If the receptor recognizes a self-antigen present in that organ and if the interaction between them is sufficiently strong, the lymphocyte will undergo activation-induced cell death (apoptosis). The death of self-reactive lymphocytes shortly after they develop

is called clonal deletion or negative selection. There are a number of conditions that must be met before a lymphocyte will kill itself. The lymphocyte has to bind the antigen with high avidity, a condition that is a combination of the strength of the receptor–antigen interaction and the number of receptor molecules engaged. This means that lymphocytes which develop with a specificity for abundant, ubiquitous self-antigens will be deleted in the bone marrow or thymus. However, if the strength of the interaction is too low or the number of receptors engaged is too few, the lymphocyte will not be deleted and development will continue.

Clonal anergy. Not all self-antigens are ubiquitous. Many antigens are restricted to particular tissues and specific stages of lymphocyte development. Lymphocytes which emerge from the thymus or the bone marrow and then encounter their self-antigen in the periphery (locations outside the thymus or the bone marrow) are dealt with in one of two ways. Either the lymphocytes are deleted or they are rendered functionally silent, a state called anergy. In both situations, tolerance is preceded by the self-reactive lymphocyte multiplying extensively before initiating a cell death program. Anergy means that some self-reactive lymphocytes persist after their encounter with autologous (self-) antigen, but they are incapable of responding to it. From recent experiments with transgenic models, it appears that anergic lymphocytes can be made functional but that this result requires a combination of strong signals. Other experiments have tended to blur the distinction between clonal anergy and deletion, suggesting that anergy is a delayed form of deletion. *See* ANERGY.

Clonal ignorance. Another fate for self-reactive lymphocytes is to never encounter the self-antigen for which they are specific and thus to be left in a state of ignorance. If that antigen is expressed at low levels (that is, if the antigen is present on only very rare cells or is restricted to sites which lymphocytes never visit—immune privileged areas such as the eye), then it is unlikely that the self-reactive lymphocyte will ever encounter the antigen in sufficient density to initiate an immune response. This is particularly true for helper T cells, which see small peptides derived from proteins only in association with the specialized proteins of the class II major histocompatibility complex (MHC). Expression of class II MHC proteins is restricted to macrophages, B cells, and dendritic cells. Consequently, only these cells can activate antigen-specific helper T cells and are thus called antigen-presenting cells (APCs). The location of the antigen, its density on the surface of the cell, and whether it is presented in a manner which T cells can "see" will all contribute to a determination of whether a self-reactive lymphocyte will respond to its self-antigen.

Lymphocyte co-stimulation. The context in which lymphocytes encounter their antigen is very important in determining the outcome of the interaction (see **illus.**). Lymphocytes will respond to an antigen only if they receive two signals: the first

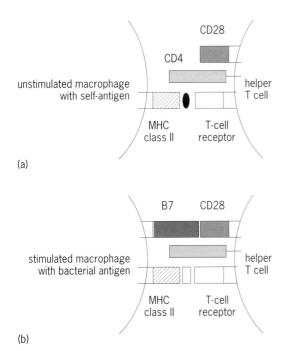

(a)

(b)

Co-stimulation is necessary for lymphocyte activation by antigen. (*a*) Normal interaction between an antigen-specific helper T cell and its antigen on an unstimulated antigen presenting cell (a macrophage) results in clonal anergy because the macrophage does not express the co-stimulatory molecule B7. (*b*) Interaction between a helper T cell specific for a bacterial antigen and a macrophage stimulates through both the T-cell receptor (TCR) and CD28, which allows the helper T cell to proliferate and differentiate into an effector cell.

through the antigen receptor and the second through a co-stimulatory molecule. Expression of co-stimulatory molecules is restricted to cells of the immune system, meaning that two different immune cells have to cooperate to produce an immune response. Usually, when a self-reactive lymphocyte meets its self-antigen, it does not receive any co-stimulation and the interaction is abortive. The recruitment of a second immune cell type is usually dependent on the antigen. Foreign antigens are effective at recruiting such cells while self-antigens are not. For example, bacteria activate macrophages, which then provide co-stimulation for T cells, which in turn provide co-stimulation for B cells, which then make antibody to kill the bacteria.

Molecular basis of tolerance. It has recently become feasible to introduce specific mutations into the genome of laboratory mice. In this way, genes have been removed and their role in the immune system evaluated in the resulting mutant animals. Such experiments have revealed a number of molecules that are critical for maintaining tolerance. For example, mutation of the *lpr* gene results in mice which develop autoantibodies and symptoms similar to systemic lupus erythematosus, an autoimmune condition in humans. The protein produced by the *lpr* gene is involved in programmed cell death which removes self-reactive lymphocytes from the periphery. Mutation of another gene, *CTLA-4,* results in a

severe autoimmune (lymphoproliferative) condition. This gene encodes a protein which is essential for turning off activated T cells; apparently, this includes self-reactive T cells. Similarly, mutation of genes encoding certain cytokines or their receptors results in the development of autoimmunity. For example, the absence of interleukin 10 results in the development of an autoimmune form of colitis (an inflammation of the colon), while deletion of the interleukin-2 receptor results in the appearance of autoantibodies. Many other examples exist; collectively these studies show that the immune system is a finely balanced system which can break down if one or more of the components are removed.

For background information *see* ANTIBODY; ANTIGEN; AUTOIMMUNITY; CELLULAR IMMUNOLOGY; GENE; IMMUNITY; IMMUNOLOGICAL DEFICIENCY; IMMUNOLOGY; IMMUNOSUPPRESSION in the McGraw-Hill Encyclopedia of Science & Technology.   David Tarlinton

Bibliography. G. J. V. Nossal, Negative selection of lymphocytes, *Cell*, 76:229–239, 1994; K. Pfeffer and T. Mak, Lymphocyte ontogeny and activation in gene targeted mutant mice, *Annu. Rev. Immunol.*, 12:367–411, 1994; K. Rajewsky, Clonal selection and learning in the antibody system, *Nature*, 381:751–754, 1996; D. M. Tarlinton, B-cell differentiation in the bone marrow and the periphery, *Immunol. Rev.*, 137:203–229, 1994.

Immunity

The usual fate of organs or tissues transplanted from one individual to another is that the graft is destroyed. Invariably, cells of the recipient's immune system recognize antigens on the foreign transplant and mount an attack that leads to graft rejection. Immune privileged sites are regions of the body where foreign tissue grafts experience extended (often indefinite) survival, compared to similar grafts placed at conventional body sites. Immune privileged tissues are organs and tissues that as grafts experience prolonged (often indefinite) survival when placed in conventional body sites (**Table 1**).

Immune privilege. Immune privilege was discovered near the end of the nineteenth century when foreign tumor cells placed in the eye of rabbits grew progressively, whereas similar cells placed below the surface

TABLE 1. Immune privileged tissues and sites

Tissues	Sites
Eye (cornea, lens)	Eye (cornea, lens, anterior chamber, vitreous cavity, subretinal space)
Cartilage	Brain
Fetal-placental unit	Uterus during pregnancy
Ovary	Ovary
Testis	Testis
Liver	Adrenal cortex
Tumors	Tumors

TABLE 2. Mechanisms responsible for immune privilege

Passive	Active
Blood-tissue barrier	Expression of Fas ligand
Deficient lymph system	Expression of enzymes CD59, MCP, and DAF
Tissue fluids drain intravascularly	Expression of class Ib molecules
Reduced expression of MHC class I and II molecules	Immunosuppressive microenvironment: transforming growth factor beta; alpha-melanocyte stimulating hormone; vasoactive intestinal peptide; free cortisol; interleukin (IL) 1 receptor antagonist

of the skin did not. The explanation for this observation emerged from studies in transplantation immunobiology by Peter Medawar and his colleagues. These investigators discovered that rejection of skin grafts placed orthotopically (on the body wall) was mediated by immunity. Medawar noted that if the lymph node draining a skin graft site was removed shortly after the graft was in place, rejection was markedly delayed. He speculated that antigens escaped from the graft via the lymphoid system and were first detected in the lymph node by antigen-reactive lymphocytes. Since skin grafts placed in the anterior chamber of the eye were not rejected, and since it was believed that neither the eye nor the brain possessed efferent (outward) lymphatic drainage, he hypothesized that immune privilege was predicated on the inability of graft-derived antigens to escape. He reasoned that foreign tissue grafts placed in the anterior chamber were tolerated by the recipient out of so-called immunologic ignorance.

In the mid-1970s three separate lines of evidence questioned the validity of the immune-ignorance hypothesis. First, certain privileged sites were found to display readily detectable lymphatic drainage pathways. Second, antigenic material placed in privileged sites was detected in the blood and in distant organs. Third, the systemic immune apparatus detects and responds to antigens placed in privileged sites, although the nature of the immune response is different from the responses to antigens placed at nonprivileged sites.

Specializations of immune privileged sites and tissues. Organs and tissues that possess immune privilege display passive features, that is, anatomical arrangements or constitutively expressed molecules that build physical and chemical barriers to immune cells and molecules. Privileged tissues and sites also possess active features that contribute to immune privilege by regulating the immune response. **Table 2** summarizes the features currently considered to be important in immune privilege.

Passive features. Blood-tissue barriers created at the level of vascular endothelium (a layer of cells lining the vessels of the circulatory system) or of an epithe-lial surface highly regulate passage of blood-borne cells and molecules into privileged sites and tissues. Many privileged sites lack lymphatic drainage pathways. However, lymph flow is only one route by which antigenic information can escape a site. In the eye and in the brain, a specialized extravascular fluid (aqueous humor and cerebrospinal fluid, respectively) drains directly into the venous circulation through anatomic specializations: the trabecular meshwork (bands of fibrous tissue) of the eye, and the arachnoid villae (the surrounding membrane) of the brain. From these privileged sites, material (such as antigen) is delivered intravenously rather than via lymph channels, thereby favoring tolerance induction.

Reduced expression of class I and class II major histocompatibility complex (MHC) molecules is a common feature of many privileged tissues and sites. In the most extreme case, the syncytiotrophoblast (a net of cytoplasm that contains irregularly scattered nuclei) of the placenta does not express conventional class I or class II MHC molecules. Instead, atypical class I (Ib) molecules are expressed on these cells, and appear to protect the cells from injury mediated by natural killer (NK) cells. The deficiency in expression of class I and II molecules on parenchymal cells of privileged tissues and sites renders these cells less vulnerable to destructive immune attack.

Active features. A number of features of privileged sites and tissues actively modify both induction and expression of immunity. Ocular cells (of the eye) constitutively express CD59 (complement receptor, or cluster of differentiation 59), MCP (membrane cofactor protein), and DAF (decay-accelerating factor)—enzymes that inactivate complement (proteins that mediate inflammation, phagocytosis, and immune defense). If antibodies against antigens on ocular cells enter the eye, ocular cells are spared certain lysis by these membrane-bound complement inhibitors. Similarly, the ligand for Fas (CD95) has recently been discovered on cells lining the anterior chamber, and on Sertoli cells of the testis. The Fas ligand, which causes T cells that express its receptor, Fas (CD95), to die (via apoptosis), may allow ocular and testicular cells to escape immune attack by T cells designed to recognize and destroy them.

Parenchymal cells within privileged sites and tissues secrete soluble molecules that create local microenvironments that are immunomodulatory. Growth factors, cytokines, and other mediators which suppress effector lymphocytes and macrophages have been found within aqueous humor, cerebrospinal fluid, and amniotic fluid.

The prolonged survival of foreign grafts in privileged sites appears to be dependent predominantly on active, rather than passive, features of these sites. Thus, immune regulation has emerged as a dominant theme in contemporary research into immune privilege.

Mechanisms that regulate immune responses. The immune response can be regulated at two different stages. In immunologically naive individuals, introduction of an antigen leads to the induction of immunity—the afferent limb of the immune reflex arc. Once an individual has been rendered immune, reintroduction of antigen leads to the expression of immunity—the efferent limb of the immune reflex arc. Immune induction and immune expression are quite different processes, and they are differentially susceptible to regulation.

Inhibition of immune expression. Although early studies indicated that grafts placed in privileged sites of preimmunized recipients were rejected, the notion that immune privilege concerns only the afferent limb of the immune response no longer holds true. Evidence is provided by tuberculin-immune mice that display delayed hypersensitivity when antigen is injected intracutaneously but fail to display inflammation when antigen is injected into the anterior chamber of the eye. In addition, mice immunized systemically to tumor-specific antigens are unable to prevent the growth of the parent tumor if it is implanted into the anterior chamber.

These findings emphasize the importance of local factors in thwarting the effectiveness of an immune response directed at antigens within privileged sites and tissues. If one considers the anterior chamber of the eye as a prototype of an immune privileged site, the blood-ocular barrier inhibits the migration of blood-borne immune effector T cells and immunoglobulin IgM antibodies into the eye. T cells that pass this barrier may fail to encounter their antigen within the eye because of insufficient expression of class I and II molecules, or they may be deleted via apoptosis or inhibited from proliferating, from secreting lymphokines, and from differentiating into functional cytolytic cells by local immunosuppressive factors. In aqueous humor, natural killer cells are inhibited from lysing their targets, macrophages cannot be activated, and antibodies are unable to cause lysis of antigen-bearing cells. *See* IMMUNE SYSTEM.

Modified induction of immunity. As discussed above, antigenic material introduced into privileged sites is not sequestered from the immune system, but escapes to impact the systemic immune apparatus. However, the immune response that is engendered is distinctly unusual. This phenomenon has been most thoroughly studied for the anterior chamber of the eye and has been termed anterior chamber associated immune deviation (ACAID). Virtually any antigen placed in the anterior chamber evokes a stereotypic, systemic immune response that is selectively deficient in T cells that mediate delayed hypersensitivity and B cells that produce complement-fixing antibodies. However, individuals with ACAID produce large amounts of non-complement-fixing antibodies, primed cytotoxic T cells, and T cells that suppress delayed hypersensitivity. This deviant response is dictated by the eye itself, where transforming growth factor beta 2 causes indigenous antigen-presenting cells to capture ocular antigens, to migrate preferentially to the spleen, and to present antigen there to immune cells in a manner that translates into ACAID.

Anterior chamber associated immune deviation plays a critical role in ocular immune privilege. For example, in the absence of ACAID, foreign grafts placed intraocularly are rejected. Alternatively, if ACAID is induced to ocular autoantigens, experimental autoimmune diseases of the eye can be prevented. Systemic immune responses similar to ACAID have been described for antigens placed in the brain and for the maternal–fetal interface.

Logic of immune privilege. All organs and tissues are susceptible to exogenous and endogenous pathogens for which immunity affords protection. The range of pathogens that threaten one tissue (such as the skin) may be quite different from the range of pathogens that affect another (such as the gastrointestinal tract). Pathogens use a diversity of virulence mechanisms to gain advantage within particular tissues, and the immune system has found it necessary to create diverse effector mechanisms to meet the challenges of pathogens. Antibodies that are particularly useful on mucosal surfaces (such as IgA) are of little value in protecting against cutaneous pathogens. Effector T cells and antibodies procure pathogen elimination either by direct binding and elimination or by enlisting nonspecific host-defense mechanisms. In the latter case, delayed hypersensitivity T cells and complement-fixing antibodies cause considerable tissue damage (immunopathology). If this nonspecific damage occurs among the cells of a vital organ, disease may result.

Certain tissues of the body (such as the eye, the brain, and the fetal-placental unit) are highly vulnerable to immunopathogenic injury. In the eye, immune privilege appears to limit intraocular expression of immunogenic inflammation. The delicate visual apparatus is exceedingly sensitive to minor adjustments, since an accurate image falls on the retina only when the apparatus is perfectly aligned. Inflammation disrupts the visual axis, and blindness is the inevitable result. In an evolutionary sense, immune privilege appears to have been created as a means by which the eye can receive certain forms of immune protection (cytotoxic T cells, non-complement-fixing antibodies) that do not disrupt vision, while suppressing the other forms that would threaten sight. The classic description of immune privilege, in which a foreign graft survives beyond expectations, reflects the important role that delayed hypersensitivity and complement-fixing antibodies play in graft rejection.

For background information *see* ANTIBODY; ANTIGEN; ANTIGEN-ANTIBODY REACTION; AUTOIMMUNITY; CELLULAR IMMUNOLOGY; EYE (VERTEBRATE); HYPERSENSITIVITY; IMMUNITY; IMMUNOLOGY; IMMUNOSUPPRESSION; TRANSPLANTATION BIOLOGY; VISION in the McGraw-Hill Encyclopedia of Science & Technology.   J. Wayne Streilein

Bibliography. C. F. Barker and R. E. Billingham, Immunologically privileged sites, *Adv. Immunol.*, 25:1–54, 1977; J. Y. Niederkorn, Immune privilege and immune regulation in the eye, *Adv. Immunol.*, 48:199–208, 1990; J. W. Streilein, Immune privilege as the result of local tissue barriers and immunosuppressive microenvironments, *Curr. Opin. Immunol.*, 5:428–432, 1993; J. W. Streilein, Unraveling immune privilege, *Science*, 270:1158–1159, 1995.

Impulse generator

Flux-compression (or magnetic cumulative) generators provide a compact and relatively inexpensive source of energy for use in experimentation at remote sites or in situations where a capacitor bank is either too bulky or too expensive to use. Essentially single-shot devices, they can generate megajoules of energy in microseconds, with terawatts of peak power.

Principles of flux compressors. The basic mechanisms that are involved may be understood by considering a closed conducting cage surrounding a region in which a magnetic field is established by either a current flowing through the cage or some external source. The process of flux compression is introduced when the volume of the cage is reduced by the action of an externally applied pressure. The compression time must be sufficiently short, of the order of microseconds, to prevent massive magnetic diffusion through the walls of the surrounding cage. In addition, the cage material and its thickness must be chosen such that there is an insignificant increase in its resistance due to Joule heating.

The force that accelerates the conducting cage to compress the magnetic field can be generated either by an explosive (solid, liquid, gaseous, or even nuclear) or by electromagnetic means. In the first case, a part (possibly up to 50%) of the chemical energy stored in the explosive is transformed into the kinetic energy of the moving cage. As this compresses the magnetic field, and does work against the internal magnetic force, a part of the kinetic energy (possibly up to about 80%) is transformed into magnetic energy. Much research effort in recent years has been directed toward minimizing the electrical loss mechanisms that are inherent in the dynamic process and that lower the magnetic flux within the cage and reduce the energy gain.

Because of their explosive nature, most, if not all, of the generator components and any associated sensors are destroyed during the generation operation, but the use of explosives in flux compression does not introduce any additional disadvantage. The magnetic flux density produced within the conducting cage may exceed 100 teslas, representing an energy density of 4 GJ/m³, with 40,000 times atmospheric pressure exerted on the conducting structure being also sufficient to destroy it even when it is protected from the effects of the explosion.

Experiments on flux compression began in the 1950s and formed part of the atomic weapons program. Since then experimental work has been undertaken in more than 10 countries, and the **table** provides a summary of the outputs obtained in the various programs. A wide variety of different geometrical arrangements has emerged from this experimental work, including cylindrical, coaxial cylindrical, helical cylindrical (considered below), plane and bellows, spherical, Archimedes-spiral-like, and disk shaped. As a general rule, flux compressors fulfil one of two distinct sets of requirements: (1) those which produce outputs of megaamperes of current and megajoules of energy by compressing the final magnetic field into an inductor forming part of an external load circuit; and (2) those which generate ultrahigh magnetic fields by concentrating the initial magnetic field into a small volume with the aid of a collapsing (or imploding) shell.

Helical flux compressors. The helical cylindrical arrangement provides a relatively simple basis for the construction of a high-energy flux compressor. The arrangement of a typical end-initiated device that produces an output of 2 MJ in 160 microseconds when using a 15-kg (33-lb) explosive charge is

Output achieved from flux-compression generators

Country or organization	Magnetic flux density, 10² T	Energy, MJ	Current, MA	Year of program start
Russia (Soviet Union)	17/25*	100	>300	1952
United States	10†/14*	50	320	1950
France	11.7	8.5	24	1961
EURATOM (Frascati, Italy)	5.4/7*	2	16	1961
United Kingdom	5	10	20	1956
Romania	5/7.5*	0.5	12	1982
Japan	5.5	‡	‡	1970
Poland	3.5	—	0.8	1973
China	‡	‡	2§	1967
Germany	—	‡	1.2§	1975

* Obtained only once.
† Inferred from x-ray pictures and a numerical simulation code.
‡ Unavailable.
§ Much higher figure was probably obtained.

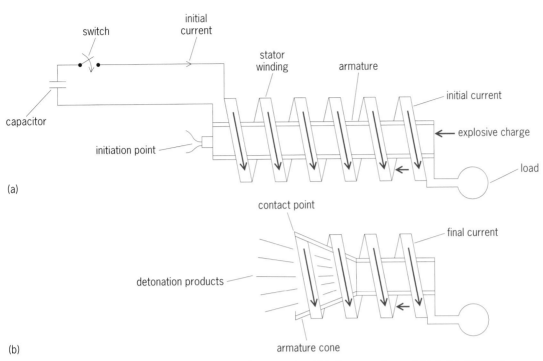

Fig. 1. Concept of a helical, end-initiated flux-compression generator, (a) before operation and (b) during operation.

shown in **Fig. 1**, and the practical details of this generator are shown in **Fig. 2**. A completed unit awaiting firing in a pit at the firing site is shown in **Fig. 3**. Similar generators have been or could be constructed on a considerably larger scale, with output energies up to 150 MJ provided at currents up to 60 MA from charges of more than 250 kg (550 lb) of special high explosive.

The energy generation process can be simply expressed by considering the flux compressor as a simple electrical circuit. Then, if the final load inductance L_1 is, say, $k_l = 10^3$ times less than the initial inductance of the generator L_i, and the final magnetic flux L_1I_1 is $k_f = 40\%$ of the initial value L_iI_i (where I_i and I_1 are the initial and final currents), then the final load energy $W_1 = L_1I_1^2/2$ and current I_1 are related to the initial energy $W_i = L_iI_i^2/2$ and current I_i by $W_1 = k_lk_f^2W_i$ and $I_1 = k_lk_fI_i$. On using the values given above, the ratio of the load energy to the initial energy (or the energy gain) is 160, with a current gain of 400. In practice, higher energy multiplications are possible, limited mainly by the need for

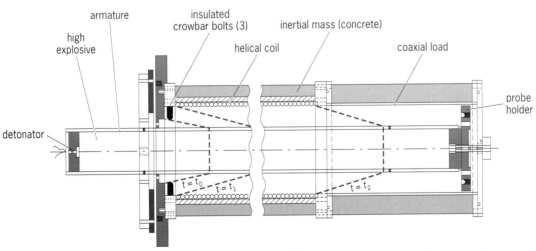

Fig. 2. Practical arrangement of a 2-megajoule helical, end-initiated flux-compression generator. The armature has extensions welded at each end. The armature, start plate, start ring, end ring, coaxial load, armature end plug, and end plate are made of aluminum. The separator plate, insulator, and probe holder are made of polyvinyl chloride. The positions of the armature cone at successive times t_0, t_1, and t_2 are shown.

Fig. 3. Helical, end-initiated flux-compression generator at the firing site.

the generator to withstand the very high electric fields that are generated within the working volume.

The action of the generator of Fig. 1 (and Fig. 2) begins with the creation of the initial magnetic field (initial flux) by the discharge of a capacitor bank through the circuit formed by the helical stator coil, the load inductor (coaxial in Fig. 2), and the tubular armature. The explosive contained within the armature is detonated, causing the armature to expand into the conical form shown in Fig. 1b and in Fig. 2 at $t = t_0$ (when the armature touches the crowbar), closing and separating the generator circuit when the current from the capacitor reaches a maximum. The point of contact made by the expanding armature with the helical coil moves progressively to the right (from $t = t_1$ in Fig. 2), and the initial magnetic flux of L_iI_i is finally compressed (at $t = t_2$) into the coaxial load inductor. A current path is completed through the helical coil, the load inductor, and the armature, and the current that circulates through this path increases progressively as the contact point moves toward the load. To accommodate this increase, the helical coil is divided into a number of sections, with both the diameter of the conductors and the number of parallel conductors in the sections increasing toward the load end. In the 2-MJ compressor mentioned above, the first section contained three parallel insulated conductors, each with seven strands of 0.85-mm-diameter copper cable, and the eighth and final section contained twenty parallel conductors, each with seven strands of 2.14-mm-diameter copper cable.

A helical generator can also be used to produce very fast rising, high voltages with predetermined waveforms. To achieve this, a special detonator is positioned inside the explosive charge to provide initiation in a linear manner along the central axis of the compressor. This results in the expanding cone of Fig. 1 extending over the whole length of the armature, thereby reducing the time needed for the compression process to that necessary for the

expanding armature to travel the radial distance between its initial position and that of the coil.

In many applications it is necessary to provide an autonomous power source with an energy gain of up to 10^6, which far exceeds that possible from a single-stage device. This requirement has led to the series coupling of generators by means of the dynamic transformer to form a chain with a single common central explosive charge that ensures the necessary firing synchronization between the individual stages.

Applications. It is readily possible to adapt the design of a helical generator to meet the requirements of many potential terrestrial and outer-space applications. This has led to their use as the basic energy source in many frontier areas of high-energy and high-current pulsed-power research, including thermonuclear fusion, electromagnetic launchers, powerful x-ray generators, powerful lasers, and ion, electron, or neutron radiation sources. However, almost all these applications require pulses with lengths shorter than the output pulse of the generator, and to obtain such pulses the output has to be conditioned by a number of switching stages, with each stage including both opening and closing switches to interrupt and transfer current to the load. Various techniques, including exploding metallic foil fuses, explosively formed fuses, and plasma switches, have been used as opening switches, while the closing switches either are activated by a fuse or an explosive or use voltage breakdown or surface tracking techniques. Very compact and powerful autonomous power sources result from the combination of a small initial energy source, possibly a permanent magnet or a battery, supplying a chain of helical generators whose output is conditioned to meet the requirements of the specific application.

For background information *see* IMPULSE GENERATOR; INDUCTANCE; MAGNETIC FIELD in the McGraw-Hill Encyclopedia of Science & Technology.

I. R. Smith; B. M. Novac; H. R. Stewardson

Bibliography. W. Baker and G. Cooperstein (eds.), *9th IEEE International Pulsed Power Conference, 1995*; M. Cowan and R. B. Speilman (eds.), *Megagauss VI Magnetic Field Generation and Pulsed Power Applications*, 1992; B. M. Novac, H. R. Stewardson, and I. R. Smith, Experimental methods with flux compression generators, *Eng. Sci. Educ. J.*, 5(5):211–222, October 1996; I. R. Smith, B. M. Novac, and H. R. Stewardson, Flux compression generators, *Eng. Sci. Educ. J.*, 4(2):52–56, April 1995; I. R. Smith, B. M. Novac, and H. R. Stewardson, High-energy pulsed-power conditioning and switching for flux compression generators, *Eng. Sci. Educ. J.*, 5(1):17–24, January 1996.

Infection

Mast cells normally are present within all tissues in humans and other mammals. They may be especially numerous beneath the surfaces of the skin, the respiratory system, and the gastrointestinal and genitourinary tracts. This distribution places mast cells near the sites of entry of parasites, bacteria, and other pathogenic organisms, as well as near surfaces that are exposed to environmental antigens (such as those derived from pollen).

Chemical mediators. Mast cells represent a source of many biologically active chemicals (mediators) that, when released by the cell, can induce changes in the function of other cells or alterations in the noncellular constituents of the tissues or the plasma. For example, in most tissues and organs, mast cells represent the major source of histamine and the only source of the anticoagulant molecule, heparin. Both of these mediators (as well as others) are stored within prominent granules in the mast-cell cytoplasm. Because mast cells often are present near blood vessels, lymphatic vessels, and peripheral nerves, these chemical mediators have to travel only short distances from the mast cell to express their effects.

At the end of the nineteenth century, the immunologist Paul Ehrlich proposed that these cells, which he termed *Mastzellen* (well-fed cells) because their cytoplasm was stuffed with prominent granules, may help to maintain the nutrition of connective tissues. Soon after, it was suggested that mast cells had a phagocytic function, and might thereby contribute to host defense against infection.

Inflammatory responses. Despite these early ideas, the mast cell was eventually recognized more as a contributor to a variety of diseases than as an important element of host defense. This view reflected the results of a large number of studies which indicated that mast cells can initiate inflammatory responses, particularly those seen in allergic individuals upon exposure to those substances which had elicited such allergies. It was found that mast cells can bind immunoglobulin E–class (IgE) antibodies to receptors on their surface, and that aggregation of these receptors by contact of cell-bound IgE with multivalent antigen stimulates the mast cells to secrete histamine, heparin, and other mediators that are stored in the cells' cytoplasmic granules, and to release other powerful proinflammatory mediators, such as leukotrienes and prostaglandins, which are synthesized by the cells upon their activation by IgE and specific antigen.

Such IgE-dependent allergic reactions can be very dramatic. Anaphylaxis, a reaction which can be triggered in sensitized individuals by such minor provocations as bee stings or doses of penicillin, can rapidly lead to death if it is untreated. Other disorders that are associated with the release of chemical mediators by IgE-dependent mast cells include hay fever and many types of asthma. As much as 20–30% of the population in the United States suffers from some type of allergic disease.

Acquired immunity. IgE-dependent immune responses, such as those responsible for allergic diseases, represent a form of acquired immunity. This type of response is expressed only in individuals that have had some previous contact with the eliciting antigen (for example, in allergic individuals a previous exposure to the offending pollen or antibiotic). Yet it was not clear why there would be in existence an effector element of acquired immunity (namely, the mast cell) that is so potent, and so widely distributed in vascularized tissues, that its activation in response to intrinsically innocuous substances such as antibiotics could result in the death of the host. One hypothesis is that the mast cell can express adaptive (that is, beneficial) functions by releasing, in appropriate settings and in submaximal amounts, many of the same mediators which are produced widely, rapidly, and in large quantities during anaphylaxis.

Mast-cell deficiency. However, in settings such as immune responses, which may involve the coordinated and potentially redundant activities of multiple cell types, it may be difficult to characterize the specific contributions of a single element, such as the mast cell. In the case of the mast cell, this problem can be addressed by using mast-cell knock-in mice. In this approach, mast cells which have been derived in the laboratory from normal mouse bone marrow cultures are used to repair selectively and locally the profound mast-cell deficiency of genetically mast-cell-deficient mice. For example, mice can be produced to contain mast cells at one site in the skin, whereas the skin site on the opposite side of the body remains essentially devoid of mast cells. The mice can also be used for laboratory experiments to compare the expression of biological responses in tissues that differ solely in whether or not they contain mast cells.

Cytokines. In a relatively recent development, mast cells were shown to be potential sources of many different types of cytokines, proteins that represent powerful signals between individual cells and can importantly influence many biological processes. At least one cytokine, tumor necrosis factor-alpha (TNF-α), can be released rapidly from preformed stores within mast cells, and more slowly from transcripts produced in response to appropriate activation of the cell. Studies in mast-cell knock-in mice showed that TNF-α represents an important mediator of mast-cell-dependent leukocyte recruitment, in response to both IgE and antigen or immune complexes, and that mast cells are required for the expression of IgE-dependent host defense against at least one species of ectoparasite.

Taken together with a large body of other evidence, derived from studies in both humans and many experimental animal species, these findings indicated that mast cells have an important role as an effector cell in acquired immune responses to parasites. This prompted some scientists to propose that, in the developed world (which, fortunately, does not bear a major burden of parasitic diseases) the mast cell may have become a rather unimportant element of the human immune system—a cell which once performed an essential role in acquired immunity to parasites but now functions primarily as an effector cell in allergic diseases.

However, recent work indicates that the mast cell may indeed retain an important role in host defense against infection, even in the developed world. Human resistance to infection with bacteria and other pathogens reflects the body's ability to express adequate natural immunity and maintain appropriate barrier function at surfaces exposed to the environment, such as the skin and the mucosa of the respiratory and gastrointestinal tracts. Natural immunity is the set of immunologically nonspecific reactions, of which many are mediated by monocytes and macrophages, granulocytes such as neutrophils, and natural killer cells, which can collectively contain and kill invading microorganisms, even in hosts which have not previously been exposed to these agents.

Acute inflammatory responses. Acute inflammation has long been regarded as a critical element of such natural immunity. Acute inflammation represents the entire group of tissue responses—including increased vascular permeability (leading to tissue swelling), increased blood flow and retention of blood within the tissues (leading to redness or erythema), and infiltration of neutrophils and other circulating granulocytes (leading to the formation of pus)—that characterizes immunologically nonspecific responses to many types of tissue injury, including those associated with splinters and cuts, and infections with microorganisms.

There are indications that acute inflammation that depends on the presence of mast cells can represent a critical component of host defense against bacterial infection. It has been demonstrated that mast cells can promote the clearance of virulent strains of the bacteria *Klebsiella pneumoniae* from the lungs or the peritoneal cavity of mice, as well as reduce the mortality associated with experimental intraperitoneal infection. Other work showed that reconstitution of genetically mast-cell-deficient mice with mast cells confers striking protection from death due to surgically induced bacterial peritonitis (an infection of the abdominal cavity), and that this effect was abolished by the administration of antibodies to TNF-α. The injection of mast-cell-deficient mice with TNF-α (instead of mast cells) was also protective, but only when the cytokine was provided in the right amount.

Functions in immunity. These findings suggest that mast cells perform two functions in host defense: they act as important components of natural immunity (for example, against bacteria) and as critical effectors in acquired immunity (for example, against certain parasites). However, much work remains to be done. Issues that need clarification include the importance of the mast cell's role in natural or acquired responses to additional types of pathogens; the signals which regulate mast-cell mediator release in these settings; whether there are quantitative or qualitative differences in the mast-cell mediators released during natural or acquired immune responses; whether mast cells can express important functions in these responses in addition to initiating inflammation (such as contributing to antigen presentation and immunoregulation or to barrier function); the extent to which the findings in mice apply to humans; and whether the function of mast cells in host defense can be manipulated for therapeutic ends.

For background information *see* ALLERGY; ANAPHYLAXIS; HISTAMINE; HYPERSENSITIVITY; IMMUNITY; IMMUNOLOGY; INFECTION; INFLAMMATION; NEUROIMMUNOLOGY in the McGraw-Hill Encyclopedia of Science & Technology.   Stephen J. Galli

Bibliography. B. Echtenacher, D. N. Männel, and L. Hültner, Critical protective role of mast cells in a model of acute septic peritonitis, *Nature*, 373:75–77, 1996; S. J. Galli, New concepts about the mast cell, *New Engl. J. Med.*, 328:257–265, 1993; R. Malaviya et al., Mast cell modulation of neutrophil influx and bacterial clearance at sites of infection through TNF-α, *Nature*, 373:77–80, 1996; H. Matsuda et al., Necessity of IgE antibodies and mast cells for manifestation of resistance against larval *Haemphysalis longicornis* ticks in mice, *J. Immunol.*, 144:259–262, 1990.

Insecta

Until the 1990s, insects were traditionally viewed as having a very poor fossil record. This perception can be attributed to the delicate nature of the living organisms, which makes them more susceptible to the vagaries of preservation. Nevertheless, it is the exceptionally delicate wings of insects that are usually represented in the fossil record, or at least the parts that are most characteristic and easily recognized. A closer examination of the fossil record indicates that insects do constitute a very high representation of extant taxa as fossils. However, there are still intervals in geologic history where fossil insects are rare—none more striking than the Triassic Period.

Isolated occurrences of Triassic insects have been reported from all continents, including Antarctica, but for a long time the only regions reporting significant numbers were in New South Wales and Queensland, Australia, and Kirghizia and Tadzhikistan, Russia. The majority of fossils reported from these areas consist of isolated wings, which, however, are found in extremely large numbers and represent a high diversity of forms, including dipterans.

Extinctions. The general paucity of Triassic insects has led to the notion that the end-Permian extinction event caused a bottleneck in insect diversity, and it was not until the Jurassic Period that diversity began to increase again. Jurassic sites are reported from Germany, including the Solnhofen quarries; England; and over 60 sites in Mongolia, Siberia, China, and Kazakhstan. Furthermore, in contrast with many of the Permian localities, the Jurassic assemblages typically include numerous groups that are well represented today. Thus, the early part of the Mesozoic Era is pivotal to understanding the evolution of modern insects. Furthermore, the Triassic–Jurassic boundary holds the key to understanding the evolution of modern terrestrial ecosystems, because the majority of modern higher vertebrate taxa also make their appearance at that time, as do many of the plants.

New site for Triassic insects. Any new site producing Triassic insects is therefore very important, but one bearing complete insects would be exceptional. The most significant advances made in recent years in the understanding of Triassic insects have resulted from excavations at a single site in North America. The Carnian (Late Triassic) sediments of the Solite Quarry on the North Carolina–Virginia border contain a wealth of fossils, including complete insects. Although the Triassic assemblage from the Solite Quarry has not yet yielded as many insects as excavations at other Triassic sites in Australia and Russia, more than 1500 individuals have been collected. More importantly, the exquisite detail of the fossils makes them unique. Preserved as silvery carbonaceous films, the specimens show such remarkable detail that microtrichia (length approximately 1 micrometer) on their bodies are frequently preserved, although such fine details are usually associated only with insects occurring in amber. Even though pre-Cretaceous amber with inclusions has not yet been discovered, resin-producing trees were undoubtedly present in the Late Triassic. The unique Solite preservation, which in some respects is remarkably reminiscent of the Burgess Shale, allows incredible distortion-free preservation of insect fossils.

Solite Quarry fossils. The deposit is composed of very fine grained shales that were deposited in a deep lake, and it is notable for the profusion of many types of fossils, including conifers, ferns, cycadeoids, bony fishes, and the prolacertiform tetrapod *Tanytrachelos*. The arthropods represented are spiders (Araneae); cockroaches (Blattodea; **illus**. *a*); eight species of beetle (Coleoptera), including one of the Lower Mesozoic family Schizoporidae, another from the living family Staphylinidae, and a probable carabid; flies (Diptera) from four extinct families and the living families Anisopodidae, Psychodidae, and Tipulidae; aquatic predacious sucking bugs (Heteroptera) in the living families Belostomatidae and Naucoridae; Sternorrhyncha "homopterans" of the extinct Permian family Archescytinidae (illus. *b*); Orthoptera; thrips (Thysanoptera); and caddisflies (Trichoptera; illus. *c*). The thrips are the oldest definitive specimens. The Coleoptera and Diptera provide the oldest records of these orders from North America. The anisopodids, belostomatids, naucorids, psychodids, staphylinids, and tipulids provide the oldest global records for these families.

Adding to the significance of the insect records from Virginia is the fact that the whole assemblage represents a significant part of an entire marginal to fully fresh-water ecosystem, represented by forms such as the water bugs, caddisflies, fish, and the diapsid reptile *Tanytrachelos*. Mixed with this assemblage are components of a fully terrestrial environment, exemplified by the phytophagous thrips

Triassic insects from the Solite Quarry, Virginia. (*a*) Roach. (*b*) Archescytinid. (*c*) Caddis fly.

and archescytinids, the detritivores such as the roaches, and the ground-dwelling predators such as the staphylinid beetle, not to mention the carnivorous theropod dinosaurs represented by tracks. The preservation of whole insects suggests limited dispersal. It can, therefore, be assumed that the terrestrial component was feeding on plants and animals that occurred very close to shore or even grew in the littoral zone, and it also suggests that a much richer, fully terrestrial insect fauna lived nearby.

Widespread extinctions. These new records of living insect families indicate that hypotheses of widespread extinctions of insects during the Triassic may be biased by the paucity of Triassic fossils. Previously it was estimated that for the Triassic only 5–20% of the families documented are extant. However, at the Solite Quarry, 15 out of approximately 21 insect families represented can be identified with certainty, and 7 of these are extant families. At the very least, one-third of the families recovered are extant, which suggests that the stasis of insect clades is much greater than previously hypothesized.

Missing flora and fauna. One particularly significant fossil in the Solite assemblage is the conifer seed *Fraxinopsis*. Although this species is well represented in Triassic sediments from the Southern Hemisphere, this is the first record of *Fraxinopsis* in the Northern Hemisphere. Coupled with the discovery of mammallike reptiles from another Carnian site near Richmond, Virginia, this record of *Fraxinopsis* distorts previously sharp distinctions between the Triassic floras and faunas of the Southern and Northern hemispheres. However, the Solite insect assemblage does not continue this trend. It seems to reaffirm the faunal provinciality of the Late Triassic, showing much greater affinity to Russian and European insect records than to Australian material. Nevertheless, certain expected groups are absent from the Solite assemblage. For example, given that the aquatic insects are the most abundant, it is surprising that no Ephemeroptera (mayflies) or Odonata (dragonflies) have been recorded so far. These orders have nymphs that live in fresh water, and their fossil record extends from the Carboniferous and the Permian respectively to the present. Another missing order is the Titanoptera, known solely from the Triassic and occurring in both the rich Russian and Australian assemblages. It is almost certain that other families and orders will be found in the Solite deposits in the future. Therefore, closer affinities with faunas from the Southern Hemisphere should not be ruled out.

Trace fossils. Although the paucity of Triassic insect localities means that each new site has a profound effect on the way insect evolution is viewed, isolated occurrences and trace fossils are also taken into consideration. A variety of trace fossils from the Triassic Chinle Formation of Arizona have been interpreted as termite nests and burrows. Since they predate the oldest known body fossils of termites (Isoptera) by 135 million years, this referral is equivocal; nevertheless, they do represent the earliest examples known of social behavior in insects, and add to the broad picture of great diversity of Triassic insects.

For background information *see* ARANEAE; COLEOPTERA; DIPTERA; EPHEMEROPTERA; EXTINCTION (BIOLOGY); FOSSIL; INSECTA; ODONATA; ORTHOPTERA; THYSANOPTERA; TRIASSIC; TRICHOPTERA in the McGraw-Hill Encyclopedia of Science & Technology.   Nicholas C. Fraser; David A. Grimaldi

Bibliography. N. C. Fraser et al., A Triassic Lagerstaette from eastern North America, *Nature*, 380:615–618, 1996; S. T. Hasiotis and R. F. Dubiel, The nonmarine Triassic, *N. Mex. Mus. Nat. Hist. Sci. Bull.*, 3:175–178, 1993; C. C. Labandeira and J. J. Sepkoski, Jr., Insect diversity in the fossil record, *Science*, 261:310–315, 1993; P. E. Olsen et al., Cyclic change in Late Triassic lacustrine communities, *Science*, 201:729–733, 1978; E. F. Riek, Fossil insects from the Triassic beds at Mt. Crosby, Queensland, *Austral. J. Zool.*, 3:654–691, 1955.

Integrated circuits

Lithography is important in making integrated circuits. Semiconductor chips are the building blocks by which advances in communications and transportation are made possible in concert with software technology. The semiconductor chip comprises a silicon wafer printed by lithography equipment, similar in operation to the common photographic camera.

Fabrication process. The integrated circuits are made by using a process that prints patterns, of up to 20 layers, on the surface of a silicon wafer. Each layer is designed to define the transistors and interconnections required to make up the circuit. A lithographic exposure process is used on every layer. The layer requiring the highest resolution (called the critical layer, most commonly the gate layer) is the most challenging element in this process.

The diameter of the wafer is between 6 and 8 in. (150 and 200 mm), depending on the product. The chips are then diced (cut out of the wafer) and mounted on chip carriers. The input and output connections are made, and the chip is sealed in a package. The chips have constantly been designed to perform more functions and to be fabricated at a lower cost per transistor. To do this, both the yield of good chips and the number of chips per wafer have been increased. Currently, the yield is very good, and the only way to continue to improve performance is to increase the number of chips per wafer.

Resolution. The performance of the chips is thus dependent on the size of the features required to make the transistors. The size of the transistor features and the space between them is dependent on the ability of the lithography equipment (the camera) to print them. This ability is commonly referred to as the resolution of the system. If a camera on the space shuttle, 190 mi (306 km) high, could see a football on Earth, its resolution would be equivalent to 0.35 micrometer on the scale of lithography equipment. The demand to improve the resolution is on the order of 10% per year (**Fig. 1**).

The lithography equipment industry has been challenged to keep up with this requirement. The result of improving resolution is an increase in number of chips per wafer (**Fig. 2**). This increase allows the semiconductor industry to maintain its record of continually providing more performance per integrated circuit at a reasonable price. The other advantage of improving resolution is the increase in chip speed due to the smaller capacitance offered by the smaller transistor features. The realization of the ben-

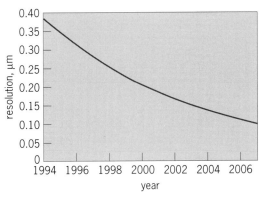

Fig. 1. Existing and forecast resolutions of lithography equipment for integrated circuits.

efits of improving resolution, resulting in increased productivity and speed of integrated circuits, is critical to the constant demand for these characteristics.

Lithography. The lithography tool is critical in producing integrated circuits with improved resolution. This tool prints the patterns onto the many layers required to make an integrated circuit. It uses optics and precision stages to print the pattern contained in a reticle (the picture to be printed) onto the silicon water (analogous to the film in a camera). The limiting resolution that the lithography tool can print is controlled by the exposing wavelength of the source and the numerical aperture (approximately $1/f$ number in the camera) of the optics. The development of these tools has progressed very rapidly, so that the modification of the exposing wavelength and the numerical aperture has kept up with the demands of the semiconductor industry. The changes in the exposing wavelength and the numerical aperture of the optics require advances in the photoresist (the photosensitive film emulsion), optical design, optical materials, optical sources, and reticle technology. The development of lithographic equipment requires around 3 years and substantial expense, assuming that the technology is evolutionary. Completely new technology takes 10 or more years and a larger order of magnitude of expenditure

to develop. The current standard lithography tool uses the I-line exposure source with a wavelength of 365 nanometers, and has numerical apertures approaching 0.6. This means that the limiting resolution is approximately 0.35 μm. The technology that will be used to fabricate the next generation of integrated circuits is deep-ultraviolet lithography equipment, which uses a wavelength of 248 nm for its light source and has numerical apertures of 0.6. The limiting resolution will be approximately 0.25 μm.

A new concept called step and scan for exposing the wafer has been developed, moving the wafer and the reticle in a very precise manner such that a smaller optical field produces a large image field. As a result, the optics can be made with higher numeric aperture while maintaining image quality. The scanning concept also provides integration that in turn results in superior linewidth control and very low distortion. Step and scan allows the resolution to be improved over the larger field size needed, to accommodate the larger chip sets planned for the near future. The next stage in improving resolution will be to use an exposing wavelength of 193 nm, also with a numerical aperture of 0.6. The limiting resolution will be extended to approximately 0.18 μm.

Major efforts are under way to explore both optics-based and nonoptics-based techniques to print features smaller than 0.18 μm. These efforts involve exposure sources such as electron beams, ion beams, and x-rays.

Although the lithography equipment is very important, the development of the photoresist and reticles is important as well. A change in wafer size from 8 to 12 in. (200 to 300 mm) is being considered to help reduce the number of semiconductor plants required to keep up with the demand for integrated circuits.

Resolution enhancement techniques. A number of techniques to enhance resolution have been under investigation and, in some cases, used to produce semiconductor chips. These techniques fall into two basic categories, relating to the illumination system and the reticle design.

The illumination-system techniques are based on the compensation of the light used to produce the higher resolution, so that more light is available for the higher-resolution features than for the lower-resolution features. A special place in the illumination system, called the pupil, is a circular area. The light in the center of the pupil illuminates the lower spatial frequencies. The spatial frequency of the illumination increases for the light at the outer radius of the pupil. By controlling the amplitude of the light, the higher-resolution features can be enhanced. Two widely used techniques block the light in the center of the pupil and let the light pass through at the outer edge of the pupil. The first method is called annular illumination and lets a ring of light pass at the outer edge of the pupil. This technique enhances the features for all orientations. The second method, called quadrupole illumination,

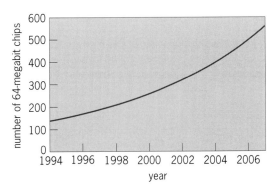

Fig. 2. Existing and forecast numbers of 64-megabit integrated-circuit chips on an 8-in. (200-mm) wafer.

lets the light pass through four holes at the outer edge of the pupil. This method provides better enhancement, but only for features at an orientation of 45°.

The second technique uses a reticle with phase shift to allow lower-resolution features to pass through the optics and interfere at the wafer in order to produce higher-resolution features on the wafer. This technique increases the cost of the reticle and is very hard to control. Variations of the phase-shift reticle have been derived to provide maximum enhancement with minimum fabrication difficulty.

Combinations of the two techniques are also being pursued. They may enable the normal resolution to be enhanced by extending the resolving power. However, this extension will be dependent on the quality of the optics and the reticles used to implement these advanced techniques.

For background information *see* INTEGRATED CIRCUITS; LENS (OPTICS); RESOLVING POWER (OPTICS) in the McGraw-Hill Encyclopedia of Science & Technology.   John J. Shamaly

Integrated services digital network (ISDN)

The integrated services digital network completes the end-to-end digital transmission capability of the public switched telephone network (PSTN). For more than a century the cable pair of copper wires has been used only for two-way analog voice communication, but as a result of digital technology it may now be used for new services. These services take advantage of the more robust digital signals, not only for digitized voice transmission but also for other bidirectional digital signals.

The technology now used for attaining this digital capability on subscriber lines is limited to bidirectional digital transmission at 144 kilobits per second. The basic service has two digital channels at 64 kilobits per second for voice and one 16-kilobit-per-second channel for call setup and low-speed data service. Extending digital service to the subscriber with 64-kilobit-per-second channels was a natural evolution from the public switched digital network, which is designed for such channels. Also available, particularly for private branch exchange (PBX) trunks, is a primary rate of 1.5 megabits per second, with twenty-three 64-kilobit-per-second voice channels and one 64-kilobit-per-second channel for data and call setup.

At somewhat high cost, ISDN technology has been added to most local time-division digital central-office switches. The service has been available from local telephone service providers in most locations for some time. The introduction of ISDN has stimulated the replacement of some earlier stored program control (SPC) space-division switching systems.

The solidification of standards, design, testing, and deployment of ISDN technology has taken more time than was anticipated. However, with the general availability of ISDN, the number of subscribers is growing rapidly. Subscribers are able to take advantage of ISDN's capabilities, and in fact have found novel applications that were not contemplated originally. Standardization of the processing of service orders has also contributed to the growing popularity of ISDN.

Standards. In the United States, test beds were established to ensure the functional interworking of equipment, particularly between the central-office and customer-premises equipment of different major manufacturers. National standard service packages were tested and demonstrated with the cooperation of the leading vendors and local telephone companies. Interface circuits, cards for personal computers, and digital telephones are generally available.

Representing increasingly complex digital services, successive national-standards phases were established for the tests among vendors. The first phase was directed toward the basic service. This included sending packetized data over the data or D channel. ISDN-capable interface cards for personal computers became known as ISDN modems, since the term modem was already familiar to most purchasers of computer equipment with analog subscriber line service. (Similarly, the term cable modem was introduced for the interface that enables digital signals to pass over cable television networks.)

The second phase (National-2-ISDN) included services such as calling line and name identification and deluxe call waiting, where the calling number is displayed while calls are in progress. Also included was the selection of multiple stations (with a maximum of eight) on the same line.

The third phase (National-3-ISDN) included more sophisticated services such as the integration of ISDN telephones with personal computer workstations, multiple (extension) ISDN telephones on a single basic rate interface (BRI) service pair, transparent feature operation between ISDN private branch exchanges, and centralized facsimile servers with ISDN access. Until the advent of ISDN, facsimile equipment operated at no more than 14.4 kilobits per second. ISDN enables the use of the much faster (64-kilobits-per-second) facsimile standard, known as Group IV. The time required to establish a facsimile connection and transmit an A4 page (210 × 297 mm or 8.3 × 11.7 in.) is then less than 30 s.

New services. As the number of ISDN subscribers has grown, new service applications and new technologies have appeared that take advantage of the ubiquitous and robust public digital network. Among the more popular of these are telecommuting, access to the Internet, and video teleconferencing.

Deploying ISDN is the first step in the process of changing the public telephone network into an all-digital switched telecommunications network. While the initial ISDN technology has been developing, standardized, and deployed, newer technologies with even greater service challenges have been explored to bring higher-speed digital transmission to the end user. This effort has been directed toward increasing the bit rate or so-called bandwidth of the digital subscriber line over copper pairs.

The high-speed digital subscriber line (HDSL) is being developed and service tested to attain channels at 1.5 megabits per second in either direction or 750 kilobits per second in both directions. For even higher speeds, the asynchronous digital subscriber line (ADSL) is being explored with the goal of achieving a rate of 45 megabits per second in one direction. The objective is to transport full-motion video and carry out higher-speed transfer of files or images over the existing subscriber distribution plant. For bit rates above this value, the use of optical fiber or coaxial cable or both must be considered for signal distribution. As a result, there are efforts to achieve higher-bit-rate ISDN on cable television.

Since the ISDN standards for basic and primary services were introduced, new technologies, such as ADSL, have begun to open capabilities that go beyond meeting the initial ISDN objectives. For identification purposes, the currently available ISDN is identified as N-ISDN for narrowband ISDN. Broadband ISDN, or B-ISDN, is expected to bring transmission at ever higher rates, such as 600 megabits per second or 1.2 gigabits per second, to end users within the next few years. For higher-speed, broadband, switched digitization of all services (including voice, data, and video) using the same optical-fiber medium for interoffice or internodal communications, a form of packetized or cell transmission known as asynchronous transfer mode (ATM) is being extensively pursued. In the meantime, N-ISDN provides an almost ubiquitous worldwide end-to-end switched digital capability.

Multirate ISDN. For private networks, a popular technique employing 24 or less 64-kilobit-per-second channels of a T1 (1.5-megabit-per-second) digital carrier line is being used for data or computer networking. This service, known as frame relay, enables subscribers to discover the usefulness of N-ISDN as a switched service. By introducing an inverse multiplexer, it is possible to couple several dialed-up switched 64-kilobit-per-second channels to obtain a higher-bit-rate ($n \times 64$ kilobits per second) connection. This service is sometimes called multirate ISDN. The connection bandwidth is determined by the value of n, which generally does not exceed 24.

Telecommuting. Many people working away from their base-office environments require simultaneous data and voice telecommunications services. ISDN has met this need by offering an increasingly popular method of computer networking. It provides a data transfer rate much higher than is possible with analog modems, as well as voice transmission over the same copper pair. Some computer modem products include not only the 28.8-kilobit-per-second digital-to-analog protocol but also the protocol for ISDN end-to-end digital transmission.

Internet access. The transmission rate for modern modem technology is 33.4 kilobits per second. By using all of the ISDN capacity for data, a fivefold improve-

ment in transmission speed may be obtained. This capability has been found particularly useful for terminal access to the Internet servers.

Video conferencing and broadcasting. The 144-kilobit-per-second two-way-transmission of ISDN makes possible reasonably high-quality conferences or lectures by using compressed digitized video. Furthermore, digital voice signals are easier to mix than analog signals in order to form multipoint dial-up voice as well as video conferences. With the quality improvement, lower cost, and greater availability provided by ISDN, switched conference calling has found greater use, and some travel requirements have been mitigated. Much software has been developed to broadcast digital video and imaging for teleconferences and school rooms as well as for individual personal video using computers.

Other applications. With readily available ISDN providing end-to-end digitized voice transmission, many other services have developed. For example, it is easier to encipher speech when it is digitized, thereby providing more secure telecommunications. Digitized voice, particularly enciphered voice, is also more difficult to wiretap.

A surprising application of switched digital transmission has been in making remote digital compact disk recordings. Artists may perform for their recording session anywhere that ISDN is available.

ISDN is a successful extension of the capabilities of local central-office switching systems. The services that it has enabled constitute the first step toward increasingly available ubiquitous worldwide switched robust digital transport in the public network. It opens the way for, and will stimulate the deployment of, newer forms of digital communications such as asynchronous transfer mode.

For background information *see* DATA COMMUNICATIONS; FACSIMILE; INTEGRATED SERVICES DIGITAL NETWORK (ISDN); SWITCHING SYSTEMS (COMMUNICATIONS); TELECONFERENCING; TELEPHONE SERVICE; VIDEO TELEPHONE in the McGraw-Hill Encyclopedia of Science & Technology.   Amos E. Joel, Jr.

Bibliography. Bell Communications Research, *National ISDN Definitions Package*, Doc. IS-796-1, 1996.

Interferometry

The phrase "to spatially resolve" refers to the ability to see structure in objects or to separate individual objects in an image. The maximum spatial resolution is given by Eq. (1), where θ is the minimum measur-

$$\theta \approx 2.4 \times 10^{-4} \frac{\lambda}{D} \qquad (1)$$

able separation in arc-seconds (an arc-second is 1/60 of an arc-minute or 1/3600 of a degree), λ is the wavelength of the light in nanometers (1 nm = 10^{-9} meter), and D is the size of the aperture through which the light travels in meters (1 m ~ 3.3 ft). This resolution limit may be explained by the diffraction of light through the aperture, a process which smears images. (The maximum spatial resolution is also called the diffraction limit.) An alternative explanation is that some spatial information from the source is lost because only a fraction of its light gets through the aperture. The human eye is most sensitive to light at $\lambda \sim 500$ nm (yellow light), and its pupil is typically $D \sim 2$ mm in diameter, which translates to a spatial resolution of approximately 60 arc-seconds (1 arc-minute). Unfortunately, this resolution is not enough to produce images of stellar surfaces or to chart the orbits of binary star systems. To overcome the limitations of the human eye, astronomers since Galileo have constructed telescopes and instruments to increase resolving power (or, to increase the effective size of the eye's pupil).

Large telescopes. Equation (1) provides some incentive for building large, single-mirror telescopes. The cost of such a telescope, however, is roughly proportional to D^3, imposing a monetary upper limit on D. The D^3 relationship arises not only from the amount of materials needed but also from the complexity of keeping a large telescope mirror in the correct shape (for high image quality) at different temperatures and orientations. The largest single-mirror telescopes as of 1996 are the two 10-m (400-in.) Keck telescopes on Mauna Kea, Hawaii, with a diffraction limit of $\theta \sim 12$ milliarc-seconds at $\lambda \sim 500$ nm. However, the typical spatial resolution obtained with these telescopes is only about 1 arc-second, no better than that obtained with a 15-cm (6-in.) telescope. The reason for this discrepancy is that light from an astronomical source must pass through the Earth's turbulent atmosphere, distorting the image. Nevertheless, such large telescopes are built because the amount of light gathered is proportional to D^2, so that larger telescopes can observe fainter objects. In fact, a 10-m (400-in.) telescope, using a detector with the same efficiency as the human eye, can view objects 2.5×10^7 times fainter than the eye can. (With more sensitive astronomical detectors, this number approaches 10^9.)

Reduction of turbulence effects. Ingenious solutions have been devised for reducing the effects of atmospheric turbulence. The Hubble Space Telescope, launched in 1990, eliminates the problem by orbiting above the Earth's atmosphere. Its diameter is $D = 2.3$ m, which translates to a resolution of $\theta \sim 48$ milliarc-seconds at $\lambda \sim 500$ nm. In the early 1970s, a passive technique called speckle interferometry was invented. When a very short exposure (a few milliseconds in duration), or snapshot, is obtained, it appears to be an ensemble of randomly placed points called speckles. When many snapshots are processed and combined, structural information (a model, not a true image) about the source is obtained. A 4-m (160-in.) telescope with a speckle camera can resolve objects separated by $\theta \sim 30$ milliarc-seconds in yellow light. Recently, an active technique called adaptive optics has been developed. Part of the light from the source is sent to a wavefront sensor, which determines how the light is distorted by the atmosphere. Signals from the wavefront sensor are then sent, in real time, to actuators mounted on the main mirror, changing its shape to focus the rest of the light into a nearly diffraction-limited image. Resolutions comparable to speckle interferometry have been obtained, but adaptive optics has the advantage of producing true images as opposed to models.

Long-baseline interferometry. The highest spatial resolutions have been produced by long-baseline optical interferometry, which takes advantage of the wave nature of light. When light is split and recombined, an alternating pattern of bright and dark bands, called fringes, is produced. The bright bands are formed in regions where the peaks of waves add (constructive interference), and the dark bands are formed where the peak of one wave is superimposed on the valley of another so that the two waves cancel (destructive interference). In a Michelson interferometer, two telescopes, separated by a baseline of length D, look at the same source (measuring the spatial coherence of different parts of the incoming waves) and send the light to a central beam-combining facility. In this facility, the light from the telescopes undergoes interference, producing fringes. The contrast of the fringes, also called the visibility, depends directly on D and the shape of the source along the telescope baseline. Larger values of D make the telescope more sensitive to smaller structures on the source, and vice versa, in accord with Eq. (1), so changing D during a night's observations provides more spatial information. Also, as the baseline rotates underneath the source (because of the Earth's rotation about its axis), spatial information can be obtained along different axes. To increase the visibility acquisition rate further, additional telescopes with different baselines and baseline orientations may be used. The number of independent baselines increases according to Eq. (2), where N_b is the

$$N_b = \frac{1}{2}N(N - 1) \qquad (2)$$

number of baselines and N is the number of telescopes.

In the 1920s, the diameters of several stars were obtained at the Mount Wilson Observatory (near Pasadena, California) by using an interferometer (two small movable telescopes) mounted on the Hooker 2.5-m (100-in.) telescope (used to view the fringes). With this instrument, the diameters of several red supergiant (highly evolved) stars were measured, typically 40–50 milliarc-seconds. At Narrabri, New South Wales, Australia, an intensity interferometer (that is, one that did not measure phases, discussed below), consisting of two movable 6.7-m (264-in.) telescopes with baselines between 10 and 188 m (33 and 617 ft), was constructed in the 1970s and used to measure more stellar diameters. In the early 1980s, several prototype interferometers were constructed on Mount Wilson, culminating in the Mark III. This instrument had five telescopes with baselines between 3 and 30 m (10 and 100 ft), but used only two simultaneously. This interferometer not only measured the diameters of 70 stars but also obtained the orbital parameters of 26 binary stars, some separated by as little as 3 milliarc-seconds. In addition, astrometry (the science of measuring stellar positions) was successfully performed on 12 stars with the two telescopes, whose baseline was known very accurately. Other optical interferometers have obtained similar results.

With only the fringe contrasts, it is possible to deduce model parameters for astronomical objects, such as the diameter of a star or the orbital elements of a binary star system. If the spatial shift of the fringes, called the visibility phase, is also obtained, true images can be created. The visibilities and phases are related to the images via a Fourier transform, which is well understood mathematically. At optical wavelengths, the phases cannot be determined directly because the fringes shift over very short time scales (because of atmospheric turbulence), typically less than 10 ms. They must be inferred iteratively from a starting model of the source and closure phases. The closure phase is the sum of the phases from three telescopes, a very useful quantity because it is independent of the effects of the Earth's atmosphere.

In 1996, the Cambridge Optical Aperture Synthesis Telescope (COAST) in England, with three telescopes, obtained the first true images with long-baseline optical interferometry, observing the binary star system Capella, whose average separation is about 50 milliarc-seconds. A few months later, the National Prototype Optical Interferometer (NPOI) in Arizona, also with three telescopes, produced images of the binary star system Mizar A, whose average separation is about 9 milliarc-seconds. Within the next few years, the NPOI will employ ten telescopes (up to six may be used simultaneously) and will be capable of resolving objects with separations of about 0.2 milliarc-second. In addition, the baselines of four NPOI telescopes will be monitored by an extensive laser metrology system, producing astrometric positions at the 1–5-milliarc-second level (a factor of about 20 better than classical methods). *See* PLANET.

Aperture synthesis. Imaging via long-baseline interferometry is also called aperture synthesis. In an optical interferometer array of several telescopes, each such telescope may be treated as a piece of an imaginary telescope whose diameter is the size of the largest baseline. As the Earth rotates underneath a source, the telescopes move as well, filling in some of the missing pieces and synthesizing a larger telescope. Aperture synthesis minimizes the problems associated with a large telescope; that is, the cost of many small telescopes is less than one large telescope, and it is easier to maintain the shape of small telescopes and remove the atmospheric effects from them. Aperture synthesis was first perfected at radio wavelengths during the 1950s–1970s, and this development has now culminated in many multitelescope radio interferometers with maximum baselines between 30 and 12,000 km (18 and 7000 mi), for resolutions (θ) between about 1 arc-second and about 1 milliarc-second. *See* RADIO TELESCOPE.

For background information *see* ADAPTIVE OPTICS; FOURIER SERIES AND TRANSFORMS; INTERFEROMETRY; OPTICAL TELESCOPE; RADIO TELESCOPE; RESOLVING POWER (OPTICS); SATELLITE ASTRONOMY; SPECKLE in the McGraw-Hill Encyclopedia of Science & Technology. Nicholas Elias

Bibliography. J. T. Armstrong et al., Stellar optical interferometry in the 1990s, *Phys. Today*, 48(5):42–49, May 1995.

Internet

The Internet is used for everything from shopping for books to keeping in touch with family members and teleoperating scanning tunneling electron microscopes. This article discusses the basic principles of data transfer on the Internet; and the Blacksburg Electronic Village, a communitywide project that provides a variety of high-bandwidth network services on the Internet.

Data Transfer

Despite the phenomenal growth in complexity and size, the Internet still contains key design features of its progenitor, DARPANET, a Department of Defense project initiated in 1969. The Department of Defense needed a robust digital communications medium to use in place of telephone networks, which are vulnerable to breakdown at switching stations. For DARPANET, the developers used an unreliable delivery model with messages transported on a per hop basis. Unreliable delivery means that while most information will be delivered successfully, any message that becomes corrupted or difficult to deliver will be discarded without warning. Today, the Internet uses the unreliable delivery model and corresponding protocols which evolved directly from the DARPANET.

Connection to Internet. To take advantage of the Internet, a user must connect the computer to it. The typical user employs a personal computer and connects via a modem. A modem is a device for turning data into sound and sound into data. The original modems sent information by changing between two tones, much like the telegraph. Modern modems use sophisticated compression and data-checking algorithms to send and receive complicated chirps, which are more analogous to the sound signals carried on modern voice telephone lines. These are the sounds that are heard when the modem is connected to another machine. The least expensive modern modems send data more than 10 times as quickly as their predecessors.

The user connects to an Internet Service Provider (ISP) by obtaining a phone number, a log-in identity, and a password from the Provider (typically for a monthly or hourly fee). The Internet Service Provider has several computers and many modem lines which serve as the first of many gateways to the Internet. The essential information that the Provider supplies to the user is an Internet address and a router name.

The Internet address allows any requests for information to be specifically labeled so that the responding computer will know where to send information. The typical Internet Service Provider applies to the international body that administers the assignment of Internet Protocol (IP) addresses. An Internet Service Provider is given a block of addresses that it can assign as it sees fit. When a user dials up, one of these addresses is assigned to the user during that session. When the user disconnects, that address is given to another user.

The router name tells the user's computer which machine to ask when it wants to find one of the more than 10 million computers connected to the Internet worldwide. A router functions as a large electronic telephone directory connected to a switching box. If the router is unable to find the address in its directory, it automatically asks a router with a larger directory.

Challenges. The Internet faces many immediate challenges. In much the same way that telephone companies have run out of telephone numbers in certain area codes, the Internet is in danger of running out of addresses. While the telephone company can simply add an area code, the problem is much more complicated for the Internet. In addition, the rapid growth in the number of computers has been accompanied by an increase in the uses of the Internet. Once a medium for transmitting text, the Internet now carries live voice and video. Several seconds of high-definition video consumes as much bandwidth as hundreds of pages of text.

The creation of browsers makes it easy to exchange information. These browsers provide a user-friendly point-and-click visual interface to what was once accessible only to the technically proficient. More people are requesting more information from more computers. With the many gigabytes of data being sent every minute between these millions of computers, the most critical challenge facing the Internet is making sure that information is delivered intact and in a timely fashion.

Protocols. Reliable data transfer on an unreliable communications medium is the major task of Internet software. This task is broken into a series of small procedures or protocols. Protocols are a collection of rules which explain a procedure for accomplishing a well-defined task. It is helpful to consider the analogy between a company sending a product and a computer transferring data.

Each company department accepts the product and whatever labels or packaging the previous department added. The department adds its own labels or packaging and sends the result to the next step. Similarly, each of the protocols used to control Internet traffic has a specific task which depends only on the actions of the previous protocol and affects only the operations of the next protocol. This lay-

Comparison of shipping a package and Internet data transfer	
Shipping a package	Internet data transfer*
Package product	TCP: break data into packet-sized pieces; label pieces
Address package	IP, DNS, PPP/Ethernet: address packets; send packets to Internet
Ship package	Routing: move packets along Internet; discard where appropriate
Receive package	PPP/Ethernet, IP: reassemble fragmented packets; receive from Internet
Examine product	TCP: check for completeness; discard duplicate or corrupt packets

* TCP = Transfer Control Protocol; IP = Internet Protocol; DNS = Domain Name Server; PPP = Point-to-Point Protocol.

ering of protocols enables small protocols to be sequentially linked to accomplish complicated tasks. Each step in the process of delivering a package has an Internet analog (see **table**). The handlers and shippers of the product are in no way involved in the creation process, nor are they concerned with the end use of the product. Similarly, almost all reliable data transfer on the Internet uses a standard set of layered protocols, no matter what the end user or end application does with the information, whether controlling scientific instruments or browsing the Web.

Organization of data. Popular phrases such as Information Superhighway or Interactive World Wide Web mystify the concrete nature of the Internet. The Internet is nothing more than a medium capable of sending 1's and 0's from one computer to another. Each digit in a string of data is called a bit. A collection of bits along the Internet is called a packet or datagram (**Fig. 1**). Any computer connected to this medium is a host or node. The medium is constructed of electrical wire, fiber-optic cables, cellular connections, and satellite uplinks. New methods of sending information through the Internet include digital radio, the coaxial cable television network,

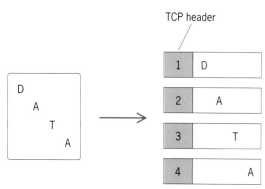

Fig. 1. Division of data into packet-sized pieces, with a Transport Control Protocol (TCP) header prefixed to each piece.

and even oscillations in alternating-current power lines. Regardless of the media along which the information travels, protocols facilitate Internet transfer by attaching a header to every data transmission. The header is analogous to the labels and packaging used in a shipping department. A header of any sort is a collection of 1's and 0's that is prefixed to a data packet (Fig. 1). For efficiency, a header should be as small as possible while still containing the information necessary to ensure reliable delivery of the packet.

Transfer mechanisms. Data transfer starts with an application attempting to send information to a remote machine. The transfer mechanisms are the same regardless of the content of the request.

The Transfer Control Protocol (TCP) divides the data into packet-sized pieces (Fig. 1). This process is analogous to the role of the packaging department in shipping a package. Each packet is given a unique number. To ensure that the contents of the packet are delivered intact, the Transfer Control Protocol software attaches a mathematical "snapshot" of the Transfer Control Protocol header and data. The receiving host can then compare this snapshot with the contents and request the retransmission of damaged parts.

Typically, a user knows a name for a remote host but not the address. Internet software relies on a Domain Name Server (DNS) to translate human-readable names, for example, www.Freesoft.org or 205.215.34.84, into computer-readable addresses. A Domain Name Server is a "telephone directory white pages" for computers. Every host has a unique address. An address consists of 4 octets of bits for a total of 32 bits. The Internet Protocol (IP) interacts with a Domain Name Server to determine an address for each packet. (This procedure is analogous to addressing a package.) To ensure that the address information remains intact, the Internet Protocol attaches a snapshot of the Internet Protocol header to the packet. This snapshot allows routers and the receiving Internet Protocol software to quickly determine if a packet has been corrupted. Because of the heterogeneous nature of Internet hardware, some packets may need to be subdivided or fragmented en route. In addition to addressing packets, the Internet Protocol controls fragmentation of packets.

Thus far in the process, no information has left the host. Typically, modems or Ethernet cards are used to transmit bits onto the Internet. These components can be thought of as dollies which carry a package from the addressing department onto the trucks of the shipping department. The Point-to-Point Protocol (PPP) is used for most modem transfer. Ethernet cards have their own protocols specific to their manufacturers.

The routing decisions for moving a packet (analogous to shipping a package) are made on a per hop basis. Each node of the Internet examines a packet and retransmits it to what is determined to be the next node. Telephone networks use a global routing scheme. If a line is cut in the middle of a phone conversation, the connection is lost. If a node is destroyed during an Internet transfer, the connection is delayed but not broken, as packets follow another route. There are a variety of algorithms and protocols dedicated solely to calculating the next hop. If a packet becomes mangled because of mechanical or electrical failure, the router can use the Internet Protocol snapshot to detect the inconsistency between the current header and the header snapshot and discard the packet. To prevent corrupt or misrouted packages from endlessly roaming the Internet, routers discard packets after a finite number of hops.

The Point-to-Point Protocol (PPP) or Ethernet translates the incoming bits into a format understandable to the receiving machine. (This operation is analogous to the unloading of transport vehicles.)

The receiving Internet Protocol software reassembles the fragments into whole packets (a process analogous to receiving packages). The received packet's Internet Protocol header is again compared with the snapshot of the header. If the packet header and the snapshot are inconsistent, the packet is discarded. The Internet Protocol software makes no effort to check the data in the packet. This is the task of the Transfer Control Protocol.

The Transfer Control Protocol examines the snapshot created by the sending machine to be sure that each packet is clean. Additionally, the Transfer Control Protocol is able to verify that all of the packets have arrived. (These operations are analogous to examining a package's contents and checking the invoice.) Intact packets are acknowledged by the receiver. Any packets that failed to arrive are retransmitted by the sender.

Transfer Control Protocol. As described above, the Transfer Control Protocol serves the important function of data quality control. Its primary tasks are to ensure complete delivery of datagrams and to control dataflow. The protocol uses a 192-bit header. When the Transfer Control Protocol software receives a packet, it need strip off and process only the first 192 1's and 0's to know how to handle the packet.

Each header has a unique sequence number and a checksum specific to that piece of data (**Fig. 2**). The checksum is a mathematical function calculated from the bits of the header (excluding the checksum bits) and the data content; it is the snapshot of the packet referred to above. The header also contains an acknowledgment number, which is the value of the next sequence number that the receiving host is expecting; and a window size, which specifies the number of packets the receiver is willing to accept.

The Transfer Control Protocol software recalculates the checksum based on the header and data in the received packet. This result is compared with the checksum value recorded in the header in order to determine if the data or header has been corrupted (**Fig. 3**). The Transfer Control Protocol uses

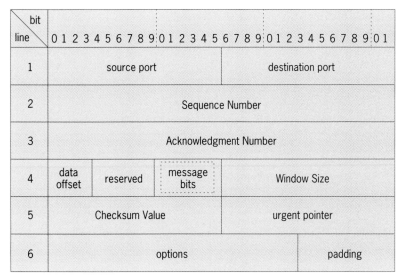

Fig. 2. Layout of a Transfer Control Protocol header in 6 lines of 32 bits each. For example, the 32d through 63d bits indicate the sequence number. The most important fields of the header (which are capitalized here) are the sequence number, acknowledgment number, window size, and checksum value.

sequence numbers to discard duplicates, request missing packets, and confirm received packets.

The other function of the Transfer Control Protocol, regulating the rate of packet transmission, is called flow control. Flow control allows very fast machines to efficiently communicate with slow machines. If the sending machine delivers packets faster than the receiver is able to process them, the receiver's memory will quickly become flooded, packets will be turned away, and the result will be unnecessary Internet traffic. To avoid this, the receiving host specifies the number of packets it is ready to accept. The sender responds by sending only as many sequence numbers as are indicated

in the window-size field and then awaits the next acknowledgment from the receiver. If the window size is zero, data transmission is suspended until the sender gets a nonzero window size from the receiving host. Then the sender commences transmission of nonacknowledged sequence numbers.

The Transfer Control Protocol also performs another sort of flow control. Sometimes both machines are capable of processing and sending packets much more quickly than the network can transport them. The Transfer Control Protocol software estimates the round-trip time between the machines. Using this estimate, the Transfer Control Protocol software establishes a reasonable rate of transmission. A good estimate of the round-trip time is essential to avoid needless network traffic (which results if the estimate is too low) and idle cycles (which result if the estimate is too high).   Joseph Libson

Blacksburg Electronic Village

The Blacksburg Electronic Village is a community-wide network project that began in 1991 with joint discussions led by Virginia Polytechnic Institute and State University with the town of Blacksburg, Virginia, and Bell Atlantic, the telephone company in the region.

The three parties agreed to work together to provide a variety of high-bandwidth network services in Blacksburg using the Internet. The university developed the software used to connect the user's computer to the Internet, based on user feedback through surveys and focus groups. It also provided authentication and authorization services. The university deployed a high-speed modem pool and provided Ethernet services to on-campus students, and Blacksburg organized its information into an online database. The telephone company agreed to administer much of the off-campus network in the town and to provide direct connections to many local schools and businesses via T1 telephone lines and bundled Internet access.

Finally, Blacksburg Electronic Village, Inc., was formed to provide community representation to the project. The board of directors of that nonprofit corporation meets several times a year to provide input to the members of the alliance and to assist in grant development.

In fall 1993, the first customers were able to sign up for Internet service via the university's modem pool using a data rate of 14,400 bits per second. In 1994 the first apartments in the town were wired for high-speed direct connection via Ethernet (1.5×10^6 bits per second). Early users of the network had access to electronic mail, a gopher server, and a Usenet news server, but by early 1994 a community Web server was also available. (A gopher server was a piece of software, now rarely used, that made it easy for Internet users to access text-based information located on various computers on the Internet. A Usenet server is a piece of software that delivers text-based discussions consisting of articles that look like pieces of e-mail. A World Wide Web server is a

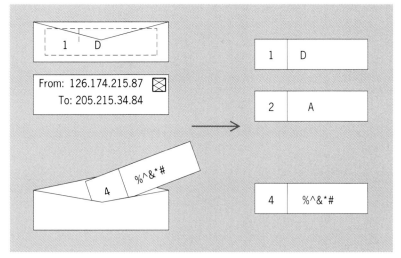

Fig. 3. Functions of Internet Protocol and Transfer Control Protocol software in receipt of packets. Internet Protocol software removes Internet Protocol header information and hands packets to Transfer Control Protocol software. Here, sequence number 4 has corrupted data, and sequence number 3 is missing.

piece of software that delivers text and graphics information to Internet users.)

Blacksburg now maintains an extensive Web site for local government information. The site is designed and administered by town employees, and any town department can be contacted via e-mail. Town officials use the Web site and e-mail as an additional channel for improving communication with local citizens.

The university provides the staff, computing resources, and much of the network resources for most of the day-to-day operational needs of the experiment, with a special focus on education and outreach. The Blacksburg Electronic Village group at the university assists other communities in southwest Virginia, helping them to design and implement their own community networks. This group also runs the Blacksburg Electronic Village Web server, a major focus of online activities.

Access. Since service started in 1993, there has been a steady increase in the number of Internet users in Blacksburg, and by summer 1996 more than 44% of the residents in the Blacksburg area used the Internet every day. High-bandwidth direct connections via local Ethernet networks (within buildings) support Internet access in the Montgomery-Floyd Regional Library, the Montgomery County public schools, six local apartment complexes, and numerous commercial office buildings in Blacksburg.

By fall 1995, there were two private Internet Service Providers in the area, and the university began moving users that were not affiliated with the university to these private service providers. This process was completed in summer 1996, and in the fall there were four local Internet Service Providers selling dial-up Internet access in the area.

The Blacksburg Electronic Village provides a standard set of software tools to users. All of its software is freeware (which can be distributed freely without fees or licenses) or has been licensed for use by members of the project.

Services. E-mail is an essential service in Blacksburg, and numerous surveys show that more than 95% of local Internet users have electronic mail. While the university was still providing e-mail for most of the community, more than 250,000 e-mail messages might be processed and delivered on busy days (in a town of 36,000 people).

World Wide Web and File Transfer Protocol (FTP) servers are also core services for local users. The Blacksburg Electronic Village and the university run the primary Web servers used by the community, but hundreds of residents and businesses also run Web servers. File Transfer Protocol servers are commonly used to allow users to download software and large files. The Blacksburg Electronic Village runs a File Transfer Protocol server to make it easy for users to download updated copies of Blacksburg Electronic Village software without making a trip to the office with a diskette. About 60% of local residents who are online use the World Wide Web regularly.

More than two-thirds of local businesses advertise on the World Wide Web, and more than 75 local community groups have home pages online. Web pages for nonprofit groups are hosted as a free service by the Blacksburg Electronic Village and by the Montgomery-Floyd Regional Library.

Listservs are also known as mailing lists. A listserv server takes a single piece of mail sent to a special e-mail address (for example, cats-L@listserv.mail.com) and resends it to a list of e-mail addresses. The Blacksburg Electronic Village seniors group uses a mailing list to keep in touch with nearly 100 members, and many other local groups use e-mail mailing lists. Some mailing lists are cross-connected with local Usenet groups so that residents can use e-mail or Usenet readers to access local public discussions. There are about a dozen Blacksburg groups (for example, bburg.general, bburg.bev.announce, and bburg.auto.repairs) where the community discusses a wide variety of topics and issues. There are also some private Usenet groups that have restricted access (only some people are allowed to read or post messages). These are most often used to support classroom activities, businesses, and community groups. The Blacksburg Electronic Village also provides several Web-based chat rooms for community use.

The local library has provided free Internet access to the community, and anyone who wants a free e-mail account can obtain one at the library. Demand for the public machines was so high that the library immediately had to initiate a sign-up system. Library patrons could sign up for 30-min sessions, and were limited to two per day. In the first year of this service, the library logged over 20,000 user sessions.

Demand has been steady since then. The Blacksburg Electronic Village has taken universal and affordable access as a serious challenge, and anyone in the community, regardless of the ability to purchase a computer or network access from home, not only can use the Internet in the library but also can have an Internet e-mail address. The ability to put an e-mail address on a resume, reply to an online job advertisement, or ask for advice online and have a reply address gives the economically disadvantaged the same power as those with network access in the home.

In providing education and training, the local library and the Blacksburg Electronic Village have cooperated on several grant initiatives to provide ''Internet librarians'' to support Internet access in the libraries and to teach the general public how to use the Internet effectively. The classes have been extremely popular.

With most businesses online and many residents as well, being ''wired'' has become commonplace in the community. There are thousands of personal home pages in Blacksburg, hosted as a free service by the Blacksburg Electronic Village (for noncommercial use) or on many private Web servers run by individuals in the community. A wide variety of local businesses provide sophisticated online ordering systems, with products ranging from groceries to textbooks and clothing.

For background information *see* DATA COMMUNICATIONS; ELECTRONIC MAIL; LOCAL-AREA NETWORKS; MODEM; PACKET SWITCHING; TELEPHONE SERVICE; WIDE-AREA NETWORKS in the McGraw-Hill Encyclopedia of Science & Technology.   Andrew Michael Cohill

Bibliography. A. Cohill and A. Kavanaugh, *Community Networks: Lessons Learned from Blacksburg, Virginia*, 1997; K. Dowd, *Getting Connected: The Internet at 56K and Up*, 1996; J. Erwitt and R. Smolan, *24 Hours in Cyberspace: Painting on the Walls of the Digital Cave Photographed in One Day by 150 of the World's Leading Photojournalists*, 1996; C. Huitema, *Routing in the Internet*, 1995; C. Hunt, *Networking Personal Computers with TCP/IP: Building TCP/IP Networks*, 1995; H. Rheingold, *The Virtual Community: Homesteading on the Electronic Frontier*, 1994; D. Schuler, *New Community Networks*, 1995.

Ion-solid interactions

The bombardment of solid surfaces by fast neutral or ionized particles has been of interest for more than 100 years. This interest stems from the many important applications that depend on these processes. For example, electron emission initiated by particle impact on a converter surface permits detection of very small particle currents or even single-particle counting for highly sensitive mass spectrometry. Sputtering, that is, the removal of atomic constituents from a solid because of particle impact, is highly important in many fields, including space flight, thermonuclear fusion research, and semiconductor technology. Radiation damage in solids, which is caused by the impact of energetic particles, can, for example, be utilized in light-ion radiotherapy for the destruction of cancerous tissue.

In most of these applications the kinetic energy of an atomic projectile is of foremost importance. This is the case in particle-induced electron emission, ion-surface scattering, and sputtering, all of which involve a series of binary collision processes. Some ion-induced phenomena, however, depend on inelastic processes, which become especially important if the internal (potential) energy of the projectile is comparable to, or even much larger than, its kinetic energy. The internal energy of ions results from their production, in which electrons have been removed from neutral atoms or molecules. As soon as such ions encounter a solid surface, they tend to regain electrons until they are again neutral. In the course of these recombination processes, the ions will deposit most of the energy which previously had been spent on their production. Internal energy also exists in the form of long-lived electronic excitation in metastable states. Ions and metastable atoms can give rise, for example, to electron emission from surfaces (potential emission) at kinetic energies so low that equally fast neutral atoms would not be able

to remove electrons by kinetic emission. Moreover, metastable excited particles are of importance in plasma chemistry and plasma technology, where they can efficiently provide additional reactivity.

However, effects depending on internal projectile energy remain of minor importance as long as the kinetic projectile energy remains dominant. This situation is drastically changed if slow multicharged ions impinge on matter, in particular a solid surface. Such multicharged ions are now abundantly available from novel ion sources, which can deliver ions up to the completely stripped limit (that is, all electrons have been removed from an atom) of virtually any element. Generally, the achievable ion flux becomes rapidly smaller with increase of the ion charge state. However, if such a multicharged ion approaches a solid surface, a situation can easily arise in which the kinetic energy of the projectile is much smaller than its internal (potential) energy. For example, a neutral thorium atom carries altogether 90 electrons. To remove 80 of them, that is, to create a very highly charged (neonlike) Th^{80+} ion, it is necessary to expend an energy of about 250 keV, which becomes available again if this ion approaches a solid surface. It is rather easy to transport very highly charged ions at a kinetic energy of not more than a few kiloelectronvolts. This kinetic energy is almost negligible in comparison to the potential energy carried by such highly charged ions. (Related experiments must be conducted under ultrahigh-vacuum conditions; otherwise the highly charged ion can easily recombine in collisions with background gas molecules.) For all potential-energy-driven effects in slow-particle–surface collisions, the state of the solid surface determines both the nature and the strength of its interaction with the ion.

Slow-ion impacts on metal surfaces. Ground-breaking studies conducted during the 1950s and 1960s showed that electron capture from a metal surface by slowly approaching ions gives rise to Auger electron emission. This process involves the mutual interaction of at least two electrons via their Coulomb repulsion, which can eject one of them into vacuum by disposing of the available potential energy (discussed above). The resulting total electron yields and observable electron energy distributions have opened the way to semiempirical explanations of the electronic transitions between target surface and projectile ion. A single ion may also capture several electrons at once, which gives rise to autoionization of the then doubly or multiply excited projectile, which is also an Auger process.

Since the mid-1980s, the increasing availability of ever more powerful multicharged ion sources and ultrahigh-vacuum equipment has remarkably extended such studies. They have involved primary ions of increasingly high charge, up to that of Th^{80+}. A single impact of such an ion on a clean metal surface can give rise to the ejection of more than 300 electrons. There have been numerous related experimental studies on the charge states and kinetic energy losses of projectiles scattered from the rather

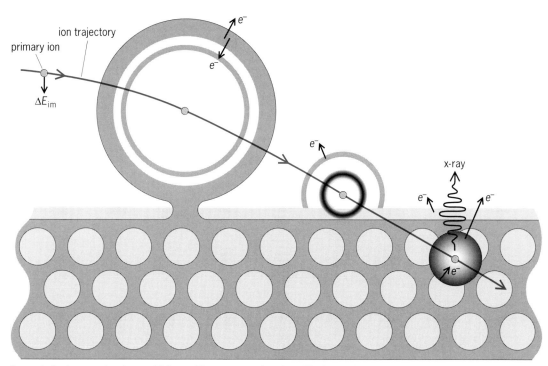

Scenario for impact of a slow multicharged ion on a metal surface. The ion is shown in four stages along its trajectory. ΔE_{im} denotes the image-charge energy gain acquired by the multicharged ion as it approaches the surface. *(After L. J. Dubé et al., eds., The Physics of Electronic and Atomic Collisions, AIP Conf. Proc. 360, 1995)*

flat surfaces of single-crystalline targets, and on the emission of characteristic soft x-rays from projectiles that are not deexcited via electron emission. The combined expertise from this work has led to a scenario for the slow-multicharged-ion–surface interaction, as shown in the **illustration**.

The process consists of four stages. In the first stage the primary multicharged ion approaches the surface. The second stage is the ion's first neutralization. After capture from the solid, many electrons temporarily reside in highly excited states of the projectile, resulting in the hollow atom state and the onset of slow electron emission due to auto-ionization. The third stage involves the ion's close surface contact, with the emission of further slow electrons and also fast Auger electrons from recombination of projectile inner-shell vacancies. In the final stage the incident ion undergoes surface penetration or is backscattered into the vacuum.

The first two stages can be satisfactorily described by the classical over-the-barrier model. While still distant from the metal surface, a highly charged ion induces a collective response of the freely moving metal electrons. This response corresponds to the classical picture of an image charge situated below the surface. This image charge accelerates the ion toward the surface until its neutralization has been completed, with the corresponding kinetic energy gain (denoted ΔE_{im} in the illustration) imposing an effective lower limit on the final ion impact energy. The image-charge acceleration can be directly observed by comparing the scattering of neutral and

multicharged particles, respectively, that are directed at grazing incidence onto a very flat surface. For the multicharged ion, the image-charge attraction will increase the angle of incidence (see illus.). The subsequently neutralized projectile is specularly reflected as at a mirror. Consequently, the reflection angle for ions that are initially highly charged will be larger than that for their neutral counterparts.

There is a potential barrier between the multicharged ion and the surface that depends on the metal work function (that is, the minimum energy required to release an electron from the solid, typically 4–5 eV for clean metals) and the instantaneous charge and distance of the ion from the surface. This potential barrier is gradually pushed down by the approaching ion. At some critical distance the barrier can be surmounted by the least tightly bound electrons from the solid, which will then be captured into highly excited (Rydberg) states of the projectile (the second stage). This electron capture continues until the projectile is completely neutralized; neutralization will be achieved after typically less than 10^{-14} s and lead to the formation of the hollow atom. Electrons in highly excited Rydberg states of the hollow atom mutually interact and thus become subject to autoionization. However, the emitted electrons are rapidly replaced by further electron capture from the surface. This replacement will restore the hollow-atom state during further approach of the projectile toward the surface (the third stage). This interplay of autoionization, electron capture, and image-charge attraction leads to the population of increasingly less excited (that is, more tightly bound) projectile states.

At sufficiently close distances to the surface, the considerable density of metal electrons will begin to screen the electrons on the hollow atom. This screening causes further shrinking of the atom and the ejection of slow electrons. At, and eventually below, the surface, deexcitation of the hollow atom proceeds via fast Auger electron emission (the fourth stage), whereby the projectile inner-shell vacancies that remain become filled by electrons captured from the solid. Alternatively, inner-shell vacancies are deexcited by emission of characteristic soft x-rays. In studies devoted to hollow-atom formation and decay, the slow-electron emission primarily delivers information on above-surface and at-surface stages (the second and third stages), whereas the spectra of fast Auger electrons and soft x-rays reveal details of hollow-atom decay at and below the surface (the third and fourth stages).

Slow ions on insulator surfaces. If slow multicharged ions are incident on insulators instead of metal or semiconductor surfaces, the fundamentally different surface properties of conducting and insulating solids come into play. (Such experiments are more complicated because the target surface may become charged, and this makes slow-electron spectroscopy, in particular, difficult.) Insulators have practically no free electrons and feature band gaps where no bound electrons are available for capture onto the approaching multicharged ions. The image-charge attraction of the ion will be weaker than for a metal surface because an insulator cannot form an image charge as effectively as a metal, and because electrons that are captured from insulators cannot be replenished as fast as for a metal. These facts have important consequences for the temporal and spatial evolution of hollow-atom formation. Under certain circumstances, complete ion neutralization is not achieved above the surface, in which case the projectile will be reflected as a hollow ion before being able to make close surface contact.

Recently, experiments studying the interaction of slow multicharged ions with particular insulator surfaces (alkali halides, in particular, lithium fluoride) have led to the discovery of potential sputtering, which comprises the rather efficient removal of (mainly neutral) target particles at such low (hyperthermal) ion kinetic energies that kinetic sputtering of metals or semiconductors would not be feasible. For example, the impact of a 600-eV Xe^{27+} ion (with a potential energy of about 10 keV) on polycrystalline lithium fluoride (LiF) gives rise to the ejection of almost 300 lithium fluoride molecules or their atomic constituents. Potential sputtering has also been observed for other, fundamentally different insulator species, such as silicon dioxide (SiO_2; quartz). The process is currently understood to be initiated by electron capture onto the approaching ion which, in turn, creates hole states and electron-hole pairs in the insulator. These holes cannot be rapidly recombined, and thus act as precursors for a specific kind of radiation damage (color-center formation), which can give rise to desorption of neutral particles from the insulator surface.

Potential sputtering has little in common with the Coulomb-explosion mechanism proposed earlier. Here, the rapid removal of electrons at and near the ion impact site leads to a local aggregation of positively charged ions in the solid, and these ions are then ablated from the surface because of their mutual Coulomb repulsion. However, it is conceivable that the Coulomb-explosion mechanism may become important for the impact of projectiles of still higher charge.

This potential-sputtering process could be of considerable interest with regard to innovative nanostructuring technologies. Such technologies would produce well-defined micropatterns on particular insulator surfaces through the impact of slow multicharged ions. The soft sputtering involved may be accompanied by less undesirable radiation damage than the conventional kinetic sputtering, which is utilized in present-day techniques for nanostructuring.

For background information *see* ATOMIC STRUCTURE AND SPECTRA; AUGER EFFECT; COULOMB EXPLOSION; ION-SOLID INTERACTIONS; ION SOURCES; RYDBERG ATOM; SPUTTERING; SURFACE PHYSICS; WORK FUNCTION (ELECTRONICS) in the McGraw-Hill Encyclopedia of Science & Technology.   Hannspeter Winter

Bibliography. L. J. Dubé et al. (eds.), *The Physics of Electronic and Atomic Collisions*, AIP Conf. Proc. 360, 1995; C. D. Lin (ed.), *Fundamental Processes and Applications of Atoms and Ions*, 1993; H. Winter, Hollow atoms make an appearance above insulators, *Phys. World*, p. 23, December 1996.

Jupiter

The *Galileo* spacecraft, which arrived at Jupiter on December 7, 1995, is conducting one of the most historic planetary exploration missions ever performed. The spacecraft consisted of two main components: an atmospheric entry probe and a planetary orbiter. The Probe plunged into the atmosphere of Jupiter on December 7, accomplishing the first direct sampling of the atmosphere of one of the outer giant planets and surviving the most difficult atmospheric entry ever attempted. On the same date the Orbiter was the first ever to be placed in orbit about Jupiter, and will return data about the Jovian system for a total of almost 2 years. Before reaching Jupiter, the spacecraft took the first closeup pictures of asteroids, discovering that one of them had a small satellite (about 1 mi or 1.6 km in diameter), and was the only platform that had a direct view of the impact of the comet Shoemaker-Levy 9 with Jupiter in July 1994.

Galileo was launched on October 18, 1989, on the space shuttle *Atlantis*. After a circuitous route that included a close flyby of Venus and two close

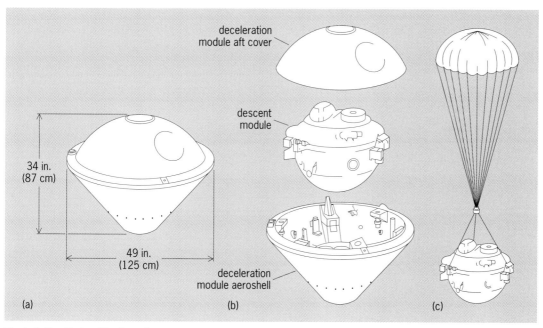

Fig. 1. *Galileo* Probe. All science instruments were housed in the descent module. (*a*) Probe prior to and during entry. (*b*) Exploded view. (*c*) Probe during descent.

flybys of Earth for gravity assists to gain energy, the Probe was separated from the Orbiter on July 12, 1995.

Galileo Probe

The success of the *Galileo* Probe was a remarkable achievement in view of the nature of its entry into the Jovian atmosphere. The Probe entered the atmosphere at 106,000 mi/h (170,000 km/h) on a flight path that was inclined 8.5° to the horizontal. An error of 1.5° too shallow, and the Probe would skip out of the atmosphere; 1.5° too steep, and it would be destroyed. Because battery power had to be conserved for the encounter with Jupiter, the Probe's trajectory could not be altered in any way after it was released from the Orbiter, 5×10^7 mi (8×10^7 km) from Jupiter; nor was there any communication with the Probe until the day of the Jupiter encounter, 5 months later. Once in the atmosphere, the Probe slowed to under 1000 mi/h (1600 km/h) in less than 2 min. This deceleration caused the Probe to experience 228 times the acceleration of gravity on Earth (228 *g*). By comparison, a jet fighter pilot in a high-speed turn experiences forces of only about 6 *g*. Due to the high-speed entry, a shock layer was set up in Jupiter's atmosphere that was located about 1 in. (2.5 cm) from the nose of the Probe. The temperature in the shock layer reached 25,000°F (14,000°C), about 2.5 times the surface temperature of the Sun. The Probe survived because of its protective heat shield.

Figure 1 shows the main components of the Probe. All science instruments were located in the descent module. **Table 1** lists the science investigations associated with the Probe and their purpose. **Figure 2** illustrates the mission profile. The Probe entered the Jovian atmosphere at 6.54° north latitude, that is, just north of the equator. The near-equatorial entry was dictated by the fact that the Probe had to take advantage of the rotational velocity of the Jovian atmosphere (which is greatest near the equator) in order to minimize the relative velocity between the atmosphere and Probe during the high-speed entry.

TABLE 1. *Galileo* Probe scientific investigations

Instrument or investigation	Objects of study
Atmospheric Structure Instrument	Pressure, temperature, and vertical winds
Energetic Particle Detector	Inner-radiation-belt parameters
Lightning/Radio Emission Detector	Optical and radio properties of lightning
Helium Abundance Detector	Relative helium abundance
Nephelometer	Cloud parameters and location
Net Flux Radiometer	Solar and planetary radiative flux
Neutral Mass Spectrometer	Composition
Orbiter Doppler Wind Experiment	Winds
Radio Scintillation Experiment	Turbulence and signal absorption
Ground-Based Doppler Tracking	Winds

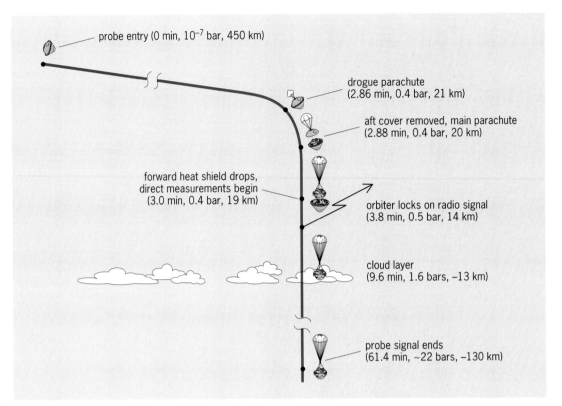

probe entry (0 min, 10^{-7} bar, 450 km)

drogue parachute
(2.86 min, 0.4 bar, 21 km)

aft cover removed, main parachute
(2.88 min, 0.4 bar, 20 km)

forward heat shield drops,
direct measurements begin
(3.0 min, 0.4 bar, 19 km)

orbiter locks on radio signal
(3.8 min, 0.5 bar, 14 km)

cloud layer
(9.6 min, 1.6 bars, –13 km)

probe signal ends
(61.4 min, ~22 bars, –130 km)

Fig. 2. Probe mission profile. Altitude is referenced to the level at which pressure is 1 bar. (1 bar equals sea-level surface pressure on Earth.) The last signal was received from the Probe at 61.4 min after entry, near 24-bar pressure. 1 km = 0.6 mi.

Motivations. One reason for the great efforts to explore Jupiter is to understand how the Earth formed and how it has evolved since. In order to do that, it is necessary to develop an understanding of how planets form in general, and what processes affect the way in which planets change over time. Jupiter is the largest planet in the solar system and influences all the other planets in one way or another. Jupiter has over twice the mass of all the other planets combined, and over 1000 Earths could fit inside Jupiter. Without understanding the processes that formed Jupiter and affected its evolution, it would be impossible to understand the solar system or the history of the Earth. Furthermore, Jupiter and its four largest moons, called the Galilean satellites, form a minisolar system. Studying the evolution of the Jovian system gives clues as to how the solar system has evolved as an ensemble of planetary bodies.

Jovian atmospheric composition. The Probe found that Jupiter's main constituents, hydrogen and helium (the two most abundant elements in the universe), are very nearly in the same proportion that they were in the Sun when it was first forming out of a gas and dust cloud. Jupiter, by mass, consists of nearly 76% hydrogen and 24% helium. Prior to the *Galileo* mission, remote-sensing estimates in 1979 from the flyby *Voyager* spacecraft had indicated that the Jovian helium abundance was considerably less than in the early Sun, and therefore less than measured by the *Galileo* Probe. Remote measurements

of helium are difficult to make and involve various assumptions. This difficulty explains the differences between the two measurements and illustrates the advantages of the direct sampling done by the *Galileo* Probe.

Although the material other than hydrogen and helium that constitutes Jupiter is only a very small fraction of the total, it provides important clues as to how Jupiter formed as a planet and what sorts of processes have affected it since its formation. Upon comparison with the Sun, the abundances of such heavier elements as carbon and oxygen, and of noble gases such as neon and xenon, as well as their isotopic ratios, can be used to trace the evolutionary history of Jupiter and compare these values with similar information that is available for the inner planets. *Galileo* found that carbon, nitrogen, and sulfur have greater relative abundances to hydrogen on Jupiter than they do in the Sun by about a factor of 2–3. However oxygen (in the form of water since Jupiter is much cooler than the Sun) is apparently scarce on Jupiter, at least where the *Galileo* Probe entered. This result is very different from expectations, since it was anticipated that thick water-ice clouds would be encountered by the *Galileo* Probe, and water would be an important constituent in the atmosphere as it is on Earth.

The meaning of the relative dryness of Jupiter has not yet been resolved, but certainly it was a great surprise. The greater abundances of carbon and other heavier elements on Jupiter as compared to the Sun imply that comets and other small bodies have impacted Jupiter over the age of the solar system (about 4.6×10^9 years) and deposited extra material. However, it is hard to understand why such impacts would not also have brought in water (and hence oxygen). The relevance of such findings to understanding the Earth has to do with the fact that impacts may have played a role in the evolution of the Earth, bringing in water that is now seen in the oceans, and other gases that are now in the atmosphere. Hence, the role of small impacting bodies in the evolution of planets is an important topic, and the information now obtained about Jupiter may provide new ways of thinking about it.

Winds. One of the main objectives of the *Galileo* Probe was to measure winds in the Jovian atmosphere. From pictures of Jupiter taken by spacecraft and by Earth-based telescopes, it was known that at cloud levels winds blew mostly in the east-west direction and reached speeds of greater than 200 mi/h (320 km/h). It was not known if these winds extended to greater depths below the clouds. The jet streams that occur in the Earth's atmosphere generally have far smaller speeds. The *Galileo* Probe encountered winds exceeding 400 mi/h (640 km/h) and showed that the winds extend deep below the visible clouds on Jupiter. It is not known exactly what produces the winds on Jupiter, but the Probe provided evidence that energy escaping from the very hot interior of Jupiter is the ultimate energy source.

Temperature and pressure. The *Galileo* Probe also measured the way in which temperature and pressure are related in the Jovian atmosphere. Temperature and pressure increase with depth. The Probe radioed signals to the Orbiter to a depth where the pressure was about 22 times sea-level surface pressure on Earth. The temperature at that point was about 311°F (155°C). At the point where the Probe first starting taking direct measurements, the pressure was about 0.4 that of sea-level surface pressure and the temperature was −225°F (−143°C). If the Probe had been able to survive to reach the very center of Jupiter (no probe could have done this), temperatures in the range of 60,000°F (33,000°C) and pressures millions of times greater than sea-level surface pressure would have been measured.

Cloud layers. Before the *Galileo* Probe mission, three cloud layers were expected to be found in the Jovian atmosphere. The Probe started direct measurements near the bottom of the top cloud layer, and detected the mostly ammonia (NH_3) ice particles by measuring how sunlight was diminished as the Probe went through this top cloud. The Probe also detected a second tenuous cloud, probably consisting of ammonium hydrosulfide (NH_4SH) ice particles. But below this, no thick water clouds were detected, in contrast to what was anticipated before the mission. The Probe entered a region which was relatively clear of clouds, as determined from Earth-based images of the Probe entry site taken by large telescopes, including the Hubble Space Telescope. Thus, the clouds encountered by the Probe may be more tenuous than is typical of the rest of Jupiter. However, the lack of detection of water-ice clouds is consistent with the fact that there was little water vapor measured in the atmosphere at the probe entry site.

Lightning. The *Voyager* spacecraft which had flown by Jupiter detected lightning events as it took pictures of the night side of Jupiter. Most of this lightning occurred near 45–50° north latitude. The *Galileo* Probe was equipped with optical sensors to detect nearby lightning flashes; however, it did not detect any lightning optically. It did pick up radio signals emitted by far-away lightning bolts. The conclusion is that Jupiter has less frequent lightning than the Earth on a per-unit-area basis, but individual lightning bolts are about 10 times more energetic than on Earth.

Radiation belts. Before it hit the atmosphere, the Probe measured high-energy charged particles trapped in Jupiter's magnetic field. These radiation belts are analogous to the Van Allen radiation belts which exist above the Earth's atmosphere. Two new radiation belts were discovered near the top of Jupiter's atmosphere.   Richard E. Young

Galileo Orbiter

The *Galileo* Orbiter is a highly sophisticated robotic spacecraft designed to orbit the giant planet for at least 2 years and provide a platform for detailed observations of the entire Jovian system. The spacecraft weighed approximately 2.5 tons (2.2 metric tons) at launch, nearly 1 ton (0.9 metric ton) of which was propellant required to perform deep-space maneuvers, place the craft in orbit about Jupiter, and adjust the orientation of the spacecraft. In flight, most of the spacecraft rotates at about three revolutions per minute to provide stability and to allow many of the scientific instruments to continuously scan in all directions. The Orbiter's scientific experiments are listed in **Table 2**.

The scientific objectives of the *Galileo* mission are (1) study of the chemistry, dynamics, and structure of Jupiter's atmosphere; (2) study of the chemical and physical state of the Galilean satellites; and (3) study of the magnetosphere (the region of space controlled by Jupiter's large magnetic field), including its composition, structure, and dynamics. The Orbiter mission relies on the technique of using gravity assists from the large satellites of Jupiter to provide a tour of the Jovian system, including 10 extremely close flybys of the satellites **(Table 3)**.

Io results. One important result of the mission was actually achieved before *Galileo* was technically in orbit, during the spacecraft's December 7, 1995, flyby of the inner Galilean satellite Io. Analysis of *Galileo's* radio signal enabled precise measurements of the speed and course of the spacecraft during its flyby to be made through use of the Doppler effect. These data, in turn, were used to measure the satellite's gravity field, and showed that the satellite

TABLE 2. *Galileo* Orbiter scientific investigations

Instrument or investigation	Objects of study
Remote sensing (despun)	
Solid-State Imaging (SSI) Camera	Galilean satellites, high-resolution imaging, and atmospheric small-scale dynamics
Near-Infrared Mapping Spectrometer (NIMS)	Surface/atmospheric composition and thermal mapping
Photopolarimeter Radiometer (PPR)	Atmospheric particles, and thermal and reflected radiation
Ultraviolet Spectrometer/Extreme Ultraviolet (UVS/EUV)	Atmospheric gases, aerosols, and so on
Fields and particles (spinning)	
Magnetometer (MAG)	Strength and fluctuations of magnetic fields
Energetic Particles Detector (EPD)	Electrons, protons, and heavy ions
Plasma Detector (PLS)	Composition, energy, and distribution of ions
Plasma Wave Spectrometer (PWS)	Electromagnetic waves and wave-particle interactions
Dust Detector (DDS)	Mass, velocity, and charge of submicrometer particles
Engineering experiment	
High-Energy Ion Counter (HIC)	Spacecraft charged-particle environment
Radio science	
Celestial mechanics	Masses and internal structures of bodies from spacecraft tracking
Propagation	Satellite radii and atmospheric structure from radio propagation

(about the size of the Earth's Moon) must have a dense central core, extending about half way from the center to the surface. This is the first planetary satellite with a confirmed core, implying that the intense heating that this moon receives as a result of varying tidal stresses under the influence of Jupiter's gravity has partially melted the interior and allowed separation of the heavy metallic materials, just as in the Earth's interior.

During the same flyby, the space-physics instruments were measuring the environment around Io. Remarkably, the spacecraft appears to have flown through a dense, cold plasma of electrons and ions at an altitude of about 550 mi (900 km), suggesting that volcanically produced gases from this active satellite may produce a patchy tenuous atmosphere. Other instruments reported a distinct drop in the magnetic field strength near Io and the presence of large electric currents flowing between the satellite and Jupiter. It is not clear whether these data mean that Io has a magnetic field of its own or whether the effects are all produced by the strong currents around Io.

Images of Io during the first three orbits (**Fig. 3**) show evidence of dramatic changes since the *Voyager* visits in 1979. There are geyser plumes active now that were not seen in 1979, while some of the eruptions seen then have ceased.

Observations of Jovian atmosphere. The first two orbits gave *Galileo* close looks at the satellite Ganymede, a rock-ice giant larger than the planet Mercury, and the first detailed observations of Jupiter's atmosphere, including the Great Red Spot (**Fig. 4**), an immense weather system which has apparently lasted for over 300 years. Pictures taken of the Great Red Spot at infrared wavelengths (beyond the range of human vision) reveal new details of cloud structure and altitudes. Some of the areas around the Great Red Spot show evidence of large-scale convective cells resembling huge thunderheads. If these are in fact thunderstorms, they are important to understanding the energy sources which drive the weather systems on Jupiter; on the Earth much of the energy which drives the global weather is generated by the exchange of heat between the atmosphere and ocean in tropical thunderstorms.

TABLE 3. *Galileo* Orbiter flybys of Galilean satellites

Orbit*	Satellite encounter	Date	Altitude, mi (km)
G1	Ganymede	June 27, 1996	519 (835)
G2	Ganymede	Sept. 6, 1996	162 (260)
C3	Callisto	Nov. 4, 1996	695 (1118)
E4	Europa	Dec. 19, 1996	430 (692)
J5	(Solar conjunction)		(No close flyby)
E6	Europa	Feb. 20, 1997	365 (587)
G7	Ganymede	Apr. 5, 1997	1901 (3059)
G8	Ganymede	May 7, 1997	985 (1585)
C9	Callisto	June 25, 1997	258 (416)
	Magnetotail, apojove	Aug. 8, 1997	
C10	Callisto	Sept. 17, 1997	326 (524)
E11	Europa	Nov. 6, 1997	699 (1125)

* Orbits are named according to MN, where M is the initial letter in the name of the satellite encountered, and N numbers the orbits from 1 to 11.

Fig. 3. *Galileo* image of Io, with Jupiter's atmosphere in the background, taken on September 7, 1996, at a distance of about 302,000 mi (487,000 km). *(Jet Propulsion Laboratory)*

Ganymede results. As with the Io flyby, the close encounters with Ganymede provided key information on its internal structure and its interactions with Jupiter's magnetic environment.

Magnetic field. Data from the Plasma Wave Spectrometer (which measures electromagnetic and electrostatic waves over a wide frequency range) during the first Ganymede flyby revealed that a large region around the satellite was filled with naturally occurring radio noise of the type generated around planets (such as the Earth) which have magnetic fields and radiation belts. This result had certainly not been expected for an icy satellite orbiting Jupiter. Magnetic field data rapidly confirmed an increase (by a factor of 5) in field strength near the satellite, suggesting strongly that Ganymede possesses its own magnetic field which, in effect, creates a little magnetosphere, immersed within the immense Jovian magnetosphere. Data from the second flyby confirmed these results and allowed scientists to calculate the geometry of the Ganymede field more precisely. It resembles that of a bar magnet (a dipole field), tipped about 10° to Ganymede's rotation axis (the Earth's dipole is tilted about 11.5°), and having a magnetic field at the equator about one-fortieth the strength of Earth's equatorial field.

Internal structure. Gravity data provided more clues to the internal makeup of Ganymede. Analysis of the radio data showed that Ganymede, like Io, must have a dense inner core, in this case surrounded by a light ice mantle about 435 mi (700 km) thick. Given the presence of a Ganymede magnetic field, it seems likely that the core itself is further separated into a rock shell and an inner metallic core, made of a mixture of iron and iron sulfide, which

is the probable source of the electrically conducting liquid thought to be required to produce a magnetic field.

Atmosphere. Ganymede does not have an appreciable atmosphere by terrestrial standards, but two instruments, the Plasma Detector and the Ultraviolet Spectrometer, provided evidence that both hydrogen and oxygen (the basic components of the satellite's water-ice surface) may be present in a thin atmosphere and on the surface. Their observations suggest that the chemistry on Ganymede's surface and in its thin atmosphere may be quite complex, and raise the question of how much oxygen is "trapped" in the surface and in what form.

Geology. Ganymede's geology is as fascinating as its internal structure. High-resolution pictures from the initial orbits show features on the surface smaller than a football field **(Fig. 5)**. Much of the surface of Ganymede seems to have been heavily modified by tectonic forces from within the satellite, producing sets of faults and cracks running for thousands of kilometers across its frozen surface. Even the older regions appear to have undergone some degree of this tectonic resurfacing, suggesting one or more events in Ganymede's geologic history which created forces within its icy mantle much like the forces which drive plate tectonics on the Earth.

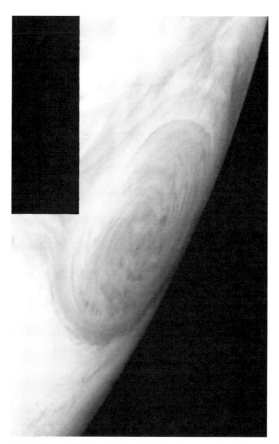

Fig. 4. View of Jupiter's Great Red Spot, a mosaic of two *Galileo* images taken on June 26, 1996. Black rectangle is an area not scanned by *Galileo*'s cameras. *(Jet Propulsion Laboratory)*

Fig. 5. Mosaic of four *Galileo* high-resolution images of the Uruk Sulcus region of Ganymede, shown within the context of an image of the region taken by *Voyager 2* in 1979. *Galileo* images were taken on June 27, 1996, at a range of 4628 mi (7448 km). Area shown is about 75 by 68 mi (120 by 110 km). *(Jet Propulsion Laboratory)*

Callisto results. Ganymede's twin sister, Callisto, was visited by *Galileo* on its third orbit. Although very similar in overall size and makeup, Callisto appears very different from Ganymede in other respects. Unlike Io and Ganymede, Callisto appears to be a relatively homogeneous mixture of ice and rock and does not seem to have a dense core. It also exhibits only small magnetic and radio signatures and probably has no magnetic field. Its heavily cratered surface is darker than most of Ganymede's surface and seems to be mostly a mixture of hydrated silicate rocks and ice, probably darkened by some carbonaceous materials similar to those found in primitive meteorites. High-resolution pictures, capable of resolving objects the size of large buildings, show that Callisto too has geological surprises. Instead of the densely cratered surface which scientists expected by analogy to old areas of the Moon and Mars, some of Callisto's surface appears to be blanketed with darker debris which has slid down the icy slopes of craters and fault scarps.

Europa results. Europa is perhaps the most enigmatic of the Galilean satellites. It is about the size of the Earth's Moon and has a rocklike density, but its surface is covered entirely by a layer of bright, relatively smooth water ice. Theoretical arguments suggest that Europa's icy crust is at least a few kilometers thick and possibly as thick as 60 mi (100 km). Heat from the decay of radioactive elements in the satellite's rocky interior combined with tidal heating similar to that occurring on Io may have produced a liquid water layer under the ice, and according to some calculations may be sufficient to maintain a liquid ocean today.

Images taken of Europa on *Galileo*'s first orbit provided a spectacular view of a region near the equator where the ice surface appears to have been pulled apart, probably by tidal stresses, much like a loose jigsaw puzzle. The first images with much better than *Voyager* resolution, about 1300 ft (400 m) per resolution element, were returned from the third orbit **(Fig. 6)**, and the first targeted Europa encounter on the fourth orbit produced some pictures with incredible detail at the 100-ft (30-m) scale. These images show a surface of astonishing complexity. In many places the surface consists, at virtually all scales, of thin, overlapping and crossing ridges of ice no more than a few hundred meters in height (Fig. 6). Older ridges and other structures appear to have lower topographic relief, suggesting that warm ice at shallow depths has allowed the viscous flow of ice (similar to glacial movement on the Earth). While these data do not prove or disprove the existence of a liquid ocean, they show clearly that this satellite has experienced a geologic history like no other planet-sized object seen so far.

Prospects. *Galileo* is scheduled to complete its primary mission in December 1997. It is estimated that there will be enough propellant and power to conduct extended operations for another 2 years, as long as the spacecraft's electronics survive the continual radiation dose they receive in Jupiter's radiation

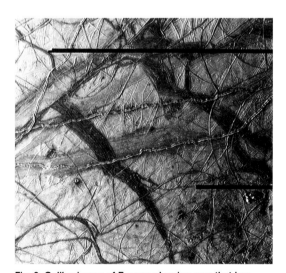

Fig. 6. *Galileo* image of Europa, showing area that has been highly disrupted by fractures and ridges. Image was taken on November 6, 1996, at a distance of 25,290 mi (40,973 km). Area shown is about 150 by 140 mi (238 by 225 km). *(Jet Propulsion Laboratory)*

belts. A proposed extension, the Galileo Europa Mission, would focus on thoroughly exploring Europa, with eight more close encounters, followed by one more close encounter with Io in 1999.

For background information *see* JUPITER; PLANETARY PHYSICS; SPACE PROBE in the McGraw-Hill Encyclopedia of Science & Technology.

Torrence V. Johnson

Bibliography. J. D. Anderson et al., Gravitational constraints on the internal structure of Ganymede, *Nature*, 384:541–544, 1996; J. D. Anderson, W. L. Sjogren, and G. Schubert, *Galileo* gravity results and the internal structure of Io, *Science*, 272:709–712, 1996; D. Gurnett et al., Evidence for a magnetosphere at Ganymede from plasma-wave observations by the *Galileo* spacecraft, *Nature*, 384:535–537, 1996; M. Kivelson et al., Discovery of Ganymede's magnetic field by the *Galileo* spacecraft, *Nature*, 384:537–541, 1996; C. T. Russell (ed.), The *Galileo* Mission, *Space Sci. Rev.*, vol. 60, no. 1–4, 1992; Special section on first results from the *Galileo* Orbiter, *Science*, 274:377–413, 1996; Special section on first results from the *Galileo* Probe, *Science*, 272:837–860, 1996.

Laser

Recent advances in laser science and technology include the development of high-power diode lasers, the use of copper vapor lasers in a wide variety of applications, and the demonstration of lasing without inversion.

High-Power Diode Lasers

Diode lasers are semiconductor devices that directly and efficiently convert electrical power into optical power. Diode lasers can achieve total electrical-to-optical conversion efficiencies greater than 60%, making them by far the most efficient light source known. Diode laser chips are also extremely small, typically only $0.1 \times 0.5 \times 0.75$ mm. The small size and high efficiency are key attributes which make a diode laser the choice in a large variety of applications, including laser printers, compact disks, laser scanners, and fiber-optic communications. This section discusses the enhancements in both crystal growth techniques and device designs that have increased the power available from diode lasers by several orders of magnitude while maintaining low cost.

Figure 1 shows the basic components of a diode laser. With techniques similar to those used in the silicon microelectronics industry, a series of planar crystalline layers are grown to form both an optical waveguide to confine the light and a *pn* junction to provide current injection into the active layer. Two mirrors are created by dividing the crystal along cleavage planes; the two mirrors and the active layer between them form a laser cavity. The facets ordinarily receive high- and low-reflectivity coatings, resulting in high-power emission from a single facet.

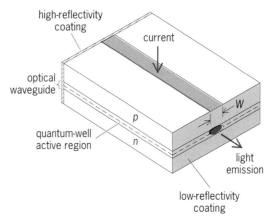

Fig. 1. Basic components of a diode laser chip.

The active layer, or light-emitting region, is composed of one or more ultrathin layers of low-band-gap material surrounded by higher-band-gap material. Since the ultrathin layers quantum-mechanically confine the carriers (electrons and holes), they are referred to as quantum wells. The thickness of a quantum well is typically 10–100 atomic layers. The reduced density of states and improved carrier confinement properties of quantum wells lead to significant performance improvements in diode lasers. For example, threshold current densities have decreased by almost two orders of magnitude from about 1500 A/cm^2 to less than 50 A/cm^2. Quantum-well diode lasers also routinely exhibit very high internal quantum efficiencies of 90–95%.

In recent years, there has been an explosion in the number of high-quality diode laser crystal material systems. The emission wavelength of a diode laser is directly dependent on the band gap of the active-region material, and with the various new material systems the wavelengths accessible directly with diode lasers at room temperature extend from 0.45 micrometer (blue) to greater than 4.5 micrometers (far-infrared).

Lasers in general can be divided into two classes: single spatial mode and multispatial mode. Applications demanding high brightness (small spot sizes) typically require single-spatial-mode lasers. Single-spatial-mode lasers allow only the fundamental mode to oscillate, resulting in a smooth single-lobed radiation pattern. For conventional single-mode lasers, the width, W, of the laser (Fig. 1) must be kept narrow, 2–5 μm, to guarantee operation in only the fundamental mode. Many applications, however, demand raw optical power rather than high brightness and use multimode lasers. In a multimode laser the width of the optical cavity is increased to 50–400 μm, which allows many modes to oscillate simultaneously. Because of the larger gain volume, multimode lasers are capable of obtaining significantly higher powers than single-mode lasers. However, the radiation pattern from a multimode laser typically contains significant structure (local peaks and valleys) and cannot be focused to as small a spot.

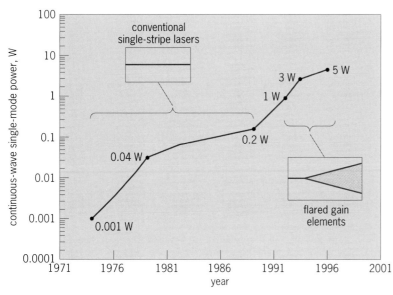

Fig. 2. Highest reported power from a single-mode diode laser versus year of fabrication of device.

the single-spatial-mode signal from a narrow single-stripe section is injected into a linearly flared gain region where the beam is allowed to expand. With current injected into the flared region, the mode experiences significant amplification (~10–50 times) due to the large gain volume, resulting in reported single-mode continuous-wave powers of 1–5 W, depending on the exact architecture. As was the case with single-stripe lasers, the 1-W continuous-wave power levels achieved with flared-gain-element diode lasers have opened up the possibility of a variety of new applications, including laser displays, spectroscopy, remote sensing of gases, laser radar (LIDAR), and genetic research. Work continues on improving the reliability and output power of these devices.

Multimode lasers. Multimode lasers are used in applications which typically require higher optical power than is available from single-mode lasers. For example, some cost-effective, high-resolution thermal printing systems use multimode diode lasers. Commercially available single-emitter multimode lasers typically have emitting apertures of 50–200 μm and produce continuous-wave optical powers of 0.5–2.0 W, while research devices have obtained over 10 W. Single-emitter multimode lasers are also commonly used for pumping compact solid-state crystal lasers which emit at wavelengths from 0.5 μm (green) to beyond 10 μm (far-infrared). For applications that require still higher continuous-wave powers, several multimode diode laser elements can be integrated into a single diode bar, 0.4 in. (1 cm) long, which produces reliable continuous-wave powers of 20 W, with research devices reaching 100 W.

Many solid-state lasers are operated only under pulsed conditions to obtain high peak powers. For example, the neodymium:yttrium-aluminum-garnet (Nd:YAG) solid-state lasers, which are useful in a variety of applications, including machining and LIDAR, approach peak pulsed powers of 1 MW (and continuous-wave powers of 2–6 kW). Historically, low-efficiency flash lamps were used exclusively to pump solid-state lasers. However, solid-state lasers pumped with diode lasers can offer higher efficiency, higher powers, longer lifetimes, and better beam quality. To satisfy pulsed solid-state laser applications, two-dimensional arrays of high-packing-density laser bars operating with pulse lengths of 0.1–1.0 millisecond are fabricated. A single diode laser bar, 0.4 in. (1 cm) long, achieves peak powers of 300 W, while a 0.4×0.8 in.^{2} (1×2 cm^2) two-dimensional stack of laser bars is commercially available at the 4.8-kW level. For ultrahigh power laser applications, manifolds containing several 0.4×0.8 in.^{2} (1×2 cm^2) stacks have been constructed. For example, a manifold assembled with sixteen 0.4×0.8 in.^{2} (1×2 cm^2) stacks has obtained 77 kW of peak optical power.   Stephen O'Brien

Copper Vapor Lasers

The copper vapor laser is a high-power laser that emits intense laser pulses in the visible spectrum.

Single-spatial-mode lasers. Some of the most significant advances in diode lasers have occurred with single-spatial-mode sources (**Fig. 2**). Significant progress occurred with a variety of single-stripe laser designs, increasing the power of research devices by two orders of magnitude from about 1 mW in 1974 to 200 mW by 1990. Achieving the 200-mW power level was significant since it enabled several important applications, including printing, free-space communications, and active fiber amplifiers. Fiber amplifiers provide an excellent example of how high-power diode lasers can assist in revolutionizing an industry, in this case the telecommunications industry. In a fiber amplifier a high-power (greater than 100 mW) single-mode diode laser operating at either 0.98 or 1.48 μm is used to pump an erbium-doped optical fiber, resulting in optical gain within the fiber at 1.55 μm, the signal wavelength. The all-optical fiber amplifier replaces expensive and noisy optical-to-electrical-to-optical repeaters used previously to amplify low optical signal levels. Fiber amplifiers are being deployed throughout the world in both terrestrial and submarine optical communications systems. *See* SOLITON.

In parallel with advances in power, the reliability of high-power diode lasers has improved dramatically. With advanced single-stripe laser designs, diode lasers operating at a variety of wavelengths have demonstrated high-confidence, extrapolated room-temperature lifetimes exceeding 10–100 years at power levels up to 150 mW.

By 1990 the power reported in technical journals from narrow single-stripe lasers began to saturate at about the 200-mW level. However, subsequent research with alternative laser designs has resulted in a further 25-fold increase in single-mode power. The device design responsible for this increase is the flared gain element (Fig. 2). In typical operation

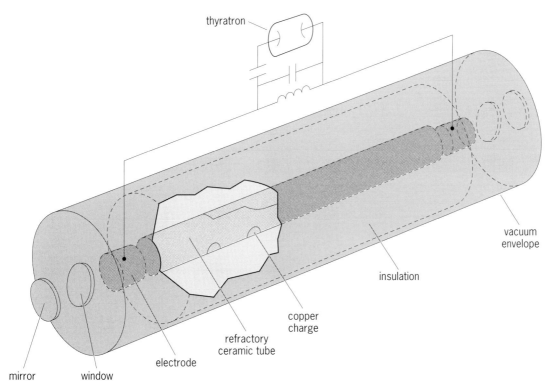

Fig. 3. Basic components of a copper vapor laser. For simplicity, mounting of mirrors outside apparatus is not shown.

The copper vapor laser was originally developed for use in the enrichment of uranium-235 for nuclear power generation. The requirements of this application demanded the world's highest-power visible laser. Single lasers can produce more than 800 W, and when lasers are used as a chain of amplifiers, powers in excess of 10 kW have been achieved (about 1000 times more powerful than the lasers used for light shows and 10 million times the power of the laser in a compact disk player). These devices find applications in micromachining, high-speed imaging, astronomy, medicine, and research.

Operation. The copper vapor laser emits light at two wavelengths, 510.6 nanometers (green) and 578.2 nanometers (yellow). The light is produced in the form of a very short pulse, typically 30 nanoseconds in duration, and these pulses are produced at a rate of 5000–50,000/s. This very high pulse rate means that the beam appears continuous to the unaided eye, and the beam itself is parallel and about 1–3 in. (25–75 mm) in diameter. In many applications the power level required is the range of 10–50 W, and these devices typically occupy a surface area of about 10 ft^2 (1 m^2).

The heart of the copper vapor laser is a refractory ceramic tube containing elemental copper and a buffer gas (usually neon). The tube is heated by a pulsed electrical discharge running in the buffer gas between electrodes at each end of the tube (**Fig. 3**). The heat generated by the discharge raises the tube temperature sufficiently to vaporize the copper. Under normal running conditions the tube temperature is approximately 2700°F (1500°C). Optimum laser performance is ensured by the thermodynamic balancing of the power input and thermal insulation of the tube. The vaporized copper atoms are excited from the ground state to the upper laser level by electrons accelerated in the discharge. This process leads to efficient excitation of the copper atoms, and this excitation combined with the large optical cross section of the copper atoms produces very high optical gain. When the electrons decay or are stimulated from the upper laser levels to the lower laser levels, photons at 510.6 nm and 578.2 nm are emitted. However, the lower laser level has a relatively long lifetime and so electrons tend to accumulate there. When more have accumulated there than can be excited to the upper level, the optical gain soon turns to optical absorption and the laser output is terminated. This whole process happens very quickly, leading to the short duration of the laser pulses (typically 30 ns). When the electrons have cleared from the lower laser level and returned to the ground state (typically after 20 microseconds), excitation can occur again, thereby facilitating the typical kilohertz pulse repetition rate. A thyratron is used to switch the high voltage required to strike a discharge in the laser tube (typically 10 kV) and to carry the large currents that are required for efficient excitation.

The very short pulse poses some interesting challenges for the optical design of the laser resonator. A typical resonator length of 5 ft (1.5 m) results in only sufficient time for a photon to make three round

Fig. 4. High-precision hole drilled by a copper vapor laser (hole diameter of 150 μm through 1-mm-thick stainless steel).

trips between the resonator mirrors. Since the initial burst of photons emitted when the laser pulse begins is incoherent, the resonator must quickly filter out photons that are not traveling parallel to the laser tube axis so that a parallel beam can be produced. This is achieved through a design called an unstable resonator. The unstable resonator uses curved mirrors that reflect photons which are not traveling parallel to the axis into the walls of the laser tube, where they are scattered and lost. Only photons that are parallel to the axis are fed back through the tube for further amplification. Some of these photons are allowed to escape through (or past) one of the resonator mirrors to form the output beam of the laser. The very high gain of the copper laser means that the feedback required to sustain oscillation is very low, so in fact most of the photons are allowed to escape through this end mirror (called the output coupler).

Micromachining. Research has shown that the short intense pulses produced by the copper vapor laser are ideal for micromachining. The visible-wavelength beam has advantages over other lasers since it is more efficiently absorbed by many materials. The short pulse efficiently vaporizes or melts the material, but the time scale for this is so fast that any melt is explosively ejected. This leads to a material removal mechanism that deposits very little heat in the bulk of the material. The very small diameters to which the beam can be focused, combined with the low pulse energy (1–20 millijoules), enable small amounts of material (10^{-5} to 10^{-7} mm^3) to be removed per pulse. The low material removal per pulse facilitates very high precision machining, and the high pulse rate produces an efficient machining process. **Figure 4** shows a high-precision hole, 150 μm in diameter, through a 1-mm-thick stainless steel sheet.

High-speed imaging. The copper vapor laser is an excellent strobe source for high-speed imaging in ballistics, impact research, production lines, machinery, and flow visualization of liquids and gases (such as in automotive engines). A typical high-speed imaging system combines the laser strobe source with a high-speed video or film camera, laser beam delivery optics or optical fibers, and software for analysis of the images. The high-speed camera receives one laser pulse in synchrony with the camera framing rate. The short laser pulse is used to effectively freeze the motion in view since camera shutters (electronic or mechanical) are too slow for high-speed objects or particles. In addition, the high intensity and single wavelength of the laser strobe allows the camera with a spectral filter to see into or through bright ambient conditions (such as flames or plasmas).

Astronomical applications. High-power copper vapor lasers are used in the adaptive optics technique that allows astronomers to correct for the atmospheric turbulence that distorts images during observations. The turbulence causes local variations in the refractive index of the air which distorts the wavefront of light from the stars (causing them to twinkle). The copper vapor laser is used to create a so-called guide star by focusing the beam to generate Rayleigh scattering at a height of 7–9 mi (12–15 km). If the scattered light (image) from the guide star is sampled with a sensor linked to a deformable mirror (the adaptive optic) in the astronomical telescope, the aberration caused by the atmosphere can be corrected. This technique greatly improves the images of real stars in the same field of view as the guide star, and results in the visible spectrum resembling those of the Hubble Space Telescope can be achieved from terrestrial-based telescopes.

Medical applications. The copper vapor laser can treat both vascular and pigmented lesions of the skin, moles, and port wine stains. The 578-nm output coincides almost exactly with an absorption peak in oxyhemoglobin, so it can selectively destroy subcutaneous blood vessels. The 511-nm output, readily absorbed by melanin, enables selective bleaching and removal of pigmented lesions. In addition, the copper vapor laser can be used to power a compact dye laser that is tuned to activate photosensitive drugs. The common application of this is in photodynamic therapy to treat cancerous tumors.

Martyn Knowles

Lasing Without Inversion

Traditional lasers can be understood by considering a group of atoms that are in one of two states (energy levels) such as those labeled A and B in **Fig. 5***a*. These two states are assumed to be separated in energy by an amount ΔE, and the interaction of the atoms with light of a specific wavelength λ, whose photons have the same energy ΔE, is considered. If a photon of this light interacts with an atom that is in the lower energy level, B, the atom can be excited to the upper energy level, A, and the light is then absorbed. If a photon of this light interacts with an

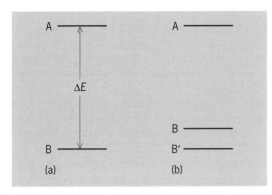

Fig. 5. Atomic energy levels for (*a*) a two-level system to describe conventional lasers and (*b*) a three-level system to describe lasing without inversion. Symbols are identified in text.

atom that is in the upper energy level, the atom can be stimulated to emit a photon (so that there are now two photons of wavelength λ) and the atom is then deexcited to the lower energy level.

Now, in the presence of light of wavelength λ, the probability that an atom in the lower energy level will absorb a photon is equal to the probability that an atom in the upper level will be stimulated to emit a photon. Consequently, if a beam of this light passes through a collection of these atoms, there are two possible outcomes:

1. The transmitted beam will be amplified (become more intense) if more than half the atoms are in the upper energy level since there will be more stimulated emission than absorption. In this case a population inversion is said to exist because there is a greater population of atoms in the upper level. (This is an unnatural, nonequilibrium situation.) It is the excess of atoms in the upper level that makes stimulated emission the dominant process and leads to laser oscillation. The stimulated emission process and resultant amplification of a beam of light is the origin of the acronym laser, from light amplification by the stimulated emission of radiation.

2. The transmitted beam will be attenuated (become less intense) if more than half the atoms are in the lower energy level because there will be more absorption than stimulated emission. Since the population of atoms in the lower level is greater in this case, there is no population inversion. It is the excess of atoms in the lower level that makes absorption the dominant process and prevents laser oscillation.

In the physical world, atoms are generally in the lowest energy level; that is, there is no population inversion. If this is not the case, and in the absence of external influences, the atom population tends to quickly adjust to that equilibrium arrangement. Previously, in order to build a laser, an inverted atomic population had to be achieved on at least some time scale. This population inversion is generally difficult to achieve, especially in the ultraviolet spectral region, and various techniques have been used to achieve it in the well-known examples of

laser oscillators, such as excimer, dye, argon ion, Nd:YAG, helium-neon (HeNe), and carbon dioxide (CO_2) oscillators.

However, an important breakthrough has occurred since 1992. Lasers have, for the first time, been made to operate without a population inversion; that is, lasing without inversion (LWI) has been accomplished. This new concept is based on quantum-mechanical coherence and interference effects. Implementing it requires consideration of an atom with at least three energy levels, such as in Fig. 5*b*, where A is an excited state and B and B' are two lower levels.

Double-slit experiment. This quantum-mechanical phenomenon can be understood by analogy with an important optical effect, the Young's double-slit experiment. In that experiment, light from a small source illuminates two slits so that on a distant screen there is a series of bright and dark regions. In a dark region, there are points at which absolutely no light can be detected; but if either slit is blocked, light will suddenly be observed at such a dark point. The absence of light at a dark point is the result of destructive interference of light waves following the two different paths from the two slits to the screen; of course, this interference can occur only when both slits are open. A crucial part of Young's experiment is the illumination of the two slits by the small source. This ensures that the light incident on both slits has the same phase, that is, that the waves incident on both slits are oscillating together; the light passing through the two slits is said to be coherent. In fact, if each slit is illuminated by a separate source, or equivalently, if an extended source illuminates the slits, then the interference effects disappear; in other words, the light passing through the slits is no longer coherent.

Interference of absorption paths. In the case of the atom with three levels in Fig. 5*b*, instead of two paths, one from each slit, to the screen as in Young's experiment, there are now two paths of absorption, one from each level, B or B', to level A; and these paths can interfere destructively. More explicitly, if light of appropriate wavelength is incident on an atom in the lower level B or B', it can be absorbed and the atom will be left in the upper level A. However, if the atom is first manipulated so that levels B and B' are coherent (in analogy with coherent illumination of the two slits in Young's experiment), then the absorption processes can interfere destructively so that the atom remains in the lower levels and no light is absorbed (in analogy with the dark regions in Young's experiment). However, an atom in the upper level A can still be stimulated to emit light and be left in the lower levels. Consequently, even if most of the atoms are in the lower levels B and B', lasing without inversion should be observable. Specifically, atoms in the lower levels cannot absorb light, and stimulated emission from just a few atoms in the upper level will then be the dominant process.

The destructive interference leading to cancellation of absorption described above is the essence of

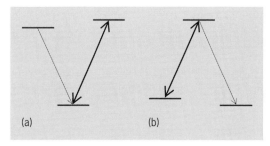

Fig. 6. Energy-level schemes for the two existing lasers that operate without population inversion. (a) V scheme. (b) Λ scheme. The heavy arrow in each part is a strong drive transition to produce the atomic coherences; the light arrow is the lasing transition.

Bibliography. D. S. Chemla, Quantum wells for photonics, *Phys. Today*, 38(5):59–64, May 1985; F. J. Duarte, *High-Power Dye Lasers*, 1991; S. E. Harris, Lasers without inversion: Interference of lifetime-broadened resonances, *Phys. Rev. Lett.*, 62:1033–1036, 1989; O. Kocharovskaya, Amplification and lasing without inversion, *Phys. Rep.*, 219:175–190, 1992; C. E. Little and N. Sabotinov, *Pulsed Metal Vapour Lasers*, 1996; J. Major et al., Single-mode InGaAs-GaAs laser diodes operating at 980 nm, *Electr. Lett.*, 27:539–541, 1991; S. O'Brien et al., Operating characteristics of a high power monolithically integrated flared amplifier master oscillator power amplifier, *IEEE J. Quant. Electr.*, 29:2052–2057, 1993; G. G. Padmabandu et al., Laser oscillation without population inversion in a sodium atomic beam, *Phys. Rev. Lett.*, 76:2053–2056, 1996; M. O. Scully and M. S. Zubairy, *Quantum Optics*, 1996; Y. Suematsu, Advances in semiconductor lasers, *Phys. Today*, 38(5):32–39, May 1985.

two more phenomena known as electromagnetically induced transparency, and ultrahigh index of refraction without absorption. Both the groups that introduced these phenomena and others have carried out detailed calculations for many schemes for lasing without inversion.

Observations. The first observation of the effect of net stimulated emission without population inversion was reported in 1992. Several other such observations were reported soon after. In these early experiments it was always possible to interpret the coherent atomic levels in terms of combinations such that there was a special combination of lower levels that did not contain any population; this was therefore considered to be lasing without inversion but with a hidden population inversion.

Only two operational lasers based on lasing without inversion have been reported so far. The first was in rubidium and was based on the V scheme, the arrangement of the energy levels as shown in **Fig. 6a**. The second, in sodium, was based on the Λ scheme illustrated in Fig. 6b. In both cases, there is a strong drive transition, indicated by a heavy double-headed arrow in each part of Fig. 6; it produces the necessary coherences. The laser transition is indicated by a light arrow. A description in terms of a hidden population inversion is not possible for either of these two lasers; consequently, they supersede previously known principles of laser operation.

Existing experimental results provide a proof of principle for the concept of lasing without inversion. Future interest will be in the development of ultraviolet or even x-ray lasers based on lasing without inversion. Obtaining a population inversion to build a conventional laser in this region of the spectrum is especially difficult, and lasing without inversion may be an excellent way to circumvent these difficulties.

For background information *see* ADAPTIVE OPTICS; COHERENCE; INTERFERENCE OF WAVES; ISOTOPE SEPARATION; LASER; OPTICAL COMMUNICATIONS; OPTICAL FIBERS; QUANTUM MECHANICS; STROBOSCOPIC PHOTOGRAPHY in the McGraw-Hill Encyclopedia of Science & Technology.   Edward S. Fry

Liquid membranes

In many areas of science there is an increasing need for chemical analysis of more complicated samples and for lower detection limits. One consequence is a growing interest in sample preparation procedures to be performed prior to chromatography as a critical step in the analytical scheme. Such procedures would typically achieve a selective removal of unwanted components in the sample and would increase the relative concentration of the target compound before the final analysis.

Sample preparation techniques. A number of sample preparation techniques are applied to practical problems.

One of the most successful techniques is solid-phase extraction. In this approach, a small column, often disposable, is packed with a solid material such as a silica adsorbent, an ion exchanger, or a reverse-phase material (alkyl chains bonded to silica support). The sample to be analyzed is drawn through the column, and the target compound is retained on the column while part of the unwanted material passes through. In a second step, the column may be washed with a suitable solvent to remove more unwanted material; and finally the target compound is eluted with another solvent and analyzed. This procedure can be applied to a large number of analytical problems; but it has limitations, especially for polar target compounds.

Supercritical-fluid extraction is a technique that utilizes supercritical carbon dioxide [pressure over 72 bars (72×10^5 Pa) and temperature over 32°C (90°F)]. It is used mainly for extraction of nonpolar compounds from solid samples. With this technique, no organic solvents are used; the efficiency of extraction compares favorably with classical methods.

The classical technique of liquid-liquid extraction can in principle solve a number of sample preparation problems; however, its application in modern

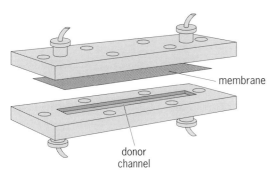

Fig. 1. Supported liquid membrane unit. The acceptor channel is hidden. The two blocks are clamped together with bolts during use. *(After J. Å. Jönsson et al., Automated system for the trace analysis of organic compounds with supported liquid membranes for sample enrichment, J. Chromatogr. A, 665:259–268, 1994)*

work has several drawbacks. These include the use of considerable amounts of organic solvents, difficulties of automation, and the formation of emulsions. Dialysis is another classical technique for selective sample pretreatment. It is very efficient in removing macromolecules, but low-molecular analytes are diluted in the dialysis process.

Supported liquid membrane extraction. The supported liquid membrane technique is a combination of liquid-liquid extraction and dialysis. With this technique, two aqueous phases, the donor phase and the acceptor phase, are separated by a membrane consisting of an organic liquid entrapped in the pores of a porous hydrophobic membrane. In this way a three-phase system (aqueous/organic/aqueous) is formed. By selecting the composition of the three phases, many types of compounds can be selectively extracted from the donor to the acceptor phase. Extraction is performed with a minimal use of organic solvents and with high selectivity; the technique has a large potential for automation and for connection to analytic instruments.

Figure 1 shows a typical membrane unit for supported liquid membrane extraction. It consists of two blocks of a polymer, such as polytetrafluoroethylene (PTFE). A groove is machined in each block, and flow connectors are provided in both ends of the grooves. By clamping the blocks together with a membrane in between, one flow-through channel is formed on each side of the membrane. The membrane is typically a porous PTFE membrane that has been soaked in an organic solvent with low volatility and low solubility in water. The organic solvent is held by capillary forces in the pores of the membrane. Typical solvents are *n*-undecane, di-*n*-hexyl ether, and tri-octyl phosphate.

An example of the supported liquid membrane technique is the extraction of amines from a sample of blood plasma (**Fig. 2**). The pH of the sample is adjusted to a sufficiently high value that the amines are uncharged. The acceptor channel is filled with an acid. The sample is pumped through the donor channel, and the uncharged amines (B) are extracted

into the organic membrane phase. An amine molecule that has diffused through the membrane will immediately be protonated at the membrane–acceptor inferface and thereby prevented from reentering the membrane. The result is a transport of amine molecules from the donor to the acceptor phase. Practically all the analyte molecules in the acceptor will be in the form of ammonium ions. Thus, the concentration gradient (and thus the rate of mass transfer) of the diffusing species (that is, the uncharged amine) will be unaffected by the total concentration of amine in the acceptor phase. This permits a high degree of concentration enrichment (several hundred times, depending on volumes) when more and more sample is pumped through the donor channel.

In the extraction in Fig. 2, any acidic compounds will be in anionic form (A^-) in the alkaline donor phase and, being charged, will be completely excluded from the membrane. This is also true for permanently charged compounds. Macromolecules, like proteins, will typically be charged and therefore rejected. They usually also have a very low solubility in the organic liquids used, even in the neutral state. Neutral compounds (N) may be extracted, but the concentration in the acceptor phase will never exceed that in the donor phase, so that no enrichment is obtained. Thus, the conditions prevailing in Fig. 2 lead to highly selective extraction of small, basic compounds. By careful selection of the organic liquid in the membrane, thereby maximizing the solubility of the target compounds, a further selectivity toward unwanted compounds can be gained.

With an acidic donor phase and an alkaline acceptor phase, acidic compounds will be extracted in a way similar to the amines in Fig. 2. Also, supported liquid membrane extraction systems for various permanently charged compounds and metal ions can be devised by incorporating ion-pairing or chelating reagents in one or several of the phases. In general, a neutral, extractable species should be formed in the donor phase, while this species should be destroyed in the acceptor phase by forming a charged, and therefore nonextractable, species.

Applications. The supported liquid membrane technique has been applied to a number of classes of compounds and types of samples for both environ-

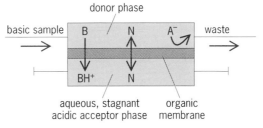

Fig. 2. Supported liquid membrane extraction, which is selective for the basic target compounds (uncharged amines; B), rejecting anions (A^-) from acidic compounds and diluting neutral compounds (N).

mental and pharmaceutical studies. Three applications that illustrate different aspects of the technique are extraction of basic drugs from blood plasma, integrating sampling of pesticides in natural waters, and extraction of metal ions.

Extraction of basic drugs from blood plasma. As an example of how supported liquid membrane extraction can be used in bioanalysis, 0.5 milliliter of alkalized blood plasma was extracted through a supported liquid membrane containing dihexyl ether into 12 microliters of an acidic acceptor buffer. The acceptor phase was subsequently analyzed by high-performance liquid chromatography. The entire flow system was completely automated by using a robotic sample handler. The resulting chromatograms from a plasma blank showed no significant peaks and were virtually indistinguishable from those from a water blank. The drug peaks from a spiked plasma sample were clearly seen and readily quantified. As virtually no components from the blood plasma were seen, a high degree of cleanup was manifested. A miniaturized extraction unit was also connected directly to a capillary electrophoresis instrument, permitting use of this high-resolution technique for direct analysis of blood plasma samples.

Integrating sampling of pesticides in natural waters. Supported liquid membrane extraction has been used for continuous field sampling of acidic herbicides. Water from a river to be studied was acidified and continuously pumped through the donor channel of a supported liquid membrane unit. The stagnant basic acceptor was removed every 24 h and analyzed, providing a true mean of the concentration during the sampling time. This technique supplies a more reliable account of the leakage of herbicides into a recipient than can be obtained with conventional grab sampling, because the leakage is very variable in time depending on weather conditions.

Extraction of metal ions. Liquid-liquid extraction of metals exploiting various complexing reactions is a classical area of inorganic and analytical chemistry, and the literature is extensive. Many of the chemical systems used can be applied to supported liquid membrane extraction. One example is the use of an extractant (Aliquat-336) in the organic membrane for extraction of several heavy metals in the form of thiocyanate complexes. In the acceptor phase a stronger complexing reagent, pentetic acid (DTPA), traps the metal ions by forming a charged complex. Other extraction systems, comprising 8-hydroxyquinoline in the donor phase or alkyl phosphonate salts in the membrane phase, can also be used. With such systems, high extraction efficiencies were obtained, permitting large enrichment factors (greater than 500) from environmental water samples and urine samples. With off-line graphite furnace atomic absorption spectrometry, detection limits well below 1 part per billion were obtained with no influences from humic acids or salts.

For background information *see* ANALYTICAL CHEMISTRY; CHROMATOGRAPHY; ELECTROPHORESIS; ION-SELECTIVE MEMBRANES AND ELECTRODES; MEMBRANE SEPARATIONS; SUPERCRITICAL-FLUID CHROMATOGRAPHY in the McGraw-Hill Encyclopedia of Science & Technology.   Jan Åke Jönsson; Lennart Mathiasson

Bibliography. J. Å. Jönsson and L. Mathiasson, Supported liquid membrane techniques for sample preparation and enrichment in environmental and biological analysis, *Trends Anal. Chem.*, 11:106–114, 1992; M. Knutsson et al., Supported liquid membranes for sampling and sample preparation of pesticides in water, *J. Chromatogr. A.*, 757:197–206, 1996; B. Lindegård et al., Automated CLC determination of a basic drug in blood plasma using the supported liquid membrane technique for sample pretreatment, *Anal. Chem.*, 66:4490–4497, 1994; M. Papantoni et al., Trace enrichment of metals using supported liquid membrane technique, *Analyst*, 120:1471–1477, 1995.

Machinability of metals

Recent developments include the use of ceramic tools in high-speed machining of superalloys. An important application is machining of nickel-based superalloys, such as Inconel 718 and Incoloy 901. Nickel-based superalloys are widely used for components with high strength at elevated temperatures, which is a typical requirement for the aircraft industry as well as for gas and steam turbines. However, these superalloys are extremely difficult to cut. The main reasons for the poor machinability of the nickel-based superalloys are their high shear strength and rate of work hardening. The poor thermal diffusivity of these alloys also results in heat concentration in the vicinity of the cutting tip, limiting tool life because of the rapid increase in wear.

Machining of nickel-based superalloys can be achieved by ceramic tools at metal removal rates four times greater than those achieved with carbide tools. The ceramic materials retain their strength up to 1200°C (2200°F), which is necessary for effective machining of these alloys.

Mixed alumina cutting tools. Alumina-based ceramic tools are used in cutting various superalloys. In spite of the fact that alumina ceramic tools have high wear resistance, they are prone to fracture because of their brittle nature. Therefore, particles of zirconia (ZrO_2) are dispersed in the alumina matrix to improve fracture toughness. Also, the addition of either titanium carbide or titanium nitride enhances the resistance of the alumina tools to thermal shock. Ceramic tools fabricated of an alumina-zirconia-tungsten mixture have been used commercially in turning Inconel 718 parts at 244 m (805 ft) per minute. Furthermore, titanium mixed ceramics have been used in turning Incology 901 at 230 m (760 ft) per minute, a feed rate of 0.2 mm (0.008 in.) per revolution, and a depth of cut of 2 mm (0.008 in.). The stability of alumina–titanium carbide tools with respect to wear was attributed to their chemical inertness at elevated temperatures.

A typical worn mixed alumina tool shows a regular abrasion on its faces. The crater wear, however, increased with discontinuous cutting. In the alumina-based ceramics, the main cause of the flank wear is plastic flow under high temperature coupled with high mechanical stresses.

Tool life can be extended through employing modifications in tool geometry. Examples include reducing the cutting-edge angle because of its influence on the chip flow direction, resulting in lowering the cutting forces. In addition, the use of specially prepared round or button inserts to ensure that the cutting edge is under compression increases the resistance to crack initiation.

The alumina–titanium carbide inserts produced values of surface roughness that are typically half of those encountered with carbide tools.

Whisker-reinforced ceramic cutting tools. Ceramic cutting tools reinforced with silicon carbide whiskers represent the most recent developments in machining nickel-based alloys. The combination of low coefficient of thermal expansion and improved thermal conductivity enhances the resistance of the composite to thermal shock as compared to that of the mixed alumina ceramic tools. This allows the use of coolant, which helps prevent workpiece distortion and facilitates the handling of the hot chips.

Reinforcing alumina with silicon carbide whiskers has led to improvements in all properties except chemical inertness. Silicon carbide is chemically reactive with nickel and iron; therefore, in many instances the wear that is observed in these composite tools when machining Inconel was attributed to the mechanism of whisker pull-out. Wear was also attributed to plastic deformation and weakening of the tool due to the diffusion process. Silicon carbide whisker–reinforced ceramic tools are recommended for surface finishing operations because they yield excellent results. In general, the use of sialon and whisker-reinforced alumina inserts in machining nickel-based superalloys offers better overall performance than either conventional or mixed alumina tools.

Tool wear mechanisms. Whisker-reinforced ceramic inserts induce several types of tool wear during end milling of Inconel 718, depending on the range of speeds and feeds. For feeds and depths of cut above 0.2 mm (0.008 in.) per tooth and 2 mm (0.08 in.) respectively, tool breakage may occur.

Round inserts improve the cutting performance in comparison with square ones, because they provide a stronger cutting edge, aiding resistance to notch wear.

Flank wear is predominant along the cutting edge at the cutting speeds, ranging 400–700 m (1300–2300 ft) per minute. It is due mainly to adhesion of the workpiece material on the tool surface followed by grains pull-out. Reducing axial depth of cut, feed, and immersion ratio causes the tool to wear faster. (The immersion ratio is the ratio of the radial width of the cut to the tool diameter.)

The average cutting force decreases with increasing cutting speeds, as well as lower axial and radial depth of cut.

The optimum performance of the whisker-reinforced ceramic inserts when milling Inconel 718 is typically obtained by using speeds of 700 m (2300 ft) per minute or higher, axial depths of cut in the range of 1–2 mm (0.04–0.08 in.), and feeds of 0.1–0.18 mm (0.004–0.007 in.) per tooth. Increasing the immersion ratio results in an improvement in tool life.

During end milling of Inconel 718, increasing the rotational speed of the spindle does not allow the cutting temperature to reach its steady state, which leads to better chip morphology and more stable cutting.

Effects of process variables on tool wear. The effects of process variables on tool wear involve cutting speed, axial depth of cut, feed, immersion ratio, cutting forces, and surface finish.

Cutting speed. Both flank wear and depth-of-cut notch wear decrease as the speed increases from 200 m (660 ft) per minute to 700 m (2300 ft) per minute. For width of cut with greater radius, the tool wear increases for the same cutting speed, because of the increase in the cutting force. Tool flaking on the rake face tends to decrease with an increase in the cutting speed. Since the temperature variation between the exit from the workpiece and another entrance depends on the cooling rate of the tool and the rotational speed of the spindle, it is expected that the higher the spindle speed, the lower the temperature difference and thermal shocks on the tool surface. This explains the improvement of the tool performance for higher cutting speeds. Experimental investigations indicate that temperature levels in interrupted cutting depend primarily on the length of the heating cycle and secondarily on the cooling time between cycles.

Axial depth of cut. Tool wear for round and square inserts is higher for lower axial depths of cut. Of the several factors, probably the most important one is that more heat is concentrated at the insert nose radius, where less cutting tool material is available to dissipate it.

Feed. In general, the increase in feed up to 0.125 mm (0.00492 in.) per tooth reduces the tool wear except for the trailing-edge wear at the speed of 350 m (1160 ft) per minute and the feed of 0.125 mm (0.00492 in.) per tooth. Increasing the feed to more than 0.125 mm (0.00492 in.) per tooth has a limited effect on tool wear. Above a feed of 0.2 mm (0.008 in.) per tooth, the cutting edge breaks because of the mechanical loads.

Immersion ratio. Tool wear is minimal for full immersion ratio. Flank wear is hardly affected by the change in the immersion ratio. However, depth of cut notch at 0.25 immersion ratio is twice that obtained at full immersion ratio. Trailing-edge wear is highest for 0.5 immersion ratio. These results are expected, since the tool is exposed to high mechanical shock load at its exit from the workpiece at low immersion ratios.

Cutting forces and surface finish. During milling of Inconel 718 using whisker-reinforced ceramic tools, the average cutting force increases with increasing feed, width, and depth of cut; however, cutting forces decrease with increasing cutting speed. The reduction of cutting forces with the increase in cutting speed is related to cutting temperature. Surface finish of 0.7 micrometer can be obtained when a cutting speed of 1000 m (3300 ft.) per minute, a feed of 0.2 mm (0.008 in.) per tooth, a depth of cut of 1.5 mm (0.059 in.), and full immersion ratio are used.

For background information *see* CERAMICS; COMPOSITE MATERIAL; HIGH-TEMPERATURE MATERIALS; MACHINABILITY OF METALS; MANUFACTURING ENGINEERING; WEAR in the McGraw-Hill Encyclopedia of Science & Technology.   M. A. Elbestawi

Bibliography. M. A. Elbestawi et al., Performance of whisker-reinforced ceramic tools in milling nickel-based superalloy, *Ann. CIRP*, 42:99–102, 1993; R. Komanduri and T. A. Schroeder, On shear instability in machining a nickel-iron base superalloy, *ASME J. Eng. Ind.*, 108:93–100, 1986; B. M. Kramer and P. D. Hartung, Theoretical considerations in the machining of nickel alloys, in F. Gorsler (ed.), *Cutting Tool Materials*, 1981; A. R. Thangaraj and K. J. Weinmann, On the wear mechanisms and cutting performance of silicon carbide whisker–reinforced alumina, *ASME J. Eng. Ind.*, 114:301–307, 1992.

Machining

Significant advances in the development of cermet materials for applications in the cutting-tool industry have been made. During the 1960s, several companies began to market cermet inserts with improved toughness with the addition of tungsten carbide (WC) and tantalum carbide (TaC). The additional elements produced a more stable cutting tool; however, the mechanical structure provided levels of toughness well below those of traditional cemented carbide materials. Applications of cermet materials was restricted to light finishing cuts with light feed rates. Such a restriction adds an expense to the manufacturing process and limits that expense to a single operation in the process. The expense would, in most cases, outweigh the value of the cermet cutting tool.

Problems and solutions. The problems with the use of cermet material were mainly in general applications. If an application involved interrupted cutting, the lack of toughness in the cermet material made it vulnerable to catastrophic failure. Because of the hardness of the cermet cutting tool and the lack of toughness in interrupted cutting, the cermet was subject to microcracking and microchipping. The extent of these cracks on the exterior of the cermet tool is not visible to the naked eye; thus, the machine operator could very well think that the tool was usable. When the tool would enter the cut in this damaged condition, a catastrophic failure resulted.

This type of failure had the potential for damaging the workpiece and the machine tool. This perception of cermet material all but eliminated its use in the machining industry; however, cermet had enough merit as a cutting-tool material because of its strength at high temperatures that cutting-tool manufacturers continued research and development to overcome its weaknesses. The merit of a cermet tool is that it leaves a surface finish that could replace an additional grinding operation. In a machining operation, cermets generally leave a smooth, mirrorlike finish; in some material a prismlike rainbow coloring is visible.

In the 1970s it was found that a denser cermet material with a considerably finer microstructure could be produced with the addition of titanium nitride (TiN). This increased density improved the cermet, making it more resistant to oxidation and increasing its strength at high temperatures (hot hardness). The finer the microstructure in a cutting-tool material, the better is the tool's ability to dissipate heat and withstand shock. Heat dissipation to the larger mass of the cutting tool helps protect the smaller mass at the cutting edge and prevents edge deformation.

Improved characteristics. The new cermet material is generally composed of titanium carbide (TiC), TiN, WC, tantalum (Ta), nickel (Ni), cobalt (Co), molybdenum (Mo), and vanadium (V). The binder material, comprising Ni and Co, holds the other elements together in a phase that enhances the strength of each element; it must be able to bond with the additional elements. Titanium carbide gives additional hardness when it is combined in a matrix, and it is added to carbide cutting tools to impart strength and wear resistance. In cermet tools, adding TiC to the matrix as the hard component achieves the same results as in carbide cutting tools and stabilizes the cermet material both chemically and thermally. The amount of TiC in cermets is generally more than that in carbide cutting tools.

In comparison with other cutting-tool materials, the new cermet material ranks well within the cemented carbide toughness area and above the high-temperature hardness area (see **illus.**); it is now positioned near the center of the group of cutting-tool materials. Cermet as a cutting-tool material offers an excellent balance, and its properties place it in the same group as carbide and coated-carbide grades. Positioning of the new cermet material in the ranges of carbide and coated carbide indicates that the material is more resistant to the failure mechanisms that hindered it in the past. Cermet material is classified as a general-purpose type of cutting-tool material, and applications probably will be extended. Cermet material can be used in interrupted cutting with fewer instances of catastrophic failure; it can be used in milling applications (continuous interrupted cutting) in which the earlier materials would surely have failed. The new cermet material maintains the ability to provide the same type of smooth surface finishes as its predecessor, thus earning a place in both the general and finishing applications in manufacturing.

Coated-cermet cutting tools. Physical vapor deposition (PVD), a low-temperature coating process used in the cutting-tool industry, has been used to improve the performance of the cermet material. The low temperatures characteristic of physical vapor deposition have eliminated the danger of changing the structure or shape of the finished cutting tool. There are several types of coating available for cermet tools, including TiN, TiC, titanium aluminum nitride (TiAlN), and a combination of TiC and TiN. Most common is the TiN coating, which is gold in color. A TiN coating on the cermet tool provides a protective barrier between the material being removed and the cutting tool. This coating provides better lubricity that reduces the contact time between the material being removed and the tool, so that the amount of heat generated in the machining process is reduced. The lower the heat in a cutting-tool operation, the better the edge integrity and the better the resistance to catastrophic failure.

Chemical vapor deposition (CVD), a process that requires higher temperatures, is still being investigated for use with the cermet material. The heat requirements of chemical vapor deposition are much higher and can cause a structural change in the matrix. A change in any portion of the matrix can lead to catastrophic failure of the entire cutting tool; however, in this type of coating process multiple layers of different coating materials can be deposited on cutting-tool material, providing additional protection to the cutting tool. It is anticipated that when this process is perfected it will impart greater protection and improved performance to the cermet material.

High-speed machining. Changes in the manufacturing environment over the past few years have changed the focus of development by cutting-tool suppliers. An increased rate of metal removal is required, accompanied by a decreased need for heavy machining resulting from the new technology of near-net-shape manufacturing. In this technology, a precision casting or forging is used to produce the finished product, and the amount of machining needed to accomplish this task is thereby reduced to light depths of cut and minimal material removal.

Requirements for increased production have led to development of cutting-tool materials that will not lose the integrity of the cutting edge. One way of increasing production is to increase the rate of metal removal. A major reason for cutting-tool failure is a high-temperature environment. When the cutting speed is increased, the temperature at the cutting edge of the tool increases rapidly. Increased temperatures will press the limits of the hot hardness of carbide cutting tools. With carbide cutting tools, the limits of temperature in the metal-removal process are much lower than those of cermet cutting tools. For example, the hardness of carbide is much lower at 900°C (1650°F) than that of cermet. The increased hardness of the cermet cutting tool will

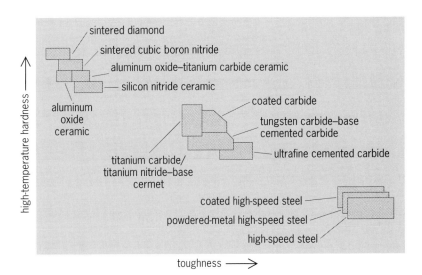

Comparison of characteristics of high-temperature hardness and toughness of various materials used to fabricate cutting tools. High-speed steel is produced by casting or forging. Powdered-metal high-speed steel is a type of processing used to produce a stronger, structured product. Cubic boron nitride was developed for hard turning, which can replace a finish grinding operation in some processes. *(After R. Biagiotti and R. Macek, The time is right for cermets, Cutting Tool Eng., pp. 24–30, March 1996)*

allow it to perform at elevated cutting speeds with better edge integrity, thus reducing the amount of time required to complete the machining process. Adding the coating to the cermet by means of physical vapor deposition will also increase the performance.

Tool life. Production can be measured by increasing the amount of time that a cutting tool will retain its cutting edge integrity. Tool-change downtime in a production run can be critical to the production requirements. In most operations the cermet cutting tool will retain a keen cutting edge long after the carbide cutting tool's edge has degraded. An additional benefit from the use of cermet cutting tools is that with a better edge integrity the finish obtained in most cases will eliminate the need for a second carbide tool for the finish requirements. An application report for machining AISI 1043 steel showed that the cermet cutting tool outperformed the carbide cutting tool by over 300%. In other words, a carbide cutting tool would have to be replaced three additional times to equal the performance of the one cermet tool. This represents a significant savings for the manufacturer.

Near-net-shape manufacturing. The amount of material to be removed from a part in a machining operation determines the amount of time needed to complete the part. With near-net-shape manufacturing, which requires minimal removal of stock, cermet material has found an increase in the number of suitable applications. Technological advances in the forging and casting industries permit production of raw products that require minimal material removal during machining operations. Near-net-shape manufacturing will benefit from the capabilities, longer tool life, and higher machining speeds of cermet cutting tools.

Research goals. Although there have been many advances in cermet material development, the investigation of new materials continues. Increased heat resistance and material toughness will remain a challenge for the cutting-tool industry. The use of cermet cutting tools is increasing but has not yet reached its potential. It is anticipated that eventually cermets will serve as primary cutting-tool materials.

For background information *see* CERMET; MACHINABILITY OF METALS; MACHINING; METAL COATINGS; VAPOR DEPOSITION in the McGraw-Hill Encyclopedia of Science & Technology.   Robert M. Macek, Jr.

Bibliography. R. Biagiotti and R. Macek, The time is right for cermets, *Cutting Tool Eng.*, pp. 24–30, March 1996; T. Kainuma, *Advanced Cermet Technology*, Training Man. DS001, Mitsubishi Materials Corp., November 1991.

Macroevolutionary dynamics

Critically important to ecology, evolutionary biology, and paleobiology is the problem of scale, that is, the idea that spatial and temporal scales of observation determine the patterns that emerge from the data. The number of trilobite species in a particular Paleozoic deposit, for example, may fluctuate in a directionless manner over short geological intervals, belying a long-term, global diversity trend. If the apparently random fluctuations observed on small spatiotemporal scales are mere noise within larger-scale, deterministic patterns, data cannot be scaled up from real-time ecological observations to predict patterns in the fossil record. If, however, at least some patterns and processes are scale independent, ecology has the potential to predict macroevolutionary dynamics.

Predation and shell form. Patterns in the evolution of skeletal morphology in marine invertebrates are strongly linked to the process of predation. Observations of fossil and modern marine faunas show that increased predation pressures from durophagous (shell-crushing) fish and crustaceans produce patterns of increased defensive architecture in hard-shelled invertebrate prey that are similar on multiple scales of time and space. Gastropod mollusks provide the best example of these scale-independent effects.

The shell-crushing abilities of living, predatory, teleostean fishes and decapod crustaceans increase from the temperate zone to the tropics and from the Caribbean to the tropical Indo-Pacific. Defensive features of gastropod shells, including heavy calcification, tight coiling, spines, ribs, low spires, and narrow apertures, increase along the same latitudinal and longitudinal gradients. In an analogous fashion, shell-breaking teleosts and decapods diversified globally on a macroevolutionary temporal scale of tens of millions of years beginning in the Mesozoic Era. The radiation of these predatory forms likely caused the accompanying macroevolutionary trend of increased defensive architecture in gastropods,

as taxonomic groups with well-defended shells replaced those less well defended. Defensive architecture also increased in a variety of other marine invertebrates, including bivalves, cephalopods, and crinoids. Smaller increases in both the capabilities of crushing predators and the defensive morphologies of prey occurred in the mid-Paleozoic, when the functionally new predators were placoderm and chondrichthyan fishes, and the skeletonized prey that responded evolutionarily included gastropods, cephalopods, crinoids, and brachiopods.

The same covariation of predation pressure and shell form can be found at smaller evolutionary scales of millions of years and hundreds of kilometers. Gastropod species endemic to Lake Tanganyika in Africa possess thick, heavily ornamented shells. These shells look like marine forms; they are very unlike other fresh-water gastropod shells, which are almost always thin and unornamented. The marine-style architectures of the Tanganyikan snails are an evolutionary consequence of predation by an endemic crab species with crushing claws that are unusually powerful compared to other fresh-water crustaceans. The evolution of defensive architecture in these snails could not have taken longer than 7 million years, the approximate age of Lake Tanganyika.

On even smaller, ecological scales of decades to centuries and kilometers to tens of kilometers, among sites along marine coastlines and within sites through time, increased development of antipredatory architectural features in gastropods is again linked to the presence of durophagous predators. Studies that include experimental, genetic, and developmental work show that the mechanism underlying the rapid appearance of defensive morphologies on such small scales is the developmental, or phenotypic, plasticity of the prey. In one set of experiments, the mere scent of predatory crabs caused growing snails to develop defensive features on their shells.

Gastropods display analogous morphological responses to increased durophagous predation on multiple spatiotemporal scales. Individuals within populations respond to small-scale, short-term predation increases through phenotypic plasticity. Larger-scale responses occur in the form of evolutionary differences at the species and genus levels, as in Lake Tanganyika, or at the largest scales, through the replacement of higher-level taxonomic groups, as in the macroevolutionary events of the Mesozoic. At all scales, however, these trends remain rooted in individual interactions between prey and predators.

Abundance of prey. Increasing predation does not guarantee the appearance of defensive architecture. A more obvious consequence of increasing predation is a decline in prey abundance. For example, dense populations of ophiuroid echinoderms are limited by predators in similar ways on multiple scales.

On small, ecological scales, predatory teleosts (particularly wrasses) and crabs limit the distribution of brittlestar beds in shallow subtidal habitats of the

Predatory seastar surrounded by suspension-feeding ophiuroids in a brittlestar bed at a depth of 90 ft (27 m) off the Isle of Man, U.K. The seastar is approximately 6 in. (15 cm) in diameter. *(Courtesy of Richard B. Aronson)*

British Isles. Brittlestar beds, in which the ophiuroids are packed in densities of hundreds to thousands per square meter, live in the open on sandy to silty sediments in certain areas around Britain and Ireland. These dense populations contrast sharply with populations in rocky reefs only meters away, where the brittlestars are relatively rare and hide in crevices.

Predatory fish and crabs are abundant in rocky reef habitats, and they rapidly attack ophiuroids tied to lengths of thread and staked out as experimental bait. Fish and crabs are responsible for 75% of attacks on tethered ophiuroids in rocky reef environments, and seastars account for the remaining 25%. The situation is completely different in brittlestar beds on nearby soft-sediment, level bottoms. First, staked-out ophiuroids are attacked at one-quarter the rate that they are on rocky reefs. Second, fish and crabs are rare and account for 25% of the few attacks on the experimental prey, with seastars accounting for 75% (see **illus.**).

Dense, well-preserved assemblages representing ancient brittlestar beds are found throughout the fossil record. Different ophiuroid species formed brittlestar beds at different times in the past, and the extreme rarity of bitten-off and regenerating arms in the fossils suggests low predation pressure in Paleozoic, Mesozoic, and Cenozoic populations. Yet, while predation remained low through time within this type of population, the number of these low-predation brittlestar beds declined dramatically in shallow water during the Mesozoic, at the same time that predatory fish and crabs were radiating. Seastars and other slow-moving invertebrates, which had been important predators in Paleozoic shallow-water communities, were supplanted by fast-moving, skeleton-crushing teleosts and decapods. Thus, living brittlestar beds provide an ecological window on individual predator–prey interactions, through which paleontologists can understand Paleozoic community structure and interpret the macroevolutionary dynamics of community replacement in the Mesozoic.

Onshore–offshore dynamics. Increased durophagous predation in the Mesozoic was part of a larger macroevolutionary pattern of faunal replacement from onshore to offshore habitats. The newly important predators of the Mesozoic, teleosts and decapods, originated in nearshore, shallow-water environments. They forced changes in gastropod shell architecture and eliminated dense populations of stalked crinoids (sea lilies) and brittlestar beds. Novel predatory groups expanded their influence offshore, progressively altering prey morphologies, and the communities in which those prey lived, along an onshore-to-offshore gradient.

There is an analogous trend in the ecological-scale consequences of human fishing activity in the Gulf of Maine off the east coast of North America. Fishing began in coastal waters several centuries ago and progressed farther offshore as nearshore stocks were depleted. The decimation of Atlantic cod and other bottom-feeding fishes nearshore reduced predation pressure on benthic invertebrates. As a result, sea urchins now live in more dense populations and are larger and more active in coastal hard-bottom habitats than in comparable less-fished habitats 60 mi (100 km) offshore. Intense grazing by urchins nearshore has produced large areas devoid of fleshy macroalgae (seaweeds); these urchin barrens are dominated by calcified algal crusts. Offshore, high-predation areas have fewer sea urchins and are more uniformly covered by kelp and other seaweeds at the same depths. Fishing people, who are novel top predators, are transforming modern teleost-macroalgae communities into postmodern human-urchin communities. This transformation is happening in ecological time, and it is occurring along the same onshore–offshore gradient that originally produced those teleost-dominated communities over macroevolutionary time, beginning in the Mesozoic. Once again, ecological processes are the basis of predation-driven macroevolutionary patterns.

Scale dependence and independence. The effects of predation are scale independent, at scales ranging from the ecological to the macroevolutionary. However, other types of processes and the patterns they produce are scale independent. Paleontologists have focused a great deal of attention on extinction.

There is substantial evidence that certain biological rules determine which taxonomic groups normally resist going extinct in a given interval of geological time. Features that promote the survivorship of a taxon include high species diversity and broad geographic range. However, these rules may not apply during mass extinctions, when survival may be governed by different rules. Whether such rule changes are a common feature of mass extinctions, or whether a mass extinction is merely the sum of many individual, normal extinctions, remains a subject of controversy in macroevolutionary studies. Regardless of the outcome of this debate, macroevolutionary trends in prey morphology and distribution, grounded in the ecology of individual predator–

prey interactions, transcended even mass extinctions and have resulted in present-day communities.

For background information *see* ADAPTATION (BIOLOGY); ANIMAL EVOLUTION; FOSSIL; MACROEVOLUTION; ORGANIC EVOLUTION; SPECIATION in the McGraw-Hill Encyclopedia of Science & Technology.

Richard B. Aronson

Bibliography. R. B. Aronson, Scale-independent biological processes in the marine environment, *Oceanography and Marine Biology: An Annual Review*, 34:435–460, 1994; M. L. McKinney (ed.), *Biodiversity Dynamics*, 1998; R. M. Ross and W. D. Allmon (eds.), *Causes of Evolution*, 1990; G. J. Vermeij, *Evolution and Escalation*, 1987.

Magnet

Since the mid-1980s, a number of magnets based on molecules and organic polymers and displaying a wide variety of magnetic phenomena have been discovered.

Principles of magnets. Lodestone, a magnetic material, has been known for 3500 years. Magnets have become increasingly important during the twentieth century, initially with the increasing use of electricity (in motors, transformers, and so forth) and subsequently with their use in information technology (in magnetic memories, and so forth). Modern metallurgically prepared magnets are based on transition metals with unpaired electrons in d orbitals [such as iron, chromium dioxide (CrO_2), and iron(III) oxide (Fe_2O_3)] and rare-earth metals with unpaired electrons in f orbitals (such as $SmCo_5$, $Co_{17}Sm_2$, and $Nd_2Fe_{14}B$).

Magnets work because of the net alignment of magnetic moments due to the spin associated with each of the unpaired electrons. (Each unpaired electron contributes a spin of ½.) If all unpaired spins in the solid align in a parallel fashion, a ferromagnet results. If they align so as to cancel, there is a net zero magnetic moment and an antiferromagnet results. Partial cancellation of spins, as in a ferrimagnet or canted ferromagnet, leaves a net magnetic moment. When the spins couple, it is nearly always through quantum-mechanical exchange. Two mechanisms of quantum-mechanical exchange have been identified as important for organic- and molecule-based magnets: spins in orthogonal orbitals (related to Hund's

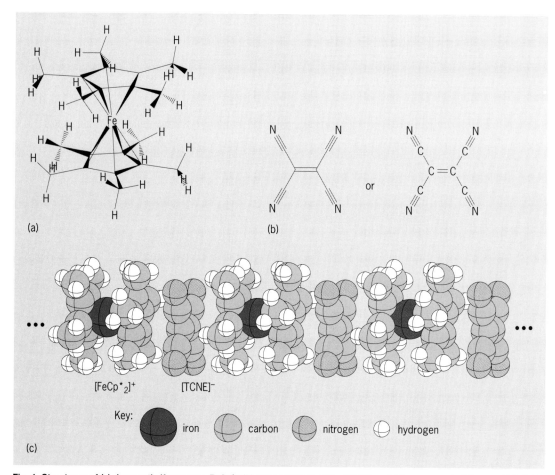

Fig. 1. Structures of (a) decamethylferrocene, $FeCp^*_2$; (b) tetracyanoethanide, TCNE; and (c) a segment of the parallel chain of decamethylferrocenium tetracyanoethanide, $[FeCp^*_2]^+[TCNE]^-$.

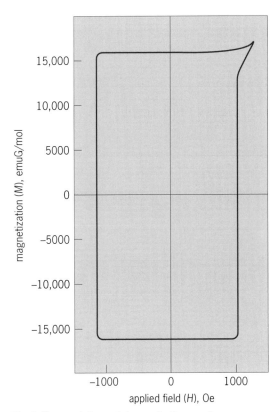

Fig. 2. Hysteresis loop of decamethylferrocenium tetracyanoethanide, [FeCp'$_2$][TCNE]. 1 oersted (Oe) = 80 A/m. 1 gauss (G) = 1.26 × 10^{-3} Wb/m^2.

transition-temperature systems. Structural disorder is sometimes present, often due to solvent incorporated into the structure. This disorder, if sufficiently strong, results in the unusual behavior associated with spin glasses, for which the relative orientations of nearby spins is nearly frozen but there is no long-range ordering of the spin orientations. Spin glasses feature a variety of time- and frequency-dependent magnetic phenomena.

The combination of magnetic properties with the customization and processing capabilities of molecules and polymers leads to the possibility of eventual use of these materials in novel technologies ranging from lightweight cores for transformers to integration with electronic and photonic applications as smart materials. For example, the polymer magnet V(TCNE)$_x$ · y(CH$_2$Cl$_2$) [TCNE = tetracyanoethanide] is as an effective magnetic shield at room temperature.

Pure organic ferromagnets. Ferromagnets composed only of elements in the first row of the periodic table can operate only at very low temperatures, less than 2 K ($-456°$F). These ferromagnets generally are based on nitroxides. The first to be reported, in 1991, was the beta (β) phase of 4-nitrophenyl nitronyl nitroxide, with an ordering temperature of 0.6 K ($-458.6°$F). The pure organic ferromagnet with the highest transition temperature, 1.48 K ($-457.0°$F), is also nitroxide based.

Mixed magnets. Molecule- and polymer-based magnets relying on spins in both the p and d orbitals have a wide variety of embodiments. One of these, reported in 1985, is the first molecule-based magnet to display a hysteresis loop: decamethylferrocenium tetracyanoethanide, [MCp'$_2$][TCNE] (M = Fe^{III}, Cp'$_2$ = dipentamethylcyclopentadienide; **Fig. 1**). Its hysteresis loop is shown in **Fig. 2**. Many modifications of this electron transfer salt have led to other magnets, for example, replacement of Fe^{III} (spin S = ½; transition temperature T_c = 4.8 K = $-451.0°$F) by Mn^{III} (S = 1, T_c = 8.8 K = $-443.8°$F), Cr^{III} (S = ³⁄₂, T_c = 3.5 K = $-453.4°$F), and Co^{III} (S = 0, with no ordering, due to interruption of the spin chain by a non-spin-containing unit). A number of analogous compounds wherein the TCNE is replaced by TCNQ (tetracyanoquinodimethanide) have been reported, especially [FeCp'$_2$][TCNQ].

While the spin-containing TCNE anion radical has a cofacial packing arrangement with the [MCp'$_2$] cation radicals (Fig. 1), it bridges between two metals in another family of molecule-based magnets, in which spin-2 (that is, with four unpaired electrons) metalloporphyrin cation radicals alternate with spin-½ TCNE. The structure of this class is typified by that of [MnTPP][TCNE] (TPP = *meso*-tetraphenylporphinato; **Fig. 3**). The spins on adjacent cation and anion radicals within a chain are strongly coupled antiferromagnetically via quantum-mechanical exchange. As a result, the isolated chains in this class of materials act as one-dimensional ferrimagnets. While this material undergoes a transition at approximately 14 K ($-435°$F) to a three-dimensional (canted)

rule) and configurational admixture of states (where the electrons associated with the important spins temporarily occupy other energy states that bring them in stronger contact with other spins). A third, classical mechanism for spin coupling has been identified: dipolar interactions between the spin-containing moieties, that is, where the weak magnetic fields generated by the individual magnetic moments cause spin alignment.

Molecule- and polymer-based magnets. Initially, organic-polymer- and molecule-based materials were not thought to be magnets. However, in the mid-1980s molecule-based magnets were discovered—the first examples containing stable organic species with unpaired spins that exhibit hysteresis (that is, are magnetic in the absence of an applied magnetic field), and also have transition temperatures to the ordered magnetic state above the boiling temperature of liquid helium. The discovery of polymer magnets, with transition temperatures above room temperature, followed in 1991. These discoveries demonstrate that nonmetallurgically prepared organic magnets can be fabricated. Since the mid-1980s, increasing numbers of families of molecule- and polymer-based magnets have been discovered. These magnets display a wide range of magnetic phenomena ranging from quasi-one-dimensional effects to the ramifications of random anisotropy and exchange in the three-dimensional coupled high-

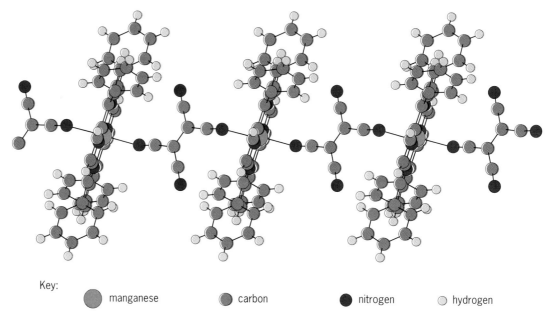

Key: manganese   carbon   nitrogen   hydrogen

Fig. 3. Segment of a one-dimensional chain of [MnTPP][TCNE].

ferromagnetic state, the similarly structured [MnOEP][HCBD] (OEP = octaethylporphyrinato; HCBD = hexacyanobutdiene), with similar one-dimensional ferrimagnetic behavior at high temperatures, is an antiferromagnet below 19.6 K (−422.4°F). The dominant role of interchain classical dipole-dipole (through space) interactions in the former compound, due to the greater separation of the chains reducing the quantum-mechanical exchange, leads to these differences.

The reactions of some nitroxides with Mn(hfac)$_2$ (hfac = hexafluoracetylacetonate) have led to ferrimagnets with transition temperatures as high as 46 K (−377°F). Meanwhile, molecule-based magnets have been prepared that solely use d orbitals of differing spin magnitudes arising from different nearest-neighbor divalent metal ions. Examples include (NBu$_4$)$_2$. Co$_2$[Cu(opbo)]$_3$ · 2H$_2$O [Bu$_4$ = tetra-n-butyl, opbo = o-phenylenebis(oxamato)], which has a transition temperature of 32 K (−402°F) and a large hysteresis effect with a coercive field of 4 × 10^5 A/m (5000 oersteds).

Raising transition temperatures. The transition temperatures of molecule- and polymer-based magnets are sensitively related to the dimensionality of the magnetic interactions. That is, materials with strong magnetic interactions along one (the chain) direction and weak magnetic interactions in the perpendicular directions have a transition temperature proportional to the geometric mean of the in-chain and out-of-chain magnetic interactions. Hence, the transition temperatures for the materials discussed above tend to be very low. A key approach to raising the transition temperature to room temperature or above is having strong magnetic coupling in all three directions between spins. For this case, the transition temperature is proportional to the strength of the

magnetic interactions, often leading to much higher transition temperatures. Such occurs in V(TCNE)$_x$ · y(CH$_2$Cl$_2$) and Prussian blue materials. Examples of each have transition temperatures above room temperature.

The V(TCNE)$_x$ · y(CH$_2$Cl$_2$) has been prepared from a variety of solvents, with that prepared from methylene chloride being the one with the highest transition temperature. The transition temperature is estimated to be 400 K (260°F), although the thermal decomposition temperature is 350 K (170°F). Disorder plays an important role in the magnetism of this class of materials.

Prussian blue, Fe^{III}$_4$[Fe^{II}(CN)$_6$ · xH$_2$O], possesses a three-dimensional structure comprising Fe^{II}-CN-Fe^{III} linkages in all three dimensions. This compound magnetically orders at 5.5 K (−450°F). Through substitution with differing metals with differing number of spins, both ferromagnets and ferrimagnets have been prepared. For example, CsNi[Cu(CN)$_6$] is a ferromagnet below 90 K (−298°F), while V[Cu(CN)$_6$]$_{0.86}$ is ferrimagnetic below 315 K (107°F).

For background information *see* MAGNETIC HYSTERESIS; MAGNETIC MATERIALS; MAGNETISM; MOLECULAR ORBITAL THEORY; ORGANIC CONDUCTOR in the McGraw-Hill Encyclopedia of Science & Technology.

Arthur J. Epstein; Joel S. Miller

Bibliography. D. Gatteschi, Molecular magnets: A basis for new material, *Adv. Mater.*, 6:635–645, 1994; O. Kahn, *Molecular Magnetism*, 1993; M. Kinoshita, Ferromagnetism in organic radical crystals, *Jap. J. Appl. Phys.*, 33:5718–5733, 1994; J. S. Miller and A. J. Epstein, Designer magnets, *Chem. Eng. News*, 73(40):30–41, 1995; J. S. Miller and A. J. Epstein, Organic and organometallic magnetic materials: Designer magnets, *Angew. Chem. Int. Ed.*, 33:385–415, 1994.

Maintainability of systems

The importance of equipment maintenance in ensuring product quality and plant safety has long been recognized. More recently, industry has begun to understand the impact of maintenance practice on broader elements of manufacturing productivity, such as factory output and product cycle time, as well as on product yield. Rising maintenance costs as a percentage of total operating costs, as well as market pressure to produce competitively priced products, have generated increased awareness and interest in equipment maintenance in industries as diverse as semiconductors and electric utilities. In such equipment-intensive industries, even small improvements in equipment utilization can result in an immediate competitive advantage. Maintenance technology has improved dramatically in the last decade with the development of sophisticated sensors that allow for on-line monitoring of critical process and equipment parameters. At the same time, new operational analysis tools that enable better decision making at the planning and operational stages have become available.

Maintenance systems and operations. The purpose of maintenance is to ensure that equipment is in proper operating condition so that it can perform its intended function when needed. Traditionally, the focus of maintenance has been on equipment availability, the ratio of total time minus downtime to the total time. Equipment maintenance was performed in either a reactive mode (that is, upon detecting an equipment failure or shutdown) or in a preventive mode (that is, at regularly scheduled intervals, without specific knowledge of the level of equipment degradation). Both modes can lead to ineffective and costly maintenance strategies: the former, to unplanned operational disturbances, and the latter, to unnecessary repairs or replacements. Recent developments in operational strategy have focused on predictive (or condition-based) maintenance, that is, on methods that assess equipment degradation, predict when failures will likely occur, and initiate appropriate maintenance activities to avert imminent failure. Predictive maintenance programs seek to control maintenance activities in order to avoid unplanned equipment outages and unnecessary maintenance and overhauls. Moreover, modern maintenance practice is focused on increasing equipment effectiveness, that is, ensuring that equipment is both available and capable of producing products of superior quality.

Equipment degradation monitoring. In order to accurately predict equipment failures, new methods for monitoring equipment degradation are being developed that combine use of sophisticated sensors with data-driven models that characterize equipment condition. The new technologies for monitoring critical parameters include acoustic sensors for vibration analysis, sensors for oil-wear analysis, infrared thermography for detecting problems in active electrical and mechanical systems, ultrasound for detecting incipient cracks and pressure or vacuum leakage, fiber optics for strain-gage measurements, high-speed video to examine wear in moving parts, and laser alignment to examine the alignment of moving parts at machining stations and installation stations. For example, a sensor has been developed to analyze transformer fault gases for power-delivery applications. This sensor, which uses a microelectronic device, is inserted directly into transformer oil and measures four key gases associated with fault currents in transformers. It allows for on-line, in-place information on present conditions as well as significant trends, and it obviates production losses necessitated by external oil sampling and analysis. On-board equipment sensors are designed for use under a variety of conditions, including extreme temperatures and vibrations. Coupled with models of equipment degradation from failure physics (which analyzes the actual physical mechanisms that lead to part or equipment failures), these sensoring techniques enable fast, safe, and accurate failure prediction as well as the identification of the root cause of a failure.

Industry initiatives. Two well-publicized industry-led initiatives have revolutionized the maintenance field over the last decades: reliability-centered maintenance and total productive maintenance.

Reliability-centered maintenance. This concept was developed in the United States in the late 1960s by the aerospace industry, and it has been widely adopted in the passenger airline industry and nuclear power industry. Reliability-centered maintenance involves the use of reliability methods (such as failure mode, effects, and criticality analysis) to identify and prioritize component failures in a systematic and structured way. Key equipment components are then selected and monitored, and appropriate maintenance actions are determined for all components based on cost, safety, and environmental and operational requirements. The rationale behind reliability-centered maintenance is that most recommendations for fixed time-in-service maintenance intervals supplied by vendors are overly conservative and are applied to individual components without consideration of the functions of the overall system. The focus of reliability-centered maintenance is on failure triage, that is, on identifying and avoiding machine conditions that are likely to be the most serious and to lead to expensive equipment failures. For critical equipment, reliability-centered maintenance replaces fixed maintenance intervals with maintenance intervals that depend on usage and condition, as determined by analysis of past performance and machine diagnostics. For noncritical equipment, cost analysis is performed to determine an appropriate balance between reactive and preventive maintenance. Maintenance tasks are not performed unless they affect system availability, safety, or overall cost. Companies that have introduced reliability-centered maintenance report significant savings in maintenance costs and improved effectiveness of equipment. Moreover, data collected as part of such a program are valuable in guiding ongoing maintenance improvements.

Total productive maintenance. This industry initiative, developed in Japan in the 1980s, has been adopted by many industries, including semiconductor manufacturers. The development of the concept stems from the recognition that production level, cost, inventory, safety, and product quality depend directly on equipment performance. The goals of total productive maintenance, namely zero unplanned downtime, zero equipment-caused product defects, and zero loss of equipment speed, are approached via a strategy of continuous improvement that involves all company workers. Managers, engineers, production operators, and maintenance personnel are organized into small autonomous groups and charged with maintaining a single machine or set of machines. Machine operators are assigned certain routine maintenance tasks, such as cleaning, lubrication, calibration, and bolt tightening; while maintenance personnel have responsibility for larger, less frequent tasks, such as overhauls and component replacements. Their activities are directed toward five main categories of equipment losses that reduce equipment effectiveness: setups and adjustments, equipment failures, idling and minor stoppages, reductions in equipment speed, and process defects. The approach of total productive maintenance prescribes a data collection strategy that ensures a consistent set of measurements and allows groups to prioritize improvements.

As the relationship between maintenance of equipment and plant productivity has become better understood, interest has focused on integrating production and maintenance decisions. While the initiatives involved in total productive maintenance have been successful in reducing variability in processing time due to unplanned equipment downtime, maintenance activities are still largely scheduled independently of the production status (the amount and location of work) of the plant. As a result, planned maintenance activities still have the potential for significant disruptions in product flow. Research is under way to develop strategies that execute maintenance activities opportunistically, that is, when production status allows, subject to achieving certain performance objectives. The performance objective may involve meeting a targeted equipment availability or minimizing total expected production and maintenance costs. The long-term goal of this approach is to develop controllers that schedule appropriate production/maintenance actions based on the combined input of a shop-floor control system and condition-monitoring sensors.

Life-cycle equipment considerations. It has been estimated that at least 50% of total life-cycle costs of equipment are attributable to operations and maintenance. As customers recognize that their options are to pay now (higher purchase price) or pay later (increased maintenance costs), maintenance considerations are becoming increasingly important at the design phase. Equipment buyers have begun to require better information on time to failure and repair. Suppliers have responded by including such things as analysis of failure modes and effects, statistical information on failure times, cost-effective maintenance procedures, and better customer training with their products. In addition, many equipment manufacturers offer optional design enhancements to improve machine maintainability; these include built-in diagnostics, greater standardization and modularity, and improved component accessibility. While these enhancements may add to the purchase price of equipment, in most cases this higher initial cost is more than offset by reduced operating and maintenance costs over the equipment's life cycle.

For background information *see* MAINTAINABILITY OF SYSTEMS; RELIABILITY (ENGINEERING); SYSTEMS ENGINEERING in the McGraw-Hill Encyclopedia of Science & Technology.   Georgia-Ann Klutke; Martin A. Wortman

Bibliography. J. Douglas, The maintenance revolution, *EPRI J.*, 20(3):6–15, May–June 1995; B. Maggard and D. Rhyne, Total productive maintenance: A timely integration of production and maintenance, *Prod. Invent. Manag. J.*, 33:6–10, 1992; S. Nakajima, *Introduction to Total Productive Maintenance*, 1988; A. M. Smith, *Reliability-Centered Maintenance*, 1993.

Manufacturing

As markets reach their saturation limits for many products and as customers grow more demanding, it becomes necessary for manufacturing industries to produce customized products of high quality at lower cost and quicker pace. Modern manufacturing industries are trying to achieve these goals with the help of technologies such as flexible manufacturing systems, computer-aided manufacturing, and computer-integrated manufacturing, and with information and communication systems. As a key component of these technologies, on-line process monitoring makes it possible to build integrated, adaptive, and high-speed manufacturing systems capable of operating with minimal human supervision to produce high-quality products at reduced cost.

Nature of process monitoring. Process monitoring (or manufacturing-process monitoring) identifies characteristic changes of a process without interrupting the normal operation by evaluating the process signatures. Process monitoring usually precedes diagnosis and control activities (see **illus**.). Diagnosis refers to the identification of causes for the change or failure of the process. Control refers to the predetermined or corrective action necessary to achieve the desired process behavior that will maintain workpiece quality, high production rate, and close dimensional tolerances.

Some of the possible sensors employed for manufacturing-process monitoring detect force, deflection, acceleration, temperature, pressure, acoustic

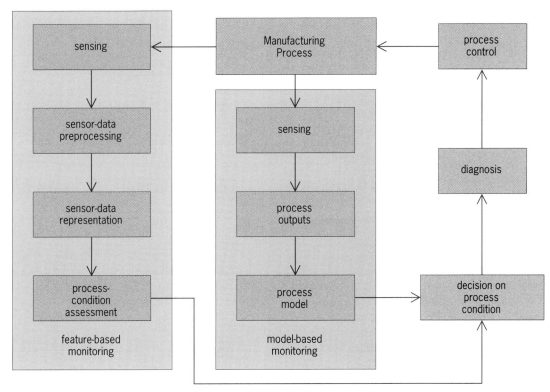

Steps in manufacturing-process monitoring.

emission, electric current or voltage, and optical signals. In practice, the major criteria for selecting sensors are noninterference with the normal operation, quick response to changing process conditions, cost, reliability, effectiveness, and signal-to-noise ratio.

Process-monitoring approaches. Process montoring works on the principle that manufacturing processes generate some dynamic signals that contain information about the changing process conditions; this information can be extracted and used for on-line process monitoring. Process-monitoring methods can be divided into model-based methods and feature-based methods.

Model-based methods. The first step in a model-based approach is to formulate a model linking the sensor outputs to the process mechanics. During the normal process-monitoring stage, the sensor outputs are fed to the model to detect or predict the change in process conditions. The model-based approach has certain limitations. It is often difficult to determine exact process models, because most manufacturing processes are nonlinear and time-variant systems. In manufacturing processes where sensor signals are affected by process working conditions such as feed and speed, it is difficult to identify whether a change in sensor output is due to the change of process working conditions or the deterioration of the process. To counter these limitations, adaptive schemes may be used in which model parameters are adaptively adjusted, keeping the model structure usually fixed.

Feature-based methods. In the feature-based approach, the first step is to select an appropriate set of sensor-signal features sensitive to the desired process conditions, on the basis of analytical studies, computer simulations, and systematic experiments. These features may belong to either the time domain or the frequency domain, depending on the application. The feature set in some applications may also include the process working conditions in order to account for their effect on process conditions. Sensor-signal features are then related to the process conditions by a suitable relationship scheme. This relationship scheme can take a time-dependent, nonlinear, or even a nonanalytic form.

The common relationship schemes used are analytical functions, pattern recognition methods, fuzzy systems, decision trees, neural networks, and expert systems. In these schemes, the relationship between sensor-signal features and process conditions is determined through a learning or training procedure from a set of sample data. The sample data are generated by human experts or from a set of experiments. Developing the relationship between the features and the process conditions is known as the learning phase. Once the relationship between the features and the process conditions is determined, the process conditions are assessed continuously or intermittently from the sensor outputs. This condition assessment is known as the classification or estimation phase.

Comparison of approaches. Feature-based methods are more practical to implement than model-based methods, but are liable to misinterpretation of sensor outputs that can lead to a wrong assessment of process conditions. This typically happens when the sensor outputs in the classification phase are drastically different from the sample data used during the learning phase. To guard against this type of misinterpretation, some procedures utilizing artificial intelligence can be incorporated in the feature-based methods.

Both model- and feature-based methods are useful for process monitoring, depending on the nature of the application. In applications where fairly accurate process models can be developed, the model-based process monitoring is desirable. However, if the manufacturing process being monitored is too complex for exact modeling, feature-based monitoring is a more viable option.

Important steps. Feature-based process monitoring consists of four main steps (see illus.): sensing, sensor-data preprocessing, sensor-data representation, and process-condition assessment. Selection of suitable sensors, mounting them at convenient locations, and collecting sensor data at appropriate sampling rates are the major components of the sensing step. The sensor-data preprocessing step enriches the sensor data by eliminating noise that is accumulated because of nonhomogeneities in the workpiece material and electric interferences during the signal transmission through cables and instrumentation, and filters out the signal frequencies that are not sensitive to the process conditions of interest. The preprocessing step increases the signal-to-noise ratio and consequently improves the accuracy of the assessment of process conditions. The main objective of the sensor-data representation step is to provide the most compact signal representation possible while preserving the structural features of interest in the sensor signals. In process-monitoring applications, the significant structural features are the sensor-signal components that are sensitive to the process conditions being monitored. It is important to select the representation scheme that correctly matches the structural features of interest in the sensor signals and that highlights these structural features from the features of no interest. The scheme of sensor-data representation is also known as the feature-extraction scheme. The process-condition assessment step involves the identification of process conditions based on the sensor-signal features. The decision, or result, from this step is passed on to the steps of diagnosis and process control.

Model-based process monitoring usually involves two steps: measurement of process outputs through sensors, and feeding them to a process model to predict or detect the process changes. As in feature-based monitoring, the outcome of the process model is fed to the diagnosis and process control steps. In feature-based monitoring, sensors usually measure dynamic changes in process variables such as force, temperature, and acoustic emission, whereas in model-based monitoring, sensors measure static values of process variables at a given instant or over a time interval.

Multisensor integration. Practical experience indicates that in many applications the output of a single sensor, such as force, temperature, acoustic emission, or vibration, is insufficient for accurate process monitoring. More reliable and consistent results are obtained with outputs from two or more sensors. Using more than one sensor for achieving the objectives of a monitoring task is known as multisensor integration. The selection of a correct set of sensors is the first step in implementing a successful sensor integration scheme. Then an algorithmic procedure is developed for combining their outputs. This technique, known as sensor-data fusion, combines output from multiple, possibly dissimilar, sensors into a single coherent result. Methods using multisensor-based process monitoring provide reliable and consistent assessment of process conditions by virtue of complementarity, redundancy, and timeliness of information.

Applications. Process monitoring plays a key role in automating many manufacturing processes. It is particularly essential in precision manufacturing, where requirements on tolerances, form, dimensions, and surface finish are very stringent. Considerable progress has been achieved in monitoring cutting-tool condition for machining processes such as turning, milling, and drilling.

The most commonly used sensors in the industrial machining environment determine force, spindle motor power, and acoustic emission. Acoustic emission is rapidly finding application in monitoring a wide variety of manufacturing processes. It is being studied for monitoring processes as well as products in manufacturing that involves material deformation, such as forging; extrusion; rolling; and wire, tube, and deep drawing. Monitoring by means of acoustic emission has also proven to be effective in welding and grinding applications. There is active interest in developing sensors for monitoring the surface quality of workpieces; current sensors are either ultrasonic or optical. Two areas closely related to process monitoring are machine-tool monitoring and critical-component monitoring.

With the availability and further development of sophisticated sensors, high-resolution data acquisition systems, effective signal-processing methods, and high-speed computers, the area of process monitoring will play a dominant role in manufacturing automation, providing customized products at reduced cost with improved quality.

For background information *see* COMPUTER-AIDED DESIGN AND MANUFACTURING; COMPUTER-INTEGRATED MANUFACTURING; FUZZY SETS AND SYSTEMS; INDUSTRIAL ENGINEERING; INSPECTION AND TESTING; MANUFACTURING ENGINEERING; MODEL THEORY in the McGraw-Hill Encyclopedia of Science & Technology.

Sagar V. Kamarthi

Bibliography. G. Byrne et al., Tool condition monitoring (TCM): The status of research and industrial applications, *Ann. CIRP* (International Institution for

Production Engineering Research), 44(2):541–567, 1995; D. Dornfeld, Application of acoustic emission techniques in manufacturing, *NDT&E Int.* (Nondestructive Testing and Evaluation), 25(6):259–269, 1992; R. Du, M. A. Elbestawi, and S. M. Wu, Automated monitoring of manufacturing processes: Part 1: Monitoring methods, *J. Eng. Ind.*, 117:121–132, 1995.

Marine sediments

Recent analyses of marine sediments have revealed much about the Earth's paleoclimates and its hidden sources of energy. First, this article discusses the analysis of sediment deposits termed Heinrich layers and how they provide clues to some of the Earth's dramatic climatic shifts. Then, the article examines natural gas hydrate in marine sediments, and its methane content both as a potential source of energy and as a potential factor in global climate change.

Heinrich Layers

There is a growing body of evidence for correlative abrupt climate changes in many parts of the globe, and the Heinrich layers represent some of the most spectacular evidence. Heinrich layers are irregular sediment deposits that formed in the northern Atlantic Ocean and contain high concentrations of very coarse detritus (rock fragments and organic sediments). Sedimentologists indicate that the only way coarse detritus could have been transported so far from the continental sources is by icebergs. The understanding of these events is important because they are related to the large climatic shifts that are indicated by polar foraminifera species assemblages associated with Heinrich layers. The causes and effects of Heinrich events pose important, unresolved questions. Identifying the sediment sources is one of the keys to understanding Heinrich events and their relation to global paleoclimates.

Origin. As glaciers thaw and freeze, they incorporate fragments of underlying bedrock into the ice. As the ice travels, these fragments are ground into very fine particles. Accordingly, the coarsest fragments will have traveled the shortest distance. Therefore the detritus in icebergs must have a large component derived very near the region of calving into the ocean. The Heinrich layers thicken toward the western North Atlantic Ocean, an indication that the source of the icebergs is the North American margin or the Labrador Sea. Among the coarse detritus transported by iceberg, limestone and dolostone fragments make up about 25–30% of the Heinrich layer sediment. Deposits of limestone and dolostone crop out over large areas in the Hudson Bay region of Canada, and line the floor of much of the Hudson Strait. These observations have been used to suggest that an ice stream through the Hudson Strait was a large sediment contributor. The other 70–75% of the detritus is composed predominantly of quartz, with lesser amounts of feldspar and minor amounts of other minerals, such as hornblende. Although their abundances are not high, minerals such as feldspar and hornblende that are unlikely to survive sedimentary recycling can provide distinctive geochemical fingerprints of the terranes that the glacier crossed.

Feldspar grains. Lead (Pb) is composed of four naturally occurring, stable isotopes: ^{204}Pb, ^{206}Pb, ^{207}Pb, and ^{208}Pb. All existed in the primordial earth in proportions that have been determined by measurements of these isotopes in meteorites. Additionally, ^{206}Pb, ^{207}Pb, and ^{208}Pb are produced by radioactive decay of uranium-238 (^{238}U; half life 4.468×10^9 years), ^{235}U (half life 7.038×10^8 years), and thorium-232 (^{232}Th; half life 1.401×10^{10} years), respectively. The bulk of the earth is estimated to have a ^{238}U/^{204}Pb ratio of about 8. The time-dependent variation of the isotope ratios ^{206}Pb/^{204}Pb and ^{207}Pb/^{204}Pb from primordial compositions from 4.55 billion years ago to the present, assuming this U/Pb ratio, indicates that the variation in lead isotope composition produced by the decay of uranium isotopes follows a systematic path with time, and thus can provide a measure of the age of feldspar samples. The initial lead isotopic composition of rocks is the composition at the time that they formed. The initial lead isotope composition of rocks can be estimated in several ways, and provides valuable clues to the age and origin of rock. Due to the characteristics of the cation sites, the U/Pb and Th/Pb ratios of feldspars, especially potassium feldspars, are very low. Therefore, in simple cases, feldspars record lead isotopic compositions that approach the initial composition of the original host rocks.

By selecting sand grains from the Heinrich layers, only portions of the sediment that were definitely of ice-rafted origin are analyzed. The lead isotopic compositions of individual feldspar grains as well as composite samples of greater than 75 grains form a well-defined linear array. The slope of the line, as well as its position in lead isotope space, requires that the feldspar grains were derived from a very ancient source. Furthermore, the data are indistinguishable from galena (lead sulfide) and feldspar data from the Churchill Province of the Canadian Shield, which forms the basement of the Hudson Strait. Such evidence may indicate that these feldspar grains were carried by icebergs from the same region of the Canadian Shield as the limestone and dolostone grains.

Relating grain analyses to bulk sediment samples. A reasonable case can be made that virtually the entire inventory of detritus in the Heinrich layers of the North Atlantic came from ancient continental sources like those surrounding the Labrador Sea. All published data from the Heinrich layer belt are consistent with derivation from the same region as the individual grains of feldspar and hornblende. In conjunction with the very high sediment flux and the pattern of thickening, it is most plausible that the detritus is from the region of the Canadian Shield termed the Churchill Province. Although there is no reason to believe that iceberg contributions from other ice sheets were reduced during Heinrich events, their contributions are hidden by the much larger contributions from the Labrador Sea, further highlighting the anomalous nature of the Heinrich events.

Climate. Heinrich layers appear to require a series of repeated, anomalous, glaciological processes within one or more ice domes (areas within an ice sheet where the center is thicker and higher than the edges) in the Canadian Shield. Several lines of evidence are consistent with the creation of Heinrich layers by massive releases of iceberg armadas into the North Atlantic Ocean. The distinct composition and large amount of the coarse lithogenic detritus point to iceberg transport. Additionally, age and sediment flux measurements (how fast the sediment was deposited) imply that these layers could have formed in less than a few hundred years, and thus the rates of sediment deposition are extremely rapid. Furthermore, oxygen isotope data from the polar foraminiferan species *Neogloboquadrina pachyderma sinistral,* have been used as evidence for decreased salinity of surface water during the interval of the Heinrich layers due to iceberg melting.

Climate changes in the North Atlantic are produced by variations in the intensity of deep-water formation there. Deep water is formed in regions where the density of surface water is raised high enough that the surface water sinks into the deep ocean. In the North Atlantic Ocean, this increase in density is a product of heat loss from the surface water, and thus the deep-water formation acts as a heat pump to warm the air over this region. Ocean models predict that abrupt additions of fresh water into the North Atlantic, such as would have been introduced by the hypothesized armadas of icebergs, should lower the surface-water density enough to stop or greatly decrease deep-water formation. This would decrease the heat-pump warming mechanism, thus cooling the region. Heinrich layers occur toward the end of cooling trends where deep-water formation was probably already dramatically reduced.

Temperature oscillations. Although temperatures were still cold immediately following Heinrich layers, shortly afterward there is a record of an abrupt transition to brief warm intervals. It seems plausible that events leading up to the Heinrich layers somehow triggered these temporary warm periods, and the gradual return to colder conditions reflected the global climate's attempt to restore normal ice-age conditions. Thus paleoecologists are attempting to determine if Heinrich events were a climate response to changes in glaciers or a glaciological response to changes in the climate. There are many oscillations in the temperature of the North Atlantic region that did not trigger massive events like the Heinrich layers. These oscillations are being documented from careful deep-sea sediment core descriptions and have an approximately 1500-year period. More than 5000 years separate any two of the Heinrich layers. It is believed that fresh-water influx into

the North Atlantic increased as the ice sheets grew during cooling cycles, thereby producing surges of icebergs periodically in the Hudson Strait, rather than the surges being triggered by every cold snap. The ice body that provided the icebergs for the Heinrich layers grew in response to this cooling until it reached an unstable condition where ice streamed through the Hudson Strait. Immediately after the event, the fresh-water content of the surface ocean would have been high and conditions would have remained cold. However, this large supply of fresh water from the glaciers would have been dramatically reduced by this surging event, so eventually the surface water would have been able to achieve the high densities needed to form deep water again. As the region warmed because of deep-water formation, more snow would have accumulated because of the increase in humidity, and the cycle would have repeated.

The origin and consequences of the Heinrich layers have not yet been confirmed. But the dominance of a source region in the Labrador Sea, quite possibly the Hudson Strait, is clearly indicated by the lead isotope data from feldspars and substantiated by all other published data.   Sidney Hemming

Natural Gas Hydrate in Marine Sediments

Earth scientists have been increasingly interested in naturally occurring gas hydrate, a solid substance composed of water and gas (predominantly methane) that is stable under conditions of low temperature and high pressure. This icelike substance occurs on Earth in geologic environments where these conditions exist and in which there is a sufficient supply of methane. Earth scientists are attempting to determine if enormous volumes of natural gas are stored as gas hydrate in marine sediments. The conditions appropriate for gas hydrate formation are very common in the upper levels (few hundred feet) of continental margin sediments, and in sediments at high latitudes where surface temperatures are very low. Current estimates suggest that about 10^{13} tons (10^{16} kg) of carbon are stored in sediments as gas hydrate. This is approximately twice the amount of carbon estimated to be stored in all other fossil fuel deposits.

Despite its abundance in marine sediments, gas hydrate is not stable on the surface of the Earth because of the low pressure, making it difficult to study. Much of what is known about its occurrence, composition, and physical characteristics comes from drilling into gas hydrate–bearing sediments and from indirect geophysical techniques. Gas hydrate deposits are believed to be common in continental margin sediments because geophysical data have indicated their presence in every ocean basin throughout the world.

Bottom simulating reflector. Sediment sequences that contain gas hydrate are usually detected by the occurrence of a bottom simulating reflector (BSR) in seismic reflection profiles. Bottom simulating reflectors occur between approximately 660 and 1980 ft (200 and 600 m) depth below the sea floor on continental rises. A common characteristic of BSRs is that they crosscut sediment bedding planes. This indicates that BSRs do not represent a depositional horizon in the sediments but are caused by a diagenetic (sediment into rock) or postdepositional change. Based on this observation and the results of experimental studies delineating the stability conditions of gas hydrate, BSRs are generally believed to represent the lower depth limit in sediments at which gas hydrate is stable. Thus, the pore spaces of sediments above the BSR are partly filled with gas hydrate, which has the effect of increasing the sediment density and seismic velocity, whereas sediments beneath the BSR probably contain bubbles of free gas. This would result in a sharp contrast in seismic velocities and a strong reflector at the base of the gas hydrate stability zone, similar to what is observed at BSRs. However, the origin of BSRs and the relationship between seismic velocities and amounts of gas hydrate and free gas are poorly constrained.

Ocean drilling. Samples of gas hydrate have been recovered by drilling in marine sediments and in permafrost in polar regions. More specifically, deep ocean drilling performed during the last several decades as part of the Deep Sea Drilling Project (DSDP) and the Ocean Drilling Program (ODP) has recovered gas hydrates off the west coasts of North, Central, and South America; off the southeastern coast of the United States; in the Gulf of Mexico; and offshore of Japan. Gas hydrate samples have also been recovered from the sea floor by using submersibles.

Methane. Natural gas hydrate has a clathrate structure, in which rigid cages of water molecules surround gas molecules. For an ideal methane hydrate, the molecular ratio of methane to water is 1 to 5.75. When solid gas hydrate decomposes, the methane (CH_3) gas that is liberated has a volume 164 times larger than the volume of the solid from which it originated. Thus, gas hydrate provides a very dense storehouse for methane. This is important both because methane (natural gas) is an important fossil fuel and because it is a strong greenhouse gas that may affect climate.

In order for gas hydrate to form in sediments, there must be an adequate supply of methane. Isotopic and compositional data for gases from gas hydrate–bearing sediments indicate that the gas is predominantly microbial methane, produced during decomposition of buried organic matter by bacteria. In sediments that are moderately rich in organic matter, sufficient methane is produced locally in the sediment for gas hydrate to form. Gas hydrate has also been found to concentrate near faults and fractures which contain methane-rich fluids that flow upward. Such occurrences have been reported from sea-floor seeps of methane in the Gulf of Mexico and off the southeastern coast of the United States, and at greater subsea-floor depths in accretionary sediments off the west coast of Central America.

Potential energy resource. Because of the widespread occurrence of gas hydrate–bearing sediments and the enormous amount of methane stored in them, gas hydrate should be viewed as a potential energy resource. Several methods have been tested by the natural gas industry and have demonstrated that methane can be recovered from naturally occurring gas hydrate. However, a practical method has yet to be developed in which the cost of the energy used to decompose hydrate at depth, and thereby release methane, is significantly less than the economic value of the methane recovered. The potential of gas hydrate as a resource has, in particular, attracted growing attention by government and industry groups in Japan. Detailed geological studies have revealed the widespread occurrence of gas hydrate–bearing sediments in the Nankai Trough off southwestern Japan. Estimates of the amount of methane stored in this region make it the largest prospective hydrocarbon reservoir in Japan. The Japanese Ministry of International Trade and Industry (MITI) has launched a research and development program focusing on methane hydrate as part of its 5-year plan (1995–1999) for development of Japan's domestic oil and gas resources.

Global climate change. Methane is present in trace concentrations in Earth's atmosphere. Because it is a strong absorber of infrared radiation (much stronger even than carbon dioxide), methane is a greenhouse gas that can affect global warming. The amount of methane stored in gas hydrate deposits worldwide is probably several thousand times larger than the amount currently in the atmosphere. Release of even a small fraction of such methane could have a significant impact on the radiative properties of Earth's atmosphere and thereby affect global climate.

Abundant geologic evidence indicates that rapid drops in sea level and deep ocean warming occurred in the past. Recently there has been considerable speculation that the release of methane from sedimentary gas hydrate played a role in past climate change, because (1) several of these time intervals of sea-level and climate change are characterized by a major input of carbon (from some unknown source) to the ocean and atmosphere, and (2) the stability of gas hydrate decreases with reduced pressure and elevated temperature. A major assumption here is that a large mass of methane exists in oceanic gas hydrate zones. Although this assumption has been confirmed by research, a link between gas hydrate and past or future climate change remains speculative because the role of gas hydrate in the current climate regime is not yet well understood.

Research on gas hydrate. In 1995, Leg 164 of the International Ocean Drilling Program was devoted to furthering understanding of gas hydrate in marine sediments and the cause of BSRs by drilling into the Blake Ridge, off the southeastern coast of the United States. The Blake Ridge gas hydrate field was targeted because it is associated with an extensive and well-developed BSR. A pressure core sampler was used to take a small sediment core from the bottom of the borehole and seal the core into a pressure housing so that it would come to the surface under its original pressure. Pressure core samples contain all their original gases until they are opened to the atmosphere. The quantity and composition of gas in a deep-sea sediment sample can be determined by releasing gas from the pressure core sampler under controlled laboratory conditions at the surface. Some samples produced as much as 3.9 gal (15 liters) of methane per each liter of sediment recovered.

Preliminary analysis of data from Leg 164 drill sites indicates that gas hydrate occupies several percent of the sediment volume in a zone that is 800 ft (250 m) thick, and that sediments beneath the BSR, where gas hydrate is not stable, contain bubbles of methane gas. Seismic reflection profiles indicate that gas hydrate and free gas zones extend over 10,000 mi^2 (26,000 km^2) of the Blake Ridge. If the Leg 164 results are representative of this entire area, then sediments on the Blake Ridge could contain as much as 5×10^{10} tons (5×10^{13} kg) of methane. If it could be extracted, this amount of methane could supply the United States with natural gas, at current levels of consumption, for the next 100 years.

For background information *see* ATLANTIC OCEAN; BIOGEOCHEMISTRY; GLACIAL GEOLOGY; ICEBERG; MARINE GEOLOGY; MARINE SEDIMENTS; METHANE; PALEOCLIMATOLOGY; PALEOCEANOGRAPHY; RADIOISOTOPE (GEOCHEMISTRY); STRATIGRAPHY in the McGraw-Hill Encyclopedia of Science & Technology.   Paul J. Wallace

Bibliography. R. B. Alley and D. R. MacAyeal, Ice-rafted debris associated with binge/purge oscillations of the Laurentide ice sheet, *Paleoceanography*, 9:503–511, 1994; W. S. Broecker, Massive iceberg discharges as triggers for global climate change, *Nature*, 372:421–424, 1994; W. S. Broecker et al., Origin of the northern Atlantic's Heinrich events, *Clim. Dyn.*, 6:265–273, 1992; G. Dickens et al., Direct measurement of *in situ* methane quantities in a large gas-hydrate reservoir, *Nature*, 385:426–428, 1997; G. Dickens et al., Dissociation of oceanic methane hydrate as a cause of the carbon isotope excursion at the end of the Paleocene, *Paleoceanography*, 10:965–971, 1995; K. Kvenvolden, Natural gas hydrate occurrence and issues, *N.Y. Acad. Sci. Ann.*, 715:232–246, 1994; C. Paull et al., *Proceedings of the Ocean Drilling Program, Initial Reports*, 1996; S. Rahmstorf, Bifurcations of the Atlantic thermohaline circulation in response to changes in the hydrological cycle, *Nature*, 378:145–149, 1995.

Mars

Martian meteorites provide an important window into the geologic history of Mars. Currently, 12 meteorites in collections throughout the world are presumed to be from Mars (see **table**); half were found in Antarctica. For many years, they have been called SNC meteorites, after three of the major types of meteorites (Shergotty, Nakhla, and Chassigny).

Links to Mars. The key data indicating that these meteorites are from Mars are of two types. First was the discovery that one of the Antarctic meteorites (EETA79001) contained gases trapped in impact-produced glass pockets that turned out to have compositions nearly identical to the gases measured in the Martian atmosphere by the *Viking* lander spacecraft in 1976. Subsequent laboratory experiments showed that shock processes which produce glass from rocks and soils can almost quantitatively trap atmospheric gases without changing the composition. Because EETA79001 reveals a close connection between the atmosphere of Mars and the group of SNC meteorites, it is sometimes called the Rosetta Stone of Martian meteorites.

The second major line of evidence connecting these meteorites is that their oxygen isotopic ratios are similar and strongly support the interpretation that they are from the same parent body. On a plot of three isotopes of oxygen (^{16}O, ^{17}O, ^{18}O), they are clearly of the same family and come from the same original planetary reservoir. These isotopes also show that these meteorites are not related to other types of meteorites and could not have come from the Earth or the Moon.

The logic, then, is that if one meteorite comes from Mars based on the trapped gas data, and the other 11 meteorites are from the same planetary body based on oxygen isotopes, then all 12 meteorites come from Mars. This conclusion is now accepted by most meteorite researchers.

Subsequent meteorite analyses have shown that several other meteorites in this group also have some trapped gases from the Martian atmosphere. Other studies have shown how meteorites can be ejected from Mars by the impacts of asteroids or comets on its surface. Isotope studies showed that the Rosetta Stone meteorite was ejected from Mars only about 600,000 years ago, and therefore it provides a fairly recent sample of Martian atmospheric composition.

Allan Hills Martian meteorite ALH84001. In 1984, an unusual greenish meteorite was found in Antarctica. Because it was mainly orthopyroxene (iron magnesium silicate), it was misidentified as a diogenite, a type of achondrite meteorite. In 1993, as part of a new diogenite study, this meteorite was found to be, in fact, an SNC meteorite. The mineralogical, chemical, and isotopic compositions were then measured in detail. The original crystallization age was found to be 4500 million years (m.y.), which is about the age of Mars itself. It was realized that this rock was part of the original Martian crust and was formed by igneous processes. At some subsequent time the meteorite was fractured by shock, and was invaded by fluids through cracks and pore spaces. Secondary minerals were formed in these spaces at times that have been variously estimated to be from 1200 m.y. to 3600 m.y. ago. Analyses of isotopes formed by cosmic rays show that the meteorite was ejected from Mars about 16 m.y. ago and it landed in Antarctica about 13,000 years ago, where it remained in and on the ice fields.

Carbonate globules. The most striking variety of the secondary minerals in meteorite ALH84001 consists

Martian meteorites				
Name	Classification*	Mass, kg (lb)	Find/fall†	Year
Shergotty	S-basalt (pyr-plag)	4.00 (8.8)	Fall	1865
Zagami	S-basalt	18.00 (39.6)	Fall	1962
EETA79001	S-basalt	7.90 (17.4)	Find-A	1980
QUE94201	S-basalt	0.012 (0.26)	Find-A	1995
ALHA77005	S-lherzolite (ol-pyr)	0.48 (1.1)	Find-A	1978
LEW88516	S-lherzolite	0.013 (0.29)	Find-A	1991
Y793605	S-lherzolite	0.018 (0.040)	Find-A	1995
Nakhla	N-clinopyroxene	40.00 (88)	Fall	1911
Lafayette	N-clinopyroxene	0.8 (1.8)	Find	1931
Gov. Valadares	N-clinopyroxene	0.16 (0.35)	Find	1958
Chassigny	C-dunite	4.00 (8.8)	Fall	1815
ALH84001	?-orthopyroxene	1.90 (4.18)	Find-A	1993

* S = shergottite, N = nakhlite, C = chassignite; ALH84001 is none of these.
† A = Antarctica.

of carbonate globules that range from a few micrometers to about a half millimeter in diameter. The carbonates are found mostly on surfaces or in depressions that are revealed when the meteorite is broken apart along old fractures. The carbonates are often pancake shaped rather than spherical. It is these carbonates that provide most of the circumstantial evidence for possible ancient life on Mars.

Textural features of these globules include complex zoning from a more calcium-rich carbonate ($CaCO_3$) in the core, to a very magnesium-rich carbonate ($MgCO_3$) rim sandwiched between two very thin iron-rich carbonate ($FeCO_3$) rims that are also enriched in sulfur. The cores also contain some manganese carbonate ($MnCO_3$). The research group found that the old fracture surfaces containing the carbonates are enriched in polycyclic aromatic hydrocarbons, common organic species found on Earth in many environments and formed by burning or decay of organic material as well as by many industrial processes. Polycyclic aromatic hydrocarbons are also common in carbonaceous chondrite meteorites, and are believed to have formed in the early solar system and even in interstellar dust clouds. In ALH84001, the polycyclic aromatic hydrocarbons are sometimes concentrated on or in the carbonate globules.

Other unusual features of the carbonate globules include the presence of very small minerals, particularly in the iron carbonate rims. These minerals include magnetite (Fe_3O_4), pyrrhotite (FeS), and a spinel form of iron sulfide tentatively identified as greigite (Fe_3S_4). The magnetites are mostly in the size range 20–100 nanometers (mean diameter about 40 nm). They consist mostly of equant crystals, although teardrop-shaped crystals and whiskers are also present. The pyrrhotite and greigite are also in the nanometer size range. Another feature associated with the carbonates, but also found on the old fracture surfaces away from the carbonates, consists of a variety of tiny spherical forms and elongate forms that range from below 50 nm in diameter to as large as 0.6 micrometer in length for some of the elongate segmented or filamentous forms.

Formation of features. These features include carbonates, polycyclic aromatic hydrocarbons, nanophase minerals, and fossillike forms.

Carbonates. The research group considered a variety of possible origins for the carbonate globules and their associated features. For example, carbonates are common in carbonaceous chondrites, and were presumably formed on asteroids by low-temperature aqueous processes. However, there are many details in which these carbonates do not resemble the ALH84001 carbonates. Carbonates also form on Earth in many types of igneous and metamorphic processes, and have been reported in impact rocks where they may have crystallized from hot impact gases. However, most carbonates on Earth are formed at relatively low temperatures as precipitates in oceans, lakes, and hot springs, and as a result of various weathering and secondary processes. In many cases, the precipitation of carbonates is associated with bacterial activity; the bacteria either form the carbonate directly or change the local environment enough to cause inorganic precipitation. The complex zonings of the Martian carbonate globules are not compatible with equilibrium high-temperature formation; the compositions within a single carbonate could never have been in thermodynamic equilibrium. It was proposed that the carbonates might also be low-temperature precipitates formed on Mars from fluids that permeated the rock through cracks and pores. It was also proposed that the carbonate precipitation may have been facilitated by life on Mars in the form of tiny organisms. Carbonate nodules found in some sedimentary locations and known to be associated with microbial activity are sometimes highly zoned, and can include relatively pure magnesium carbonate.

Polycyclic aromatic hydrocarbons. It was proposed that the associated polycyclic aromatic hydrocarbons (PAHs) may result from decay of these microorganisms. Two other major hypotheses have been proposed for the formation of these PAHs. It is known that PAHs can be made in nature by strictly inorganic processes such as the Fischer-Tropsch process, in which carbon monoxide reacts with hydrogen to form methane and higher hydrogen-carbon compounds (hydrocarbons), including PAHs. This reaction can take place at relatively low temperatures, and it has been suggested to explain hydrocarbons found in terrestrial settings such as pegmatites and granites. If the PAHs in meteorite ALH84001 formed on Mars, it is possible that they were formed by such inorganic processes. The pattern of compounds in ALH84001 does not closely resemble that of PAHs of known inorganic origin; however, fractionation processes associated with Martian ground water could have changed the composition of the compounds. Another hypothesis suggests that the origin of the PAHs in ALH84001 is contamination from Antarctic melted ice. However, the profile of PAHs in ALH84001 shows low values at the exterior and higher values in the interior, suggesting that these PAHs are not contaminants that filtered in from melted ice. The measured abundances of PAHs in the meteorite are much higher than those in the ice, so that an efficient concentration mechanism is necessary. In summary, the detected PAHs in the meteorite can possibly be explained by Antarctic contamination or by inorganic processes, but the PAHs are also compatible with a biogenic origin.

Nanophase minerals. The presence of nanophase magnetite, pyrrhotite, and an iron monosulfide such as greigite can also be explained by either inorganic or organic processes. All of these minerals can be formed by low-temperature precipitation from aqueous solutions under proper conditions. It has been proposed that some of the magnetite may have been formed by vapor deposition at higher temperatures. Some of the ALH84001 magnetite crystals are elongated (whiskers) and twinned, and some have screw dislocations. However, the magnetite in meteorite ALH84001 is very similar to magnetite produced by a number of different kinds of bacteria on Earth. Some of the bacteria produce the magnetite internally and use it for orientation and locomotion along magnetic lines of force. Other kinds of bacteria produce magnetite on the exterior of their cell surfaces, and they may use it as an "energy battery" when going back and forth between an aerobic and an anaerobic environment. Whisker magnetite has been found within bacteria cells as well as twinned magnetite, although magnetite with screw dislocations has not yet been reported. Bacteria also produce greigite within their cells, which they can use similarly for orientation or locomotion. These minerals are often left behind as the bacteria decay, and then constitute magnetofossils (fossils of magnetite) in some sediments. It may be that the Martian magnetite and sulfides have similarly been produced by Martian bacterialike organisms.

Fossillike forms. A research group proposed that some of the ovoid, spherical, and elongated forms that are found particularly in the rims of the carbonate globules may be fossilized microorganisms from Mars. Some of these forms resemble identified fossils from ancient Earth sediments, but are much smaller. Controversy has developed over the smallest possi-

High-resolution transmission electron microscope image of a cast, or replica, from a chip of Martian meteorite ALH84001 that shows the outline of what may be microscopic fossils of bacterialike organisms that may have lived on Mars more than 3.6 billion years ago. (*NASA*)

ble size for a microorganism; the smallest known bacteria is about 100 nm in diameter. Some of the forms described by the research group are as small as 20 nm in diameter, but most are 50 nm or larger. Other research groups have interpreted small forms 20–100 μm in diameter that have been found in Earth environments and sediments as microorganisms and microfossils. Still others have suggested that larger bacteria may shrink to much smaller dimensions when deprived of nutrients or water. Other suggestions for the forms include additional types of nanominerals, reproduction spores from larger bacteria, or even viruses.

Other evidence. The study of meteorite ALH84001 is continuing with new instruments and techniques. Recent data on carbon and oxygen isotopes seem to support a low-temperature origin for the carbonates and even a biogenic origin for some of the carbon-rich material (very light carbon has been found). Measurements of magnetic properties of the ALH84001 magnetite are consistent with a biogenic origin. However, recent data on sulfur isotopes in associated pyrites do not support a biogenic origin of these pyrites, although the extremely fine-grained sulfides that have been described have not yet been analyzed.

Implications. A number of observations of the meteorite ALH84001 have suggested that the activity of past microorganisms on Mars is a reasonable explanation (see **illus.**). This interpretation has stimulated some major controversies over the origin of the PAHs, the temperature at which the carbonates formed, the origin of the nanominerals, the interpretation of the fossillike forms, and the possibilities of terrestrial contamination. None of these controversies will be easily settled, but they are promoting considerable research in a variety of fields, including micropaleontology, geochemistry, biology, and isotope analysis. Whether these controversies can be settled with the other 11 Martian meteorite samples remains to be seen. It may be necessary to await the return of actual Mars samples by future space missions.

For background information *see* CARBONATE MINERALS; INTERSTELLAR MATTER; MARS; METEORITE in the McGraw-Hill Encyclopedia of Science & Technology.   David S. McKay

Bibliography. P. H. Benoit, Meteorites as surface exposure time markers on the blue ice fields of Antarctica: Episodic ice flow in Victoria Land over the last 300,000 years, *Quat. Sci. Rev.*, 14:531–540, 1995; J. Bradley et al., Magnetite whiskers and platelets in the ALH84001 Martian meteorite: Evidence of vapor phase growth, *Geochim. Cosmochim. Acta*, 60:5149–5155, 1996; D. S. McKay et al., Search for past life on Mars: Possible relic biogenic activity in Martian meteorite ALH84001, *Science*, 273:924–930, 1996; R. O. Pepin, Evolution of the Martian atmosphere, *Icarus*, 304:289–304, 1994.

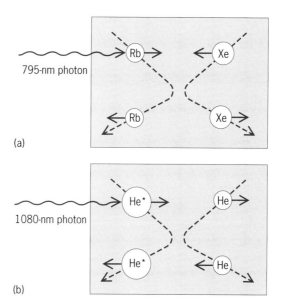

Fig. 1. Hyperpolarization (*a*) of xenon-129 (^{129}Xe) nuclei by spin exchange with optically pumped rubidium atoms, and (*b*) of helium-3 (^{3}He) atoms by exchange with optically pumped metastable atoms (designated He*).

Medical imaging

The signals for conventional magnetic resonance imaging (MRI) come from the nuclei of ordinary hydrogen (^{1}H) atoms in the human body. It is difficult to image body parts such as the lungs which contain relatively few hydrogen atoms. However, excellent magnetic resonance images of the lungs can be obtained if the patient first takes a deep breath of the gases helium-3 (^{3}He) or xenon-129 (^{129}Xe) which have had their nuclear spin polarizations increased to tens of percent by laser optical pumping. The magnetic resonance images from the hyperpolarized nuclei of these gases are as good as, and often better than, conventional images from hydrogen in tissue. Because of its high solubility in human tissue, hyperpolarized xenon-129 is useful for imaging other body parts as well as the lungs.

Nuclear magnetization. A nucleus with a nonzero spin has a magnetic moment, μ, and it can therefore interact with magnetic fields. In equilibrium, a spin-½ nucleus can have its spin point along an applied magnetic field H (spin-up) or opposite (spin-down). It is convenient to express the difference, ΔN, in the number density of spin-up and spin-down nuclei in terms of the spin polarization defined by Eq. (1),

$$P = \frac{\Delta N}{N} \qquad (1)$$

where N is the total number density of nuclei. The tendency of the nuclear moments to orient parallel to the field direction is counterbalanced by thermal agitation, so in thermal equilibrium at a temperature T the nuclear polarization is very nearly given by Eq. (2), where k is Boltzmann's constant. At human

$$P = \frac{\mu H}{kT} \qquad (2)$$

body temperature, the thermal agitation energy, kT, of the nuclear spins is much larger than the magnetic alignment energy, μH. Therefore, the polarization is very small, typically of the order of 3×10^{-6}, even

in the relatively large fields, of the order of 1 tesla, used in magnetic resonance imaging facilities.

Magnetic resonance imaging works by using pulses of magnetic field, oscillating at radio frequencies, to tip the nuclear spins away from the direction of the static magnetic field. The tipped spins then precess about the field. For magnetic fields of the order of 1 tesla, which are typical for magnetic resonance imaging applications, the precession frequencies are comparable to those of frequency-modulation (FM) radio or television broadcasts. The radio waves that the nuclei emit at these frequencies can be detected and processed with great sensitivity. By adding controlled inhomogeneities to the static magnetic field, the nuclei from different parts of the body can be made to precess at different frequencies, and the resulting radio signals can be used to image the spin densities in the body.

The signal-to-noise ratio for magnetic resonance imaging is proportional to the product of the polarization and the spin number density. In conventional magnetic resonance imaging, the relatively small thermal polarization is compensated by the high number density of hydrogen atoms in human tissue. The number density of spins in a gas is several orders of magnitude smaller, so magnetic resonance imaging with the nuclei of gases is normally quite difficult. However, by laser optical pumping, the nuclear spin polarization of the noble gases helium-3 and xenon-129 can be increased by five or six orders of magnitude, thereby more than compensating for the low number density of nuclear spins in the gas. The two methods currently used to hyperpolarize helium-3 and xenon-129 nuclei, spin exchange and metastability exchange, are shown in **Fig. 1**.

Spin exchange. In this method (Fig. 1a), helium-3 or xenon-129 gas is mixed with the vapors of one of the alkali metals, say rubidium, and pumped with circularly polarized laser light that is suitably tuned to be absorbed by the rubidium atoms. Each absorbed photon deposits a half unit of spin. A small fraction of the rubidium spin is then transferred to the noble-gas nuclei during a collision, with the remaining fraction dissipated to the rotational angular momentum of the colliding atoms about each other. For currently available laser powers (tens of absorbed watts), the gas must be optically pumped for several hours to produce the liter or so of highly polarized noble gas needed to image a human lung. In the case of helium-3 this is best done by pumping the entire batch of gas in a sealed-off cell, but for xenon-129 it is more expedient for the gas to flow through the optical pumping cell to a cold trap, where the xenon is condensed and accumulated as a spin-polarized solid. For flowing systems, a xenon-129 atom spends only a few minutes in the optical pumping cell.

Metastability exchange. In this method (Fig. 1b), a weak electrical discharge is maintained in low-pressure helium-3 gas to produce metastable atoms. The metastable atoms absorb circularly polarized laser light. Each absorbed photon deposits one unit of spin. The spin polarization of the metastable helium atoms is passed on with nearly 100% efficiency to ground-state atoms in metastability exchange collisions. Once polarized, the low-pressure helium gas must be compressed to atmospheric pressure for use in magnetic resonance imaging. Both spin exchange and metastability exchange can be used to hyperpolarize helium-3. Hyperpolarized xenon-129 can be produced only by spin-exchange pumping, since the collisional depolarization rates of metastable xenon atoms are too fast to permit optical pumping.

To get the same production rate of gas, about 40 times as much laser light must be absorbed for spin-exchange pumping as for metastable pumping. However, because of the much less demanding tolerance on the spectral linewidth of lasers for spin-exchange pumping, they are often less expensive than the lower-power lasers used for metastability-exchange pumping.

Hyperpolarized helium-3 gas can be stored in the Earth's magnetic field without significant loss of polarization for a day or more in carefully prepared glass cells. Useful storage times for hyperpolarized xenon-129 gas seldom exceed 1 h because of much more potent relaxation mechanisms at the cell walls. However, when xenon-129 is condensed in a holding field of 0.1 tesla or more, the spin-polarized solid can be stored for a few hours at liquid nitrogen temperatures and for a week or more at liquid helium temperatures.

Imaging of living subjects. Conventional magnetic resonance imaging systems can be used for imaging with hyperpolarized helium-3 or xenon-129. The radio-frequency coils and the associated electronics must be capable of handling the resonant frequen-

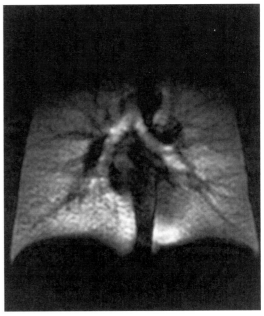

Fig. 2. Magnetic resonance image of a human lung obtained with hyperpolarized helium-3. *(From J. R. MacFall et al., Human lung air spaces: Potential for MR imaging with hyperpolarized He-3, Radiology, 200:553–558, 1996)*

cies of helium-3 or xenon-129, which are lower than those of hydrogen. Since the gas polarization is independent of the magnetic field, hyperpolarized gas imaging can work well at much smaller magnetic fields than those needed for conventional magnetic resonance imaging, where a large field is needed to provide adequate thermal polarization. The thermal polarization of the hydrogen nuclei recovers in 1 s or less. The spin polarization of hyperpolarized gas, once expended to make an image, never recovers. A new charge of hyperpolarized gas is needed to take another image. The spatial diffusion rates of spins in the gas are about a million times faster than the diffusion rates of hydrogen spins in tissue, limiting the practical resolution of the gas image to a fraction of a millimeter. Diffusion of the gaseous spins through a magnetic field gradient also causes relaxation of the spin polarization, so ambient field gradients must be minimized to permit hyperpolarized gases to be stored for hours or more.

The first image of a human lung obtained with hyperpolarized helium-3 gas is reproduced in **Fig. 2**. To acquire this image, hyperpolarized helium gas was produced by rubidium spin-exchange pumping with a diode laser array for about 11 h in a sealed-off glass cell at high temperature and pressure. The gas was then cooled to room temperature to condense out the rubidium vapor, and expanded into a polyethylene bag at atmospheric pressure. The gas was inhaled by a human volunteer in a conventional magnetic resonance imaging system. The pulse sequences and radio frequencies of the imager had been modified to be appropriate for hyperpolarized gas imaging. Excellent images, similar to those of Fig. 2, have also been obtained with hyperpolarized helium-3 prepared by metastability-exchange pumping.

Hyperpolarized xenon-129 was the first gas used for magnetic resonance imaging (of a mouse), and successful preliminary images of human volunteers have now been done with xenon-129. Substantial amounts of inhaled xenon dissolve in blood, which carries it rapidly from the lung to other body parts. Xenon-129 retains its polarization in tissue for tens of seconds, so it is possible to use hyperpolarized xenon-129 for magnetic resonance imaging of tissue as well as the lungs. The anesthetic effects of xenon gas must be considered when xenon-129 is used for magnetic resonance imaging of living subjects. Hyperpolarized xenon-129 gas has been used to image the brains of laboratory animals, and human brain imaging will be possible as well.

For background information *see* MEDICAL IMAGING; NUCLEAR MAGNETIC RESONANCE (NMR); OPTICAL PUMPING in the McGraw-Hill Encyclopedia of Science & Technology.   William Happer

Bibliography. M. S. Albert et al., Biological magnetic resonance imaging using laser-polarized ^{129}Xe, *Nature*, 370:199–201, 1994; N. D. Bhaskar, W. Happer, and T. McClelland, Efficiency of spin exchange between rubidium spins and ^{129}Xe nuclei in a gas, *Phys. Rev. Lett.*, 49:25–28, 1982; J. R. MacFall et al., Human lung air spaces: Potential for MR imaging with hyperpolarized He-3, *Radiology*, 200:553–558, 1996; H. Paetz gen. Schieck and L. Sydow (eds.), *Polarized Beams and Polarized Gas Targets*, 1996.

Medical mycology

Medical mycology is the study of fungi that are pathogens of humans and animals. Most fungi are filamentous, growing as tubelike structures or hyphae, but one important group are single celled and are referred to as the yeasts. The fungi are classified in a separate kingdom; they are eukaryotic like mammalian cells but different from bacteria, which are prokaryotic. Most fungi form spores asexually and reproduce sexually, although some lack a mechanism for sexual reproduction. They are very diverse in appearance microscopically. Therefore, the diagnostic laboratory technician is often confronted with problems in identification of fungal samples.

Fungi cause a variety of human diseases. These diseases can occur on or below the skin surface, in mucosal membranes, or at deep-seated locations, invading via the bloodstream to many sites in the body. They can cause endogenous disease, residing on the mucous membranes of healthy individuals until sparked into activity (as does *Candida albicans*). They can initiate infection from an exogenous source following the inhalation of spores from the environment (for example, histoplasmosis, aspergillosis, cryptococcosis, and coccidioidomycosis). Fungal pathogens can be opportunistic and invade the immune-compromised host, or can infect the healthy individual.

Historical diseases. Many fungal diseases have been known since the beginning of recorded history. Thrush, an oral or genital infection caused by the yeast *C. albicans,* was described by Hippocrates in the 4th century B.C. Ringworm, or tinea, has been known since the time of the Roman Empire. It is estimated that, of the approximately 100,000 species of fungi, only about 150 are known to cause human disease. Unfortunately, the medically important fungi received less attention in research than did the fungal pathogens of food crops or the human pathogenic bacteria. Even by the mid-1960s, these medically important organisms were not studied extensively by the scientific community, probably because they were perceived as being less important causes of human disease. Interestingly, research on the nonpathogenic fungi such as *Aspergillus* and baker's yeast, *Saccharomyces cerevisiae,* was extensive, but the application of such knowledge was not easily translated to research on the medically important fungi because of the complexity of pathogens such as *C. albicans* and *Histoplasma capsulatum,* whose genetic systems were unknown.

Early study. The impetus for the study of medically important fungi was initiated by the changes in the management of human diseases. Treatment of various forms of cancer demanded the use of cytotoxic drugs, and failing organs fostered the science of transplantation. A concomitant outcome of these advances was the development of immunosuppression in treated persons. Further, the excessive use of antibacterial drugs resulted not only in the development of drug-resistant bacteria but also in the elimination of specific bacterial competitors of normally benign, resident fungi, such as *C. albicans*. Left unchecked by specific bacterial competitors in the gut, this yeast became a formidable pathogen, one of many microorganisms capable of causing opportunistic infections. Nosocomial, or hospital-acquired, infections created a new management problem for physicians. Human immunodeficiency virus (HIV) infection has resulted in large numbers of infections caused by fungi such as *C. albicans, Cryptococcus neoformans,* and others capable of unchecked invasion in the immune-weakened individual.

The medical advances and the spread of HIV and acquired immune deficiency syndrome (AIDS), then, have pushed fungal infections into prominence. For example, candidiasis now ranks fourth among all nosocomial infections. Oral and vaginal candidiasis has been estimated to occur in over 70% of people with HIV and AIDS. In the normal host, respiratory infections such as histoplasmosis and coccidioidomycosis are common but usually self limiting; however, in persons with AIDS, they can be devastating.

Fungal pathogens. The judgment might be made, based upon the discussion above, that the fungi are rather weak pathogens of humans and capable of causing serious disease only in the immune-compromised individual. In fact, some of these organisms have evolved as part of the normal biota of humans, such as *C. albicans*. Some fungi, such as *H. capsulatum*, can escape destruction by white blood cells and so resist immune elimination. The surface capsule of *C. neoformans* prevents destruction by host phagocytic cells. *Candida albicans* invades systemically when the host becomes neutropenic (develops an abnormally low number of white blood cells), but in the CD4+ T-lymphocyte-depleted individual with HIV or AIDS, a mucosal infection develops. The ability of *C. albicans* to invade tissues according to host's type of debilitating condition is evidence of its adaptability. Fungi must be able to grow in the harsh environment of the oral cavity or the gut, where competition for food, pH extremes, and high temperature eliminate other, more benign microorganisms. *Aspergillus* species and *C. neoformans* cells, common in the environment, become airborne and can enter the ventilation systems of health care facilities.

Transfer of technology. Since the 1960s, much has been learned about many fungal pathogens. These scientific advances have had a major translational impact in the development of better diagnostic tests and treatments for these diseases. Molecular biology studies have yielded assays for typing strains of fungal pathogens for epidemiologic studies. Gene manipulation has resulted in the identification of virulence factors. From these types of studies, candidate vaccines may be identified.

Antigens of pathogenic fungi. The identification of major antigens of pathogenic fungi has led to studies of their role in protection against infections. This approach has been a major focus of study with fungi such as *Histoplasma capsulatum, Coccidioides immitis,* and *Blastomyces dermatitidis*. These fungi are acquired from the environment by inhalation of spores; they cause endemic disease because of their geographical restriction to certain parts of the United States. Coccidioidomycosis is found exclusively in the southwestern United States, while the others are concentrated in the eastern half of the country. These organisms can infect the normal host (in the case of *H. capsulatum* and *C. immitis*) and are increasingly observed in the immunocompromised individual.

Promising data. The study of the major antigens of these fungi offers two approaches to the treatment and prevention of fungal diseases. First, the identification of protective antigens will allow the development of vaccines. Second, in individuals who are immunocompromised and susceptible to opportunistic pathogens, prophylactic, passive treatment with protective antibodies made against key antigens may be useful in combination with antifungal drugs to prevent disease. Promising data have emerged on the treatment of cryptococcosis with anti–*C. neoformans* antibody in high-risk individuals. Studies on the immune response to fungi have also identified cytokines (biological modulators produced by immune cells) which help protect against fungal diseases. Again, in the immunocompromised individual with inadequate levels of cytokines, adjunct therapy

with these compounds may be useful in preventing or eliminating diseases.

Diagnosis. The number of fungi which cause disease in the immunocompromised host is increasing. Their identification in clinical specimens is critical to early diagnosis and successful treatment. Ultimately, culture of the organism from the individual and its identification is the most reliable means of verifying a specific infection. In modern clinical laboratories, this may be impossible because many of these organisms take considerable time to grow; also, isolation of the pathogen from the individual is not always successful. For example, in individuals who have undergone transplant surgery or have leukemia, aspergillosis is rarely identified before death. Serology (finding antibody to the pathogen) has been useful in the diagnosis of histoplasmosis and coccidioidomycosis, but is not particularly revealing in the individual with invasive candidiasis who may not be making antibody. Or, since the organism is a normal resident of humans, who usually have anti-*Candida* antibody, finding antibody may be inconsequential. For these diseases, antigen capture assays are useful, but not fully developed except in the case of the polysaccharide capsule of *C. neoformans,* which is detected in individuals by using an antibody-conjugated latex agglutination test (latex beads coated with antibody to capsule agglutinate if the capsule is present in body fluids). Similar approaches are being tested experimentally in the diagnosis of aspergillosis, and the development of assays seems promising. A successful diagnosis should be more likely once the physician is alerted to the likelihood of fungal disease in high-risk individuals.

Treatment. The high impact of fungal infections in modern medicine came rather quickly and, at least for the invasive infections, the choice of an antifungal drug for treatment was limited to a few compounds.

Amphotericin. Amphotericin B is still the drug of choice for many invasive fungal infections, but does have disadvantages. The similarity of fungal and mammalian cells often means that antifungal drugs cannot discriminate between host and fungus; hence, amphotericin B therapy often causes toxicity in the individual. Nevertheless, the drug remains the primary treatment for many invasive fungal diseases.

Less toxic compounds. The impact of fungal infections has led to the development of new, less toxic compounds. The imidazoles and triazoles, inhibitors of fungal ergosterol synthesis, look promising and may eventually replace amphotericin B. Intraconazole and fluconazole are newer triazoles that have undergone considerable study and testing. Fluconazole is used frequently in individuals with HIV or AIDS for the treatment of oral or vaginal candidiasis.

Drug resistance. Just as drug-resistant bacteria are observed as a consequence of the widespread nature of antibiotics, drug-resistant fungi, once thought to be uncommon, are now observed more frequently as physicians treat chronic fungal infections in the immunocompromised individual. For example, resistance to fluconazole by *Candida* is frequently reported in individuals with HIV or AIDS with recurrent oral or vaginal disease. Therefore, in certain instances fungi react to antibiotics in the same way as bacteria. The extended use of antifungals selects for resistant strains of specific organisms. The mechanisms of resistance, while studied less than in their bacterial counterparts, appear to be similar phenotypically but probably different molecularly. The other approach to the treatment of fungi infections is to modify the toxicity and availability of amphotericin B by incorporating the drug in lipid vesicles. Finally, combination drug therapy or, still to be tested, the use of drugs and antibodies or cytokines in combination may offer promise.

For background information *see* ACQUIRED IMMUNE DEFICIENCY SYNDROME (AIDS); ANTIBODY; ANTIGEN; FUNGI; IMMUNITY; IMMUNOLOGICAL DEFICIENCY; MEDICAL MYCOLOGY; MYCOLOGY; OPPORTUNISTIC INFECTIONS; TRANSPLANTATION BIOLOGY in the McGraw-Hill Encyclopedia of Science & Technology.

Richard Calderone

Bibliography. A. Casadevall, Antibody-based therapies for emerging infectious diseases, *Emerging Infect. Dis.*, 2:200–208, 1996; N. Georgopapadakou and T. Walsh, Human mycoses: Drugs and targets for emerging pathogens, *Science*, 264:371–372, 1994; S. Sternberg, The emerging fungal threat, *Science*, 266:1632–1634, 1994.

Meteorite

New interest in the details of condensation of liquids and solids from cosmic gas and dust mixtures has been stimulated by dramatic images captured by the Hubble Space Telescope. These views show partially opaque (dusty) nebulae in which stars are forming, and there is evidence that suggests the existence of planets in several other solar systems. The most primitive materials found in the Earth's solar system are meteorites containing objects called chondrules: small spherical objects that probably formed early in the history of the solar system by coagulation and sedimentation in a dust-enriched inner accretionary disk, subject to melting or revaporization by the early Sun. Since the beginning of mineralogical science in the nineteenth century, great efforts have been made to understand the formation of these objects. They have been examined with the best analytical apparatus. Some workers have statistically cataloged the objects' textures, chemical compositions, and other properties. Others have duplicated some of their features under controlled laboratory conditions. Still others have attempted to calculate how the chondrules could form from a vapor, based on thermodynamic principles and present understanding of the chemistry of minerals, liquids, and gases.

Chondrite meteorites. The carbonaceous chondrite meteorites contain nonvolatile elements in proportions very similar to those in the solar photosphere. Chondrites offer the best physical evidence for the state of matter before and during formation of planet-sized bodies in the solar system. They basically consist of iron-nickel (Fe-Ni) metal, metal sulfide, chondrules, and calcium + aluminum–rich inclusions (CAIs), with chondrules dominant by volume, enclosed in a silicate matrix. The chondrules and CAIs are generally spherical, ranging up to 1.2 mm (0.047 in.) and 2 cm (0.78 in.) in diameter, respectively. Their shapes are characteristic of liquid droplets solidified in space.

The most refractory (high condensation temperature) objects are the CAIs. Type A are dominated by the minerals melilite ($Ca_2Al_2SiO_7$–$Ca_2MgSi_2O_7$, solid solution), hibonite ($CaAl_{12}O_{19}$), and spinel ($MgAl_2O_4$), with minor amounts of perovskite ($CaTiO_3$). Type B include melilite, spinel, fassaitic pyroxene ($Ca[Al,Ti,Mg][Al,Si]_2O_6$, solid solution), and anorthite feldspar ($CaAl_2Si_2O_8$). Most chondrules have major-element compositions similar to terrestrial mantle-derived lavas (basalt to komatiite), and many contain silicate glass, a quenched liquid. Their mineralogy includes Fe-bearing pyroxene and olivine ($[Fe,Mg]_2SiO_4$), sodium (Na)-bearing feldspar, and Fe-, chromium (Cr)-bearing spinels.

It was demonstrated in 1972 that oxide minerals rich in aluminum (Al), titanium (Ti), and calcium (Ca) condense first from a solar gas, followed by magnesium (Mg)–rich silicates. The compositions of CAI type A, then B, followed by metal and chondrules, define a sequence that tracks the expected order of solid condensation from such a gas under equilibrium conditions. Therefore, the compositions and mineralogies of the CAIs and chondrules could originate by accretion of solid mineral grains into nodules at temperatures defined by the phase with the lowest condensation point of those present in the nodule. The textures of many of the inclusion types, particularly the ferromagnesian chondrules, indicate their formation by rapid quenching from independent, partially liquid droplets prior to incorporation into their host meteorites. Current theory calls upon a heating event to melt the precursors of chondrules, and some CAIs, before incorporation into meteorite parent bodies. An alternative hypothesis is that some liquids condensed directly from the nebular gas under equilibrium conditions.

Condensation calculations. The thermodynamic principles most often applied to calculations of solar condensation at temperatures between 1000 and 2000 K (1340 and 3140°F) are time independent. The equilibrium states for a particular fixed bulk (total) composition and total pressure are calculated at a series of successively cooler temperatures by solution of systems of simultaneous equations. The assumption of thermodynamic equilibrium can be questioned on many grounds; however, it serves to constrain what is possible. Even when making the assumption that all constituents reach equilibrium at each temperature step, such calculations are close to the limit of what existing data can constrain.

In all condensation calculations, the gas is treated as a mixture of ideal chemical species, probably a valid assumption at the very low pressures of interest. Over 100 gas species must be considered, and nearly as many pure solid phases. The most difficult minerals to characterize are the solid solutions, those solids incorporating variable proportions of two or more cations on a particular crystalline lattice site. The stabilities, or energetics, of such phases are usually nonlinear functions of composition. Liquids are a more formidable challenge, because they incorporate more elements and their thermodynamic properties are even more nonlinear. All present knowledge of the thermodynamic properties of gas, solid, and liquid phases derives from painstaking experimental work carried out by a myriad of researchers over the past century. Recently, geochemists utilizing computers have been able to synthesize the entirety of the experimental database into optimized formulations capable of describing the minerals and their solid solutions with a single set of internally consistent data and equations.

A good thermodynamic model for liquids in the CaO-MgO-Al$_2$O$_3$-SiO$_2$ system, suitable to model CAIs, was developed in 1983; but it has not been improved since. A much more complete model, called MELTS, including the 15 most abundant oxide components in silicate melts, appeared in 1995. Devised for the study of terrestrial silicate lavas, this model cannot address the compositions low in silicon dioxide (SiO$_2$) content that are typical of CAIs; however, MELTS correctly calculates the inferred crystallization histories and observed final products, given chondrule bulk compositions as input. Currently, no single model for liquids enables exploration of the entire composition range of interest with any accuracy; nevertheless, the existing models nearly overlap, and are compatible with each other in many respects.

The choice of initial bulk composition of gas strongly influences the results of any condensation calculation. The accretionary disk in which chondrules formed was heterogeneous, with parts enriched in dust (preexisting micrometer-scale grains) relative to solar abundances. It has proved instructive to explore the effects of total vaporization of interstellar dust + gas mixtures, followed by condensation upon cooling. Calculations from a variety of initial compositions enable consideration of condensation sequences in different regions of the solar nebula. The composition of the dust that enriches the solar gas is a further variable. A system enriched in a refractory dust component will condense matter at higher temperatures, and consequently will stabilize liquids (see **illus.**). Addition of more oxygen-rich dust will further enhance condensation. Condensation of solids or liquids from gas is also enhanced by increasing total pressure (P^{total}). A total pressure of 10^{-3} bar (10^2 pascals) is considered a maximum, while $P^{\text{total}} = 10^{-6}$ bar (10^{-1} Pa) is more realistic for solar nebular conditions.

Recent results. In 1995, condensation-sequence calculations applicable to CAI condensation were performed, using the 1983 model for CaO-MgO-Al$_2$O$_3$-

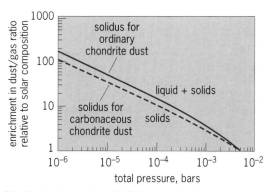

Plot of pressure versus dust/gas ratio, showing the solidus (the line below which liquids are no longer stable) for two possible compositions of dust, enriching a gas of solar composition. At any particular total pressure, increasing the amount of condensable matter, or dust, increases the temperature at which solids condense, and therefore it increases the possibility that liquids will form. 1 bar = 10^5 Pa.

SiO$_2$ liquids. These calculations indicated that only minor enhancements in the dust/gas ratio (relative to solar abundances) or slight increases in P^{total} are sufficient to produce condensate liquids of composition similar to type A CAI. Upon cooling, the resulting liquid would dissolve magnesium oxide (MgO) and SiO$_2$ from the gas, attaining a bulk composition similar to type B CAI. Liquids were found to be stable in a solar gas (dust/gas ratio = 1.0) at $P^{\text{total}} \approx 5 \times 10^{-2}$ bar (5×10^3 Pa), and at $P^{\text{total}} = 10^{-3}$ bar (10^2 Pa) if the dust/gas enrichment ratio is as low as 16. Liquids were found that coexist with olivine, which occurs in nearly all chondrules and also as isolated mineral grains in chondrites. Yet condensation of olivine with a liquid of its own composition at pressures less than ~100 bars (10^7 Pa) had been ruled out in the mid-1960s. The new results illustrate the importance of considering colligative properties in even a simple silicate liquid. (Colligative properties are those that depend on the number of molecules but not their nature.) They also point to the need for considering nonlinear composition effects, or nonideality, in modeling thermodynamically the more complex liquid and solid solutions required for condensation of chondrule compositions.

In 1996, calculations were performed to delineate the minimum P^{total} and dust/gas enrichment ratio at which liquids of any composition are stable (the solidus). The minimum temperature was found to be near 1500 K (2650°F) in all cases. For solar gas, the solidus was at $P^{\text{total}} \approx 3.5 \times 10^{-3}$ bar (3.5×10^2 Pa). At $P^{\text{total}} = 10^{-6}$ bar (10^{-1} Pa), a dust/gas ratio of ~165 defined the solidus. That is, no liquids were found to be stable at a lower dust/gas ratio at this P^{total}. These calculations are for dust of the composition of ordinary chondrites, which are strongly depleted in the volatile elements carbon (C), nitrogen (N), oxygen (O), and hydrogen (H) relative to solar abundances. If a dust of carbonaceous chondrite composition is used instead, the solidus dust/gas ratio at $P^{\text{total}} = 10^{-6}$ bar (10^{-1} Pa) is only ~105.

The liquids found to exist at the solidus in these calculations are not of chondrule composition. They contain too little ferrous oxide (FeO) and sodium monoxide (Na_2O). However, only modest increases in P^{total} or the dust/gas ratio should be sufficient to make chondrule liquids thermodynamically stable, especially if carbonaceous chondrite dust enriches the gas. Such a dust adds more oxygen than ordinary chondrite dust, in combination with other volatile elements. Upon conversion to gas with increased temperature, this oxygen becomes available to the refractory elements. The concentration of FeO in the condensed fraction is particularly enhanced by this mechanism, further promoting the stability of liquids. Ongoing work is focused on incorporating existing models for gas and solid phases with the MELTS liquid model to explore more accurately the condensation of liquids from cosmic gases, and the implication for the origin of chondrules.

For background information *see* COSMOCHEMISTRY; INTERSTELLAR MATTER; METEORITE; MINERALOGY; SOLID SOLUTION; THERMODYNAMIC PRINCIPLES in the McGraw-Hill Encyclopedia of Science & Technology.   Denton S. Ebel

Bibliography. M. S. Ghiorso and R. O. Sack, Chemical mass transfer in magmatic processes: IV. A revised and internally consistent thermodynamic model for the interpolation and extrapolation of liquid-solid equilibria in magmatic systems at elevated temperatures and pressures, *Contrib. Mineral. Petrol.*, 119:197–212, 1995; L. Grossman, Condensation in the primitive solar nebula, *Geochim. Cosmochim. Acta*, 36:597–619, 1972; J. A. Wood, On the origin of chondrules and chondrites, *Icarus*, 2:152–180, 1963; S. Yoneda and L. Grossman, Condensation of CaO-MgO-Al_2O_3-SiO_2 liquids from cosmic gases, *Geochim. Cosmochim. Acta*, 59:3413–3444, 1995.

Fig. 1. Hot acid solfataric fields (low pH and high temperature), Naples, Italy. *(R. Koch and G. Antranikian)*

are usually associated with tectonically active zones. Terrestrial biotopes are mainly solfataric (relating to a volcanic area) fields which consist of soils, mud holes, and surface waters heated by volcanic exhaust from magma chambers below (**Fig. 1**). The surface of solfataras is rich in sulfate and is very acidic (pH 0.5–4.0). Marine hydrothermal systems are situated in shallow and abyssal depths, and consist of hot fumaroles, springs, and deep-sea vents (''black smokers'') with temperatures up to 750°F or 400°C (**Fig. 2**). Due to the presence of reducing gases, most biotopes of thermophiles do not contain oxygen and are anaerobic. However, the surface of the terrestrial solfataric fields contains high amounts of oxygen and therefore harbors aerobic microorganisms.

Additional communities of thermophiles have been discovered in artificial high-temperature biotopes, such as smoldering coal refuse piles and geo-

Microbiology

Extremophiles are microorganisms that can live in environments which are considered by humans to be extreme. These microorganisms, mostly prokaryotes belonging to the domains Bacteria and Archaea, are adapted to survive in ecological niches such as at high or low temperatures (160–230°F or 70–110°C; 32–50°F or 0–10°C), extremes of pH (acidic or alkaline conditions; pH below 4 and above 9), high salt concentrations (above 20% salt), and high pressure (above 300 bars or 290 atmospheres). Extensive work since the mid-1980s has clearly demonstrated that extremophiles are interesting objects for basic research and innovative biotechnology. *See* BIOTECHNOLOGY.

Biotopes of extremophiles. The extremophiles include three groups: thermophiles and psychrophiles, alkaliphiles and halophiles, and acidophiles and barophiles.

Thermophiles and psychrophiles. The most common biotopes (environments) in which thermophiles (heat lovers) grow optimally are of geothermal origin, and

Fig. 2. Black smoker, a hot environment (750°F or 400°C) in the ocean approximately 6600 ft (2000 m) deep. The superheated saline water is highly mineralized. The precipitation of insoluble metal salts forms the black smoker. *(G. Antranikian)*

thermal power plants. They also have been isolated from oil-bearing, deep, geothermally heated soils. Most psychrophilic (cold-loving) microorganisms thrive in polar regions, especially in Antarctic sea water and deep-sea sediments which represent permanently cold habitats ranging from 28 to 39°F (−2 to 4°C). *See* HYDROTHERMAL VENT.

Alkaliphiles and halophiles. Alkaline ground water, soda lakes, and deserts harbor microorganisms that are able to grow at pH values above 9 (alkaliphiles) and at high salt concentrations (halophiles). Halophilic microorganisms colonize saline environments, where the salinity may range from that of sea water (3%) to saturated solutions (35%, as in dead-sea, hypersaline lakes and in saline soils). Halophilic microbes can also be isolated from artificial environments such as salted food.

Acidophiles and barophiles. Extremely acidophilic microorganisms are usually found in terrestrial and marine solfataric fields. Many microbial communities in these environments are adapted to grow at low pH values (about 1.0–3.5) and high temperatures (158–185°F or 70–85°C) and are designated thermoacidophiles. Microorganisms that are able to survive only at high pressure in the deep sea are known as strict barophiles. Such microbes have been discovered in deep-sea sediments collected from depths of 9800–32,800 ft (3000–10,000 m).

Sampling and isolation of extremophiles. Samples such as sediments, waters, and rocks can be obtained from various sources and transported to the laboratory. To isolate microorganisms that live in environments without oxygen (anaerobes), the samples should be kept in bottles under reduced conditions. To support growth of microbes in the laboratory, samples are then incubated with media containing various substrates (carbohydrates, proteins, peptides, amino acids), vitamins, and minerals. The organisms that are able to grow in the enrichment cultures are transferred to plates containing solidified media. In most cases, media can be solidified by the addition of 1.5% agar. To solidify media at high temperatures, agar should be replaced by more heat-stable polymers such as gellan gum. Small colonies (0.08–0.16 in. or 2–4 mm in diameter) that appear on plates after 2–5 days of incubation can be isolated and studied in detail. By applying various techniques [such as ribosomal ribonucleic acid (rRNA), deoxyribonucleic acid (DNA)–DNA hybridization], the phylogenetic position of extremophiles can be determined. It has been estimated that so far only 1% of extremophilic microorganisms have been isolated from nature. Therefore, further efforts should be made to isolate uncultivated microorganisms that thrive in nature.

Thermostability. The isolation of various groups of extremophilic microorganisms from extreme environments demonstrates an unexpected complexity of these unexplored ecosystems. The existence of Archaea (such as *Pyrococcus* and *Pyrodictium*) that live at temperatures near the boiling point of water demonstrates that still unrecognized thermostabiliz-

ing principles must exist. The principles of heat stabilization of cellular components such as proteins, DNA, RNA, and adenosine triphosphate (ATP) are still unknown. Most hyperthermophiles (microorganisms growing above 176°F or 80°C) grow in the absence of oxygen (anaerobic) and belong to Archaea. Since the stability of biomolecules above 212°F (100°C) decreases rapidly, the maximal growth temperature at which life can exist may be 266°F (130°C). Unlike the enzymes from microorganisms living at 99°F (37°C; mesophiles), the enzymes from thermophiles are adapted to high temperatures and are even active at 266°F (130°C), for example, amylase from *Pyrococcus woesei.* Experimental evidence indicates that small and subtle changes (for example, increased salt bridges) account for the increased thermostability of enzymes. Another group of extremophiles that prefer to live at low temperatures are the psychrophiles. Microorganisms have been identified in deep-sea sediments and Antarctic sea water at temperatures lower than 37°F (3°C).

Because of their requirement of acidic pH (0.5–4.0), acidophilic thermophiles can survive at low pH and high temperatures. Acidophilic microorganisms belong to Archaea (for example, *Sulfolobus* and *Acidianus*) and can grow under aerobic and anaerobic conditions. Alkaliphilic microorganisms (growth above pH 9.0) are widely distributed throughout the world in soil and alkaline lakes. Besides bacteria, various kinds of microorganisms, including yeast and fungi, have been isolated. Thermophilic alkaliphiles that grow above 140°F (60°C) and pH 9 have been isolated recently from alkaline lakes such as Bogoria in Kenya. So far, microorganisms which are able to grow above 194°F (90°C) and pH 10 have not been isolated. A variety of organisms (halophiles) inhabit hypersaline environments, ranging from higher organisms such as the brine shrimp to microorganisms including Archaea (*Methanohalobium*) and Bacteria (*Halobacterium*).

Biotechnological applications. The application of thermophilic, acidophilic, alkaliphilic, halophilic, and barophilic microorganisms in industrial processes opens up a new era in biotechnology. Each group has unique biochemical features which can be exploited for use in biotechnological industries (see **table**). The predominant reasons for selecting enzymes from extremophiles are the high stability and reduced risk of contamination. The additional benefits in performing industrial processes at high temperature include improved transfer rates, lower viscosity, and higher solubility of substrates. Because of the unusual properties of these enzymes, they are expected to fill the gap between biological and chemical processes. The commercial importance of enzymes from thermophiles is illustrated by the application of the polymerase chain reaction (PCR), a rapid and efficient method of amplifying specific DNA sequences. For this application, a heat-stable DNA-polymerase from the hyperthermophile *Pyrococcus* sp. (optimal growth temperature of 212°F or 100°C) is used.

Biotechnological applications of extremophiles

Microorganisms	Temperature, pH	Enzymes, endogene compounds	Applications, products
Thermophiles	50–110°C	Amylases	Glucose, fructose for sweeteners
		Xylanases	Paper bleaching
		Proteases	Amino acid production from keratins, food processing, baking, brewing, detergents
		DNA-polymerases	Genetic engineering
Psychrophiles	5–20°C	Neutral proteases	Cheese maturation, dairy production
		Proteases	
		Amylases	
		Lipases	Detergents
		Polyunsaturated fatty acids	Pharmaceuticals
		Ice-protein	Artificial snow
Acidophiles	pH < 2	Sulfur oxidation	Desulfurization of coal
Alkaliphiles	pH > 9	Proteases	
		Amylases	
		Lipases	Detergents
		Cyclodextrins	Stabilization of volatile substances
		Antibiotics	Pharmaceuticals
Halophiles (3–20% salt)		Carotene	Food coloring
		Glycerol	Pharmaceuticals
		Compatible solutes	Pharmaceuticals
		Membranes	Surfactants for pharmaceuticals

Catalysts. The recent developments clearly show that extremophiles are an attractive source of novel catalysts for industrial use. Some of these enzymes have been isolated and their genes successfully cloned and expressed in mesophilic microorganisms (growth at 99°F or 37°C) such as *Escherichia coli.* Various polymer-degrading enzymes from extremophiles such as amylases, xylanases, cellulases, lipases, and proteases will play an important role in food, detergent, paper and pulp, and pharmaceutical industries. Starch-degrading enzymes from thermophiles are attractive candidates for application in the starch industry for the production of high fructose syrup at elevated temperatures.

Detergents and decontaminants. Recently, the effects of enzymatic debleaching in pulp-milling industries has been studied by using thermostable alkaline xylanases. The main industrial application of alkaliphilic enzymes such as proteinases and cellulases is in detergents. Halophilic microorganisms are already used for the production of bacteriorhodopsin and β-carotene. It has been shown that a number of thermophilic microorganisms are able to degrade toxic substances that are hydrophobic and insoluble at low temperatures. Soils that are contaminated with linear and branched alkanes (C20–C40) and polyaromatic hydrocarbons (for example, pyren, phenanthren, and anthracen) can be decontaminated with newly isolated thermophilic microorganisms. The biotechnology of extremophiles is still in its infancy but clearly has important and far-reaching implications. As the enzymology of this group of microorganisms is better understood, protein engineering methods will also be applied to tailor more efficient enzymes for specific biotechnological applications.

For background information *see* ARCHAEBACTERIA; BACTERIAL PHYSIOLOGY AND METABOLISM; BIOTECHNOLOGY; DEOXYRIBONUCLEIC ACID (DNA); ESCHERICHIA; HALOPHILISM (MICROBIOLOGY); MICROBIAL ECOLOGY; MICROBIOLOGY in the McGraw-Hill Encyclopedia of Science & Technology.   G. Antranikian

Bibliography. G. Antranikian, W. N. Konings, and W. M. de Vos (eds.), Extremophiles, *FEMS Microbiol. Rev.*, 18:89–285, 1996; C. Ratledge (ed.), Special topic review: Biotechnology of extremophiles, *World J. Microbiol. Biotechnol.*, 11(1):7–131, 1995.

Microsensor

A sensor is a device that converts a nonelectrical physical or chemical quantity, such as pressure, acceleration, temperature, or gas concentration, into an electrical signal. Sensors are an essential element in many measurement, process, and control systems, with countless applications in the automotive, aerospace, biomedical, telecommunications, environmental, agricultural, and other industries. Technological advances in the semiconductor industry since the early 1980s have led to the production of very small sensors, or microsensors, with physical dimensions in the submicrometer to millimeter range. The stimulus to miniaturize sensors lies in the enormous cost benefits that are gained by using semiconductor processing technology, and in the fact that microsensors are generally able to offer a better sensitivity, accuracy, dynamic range, and reliability, as well as lower power consumption, than their larger counterparts.

Technology and materials. Microsensors are typically batch fabricated from silicon wafers by using stan-

Fig. 1. Scanning electron micrograph of a polysilicon resonant structure made with a surface micromachining technique. *(From C. J. Welham et al., A laterally driven micromachined resonant pressure sensor, Sensors Actuat. A, 52:86–91, 1996)*

dard semiconductor process technologies in combination with specially developed processes. This technology, known as micromachining, allows hundreds of complex microsensors (or microstructures) to be produced on a single silicon wafer, resulting in a very low unit cost. This process technology is accurate and repeatable. A wide variety of passive and active materials are used to make microsensors. Single-crystal silicon is commonly used as a passive substrate material, although other substrate materials, such as gallium arsenide, quartz, and silicon carbide, are used for more specialized microsensor applications. These base materials are used in combination with a variety of thin films, including polysilicon, oxides, nitrides, and metals. It is often necessary to deposit an active material which is essential to the operation of the microsensor. Examples include zinc oxide, which is commonly used as the active material in piezoelectric sensors, and conducting polymers, which are commonly used in chemical microsensors.

Engineering a microsensor (sometimes called microengineering) requires the use of a number of micromachining processes, together with an understanding of micromechanics and microelectronics. The term bulk micromachining is often used for a set of processes, such as anisotropic etching, laser ablation, silicon-wafer bonding, and deep reactive ion etching, that enable the three-dimensional sculpting of single-crystal silicon to make a small structure, such as a diaphragm for a pressure sensor or a pressure switch, or a suspended mass structure for an accelerometer and flow sensor. Surface micromachining refers to a set of processes based upon deposition, patterning, and selective etching of thin films to form a free-standing microstructure on the surface of a silicon wafer. **Figure 1** shows a surface-micromachined microflexural resonator that has been integrated onto a thin silicon diaphragm to form a pressure sensor. The resonator is electrostatically driven into a lateral resonant mode of oscillation (around 50 kHz) by using a comb capacitor,

and its motion is sensed by using another comb capacitor. The deflection of the diaphragm under an applied pressure stretches the microresonator, thus changing its spring rate and its fundamental resonant frequency.

Mechanical microsensors. Mechanical microsensors form perhaps the largest family of microsensors because of their widespread availability. Microsensors have been produced to measure a wide range of mechanical measurands, including force, pressure, displacement, acceleration, rotation, and mass flow. Force sensors generally use a sensing element that converts the applied force into the deformation of the elastic element. **Figure 2** shows a scanning electron micrograph of a simple microcantilever stylus structure that is used in an atomic force microscope for analyzing surfaces at an atomic scale of resolution. The tip of a flexible force-sensing cantilever stylus (with a tip curvature radius of a few nanometers) is scanned over the sample surface. The forces acting between the tip and sample cause minute deflections of the cantilever, which can be detected optically, capacitively, or with integrated piezoresistors to produce an atomistic surface image.

Solid-state microsensors. Some microsensors can be made by using conventional bipolar or metal oxide semiconductor (MOS) technology, and a variety of radiation sensors are available that can detect visible and near-infrared radiation by using, for example, silicon *pn* diodes, avalanche photodiodes, and pyroelectric devices. Digital thermometers can be engineered by using a thermally sensitive device, such as a resistor, *pn*-junction diode, or transistor. A variety of magnetic microsensors, based upon Hall-plate devices (which are sensors that measure the magnetic field strength through the Hall effect), magnetoresistors, magnetodiodes, and magnetotransistors, are also available.

Chemical and biochemical microsensors. Two applications for chemical and biochemical microsensors are environmental monitoring and medicine. Both are

Fig. 2. Micrograph of a silicon microcantilever above an integrated circuit. *(From J. Brugger et al., Silicon cantilevers and tips for scanning force microscopy, Sensors Actuat. A, 34:193–200, 1992)*

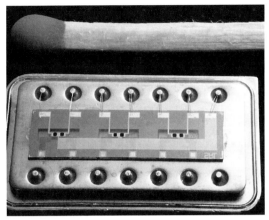

Fig. 3. Micromachined six-element chemical microsensor array in a dual in-line package. *(From J. W. Gardner et al., Integrated sensor array for detecting organic solvents, Sensors Actuat. B, 26-27:135–139, 1995)*

relatively undeveloped, but should be of great importance in the future.

Unfortunately, chemical microsensors are the least well developed, for many reasons. For example, it has proved difficult to make a stable sensor that is sensitive to just the one gas of interest. One approach successfully applied to overcome this problem has been to employ an array of nonspecific gas sensors in a microprocessor-based instrument. The microprocessor is able to apply an appropriate pattern-recognition technique to extract the required information from the output of the microsensor array. This approach has opened up an enormous range of applications for gas and odor measurement in many different areas, such as the automotive, medical, and food–beverage industries and environmental monitoring. In addition, there have been advances in the design and synthesis of new materials such as conducting polymers, which coupled with the above approach make high-quality gas microsensors a distinct possibility in the near future. Microsensor array devices are also able to measure multiple measurands and compensate for interfering variables, such as temperature. For example, one chemical microsensor **(Fig. 3)** is employed as the sensing element in an electronic-nose instrument.

Applications in the medical industry may involve monitoring blood, urine, and breath, which contain a wealth of information about the patient's state of health. Only a few such devices now exist. Examples include a glucose biochemical microsensor and ion-selective field-effect devices used to measure blood pH. The use of microsensors to gather medical diagnostic information is an attractive proposition, and eventually there may even be implanted microsensors to diagnose health problems, using smell-sensitive array devices.

Smart sensors. Since the processes used to fabricate microsensors are similar to the ones used to produce conventional integrated circuits, signal-conditioning electronics such as analog-to-digital converters can be integrated on the same chip, resulting in an integrated microelectromechanical component. Such a smart or intelligent sensor is able either to process information itself or to communicate with an embedded microprocessor. One notable example is a microaccelerometer developed for the automotive industry to control air-bag inflation. This inexpensive monolithic accelerometer comprises a surface-micromachined polysilicon sensor and bipolar–MOS interface circuitry on a single silicon chip. The sensor has been developed for a measurement range of $\pm 5\ g$ (where g is the acceleration of gravity, 32 ft/s^2 or 9.8 m/s^2), operates on a single 5-V supply, and has a power consumption of 40 mW. As the level of integration increases, a single-chip device incorporating microsensors, signal conditioning, and a microprocessor can be expected.

For background information *see* ACCELEROMETER; BIOELECTRONICS; INTEGRATED CIRCUITS; MICROPROCESSOR; OPTICAL DETECTORS; PRESSURE TRANSDUCER; SCANNING TUNNELING MICROSCOPE; SEMICONDUCTOR in the McGraw-Hill Encyclopedia of Science & Technology.   Andrew C. Pike; Chris J. Welham; Julian W. Gardner

Bibliography. J. W. Gardner, Microsensations, *IEE Rev.*, pp. 185–188, September 1995; J. W. Gardner, *Microsensors: Principles and Applications*, 1994; E. Kress-Rogers (ed.), *Handbook of Biosensors and Electronic Noses*, 1997; S. Middlehoek and S. A. Audet, *Silicon Sensors*, 1989.

Mineral

Repetitive banding in rocks frequently may be traced directly or indirectly to astronomical cycles that controlled the deposition of the minerals constituting the rock. Varves, for example, are laminated shales from glacial lake beds and exhibit oscillations between dark winter clays and lighter summer silts. By contrast, some instances of petrologic zonation are explicable only in terms of chemical fluctuations in the magmas or aqueous liquids from which the rocks crystallized. This kind of chemical banding reflects steady-state reactions that are very far removed from thermodynamic equilibrium. The Belousov-Zhabotinskii reaction is an (incompletely understood) example of a chemical feedback system that exhibits periodic behavior arising from dynamical disequilibrium. During Belousov-Zhabotinskii reactions, concentrations of certain chemical species repeatedly rise and fall as reactants enter the system at constant influx rates. The production of repetitive sequences in rocks by such disequilibrium reactions is remarkable, as it implies the persistence of metastable chemical conditions over geological time scales.

Provenance of agates. Agates are among the most beautiful and enigmatic of naturally banded rocks. Recent investigations suggest that their oscillatory layering arises from self-sustaining fluctuations in the fluids from which they formed. Agates typically are found as spherical or amygdaloidal (almond-shaped) nodules within basaltic hosts, and can range from less than a millimeter to several meters in diameter. Most geoscientists agree that agates occupy cavities that originated as gas bubbles within cooling, depressurized magmas. As waters laden with dissolved silica (SiO_2) percolated through these cavities, crystalline silica precipitated onto the cavity walls, depositing layer after layer until either the dissolved silica was spent or the vug (cavity) was completely filled. The source of these fluids remains controversial, with some evidence implicating cold, surface waters and some suggestive of deep, hydrothermal reservoirs. Regardless of its origin, silica is by far the most abundant constituent of all agates, generally accounting for more than 99% by weight.

The successive deposition of silica layers that underlies the formation of agate is the basis for a richly diverse array of repetitive sequences. With their colorful zonation, agates with dazzling pigmentation have been valued as gems since biblical times. When colored bands are present, they can range from a micrometer to more than a centimeter in thickness (**illus.** *a*). Examination of such specimens by light optical and transmission electron microscopy suggests that the coloration usually is due to the inclusion of metal oxide crystallites, particularly hematite (Fe_2O_3). The distribution of these crystallites and the sequence of pigmented zones usually are aperiodic in agates; thus, rather than representing an oscillatory chemical system, these colored bands likely reflect random changes in the trace-metal compositions of the depositional fluids.

Chalcedony fiber twisting. More interesting than the intergrowth of silica with metal oxides are the fluctuations that occur within the crystalline silica itself. The precipitation of silica on the cavity walls occurs by the nucleation and growth of siliceous fibers known as chalcedony. These whiskers grow perpendicular to the vug wall, and they twist about the fiber axis with periodicities of tens to hundreds of micrometers. The twisting of these fibers indicates the existence of a residual torque during the growth process, probably as a result of numerous screw dislocations (crystallographic missteps that can activate rapid, spiraling fiber growth). In images of chalcedony bundles taken with cross-polarized light (illus. *b*), the twisting of the fibers is revealed by jagged black stripes called zebraic extinction bands.

Iris banding. In addition to, and probably independent from, the helical fiber twisting is an oscillation known as iris banding. In 1813 the Scottish scientist David Brewster noted that thin slices of agate can yield spectral colors when held against a light, but not until 30 years later did he realize that this iridescence arises from periodic, concentric striations that act as a diffraction grating for visible light (like the grooves on a compact disc). This phenomenon indicated that the periodicities of these striations must be close to the wavelength of visible light (380–760 nanometers) and must be regular and extensive.

Since then, repetitions of more than 8000 bands

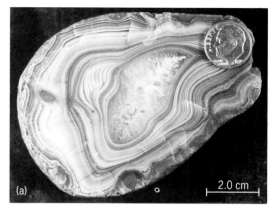
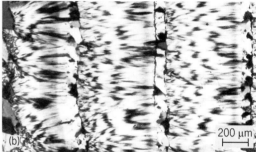
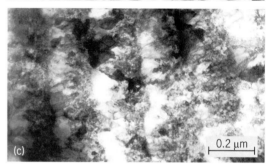

Banding at three length scales in a single iris agate from Minas Gerais, Brazil (U.S. National Museum #83325). (*a*) A polished cross section of the agate reveals a chalcedony rim around an interior quartz druse. Within the chalcedony rim, white bands alternate with darker bands pigmented by iron oxide crystallites. (*b*) A cross-polarized light micrograph reveals two independent oscillations. First, jagged black extinction bands reflect periodic twisting of chalcedony fibers. Second, thin bands of equigranular quartz alternate with thick bands of fibrous chalcedony. (*c*) Vertical iris bands are seen as light coarse-grained zones (L bands) that alternate with darkly mottled fine-grained zones (H bands) in this transmission electron micrograph. *(From P. J. Heaney and A. M. Davis, Observation and origin of self-organized textures in agates, Science, 269:1562–1565, 1995)*

have been found within single agates. Optical studies using visible light have demonstrated that iris banding is restricted to the fibrous chaldecony layers and the bands are oriented perpendicular to the twisted fiber growth axes. Although iridescence is not obvious in all agates (especially the heavily pigmented), most specimens reveal iris striations on close examination. Nevertheless, the true nature of these striations remained a mystery for the 150 years following Brewster's work. Some authors attributed them to the rhythmic segregation of amorphous opal within chalcedony, and others to regular changes in fiber orientation; however, subsequent examination of iris agate by x-ray diffraction and electron microscopy disproved both hypotheses.

The model that gained the greatest currency was proposed by C. Frondel. He etched sawed sections of iris agate with hydrofluoric acid, and on investigating the partially dissolved surfaces with a scanning electron microscope he discovered that the concentric striations constituting iris banding actually consist of alternating zones with different physical properties. One zone (the L band) is highly resistant to acid etching, whereas the neighboring zone (the H band) is highly susceptible. In the simplest case, iris agate consists of rigorous oscillations between H and L bands. Knowing that even trace quantities of interstitial hydrogen can dramatically weaken the structure of quartz, Frondel proposed that the difference in etching rates between the H and L bands arises from modulations in hydrogen concentration parallel to the fiber direction.

In order to test this hypothesis, the concentration of hydrogen and other impurity elements within iris agates was investigated with secondary ion mass spectrometry (SIMS). It was possible to tunnel through a series of iris L and H layers by bombarding an agate surface with oxygen-16 ions, and the ablated material was analyzed as it was being produced. The results obtained in this study did not support Frondel's model. Although SIMS is capable of detecting small changes in hydrogen concentration, no fluctuations in hydrogen content between the L and H zones were observed.

Brazil twins and moganite. In the absence of evidence for compositional variations, researchers next studied modulations in structure and found that these occur in abundance. Images of iris bands obtained by transmission electron microscopy (TEM; illus. *c*) reveal that the striations correspond to alternations in average crystal size. The H bands, which are easily etched with acid, are very fine grained (with crystal diameters of 5–10 nm), whereas the acid-resistant L bands contain coarse crystals (100–1000 nm in diameter). Moreover, the fine-grained H bands differ from the coarse-grained L bands in other important ways. Quartz, like many organic molecules, has a structural handedness. Crystals that contain right-handed zones intergrown with left-handed zones are described as Brazil twinned. The study with the transmission electron microscope revealed that the H bands contain a profusion of Brazil twin boundaries, whereas these structural defects are rare in the L bands.

When right- and left-handed quartz alternate regularly at the atomic scale, they create a mineralogical hybrid called moganite, and it now is known that virtually all chalcedony fibers contain intimate intergrowths of quartz and moganite. A recent study indicates that the moganite within the chalcedony fibers of iris agate is segregated entirely into the fine-grained, Brazil-twinned H bands. In summary, the striations that correspond to iris banding represent oscillations in grain size, Brazil-twin defect concentration, and moganite content.

Chemical origin of iris bands. The cause of the extreme regularity in iris banding remains unclear. It has been proposed that pure quartz is produced when dissolved silica is present in solution as individual units, or monomers, of $Si(OH)_4$. By contrast, when the concentration of dissolved silica becomes so great that the silica monomers link to form extended chains, or polymers, Brazil twinning is induced and quartz-moganite mixtures precipitate. This model implies that the modulation between defect-free quartz L bands and highly defective quartz-moganite H bands results from oscillations in the dissolved silica content in the depositional fluid.

This hypothesis suggests a chemical feedback that may explain the periodic quality of the iris bands. Field and laboratory evidence indicates that the spiral growth of chalcedony from polymeric silica occurs much more rapidly than the slow growth of defect-free quartz crystals from monomeric silica. If the influx of dissolved silica monomer into an agate cavity outpaces the loss of silica upon precipitation, the concentration of silica within the vug will gradually increase. At some critical concentration of silica, polymerization will ensue, and fibrous chalcedony will precipitate. As this is a rapid reaction, the deposition of silica now outpaces the influx into the agate cavity, and dissolved silica concentrations decrease. When levels are sufficiently low, polymerization cannot be sustained, and defect-free quartz again precipitates from monomeric silica. Thus, iris textures in agates may reflect a steady-state disequilibrium in the chemistry of the system.

In actuality, structural banding within agates occurs not only at the hundred-nanometer scale. Examination of agates with light optical microscopy reveals banding between defective chalcedony fibers and defect-free quartz over wavelengths of hundreds of micrometers (illus. *b*). Moreover, agates in hand specimens contain modulations from chalcedony to coarse-grained quartz over centimeter scales (illus. *a*). Just as fossil corals may record yearly rhythms superposed upon diurnal bands, agates can register chemical modulations of different periodicities in a hierarchical fashion. The explanation for this self-similarity in defect textures remains an exciting area for future exploration.

For background information *see* AGATE; CHALCEDONY; MAGMA; MINERAL; SECONDARY ION MASS SPECTROMETRY (SIMS); SILICA MINERALS; VARVE in the McGraw-Hill Encyclopedia of Science & Technology.   Peter J. Heaney

Bibliography. C. Frondel, *The System of Mineralogy*, 7th ed., vol. 3, 1962; P. J. Heaney and A. M. Davis, Observation and origin of self-organized textures in agates, *Science*, 269:1562–1565, 1995; P. J. Heaney and J. E. Post, The widespread distribution of a novel silica polymorph in microcrystalline quartz varieties, *Science*, 255:441–443, 1992; Y. Wang and E. Merino, Origin of fibrosity and banding in agates from flood basalts, *Amer. J. Sci.*, 295:49–77, 1995.

Mineralogy

Important evidence for microbial mediation in rock–water interactions has been recognized since the mid-1980s. Bacterial activity has been shown to contribute not only to the weathering and leaching of rocks but also to the deposition of new minerals. One example is the concentration of large amounts of iron and silica in sediments, such as the Banded Iron Formations, which had been difficult to explain by abiogenic physicochemical means alone. Since the discovery in 1954 of coccoid and filamentous microfossils in the Early Proterozoic Gunflint chert formation of Canada and the United States, it has been realized that microorganisms would likely have influenced the formation of such deposits. Today many of the reactions involved in the deposition of sediments under natural conditions are thought to be microbially mediated.

Bacterially mediated mineralogical change must be thermodynamically possible. This mediation, which generally occurs at the rock–water interface, can be likened to that of a living catalyst that greatly decreases reaction times. There are three important areas of bacterial mediation: absorption of nutrients, metabolism to obtain energy, and physical biosorption. These processes can also interact with the local environment.

Bacterial growth. Bacteria are microscopic organisms commonly consisting of a single cell, with a membrane that separates the metabolically active unit from the external medium. Bacteria can grow wherever water is present in the liquid state, and they are able to exist in such extreme environments as deep-sea thermal vents, hot springs, saline lakes, and deep oil wells, as well as in the nutrient-poor oceans and in deeply buried sedimentary rocks. The two most important bacterial reactions in the natural environment are biosorption and metabolic biomineralization.

It is difficult to show that bacteria are instrumental in the deposition of sediments, since few bacterial remains or indirect chemical evidence have been found within sediments. In general, bacteria preferentially metabolize the lightest isotope available, so that carbonates and kerogen that are formed biologically will contain greater amounts of light carbon (more ^{12}C compared to ^{13}C). Bacteria are also important as participants in inorganic chemical reactions, for they are able to selectively gather nutrients into the cell, and to modify them for use either as cellular building blocks or as an energy source for growth, thus transforming the surrounding natural environment.

Biosorption. Since bacteria have surface charges, they make excellent nucleation sites for mineral accumulation. Wherever nutrients are limited in marine and terrestrial ground waters, bacteria survive by the formation of a biofilm or mat that contains a symbiotic consortium of microorganisms surrounded by extracellular slime (polymer formation outside the cell wall). This slime mediates the surrounding environment, but it is also a site for mineral nucleation since both the slime and the cell walls have unsatisfied charges that sorb metals and silicates. Most of the surface charges are negative, allowing the positively charged metal cations in the surrounding aqueous medium to react rapidly. However, since silica is not metabolized by bacteria and since physical biosorption is the only type of interaction, positively charged aluminum and iron are needed not only to act as a bridge between the negatively charged biofilm and the silicate anion but also to form biogenic aluminosilicate precipitates.

An example of highly mineralized microbial biofilm communities are those found on submerged riverine rocks. In these biofilms, remains of individual bacterial cells indicate that cells act as template for the deposition of amorphous-to-microcrystalline metal sulfides, iron oxides, silica crystallites, and limonitic clays. Initially the grain size of the precipitate is extremely small (a few nanometers). The size increases over time, and the precipitate may become ordered and crystalline. In some cases, completely mineralized microbial cellular forms are preserved as fossils. Iron and aluminum silicates are always formed in riverine biofilms, indicating that the biomineralization process is associated with the anionic nature of the cell wall rather than with the nature of the rock they are attached to.

Metabolic biomineralization. Life on Earth is based on carbon, comprising half the cell mass. The source of this carbon is ultimately atmospheric carbon dioxide, which may be precipitated as sedimentary carbonates or converted into organic carbon by plants or bacteria. In addition, carbon is commonly the source of energy for the cell. However, in most natural environments carbon is sufficiently scarce that bacteria must utilize an alternative energy source. The maintenance of the complex biological structures of the living cell is thermodynamically unfavorable, so to sustain themselves and to grow, bacteria couple energy-producing oxidation-reduction (redox) reactions to their endergonic biosynthetic ones. Redox reactions are able to transfer electrons from a donor (reducing potential), which becomes oxidized, to an acceptor, which is then reduced. It is these energy-producing reactions that transform the natural environment. The redox reaction that occurs most often is the reduction of iron, but reduction of nitrate, manganese, uranium, sulfate, and carbon dioxide occurs where they are available. Evidence of this bacterial reduction can be seen in locally bleached spots of otherwise red, ferric-rich rocks (ferric iron being in the highest oxidation state).

Iron reduction. Since iron is extremely common in the Earth's crust, it is readily available as a source of energy in nutrient-poor environments, even though the amount of energy that can be extracted from it is small. The reduction of ferric to ferrous iron is commonly coupled to the oxidation of organic matter. This reaction is one of the most important

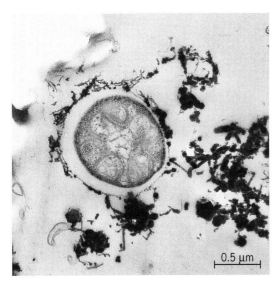

Transmission electron micrograph of a cell surrounded by precipitated iron from a microbial biofilm that produces both hematite and siderite. *(Courtesy of Danielle Fortine)*

geochemical processes in anaerobic sedimentary environments. It releases highly insoluble ferric iron from the rock mass into the ground water by reducing it to the more soluble ferrous form, and is responsible for the generally reduced state of deep ground water.

When the reduction of ferric iron is coupled to the oxidation of organic matter, bacteria can produce large quantities of ultrafine-grained magnetite ($FeO \cdot Fe_2O_3$), the possible source of the large masses found in the Banded Iron Formations. Recently, it has been established that reduced iron (Fe^{2+}) can also react with carbonate ion (CO_3^{2-}) to form siderite ($FeCO_3$), another component of iron ores (see **illus.**). Another common process in the reprecipitation of dissolved iron is chemical reaction with bacterially produced sulfide to form amorphous iron sulfides that mature to pyrite (FeS_2).

Iron oxidation. Bacterial oxidation of iron at neutral pH initially forms a gel of iron oxides and hydroxides, which alters first to ferrihydrite ($5Fe_2O_3 \cdot 9H_2O$) and then to hematite (Fe_2O_3). However, when reduced iron minerals such as pyrite come into contact with an oxidizing atmosphere, either by erosion or by mining, they can be used by bacteria as a substrate. An example is the formation of acid mine drainage by *Thiobacillus ferrooxidans*, which can utilize pyrite oxidation as a source of energy. Since this reaction is thermodynamically inefficient, a large amount of pyrite must be degraded to provide sufficient energy for cell growth, with the concomitant production of sulfuric acid (H_2SO_4).

Biogeochemistry. As noted earlier, bacteria mediate the environment through absorption of nutrients, metabolism, and physical biosorption. Physical biosorption is responsible for the formation of many minerals, from laterite soils and rock varnish to the in-place weathering of granite to clay. The mineral formed is often determined by the local environment. In some cases, biofilms in fresh-water lakes may initially accumulate gypsum (calcium sulfate; $CaSO_4$) through physicochemical binding; but when metabolic activity increases the pH of the environment, the mineral precipitated changes to calcite (calcium carbonate; $CaCO_3$). Active cells are able to shed this mineralized layer, but as the cells age they become entombed and form calcite bioherms. The physical presence of bacteria can also affect the actual crystal form. Modern stromatolites growing in the Great Salt Lake, Utah, contain bacteria that appear to act as catalysts in the formation of aragonite rather than calcite. Rapid precipitation again results in entombment, which interferes with the crystal growth, producing less defined aragonite faces compared to faces that are formed abiotically. Another example of bacterial mediation is thought to be supergene chalcocite enrichment during weathering of primary sulfides, as many bacterialike structures have been found near the replacement interfaces.

The accumulation of Banded Iron Formations appears to be more complex. It is likely that several bacterially mediated reactions were involved, and that the natural environment (possibly biologically mediated) at the time of deposition determined the form of precipitation. Silica would have accumulated on the bacterial biomass by physical biosorption, whereas iron is more likely to have been metabolized as a source of energy before being sorbed. Depending on the redox state, iron could have been deposited in the ferrous state (as siderite or iron silicates), in the ferric state (as ferrihydrite maturing to hematite), or as magnetite with mixed oxidation states. Microbial mediation appears to be essential for the typical deposition in narrow interbedded layers of the oxides, silicates, and carbonates that are found in banded iron lithofacies.

The range of microbial mediation affecting mineral formation and alteration is only now becoming appreciated. Many mineralogical reactions that occur in the aqueous environment can be bacterially mediated or induced. Much more work is needed to describe the different microbial communities and their geochemical abilities.

For background information *see* BIOFILM; MICROBIAL ECOLOGY; MINERALOGY; OXIDATION-REDUCTION in the McGraw-Hill Encyclopedia of Science & Technology.   D. Ann Brown

Bibliography. T. D. Brock and M. T. Madigan, *Biology of Microorganisms*, 1996; D. A. Brown, G. A. Gross, and J. A. Sawicki, A review of the microbial geochemistry of Banded Iron-Formations, *Can. Mineral.*, 33:1321–1333, 1995; G. F. Ferris, Microbes to minerals, *Geotimes*, 42:19–21, September 1995; W. E. Krumbein (ed.), *Microbial Geochemistry*, 1983; J. S. Poindexter and E. R. Leadbetter (eds.), *Bacteria in Nature*, 1989; S. A. Tyler and E. S. Barghoorn, Occurrence of structurally preserved plants in Precambrian rocks of the Canadian Shield, *Science*, 119:606–608, 1954.

Mining

As mining projects encroach on civilization, the impacts on water quality are becoming more important. As mineral extraction policies with regard to the environment have evolved, it has become easier and less expensive for a mining company to design mitigation features into a mine plan rather than to try cleaning up environmental degradation after the fact.

Mining engineers now must consider more than just the volumes and tonnages of ore and waste that must be safely and economically extracted to meet the demand for raw materials by modern industry. The engineers also need to consider the chemical characteristics of the ores and wastes in their design and the interaction of those wastes with atmospheric conditions, ground water, and surface water in order to anticipate potential pollution sources. After mitigation measures have been implemented, some mines are adopting natural cleanup processes, such as constructed wetlands to treat noncompliant mine water.

Primary sources of water degradation. Water-quality problems associated with mining are not new. The pioneering hard-rock miners in the Colorado Rockies recognized the impacts from poor water quality; the water entering some of their underground workings was so metal laden and acidic that they could not use it to feed the steam boilers. In some mines, acid water dissolved iron rails, plating out dissolved copper in the process. Water pollution from mining was documented in the early sixteenth century. G. Agricola, in *De Re Metallica*, noted that water contaminated by mining activities "poisons the brooks and streams, and either destroys the fish or drives them away."

Both coal and metal or hard-rock miners now face much the same problem. But with the large volumes of ore and waste rock being moved, the problem can be magnified a thousandfold. Typical sources of acid or metal drainage at mining sites include drainage from underground mines, leachates from waste rock and from mill tailings or refuse facilities, and water accumulating in abandoned open pits.

The formation of acidic or dissolved-metal drainage or acid rock drainage (ARD) is a natural weathering process. In the presence of air, water, and bacteria, sulfide minerals such as pyrite (fool's gold) are oxidized and produce sulfuric acid; concurrently, iron and other metals are released into the water. This can also be a problem in both coal and hard-rock or metal mines where previously buried sulfide minerals are exposed to oxygen and water.

In mines where the pH of water is 4.5, some ARD-producing reactions are catalyzed by the action of natural bacteria, *Thiobacillus ferrooxidans*, which accelerates the pyrite oxidation process and lowers the pH even further.

Hydrogen ions (H^+) and ferric ions (Fe^{3+}) catalyze the oxidation of other metal sulfide minerals that may be present with the pyrite, releasing additional metals such as copper, lead, zinc, and manganese into mine water. It appears that if the other sulfides are present and pyrite is absent, the predominance of bacteria-assisted oxidation may be usurped by chemical oxidation; this may be a slower process, depending on the oxidation conditions present.

Just as air, heat, and a fuel source are necessary to have a fire, to have ARD the requirements are air (oxygen), water, a pyrite source, and the bacteria to speed reactions that would otherwise occur slowly. These four components of ARD formation are codependent (and are sometimes referred to as the ARD tetrahedron). As with fire, if any of the primary ingredients are missing, ARD will not form. Prevention of mine-water pollution at its source essentially begins with the disruption of one or more of these four components or isolation from each other or the environment.

Geologists and geochemists have observed that some deposits containing pyrite have not resulted in the formation of ARD after mining. There are many methods available for prediction of waste behavior that can approximate the risks of ARD's formation in a given mining situation. These include geologic models; experience or geographical comparisons; chemical, physical, and mineralogic analyses; tests for readily extractable metals; static acid-base geochemical tests; and laboratory and field kinetic tests.

Prevention of mine-water problems. Each potential mining situation is unique, and the methods used to prevent ARD formation must be tailored to site-specific conditions. The methods rely on disrupting the ARD components as much as possible. Eliminating water seems to be the easiest general approach; thus prevention methods focus primarily on a general design philosophy that mining-waste storage facilities should store solids, not liquids. A number of tools are utilized in implementing this approach: Impervious caps (typically mine-site soils) with lateral drainage layers above the impervious zone reduce hydraulic head (and thus infiltration) on the cap. Drained (managed deposition) tailings facilities reduce pore water volumes, improve placed (consolidated) density, and lower final placed tailing permeability. Horizontal drains and wick drains are installed either during construction or after the fact. (A wick drain comprises a geotextile "sock" surrounding a corrugated plastic core.) Surface water diversions are included in the design to prevent run-on infiltration. Reclamation plans promote evapotranspiration with strong permanent vegetation covers.

If controlling the water aspect of the ARD components is not practical, engineers eliminate the effects of pyrite (or mineral acidity) with one or more of the following methods: selective placement of acid-neutralizing materials (such as limestone or fly ash), placed on top of acid-forming wastes or intimately mixed with the wastes; modification of milling process flowsheets to produce a concentrated sulfide mineral by-product that might be easier to manage

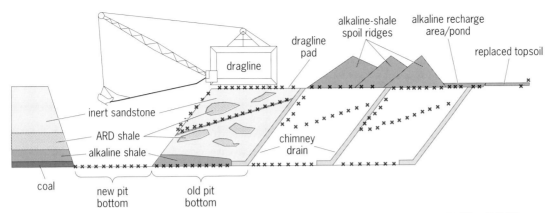

Cross section through a mine pit, showing locations (X) where limestone was added to the mine spoil. *(After V. P. Wiram and H. Naumann, Improving ground water quality in the backfill with alkaline additions, Annual Meeting of the American Society of Surface Mining and Reclamation, Knoxville, May 1996)*

by engineered burial or perhaps to sell for a marginal profit; and promotion of strong vegetative growth above ARD-prone materials (plants naturally provide neutralizing ions through their roots that slow the process of pyrite oxidation).

If the exclusion of air appears to be appropriate for controlling ARD formation, there are several methods that might be considered: Impervious caps can be made with clay or plastic geomembranes such as polyvinyl chloride, or high-density polyethylene. Another method involves flooding; this may apply to open pits, tailings, waste rock, and underground mine workings. In addition, oxygen-consuming (organic) materials can be included within waste-disposal schemes; oxygen depletion in sanitary landfills is well documented.

The disruption of bacterial activity can be accomplished with commercially available bactericides containing the active ingredient sodium lauryl sulfate, a common detergent, or phosphates. Such products may be appropriate in controlling ARD formation in active mines until reclamation activities are completed.

Example. In 1987, the Skyline Coal Company opened the Glady Fork Mine and initiated open-pit mining of the Sewanee coal seam in Sequatchie County, Tennessee. The mine moved the overburden that covered the coal with a combination of blast casting (using explosives to enhance overburden stripping), bulldozer pushing, and dragline placement. Data from baseline geochemical and hydrologic tests at the time indicated that the ARD potential at the site was minimal; however, in 1990 ARD-type seeps were found affecting a nearby creek. New studies showed that the presence of the mineral siderite (iron carbonate) had masked the potential of the mine wastes to form ARD.

Further geochemical investigations culminated in the development of a specific Toxic Materials Handling Plan, which was Skyline's approach to disruption of the ARD components. Four overburden horizons were identified as ARD forming; the horizons became moderately mixed in the blast or casting

dragline placement mining method. Limestone was added in different zones of the backfill (see **illus.**) in proportion to requirements indicated by laboratory test results. Chimney drains, consisting of inert coarse mine rock, were also installed to foster water percolation through the overburden along preferred pathways deficient in ARD-producing rock. The Toxic Materials Handling Plan was implemented from 1992 to 1995, when the pit was closed.

To further mitigate ARD at the site, the mine installed crushed limestone-filled alkaline recharge ponds over some of the mined-out areas. The purpose was to inject ARD-neutralizing solutions into reactive zones and improve ground-water quality by inducing metal precipitation within the backfill.

As of 1996, it appears that the Skyline plan, particularly the limestone additions, has resulted in measurable benefits in ground-water quality. Monitoring data reflect that design criteria and project objectives are being met. The mine plans to install constructed wetlands to further improve the quality of surface water discharging from the site.

Prospects. While the geochemistry of mine-water quality problems is well understood, site-specific hydrologic and material-handling situations can present significant challenges to the mine engineer. Control of mine-water pollution can be implemented in many ways that disrupt the geochemical relationship of the ARD components. Prevention of mine-water pollution is now being integrated into typical mine operation activities in lieu of reacting to postproject problems of ground- and surface-water quality.

For background information *see* BIODEGRADATION; LAND RECLAMATION; MICROBIAL ECOLOGY; MINING; PLACER MINING; SURFACE MINING; WETLANDS in the McGraw-Hill Encyclopedia of Science & Technology.   James J. Gusek

Bibliography. G. Agricola, *De Re Metallica* (1556), trans. by H. C. Hoover and L. H. Hoover, 1950; J. D. Hem, *Study and Interpretation of the Chemical Characteristics of Natural Water*, 3d ed., USGS Water Supply Pap. 2254, 1992; *Mine Drainage and Surface Mine Reclamation*, vol. 1: *Mine Water and*

Mine Waste, U.S. Bur. Mines Inform. Circ. IC 9183, 1988; V. Rastogi, Water quality and reclamation management in mining using bactericides, *Min. Eng.,* 48(4):71–76, 1996.

Molybdenum

Transition-metal complexes, molecules in which a small number (usually four to six) of ligating groups surround a central transition-metal atom (for example, chromium, molybdenum, and tungsten), play critical roles in biology and in the chemical industry. Two examples from biology include the oxygen-evolving complex (a manganese-containing assemblage) in photosynthetic organisms; and the iron-molybdenum cofactor thought to be the site of dinitrogen binding, reduction, and cleavage by enzymes in nitrogen-fixing bacteria. Both of these processes, the production of the dioxygen (O_2) molecule and the destruction of the dinitrogen (N_2) molecule, involve the manipulation of ubiquitous small molecules. Transition metals also catalyze transformations of small molecules in processes optimized for the modern chemical industry. Biological nitrogen fixation has an industrial counterpart, the Haber-Bosch process, by which ammonia (NH_3) is synthesized. Another example of industrial utilization of small molecules is the production of hydrocarbon fuels from a mixture of carbon monoxide (CO) and hydrogen (H_2), termed Fischer-Tropsch chemistry.

Dinitrogen reduction. Despite the utilization of transition metals as catalysts by the chemical industry and by nature, it is often the case that the most efficient systems are extremely mysterious with respect to their mechanisms of action. ''How do these things work?'' is asked whenever the goal is to mimic or optimize these processes or to create new ones. The chemistry described here relates to the question of dinitrogen reduction at a molybdenum metal center. For the first time a well-defined metal complex has been shown to dissect the dinitrogen molecule under mild conditions of temperature and pressure in a manner amenable to close mechanistic scrutiny.

In this article the design, synthesis, and characterization of the first dinitrogen-splitting complex is described. The chemical reactivity of this complex, including dinitrogen cleavage, is discussed in the context of element-element triple bonds and free-radical chemistry.

Coordination number. Alfred Werner, 1913 Nobel laureate, introduced the concept of the coordination number essentially as the number of atoms within primary bonding distance of a central (metal) atom. The most common coordination numbers found for isolable transition-metal complexes are 4, 5, and 6.

A relatively small number of isolable complexes exhibit a coordination number higher or lower than the standard values. For the element molybdenum (Mo), the coordination number 3 was unknown prior to 1995. The research project described here, which was initiated in 1994, had as an initial objective the production of an isolable three-coordinate molybdenum(III) complex. The potential accessibility of this molecular class was suggested by the synthesis of related three-coordinate chromium(III) complexes [structure (**1**)] first described in the late

$$[(CH_3)_3Si]_2N \longrightarrow Cr \underset{\displaystyle N[Si(CH_3)_3]_2}{\overset{\displaystyle N[Si(CH_3)_3]_2}{\diagdown}}$$

(**1**)

1960s and early 1970s. Although the latter studies produced textbook examples of coordination number 3, accompanying reactivity studies indicated that three-coordinate complexes of 3*d* elements (atomic number 13–23) were not exceptionally reactive toward N_2 or H_2, although they did tend to be sensitive to dioxygen.

More recently a three-coordinate complex of tantalum, a 5*d* element, was prepared and shown to exhibit significant patterns of reactivity toward small molecules, especially carbon monoxide. Reductive scission of carbon monoxide by the three-coordinate tantalum(III) complex led to four-coordinate species featuring multiple tantalum-carbon and tantalum-oxygen bonds (**Fig. 1**). A propensity toward multiple bonding was thereby established phenomenologically, in conjunction with a low coordination number, as a criterion for high reactivity. That the 3*d* elements such as chromium, in contrast with their heavier 4*d* and 5*d* congeners, provide relatively un-

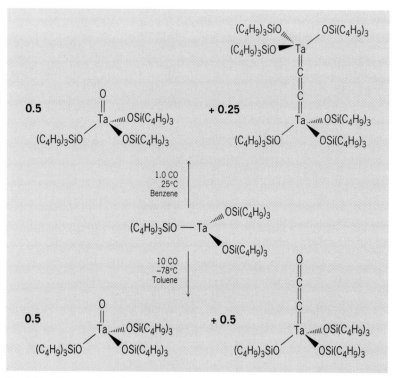

Fig. 1. Reactions of a tantalum(III) complex.

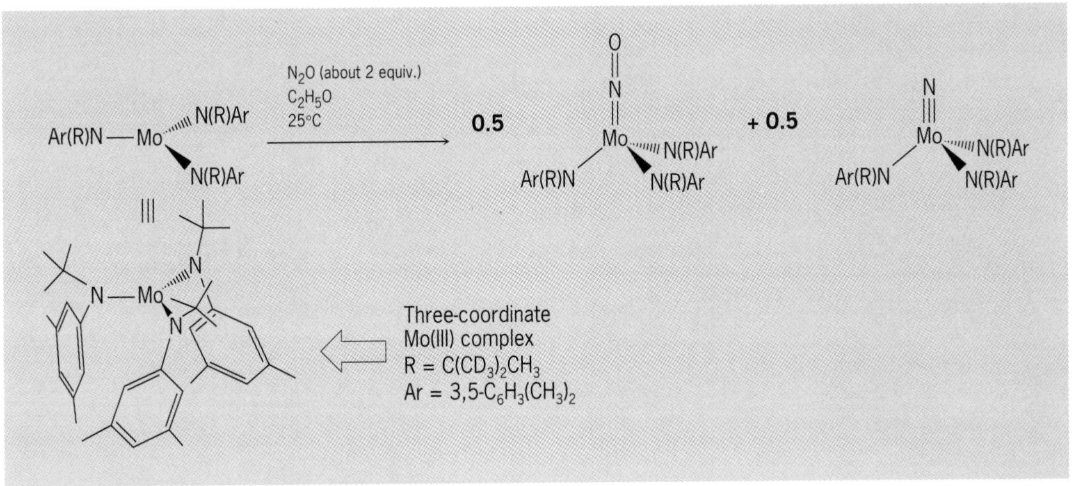

Fig. 2. Conversion of molybdenum(III) complex.

reactive examples of coordination number 3 is consistent with their known lack of propensity for multiple bonding.

A formula for high reactivity in a three-coordinate complex can be given as follows: corresponding four-coordinate species should be known or predicted to be stable, where the fourth (new) substituent is multiply bonded to the central metal atom.

Three-coordinate complex. Molybdenum in its +3 oxidation state, designated molybdenum(III), is known for its propensity toward metal-metal triple bonding. A well-known class of compounds has the formula Mo_2X_6 (X = alkyl, amide, alkoxide, thiolate, or mixtures thereof), and exhibits an ethanelike structure about a central Mo-Mo triple bond (**2**). Correspond-

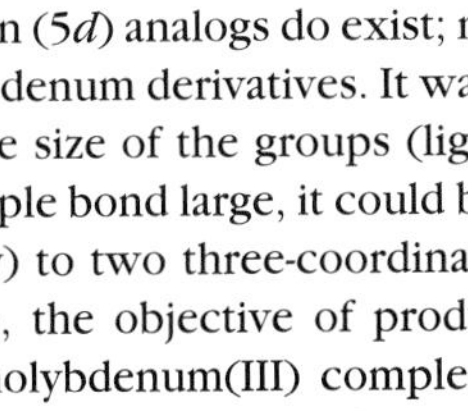

(**2**)

ing chromium (3d) compounds are not known, while tungsten (5d) analogs do exist; most prevalent are the molybdenum derivatives. It was thought that by making the size of the groups (ligands) flanking the Mo-Mo triple bond large, it could be pulled apart (conceptually) to two three-coordinate fragments.

In practice, the objective of producing a three-coordinate molybdenum(III) complex was met by treating a common six-coordinate molybdenum starting material, $MoCl_3(OC_4H_8)_3$ [OC_4H_8 = tetrahydrofuran], with three equivalents of the lithium salt of a voluminous N-donor ligand, Li(N[R]Ar) [R = $C(CD_3)_2CH_3$, Ar = $3,5-C_6H_3(CH_3)_2$]. Lithium chloride was expelled, the weakly bound tetrahydrofuran molecules were lost, and the orange-red three-coordinate complex $Mo(N[R]Ar)_3$ was obtained (**Fig. 2**).

The above successful methodology is quite similar to that used previously for the preparation of three-

coordinate chromium(III) complexes, with the exception that the N[R]Ar ligand type had not been utilized previously. The choice of N[R]Ar can therefore be regarded as key. N[R]Ar was chosen to meet, at a minimum, the following criteria: it is not subject to low-energy decomposition routes such as β-hydrogen elimination (it lacks β-hydrogens), nitrogen-silicon (N-Si) bond cleavage (no N-Si bonds), and ligand dissociation in low-dielectric media (N[R]Ar is a strong nucleophile); and it is sufficiently voluminous to prevent metal-metal bond formation by association. In addition, N[R]Ar is readily prepared in partially deuterated form, facilitating detection of paramagnetic derivatives such as the desired $Mo(N[R]Ar)_3$, using hydrogen-2 nuclear magnetic resonance (^2H NMR). An important consideration is that electronic and steric variation on the N[R]Ar ligand type is straightforward synthetically, permitting fine-tuning of reactivity and other physical properties. Finally, the introduction of the N[R]Ar class of ligands opens up a host of new possibilities for three-coordinate complexes.

Nitrous oxide. The isolation of the complex $Mo(N[R]Ar)_3$, and the demonstration by x-ray crystallography that it contains three-coordinate molybdenum centers, was an exciting development in view of the existence of a related, well-characterized four-coordinate molybdenum nitrido species, $NMo(NPh_2)_3$ [Ph = C_5H_6]. It was reasoned that the stored energy in $Mo(N[R]Ar)_3$ could be harnessed for N-atom abstraction from a suitable substrate. Since nitrogen in most of its compounds forms three bonds, the same number of bonds are subject to cleavage during an N-atom abstraction event. The question arose as to what molecule could conceivably give up a nitrogen atom while simultaneously releasing a stable molecule. Answering it led to the choice of nitrous oxide (N_2O) as a substrate for $Mo(N[R]Ar)_3$. Nitrous oxide could conceivably donate a nitrogen atom to $Mo(N[R]Ar)_3$ with the release of nitric oxide (NO).

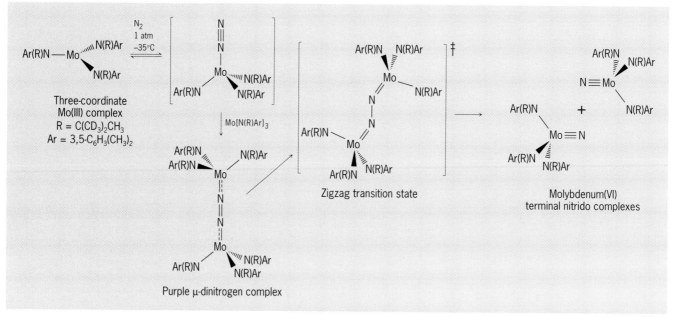

Fig. 3. Cleavage of dinitrogen by molybdenum(III) complex.

In practice, treatment of Mo(N[R]Ar)$_3$ in an ethyl ether solution with a moderate excess of N$_2$O added by injection did in fact lead to smooth cleavage of the N-N bond. The nitric oxide formally generated, rather than being expelled from the system, was trapped to produce one molecule of the nitrosyl complex (ON)Mo(N[R]Ar)$_3$ for every molecule of the nitrido complex NMo(N[R]Ar)$_3$ (Fig. 2).

The reaction between nitrous oxide and Mo(N[R]Ar)$_3$ represents a surprising and previously unknown mode of reductive scission of N$_2$O. Typically, reactions of N$_2$O result in scission of the weaker (by about 75 kcal/mol) N-O bond, with release of the very stable dinitrogen molecule. For this reason, nitrous oxide is usually considered a clean oxidant, and is sometimes used in place of dioxygen in combustion mixtures. In the case of nitrous oxide N-N bond cleavage by Mo(N[R]Ar)$_3$, the dramatic selectivity is presumably dictated by the unusual electronic structure of the three-coordinate molybdenum(III) complex. A remaining question concerns the mechanism of nitrous oxide N-N bond cleavage by Mo(N[R]Ar)$_3$: are two Mo(N[R]Ar)$_3$ molecules per N$_2$O molecule required for cleavage as suggested by the reaction stoichiometry? A conclusive answer is required in order to illuminate fully the origin of the remarkable selectivity.

Dinitrogen. As noted above, the reaction of nitrous oxide with Mo(N[R]Ar)$_3$ demonstrated the propensity of Mo(N[R]Ar)$_3$ to act as a nitrogen-atom acceptor. Dinitrogen, with its extremely strong (about 225 kcal/mol) N-N triple bond, is the simplest conceivable N-atom donor. Purification of Mo(N[R]Ar)$_3$ involves recrystallization under an inert atmosphere at −35°C (−31°F) from ethyl ether solutions. Dinitrogen gas, typically assumed to be unreactive and

therefore employed as an inert atmosphere in the synthesis of air-sensitive molecules, was found to react slowly with Mo(N[R]Ar)$_3$ when attempts were made to purify it. Cooling red-orange solutions of Mo(N[R]Ar)$_3$ under 1 atm (10^5 pascals) of N$_2$ led to the development of a purple color, attributed to the formation of a dinuclear dinitrogen complex, (μ-N$_2$){Mo(N[R]Ar)$_3$}$_2$ (**Fig. 3**). The latter paramagnetic dinitrogen complex has been characterized by a variety of techniques, including extended x-ray absorption fine structure (EXAFS) spectroscopy, which indicated a linear MoNNMo structure.

Purple (μ-N$_2$){Mo(N[R]Ar)$_3$}$_2$ persists for long periods of time (weeks) at −35°C (−31°F), but when solutions containing it are warmed to about 25°C (77°F) the purple color is lost with a half-life of about 30 min, the final color being amber-gold. The final product of the reaction is the nitrido complex NMo(N[R]Ar)$_3$, known from the nitric oxide study. Use of isotopically labeled dinitrogen (^{15}N$_2$, 1 atm or 10^5 Pa) in conjunction with Mo(N[R]Ar)$_3$ resulted in clean formation of ^{15}NMo(N[R]Ar)$_3$, establishing dinitrogen as the origin of the nitrido nitrogen atom. The reaction of Mo(N[R]Ar)$_3$ with dinitrogen gas represents the first example of N$_2$ reductive scission by a well-characterized transition-metal complex, in the absence of any added reagents. That the reaction proceeds in solution and under mild conditions makes the discovery all the more remarkable since it allows detailed mechanistic investigations to be carried out with a variety of sophisticated techniques.

Dinitrogen cleavage by three-coordinate molybdenum complexes has provided chemists with a detailed picture of one possible mode of nitrogen fixation at a reducing metal center. This reaction provides a starting point for the rational development of new nitrogen-fixing catalysts.

For background information *see* CHEMICAL BONDING; COORDINATION CHEMISTRY; COORDINATION COMPLEXES; LIGAND; MOLYBDENUM; TRANSITION ELEMENTS in the McGraw-Hill Encyclopedia of Science & Technology.   Christopher C. Cummins

Bibliography. F. A. Cotton and G. Wilkinson, *Advanced Inorganic Chemistry*, 5th ed., 1988; C. E. Laplaza et al., *J. Amer. Chem. Soc.*, 118:8623–8638, 1996; D. R. Neithamer et al., *J. Amer. Chem. Soc.*, 111:9056–9072, 1989; T. Travis, *Chem. Ind.*, 15:581–585, 1993.

Nanostructure

Carbon nanotubes are composed entirely of carbon atoms, and have diameters as small as 0.7 nm (the same diameter as the closed-cage C_{60} molecule, buckminsterfullerene). These tubes can be prepared either as single-wall structures consisting of a tube one atom in thickness or as multiwall nanotubes containing coaxial shells of the elemental single-wall nanotubes.

Research on carbon nanotubes is of particular interest to the rapidly developing field of nanostructures. Because of the very small diameters, it is expected that new techniques will be introduced to facilitate purification, characterization, and manipulation of nanotubes, and that these new techniques will further enrich experimental capabilities in the broader nanostructure field.

Discovery of carbon nanotubes. Multiwall carbon nanotubes less than 6 nm in diameter were grown by an arc-discharge method and were first identified experimentally in 1991 in relation to fullerene nanostructures. Actually, observations of similar structures had been reported 15 years earlier on carbon fibers with diameters as small as 7 nm grown from carbon atoms in the vapor phase. Much interest in carbon nanotubes was aroused by theoretical predictions in 1992 that single-wall nanotubes could have either metallic or semiconducting electrical behavior, depending purely on geometrical considerations. This discovery attracted much attention because materials typically are made to conduct electricity by suitable doping with impurities. For example, a pure semiconductor such as silicon in column 14 of the periodic table would have a very

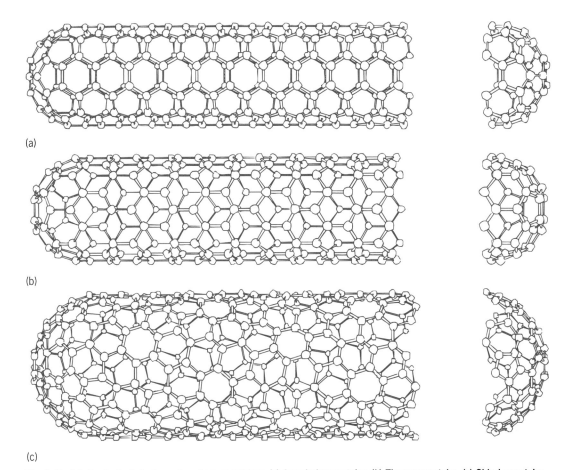

(a)

(b)

(c)

Fig. 1. Models for typical single-wall carbon nanotubes. (*a*) Armchair nanotube. (*b*) Zigzag nanotube. (*c*) Chiral nanotube. By rolling up a single layer from a three-dimensional graphite crystal into a cylinder and capping each end of the cylinder with half of a fullerene molecule, a single-wall carbon nanotube is formed.

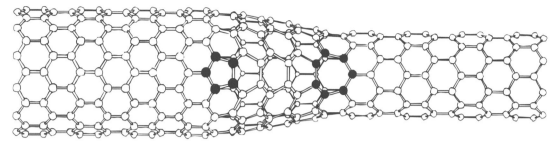

Fig. 2. Junction between two zigzag nanotubes, the larger-diameter tube being conducting and the smaller-diameter tube being semiconducting. In the junction region the carbon atoms of the pentagon and the heptagon needed to form the junction between the two zigzag nanotubes are indicated by filled circles. For each nanotube, eight circumferential zigzag chains are shown.

low conductivity at room temperature, but doping silicon with arsenic in column 15 or with gallium in column 13 greatly enhances the conductivity of silicon. Soon after these theoretical predictions about their remarkable electronic properties, interest in carbon nanotubes was fueled by the successful synthesis in 1993 of single-wall carbon nanotubes, in a catalytic process employing a small concentration of transition-metal atoms. These single-wall nanotubes were one atomic layer thick, with diameters predominantly in the 1.0–1.3-nm range and lengths in excess of 1 micrometer. The experimental discovery of single-wall nanotubes was very important to the development of the field since almost all of the theoretical work on carbon nanotubes has been done on single-wall nanotubes. Since 1992, the investigation of carbon nanotubes has been a very active research field.

Structure and properties. **Figure 1** shows examples of typical carbon nanotubes. The cylinders of these nanotubes are formed by rolling up a single layer of a graphite crystal which consists of sheets of carbon atoms on a hexagonal open network. The nearest-neighbor carbon-carbon distance in graphite and in a carbon nanotube is 0.142 nm, while the interlayer separation between graphite sheets is 0.335 nm. Because of the curvature of the cylinders in Fig. 1 and the different diameters of adjacent cylindrical shells, there is no well-established interlayer site correlation in multiwall carbon nanotubes; therefore the interlayer separation in the nanotubes is over 0.34 nm.

The graphite sheet can be rolled up so that the edge is in the shape of armchairs (Fig. 1*a*), or if the sheet is rolled up at right angles to Fig. 1*a*, a zigzag edge (Fig. 1*b*) is obtained. If the sheet is rolled up in other directions but still makes a seamless joint, a chiral nanotube is obtained (Fig. 1*c*). In the chiral nanotube, the succession of hexagons on a particular cylinder makes an angle with the axis of the nanotube, giving rise to chirality. The remarkable property of these nanotubes is that all of the armchair nanotubes (Fig. 1*a*) show metallic conductivity. This, however, is not the case for the zigzag nanotubes (Fig. 1*b*), where metallic conductivity occurs

only when the number of carbon atoms around the nanotube diameter is a multiple of 3. Otherwise the zigzag nanotube exhibits semiconducting properties. Thus two-thirds of the zigzag nanotubes are expected to be semiconducting and one-third are metallic. The chiral nanotubes (Fig. 1*c*) also exhibit either metallic or semiconducting behavior. A simple model can predict the expected electronic properties of a given nanotube from its diameter and the angle of the hexagon rows with respect to the axis of the nanotube. Also shown in Fig. 1 are the caps to the nanotubes, which are hemispheres of appropriate fullerenes which fit perfectly on the cylinders. The caps are half of a C_{60} molecule cut to be normal to a pentagon face (Fig. 1*a*) or to a hexagon face (Fig. 1*b*).

Because of the special atomic arrangement of the carbon atoms in a carbon nanotube, it is difficult for impurities to enter the nanotube structure. This follows from the small size of the carbon atoms relative to the size of potential impurity atoms that may wish to substitute for carbon atoms, similar to what was discussed above for an arsenic atom substituting for a silicon atom. Therefore, carbon nanotubes tend to display highly crystalline well-ordered structures. These structures have been studied in some detail with transmission electron microscopy techniques. The more recently discovered scanning tunneling microscope technique has also been utilized extensively to study individual carbon nanotubes. Results of such studies on individual multilayer nanotubes have shown metallic, ohmic behavior in some nanotubes and semiconducting, tunneling behavior in others. For nanotubes that were semiconducting, an energy gap could be measured, and this energy gap was found to have the same dependence on the diameter of the nanotube as the theoretical predictions.

Heterojunctions. Having demonstrated the existence of semiconducting and metallic nanotubes experimentally, researchers started to investigate the electronic properties of a heterojunction of nanometer size formed by joining a semiconducting and a metallic nanotube. Such junctions have now been studied theoretically but await experimental elucidation. The results for a junction less than 1 nm in diameter indicate that the electronic heterostructure behavior is confined to about three rows of hexagons on either side of the physical junction (**Fig. 2**). Such a heterojunction would thus constitute the smallest electronic device produced so far if it could be prepared under controlled conditions, with an active area of 1 nm both longitudinal and transverse to the junction.

Measurements. These encouraging results led to attempts by researchers in 1995 to make measurements of the temperature and magnetic-field dependence of the electrical resistivity of individual multiwall carbon nanotubes. Several successful transport measurements have been reported on individual multiwall nanotubes less than 20 nm in diameter. Unexpected new physical phenomena have already been observed in these structures at very low temperature (below 1 K) and high magnetic field (5 teslas). Measurements of the mechanical properties show that single-wall nanotubes can be flexible but have a very high Young's modulus, exceeding that of graphite. Such unique properties offer attractive possibilities for applications of these structures in practical devices.

High-yield synthesis. In 1996 a high-yield synthesis method was reported for preparing bundles of single-wall carbon nanotubes using a laser to vaporize carbon containing a small amount of a transition-metal catalyst. The laser vaporization takes place within a furnace operating at 1200°C (2200°F). The carbon nanotubes thus produced are of small diameter (1.2–1.3 nm) and have a narrow diameter distribution ($\pm 10\%$). X-ray diffraction measurements and theoretical arguments have been offered in support of the preponderance of armchair nanotubes (Fig. 1a) of a single diameter about twice that of C_{60}. The single-wall nanotubes within the bundle are arranged in ordered triangular arrays. Electrical measurements have shown metallic conductivity, as expected from theoretical considerations. A unique Raman spectrum is expected from each type of carbon nanotube, including a dependence on diameter. The observed Raman spectra are consistent with the structural identification of single-wall nanotubes, and support a model which relates the vibrations in a nanotube to vibrations on the graphite sheet which forms the basis of the nanotube (Fig. 1).

For background information *see* FULLERENE; GRAPHITE; NANOSTRUCTURE; RAMAN EFFECT; SCANNING TUNNELING MICROSCOPE; SEMICONDUCTOR in the McGraw-Hill Encyclopedia of Science & Technology.

M. S. Dresselhaus; R. Saito

Bibliography. M. S. Dresselhaus, G. Dresselhaus, and P. C. Eklund, *Science of Fullerenes and Carbon Nanotubes*, 1996; S. Iijima, Helical microtubules of graphitic carbon, *Nature*, 354:56–58, 1991; J.-P. Issi et al., Electronic properties of carbon nanotubes: Experimental results, *Carbon*, 33:941–948, 1995; A. Thess et al., Crystalline ropes of metallic carbon nanotubes, *Science*, 273:483–487, 1996.

Negative ion

The independent-particle model forms the basis of the theoretical description of multiparticle systems in many areas of microscopic physics. In such systems, the sheer number of particles involved often precludes a description encompassing the interactions among all the individual particles. For example, the independent-electron model has been the cornerstone of atomic theory since its inception. In this approximation, each electron in an atom is considered to move independently in a potential created by the nucleus and the time average of the other electrons. In this relatively simple model, the individual interactions between pairs of electrons are, to a first approximation, neglected. Any correlated motion among the electrons, that is, the direct interaction between two electrons, is treated as a perturbation, and several techniques have been developed to handle this normally small correction.

In certain cases, however, the interaction between pairs of electrons can no longer be neglected relative to the usually dominant interaction between the electrons and the nucleus. Under such a condition, the independent-particle description becomes inadequate. The independent-electron model must then be supplanted by one that incorporates individual interelectronic interactions.

In negative ions, for instance, the positive nuclear charge is smaller than the total negative charge of the electron cloud, making the electron–electron interaction inherently more important. In particular, in doubly excited states of these ions two electrons are simultaneously excited from their lowest allowed orbits. In this case, the interaction between the two excited electrons is frequently of comparable strength to the interaction between each excited electron and the rest of the system. The pair of excited electrons strongly correlate their motion in order to reduce their interelectronic repulsion energy while enhancing their attraction to the nucleus. These states can be produced by electron or photon impact, but the probabilities for these processes are rather small, and experimental information is therefore scarce. In recent years, however, new methods have been developed based on the photodetachment process, in which a negative ion absorbs a photon and breaks up into an electron and a residual atom.

Resonances in photodetachment spectra. Generally, if an amount of energy greater than the binding energy is deposited in an atomic system, a single electron will be ejected. If a doubly excited state is produced, however, the energy can be shared for a short time by two electrons in such a way that neither has sufficient energy to be immediately emitted. Eventually, the excitation energy becomes concentrated on one of the electrons of the pair, and this electron is ejected while its partner drops back to a lower energy level. This process is called autodetachment.

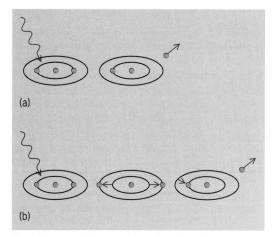

Fig. 1. Two possible mechanisms for photodetachment of a negative ion. (*a*) Direct process. (*b*) Process involving the excitation of an intermediate doubly excited state followed by its autodetaching decay.

According to quantum theory, the emitted electron can be described as a (matter) wave whose amplitude reflects the probability of finding the electron at a specific time and location. The waves produced by the photodetachment process or indirectly via the excitation and subsequent autodetachment of doubly excited states will, except for the phase, be identical; and they do, further, originate from the same initial state, namely the ground state of the negative ion (**Fig. 1**). The two waves will, in analogy to Young's double-slit experiment in optics, interfere constructively or destructively. The probability for photodetachment will consequently increase or decrease when the photon energy is tuned to excite a doubly excited state. These variations in rate of photodetachment are called resonances. They provide a means of observing doubly excited states. The range over which the resonance structure appears, called the width of the resonance, is a measure of the lifetime of the doubly excited state.

Laser-ion beam apparatus. **Figure 2** shows the experimental arrangement used to investigate doubly excited states of negative ions via photodetachment. Positive ions are produced in an ion source from which they are extracted and accelerated to form a beam of selected energy. This beam is then passed through a cesium vapor in a charge-exchange cell where some of the positive ions sequentially capture two electrons from the cesium atoms to form negative ions. The negative ions thus produced are bent twice by 90° in a pair of electrostatic charge-state analyzers. In the 50-cm-long (20-in.) path region between the two analyzers, the ions are merged with one or two laser beams. After the second analyzer, three detectors are placed to measure the number of negative ions, neutral atoms, and positive ions in the beam after the interaction with the laser light. The number of negative ions detected after the passage between the two analyzers depends upon the number of ions produced in the ion source and the

rate of detachment due to collisions with the residual gas. When a laser beam is superimposed on the ion beam, the detachment rate will be enhanced by the process of photodetachment. Instead of monitoring the number of surviving negative ions, it is preferable from an experimental point of view to monitor the yield of natural atoms produced in the photodetachment process. A gated detection procedure is then used to distinguish between those atoms produced by photodetachment and those produced by collisional detachment. In this case, the detectors are gated on or off in coincidence with the time structure of the pulsed laser.

To gain further information about the photodetachment process, the residual neutral atoms must be interrogated to find out their excitation state since this defines the detachment channel through which the negative ions have decayed. It has been found that autodetachment preferentially leaves the residual atom in excited states whereas the direct photodetachment process tends to leave it in its ground state. State-selective detection of the residual atoms can be achieved by the use of a second laser to resonantly excite the atom from the excited state of interest to a Rydberg state, a weakly bound state situated just below the ionization limit. Atoms in the Rydberg states are easily ionized in an electric field which, in the present experiment, is provided in the second charge-state analyzer. The subsequent positive ions are then detected in the presence of very little background. This sensitive and selective detection scheme is called resonance ionization spectroscopy.

Doubly excited states of Li⁻. Recent studies of double excitation in the negative lithium (Li⁻) ion illustrate the potential of this method. **Figure 3** shows a portion of the photodetachment cross section for the negative lithium ion in the ultraviolet spectral region. The pair of prominent dips in the cross section, called window resonances, are associated with two

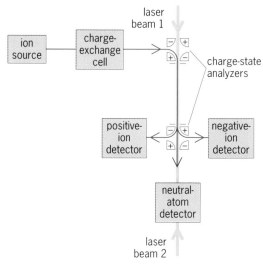

Fig. 2. Collinear laser–ion beam apparatus.

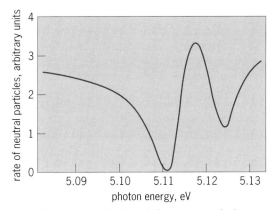

Fig. 3. Changes, as a function of photon energy, in the rate for the process that photodetaches an electron from a negative lithium (Li⁻) ion and leaves the residual atom in the 3s state.

doubly excited states in the negative lithium ion that are weakly bound with respect to the excited lithium atom in the $4p$ state. Studies of high-lying doubly excited states of negative ions present a challenge to the experimentalist since photodetachment cross sections in the ultraviolet are typically small and the associated resonances are usually closed spaced. The present measurement was made possible only by the high sensitivity and energy resolution associated with the collinear beam method plus the added selectivity and sensitivity of the detection scheme based on resonance ionization spectroscopy.

Electron correlation. Electron correlation plays a major role in determining the energy positions and widths of resonances, and so experimental studies, such as the one described here, provide sensitive tests of the ability of theory to go beyond the independent-particle model. For high-lying doubly excited states, the independent-electron model breaks down completely. It is then impossible to label the electrons with individual quantum numbers. Instead, the two outermost electrons have to be treated as a correlated pair of electrons with a common set of quantum numbers that describes their collective motion. If the excitation energy is shared approximately equally by the two electrons, for example, their motion tends to be correlated in such a way that each electron avoids the other by staying on opposite sides of the nucleus at roughly equal distances from it. Correlation becomes increasingly important in negative ions as the energy of the doubly excited state approaches the double detachment limit, that is, the threshold for the simultaneous ejection of two electrons. Here, the slow motion of the two detached electrons in the field of the positive ion core is highly correlated.

There exists a great deal of experimental data on systems that are adequately described by the independent-particle model. Far less data, however, are available for correlation-sensitive systems such as negative ions. The method described here has the potential of providing accurate data on such systems, which in turn should better the understanding of atomic theory.

For background information *see* ATOMIC STRUCTURE AND SPECTRA; INTERFERENCE OF WAVES; NEGATIVE ION; PHOTOIONIZATION; QUANTUM MECHANICS; RESONANCE (QUANTUM MECHANICS); RESONANCE IONIZATION SPECTROSCOPY; RYDBERG ATOM in the McGraw-Hill Encyclopedia of Science & Technology.

Dag Hanstorp

Bibliography. P. G. Harris et al., Observation of high-lying resonances in the H⁻ ion, *Phys. Rev. Lett.*, 65:309–312, 1990; C. D. Lin, Doubly excited states, including new classification schemes, *Adv. Atom. Mol. Phys.*, 22:77–142, 1986; U. Ljungblad et al., Observation of doubly excited states in Li⁻, *Phys. Rev. Lett.*, 77:3751–3754, 1996; C. Pan, A. F. Starace, and C. H. Greene, Photodetachment of Li⁻ from the Li 3s threshold to the Li 6s threshold, *Phys. Rev. A*, 53:840–852, 1996.

Nobel prizes

The Nobel prizes for 1996 included the following awards for scientific disciplines.

Physics. Professors David M. Lee and Robert C. Richardson of Cornell University and Professor Douglas M. Osheroff of Stanford University were awarded the prize for their 1972 discovery of superfluidity in the rare isotope helium-3.

Superfluidity was produced in helium-4, the common isotope of helium, in 1938 by cooling it to a temperature of 2.2 kelvins above absolute zero. Helium-4 atoms are bosons, and at very low temperature they undergo Bose-Einstein condensation, in which a large proportion of them drop into a single quantum-mechanical state and move in a coordinated manner. This phenomenon explains the unusual properties of superfluids, such as frictionless flow and flow up the sides of containers.

Helium-3 atoms are fermions, which obey the exclusion principle forbidding any two of them from occupying the same quantum-mechanical state. However, in 1957 J. Bardeen, L. Cooper, and R. Schrieffer showed that, in a liquid of fermions, any small attraction between particles can bind them into bosonlike pairs, which can then undergo Bose-Einstein condensation. The Bardeen-Cooper-Schrieffer (BCS) theory explained the phenomenon of superconductivity in low-temperature metals and alloys. Numerous predictions of superfluidity in helium-3 followed, but initial efforts to detect it were unsuccessful.

In 1971, Lee, Richardson, and Osheroff (then a graduate student at Cornell University) were investigating nuclear antiferromagnetism in solid helium-3. While testing a supercooled cell containing a mixture of liquid and solid helium-3, Osheroff observed unexpected pressure fluctuations as its volume was gradually changed at temperatures near 2.7 millikelvins. Initially, the group mistakenly reported a phase transition in solid helium-3. It took months of experiments using nuclear magnetic resonance techniques to establish the existence of superfluidity in liquid helium-3 and to determine its nature. Three distinct superfluid phases were eventually discovered. The bosonlike pairs of helium-3 atoms have complex structures and can be aligned in a magnetic field. Moreover, helium-3 displays some unique anisotropic properties.

Interest in superfluid helium-3 has increased in recent years. Study of its complex pairing structures may help in understanding those of high-temperature superconductors, which are also anisotropic. Phase transitions of helium-3 may be analogous to the formation of cosmic strings in the early universe, and study of superfluid helium-3 may also help to elucidate the rotational behavior of neutron stars.

Chemistry. Robert F. Curl, Jr., Harold W. Kroto, and Richard E. Smalley shared the prize for their discovery of buckminsterfullerene, a new form of carbon in which the atoms are arranged as a closed cagelike sphere, a C-60 buckyball. Curl is professor of chemistry at Rice University, Houston, Texas; Kroto is Royal Society Research Professor at the University of Sussex, Brighton, England; and Smalley is the Hackerman Professor of Chemistry and a physics professor at Rice University.

Prior to 1985, elemental carbon was known to exist in two well-defined, common, allotropic crystalline forms: sheetlike diamond and tetrahedral graphite. However, under certain unusual conditions, carbon atoms had been shown to cluster in soccerball-like groups—usually groups of 60, but sometimes 70. Such groups are in the shape of polyhedra having pentagonal and hexagonal faces.

In 1985, initiated by interest in reproducing the long carbon chains of the carbon-rich atmosphere of red giant stars, Kroto consulted with Curl and Smalley at Rice University. Using intense laser light, they blasted carbon atoms from a sheet of graphite, forming a plasma, which was then cooled with a stream of helium and injected into a vacuum chamber. Carbon clusters formed as the mixture was cooled down to nearly absolute zero. The molecules formed were studied in a mass spectrometer. The researchers noticed a strong spike in the mass spectrometer readings, indicating that molecules with a mass equal to 60 carbon atoms were forming in the vapor. The carbon clusters were exposed to other molecules at high temperatures, but they remained unreactive. Knowing that carbon is stable only when joined in large continuous structures, they proposed a molecule constructed of 60 carbon atoms having no loose edges (because the presence of such edges would allow the structure sites to be susceptible to reaction with other molecules). The 60 carbon atoms were arranged in a sphere, as a truncated icosahedron, consisting of 12 pentagons and 20 hexagons, forming closed cagelike structures with a minimum of 32 atoms.

This work by Curl, Kroto, and Smalley has led to an explosion of fullerene research. These spherical molecules have numerous physical and chemical properties which enable fullerenes and their derivatives to have a wide range of uses. They show promise for synthetic, pharmaceutical, and industrial materials, including use as lubricants, drug delivery vehicles, catalysts, and superconductors.

Physiology or medicine. Peter C. Doherty, adjunct professor of pathology and pediatrics at the University of Tennessee College of Medicine, and Rolf M. Zingernagel, head of the Institute of Experimental Immunology at the University of Zurich in Switzerland, were awarded the prize for describing how T cells recognize, attack, and destroy virus-infected cells without causing injury to healthy cells.

During the 1970s at the John Curtain School of Medical Research in Canberra, Australia, Doherty and Zingernagel performed a series of experiments (assays) on a specific strain of mice infected with lymphocytic choriomeningitis virus (LCMV), which causes a disease of the central nervous system. They wanted to explain why some mice develop a more virulent strain of the disease than others and to determine if T cells were responsible for damaging healthy brain tissue while fighting off the infection. The assays were performed by combining cerebral fluid containing T cells with LCMV-infected cells from the same strain of mice. Considering the previous immunological theory, the results of the assays were predictable: the T cells recognized the LCMV-infected cells and attacked them.

In previous transplant rejection studies, researchers discovered that T cells attack the cells of transplanted organs if they do not have a matching group of self-recognition molecules [also called major histocompatibility complex (MHC) antigens]. Doherty and Zingernagel decided to apply the same principles to viral infections by combining the T cells and LCMV-infected cells from different strains of mice. To their surprise, this combination did not trigger an immune response. They concluded that T-cell receptors require two elements to recognize a pathogen and mount a successful immune response: a recognizable viral antigen produced by the infected target cell, and an identifiable set of MHC antigens. Additional molecular evidence confirmed their theory. When a virus attacks a healthy cell, a small segment (a peptide) of the virus binds to the cell's own histocompatibility antigens. Upon recognition of this complex, the T cell binds to and begins to destroy the infected cell.

The work of Doherty and Zingernagel encouraged other researchers to pursue the development of vaccines to eradicate the human immunodeficiency virus, eliminate certain types of cancer, and prevent the rejection of transplanted organs. Their discovery may also allow immunologists to eventually develop techniques to alter or modify immune reactions to microorganisms and self-tissues and to many inflammatory diseases.

For background information *see* CARBON; CELLULAR IMMUNOLOGY; DIAMOND; FULLERENE; GRAPHITE; IMMUNOLOGY; LIQUID HELIUM; TRANSPLANTATION BIOLOGY in the McGraw-Hill Encyclopedia of Science & Technology.

Nonlinear optics

Nonlinear optics became an important field of research with the invention of the laser in the early 1960s. In nonlinear optics, because of the interaction between light and matter, new wavelengths (new frequencies) are created. For this reason, nonlinear optics has become an important tool for generating new coherent wavelengths. To generate new wavelengths efficiently, it is important to have high light intensities and large nonlinear coefficients. These coefficients (also known as electric susceptibilities) describe how well the light can interact with the particular piece of matter involved. Many crystals, both inorganic and organic, have large coefficients. Glass, however, which is the material of choice in optical fibers, has very small nonlinear coefficients. This is related to the fact that glass has very low absorption. Nevertheless, it is easy to observe optical nonlinear effects in glass fibers for two reasons. First, because of the smallness of the optical fiber (approximately 10 micrometers in diameter for single-mode fibers), the light is necessarily tightly confined, causing large light intensities for moderate input powers. Second, since optical fibers can be very long, the light can interact with the glass for many kilometers, rather than for only a few millimeters or at most centimeters, as is the case for crystal materials. These factors compensate for the small nonlinear coefficients.

Historically, nonlinearities in fibers were studied because of their detrimental effects in wavelength-division-multiplexed (WDM) systems and their nonlinear attenuation of signals. As the understanding of the optical nonlinearities in fiber grew, it was realized that the nonlinearities might even be useful. Examples are soliton propagation in optical fibers, four-wave mixing for wavelength multiplexing, and the nonlinear optical loop mirror for extracting clock rates from large-bandwidth digital signals. The optical nonlinearities in optical fibers make it possible to do signal processing directly in the fiber. Since optical nonlinearities in glasses react extremely rapidly (10–20 femtoseconds; 1 fs = 10^{-15} s), it is possible to avoid the electronic bottleneck typically degrading the performance of existing fiber-optic systems.

Soliton propagation. Optical fibers have inherently a very large bandwidth, of the order of 25 terahertz (1 THz = 10^{12} Hz). To fully exploit this bandwidth, it is necessary to defeat dispersion. Dispersion, that is, the wavelength dependence of the refractive index, causes a pulse to broaden in time (**Fig. 1a**), and this temporal broadening forces the system engineer to separate the pulses in time to guarantee error-free detection at the receiver. This separation causes dead times in the fiber and therefore a less efficient utilization of the bandwidth. An optical soliton, by contrast, can propagate for thousands of kilometers without broadening. The soliton is generated from an ordinary optical pulse in the fiber when the optical Kerr effect on the phase is exactly counterbalanced by dispersion (Fig. 1b). Recent experi-

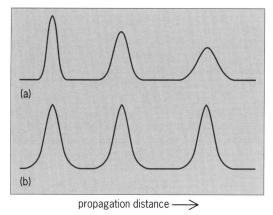

Fig. 1. Pulses in an optical fiber. (a) Ordinary dispersive pulse, which is broadened in time as it propagates along the fiber. (b) Soliton pulse, which does not alter its temporal width.

ments have shown that bit rates of 80 gigabits per second over 300 mi (500 km) can be obtained. *See* SOLITON.

Nonlinear optical loop mirror. The optical Kerr effect can also be used to carry out ultrafast optical switching for signal processing directly in the fiber. Additionally, this effect can be used to passively mode-lock lasers. The device to perform all these functions is named the nonlinear optical loop mirror (NOLM). Its principle is as follows (**Fig. 2**). The input light is split into two counterpropagating waves by a coupler. If the coupler divides the light equally in the two directions, the device operates as a mirror. The trick is to have a coupler that sends more light in one direction than in the other direction. Nonlinearly the two counterpropagating waves will obtain different amounts of phase. For certain phase differences the output power will be very low because of destructive interference, and similarly for other phase differences the output will be equal to the input power. A very important use of the nonlinear optical loop mirror is to optically recover the clock frequency in high-bit-rate communication systems.

Wavelength conversion. To use the available bandwidth efficiently in optical fibers, wavelength multiplexing is employed. This means that it is necessary to convert optical signals into different wavelengths for propagation. One way is through a nonlinear effect called four-wave mixing. A wavelength con-

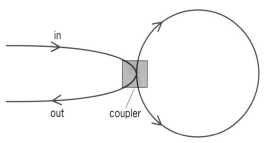

Fig. 2. Nonlinear optical loop mirror.

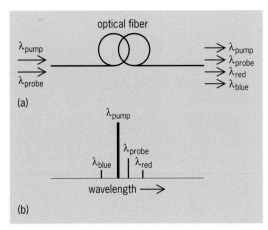

Fig. 3. Wavelength conversion. (a) Apparatus. (b) Spectrum of wavelengths.

version system can be set up by generating a very strong pump wave in the fiber at all times. This pump wave interacts with a relatively weak probe wave that carries the information. These two waves can interact nonlinearly to convert all the energy in the probe wave to a new wavelength, although 100% conversion efficiency is never achieved in reality (**Fig. 3**). Since this process needs to be phase matched (that is, a strict relationship must be satisfied between the phases of the interacting waves), it is necessary to carefully choose the pump and probe wavelengths in relation to the particular optical fiber properties to obtain optimum wavelength conversion. Four-wave mixing is also responsible for a phenomenon called phase conjugation, which can be used to compensate for dispersion in the fiber.

Nonlinear crosstalk. A problem in wavelength-division-multiplexed systems is that channels occupying different wavelengths interact with each other because of optical nonlinearities. This crosstalk severely limits the number of possible channels in the optical fiber. One nonlinearity that causes this crosstalk is the Raman effect. The Raman effect is due to optical phonons, which are created when the atoms in the glass (silicon and oxygen) move in opposite directions to each other. These vibrations can transfer energy from one optical wavelength to another, causing crosstalk. If the energy difference between two channel wavelengths exactly corresponds to the vibration energy, the crosstalk will grow exponentially with distance. Since the Raman vibrations are spectrally very broad, it is difficult to find a channel separation which is not affected by Raman-induced crosstalk.

Nonlinear photoinduced effects. All of the nonlinearities discussed so far have been third-order nonlinearities. This means that the nonlinear effect is proportional to the cube of the amplitude of the incoming light. For optical signal processing, these third-order nonlinearities are the most important; furthermore, third-order nonlinearities can occur in any type of material regardless of its atomic structure. This is not the case for second-order nonlinearities, which depend on the square of the amplitude of the incoming light. In this case the material structure has to obey certain criteria in order for the second-order effect to occur. Many materials obey these rules, for example, most inorganic crystals. Gases, liquids, and glasses do not. This means that nonlinear effects such as second-harmonic generation (the frequency of the outgoing light is twice that of the incoming) or the Pockels effect (the refractive index is altered by the light) should not occur in an optical fiber. Nevertheless, it has been found that the glass can be modified by light itself in such a way as to acquire the necessary properties to exhibit second-order nonlinearities. The phenomenon is referred to as self-organization. What is believed to happen is that the light interacts with particular defects in the glass and this interaction causes the defects to change, thereby altering the dielectric properties of the glass. The self-organization arises through the fact that these interactions occur at regularly spaced intervals to facilitate phase matching.

The same defects that are important for providing second-order effects in the glass are also believed to interact with light to provide a refractive-index grating along the core of the fiber. These gratings are referred to as Bragg gratings because their periodicity is half the wavelength of the light used.

These photoinduced effects are important for many active and passive devices in optical communication systems. The second-order nonlinearity through the electrooptic effect can be used to transfer information from an electrical signal to an optical signal. The Bragg gratings are useful in wavelength-division-multiplexed systems to route information channels at different wavelengths to their correct destination.

For background information *see* CROSSTALK; ELECTROOPTICS; NONLINEAR OPTICAL DEVICES; NONLINEAR OPTICS; OPTICAL COMMUNICATIONS; OPTICAL FIBERS; RAMAN EFFECT in the McGraw-Hill Encyclopedia of Science & Technology.   Ulf L. Österberg

Bibliography. G. P. Agrawal, *Nonlinear Fiber Optics*, 1989; N. Doran and D. Wood, Nonlinear fiber loop, *Opt. Lett.*, 13:56–58, 1988; S. E. Miller and I. P. Kaminow (eds.), *Optical Fiber Communications II*, 1988; W. W. Morey, G. A. Ball, and G. Meltz, Photoinduced Bragg gratings in optical fibers, *Opt. Photonics News*, 5:8–14, 1994; U. Österberg and W. Margulis, Dye laser pumped by Nd:Yag laser pulses frequency doubled in a glass fiber, *Opt. Lett.*, 11:516–518, 1986.

Nuclear physics

Nuclear physics is experiencing a renaissance based on important fundamental discoveries and on the construction of major experimental facilities to explore their consequences.

The discovery that quarks are the fundamental building blocks of nuclear matter has led to an in-

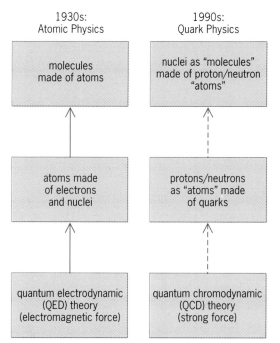

1930s:
Atomic Physics

1990s:
Quark Physics

molecules
made of atoms

nuclei as "molecules"
made of proton/neutron
"atoms"

atoms made
of electrons
and nuclei

protons/neutrons
as "atoms" made
of quarks

quantum electrodynamic
(QED) theory
(electromagnetic force)

quantum chromodynamic
(QCD) theory
(strong force)

Fig. 1. Analogy of quantum chromodynamics to quantum electrodynamics.

creasing focus on the structure of the proton and neutron as "atoms" of quarks, and on understanding how the residual forces between these "quark atoms" lead to the ordinary nuclear forces that bind the protons and neutrons into nuclei. A description of these systems at the quark level is required to explain the origin of the laws of standard nuclear physics. At stake is a fundamental understanding of how protons and neutrons combine into the myriad varieties of nuclei ("quark molecules") which in turn create the diversity of chemical elements.

Quantum chromodynamics. The analogy (**Fig. 1**) between ordinary atoms and molecules, which are governed by the laws of quantum electrodynamics (QED), and "quark atoms" and "quark molecules," which are governed by the newly discovered laws of quantum chromodynamics (QCD), is very powerful. Electrons are held in their atomic orbits (typically of size 10^{-10} m) by the electric and magnetic forces created by the exchange of the familiar light quantum, the photon. Quarks are held in their much smaller orbits inside the proton and neutron (typically of size 10^{-15} m) by the analogous chromoelectric and chromomagnetic forces created by the exchange of a new quantum named the gluon. These superstrong forces, roughly 10^9 times stronger than their electromagnetic counterparts, are responsible for binding the quarks into such small regions of space. Moreover, it is the leakage of these force fields into the region outside the proton and neutron which creates the much weaker (but still very powerful) ordinary nuclear forces between protons and neutrons, in the same way that the fringe fields around ordinary atoms create the considerably weaker forces which bind them into molecules.

Quark confinement. As powerful as this analogy is, it ignores one outstanding difference between the electronic structure of atoms and the quark structure of nuclei: Electrons can be individually knocked out of atoms, but an individual quark cannot be removed from the nucleus. A great triumph of quantum chromodynamics is that it predicts this phenomenon, known as quark confinement.

The key difference between quantum electrodynamics and quantum chromodynamics, which is responsible for this dramatic contrast, is that there is no electric force between two photons (since they have no electric charge) but there is a strong chromoelectric force between two gluons. (This feature arises because the analog of electric charge in quantum chromodynamics is not a simple number as in quantum electrodynamics, but a three-dimensional vector charge. Chromodynamics gets its name from the mathematics of color vision, where any visible color can be described as a three-dimensional vector in a coordinate system whose axes are the three primary colors.)

The quantum-electrodynamic force field between positive and negative electrical charges has the familiar dipole shape (**Fig. 2**a). In quantum chromodynamics, the intergluonic forces distort the force field that the gluons make between quarks into a flux tube of constant cross section (Fig. 2b). Since a unit length of this flux tube requires a fixed amount of energy for its creation, any energy that might be expended "pulling on a quark" just has the effect of creating a longer flux tube, and not of separating the quark from its partner. The resulting constant

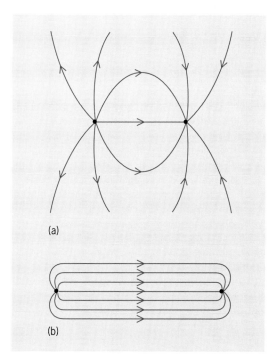

(a)

(b)

Fig. 2. Comparison of field lines. (a) Dipole electric field lines of quantum electrodynamics. (b) Flux tube in quantum chromodynamics.

restoring force has a strength of about 20 metric tons.

To be precise, this description of flux tubes is a simplification which would apply if quarks were very massive. In nature, the quarks are light enough that stretching the flux tube by many proton radii stores enough energy to create new quark–antiquark pairs (that is, to create mass from energy). When this happens, the flux tube can break, and the pulled quark can escape with the newly created antiquark glued to the end of its flux tube. This process is called meson production, since quark–antiquark atoms are called mesons, and is very common in practice.

Unanswered questions. While it is known that nuclear matter is made of quarks and gluons, and while the most notable property of such systems, quark confinement, has been demonstrated by very large scale numerical simulations of quantum chromodynamics using supercomputers, little more is understood about the quark structure of matter. For example, at the level of the "quark atoms," aside from the knowledge that the intergluonic forces are responsible, the mechanisms behind the formation of the flux tubes are not understood.

Additional puzzles include the unexpected pattern of excited states of the proton and neutron, understanding how the spatial distributions of electric charge and magnetization of the proton and neutron arise from the distribution of quarks inside them, and at an even more basic level, understanding why the only stable quark atoms, the proton and neutron, are each made of three quarks (one of each "primary color"). This latter question is closely related to a key open question at the level of "quark molecules," that of understanding why a nucleus such as helium behaves like a system of two protons and two neutrons instead of like a 12-quark "atom." Given that it does behave like a system of protons and neutrons, there is the question of understanding how to derive from quantum chromodynamics the standard nuclear force between protons and neutrons determined experimentally and used to describe the observed nuclei and their properties.

In the early universe, matter was very hot and very dense. Just as ordinary atoms make a phase transition into an electron plasma at high temperature and densities, quantum chromodynamics predicts that under extreme conditions nuclear matter will dissociate into a quark-gluon plasma. It is a challenge to test this prediction by observing a quark-gluon plasma in the laboratory, thereby reproducing conditions which existed in the first few hundred microseconds after the big bang.

New accelerator facilities. Since the mid-1980s, nuclear physicists have been building two large accelerator facilities to allow the experimental study of many outstanding questions in the field.

The Continuous Electron Beam Accelerator Facility (CEBAF) at Jefferson Laboratory in Newport News, Virginia, was completed in 1996 and has begun its program of studying nuclear matter with its very precise high-energy electron beam (**Fig. 3**).

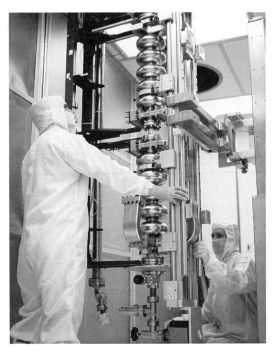

Fig. 3. Novel superconducting accelerator cavities which produce the continuous electron beam in the Continuous Electron Beam Accelerator at Jefferson Laboratory.

CEBAF may be thought of as a giant electron microscope capable of resolving the structure of nuclear matter at distances of the order of 1% of the size of the proton itself. Its unique beam, which illuminate the nucleus with a steady continuous beam rather than a flashing one (the only type available in the past), will allow experimentalists to see the internal structure of the nucleus with unprecedented clarity. When combined with modern particle detector technology, high-speed data-acquisition techniques, and high-speed computers, CEBAF is expected to revolutionize the understanding of the quark structure of matter.

Before the year 2000, CEBAF is expected to be joined by the Relativistic Heavy Ion Collider (RHIC), a facility under construction at Brookhaven National Laboratory on Long Island, New York. RHIC is designed to create a quark–gluon plasma in the aftermath of high-speed head-on collisions of two massive charged nuclei (ions). The technical challenges facing the RHIC experimentalists are formidable: A typical head-on collision at RHIC will lead to an event with thousands of particles, because the energy of the impact not only disintegrates the original nuclei but is also converted to a multitude of additional quark–antiquark pairs. It is essential to see and measure the properties of these particles to accumulate experimental evidence that they had their origin in the creation of a quark-gluon plasma.

For background information *see* ATOMIC STRUCTURE AND SPECTRA; FUNDAMENTAL INTERACTIONS; GLUON; NUCLEAR STRUCTURE; PARTICLE ACCELERATOR; PARTICLE DETECTOR; QUANTUM CHROMODYNAMICS;

QUANTUM MECHANICS; QUARKS; STANDARD MODEL in the McGraw-Hill Encyclopedia of Science & Technology. Nathan Isgur

Bibliography. F. Flam, A new accelerator explores the social life of quarks, *Science*, 267:1266–1267, 1995; M. Mukeerjee, CEBAF readies its electron beam for studies of nucleons and nuclei, *Phys. Today*, 46(8):17–19, August 1993; I. Peterson, The stuff of protons: Gluing quarks to make protons, neutrons, and atomic nuclei. *Sci. News*, 146:140–141, 1994; H. C. Von Bayer, Sweet little accelerator (Continuous Electron Beam Accelerator Facility), *Discover*, 15(8):52–58, August 1994.

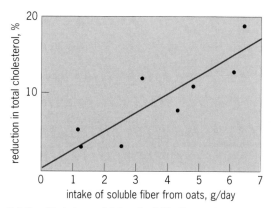

Relationship between intake of soluble fiber and reduction in blood cholesterol from recent American studies. *(After I. T. Johnson and E. Lund, Soluble fibre, Nutrit. Food Sci., 2:7–9, 1990)*

Nutraceuticals

Nutraceuticals comprise any food or ingredient that provides medical and health benefits, particularly in the prevention or treatment of diseases. Current research focuses on antioxidants, vitamins, dietary fiber, and fatty acids. This article reviews selected foods that are sources of these nutraceuticals.

Fruits and vegetables. Increased consumption of fruits and vegetables has been reported to decrease the risks of cancers. Data suggest that a reduction in the risk of pancreatic cancer is associated with increased consumption of cruciferous vegetables (broccoli, cabbage, cauliflower), and that a decrease in the risk of lung and pancreatic cancers is related to increased consumption of a variety of fruits and vegetables. The relative reduction of risks associated with the consumption of green vegetables is summarized in the **table**.

Phytochemicals, the plant ingredients responsible for these effects, include a wide range of compounds diverse in structure and function. For example, antioxidant vitamins in fruits and vegetables quench free radicals and prevent them from interacting with cellular proteins, lipids, and deoxyribonucleic acid (DNA). If unchecked, these reactions can result in the formation of cancerous cells. For example, the

intake of plant-derived antioxidant vitamin A or β-carotene appears to be inversely related to the incidence of lung cancer. Other plant antioxidant vitamins such as C and E are under extensive investigation because of their possible roles in preventing diseases and cancers.

Another group of compounds exhibiting antioxidant properties in fruits and vegetables are the flavonoids. Studies suggest that these polyphenolic compounds prevent the oxidation of low-density lipoprotein (LDL), an important contributing factor in atherosclerosis.

Cereals and oilseeds. Cereals and oilseeds contain a wide range of compounds that have considerable potential as nutraceuticals. Dietary fiber, flax, canola oil, and palm oil are examples of sources.

Dietary fiber. Dietary fiber, a group of complex carbohydrates derived from plant cell walls, is important in protecting against a range of diseases of the alimentary canal, such as constipation, hemorrhoids, and diverticulosis. Sources of dietary fiber include cereals, fruits, and vegetables. Specific components within dietary fiber appear to significantly reduce serum cholesterol and hence the risk of cardiovascular disease. For example, oat fiber is composed predominantly of a linear polysaccharide, β-glucan. Recent human metabolic studies found β-glucan to be hypocholesterolemic (reducing the level of cholesterol in the blood; see **illus.**). Because of this polysaccharide, oat fiber has been incorporated into a wide range of food products. β-Glucan was the first polysaccharide to receive approval from the U.S. Food and Drug Administration to include its health benefits on the food package.

Flax. A number of components in flax (*Linum usitatissimum*), including linolenic acid (LA) and lignan precursors, appear to have significant health benefits. The importance of lignan precursors was identified in a Finnish study which showed that increased risks for breast and colon cancer were associated with women having low levels of urinary lignans. These results indicated the importance of

Reduction in relative risks of cancers associated with consumption of green vegetables*		
	Consumption		
Type of cancer	Low	Medium	High
Esophagus	1	0.5	0.2
Stomach	1	0.8	0.4
Colon	1	1.0	0.5
Rectum	1	1.0	0.6
Liver	1	0.8	0.2
Pancreas	1	0.7	0.4
Breast	1	0.9	0.7
Prostate	1	0.8	0.3
Thyroid	1	0.7	0.5

* All risks are relative to a value of 1, the maximum risk.
SOURCE: G. Williamson, Protective effects of fruits and vegetables in the diet, *Nutrit. Sci.*, 1:6–10, 1996.

dietary lignan precursors, which are converted into mammalian lignans by colonic bacteria. Flax was found to be the richest source of the lignan precursors secoisolariciresinol diglucoside (SD) and matairesinol. Secoisolariciresinol diglucoside was shown to be particularly effective against new tumors. This was subsequently confirmed using a colon cancer model which showed that purified lignan precursors reduced both the number and size of colonic crypts developed in rats in response to known carcinogens. The antitumor properties of flax are presently being investigated as a possible dietary treatment for breast cancer between times of diagnosis and surgery. The protective effects of linolenic acid against cardiovascular disease and cancer also make flaxseed particularly attractive, as it is a rich source of linolenic acid. The lowering effect of linolenic acid on total cholesterol is comparable to that obtained with drug therapy.

Canola oil. Canola oil, developed in Canada from rapeseed, can lower serum total and LDL cholesterol. The oil is very low in saturated fatty acids (less than 7%), very high in monosaturated fatty acids (60%), and high in polyunsaturated fatty acids (30%). Studies with normolipidemic individuals (with normal cholesterol levels) found canola oil to be as effective as fat sources containing higher levels of polyunsaturated fatty acids in lowering plasma total and LDL cholesterol. Elevated LDL cholesterol is a major risk factor in the development of coronary heart disease. The reduction of cholesterol is due in large part to the monounsaturated fatty acid, oleic acid, which has been demonstrated to be as effective as the polyunsaturated fatty acid, linoleic acid, in lowering plasma total and LDL cholesterol. Similar cholesterol-lowering effects were reported when using canola oil on hypercholesterolemic individuals.

Palm oil. Tocotrienols are vitamin E derivatives characterized by an unsaturated side chain with three double bonds, as distinguished from the saturated side chain of normal tocopherols. Palm oil is the only commercial vegetable oil that is a rich source of tocotrienols. Considerable interest has been focused on its anticancer properties. Palm oil tocotrienols appear to exert both cholesterol- and tumor-suppressive effects.

Miscellaneous sources. Certain beverages and spices also contain important nutraceuticals. Tea and garlic are examples.

Recent epidemiological and experimental studies show tea consumption to be beneficial to human health. Inhibition of carcinogenesis and mutagenesis, atherosclerosis prevention, serum cholesterol reduction, and inhibition of platelet aggregation are some beneficial effects cited in recent literature. Polyphenolic compounds with antioxidative and free-radical scavenging properties are responsible for these effects. Considerable research on green tea attributed its beneficial effects to the presence of polyphenols and catechins, particularly epigalocatechin (EGCG). Similar effects were also found with fermented black or semifermented oolong teas. A recent Canadian study reported comparable antimutagenic effects for the different brands of green, black, oolong, decaffeinated, and instant teas examined, with some variability among herbal teas.

Garlic (*Allium sativum*), a perennial bulbous plant closely related to the onion, remains an invaluable flavoring agent in foods. The sulfur compound allicin (structure **1**), the major component of garlic odor, is formed from the precursor alliin (**2**) by the enzyme alliin lyase. Allicin can also be converted to ajoene (**3**) which is associated with the health effects

of garlic. Ajoene has strong antiplatelet activity in blood cells and interferes with dietary fat absorption through inhibition of gastric lipase. Other pharmacological properties have also been associated with garlic, including antimicrobial and antitumor activities.

For background information *see* ANTIOXIDANT; CANCER (MEDICINE); CHOLESTEROL; FLAX; FOOD; FRUIT; GARLIC; LIPID; NUTRITION; TEA; VITAMIN E in the McGraw-Hill Encyclopedia of Science & Technology.   N. A. Michael Eskin

Bibliography. J. A. Ahmad, Garlic: A panacea for health and good taste, *Nutrit. Food Sci.*, 1:32–35, 1996; I. T. Johnson and E. Lund, Soluble fibre, *Nutrit. Food Sci.*, 2:7–9, 1990; D. Mackerras, Antioxidants and health: Fruits and vegetables or supplements?, *Food Austral. Suppl.*, 47(1):S3–S23, 1995; A. S. H. Ong, E. Niki, and L. Packers (eds.), *Nutrition, Lipids, Health and Disease*, 1995; B. Stavric et al., The effect of teas on the *in vivo* mutagenic potential of heterocyclic aromatic amines, *Food Chem. Toxicol.*, 34:515–523, 1996; L. Thompson and S. Kunane, *Flaxseed in Human Nutrition*, 1995; G. Williamson, Protective effects of fruits and vegetables in the diet, *Nutrit. Food Sci.*, 1:6–10, 1996.

Oceanography

Oceanic thermohaline circulation is driven by differences in seawater density. As warm Atlantic Drift Water is carried from the tropics into the Norwegian and Greenland seas, it cools, becomes denser, and sinks. Thus is initiated a southward-flowing deep-

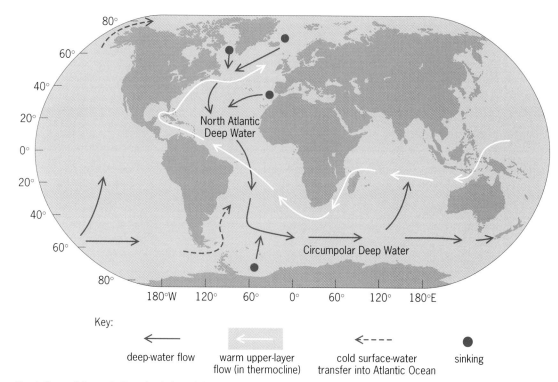

Fig. 1. General thermohaline circulation of the world ocean. *(After E. K. Berner and R. A. Berner, Global Environment: Water, Air, and Geochemical Cycles, Prentice Hall, 1995)*

water mass that joins deep water produced by winter convection in the Labrador Sea to produce the North Atlantic Deep Water. The flow of the North Atlantic Deep Water is about $2 \cdot 10^7$ m^3/s (or 20 sverdrups), which is about 10 times the combined flow of all the world's rivers. After reaching the southern ocean, the North Atlantic Deep Water mixes with southern ocean deep water, forming the Circumpolar Deep Water, which flows eastward around Antarctica and supplies deep water to the Indian and Pacific oceans. Deep-water transport from the Atlantic to the Pacific Ocean must then be compensated by a return flow of shallower water. Following a slow (about 1 meter per year), diffuse, basinwide-scale upwelling of deep Pacific and Indian ocean water, this return flow into the Atlantic is accomplished through the Bering Strait and the Arctic Ocean, around the tip of South America, and around the southern tip of Africa (**Fig. 1**).

Conveyor-belt circulation. The term conveyor-belt circulation has been coined to describe this general oceanic circulation pattern. As the mean temperature of the water entering the nordic seas is around 10°C (50°F) while that of the southward-flowing North Atlantic Deep Water is 2°C (36°F), around 8 calories are released per cubic centimeter of deep water formed, amounting to a total of (8,000,000 cal × 20 Sv) = 1.6 10^{14} cal/s. This is equivalent to one-third of the energy provided by direct solar radiation to the troposphere over the North Atlantic, pointing to the importance of the formation of the North Atlantic Deep Water for regulating climate in

this region. Consequently, changes in the rate of formation of the North Atlantic Deep Water may have played a key role in controlling the waxing and waning of the northern polar ice cap during the Quaternary. A reduced formation of the North Atlantic Deep Water during glacial periods could have promoted glaciation by reducing heat transport to high northern latitudes in the Atlantic.

Sedimentary record. As a means of evaluating whether there was a dramatic change in the rate of formation of the North Atlantic Deep Water during the last glacial maximum, when large ice caps covered North America and Eurasia, a research group used the sedimentary record of two naturally occurring radioisotopes, protactinium-231 (^{231}Pa; half-life 32,000 years) and thorium-230 (^{230}Th; half-life 75,200 years). These nuclides are produced in seawater by the decay of two uranium isotopes (^{235}U and ^{234}U, respectively) that are weathered from continental rocks and added into the ocean by rivers. Because uranium is relatively soluble in seawater, it resides in the water column for several hundred thousands of years before removal into sediments and at hydrothermal vents. Uranium is thus well mixed within the ocean, and its concentration and isotopic ratio (^{235}U/^{234}U) is uniform. Consequently, ^{231}Pa and ^{230}Th are produced at the constant activity ratio of 0.093 throughout the water column. The activity of a radionuclide is the number of atoms disintegrating per unit time (dN/dt; disintegrations per minute, or dpm); and it is inversely proportional to the number of atoms present (N): $dN/dt = -\lambda N$, where λ is the

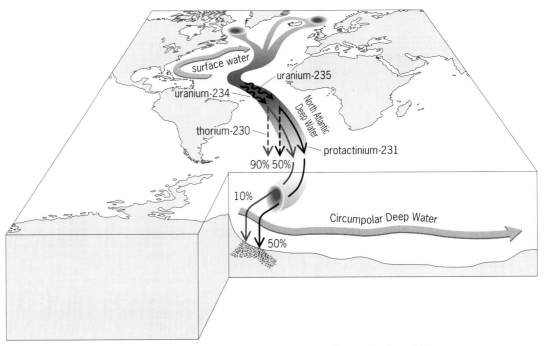

Fig. 2. Conveyor-belt circulation from the North Atlantic Deep Water to the Circumpolar Deep Water.

decay constant of the radionuclide. Unlike uranium, protactinium and thorium are highly insoluble in seawater and are rapidly removed from the water column into the underlying sediment by adsorption on sinking particles (a process called scavenging). Thorium is the more insoluble (or particle-reactive) of the two and, on average, resides in the water column for only 20–40 years after its formation. In contrast, protactinium is slightly less particle reactive and stays in the water column for an average of around 200 years before removal by scavenging.

A new approach to estimating the rate of deep-water formation in the North Atlantic or the strength of the conveyor-belt circulation is based on the fraction of ^{231}Pa produced in the Atlantic water column that is exported with the North Atlantic Deep Water into the Circumpolar Deep Water. This export creates a deficit of ^{231}Pa compared to ^{230}Th in Atlantic sediments and an excess in sediments around Antarctica, both of which are preserved in the sedimentary record. The stronger the conveyor-belt circulation, the larger the contrast between the ^{231}Pa/^{230}Th activity ratio of sediments deposited in both ocean basins.

It takes about 200 years for the North Atlantic Deep Water to renew deep water in the Atlantic. Because this is similar to the mean residence time of ^{231}Pa in the water column, half the ^{231}Pa produced in the North Atlantic Deep Water as it transits across the Atlantic Ocean is advected with the water into the southern ocean around Antarctica (**Fig. 2**). In contrast, ^{230}Th resides in the water column for a period much shorter than the mean transit time of the North Atlantic Deep Water through the Atlantic, and less than 10% of the ^{230}Th produced in North

Atlantic Deep Water is similarly removed from the Atlantic basin. Because 90% of the ^{230}Th but only 50% of the ^{231}Pa produced in the transiting North Atlantic Deep Water is scavenged within the Atlantic basin, Atlantic sediments have a mean ^{231}Pa/^{230}Th activity ratio smaller than the production rate ratio (0.093), while sediments deposited around Antarctica have a ratio greater than 0.093. This has been confirmed by analyzing the distribution of ^{231}Pa/^{230}Th in modern surficial sediments, which shows an abrupt increase in ^{231}Pa/^{230}Th at 50°S latitude.

Scavenging intensity. The difference in mean ^{231}Pa/^{230}Th activity ratio between Atlantic and Antarctic sediments depends on the rate of the formation of the North Atlantic Deep Water and the scavenging intensity in the Atlantic Ocean. If the formation rate of the North Atlantic Deep Water increases and scavenging intensity does not change, a higher fraction of ^{231}Pa will be flushed from the Atlantic into the southern ocean, while ^{230}Th will be relatively unaffected, as long as the mean residence time of deep water in the Atlantic Ocean remains significantly longer than 20–40 years. This would increase the contrast in the ^{231}Pa/^{230}Th ratio of the sediments deposited in the two oceans. Therefore, if past changes in scavenging intensity can be evaluated, past changes in the rate of formation of the North Atlantic Deep Water could be deduced from the sedimentary record of ^{231}Pa/^{230}Th ratio.

Scavenging intensity is primarily a function of particle-settling flux. The latter can also be reconstructed from the sedimentary record of ^{230}Th. Because of the very short residence time in the water column, the flux of ^{230}Th scavenged to the sediment is always within 30% of the rate at which it is produced in the overlying water column. Thorium-230 is produced uniformly throughout the water column at a rate of 0.00258 dpm/m³; therefore, the flux of ²³⁰Th to the sediment is known and is a simple linear function of water column depth. Consequently, there is a predictable inverse relationship between the particle-rain rate and the ²³⁰Th concentration (or activity) of settling particles. As the particle-rain rate increases, ²³⁰Th concentration decreases. Past changes in the particle-rain rate can thus be estimated from variations down along drilling cores in decay-corrected ²³⁰Th concentration in sediments. By applying this approach to Atlantic sediments deposited during the last glacial maximum, it can be deduced that the mean particle-rain rate was not dramatically different from that of the modern ocean. Thus, any changes in the ²³¹Pa/²³⁰Th activity ratio of Atlantic sediments deposited during the last glacial maximum primarily reflect past changes in the rate of formation of the North Atlantic Deep Water.

Heat budget of glacial ocean. The mean decay-corrected ²³¹Pa/²³⁰Th ratio of glacial Atlantic and Antarctic sediments has been found to be essentially identical to that of the modern sediment, indicating essentially no changes in the rate of thermohaline circulation. This is an important piece of information toward a better understanding of the heat budget of the glacial ocean. Although the rate of formation of the North Atlantic Deep Water does not appear to have changed significantly, other proxies (cadmium/calcium and carbon isotopic composition of the calcite skeletons of foraminifera living on the sea floor) have shown that the glacial North Atlantic Deep Water, known as the Glacial North Atlantic Deep/Intermediate Water, was a shallower water mass; thus the Glacial North Atlantic Deep/Intermediate Water must have been produced at a rate similar to that of the modern North Atlantic Deep Water. Based on current estimates from the oxygen isotopic composition of foraminifera, the difference in temperature between northward-flowing glacial upper water and southward-flowing Glacial North Atlantic Deep/Intermediate Water was only slightly less than that between modern Atlantic Drift Water and North Atlantic Deep Water. Consequently, the oceanic heat transport to the North Atlantic was only slightly lower during the last glacial period. This surprising result contradicts the view that the growth of continental ice sheets in the Northern Hemisphere was directly linked to North Atlantic heat transport and the strength of the conveyor-belt circulation. It has been suggested that a vigorous conveyor-belt circulation during glacial times would have affected global climate by cooling the tropics. Ongoing modeling efforts that take into account this new insight gained from the sedimentary record of ²³¹Pa and ²³⁰Th should provide a clearer picture of the cause-and-effect relationship between changes in ocean circulation, the growth of continental ice sheets, and global climate.

For background information *see* CLIMATOLOGY; OCEAN CIRCULATION; OCEANOGRAPHY; QUATERNARY;

SEAWATER; UPWELLING in the McGraw-Hill Encyclopedia of Science & Technology.　　Roger François; Michael P. Bacon; Daniel M. Sigman; Ein-Fen Yu

Bibliography. E. K. Berner and R. A. Berner, *Global Environment: Water, Air, and Geochemical Cycles*, 1995; W. S. Broecker, The great ocean conveyor, *Oceanography*, 4:79–89, 1991; W. S. Broecker and G. H. Denton, The role of ocean-atmosphere reorganizations in glacial cycles, *Quat. Sci. Rev.*, 9:305–341, 1990; L. D. Labeyrie et al., Changes in the vertical structure of the North Atlantic Ocean between glacial and modern times, *Quat. Sci. Rev.*, 11:401–413, 1992; R. S. Webb et al., Increased ocean heat transport and glacial cooling, *Nature*, 385:695–699; E.-F. Yu, R. François, and M. P. Bacon, Similar rates of modern and last-glacial ocean thermohaline circulation inferred from radiochemical data, *Nature*, 379:689–694, 1996.

Open pit mining

An open pit mine or surface mine is an excavation or cut made at the surface of the ground for the purpose of extracting ore minerals or construction materials. The cut is open to the surface for the duration of the mine's life. In the United States, the term open pit mine is used for mines that extract metals such as copper, gold, molybdenum, or iron. Quarry is the general term for the open pit mines that extract materials such as sand, gravel, or limestone for use as construction materials. In all of these surface-extraction techniques, highwalls are created as the excavated pit grows deeper.

Pit slope design. Slope design involves a significant economic consideration in planning a large open pit mine. A realistic determination of future pit slope angles is an essential element in the design of an open pit mine. Defining a practical and safe slope angle must take into account the economic return to investors. The evolution of powerful personal computers and slope-stability software has enabled mining engineers to account for multiple variables in slope design and to provide economic sensitivity analyses. This has led to a technique known as total slope analysis, in which major parameters affecting overall slope performance are analyzed at each stage of pit excavation and benching. Structural geology, rock strengths, pit blasting, bench geometry, pit depth, haul roads, surface drainage, in-pit drainage, pit economics, and pit hydrology are some of the major parameters for total slope analysis. Large open pit mines such as Bingham Canyon in Utah, Sierrita in Arizona, and Gold Strike in Nevada divide the pits into several vertical and horizontal pit slope sectors. Each sector is assigned a range of slope angles derived from slope-stability analysis, slope life, slope safety, and economic considerations. The impact of slope angles and the risk of slope failure on pit geometry are further analyzed with three-dimensional imaging, using mine-planning software such as MEDSYSTEM and SURPAC. Slope-stability software such as PCSTABL and GEOSLOPEs have become the main staples of rock mechanics and the mining engineer community worldwide.

Pit-slope stability and slope monitoring. Monitoring of moving slopes and the inspection of all slopes are the most important activities for establishing slope safety and pit-slope stability. Each slope is unique and thus requires a different level of monitoring. A good slope-monitoring program, which provides well-designed and -maintained slopes, can improve a mine's safety and economics. Advances in technology have dramatically expanded the scope of mining and highwall stability. These include use of microchips, the Global Positioning System, communication by cellular phones and the Internet, digital imaging, and applications of robotics in mining, especially in open-pit-slope stability and surveying. Large copper mines in Arizona, New Mexico, Utah, and Chile; gold mines in Nevada and Canada; and iron mines in North America and Australia are leaders in the introduction of new technology.

The Global Positioning System (GPS) has become a major surveying tool for slope-failure mapping, slope monitoring, and tension crack mapping. GPS is basic to pit and dump progress updates, leach pad construction, and geologic field mapping. Rather than applying traditional methods of turning angles, taking solar observations, and using electronic distance measurement, mines utilize GPS real-time positioning to locate blast hole patterns and to monitor the positions of haul trucks and loading shovels. The work with GPS in most places in the United States and other countries is enhanced by the differential GPS (DGPS) technique, which can achieve a measurement accuracy that is improved to 0.75–5 m (2.5–17 ft) coarse acquisition (C/A) data transmission, and to 0.3 m (1 ft) or less with L1 carrier-wave data transmission (L1 carrier wave operates at 1575.42 MHz). Line-of-sight limitation between the GPS antenna and the satellite is minimal in open pit mines. If current trends continue, GPS units will find widespread application in all phases of open pit mining and pit-slope stability.

Automated robotic theodolite systems. These are used for open pit highwall monitoring for copper and gold mines. The main component of such a system consists of a motorized precision theodolite equipped with an electronic distance measuring (EDM) device. Different types of motorized theodolites are available for different applications. If the measurement distances are short, a standard theodolite with attached EDM is used. For longer distances, advanced theodolites equipped with video cameras and distance meters can be used. All measuring procedures are controlled by personnel, with or without software, or by the system in conjunction with a desktop or laptop computer.

In systems being used in copper mines in Arizona and Chile and in gold mines in Nevada, computer software basically consists of two main components. One component drives the measurement stations, takes the readings, and transfers the data from the station to the computer. The other component is application related; as such, the monitoring for slope movements on open pit highwalls is automatic and continuous, 24 h a day. The theodolite itself is equipped with servo motor drives and a motorized focus drive, and is controlled by local or remote software. This type of system offers a unique advantage wherever permanent or temporary monitoring is required to assure the safety of open pit highwalls and areas subject to slope failures. The monitored survey data are not dependent on the subjectivity of the observer. With the system, a real-time 24-h monitoring of large survey targets located at multiple bench levels is possible. A survey target consists of reflecting prisms that are located on the pit slope. The technology allows surveying of a large array of prism targets scattered over the unstable pit slopes.

Another type of slope monitoring, a solar powered electronic extensometer, is becoming popular in open pit mines. The magnitudes of slope movement are conveyed by telemetry to the base station or the centralized dispatch facility. The movement data are processed and are available for immediate review. The units are very useful for monitoring short-term stability, haulage roads, and electric shovel workbenches.

Slope management and slope reinforcement. In the 1990s, open pit mining experienced a dramatic shift toward total slope analysis and pit slope management. Pit slope management is an integral part of mine operation for almost all major open pit mines worldwide. Operators would rather determine the parameters that affect their slopes than combat the geologic settings.

The presence of water and the patterns of blasting are the most significant parameters over which the operator has direct control. Open pit mine design and operation are becoming sensitive, and proactive approaches are being adopted in this area. Closely spaced horizontal drain drilling as was practiced in the 1970s is a current practice. Drains for this purpose, fabricated of preslotted polyvinyl chloride screening, are spaced 60–120 m (200–400 ft) apart and within 90–150 m (300–500 ft) in horizontal length into the slope face. They have been successful in reducing effective normal stress along the failure planes and in inducing free and open drains at the base or toe of the slope face. The Gold Strike Mine in Nevada, the Cyprus Miami operation in Arizona, and many operations in South America have effectively used this technique to stabilize failing slopes and to continue operation in a safe environment.

Whether the slope-failure mass is large or small, slope terracing and surface water management are being practiced by all open pit operators worldwide. Terracing of slopes, with provision for upslope, mid-slope, and downslope drainage, has resulted in a yielding failure for which behavior can be monitored and made predictable. A major unstable slope in an Arizona mine was stabilized with a combination of slope terracing, surface drainage, and horizontal drains.

Open pit mine operations are becoming proactive in slope-control pit blasting. As mine operators choose steeper slope angles, slope-control blasting is finding its way into the production process as a remedial step in controlling overbreak, tension cracks, and slope failure. A recent trend is to isolate production blasting rows from slope-control blasting. The last three or four rows of holes in such a pattern are left for slope-control blasting. Rather than drilling a small-diameter hole in the backmost row for presplitting a clean bench face, large-diameter blast hole drills are used for drilling a series of trim row holes, followed by designing a crest line for a given slope geometry. These trim row holes are lightly loaded with explosives and shot in advance of the production row holes. The interim row between the trim and production holes is a buffer row. Blast holes in the buffer row are also lightly loaded with explosives, but they are decked in two separate columns. All three rows of holes are shot in echelon, and are delayed hole by hole. However, the actual geometry of the blast drilling pattern is geologic site specific. The main difference between a slope control blast pattern and earlier patterns is the shortening of both the trim-row and buffer-row blast holes in soft, medium, and hard rock conditions. This provides slope toe integrity and controls bench and slope failure. If water control and blasting are managed properly, residual slope problems are minimal and can be managed with proper mine sequencing, slope terracing, and proactive pit slope monitoring.

For background information *see* MINING; OPEN PIT MINING; SATELLITE NAVIGATION SYSTEMS in the McGraw-Hill Encyclopedia of Science & Technology.   Rohini P. Sharma

Optoelectronic integration

Optoelectronic integration combines optical components—devices that emit, modulate, or detect light—with electronic components such as transistors on a single wafer to obtain highly functional circuits. Typically, the optical components convert light to and from electrical signals, while the electronics perform various signal-processing functions on these signals. There is a critical difference between purely electronic integration (such as in microprocessors) and optoelectronic integration. For electronics, one type of device (transistors) is replicated many times on a substrate and then interconnected. For optoelectronics, at least two types of device, an optical and an electronic device, need to be replicated and interconnected. Generally, optical and electronic devices are very dissimilar, and making both on the same wafer often requires compromising each. Much of the effort in optoelectronic integration is in finding ways to make the necessary devices on the same wafer while keeping cost low and improving on the circuit performance.

Both optical and electronic components are made from semiconductors. Silicon, commonly used for electronics, has physical properties inappropriate for many of the optical devices of interest. A notable exception is the silicon photodetector for sensing visible and near-infrared radiation, which, when integrated with transistors, is the basis for the solid-state video camera. But for applications involving other optical components or for longer-wavelength radiation, group 13-15 compound semiconductors (consisting of elements from groups 13 and 15 of the periodic table), such as gallium arsenide and indium phosphide, and to a lesser extent group 12-16 compound semiconductors, such as mercury cadmium telluride, are used because they have desirable optical and electronic properties.

Motivation. As in electronics, the motivations for integration are lower cost and better performance. Fabrication of integrated circuits is a highly automated process where many circuits are made simultaneously on the same wafer to give a low cost per circuit. The alternative to integration is making each component separately and connecting them externally in a hybrid circuit. Hybrids are relatively expensive because extra steps are needed to house the components in protective packages and to assemble the circuit, and because the circuits are made one at a time.

By combining the components into a single integrated circuit, the assembly steps become part of the process of integrated circuit fabrication. The lower cost can come both in the reduction in the number of manufacturing steps and in the number of components that have to be individually packaged. But this savings needs to be compared with the cost of including the additional devices in the circuit. For example, if adding an optical device to an integrated circuit halves the fraction of good circuits on a wafer, the cost for each good circuit will approximately double. If this cost is more than for a hybrid, the economic motivation for integration is lost.

The performance advantage of an integrated circuit comes primarily from having the components closer together. For an image-processing circuit, close packing of components will improve the spatial resolution. But there is a less obvious reason for close packing. Wires connecting the components are far from ideal, since they have resistance, inductance, and capacitance that degrade the electrical signal that they carry. Since the degradation is worse for longer wires, it is imperative that the wires, and hence the distance between the components they connect, be kept short. How short the wires need to be depends on the frequency of the signal; the higher the frequency, the shorter they need to be. Thus, for applications involving high frequencies, such as optical communication networks, optoelectronic integration is attractive because of the lower signal degradation in the wiring.

Material and fabrication problems. The devices used for optoelectronics are made with layers of semiconductor of various composition and doping. The layers used are highly dependent on the particular de-

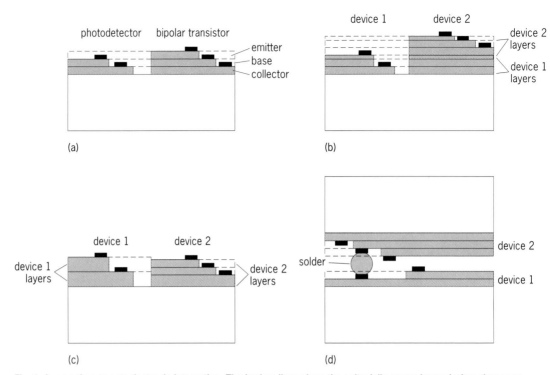

Fig. 1. Approaches to optoelectronic integration. The broken lines show the epitaxially grown layers before they were etched. For simplicity, connections between electrical contacts (black rectangles) are not shown. (*a*) Two devices are made from common epitaxial layers. (*b*) The layers for device 2 are stacked on those for device 1. (*c*) With regrowth, the layers for device 1 are removed in some areas and replaced with layers for device 2. (*d*) Chips from two wafers are flip-chip-bonded together with solder. Only one of the many bonds between chips is shown.

vice. A universal requirement is that these layers be added, or grown, on a single crystal wafer (one which has its constituent atoms in a perfect lattice) in such a way that the added atoms continue forming a perfect lattice. This is known as epitaxial growth. Crystalline imperfections, which are to be avoided, are sites where electrons in the semiconductor are trapped, resulting in device degradation. Epitaxial growth techniques include molecular beam epitaxy and organometallic chemical vapor deposition.

Once the appropriate layers are epitaxially grown, they are made into devices by using techniques similar to those used in making electronic integrated circuits. Unneeded material is removed, and insulators and metals are deposited and patterned on the semiconductor. Also, the doping may be changed locally by diffusion or ion implantation. Typically, lateral device dimensions range from a few tens of micrometers to a few tenths of a micrometer, while layer thicknesses range from about a micrometer to about a nanometer.

Approaches to integration. Several approaches are used to integrate optical and electronic devices, the easiest being to connect devices that are made side by side (**Fig. 1***a*). This approach can be used only in the exceptional cases where each type of device in the circuit can be made from the same starting layers, perhaps with differences in doping that can be accomplished by diffusion or ion implantation. An example is the integration of bipolar transistors

with photodiodes which are made from the base and collector layers of the transistor.

Another approach is to grow the layers for each device as one on top of another (Fig. 1*b*). To make a particular device at a particular location, all of the unwanted upper layers are etched away so that the desired layer is at top. While this method is more general than the first, it has several drawbacks. First, a device may interact with the layers below it through capacitive coupling or leakage currents. Hence, the design of the layer sequence is important, and extra layers may be required for isolation. Second, the etch needs to stop at precisely the correct depth. Because both the layer thicknesses and the etch rates have variations across the wafer as well as from wafer to wafer, simply etching for a fixed time period is too imprecise. The usual approach is to take advantage of chemical solutions that etch layers at a rate determined by the layer's composition. The layers in the stack differ in composition, and it is often possible to design the devices so that their layers can be removed one at a time by using a chemical solution that etches a layer of one composition much more quickly than the one below it. In this way, each layer can be etched for longer than the expected time to ensure its complete removal. To reach a particular depth in the layer sequence, the correct sequence of chemical solutions must be used. A third drawback is that the chemical etches tend to etch laterally (known as undercut) as well

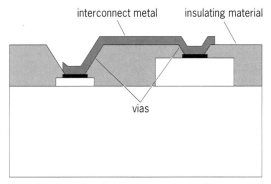

Fig. 2. Cross section showing the connection between two devices. A layer of insulating material is deposited and vias opened down to the device contacts. The via walls are sloped so that the interconnect metal can be deposited without breaks.

as vertically. To accommodate the undercut, the devices in the optoelectronic integrated circuit must be spaced at a greater distance. Etch methods that are based on plasmas (for example, reactive ion etching and ion beam etching) have less undercut than liquid-based etch methods, but their selectivity tends to be lower. A final drawback is that after all of these etches the wafer surface is terraced. It is difficult to deposit the interconnect metal over these steps, particularly the steep ones. One solution is to coat the entire wafer with an insulator (polyimide is a common choice) that provides a planar surface, and then etch vias (pathways) to allow the metal to contact the device (**Fig. 2**). While the metal still has to go down the vias, these can be etched to give a gentle slope that makes it easier to form the connection.

A third approach is to do selective epitaxial growth (Fig. 1*c*). Layers for one device are grown, then completely etched away in certain areas, and the layers for a second device are grown in these openings. This avoids many of the problems of the stacked device approach, including unwanted device coupling, complex selective etch sequences, and high terraces on the surface. Complications in the growth near the edge of the openings lowers the quality of the semiconductor that is selectively grown compared to that grown over the entire substrate.

Because of the difficulties in integrating several device types on the same wafer, a different strategy is to find ways that will give the desired results of improved performance and lower cost without the need to integrate on the same wafer. With this approach, each type of device is made on a separate wafer under the conditions that yield the best devices. Depending on the technique, either chips from these wafers or the wafers themselves are bonded together and electrical connections made between them. The key difference from the hybrid approach is that the connections are kept as short as they would be on an integrated circuit. The most widely used of these techniques is flip-chip bonding (Fig. 1*d*), where balls of solder 25–50 micrometers

in diameter bond two chips and serve as electrical connections between them.

For background information *see* CRYSTAL GROWTH; INTEGRATED CIRCUITS; INTEGRATED OPTICS; PERIODIC TABLE; SEMICONDUCTOR; SEMICONDUCTOR HETEROSTRUCTURES in the McGraw-Hill Encyclopedia of Science & Technology.   Winston Chan

Bibliography. M. Dagenais, R. F. Leheny, and J. Crow (eds.), *Integrated Optoelectronics*, 1995; S. E. Miller and I. P. Kaminow (eds.), *Optical Fiber Telecommunications II*, 1988.

Paleoecology

The fossil record provides the most concrete insight into the history of life. Increasingly, scientists are turning to this record to gain insight into the ecology of organisms and the dynamics of their communities. The recent emphasis on such studies is due to a number of landmark projects whose findings seem to contradict widely held expectations. The composition and structure of paleocommunities (recurrent time-averaged associations of fossils, not necessarily directly analogous to modern communities) have been shown to remain relatively stable over periods of 2–7 million years. This stands in marked contrast to the impressions gained by most neoecologists from the study of modern communities, which appear as ephemeral associations of organisms formed by the chance conjunction of their environmental tolerances. Mapping the relative frequency of faunal patterns displaying such stability (coordinated stasis) in the fossil record, and exploring this pattern's relationship with a large body of contradictory ecological research and theory, forms a critical agenda for paleontology and ecology.

Geologic divisions. Recurrent assemblages of organisms have been recognized in the fossil record from the beginning of paleontology. At the broadest scale, the observation that paleocommunities from different time periods look different provides the basis for the subdivisions of the geologic time scale. Such geologic divisions (eras, periods) are global in their extent and encompass many hundreds of millions of years, thus marking very long periods of Earth history during which consistently recognizable associations of organisms, or organism types (orders, families), were dominant.

However, the stability of paleoecological systems is not solely established at this broadest scale of space and time. The ability to recognize recurrent assemblages of fossil organisms at more restricted scales has allowed researchers to subdivide the geologic column more finely, and provided them with a powerful tool to correlate strata and to gain new insight into the environments in which these fossil paleocommunities existed. The scale at which this tradition of research has been carried out is the biofacies, a suite of specific organisms (a paleocommunity) of usually local-to-regional extent that is

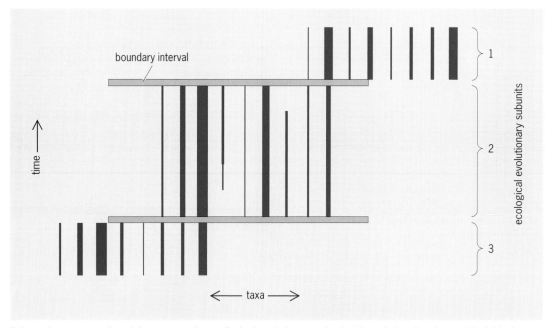

Schematic representation of three successive ecological-evolutionary subunits. The relative abundance of individual taxa is represented by the width of each vertical range bar, where greater width represents greater abundance. The full stratigraphic extent is depicted only for subunit 2. Within this subunit, most taxa range from top to bottom, maintaining consistent abundances. Natural causes for speciation and extinction are rare but are depicted here. Boundary intervals are relatively brief and represent times of extinction (range termination), speciation (range origination), immigration and emigration, and ecological reorganization—represented here as changes in the relative abundance of surviving taxa.

characteristic of a given paleoenvironmental setting and recognizable for upward of a million years.

Punctuated equilibria. Individual lineages within the fossil record remain stable for many millions of years and appear to speciate relatively quickly (geologically speaking). This phenomenon forms the paleontological basis for punctuated equilibria. Niles Eldredge and Stephen Gould drew upon the common experience of generations of field geologists and biostratigraphers, and coupling it with well-established biological ideas (specifically allopatric speciation), were able to resurrect what had always been viewed as an imperfect fossil record as the expected outcome of the evolutionary process. The evolution of a lineage was not continuous but proceeded in a stair-step fashion, with long periods of stasis periodically punctuated by speciation.

The realization that this mechanism proposed for the evolution of individual taxa might be expanded to encompass similar patterns observed across many lineages living together soon followed. Several studies utilizing a terrestrial mammalian database prompted the turnover-pulse hypothesis. This concept invoked similar causes to explain the long periods of evolutionary stasis as punctuated equilibria (developmental constraints and stabilizing selection), but looked to extreme environmental changes that might affect many members of a fauna concurrently to provide the impetus for the observed simultaneous evolutionary punctuations in many lineages.

Coordinated stasis. The recent documentation of a similar pattern in marine rocks, with the added density of sampling and more reliable stratigraphic

framework that they afford, has sparked renewed interest in the stability of fossil faunas. This interest has included a new emphasis on the apparent stability of the paleoecological associations of fossil organisms, as well as the usual attention paid to the evolutionary dynamics of individual taxa.

The documentation of coordinated stasis by researchers working in the Paleozoic rocks of the Appalachian Basin of the United States relies on two decades of exhaustive field work. This research program has come to the forefront of the discussion of long-term faunal dynamics because of its scope (Silurian–Devonian periods), resolution (the research has been carried out within a well-defined sequence stratigraphic framework that can resolve events to tens to hundreds of thousands of years), and the fact that it has identified faunal patterns at the species level.

Faunal patterns. Coordinated stasis consists of two distinct faunal patterns. First, the evolutionary side of the pattern may be represented as synchronized stasis and change across the entire fauna encompassed in a particular biofacies. This is demonstrated by the stratigraphic overlap of species ranges (see **illus.**). Only in rare instances has evolution or extinction of a biofacies member been observed within the geographic and stratigraphic range of that biofacies. Second, ecological measures describing the structure of the biofacies (such as guild membership, relative and rank abundance, and dominance diversity) display a similar bimodal pattern of stability, with periods of reorganization occurring during the same relatively brief windows of time that witness evolutionary change within individual lineages. It is during these intervals of reorganization that migration of exotic taxa into otherwise resistant biofacies has been shown to be widespread.

Ecological-evolutionary subunits. Biofacies that display patterns of coordinated stasis have been observed to be temporarily linked such that entire environmental gradients share periods of relative ecological and morphological stability and change. This broader pattern has led to the recognition of 12 ecological-evolutionary subunits within the Appalachian Basin. Each of these subunits represents a period lasting 2–7 million years during which all the biofacies in the basin display concurrent coordinated stasis. The boundaries of these subunits are recognized across all biofacies, and some appear to correspond to well-documented global bioevents.

Other patterns in fossil record. Apart from the large body of existing literature on biofacies, there are many patterns similar to coordinated stasis that have been reported in the fossil record. These include the Cambrian trilobite biomeres, vertebrate chronofaunas, and analogous findings from Quaternary coral reefs and Pennsylvanian coal swamp environments.

Causes of coordinated stasis. The four distinct aspects of coordinated stasis are (1) the morphological stasis of individual taxa; (2) the stability of the ecological structure of biofacies through time and space—which can often be seen to track preferred environments in the fossil record over hundreds of miles while maintaining ecological coherence; (3) the lack of new additions (through speciation and immigration) and the rarity of extinction within stable units; and (4) the coordination of evolutionary and ecological change.

The most basic explanations seek to rule out the possibility that coordinated stasis is some type of artifact of the fossil record. Two mechanisms have emerged within this general approach that warrant attention. The first mechanism is the cautionary demonstration that sequence stratigraphic boundaries might be mistakenly interpreted as periods of faunawide reorganization relative to some prior condition if studies are carried out at only a single outcrop. Second, it has been pointed out that a coordinated stasislike pattern might be produced if only evolutionary long-lived, robust taxa were preserved among a much more diverse fauna whose dynamic nature was lost. Both of these scenarios appear to be demonstrably false in the Appalachian Basin case; however, it is important to pay attention to sequence stratigraphic context and account for preservational biases during the first steps of any such faunal study.

The next level of explanation invokes a lack of external stimulus for ecological and evolutionary change. Much like the turnover-pulse hypothesis, change is seen as externally driven. Thus, periods during which faunas display coordinated stasis are interpreted as environmentally quiescent. Again, this mode of explanation relies on the same (primarily internal) mechanisms acting on individual lineages to explain evolutionary stasis as proposed by punctu-

ated equilibria. Paleoecological stability may be explained as the outcome of each lineage individually coming to some equilibrium with its functional environment. In order to explain the lack of additions and immigrations to stable biofacies, this theory is often coupled with a notion of incumbency. Those organisms that colonize a given region, or are the first to utilize a resource, have the upper hand in any struggle that might ensue with a competitor (either by virtue of their greater numbers or entrenched lines of supply).

The most controversial mode of explanation sparked by the unique combination of details in coordinated stasis is the notion that the method by which ecosystems and communities are organized might contribute not only to their compositional and structural stability but also to the evolutionary inertness of their constituent taxa. This explanation has taken several forms. Most formulations such as ecological locking, or the idea of structural hubs and system breakpoints, are built around the tenet that hierarchical principles which govern the structure of ecosystems both limit their dynamics (due to the rates and frequencies of processes grouped at different hierarchical levels) and introduce a degree of redundancy (due to the compartmentalized nature of ecosystems).

For background information *see* ANIMAL EVOLUTION; FOSSIL; ORGANIC EVOLUTION; PALEOECOLOGY; SPECIATION; SPECIES CONCEPT in the McGraw-Hill Encyclopedia of Science & Technology.

Kenneth M. Schopf

Bibliography. A. J. Boucot, *Evolutionary Paleobiology of Behavior and Coevolution*, 1990; D. H. Erwin and R. L. Anstey (eds.), *New Approaches to Speciation in the Fossil Record*, 1995; L. C. Ivany and K. M. Schopf (eds.). Theme issue on coordinated stasis, *Palaeogeog. Palaeoclimatol. Palaeoecol.*, vol. 127, no. 1–4, 1996; T. J. M. Schopf (ed.), *Models in Paleobiology*, 1972; E. S. Vrba, Environment and evolution: Alternative causes of the temporal distribution of evolutionary events, *S. Afr. J. Sci.*, 81:229–236, 1985.

Parachute

Parachutes have continued to be the simplest and cheapest devices for the deceleration of loads, people, and vehicles since their first recorded use in 1797. Comprising cloth and suspension lines, their construction is far simpler than that of aircraft. However, their very simplicity makes their aerodynamics much more complicated. Indeed, unlike aircraft, which are solid structures that deflect air around them, inflating parachutes not only deflect surrounding air but also adopt shapes that are dictated by the airflow they generate. Such shapes can be very complex (**Fig. 1**) and, since they change continuously, the process is intrinsically unsteady. Finally, given the lack of streamlining during inflation (Fig.

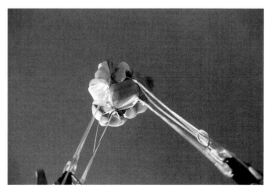

Fig. 1. Initial inflation stage of a square parachute.
(Photograph by Gary Peek)

1), turbulent flow rather than laminar flow dominates the aerodynamics.

Types. Parachutes come in many shapes and fab-
rics. Modern parachutes are termed either circular
or square, and most are made of nylon fabric. When
fully inflated, circular parachutes have the shape of
a hemispherical cap. In some cases the cap is slightly
conical, and in others slots or concentric gaps are
cut to enhance stability. Square parachutes are
shaped like an aircraft wing. They are able to maintain such a shape because the wing is built out of
cells which are inflated by ram-air action, that is, by
the relative wind created by the parachute's motion;
this wind enters through inlets cut on the leading
edge of the parachute.

Deployment stages. The initial deployment stages of
square and circular parachutes are similar. Both begin with the extraction from the harness (or vehicle)
of the bag containing the folded parachute, usually
by a drogue chute. As the bag separates from the
harness, the suspension lines unfold. Only when the
lines are fully stretched is the parachute allowed to
unfurl out of the bag and begin inflating. Inflation
is typically characterized by several stages whose
duration depends on the design of the parachute.
For sport square parachutes, the deployment of the
suspension lines typically lasts about 1 s, while inflation lasts 2–5 s.

Inflation. The inflation stages of a circular parachute are shown in **Fig. 2**. After unfurling out of
the bag, the parachute adopts a rather elongated
shape, resembling a vertical tube opened at its lower
end (Fig. 2*a*). Because of the system's rapid descent,
air rushes in through the tube's opening and accumulates at the apex of the canopy to create a high-
pressure air "bubble." Steady inflow continues to
build up internal pressure, allowing the bubble's
volume to expand horizontally as well as vertically.
However, there is more expansion in the vertical
direction than in the horizontal direction. This process continues until the bubble is large enough to
occupy the entire design volume of the parachute.

The shape of the expanding air bubble trapped
inside the canopy is dictated by the balance of aerodynamic forces that act in opposite directions along

the boundary defined by the parachute's fabric. The
bubble's expansion rate first depends on the pressure differential between the outside and inside of
the parachute. As for any blunt object moving
through air, wake turbulence generated on the
downwind side of the parachute causes the external
pressure to be lower than the internal pressure near
the apex (Fig. 2*b*). The faster the parachute, the
larger the pressure differential, the faster the inflow
into the parachute, and the faster the bubble's
expansion. However, rapid expansion generates a
large external pressure which squeezes the bubble
on its upwind side, slowing down the expansion.
This pressure arises because the bubble deflects outside air outward and increases air resistance in the
process. Again, this squeezing effect increases with
the parachute's descent speed. The balance between
these factors (rapid inflow, low external pressures
near the apex, and the squeezing force at the lower
end) is achieved by the bubble adopting the "optimal" shape, and leads to faster expansion along the
vertical than the horizontal. Because the parachute-
load system decelerates during inflation, this balance
of pressure inside and outside the parachute is continuously readjusted as the inflow becomes slower
(Fig. 2*c*).

The inflation of square parachutes (Fig. 1) evolves
similarly, at least during the first seconds of the inflation process. Here, inflation takes place in the cells
of the parachute, which later define the shape of
the canopy. Initially, these cells are shaped like horizontal tubes with inlets at one end, which scoop up
the air coming from below. The walls between the
tubes have holes which allow the internal pressure
to spread more evenly over the entire parachute.
The inflation process is regulated by the pressure
differentials between the outside and inside of the
cells in a manner similar to the processes in Fig. 2.

Computer simulations. The availability of supercomputers has led to the development of computer models in which the interactions of the airflow with the
fabric structures are demonstrated with increasing
fidelity. In a typical simulation, such models first
map the internal and external airflow (and pressures)
on grids which are adapted to the various scales and
shapes of the flow near and far from the parachute.
Second, the fabric and suspension lines are allowed
to move according to the pressures acting on them.
Finally, the flow is recalculated according to the new
positions and shapes adopted by the fabric and lines.
Over many such iterations, the evolution of the inflation process is simulated (**Fig. 3**). Current simulations are quite realistic in some respects. For example, at the midpoint of the time history shown in
Fig. 3, the top of the canopy flattens out as a result
of a phenomenon called wake recontact, where the
air sucked in the apex's wake is catching up with
the continuously decelerating parachute.

Tensed versus crumpled fabric. Such computer simulations leave out several important aspects, such as
fabric flutter. Fabric is a medium that is difficult to
simulate because it can exist in two very different

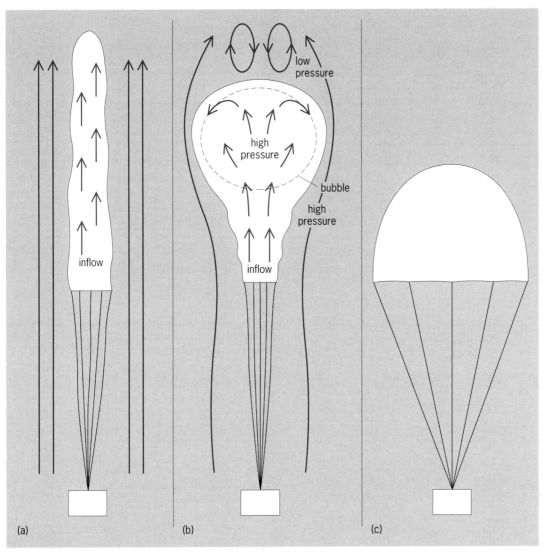

Fig. 2. External and internal airflows and pressures acting on an inflating circular parachute (*a*) after bag unfurling, (*b*) at midpoint, and (*c*) at a late stage.

states: tensile and crumpled. In the former, the fabric is stretched and adopts flat or curved shapes according to the average pressure differential across its surface and the local fabric stretching tension. In the latter, the fabric sustains no tension and is irregularly shaped according to the pressure differences existing at all length scales on either side. Moreover, the dynamics of crumpled fabric are highly unstable and involve small-scale motions which can be much faster than the large-scale motions of the entire fabric surface. An accurate simulation of the motion of crumpled fabric is beyond current computer capabilities. Current models leave out small-scale turbulence and reduce crumpled-fabric instabilities by reducing the number of crumpled configurations and by damping their motions. This is why the simulated canopy shapes in Fig. 3 look so smooth. These models, however, are reasonably accurate for describing the time-averaged motions and forces that occur during the inflation of highly

stressed parachutes, since large fabric tension forces tend to reduce the effects of small- and large-scale transient and turbulent flows that shape the surface. Also, parachutes released at very low speeds are well described by these simulations (such as Fig. 3) since there are no high-frequency oscillations of crumpled fabric involved.

Porosity and permeability. Current computer models are sophisticated enough to begin including the effects of other factors which affect real deployments, such as fabric permeability and canopy (geometric) porosity. The former measures the property of fabric to let a certain quantity of air escape through its strands and pores. The latter characterizes round canopies that have gaps and holes cut into them. Greater porosity increases the stability of a parachute during its inflation and subsequent glide. Circular canopies with zero geometric porosity tend to undergo wild oscillations, which are caused by air spilling out of one side of the canopy opening (the lower

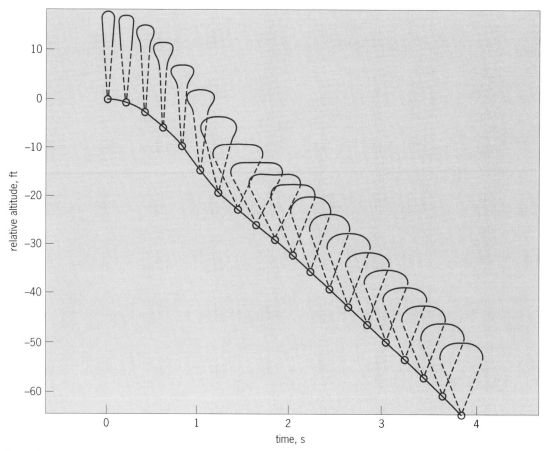

Fig. 3. Computer simulation of an inflating circular parachute released from rest. The process is superimposed on the altitude loss incurred during the process. 1 ft = 0.3 m. *(After R. J. Benney and K. R. Stein, U.S. Army Natick Research, Development and Engineering Center)*

part of the tube in Fig. 2b). These oscillations are eliminated if air is allowed to escape in a symmetric manner, for example, through a hole cut at the apex of the canopy, or through concentric, circular gaps cut along its skirt. However, introducing geometric porosity tends to slow down inflation, since allowing air to flow through gaps reduces the rate of internal pressure buildup.

Fabric permeability and canopy porosity also change the nature of the airflow in and out of the canopy since they generate large- and small-scale vortices near the parachute. Therefore, the shape and relative location of the gaps and slots must be carefully simulated in order to ensure optimal stability.

Reefing systems. If not limited, opening forces may be large enough to destroy the parachute or damage its load. Typically, if a parachute inflates too quickly, little or no deceleration occurs. The parachute and its load still travel at very large speeds by the time the parachute has opened, generating a very large amount of drag and consequently a large force on the parachute structures. Reefing devices limit the rate of canopy expansion during the early phase of inflation. With some circular parachutes, for example, a line routed around the skirt limits the inflow

and permits early deceleration, until a pyrotechnic device cuts the line at a preset time or altitude. On sport square parachutes, a slider is used to achieve the same result. Here, a square of nylon fabric is allowed to slide freely and slowly down the suspension lines, at a rate which is controlled by the slider's own drag and the tension created by the suspension lines, which fan out of the slider.

Introducing the effects of reefing devices actually simplifies some computer simulations. For slider-reefed parachutes (round or square), since the location of the slider directly controls the amount of expanded canopy surface area, it is possible to use simpler, yet quite informative equations to describe the inflation process. On skirt-reefed circular canopies, the fabric has acquired more tension by the time the pyrotechnic cutting device is activated, thus making simulation models more reliable.

Forebody wake. Another element currently left out of the computer simulations is the accurate description of the interaction between the wake generated by the suspended load (or forebody wake) and the inflating parachute inflow. In certain high-speed flight regimes, such wakes are wide and long enough to reach the opening of the canopy. Because the wakes are characterized by turbulent flow rather than laminar flow, they cause air rushing into the canopy to become more turbulent, thus creating more fabric flutter and much crumpled fabric—all processes that are impossible to simulate accurately. On average, forebody wake decreases the pressure differentials across canopy fabric and leads to a slower and more unstable inflation process.

For background information *see* AERODYNAMIC FORCE; COMPUTATIONAL FLUID DYNAMICS; PARACHUTE; TURBULENT FLOW in the McGraw-Hill Encyclopedia of Science & Technology.   Jean Potvin

Bibliography. C. W. Peterson, High performance parachutes, *Sci. Amer.*, 262(5):108–115, May 1990; D. Poynter, *Parachuting: The Skydiver's Handbook*, 5th ed., 1990; J. H. Strickland and H. Higuchi, Parachute aerodynamics: An assessment of prediction capability, *J. Aircraft*, 33:241–252, 1996.

Particle accelerator

For many years, experiments studying the scattering of electrons, muons, and neutrinos on nucleons have provided abundant information on the structure of the nucleon and on the properties of the basic forces acting between elementary constituents. With the opening of the electron-proton collider HERA (Hadron-Electron-Ring Accelerator) at DESY (Deutsches Elektronen-Synchrotron) in Hamburg, Germany, the available center-of-mass energy in the lepton-nucleon system increased by more than an order of magnitude, corresponding to a tenfold improvement in spatial resolution compared to that provided in fixed-target experiments.

HERA is the first accelerator constructed through a global collaboration, with institutions from 10 countries either contributing components or delegating skilled personnel to the project. The project was approved in 1984 and completed on schedule at the end of 1990, with roughly a third of the total effort contributed by non-German institutions.

HERA facility. HERA consists of two independent accelerators designed to store 820-GeV protons and 30-GeV electrons, respectively, and to collide the two counterrotating beams head-on in four interaction regions spaced equidistantly around its 6.3-km (3.9-mi) circumference. **Figure 1** shows an aerial view of the DESY site with the HERA ring indicated. The HERA tunnel, located 15–30 m (50–100 ft) below the surface, consists of four 90° arcs joined by straight sections 360 m (1180 ft) in length. The beams collide in the middle of the straight sections. At these locations, large subterranean buildings seven stories deep have been erected to house the experiments as well as accelerator utilities. Superconducting dipole and quadrupole magnets are used to guide the protons in their arcs. The proton beam energy is limited to 820 GeV by the 4.68-tesla bending field of the dipole magnets. These magnets are cooled with liquid and gaseous helium, which is distributed around the ring from a central plant installed on the DESY site. Over a distance of 60 m (200 ft) centered at the interaction point, the protons travel in the plane of the electron beam, and there they are guided by conventional iron magnets. The electron energy is limited to 30 GeV by the energy loss due to the synchrotron radiation. This energy is replenished by a large radio-frequency system that provides a total voltage of 200 MV. This system consists of warm 500-MHz cavities augmented by 16 four-cell, 500-MHz superconducting cavities. The magnetic bending field of 0.165 T is rather modest and is provided by conventional iron magnets.

The peak luminosity and the integrated weekly luminosity are approaching the design values of

Fig. 1. Aerial view of the **DESY** site with the **HERA** ring included. The **HERA-B** experiment is located in Hall West (labeled W), the collider experiments H1 and ZEUS in Hall North and Hall South (N and S) respectively, and the **HERMES** experiment in Hall East (E).

1.5×10^{31} cm^{-2} s^{-1} and 1.5 pb^{-1} week^{-1} (1 picobarn equals 10^{-36} cm^2), respectively. That is, for a given scattering process, when the accelerator is operating at peak level, the number of scattering events per second will be 1.5×10^{31} times the total cross section for that process, measured in square centimeters. For a scattering process whose total cross section is 1 pb, scattering events will be observed at an average rate of 1.5 events per week. The observed beam-beam effects which limit the ultimate value of the luminosity are in agreement with the design values. The superconducting magnet system and the cryogenic system have been quite reliable, and as of December 1996 none of the superconducting magnets had failed.

Unexpected opportunities have been provided by the surprising observation of a large transverse electron polarization with values up to 70%. A spin-rotation system has been installed in Hall East for the HERMES experiment, and high and reproducible longitudinal beam polarizations of order 60% are routinely achieved. Similar systems that will provide the H1 and ZEUS experiments with longitudinally polarized electrons will be installed during the planned 1999–2000 shutdown of the accelerator.

HERA research program. HERA is the basis for a rich and varied program in particle physics. Some 1300 physicists, of whom nearly 900 are from non-German institutes, are engaged in the experimental program. At present, two of the straight sections are occupied by the general-purpose detectors, H1 and ZEUS, which are used to study electron-proton collisions. In the two remaining straight sections the two beams are separate and are used for high-luminosity, high-duty-cycle, fixed-target experiments. HERMES is designed to study the spin structure of the nucleon, using the interaction of longitudinally polarized electrons with polarized hydrogen, deuterium, or helium-3 nuclei. HERA-B uses an internal wire target which intercepts protons leaving the radio-frequency bucket to produce $b\bar{b}$ quark-antiquark pairs for a measurement of CP-violating effects in the decay of neutral B mesons. The experiment is under construction and is expected to provide its first data by the end of 1999.

Quark structure of protons. Earlier experiments on lepton-nucleon interactions had shown that the proton is made of colored quarks bound by the exchange of colored, massless gluons. Free quarks have never been observed, and there is evidence that quantum chromodynamics (QCD), the underlying nonabelian field theory of strong interactions, gives rise to the color confinement of quarks and gluons inside the nucleon. That means that a quark will not appear as a free particle regardless of how hard it is hit, but will materialize as a well-collimated stream of colorless hadrons.

According to this picture, the spacelike current, neutral or charged, emitted by the incident electron interacts with one of the quarks in the proton, and the struck quark materializes as one or several jets of hadrons, called current jets, with momentum components transverse to the beam axis that are balanced by the transverse momentum of the outgoing lepton. The remains of the proton appear as a sharply collimated jet of hadrons that travel along the initial proton direction, that is, along the beam axis, where in general they cannot be observed. In a neutral-current event the electron scatters inelastically off the struck quark through exchange of a photon or a neutral vector boson (Z^0). In a charged-current event a charge vector boson (W^+ or W^-) is exchanged and the electron is transformed into a neutrino. In the case of charged-current events, only the hadrons in the current jet can be observed since the neutrino is, for all practical purposes, unobservable. The predicted event topology is observed directly at HERA (**Fig. 2**).

Structure functions. The momentum distributions of the quarks in the proton are described by three structure functions. These functions depend on two variables, labeled x and Q^2. The Bjorken parameter, x, is the relative momentum carried by the struck quark, that is, the momentum of the struck quark divided by the momentum of the nucleon. The variable Q^2 is the square of the momentum transfer between the electron and the proton. The structure functions are measured at HERA as functions of these variables by observing either the scattered electron or the hadron jet in neutral-current events or the hadron jet only in charged-current events. The kinematic region in Q^2 and x which can be probed in deep inelastic scattering experiments at HERA is shown in **Fig. 3.** The kinematic range covered by earlier fixed-target data is also indicated.

At low values of Q^2, neutral-current events are copiously produced compared to charged-current events. At large values of Q^2 (equal to or greater than 5000 GeV^2), the collider experiments at HERA have shown that these two cross sections are of similar size.

With increasing values of Q, it is possible to resolve ever finer structures. This phenomenon is due to the facts that the observation of an object of a given size requires viewing by "light" whose wavelength is smaller than the dimensions of the object, and in an inelastic process the wavelength of the exchanged particle is proportional to $1/Q$. The increase in spatial resolution of order $1/Q$ achieved at HERA has so far not revealed yet another level of substructure of matter. Indeed, the behavior of the structure functions as measured over a wide range in Q^2 and x is in excellent agreement with predictions based on quantum chromodynamics and provides a stringent test of the understanding of strong interactions. The strong coupling constant, α_s, actually depends on Q^2, and from this analysis its value is found to decrease with increasing values of Q^2 as predicted by quantum chromodynamics. For Q^2 equal to M_Z^2, where M_Z is the mass of the Z^0 boson, the value of α_s as found in deep inelastic scattering processes at HERA, is in agreement with other independent determinations. *See* SUPERSYMMETRY.

The HERA data extend the measurements of the structure functions by two orders of magnitude in x down to values of 10^{-6}. The structure functions

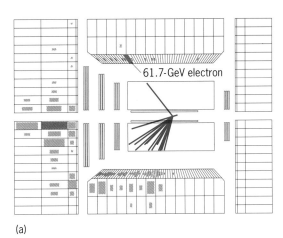

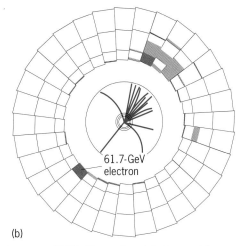

(a)

(b)

Fig. 2. Example of a deep inelastic event as observed with the **ZEUS** detector at HERA. The scattered electron and the hadrons in the current jet are detected by electromagnetic and hadronic calorimeters. (*a*) Event viewed from the side. A 27-GeV electron from the left collides with a 820-GeV proton from the right. Remnants of the proton are seen in the forward calorimeter along the incident proton direction. (*b*) Event viewed along the beam direction.

at moderate values of Q^2 increase steeply with decreasing values of x. This increase reflects a strong increase in the gluon density of the proton as x decreases, which in turn reflects the fact that there are many more gluons with low values of x (that is, with a low momentum fraction). At very low values of x, due to the high gluon density, interactions of the gluons become important and perturbative quantum chromodynamics breaks down. Thus, nonperturbative quantum chromodynamic effects such as gluon recombination leading to saturation may become detectable at HERA. Measurements of outgoing hadrons, produced in deep inelastic interactions, are particularly sensitive to a breakdown of standard perturbative quantum chromodynamics.

Evidence for the pomeron. In general, hadrons populate the gap in production angle between the current jet and the forward direction as a result of color transfer between the struck quark and the proton remains. Experiments at HERA have observed a new class of events characterized by a large gap in production angle between the proton direction and the nearest hadron. This behavior signals the exchange of a colorless object, the pomeron. First measurements have demonstrated that the pomeron is dominated by its gluonic component. Diffractive processes and their underlying mechanism can be studied at HERA in a controlled manner and in great detail for the first time.

Exotic particles. HERA is well suited to search for excited electronlike heavy leptons, scalar quarks, and particles with combined lepton-baryon numbers. An analysis of early HERA data shows no evidence for such particles at a mass up to roughly 200–280 GeV with a coupling strength less than or equal to $\sqrt{4\pi\alpha_{\mathrm{em}}}$, where α_{em} is the electromagnetic coupling constant.

Nucleon spin structure. Early experiments on deep inelastic polarized lepton-nucleon scattering showed that only roughly a third of the nucleon spin is carried by quarks. The HERMES experiment is designed to measure the outgoing hadrons in coincidence with the scattered electron. First data obtained using polarized hydrogen and helium-3 targets are promising, and it is expected that the observation of the final state will help reveal the basic mechanism of the nucleon spin.

HERA luminosity upgrade. To exploit the physics potential of HERA, in particular to search for subtle nonperturbative quantum chromodynamic effects such as instantons, to provide high-precision tests of the electroweak sector, and to search for new physics outside the standard model such as right-

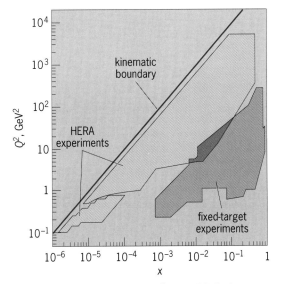

Fig. 3. Kinematic region in x and Q^2 accessible in deep inelastic experiments at HERA. The region so far probed by the HERA experiments is compared to the x and Q^2 values probed by earlier fixed-target experiments. The kinematics of the experiment makes values of x and Q^2 above and to the left of the kinematic boundary impossible.

handed charged currents or leptoquarks, will require an integrated luminosity (integrated, that is, over the duration of the experiment) of 1 femtobarn^{-1} or more. It has been shown that the HERA electron-proton luminosity can be increased by a factor of 5 by moving the low-beta focusing quarupole magnets closer to the interaction region. The reconfiguration of the interaction region will be done during the 1999–2000 maintenance shutdown and should yield a yearly integrated luminosity of the order of 150 pb^{-1} year^{-1}.

For background information *see* ELEMENTARY PARTICLE; MESON; PARTICLE ACCELERATOR; PARTICLE DETECTOR; QUANTUM CHROMODYNAMICS; QUARKS; STANDARD MODEL in the McGraw-Hill Encyclopedia of Science & Technology.   Björn H. Wiik

Bibliography. G.-A. Voss and B. H. Wiik, The electron-proton collider HERA, *Annu. Rev. Nucl. Part. Sci.*, 44:413–452, 1994; Z. Zhi-Peng and C. He-Sheng (eds.), *Proceedings of the 17th International Symposium on Lepton-Photon Interactions, Beijing 1995*, 1996.

Peptide

Polypeptides and proteins exhibit enormous structural diversity. The diverse three-dimensional organization of atoms within these macromolecules underlies the remarkable versatility in protein function. The protein biopolymer plays both structural and functional roles in living systems. For example, the molecular machines that convert light energy into chemical energy, coordinate the movement of muscle, transport oxygen in the blood, and catalyze chemical reactions are based primarily on protein biopolymers. In addition, the distinctive physical properties of proteins such as collagen and silk arise from their unique molecular structure.

Forces governing protein structure. The primary sequence (or order) of α-amino-acid building blocks uniquely defines the precise three-dimensional architecture of a particular protein. In nature, a limited library of 20 amino acids, differing only in the chemical identity of the residue side chains [R in structures (**1**) and (**2**)], provides for the limitless permutations

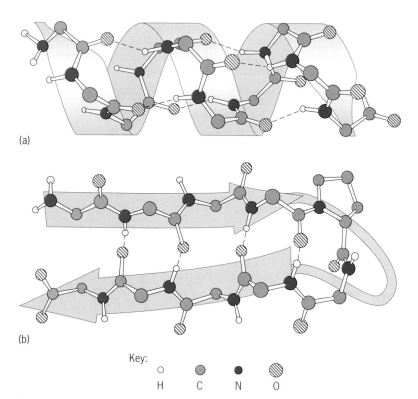

of primary sequences that form the basis of this multitalented biopolymer. The polymeric structure results from the condensation of the α-amino group ($-NH_2$) of one amino acid with the carboxyl group ($-CO_2H$) of another to form an amide bond.

This covalent bond (also termed a peptide bond) has unique properties that make it eminently suited for its role in the protein structure. The peptide bond has limited rotational freedom because of the partial double-bond character of the amide moiety, while the remaining bonds in the polymer main chain can rotate freely (**2**). Therefore, one in every three bonds in the chain is fixed. Because of this constraint, the polypeptide backbone is significantly more structured than typical synthetic polymers, such as polyethylenes and polyesters, which lack such rigidifying elements. Furthermore, the amide bond is equipped with both hydrogen-bond donors and acceptors, allowing the peptide backbone to engage in intricate networks of noncovalent hydrogen-bonding interactions. While these forces are inherently far weaker than the covalent bonds of the polymer chain, the additive effect of a large number of such interactions makes the folded, three-dimensional structures of proteins relatively stable. In fact, some proteins have evolved an unusual stability that allows them to retain their structure under the extreme heat and pressure conditions of deep-sea geothermal vents.

Structural studies have revealed that the hydrogen-bonding interactions of proteins fall into specific patterns (termed secondary structures). The most

Fig. 1. Protein secondary structure. The hydrogen bonding network is shown in the ball-and-stick diagram, and the backbone chain in the ribbon diagram. (a) Structure of α-helix. (b) Structure of β-hairpin.

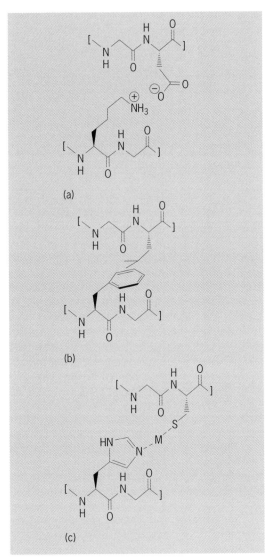

Fig. 2. Protein tertiary structure interactions. (*a*) Electrostatic interactions. (*b*) Hydrophobic interactions. (*c*) Metal coordination sites. M = metal.

common elements of secondary structure are the extended β-sheet, composed of individual β-strands (with an extended polypeptide backbone) arranged parallel or antiparallel and the coiled α-helix (**Fig. 1**). Reverse turns are also important structural elements that frequently connect α-helix and β-sheet domains. While protein secondary structure is dominated by interactions between the amide groups within the peptide backbone, higher-order or tertiary structure arises from a variety of long-range interactions between the functional groups on both the backbone and side chains. Prime determinants of tertiary structure include electrostatic and hydrophobic interactions, as well as coordination between metal ions and particular amino-acid side chains (**Fig. 2**).

Thus, protein structure is hierarchical; a linear sequence of amino acids can form ordered elements of secondary structure, which may cooperatively interact to create defined tertiary structures. Understanding how a linear sequence of amino acids dic-

tates the precise fold of a protein (the protein folding problem) remains one of the central issues in the biological sciences. An aspect of this problem is addressed in the field of protein design, which seeks to exploit the protein architecture and a knowledge of protein-folding events in the construction of artificial proteins.

Polypeptides and proteins. In nature, proteins are assembled by a complex biochemical machinery that uses information contained in the genetic code &deoxyribonucleic acid (DNA) or ribonucleic acid (RNA); to program the primary sequence of amino acids. Although this biological process is typically limited to the 20 naturally occurring amino acids, it is also possible to build polypeptides and proteins with an expanded set of building blocks in the laboratory by using chemical synthesis. The most powerful technique for the chemical synthesis of peptides involves the assembly of a polypeptide in a stepwise fashion on a solid polymeric support. The method has been automated to maximize the efficiency of the process and has been applied to the synthesis of peptides up to about 100 residues in length.

Although the basic chemical structures of polypeptides and proteins are the same, the distinction between them is one of size; polypeptides are typically small (10–40 amino acids in length), while most proteins exceed 100 residues. In general, the minimum size of a structured, functional protein is considered to be 70 residues. Small folded proteins or protein domains do exist, but these usually require cross-links such as disulfide bridges (bonds between two amino acids containing sulfur) or metal coordination sites for stability (**Fig. 3**). These small proteins tend to be functionally more limited than their large counterparts.

Protein design. Key goals in the field of protein design are to define the minimum size of a polypeptide required to achieve a compact and organized architecture, and to utilize these miniproteins as structural templates for the design of novel functional polypeptides. A recent protein design effort showed that a uniquely folded structure could be achieved in a 23-residue polypeptide in the absence of any strong cross-links. The peptide sequence was obtained through an iterative design process starting from a naturally occurring family of polypeptides known as the zinc-finger domains. Zinc fingers are small (23–28 amino acids), conserved polypeptide units that adopt a folded ββα topology, comprising a β-hairpin connected to an α-helix (**Fig. 4***a*). A β-hairpin is a structure in which two β-strands are joined by a reverse turn (Fig. 1*b*). The zinc-finger structure is stable only if divalent zinc cations are present and engaged in cross-linking the β-hairpin and α-helix portions of the polypeptide sequence together to form a fingerlike structure. The metal ion provides the cement that holds the structure together in three dimensions. These polypeptides also incorporate key hydrophobic residues at specific locations within the sequence to stabilize the

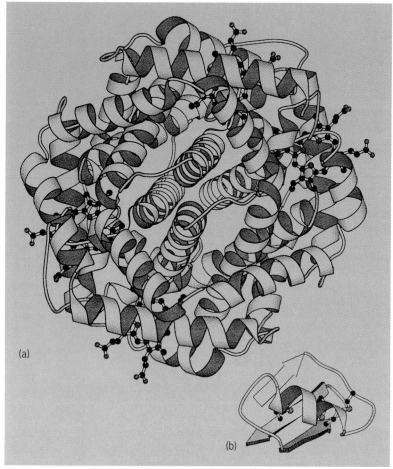

(a)

(b)

Fig. 3. Protein structures. (*a*) Hemoglobin; 574 amino acids. (*b*) Charybdotoxin; 37 amino acids; 3 disulfide bridges.

fold. However, the long-range, noncovalent interactions between these residues are insufficient to stabilize the structure without metal coordination.

The problem is to redesign this structure to eliminate the requirement for metal-coordinating crosslinks and thus to create a template for protein design. To generate a zincless zinc finger, the sequence of the polypeptide must be modified to compensate for the absence of metal coordination. Here it is possible to take advantage of the fact that certain residues have a conformational preference for a specific type of secondary structure (β-sheet or α-helix). Therefore, in the redesigned peptide (termed BBA1), appropriate residues were incorporated into the primary sequence to enhance the formation of each individual element of secondary structure. Of additional concern was the fact that the β-hairpin (strand–turn–strand) structure in the native zinc-finger domains is intrinsically disordered in the absence of metal coordination. Therefore, in BBA1 a tetrapeptide sequence that adopts a defined reverse-turn conformation was employed to connect the strands and fold the hairpin in the absence of metal (Fig. 4*b*).

The three-dimensional structures of complex molecules may be determined either by x-ray crystallography or by nuclear magnetic resonance (NMR). The small size of the peptide BBA1 made it amenable to proton NMR analysis. BBA1 includes 191 protons, 127 carbons, 35 nitrogens, and 34 oxygens. The NMR experiments indicate which protons lie within 0.5 nanometer of each other. For comparison, the length of a typical carbon–carbon single bond is 0.15 nm. The proton-to-proton distance information then permits deduction of the relative location of all the protons in three dimensions. The placement of the remaining atoms can be inferred from what is known about the chemical structure of the polypeptide.

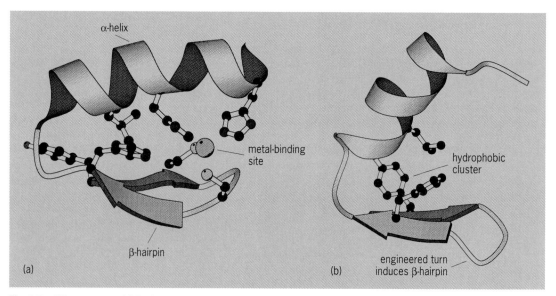

Fig. 4. Peptide structures. (*a*) Typical zinc-finger domain; 26 amino acids. (*b*) Redesigned peptide BBA1 motif; 23 amino acids.

Solving a structure by NMR is a lot like solving a three-dimensional jigsaw puzzle. The pieces of information are the interproton distances; when these distances are all put together (computationally), a picture of the chemical structure emerges. The structure of BBA1 as derived from the NMR analysis agreed well with the design; the α-helix and β-hairpin are held together by defined hydrophobic interactions between residues at positions 1, 8, and 14 in the sequence. While BBA1 no longer includes metal-coordinating cross-links, the presence of a defined tertiary structure is clear.

The structure of BBA1 reveals that the motif is more open than the natural zinc fingers. In the latter peptides, a metal-binding site cross-links the end of the α-helix to the β-hairpin, and this interaction defines the angle between the helix and the sheet. It is therefore not surprising that the removal of such a cross-link would result in a more open structure. It may be possible in the future to exploit the openness of the motif in the design of enzyme mimics that utilize the groove created between the helix and sheet as a substrate-binding site. Thus, BBA1 represents a versatile template for the future design of functional polypeptides.

This design effort demonstrates that, although the rules for protein folding are still poorly understood when it comes to predicting a folded structure from a linear sequence from first principles, the guidelines are emerging for the redesign and restructuring of existing folded structures. This small folded motif may be useful as a simple model system to provide insight into the protein-folding problem.

For background information *see* AMINO ACIDS; CHEMICAL BONDING; COORDINATION CHEMISTRY; NUCLEAR MAGNETIC RESONANCE; PEPTIDE; PROTEIN in the McGraw-Hill Encyclopedia of Science & Technology.   Barbara Imperiali

Bibliography. F. M. Richards, The protein folding problem, *Sci. Amer.*, 264(1): 54–63, January 1991; J. S. Richardson, Protein anatomy, *Adv. Protein Chem.*, 34:167–339, 1981; M. D. Struthers, R. P. Cheng, and B. Imperiali, Design of a monomeric 23-residue polypeptide with defined tertiary structure, *Science*, 271:342–345, 1996.

Perception

Attempting to listen to another person talking in a crowded party room is more difficult than in a quiet room. The ability of the listener to understand the speaker (or of even recognizing that someone is speaking) is clearly diminished by the noise of the party. Because it is generally believed that noise degrades signals, much effort has been put into filtering out the noise. However, recent research has unveiled a wide class of systems that use noise to detect weak signals via a mechanism known as stochastic resonance (SR). Stochastic resonance was discovered in an electronic circuit and then in other physical systems. It is currently of interest in sensory biology and in psychophysics as a method of enhancing the ability of individuals to detect weak signals received from the environment.

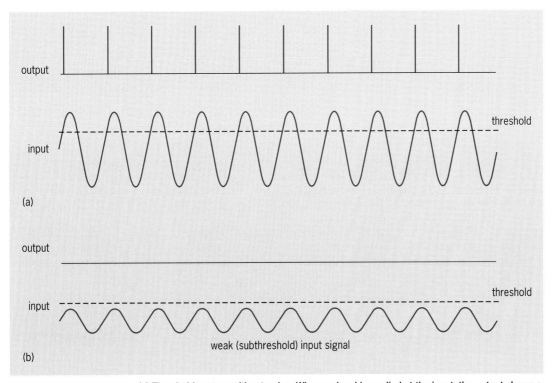

Fig. 1. Stochastic resonance. (a) Threshold system without noise. When a signal is applied at the input, the output shows a pulse every time the threshold is crossed from below. (b) If the input is a subthreshold signal, no pulses can be seen at the output.

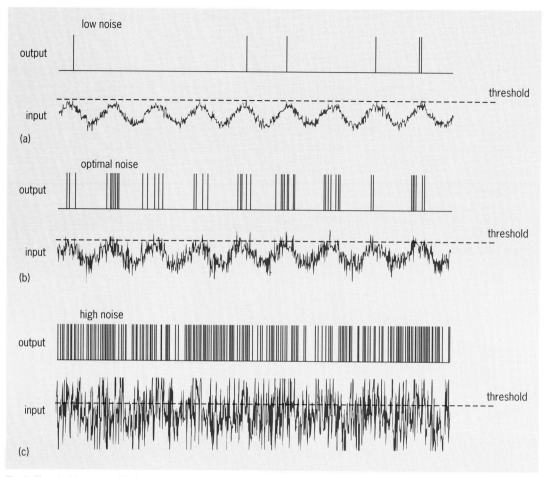

Fig. 2. Threshold system with (a–c) low, optimal, and high noise. (a) For noise levels that are too low or (c) too high, the pulses are observed at the output, but their statistical distribution does not give enough information on the input. (b) At the optimal noise level, which is intermediate, the probability of observing pulses is clearly modulated by the input.

Stochastic resonance. Stochastic resonance can be illustrated in a simple but highly nonlinear system that can transmit a single bit of information. For example, a system can be used in which a pulse is generated every time an input signal crosses a fixed threshold value. Some information about the input signal may be obtained by looking at the output of this threshold device. Consider a periodic input signal. By analyzing the time-distribution pattern of the threshold crossings, it becomes easy to detect the period of the input (but not the shape or the amplitude). Weak signals that are completely below the threshold do not trigger a pulse, so no information is transmitted. It has been determined that the poor performance from this type of system can be enhanced by adding noise. The addition of a small amount of noise at the subthreshold input signal can produce observable threshold crossings. The number of threshold crossings could be quite low, allowing only a very small amount of information to be transmitted per unit of time. An increase in the amount of noise added to the input signal would also increase the number of threshold crossings. By observing the time distribution of the output, it can be observed that the pulses are randomly distributed with time but their frequency is clearly modulated by the input signal. Thus, more information is trans-mitted through the threshold, and the input sensitiv-ity of the system increases with increasing noise, but only up to a point called the optimal noise level. When the noise level is raised above the optimal, the amount of information transmitted by the system decreases until, for amounts of noise very much larger than the optimal noise level, no information is transmitted (**Figs. 1** and **2**).

The threshold system bears some resemblance to some sensory neurons that encode the input signals in trains of spikes. By using the mechanism of sto-chastic resonance, these systems can detect sub-threshold information-carrying signals. This en-hancement of input sensibility has been observed in the peripheral nervous system of crayfish, crick-ets, rats, and humans. In all the experiments involv-ing these organisms, external noise has been added to a weak signal in the same manner as in the thresh-old system previously described. Internal noise has long been associated with the nervous system, and it is possible that both internal and external noise may serve a useful role in brain functions. In an attempt to address this problem, researchers are us-ing the human visual system.

Images, thresholds, and noise. In much the same way that a periodic input signal is processed with noise and filtered with a threshold system, noise can be

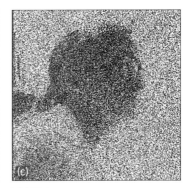

Fig. 3. Static effect of three noise levels after threshold filtering. (*a*) Low. (*b*) Optimal. (*c*) High.

added to an image and then filtered with a threshold system by using a digital computer. When a gray-scale image is digitized in the computer memory, it is first divided in a discrete grid and then a number is assigned to every point (pixel) of the grid, representing the brightness of the image at that point; the higher the number the brighter the pixel. The threshold system can be easily applied to every pixel by comparing the gray-scale entry with a fixed value termed the threshold value. Pixels whose gray-scale values are above the threshold are painted white on the screen, and all the others are painted black. Noise can be added to the image before the threshold filtering by adding random numbers to the gray-scale level of the pixels. These random numbers follow the same statistical distribution for every pixel of the image with zero mean and a fixed value for the standard deviation that can be called the intensity of the noise. After filtering with the threshold, each pixel can assume only one of two values (white if above threshold, black otherwise), thus carrying only one bit of information. The result of adding noise to a subthreshold image can be shown for three different levels of noise: low, optimal, and high (**Fig. 3**). As with the threshold system described in the previous paragraph, there is an optimal noise intensity which maximizes the image quality. Only the static effect of noise is represented in Fig. 2; that is, for every pixel, a random number (within a specific noise-level gaussian distribution) has been picked up, added to the original gray scale, and then filtered with the threshold. Striking effects can be observed when images are processed as noise changes with time. Animations can be generated in which the noisy component is different from frame to frame, while the subthreshold image is kept constant.

Dynamical noise has a profound influence on image recognition, with the perceived image quality improving drastically as the time between refresh frames decreases. Two noise parameters affect the observer's perception (at a fixed value of the threshold): the noise intensity and the speed at which noise changes. While the noise intensity effect can be ascribed to the stochastic resonance mechanism, the speed effect depends on the ability of the brain to detect information among noisy stimuli. Fast, local fluctuations of noise are easier

to filter out, and more details of the original image can be recognized.

In the experiments performed, the visual acuity of all the observers has been higher (that is, the ability of perceiving finer details of the noisy image) with faster noise; and for the fastest changing noise, it was possible to establish an optimal noise level for which the best perception of all noise levels and refresh rates could be obtained. There were substantial differences between the subjects only for the perceptive threshold at the optimal noise, showing that some individuals are more capable of dealing with noisy images than others.

For background information *see* ACOUSTIC NOISE; ACOUSTIC SIGNAL PROCESSING; HEARING (HUMAN); IMAGE PROCESSING; PERCEPTION; PSYCHOLOGY; VISION in the McGraw-Hill Encyclopedia of Science & Technology.   Enrico Simonotto

Bibliography. W. R. Hendee and P. Wells, *The Perception of Visual Information*, 1993; W. Prinz and B. Bridgeman, *Handbook of Perception and Action*, vols. 1 and 2, 1996; K. Wiesenfeld and F. Moss, Stochastic resonance and the benefits of noise: From ice ages to crayfish and SQUIDs, *Nature*, 373:33–36, 1995.

Petroleum exploration

A unique true distribution of rock properties exists in each petroleum reservoir as a result of a complex sequence of physical, chemical, and biological processes. Although some of these depositional and diagenetic processes are well understood, it is impossible to define the initial and boundary conditions in sufficient detail to provide a deterministic picture of the reservoir. There is unavoidable uncertainty in the distribution of rock properties. Geostatistical techniques are increasingly used to quantify this uncertainty and to create numerical models that mimic the physically significant features of rock property variation.

Uncertainties. The uncertainty is assessed by creating alternative realizations or descriptive models of the reservoir. The response variables of interest, such as peak production rate and recoverable reserves, are computed with each realization. A distribution or histogram of uncertainty can be assembled by pooling the results from a number of realizations. This assessment of uncertainty leads to more informed reservoir-management decisions.

To be reliable, each geostatistical realization must be constrained to all available information about the reservoir. This information comes from a variety of sources that represent different volume scales with different precision. The challenge in geostatistics is to honor the variety of available data without assuming information that is not available.

Hand contouring has been used to model reservoirs. Although this method allows the main geological trends in rock properties to be accounted for, there are some problems. It is not possible to construct a model with the short-scale variability or heterogeneity encountered in real geological formations; complex data from seismic and historical production cannot be accounted for; and there is no assessment of uncertainty in the reservoir response. A number of computer-aided mapping techniques based on inverse distance, splines, kriging, or local smoothing are available. These methods suffer the same problems as hand contouring. They are appropriate for visualizing large-scale trends but inappropriate for representing detailed heterogeneities and assessing uncertainty.

Data integration. Information about the reservoir comes from a wide variety of sources, in a wide variety of formats, covering a huge range of scales from the pore to the basin. Moreover, many data sources measure rock properties indirectly related to the rock properties of interest. For example, seismic data measure sonic properties that are imprecisely related to the properties being mapped such as fluid saturations and porosity. Accounting for information from different scales with different precision is a fundamental challenge faced by reservoir geologists and engineers and, more generally, earth scientists.

A second challenge for earth scientists is that some data are global and some are local. A correlation between porosity and permeability of 0.6, and knowledge of coarsening upward trends in porosity are global data. A porosity measurement of 22.5% and a density of 2.14 g/cm³ are local data.

Core data and well-log-derived properties provide the only hard data on the reservoir rock properties (lithofacies, porosity, permeability, and fluid saturations) that must be mapped. Historically, only vertical wells were drilled, providing detailed information on only the vertical succession of rock properties. It is increasingly common, however, to drill horizontally through the reservoir interval. Horizontal well data provide valuable additional information on lateral changes in rock properties.

In addition to hard well data, there are many indirect sources of information; these indirect data, such as seismic, are called soft data. One source of soft data is a conceptual geological understanding of the reservoir depositional system. For example, knowing the direction of sediment transport at the time

of deposition provides information on large-scale trends, stacking patterns, and the spatial continuity of the rock properties. Analog data from outcrops or densely drilled similar fields also provide conceptual geological data.

Seismic data provide a wealth of information about a reservoir. Often, structural surfaces such as the top of the reservoir, intermediate stratigraphic surfaces, and important faults can be interpreted from the seismic data. In addition to large-scale geometric information, seismic data are increasingly used to provide information on internal reservoir heterogeneities. Seismic attributes, derived from an inversion or statistical analysis of the seismic traces, are calibrated to the vertically averaged rock-type proportions and porosity.

Only recently have geostatisticians been faced with repeated three-dimensional or four-dimensional seismic surveys. The four-dimensional seismic survey provides additional information that further reduces uncertainty and improves the predictions from flow simulation. The information may include fault transmissibility, for example, the presence of sealing faults or fluid flow across nonsealing faults; the presence of high-permeability regions, that is, regions of preferential fluid movement; and more reliable measurements of lithofacies proportions and porosity. The value of a four-dimensional seismic survey is that it allows the reservoir geologist to subtract much of the geologic uncertainty in the reservoir and overburden, while highlighting temporal changes in reservoir conditions related to production.

Another important source of soft data is from well tests and historical production data. This data can be interpreted to yield information on reservoir boundaries, connectivity between wells, and volume-averaged properties around wells. In general, honoring dynamic pressure or historical production data is quite difficult; in practice, trial-and-error history matching is still the most common approach at the final stage of modeling. The most promising approach to production data integration is interpretation of the dynamic data to yield spatial data at some coarse resolution. Detailed geostatistical realizations are then constrained to that intermediate-scale soft spatial data.

The challenge addressed by geostatistics is to create detailed three-dimensional realizations of the reservoir rock properties that are consistent with all of the relevant hard and soft data.

Geostatistical modeling. The major obstacle in data integration is the difference in the format (not only measurement units) under which each data type is presented. Some information types are interpretive in nature yet could be critical in the understanding of the reservoir; some data are categorical (for example, facies types); some are continuous (for example, porosity and permeability); some appear as constraints (for example, intervals derived from seismic data); and some are prior probability distributions. A standardization of formats and units that does not tamper with the information content is an important first step toward data integration. The approach taken in geostatistics is to code as much data as possible as probabilities.

The probabilistic language and methodology are unique and universal in that they are not linked to any particular field of application or data type. Probability values are unit free; in their cumulative form they are standardized in the ultimate standard interval [0, 1]. The concept of priority probability distributions allows a common coding of diverse data related to the same goal. The concept of Bayesian updating provides a methodology for merging prior distributions into a single posterior probability distribution. All original hard and soft data and the final posterior probability distribution are coded as cumulative probability values in the interval [0, 1]. From the posterior cumulative distribution function, probability intervals can be derived, simulated values can be drawn for the unsampled value, or a best estimated value can be retained for any given optimality criterion.

Staged modeling. There is a vast interwell region to be filled in by geostatistical techniques (**illus.** *a*). Most often, the hard well data are hundreds to thousands of feet apart. More extensive soft data provide imprecise information at a relatively large scale. For example, the intervals at wells 1 and 2 in illus. *a* are shale and sandstone. The task is to establish the distribution of shale and sandstone in the interwell region. Seismic or production-related soft data may provide information on the proportion of shale and sandstone between the wells; it cannot identify the precise location of the shales.

A staged modeling approach will be considered whereby the main structural control (gridding) will be established first; then the sandstone and shale (lithofacies); and finally, continuous rock properties such as porosity and permeability are assigned within each lithofacies.

Most reservoirs are hosted in sedimentary rock formations. An essential character of these formations is a stratigraphic layering that can often be correlated between wells. These surfaces correspond to some specific time (where the sedimentary process changed) in the depositional history of the reservoir. For example, the sea level could have dropped, causing changes in the locations of erosion and deposition. An important first step in petroleum reservoir modeling is to establish this gridding. Illustration *b* shows five stratigraphic surfaces defining four layers. Each layer has its own gridding style, depending on whether erosion occurred at the bounding surface. After establishing a geologically realistic gridding style, the lithofacies, porosity, and permeability are modeled sequentially.

Stochastic modeling. The distribution of rock properties in a reservoir is not random, nor is it smoothly varying between the available well data. Geostatistical techniques allow multiple realizations to be created that are constrained to different types of data. One important piece of information is a global mea-

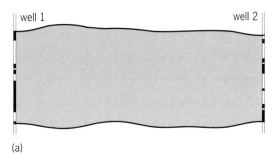

(a)

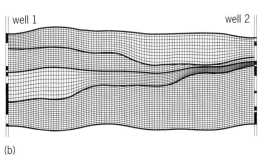

(b)

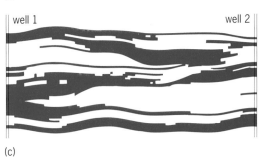

(c)

(d)

Geostatistical modeling; the problem being addressed is to map the vast (*a*) interwell region. (*b*) Geological gridding. (*c*) First and (*d*) second geostatistical realizations; shale layers are colored and sandstone layers white.

sure of spatial correlation. Geostatistical modeling techniques are based on the theory of random variables, and the concept of Bayesian updating that provides a methodology for merging prior probability distributions into a single posterior probability distribution. Geostatistics may also be seen as a branch of applied statistics that emphasizes multiple data types and the spatial context of those data.

Illustration *c* and *d* show two geostatistical realizations. Both realizations honor the well data and have the same appearance; that is, the shales have about the same size and aspect ratio. The differences between multiple realizations quantify the uncertainty in the sandstone and shale distribution.

In practice, geostatistical realizations of continuous rock properties such as porosity and permeability are constructed after the lithofacies (shale and sandstone) has been modeled. The resulting complete realizations are then used to address reservoir-management questions. For example, the hydrocarbon recovery could be established for a particular realization by fluid flow simulation. Repeating the flow simulation with multiple geostatistical realizations allows building a histogram of uncertainty. The 5% low and 95% high values on this histogram indicate the 90% probability interval for the recoverable reserves.

For background information *see* DECISION THEORY; ESTIMATION THEORY; GEOPHYSICAL EXPLORATION; MODEL THEORY; PETROLEUM ENGINEERING; SEDIMENTOLOGY in the McGraw-Hill Encyclopedia of Science & Technology.   Clayton V. Deutsch

Bibliography. C. V. Deutsch and A. G. Journel, *Geostatistical Software Library and User's Guide*, 1992; H. H. Haldorsen and E. Damsleth, Stochastic modeling, *J. Petrol. Technol.*, 42:404–412, April 1990; E. Issacs and R. M. Srivastava, *An Introduction to Applied Geostatistics*, 1989.

Petroleum reservoir production

Hydrocarbon deposits (oil and gas) are typically trapped in permeable rock formations, often at high temperature and pressure. Present-day production techniques usually recover only 35–40% of the oil in place.

Increasing recovery. The continuing imperative for improving yields is driving the petroleum industry to look for ways to increase recovery to 60–70%. Much more data on the response of the reservoir to various engineering interventions will be required. One approach is to make use of information collected from inside the reservoirs on a permanent basis with optical-fiber sensors to measure temperature profiles, pressure changes, and acoustic energy. This will allow better prediction and control of the distribution and movement of oil, gas, and water inside the reservoir.

Today reservoir engineers must form the strategy for extraction based on sparse and mostly static information. Typically, this information includes a seismic survey, carried out before the field is developed, and geological samples acquired during drilling. These data provide the basic physical description of the reservoir formations, but often lack the detail to allow reliable prediction of fluid-flow patterns. Such static data are supplemented by occasional tests of the pressure response of individual wells and of sampled production fluids. Using this information, engineers form their judgments for anticipating reservoir fluid movements such as breakthrough of water, which will result in oil reserves being cut off from the wells irretrievably.

Oil companies have occasionally repeated seismic surveys and analyzed them for differences in order to show up movements of fluids in reservoirs. This technique shows good promise but is too expensive to be used routinely.

Recent developments in optical-fiber sensors, and in particular, their deployment to remote measurement locations in reservoirs, promise to change all this. A range of different sensors can now be deployed into the heart of the reservoir, without tying up valuable wells, to continuously collect extensive temperature, pressure, and acoustic information at a much lower cost.

The expectations are that the changes in temperature distribution and the acoustic and hydrostatic pressure response of the reservoirs can be monitored and analyzed in real time to create dynamic pictures of the fluids moving inside the reservoir. The data can then be used by the engineers to control and improve extraction yield.

Optical fibers in telecommunications. Optical-fiber technology has already caused a radical transformation in telecommunications by offering essentially unlimited transmission capacity over unlimited distances, with practically no degradation in information quality.

Single-mode optical fibers, now used in all long-distance telephone links, are made of silica glass. They consist of an outer (cladding) region that has a diameter of 125 micrometers and an inner (core) region with a diameter of 5–10 μm. Light is guided and transmitted in the core region. A thin polymer coating (typically 60 μm) is applied during manufacture to protect the surface of the silica fiber against abrasion. These fiber structures, although thinner than a human hair, are extremely strong and are manufactured in lengths of hundreds of kilometers. Sections of optical fiber can be joined with connectors similar to electrical connectors, fusion-spliced by melting the silica at temperatures around 1600°C (2900°F), rather like high-temperature welding. Over 10^8 km (6×10^7 mi) of optical fibers have been installed in the world's communications systems, and installation is continuing at the rate of 10^7 km (6×10^6 mi) per year.

Optical fibers in sensing applications. As well as transmitting the information efficiently, this same optical-fiber technology now offers sensors that promise dramatic improvements in gathering information.

Optical fibers have unusual properties that make them attractive for sensing applications. Silica is a highly inert material that can survive long exposure at high temperatures and pressures. Information is transmitted with very little loss (less than 4%/km) and is not degraded at all by electromagnetic interference. This means that the information generated by an optical-fiber sensor can be transmitted over many kilometers without concern about loss of information quality.

An important advantage of optical-fiber sensors is that they utilize light in the fiber, doing away with the need for electronics deep in the oil well. Examples of optical-fiber sensors relevant to petroleum production include distributed-temperature sensors, pressure sensors, and acoustic sensor arrays.

Distributed-temperature sensors. Optical-fiber distributed-temperature sensors are commercially available that can measure the temperature profile along a 30-km (18-mi) length of fiber, with a temperature resolution of 0.1°C (0.18°F) and with a spatial resolution of 1 m (3 ft). The full profile is updated every few seconds. The sensor is thinner than a human hair and retains full performance at temperatures up to 350°C (660°F).

Distributed-temperature sensors have already been used to measure temperature profiles in oil reservoirs. They have provided high-resolution, dynamic pictures of steam fronts moving through heavy-oil reservoir formations in California and in Canada, where the oil flows only after it has been heated to reduce viscosity. This information is essential for process control inside the reservoir.

Pressure sensors. Accurate knowledge of reservoir pressure and the dynamic behavior of pressure during well shut-down tests is important for controlling extraction. Optical-fiber pressure sensors have been proven in field tests at 150°C (300°F) in steam-flood fields in California and at 350°C (660°F) in the laboratory. These sensors can measure very small pressure changes (less than 69 pascals or 0.1 lb/in.²) and guarantee accurate measurements over the life of the reservoir. By contrast, electronic sensors become unreliable at 125°C (257°F), and fail after relatively short exposures.

Acoustic sensor arrays. Acoustic sensor arrays (hydrophones) constructed from optical fibers are in use today, detecting and tracking submarines and ships at distances of many hundreds of kilometers. This military hydrophone technology is being adapted for use in oil wells, where it can serve as a permanent seismic array and in detecting the microseismic events in oil-bearing formations. The same sensors can monitor other "sounds" of oil production, such as flow noises in different sections of a well, vibration signatures of downhole pumps, and the potentially disastrous production of abrasive sand.

Deployment of sensors into the reservoir. Optical fibers become even more attractive when combined with a system for implanting (and retrieving or replacing) the hairlike sensors and their equally thin optical cables into the heart of a reservoir many kilometers inside the Earth's crust. This remarkable technique has been developed commercially, and it has been used to pump optical-fiber sensors along many kilometers of conventional 6-mm-diameter (0.2-in.), hydraulic control lines. Such a sensor highway has already been used to inject sensors into high-temperature oil wells in California and in Canada.

Space is at a premium inside oil wells, and the economics of intervention makes it totally impractical to consider installing hundreds or thousands of conventional electronic sensors in a single well. The sensor highway, and the optical-fiber sensors that travel along it, fundamentally changes the economics of permanent down-hole information. The equivalent of many thousands of conventional sensors can now be accommodated and moved along within a single small-bore control line.

Intervention costs are slashed, and sensors can easily be retrieved and replaced at very modest costs in the unlikely event of a failure—all without interrupting normal production. For the first time it has become economically feasible to guarantee access to high-quality information over the lifetime of an oil field.

The small-bore hydraulic control lines that make up the sensor highways become part of the modern oil well—the intelligent well. By using modern slimline, coiled-tubing drilling techniques, the sensor highway can even become a separate and dedicated information-gathering network—the nervous system of oil production.

Real-time, global reservoir management that uses the full power of modern computer and telecommunications technology is the challenge. Doubling of oil yields is the goal.

Information and visualization. A single sensor highway can accommodate the distributed-temperature sensor, the acoustic sensor array, and one or more hydrostatic pressure sensors.

The temperature sensor establishes the thermal profiles in the reservoir. It can locate where oil is entering the well along the extended horizontal producing sections, or it can identify the direction of movement of fluids inside the reservoir by relating the temperature of the fluid to its depth of origin.

The pressure sensor records the changes in the reservoir pressure at key sampling points. This has been the main source of information available to the reservoir engineer in the past, and it will continue to play an important role in testing the reservoir computer models.

Acoustic arrays, the future "ears" of the reservoir, establish the basis for dynamic three-dimensional images of the sounds and vibrations in the reservoir. They will make it possible to track fluid movements and identify reserves that have been left behind. The acoustic signature will identify where the oil and gas enter the well, when sand enters the well, and what the flow rates are in different parts of the well. In other words, acoustic sensors allow the production engineer to monitor the condition of the well.

Passive acoustic systems can also be adapted for active imaging by using sound or vibration sources adapted from existing seismic survey systems. Permanent seismic arrays deployed inside the reservoir will improve the spatial resolution and the image quality. The seismic information traverses much shorter distances to reach the sensors, and therefore it is of much higher quality than when seismic-sensor arrays are located at the surface or in the ocean.

The vast stream of data that will flow from the reservoir presents a major opportunity for specialists in information processing and visualization. The wealth of new information, updated continuously throughout the life of an oil field, will be combined with the existing geophysical and geochemical database. This data fusion will form a "multicolor," dynamic picture of the reservoir that can be readily used by teams of reservoir engineers to gain control over petroleum reservoirs. This information explosion is expected to produce a major increase in the yield from reservoirs, possibly doubling the extraction rate.

For background information *see* FIBER-OPTIC SENSOR; OPTICAL FIBERS; PETROLEUM RESERVOIR ENGINEERING; SEISMIC EXPLORATION FOR OIL AND GAS; WELL LOGGING in the McGraw-Hill Encyclopedia of Science & Technology.   Edward Kluth

Petrophysical logging

Boreholes or wells are one of the most expensive investments made by oil and gas companies. Petrophysics, in the oil and gas business, is the study of the physical properties of rocks, and it attempts to measure the mineralogy and amount of open pore space available to store hydrocarbons. Well logs, which are paper or digital plots of depth versus various petrophysical measurements taken from a borehole, are often the only record of the rocks and fluids encountered in drilling a well.

Borehole images. Borehole images, a special kind of well log, are electronic pictures of the rocks and fluids encountered by a wellbore. Such images are made by acoustic, electrical, or video devices lowered on a wireline into the well. Images are oriented; have high vertical and lateral resolution; and provide critical information about the presence or absence of porosity, thinly bedded rock layers, and fracture frequency and orientation.

Cores are cylinders of rock drilled by hollow diamond bits and extracted from the earth at great depth. Core materials provide the best samples of rock material from the subsurface, and prior to borehole imaging were the leading source of petrophysical data; however, because of high expense and risk, fewer wells are now being cored. The cores are generally short, so they may miss all or part of the target formation. In exploration wells, the depth to the target formation may even be unknown. All of these factors have led to the increased use of borehole images to characterize subsurface sedimentary rocks.

Televiewers. Some of the first borehole images came from borehole televiewers developed for commercial use in the late 1960s (**Fig. 1**). These tools used a rotating transducer that emitted a pulse of acoustic energy. This energy bounced off the borehole wall, and the reflected wave's amplitude and travel time were recorded by the same transducer. Modern borehole televiewers use the same physical principle, but with better transducers and detectors. Televiewers work best in boreholes filled with fluids that do not conduct electricity, such as oil. They do not

work well in very dense fluids or in very large- or small-diameter boreholes. Also, the tool must be centered in the borehole for the best results.

Electrical logs. Dipmeters are electrical well logs that have 3, 4, 6, or 8 measurement electrodes placed around the well on pads that contact the borehole wall. Dipmeter logs have been available since the early 1960s. Electrical borehole-imaging logs, which are based upon the dipmeter concept, have complex arrays of measurement electrodes that are placed on 2, 4, 6, or 8 pads around the well. Such logs became available in the mid-1980s, and their usage has increased rapidly. Electrical borehole images (**Fig. 2**) work well in the common situation where the borehole is filled by an electrically conductive fluid, such as salt water or water-based mud.

Video logs. Video imaging logs, available since the 1960s, allow the operator to view rocks, steel pipe, and moving fluids as the well is being logged. Their usage has been restricted by the high temperatures often encountered in oil and gas wells and by the fact that borehole fluids are commonly opaque. Better tools and digitally processed images are being developed that are increasing the use of video imaging logs.

Processing and interpretation. Before borehole images can be interpreted, a number of quality-control and processing steps are needed. These steps involve correcting the log for magnetic declination, applying accelerometer corrections for jerky movement of the tool, and filtering erroneous data spikes. Subsequently, two main types of images are created: static

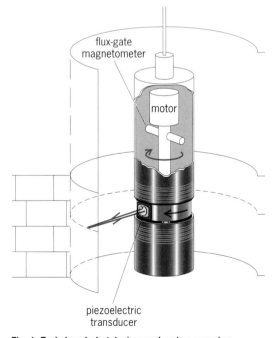

Fig. 1. Early borehole televiewer showing scanning acoustic transducer, magnetic north–sensing magnetometer, and driving motor. *(After F. Paillet et al., Borehole Images: Society of Petroleum Well Log Analysts Reprint Volume, 1990)*

Fig. 2. Electrical borehole image of a normal fault in the Niobrara Formation (Upper Cretaceous), Colorado. *(Courtesy of D. Thorn, Schlumberger, Inc., and R. Gries)*

and dynamic. Static images have had one contrast setting applied to the entire well; they provide useful views of relative changes in rock conductivity or acoustic properties. Dynamic images have had variable contrast applied in a moving window; they provide enhanced views of features such as vugs (coarse, irregular pores), fractures (cracks in the rocks), and bed boundaries. Dynamic images bring out subtle features in rocks that have very low resistivities, such as shales, or very high resistivities, such as carbonates and crystalline rocks.

In practice, the interpreter scrolls through the logged interval on a desktop computer or a computer workstation, viewing static or dynamic images and orienting bed boundaries and fractures. Bed boundaries and fractures are planar features that have a particular orientation in space. Important values to record are the dip magnitude (amount of dip) and dip direction (compass orientation of the dip vector). The computer does calculations of dip and of dip direction based upon its stored information about the borehole deviation or inclination, azimuth (compass direction), and diameter. These orientations are known because magnetometers are run with the well log. Each bed boundary and fracture that intersects the borehole wall appears as a sinusoidal curve on the flattened two-dimensional image. Once all features have been described and tabulated, the operator can perform the structural, stratigraphic, and sedimentologic interpretation.

Structural interpretations. These involve looking up and down the borehole for folds, faults, and fractures. Gradational changes in bed orientation may be related to the fact that the borehole has crossed a fold axis. Graphical techniques that look at changes in

bed orientation are available to help orient the fold. Dip domains are groups of bedding-plane dips that may be structural blocks or sequences at a large scale, for example, tens to hundreds of meters. Boundaries between dip domains may be faults. In some cases, the fault plane itself is imaged and can be oriented. In other cases, fractures are observed that are related to fault orientation.

A borehole breakout is an oval elongation of the wellbore that is caused by differential subsurface stress. Borehole breakouts, which can be measured with various borehole-imaging devices, are related to the orientation of present-day stress. When combined with the orientation of inferred natural and induced fracture sets, they may be related to directional permeability in the subsurface. Permeability is a measure of the ability of fluids to flow through a rock. This information has considerable application to optimizing the orientation of horizontal wells and configuring injection patterns in water and gas injection schemes. Also, the orientation of present-day maximum horizontal stress may be parallel to the orientation of artificially created fractures.

Stratigraphic interpretations. These involve looking at the orientation and frequency of bed boundaries. In some cases, the boundaries between dip domains can be angular discordances between overlying and underlying sediments. These discordances, called unconformities, can be important traps in oil fields, or they can be sequence boundaries that equate to considerable periods of time in which no sediments were deposited.

Sedimentologic interpretations. Such interpretations are aided by the fact that the intrinsic lower resolution limit of borehole images is on the order of a few millimeters. Conductive fractures in resistive rock that are fractions of a millimeter in width can be imaged, because conductive drilling mud invades the fractures and focuses the electric current. Although individual grain types generally cannot be discerned, features such as burrows, broken rock fragments, and vugs are common. Vug shape may be diagnostic for certain fossils. Baseline color shifts in static images can indicate changes in pore type. Cemented versus open fracture and breccia porosity can commonly be imaged. Sedimentary structures, such as fluid-escape features, ripples, cross beds, and imbricated clasts, may be apparent; and the observation of such structures may yield paleocurrent interpretations. These sedimentary structures are critical to the interpretation of the ancient environment in which rocks were deposited. For example, directional paleocurrents can yield valuable information concerning the direction in which an ancient river flowed. This can be very important in selecting the next well location if it is decided to drill more wells in the same channel deposit.

Outlook. The outlook is bright for the use of borehole images. Image quality is being enhanced by mechanical improvements to the tools and better digital acquisition and processing methods. As more and more images are analyzed by geologists who have experience in examining rock outcrops on the surface of the Earth and cores, previously unseen features in boreholes are discovered. The ultimate goal of all approaches is to improve well completions, to allow operators to improve placement of horizontal wells, to properly design gas and water injection patterns, and to better manage reservoirs for improved recovery of oil and gas.

For background information *see* FAULT AND FAULT STRUCTURES; PETROLEUM ENGINEERING; SEDIMENTOLOGY; WELL LOGGING in the McGraw-Hill Encyclopedia of Science & Technology.   Neil F. Hurley

Bibliography. F. Paillet et al. (eds.), *Borehole Images: Society of Petroleum Well Log Analysts Reprint Volume*, 1990.

Pharmacology

High throughput screening (HTS) is used in the identification of small molecules that may lead to new therapeutic or commercial developments. Utilizing robotics, miniaturization, computer-aided molecular design (CAMD), combinatorial organic synthesis (COS), and screening information managing systems (SIMS), it provides a rapid, cost-effective method to screen potential sources for novel molecules often used in drug development. Unlike conventional screening, high throughput screening deals with a wide variety of input samples. Each sample has its own set of data and tracking issues, and produces new experimental results.

Interpretation of data. High throughput screening provides massive quantities of information that must be analyzed and separated. Data management is usually accomplished through the use of computers with software specifically designed for process screening. Then, the data can be interpreted by humans while all other processes are managed by robotics. In high throughput screening, standard biochemical experiments for drug discovery are performed by robots. Computer-aided molecular design, which is usually directed toward specific targets for drug action, attempts to discover new drugs in a rational manner with the help of modern technology. This type of design tests a small number of particular molecules that might have the potential to interact with the specific target. A difficulty is that computer-aided molecular design requires background knowledge of the structure of the molecule or base that is being studied.

Workgroups. The screening information management system works with high throughput screening to provide an effective means to analyze information and data. Each screening workgroup has its own laboratory-based server and can operate independently. A central laboratory server obtains all information from all the independent workgroup servers. Sometimes located in many different facilities, the central laboratory server tracks and coordinates activities through all workgroups to increase productivity and prevent loss of samples. The benefit of such operation is that workgroups can be added and dropped without interfering with other experiments. Each workgroup can optimize its own resources to provide the central laboratory server with information that needs to be distributed to other workgroups.

Robotics. With implementation still largely in the future, robotics enables experiments to be carried out with minimal human intervention. The technology is expensive to install; but considering the long-term effects, the speed and accuracy of each screening, and the minimal required operators, it is very cost effective. For example, a fully automated robot incubator can load itself, pipet, incubate, wash, separate, measure, and so on.

Combinatorial chemistry. Potential sources of novel molecules for drug development include both natural substances, as in traditional herbal medicine, and modern technology, as in computer-aided molecular design. The latter, coupled with combinatorial chemistry technology, can create a large number of molecules, or libraries, for high throughput screening to search for therapeutic and commercial possibilities.

Initially, the libraries may not contain compounds with known novelty. They may consist of a collection of lead compounds or of structures or building blocks of set proven activity or utility. However, with each modification, derivation, or addition of chemical functional groups to the lead structures, the library can grow into a great quantity of new compounds, and subsequently new lead compound stories (multilevels), after series of screens. Combinatorial chemistry, or combinatorial organic synthesis, is used to establish and manage these extensive libraries through miniaturization and automation, necessary to reduce time, space, labor, and materials to a cost-effective level.

Combinatorial organic synthesis uses chemical building locks to form a set of different molecules. It uses a logical systematic method to synthesize and search for active compounds. There are usually three approaches to combinatorial organic synthesis: (1) Formation of separated, discrete molecular derivatives: the building blocks are reacted in a systematic order at individual positions on a grid array, which is later used to locate active compounds. (2) Encoded mixture synthesis: active compounds are identified by using inert chemical tags such as nucleotides and peptides. (3) Stepwise modification or addition of functional groups to lead structures: each step adds a specific structural feature that may be randomly placed at multiple chemically similar sites. The combinatorial synthesis procedure produces a series of compound mixtures for further screening and searching. The three approaches are used to scale down individual reaction volume and then to scale up the number of resulting compounds. Hence, they expand the conventional trial-and-error method of creating and testing by several orders of magnitude without proportional cost increases.

A great deal of information is generated by combinatorial chemistry. Combinatorial chemists often search for the best method to obtain the new molecule, along with determining the availability and price of the building blocks. Information from prior tests may be useful in experiments that are planned. Researchers can quickly access recent information on the reaction of certain molecules and avoid errors that were made in the past. They may also be able to modify the experiment to obtain slightly different results in order to determine which method is most efficient.

Combinatorial libraries. Combinatorial libraries can produce large quantities of information. The results of each compound must be cataloged and stored for future reference. All biological, structural, and screening data must be integrated. This library is often used in cost-effective research and development of other potential drugs, and allows the managers to justify their costs in combinatorial research.

Some biotechnological companies have developed products to help meet the needs of combinatorial chemists. For example, commercial software packages created to handle the excessive amount of data generated from combinatorial chemistry are now available.

Mass screening. Chemical libraries provide the largest source of chemicals for mass screening. Chemical structures in these libraries often reflect the research that is being conducted in each of various companies. However, it is often difficult for high throughput screening operators to gain access to different chemical libraries. Some companies have established bilateral agreements that allow each company to access each other's library. An advantage of chemical libraries is that if one of the molecules of the library has therapeutic value, other molecules within the library may also possess potential pharmaceutical or industrial uses.

For background information *see* CHEMICAL ENGINEERING; DATABASE MANAGEMENT SYSTEMS; MOLECULAR BIOLOGY; ORGANIC SYNTHESIS; PHARMACEUTICAL CHEMISTRY; PHARMACOLOGY; ROBOTICS in the McGraw-Hill Encyclopedia of Science & Technology.

Duen-Gang Mou

Bibliography. D. Bevan et al., Identifying small-molecule lead compounds: The screening approach to drug discovery, *Trends Biotechnol.*, 13(3):115–121, 1995; S. Brenner and R. A. Lerner, Encoded combinatorial chemistry, *Proc. Nat. Acad. Sci. USA*, 89:5381–5383, 1992; M. A. Gallop et al., Applications of combinatorial technologies to drug discovery (I), *J. Med. Chem.*, 37(9):1233–1251, 1994; E. M. Gordon et al., Applications of combinatorial technologies to drug discovery (II), *J. Med. Chem.*, 37(10):1385–1401, 1994.

Phylogeny

Since life first appeared on Earth some 3500 million years ago, species have been evolving from one another to create the diversity that is seen today. An analogy between this branching pattern of evolution and the growth of a tree has been popular since the time of C. Darwin—thus the "tree of life" metaphor. Phylogeny is the study of how species are related to one another and involves reconstructing or mapping the historical pattern of branching that best explains the patterns of similarity among organisms.

Branching process. Although there is no direct way of discovering exactly what happened in the geological past, the branching process has provided distinctive evidence of the way in which species resemble one another. Two species that recently separated from one another will share a number of unique features not found in a third species that branched off much earlier, because the species that diverged more recently shares a longer common history (branch on the "tree of life") along which additional novel characteristics have had time to arise. Phylogenetic relationships among species are therefore determined through analysis of their shared similarities.

Traditionally, phylogeny has been reconstructed on the basis of morphological (structural) similarity, but more recently molecular data (particularly gene sequence data) are becoming an important part of the reconstruction process. The sequence of nucleotides along a deoxyribonucleic acid (DNA) strand provides a vast source of phylogenetically informative characteristics easily comparable among taxa. Where morphological characteristics are hard to find, either because taxa are very distantly related and morphologically very dissimilar, or because divergence has been relatively recent and few distinct morphological characteristics have arisen, molecular data are unrivaled.

Characteristics and cladograms. Characteristics that arise through the process of evolution fall naturally into a nested arrangement. Some features, such as the presence of DNA, are common to all organisms, whereas having DNA bound within a nucleus is true for only a subset (eukaryotes). Only a small subset of eukaryotes have a backbone (vertebrates), with a further subset having features (birds). Each of these characteristics acts as a marker that identifies a group (clade) of related organisms. It is this pattern of nested characteristics that helps to reconstruct the order in which taxa have branched. These diagnostic characteristics are termed synapomorphies for the group they identify.

If every novel feature evolved only once in the "tree of life" and was retained in all descendants, the reconstruction of phylogeny would be straightforward. Feathers, for example, are found in all members of the clade Aves and appear to have evolved only once; there is no possible ambiguity. However, conflicting patterns of distribution of characteristics are often encountered because reversal or convergent evolution has occurred. Convergent evolution arises when structures of similar appearance have originated two or more times independently, while in reversal a common inherited structure has been subsequently lost from a subset of taxa. Therefore,

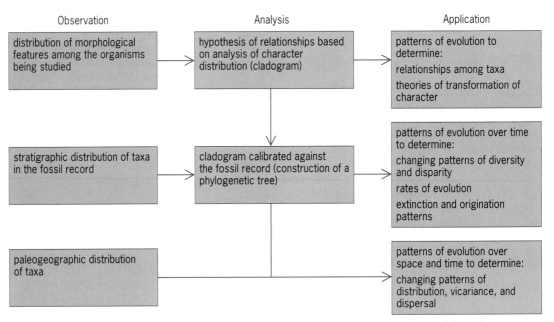

Three types of observations that can be made from the fossil record, and how they can be integrated through phylogenetic analysis to study broad patterns of evolution.

the branching diagram that explains the distribution of the greatest number of observed characteristics (with the minority of inconsistent distributions of characteristics being attributed to convergence or reversal) is sought. Since taxonomists often work with large numbers of different characteristics, computers are widely used to conduct this search. The resulting branching diagram is called a cladogram and represents the best-supported hypothesis of phylogenetic relationships.

Rooting. Cladograms, in their simplest form, are branching networks that describe taxon relationships based on similarity. These networks allow distinguishing between characteristics that have evolved once and those that show multiple independent origins. However, to discover the direction in which evolutionary change has occurred, it is necessary to root the cladogram. Rooting determines the order in which taxa have branched over time by identifying which of two or more possible states shown by a characteristic represents the primitive condition from which the other(s) arose. Again, this is accomplished by referring to the distribution of characteristics among taxa and discovering what form the characteristics take in groups more distantly related to the species of immediate interest.

Among mammals, for example, some taxa give birth to live young (placentals and marsupials) while others lay eggs (monotremes). Since species of all other vertebrate groups lay eggs, the obvious deduction is that egg laying represents the primitive condition and that giving birth to live young is an evolutionary innovation. Once again, though, the cladogram is rooted by reference not to a single characteristic but to the total amassed evidence. The rooting method is usually implemented by including

a number of additional taxa that lie close to the group being studied and that can separate characteristics for rooting purposes.

Phylogenetic trees. A rooted cladogram, depicting the order in which taxa have evolved from one another, is widely used in the study of evolutionary patterns. However, it provides only a relative framework and lacks information about the absolute time separating branching events. The most informative way of presenting phylogenetic information is in the form of a tree. Trees differ from cladograms in that the branch lengths are calibrated against absolute time.

The calibration of a cladogram and its transformation into an evolutionary tree can be achieved in one of two ways. First, if molecular data are available the relative degree of dissimilarity can be quantified and used as a molecular clock. Thus, the further back in time two taxa separated, the less similar they will be. However, because of the inexactness of the process involved, a molecular clock provides only a crude measure of elapsed time.

A second approach is to use the fossil record of first appearances to provide estimates for the time of divergence of a clade. Thus, the oldest fossil that unambiguously belongs to a clade fixes the latest possible time at which that branch must have existed, and provides a minimum estimate for the date of the branch point in the cladogram. If the fossil record is reasonably well known, estimated divergence dates will be close to the actual dates. There should also be good correspondence between the stratigraphical order that taxa appear in the fossil record and the order of branching in the rooted cladogram. Indeed, just such a statistical correspondence has been demonstrated for a majority of cases.

In these cases, stratigraphic data provide independent support for a phylogenetic hypothesis by demonstrating that appearance in the fossil record and branching order of the cladogram agree within acceptable statistical limits of error. This approach requires that stratigraphic data be kept completely separate from the phylogenetic analysis of morphological or molecular characteristics (see **illus.**). Some workers, however, argue that stratigraphic data should play an integral part in cladogram construction right from the start.

Phylogenies and paleobiology. Cladograms calibrated against the fossil record provide the best estimate of the "tree of life" for many groups. As such, they are being put to many different uses in paleobiology.

Diversity studies. Phylogenetic trees provide the best means of compensating for gaps in a patchy fossil record since they predict when branches must have existed in the geological past irrespective of whether fossil representatives of that branch have been discovered. They therefore provide estimates of past diversity that are only minimally affected by sampling biases.

Extinction studies. Phylogenetic data are crucial in extinction studies to discriminate between disappearances in the fossil record due to biological extinction and those due to taxonomic convention (where a systematist imposes a change of name along a branch of the tree purely for bookkeeping purposes). Cladograms calibrated against the fossil record provide the most accurate estimates of extinction patterns in the geological past, and these have been shown to differ significantly from those generated from raw taxonomic counts.

Origination patterns. Because a branch point is dated by reference to the earliest appearance in the fossil record of a member of either branch, a cladogram may often predict a longer history for the other branch than the fossil record indicates. Thus, phylogenetic trees provide evidence for earlier originations than does a direct reading of the fossil record.

Disparity studies. Cladograms provide information on the number of changes in characteristics that have occurred along each branch, and so long as the original compilation of characteristics is not significantly biased in any way, cladograms can be used to determine evolutionary distances separating taxa. This complements studies based on raw dissimilarity and provides an alternative measure of how biodiversity has evolved over time.

Rates of evolution. A cladogram calibrated against the fossil record provides the most accurate means of estimating rates of evolution over time for both molecular and morphological data. Phylogenetic trees provide estimates of the length of time that each branch represents and the number of transformations of characteristics that have occurred along that branch. This approach allows for the recognition of periods of rapid morphological innovation or clades that have undergone unusually slow rates of change.

Biogeographic history. Analysis of past distributions of organisms in a phylogenetic context is used to test ideas about where taxa originated and how they dispersed geographically over time.

For background information *see* ANIMAL EVOLUTION; FOSSIL; GENETIC ENGINEERING; PHYLOGENY; TAXONOMY in the McGraw-Hill Encyclopedia of Science & Technology.   Andrew B. Smith

Bibliography. M. J. Novacek and Q. D. Wheeler (eds.), *Extinction and Phylogeny*, 1992; A. B. Smith, *Systematics and the Fossil Record*, 1994.

Physiological acoustics

Correlations between acoustical features of sound (for example, the frequency of a pure tone) and physiological activity in the brain (for example, the number and timing of electrical pulses fired by a neuron) are being investigated with scientific methods that incorporate recent technological advances in the processing of bioelectric, biomagnetic, and biochemical signals. On the rich foundation of basic knowledge accrued since the 1950s about how the brain and ear process acoustical features of simple synthetic stimuli such as pure tones, scientists are analyzing the relationship of acoustical-physiological correlates to perceptual attributes of sounds (for example, the pitch of one's voice). This effort has been accelerated in recent years by rapid developments in human brain imaging.

One area of active research in physiological acoustics is how music is perceived by the brain. Psychoacoustic phenomena involved in music perception (for example, pitch, tonal consonance, and octave equivalence) have inspired scientific inquiry since at least the time of Pythagoras. The results of psychoacoustic experiments have motivated compelling and, to some extent, conflicting theories about how the brain processes tonal information. Modern scientific methods render some of these theories testable. These methods include (1) analyzing auditory behavioral deficits (for example, impaired pitch perception) in relation to the neuroanatomical location of brain lesions (for example, strokes involving the auditory cortex); (2) analyzing auditory behavior in relation to the neuroanatomical location of associated changes in blood flow and metabolism; (3) analyzing sound-evoked changes in the location, magnitude, and timing of bioelectric and biomagnetic field potentials generated by the brain; (4) analyzing sound-evoked changes in the location, number, and timing of electrical discharges fired by individual neurons and neuron clusters in the brain; and (5) mapping the gross anatomy, microanatomy, histochemistry, and connections of auditory structures in the brain.

Pitch, melody, and harmony. The brain and inner ear are physiologically and anatomically specialized to process sound frequency. In general, a pure tone's frequency is the acoustical feature that determines its pitch. The pitch associated with a pure tone vibrating at 440 Hz is the international standard for the musical note A_4.

Tonal music is structured in relation to the various scales used by different cultures to define the orderly spacing of discrete pitch elements within an octave set (for example, the seven elements of the diatonic scale of A major). Since melody derives from a succession of pitches and harmony a simultaneous combination of pitches, the neural substrate of melody and harmony perception and the neural substrate of sound-frequency processing probably share common elements within the peripheral and central auditory systems.

Acoustic signals with different physical features can have the same pitch. For example, the pitch corresponding to a pure tone whose frequency is 220 Hz (A_3) has the same pitch as three simultaneous pure tones with frequencies at 660, 880, and 1100 Hz (the second inversion of an A major triad with its root position at A_4). This perceptual phenomenon has a number of terms associated with it, including the pitch of the missing fundamental (that is, the 220-Hz fundamental frequency of the 660-880-1100-Hz series is missing), virtual pitch (that is, there is no physical energy at 220 Hz), and periodicity pitch (that is, in the time domain, the fundamental period of both stimuli, 4.5 ms, is the same). There are two general categories of theories about how the brain derives a unitary pitch from a sound composed of multiple, simultaneous frequencies. Place theory holds that the location of the active neurons determines pitch. Temporal theory holds that the timing of neuronal activity determines pitch. There is now sufficient physiological evidence to conclude that both types of neural coding schemes are operative to different degrees at different neuroanatomical levels of the auditory system.

Functional neuroanatomy. With the advent of human brain imaging via high-resolution computerized x-ray tomography (CT), magnetic resonance imaging (MRI), and positron emission tomography (PET), scientists have been able to design experiments that test specific hypotheses about the role of neuroanatomical structures in auditory perception. In contrast, when the seminal observations about the effects of brain lesions on auditory functions were made in the nineteenth century, the specific location of neuroanatomical lesions had to be deduced imprecisely from the neurological history and examination; occasionally, lesion localization was confirmed by autopsy, sometimes years later. From this remarkable body of work, along with selective ablation studies (involving removal of a part of the brain) in nonhuman primates and other mammals, emerged the still prevalent view that lesions of the auditory forebrain severely and irreversibly impair the discrimination or recognition of music, speech, voice, and environmental sounds; these same lesions do not severely and irreversibly impair the detection of sounds. These differential effects, or functional dissociations, lend empirical support to the view that the auditory forebrain is not only anatomically but also functionally the highest brain center in the auditory nervous system. Implicit in this view is the concept of serial, hierarchical processing of acoustic information from the ear to the brainstem and up to the forebrain, wherein the auditory cortex and related structures generate the conscious realization of sound and the intellectual and esthetic experience of music.

It has been established recently that lesions of the auditory cortex impair pitch perception, especially virtual pitch perception. Within the auditory cortex, damage to the primary area appears to be necessary to produce the impairment, but it is not clear whether it alone is sufficient or whether additional damage to higher, secondary areas (also known as the auditory association cortex) is necessary. The damage to the auditory cortex is typically bilateral or lateralized to the right hemisphere (the nondominant or nonverbal hemisphere in about 90% of the general population). Positron emission tomography measurements of changes in brain blood flow when healthy listeners are engaged in pitch perception tasks generally corroborate the results of lesion effect studies. Magnetoencephalography (MEG) data in healthy listeners suggest that pitch is represented by the place of neural activity within the primary auditory cortex.

The organization of cortical mechanisms for pitch perception in most individuals is probably different from that for perfect pitch (also known as absolute pitch), the ability to name the musical note corresponding to an isolated pitch. This uncommon skill, typically possessed by musicians trained early in childhood, requires lexical processing as well as pitch processing. Recent magnetic resonance imaging measurements of auditory cortex size suggest that perfect pitch is associated with anatomical specialization in the posterior auditory association cortex of the left hemisphere (the dominant or verbal hemisphere).

Bilateral, right-sided, or left-sided lesions of auditory cortex can impair melody perception, even when the primary auditory cortex is spared and there is no discernible deficit in pitch perception or rhythm perception. The observed dissociation between pitch perception and melody perception implies a hierarchical processing scheme within the auditory cortex: Individual pitches are derived at the level of the primary auditory cortex, and successive pitches are integrated into melodies at the level of the auditory association cortex.

Bilateral lesions of the auditory cortex can impair harmony perception. One theory holds that the primary auditory cortex carries out a harmonic analysis of simultaneous pitches (for example, a musical chord) and the auditory association cortex integrates these analyses over time (for example, the integration of individual chords into chord progressions and, ultimately, the harmonic structure of a composition). There appears to be some degree of specialization for harmony perception in the right hemisphere.

Taken together, clinical and experimental analyses reveal a remarkable degree of neuroanatomical specialization within the human cerebral cortex for auditory and cognitive functions involved in music perception. The location of a lesion in the cortex will influence not only whether music perception is affected, but which auditory and cognitive functions involved in music perception are affected.

Single-neuron physiology. Since the 1950s, it has been possible to record the electrical activity of individual neurons in the brain by using metal or glass electrodes with tips smaller than 1 micrometer. Applied to physiological acoustics, this technique makes it possible to determine how sound affects the number and timing of electrical discharges fired by a neuron. In this way, it was discovered that (1) neurons in certain parts of the auditory system will increase their firing rate only when a pure tone's frequency falls within a narrow range; and (2) the neuroanatomical arrangement of frequency-selective neurons is orderly (for example, low-to-high frequencies are represented in neurons going from front to back in the primary auditory cortex of nonhuman primates).

Most single-neuron experiments germane to physiological acoustics have been carried out in the auditory nerve. In order for information about sound to reach the brain, acoustic signals striking the ears must be transduced to neural signals by the 7000 hair-cell receptors and 60,000 spiral ganglion cells that are located deep inside the temporal bone of the skull. Each spiral ganglion cell has two arms, one that reaches outward to communicate with a hair cell and one that reaches inward to communicate with the brain (specifically, neurons in the cochlear nucleus of the lower brainstem). The 30,000 fibers that pass information from each ear into the brain are collectively called the auditory nerve.

At the level of the auditory nerve, the neural code for pitch appears to work in the following way: (1) which frequencies are present in the acoustic signal determines which fibers fire; and (2) when the fibers fire determines the pitch. In the example above, the 220-Hz pure tone and the 660-880-1100-Hz complex tone would excite different ensembles of auditory nerve fibers; however, for both stimuli and both ensembles, most firing would occur at time intervals of 4.5 ms (1/220 Hz). The two acoustic signals are physically different, but their neural representations in the time domain and their pitches are the same. A similar combination of place and temporal coding schemes may be used to represent tonal harmony. In everyday musical experience, melody and harmony perception are derived from the integration of these neural representations over time by the auditory cortex and connected memory areas, wherein knowledge about music influences perception in a top-down fashion.

For background information *see* BRAIN; ELECTROENCEPHALOGRAPHY; HEARING (HUMAN); HEMISPHERIC LATERALITY; LEARNING MECHANISMS; PHYSIOLOGICAL ACOUSTICS; PITCH; PSYCHOACOUSTICS in the McGraw-Hill Encyclopedia of Science & Technology.

Mark Jude Tramo

Bibliography. P. A. Cariani and B. Delgutte, Neural correlates of the pitch of complex tones I. Pitch and pitch salience, *J. Neurophysiol.*, 76:1698–1716, 1996; A. N. Popper and R. R. Fay (eds.), *The Mammalian Auditory Pathway: Physiology*, 1992; M. J. Tramo, J. J. Bharucha, and F. E. Musiek, Music perception and cognition following bilateral lesions of auditory cortex, *J. Cognit. Neurosci.*, 2:195–212, 1990; R. J. Zatorre, A. C. Evans, and E. Meyer, Neural mechanisms underlying melodic perception and memory for pitch, *J. Neurosci.*, 14:1908–1919, 1994.

Planet

Interest in the possibility that planets might orbit stars originated with the revolutionary proposal by N. Copernicus in 1543 that the Earth itself is a planet and the Sun is a star. However, it is technically very challenging to detect planets orbiting Sun-like stars. Planets do not generate light; they shine only by the reflected light of the host star. The largest (and easiest to detect) planet in the solar system, Jupiter, has a diameter 11 times larger than that of the Earth, and is more massive than 300 Earths. Jupiter is nonetheless 10^9 times fainter than the Sun. Taking a picture of a Jupiter-like planet orbiting another star would be similar to taking a photograph of a firefly near a megawatt searchlight. Although such a photograph is beyond current technology, a number of strategies have been suggested that might be capable of taking direct images of Jupiter-like, or even Earth-like, planets within the next 20–50 years.

In the meantime, the presence of extrasolar planets must be deduced by indirect means. The two most developed techniques, astrometry and Doppler spectroscopy, rely on the gravitational perturbations that giant planets impose on their host stars. For example, the Sun and Jupiter jointly orbit a common center of mass, which lies on the line connecting the Sun and Jupiter, just outside the surface of the Sun. A hypothetical alien astronomer could detect the presence of Jupiter either by noting that the position of the Sun is periodically wobbling about against the background stars (astrometry), or by measuring the periodic velocity variation of the Sun (Doppler spectroscopy) as Jupiter pulls the Sun about their common center of mass. Such measurements would reveal the orbital period and the magnitude of the wobble. From these data, the orbital radius and mass of the unseen Jupiter could be calculated from Kepler's third law of planetary motion and the principle of momentum conservation.

Doppler technique. The Doppler spectroscopy method provided the first definitive detections of extrasolar planets (**Fig. 1**). Any wave (sound or light) emitted by a moving object will be shifted as observed by a stationary observer. An object moving toward an observer will emit light or sound that is shifted toward higher frequencies (blueshift), while an object moving away from an observer will emit waves shifted to lower frequencies (redshift). The

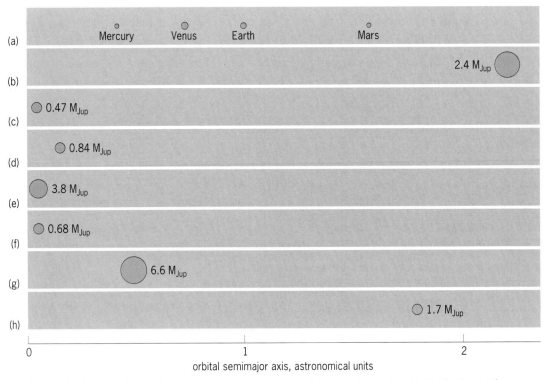

Fig. 1. Recently discovered extrasolar planets. The orbital distances of the planets are compared to the inner solar system. (1 astronomical unit = 9.3 × 10⁷ mi = 1.5 × 10⁸ km.) The masses of the planets are indicated in units of Jupiter mass (M_{Jup}). All of these planets have been found by the Doppler spectroscopy technique. (*a*) **Inner solar system.** (*b*) **47 Ursae Majoris.** (*c*) **51 Pegasi.** (*d*) **55 Cancri.** (*e*) **τ Boötis.** (*f*) **υ Andromedae.** (*g*) **70 Virginis.** (*h*) **16 Cygni B.**

velocity of the emitting object can be directly deduced from the magnitude of this frequency (or wavelength) shift. Common examples of the Doppler effect include the changing pitch of a train whistle as a train passes a stationary observer, and the so-called radar guns used to measure the speed of cars and baseballs.

Stellar Doppler shifts are measured by collecting a star's light with a telescope and passing this light into a spectrometer. A spectrometer is basically a sophisticated prism that spreads the starlight into its component colors, showing the rainbow of red, orange, yellow, green, blue, and finally violet. A more detailed inspection of a star's spectrum reveals that a number of dark lines interrupt this continuous rainbow (**Fig. 2**). Each of these lines is due to the absorption of light, at a specific wavelength, by atoms in the outer atmosphere (photosphere) of the star. In addition to revealing the chemical composition of stars, these lines serve as wavelength markers. Doppler velocities are calculated based on the measured wavelength shift of these absorption lines.

51 Pegasi planets. After four centuries of frustration and increasingly sophisticated searches, the first confirmed discovery of an extrasolar planet was announced in October 1995 by M. Mayor and D. Queloz. They had been surveying 140 starts for 1½ years with a specialized Doppler-velocity spectrometer, with which they had achieved a precision of 50 ft/s (15 m/s). With this instrument, Mayor and Que-

loz discovered that the Sun-like star 51 Pegasi periodically changes in its velocity by 187 ft/s (57 m/s) every 4.2 days (**Fig. 3**). This implies that a planet with about one-half of a Jupiter mass orbits the star at a distance of only 5 × 10⁶ mi (8 × 10⁶ km), 20 times closer than the Earth is to the Sun (1 Earth–Sun distance is 9.3 × 10⁷ mi or 1.5 × 10⁸ km, and is defined to be 1 astronomical unit). This result was completely unexpected. Virtually all theoretical predictions of planet formation suggested that giant Jupiter-like planets should form more than 3 astronomical units away from their host stars.

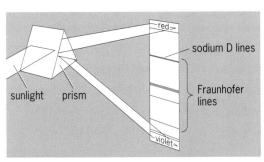

Fig. 2. Fraunhofer lines in the solar spectrum. Passing sunlight (or starlight) through a prism (or spectrometer) reveals the presence of a number of dark lines that interrupt the continuous (rainbow) spectrum. These lines act as wavelength markers, needed to measure Doppler shifts. *(After J. M. Pasachoff, Contemporary Astronomy, 4th ed., Saunders College Publishing, 1989)*

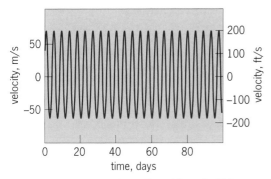

Fig. 3. Doppler velocities for 51 Pegasi from the Lick Observatory Planet Search program. These observations were made between October 1995 and January 1996. The curve is an orbital fit to the data that yields a period of 4.23 days. The semiamplitude of the velocity variations is 187 ft/s (57 m/s), indicating a companion having a mass one-half that of Jupiter at an orbital distance of only 5 × 10^6 mi (8 × 10^6 km), 1/20 of the Earth–Sun distance.

Though this odd planet around 51 Pegasi was theoretically troublesome, observationally it is the easiest type of planet to detect. Its close orbital distance increases the gravitational tug on the host star, thus increasing the Doppler-velocity signal relative to more distant planets. The 4-day period allows many orbits to be followed in just a few months. In contrast, Jupiter induces a meager 43-ft/s (13-m/s) velocity on the Sun, and 12 years are required to follow a single orbit.

51 Pegasi's planet orbits at a distance of only 9 stellar radii from the star. At this close distance, the planet is heated to 2400°F (1300°C). In the solar system, the giant planets (Jupiter and Saturn) are composed primarily of hydrogen and helium gas. Assuming that this is also true of 51 Pegasi's planet, there were initial concerns that the high temperature might give the gas enough kinetic energy to achieve escape velocity, literally boiling the planet away. Numerous calculations have now demonstrated that even at this high temperature hydrogen gas will not escape from the planet. It is also possible that the planet is primarily rocky, like a giant nickel-iron bowling ball, unlike anything seen in the solar system.

By August 1996, three more 51 Pegasi–like planets were discovered by G. Marcy and P. Butler. They had been surveying 120 Sun-like stars for 9 years and had achieved a precision of 10 ft/s (3 m/s). These three additional short-period planets were found orbiting the stars 55 Cancri, τ Boötis, and υ Andromedae. Their orbital periods ranged from 3.3 to 15 days, and their masses ranged from 0.6 to 4 Jupiter masses. The immediate implication is that 51 Pegasi–like planets are not rare freaks and that at least several percent of all stars have extremely close planetary companions.

Many kinds of planets. Though Marcy and Butler had been collecting data for several years, a lack of computational power had prevented them from analyzing their data. After 51 Pegasi's discovery, they ob-tained sufficient computer resources to quickly analyze all the observations of their 120 target stars. Beyond the three 51 Pegasi-type planets, they discovered three additional planets with very different characteristics.

The star 70 Virginis was found to have a companion with a minimum mass of 6.6 Jupiters, completing an orbit every 117 days, at a distance of about 0.5 astronomical unit. At this distance, the temperature of the planet is calculated to be 194°F (90°C), or slightly cooler than boiling water. Thus, liquid water, in the form of aerosollike droplets, might exist in the atmosphere of this object. Unlike the 51 Pegasi planets and the planets in the solar system, the orbit of this object is not circular, but oval or egg shaped. The large mass and eccentric orbit of this object have led many astronomers to suggest that it is a brown dwarf and not a planet. Brown dwarfs are intermediate objects between stars and planets. Brown dwarfs form in a manner similar to stars but differ from stars in that they lack sufficient mass to ignite nuclear fusion in their central cores. Brown dwarfs are usually thought to have masses 20–80 times larger than Jupiter. Marcy and Butler's survey of 120 stars did not find any companion objects in this mass range. *See* BROWN DWARF.

The first extrasolar planet with characteristics similar to the planets in the solar system was found orbiting the star 47 Ursae Majoris. This planet has a mass 2.4 times greater than Jupiter, a 3-year orbital period, a circular orbit, and an orbital distance of 2 astronomical units. Such a planet in the solar system would share a number of characteristics with Jupiter and look like Jupiter's big brother.

Another planet with strange characteristics, which was independently discovered by Marcy and Butler and by W. Cochran and A. Hatzes, orbits the star 16 Cygni B. This star is the fainter of two Sun-like stars that orbit each other with a separation of about 1000 astronomical units. The planet lies 1.7 astronomical units from 16 Cygni B, has a mass that is 1.7 times larger than Jupiter, and requires about 800 days to complete one orbit. Like that of the companion to 70 Virginis, the shape of this planet's orbit is noncircular (elliptical), but its low mass strongly argues against the designation of brown dwarf.

Prospects. In 1996 Marcy and Butler began a survey of 400 Sun-like stars with the world's largest telescope, the 400-in.-diameter (10-m) Keck telescope in Hawaii. In 1997 Mayor and Queloz started a survey of a different set of 500 stars with a dedicated telescope in Chile. During 1998 Cochran and Hatzes planned to begin a survey of an additional 400 stars with the 360-in. (9-m) Hobby-Eberly telescope in Texas. By the year 2000 it is expected that more than 2000 stars will be under Doppler surveillance for planetary companions. Before the year 2000 optical interferometry systems with the capability to astrometrically detect extrasolar planets will also be in operation. By 2010 many hundreds of planets will probably be discovered. *See* INTERFEROMETRY.

Prior to the discovery of extrasolar planets, most theories of planetary formation suggested that extrasolar planetary systems would resemble the solar system. They predicted that planets would have circular orbits and that giant planets such as Jupiter would orbit several astronomical units away from their host stars. These theories must now be seriously reconsidered.

By 2010 it should be possible to determine what fraction of Sun-like stars have planets, what fraction of planetary systems are similar to the solar system, and what processes lead to the various types of planetary systems, including the solar system and the 70 Virginis and 51 Pegasi systems. The impetus for these questions is to learn how common or unique the solar system and the Earth are, how common or unique life is, and finally how common or unique humans are in a universe with at least 10^{10} galaxies, each of which contains 10^{10} stars.

For background information *see* ASTROMETRY; ASTRONOMICAL SPECTROSCOPY; BROWN DWARF; CELESTIAL MECHANICS; DOPPLER EFFECT; FRAUNHOFER LINES; KEPLER'S LAWS; PLANET in the McGraw-Hill Encyclopedia of Science & Technology.   R. Paul Butler

Bibliography. R. Angel and N. Woolf, Searching for life on other planets, *Sci. Amer.*, 274(4):60–72, April 1996; A. Boss, Proximity of Jupiter-like planets to low-mass stars, *Science*, 267:360–368, 1995; M. Mayor and D. Queloz, A Jupiter-mass companion to a solar-type star, *Nature*, 378:355–363, 1995.

Polymer

A new family of cationic late-transition metal α-diimine complexes that function as catalysts for polymerization of ethylene and α-olefins has recently been identified. The unprecedented tendency of these catalysts both to produce branched polymers from ethylene and to tolerate polar functionality has opened the door to a new class of polymers with unique microstructure and properties. (Such a complex is a compound of an α-diimine and a metal atom bound through the electron lone pairs on the nitrogen atoms.)

Catalysts. The late transition metals are in the transition series on the right side of the periodic table. Compounds (**1**) and (**2**) (Et = C_2H_5; R = a substitu-

(**1**)

(**2**)

ent, such as hydrogen (H), or an alkyl or phenyl group) are examples of this family of catalysts. The most effective catalysts identified thus far are palladium(II) or nickel(II) complexes with noncoordinating counterions such as hexafluoroantimoniate (SbF_6^-) or polymethylalumoxane ($[CH_3AlO]_n$), although other metals such as cobalt and iron have shown promise. While cationic palladium (Pd) catalysts can be synthesized directly and stored in a bottle, the more reactive nickel (Ni) cations are typically generated in place by activating the corresponding nickel dihalide complexes with aluminum alkyls (for example, polymethylalumoxane). These Lewis-acid metal alkyl–diimine complexes polymerize olefins via a coordination-insertion mechanism. This is similar to the Ziegler-Natta polymerization catalysts, which are based on early transition metals, on the left side of the periodic table, such as titanium, zirconium, or vanadium. In the process of polymerization, the olefin monomer first complexes to the metal and then inserts into the metal-carbon bond of the metal alkyl group, lengthening the metal alkyl chain by two carbons.

Previously known late-transition-metal catalysts generally give oligomers or low-molecular-weight polymers with ethylene, and none were known to polymerize α-olefins (for example, propylene). The ability of these new catalysts to yield high-molecular-weight polymers is attributed to steric hindrance about the metal center in the square planar complex. This hindrance derives from the substituents in the phenyl 2 and 6 positions [isopropyl groups in the case of compounds (**1**) and (**2**)]. X-ray crystal structures have shown the phenyls [shown as benzene rings in (**1**) and (**2**)] to be rotated nearly 90° to the plane of the metal-diimine metallacycle (the ring containing the metal atom), so that the isopropyl substituents lie above and below the plane, blocking the axial orbitals of the metal (that is, those orbitals normal to the plane of the metallacycle). This steric hindrance inhibits complexation of olefin monomer at the axial orbitals. Such monomer complexation would normally result in displacement of the growing polymer chain, terminating its growth and resulting in the beginning of a new chain bound to the metal (that is, a chain-transfer event). Minor alterations to the phenyl substituents (for example, methyl groups rather than isopropyls) can significantly relax the steric congestion about the metal

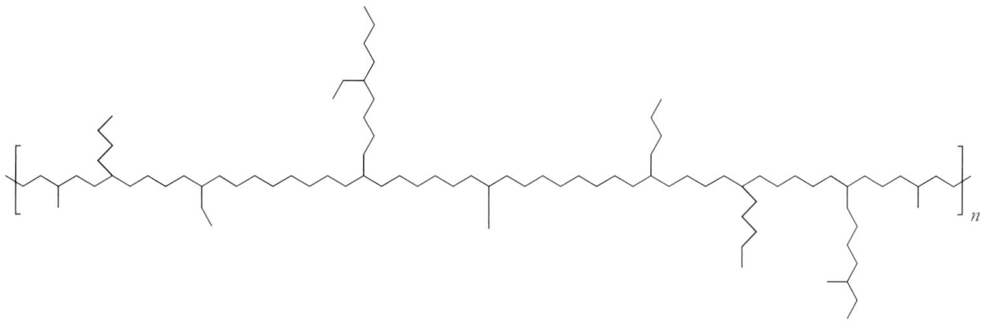

(3)

center, resulting in the production of oligomers rather than high polymers.

Branched polymers. A remarkable property of this class of catalysts is their ability to make branched polymers from ethylene, as well as to make olefin homopolymers containing linear polymethylene sequences from α-olefins such as propylene or butene. The palladium complexes are more pronounced in this respect than their nickel analogs. Polyethylenes made by palladium(II) diimine catalysts are so branched as to be nearly dendritic, with up to 30% of the side chains being amyl or greater in length. A typical branching distribution is methyl (CH_3; 42%), ethyl (C_2H_5; 16%), propyl (C_3H_7; 7%), butyl (C_4H_9; 8%), and amyl (C_5H_{11}) and greater (27%), with an overall branching level expressed as 90–100 methyls per 1000 methylenes (CH_2). Structure (**3**) is such a polymer. In contrast, the nickel catalysts tend to give a larger proportion of methyl branches and fewer branches greater than butyl. *See* DENDRITIC MACROMOLECULES.

A common branching mechanism is attributed to both catalysts. The metal is considered to migrate along the polymer chain via a series of β-hydride eliminations and readditions [reaction (1), where

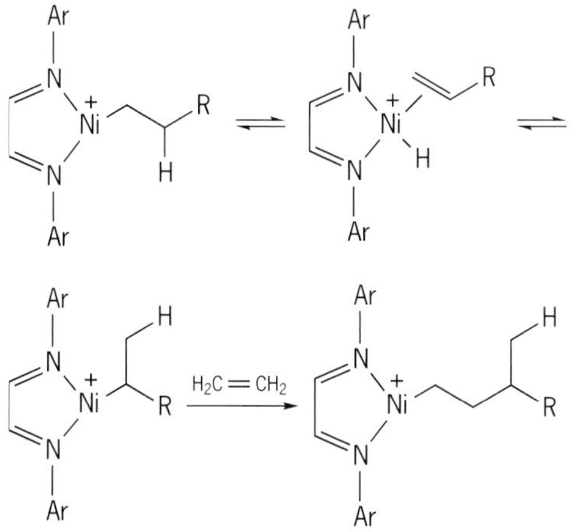

(1)

Ar = substituted phenyl; R = alkyl group]. In the case of palladium, some of the branches are formed on other branches, showing that the catalyst is capable of "walking" backward along the polymer chain past tertiary carbons before the next monomer addi-

tion step. Nickel diimine catalysts tend to yield polymers with fewer branches than palladium catalysts, because the rate of monomer addition and insertion with the nickel catalyst is faster and more competitive with rearrangement. Branching can be controlled by the polymerization conditions: raising the ethylene pressure gives less branching, reflecting the competition between monomer addition and the rearrangement, while higher temperature gives more branching. Polymer molecular weight can be similarly influenced: higher ethylene pressure raises molecular weight, while higher polymerization temperature lowers molecular weight.

Polymer properties. The highly branched palladium-derived ethylene homopolymers range from viscous fluids to elastomeric solids. The higher-molecular-weight elastomers have high extension and good recovery in the uncured state. Their strength presumably derives from chain entanglements, since they show no tendency toward stress crystallization. Such polyethylenes are totally amorphous, with densities below 0.9 g/cm³ (15 g/in.³) and glass transition temperatures (T_g; the temperature below which a rubber becomes a glassy solid) below $-60°C$ ($-76°F$). Because of their extreme branching, the fluids tend to have low viscosity for a given molecular weight, and they dissolve easily at room temperature in organic solvents such as toluene, chloroform, or tetrahydrofuran.

A wider variety of polyethylenes can be made with the nickel diimine catalysts. Minor variations in catalyst ligand structure (such as changing the substitution on the diimine backbone) can yield ethylene homopolymers ranging from essentially linear, crystalline, high-density polyethylene [catalyst (**2**), R = H] to amorphous, branched polyethylene resembling an ethylene-propylene copolymer [catalyst (**2**), R = CH_3]. In the latter case, about a quarter of the branches are longer than methyl. A typical branching distribution is methyl (72%), ethyl (6%), propyl (3%), butyl (3%), and amyl and greater (15%), with an overall branching level expressed as 110–120 methyls per 1000 methylenes. While this branching level is actually higher than the palladium-derived polyethylene described above, the preponderance of methyl branches makes the nickel-derived elastomeric polyethylenes quite distinct; they have much higher tensile strength and are nontacky solids at molecular weights where palladium-derived polyethylenes are liquids. The properties of these elastomeric polyethylenes, such as modulus, tensile strength, density, glass transition temperature, and crystallinity, can be tailored by modest changes in the polymerization conditions of temperature and ethylene pressure. The molecular weight distributions for all these polymers are relatively narrow (typical ratio $M_w/M_n = 1.6$–2.5, where M_w is the weight average molecular weight and M_n is the number average molecular weight), reflecting a single-site type of catalyst.

Polypropylene prepared with either nickel or palladium catalysts is completely amorphous, with glass transition temperatures ranging from $-25°$ to $-46°C$ (-13 to $-51°F$), depending on the catalyst used; the glass transition temperature of normal amorphous polypropylene is $-17°C$ ($1.4°F$). These propylene homopolymers can have a CH_3:CH_2 ratio below 0.4 versus the expected value of 1.0, demonstrating chain straightening by the catalyst. As a result, they resemble ethylene-propylene copolymers of high propylene content. Higher olefins (1-butene, 1-pentene, 1-hexene, and so forth) also homopolymerize to give polymers that have a lower-than-expected number of branches due to chain straightening.

Cyclopentene [(**3**) in reaction (2)] yields a high-

(3)

(2)

(4)

molecular-weight, stiff, semicrystalline homopolymer with a weak glass transition temperature around 100°C (212°F) and a broad melting point ranging from 250 to 350°C (480 to 660°F). This polycyclopentene (**4**) has exclusively 1,3 enchainment, due to catalyst rearrangement before each subsequent monomer insertion. The polymer can be formed into tough, clear films by using melt-pressing, which involves heating of the polymer above its melting point and pressing it between two flat metal plates; the polymer spreads into a uniform thin film, which is then cooled and removed.

Even internal olefins (alkenes with carbon substituents on both ends of the double bond) such as 2-butene ($H_3CCH{=}CHCH_3$) can be polymerized. That the 2-butene does not first isomerize to 1-butene ($H_2C{=}CHCH_2CH_3$) before polymerizing is shown by the branching analysis of the polymer. Poly(2-butene) has a much higher proportion of methyl to ethyl branches than an authentic sample (a sample whose origin and structure are positively known) of the corresponding poly(1-butene) made with the same nickel catalyst.

Another remarkable feature of the palladium(II) diimine complexes is their ability to effect the random copolymerization of ethylene with polar vinyl monomers such as methyl acrylate ($H_2C{=}CH\text{-}COOCH_3$) and even acrylic acid ($H_2C{=}CHCOOH$). Ethylene copolymers with over 50 wt % methyl acrylate have been made. Such tolerance of polar monomers is completely lacking with early-transition-metal Ziegler-Natta polymerization catalysts, which are irreversibly poisoned by oxygenated monomers. Polymerization rate and polymer molecular weight are generally inversely related to acrylate incorporation; copolymers with 1–2 wt % methyl acrylate are high-molecular-weight elastomers, while copolymers with more than a few weight percent acrylate are fluids. Many of the ester groups in these copolymers are located at the ends of hydrocarbon branches; this is due to the catalyst migrating back down the polymer chain after methyl acrylate insertion and before the next ethylene insertion, as previously described for ethylene homopolymerization. Higher ethylene pressure increases the rate of monomer insertion, so polymers produced at higher ethylene pressure tend to have a higher proportion of ester groups bound directly to the main polymer backbone. These ethylene–methyl acrylate copolymers are completely amorphous, and they are soluble in a variety of organic solvents. The ester groups can be hydrolyzed with aqueous base to produce elastomers with ionic cross-links (ionomers).

Methyl 4-pentenoate ($H_2C{=}CHCH_2CH_2COOCH_3$) incorporates more easily than the acrylic monomers, because its double bond is not deactivated by conjugation with the ester group. However, chain walking by the catalyst causes straightening of some of the branches, so that the structure of the ethylene–methyl 4-pentenoate copolymer strongly resembles ethylene–methyl acrylate copolymer. The same situation obtains even with methyl 10-undecenoate ($H_2C{=}CH(CH_2)_8COOCH_3$); its ethylene copolymer still looks much like ethylene–methyl acrylate.

The more highly-branched polymers will possess interesting rheological (flow) properties; accordingly, they should find uses in films and coatings, as viscosity modifiers for lubricants, and as impact modifiers for engineering plastics.

For background information *see* CATALYSIS; CATALYSIS COORDINATION CHEMISTRY; ORGANIC REACTION MECHANISM; POLYMER; POLYMERIZATION; TRANSITION ELEMENTS in the McGraw-Hill Encyclopedia of Science & Technology.　　　　Samuel Arthur

Bibliography. L. K. Johnson, C. M. Killian, and M. Brookhart, New Pd(II)- and Ni(II)-based catalysts for polymerization of ethylene and α-olefins, *J. Amer. Chem. Soc.*, 117:6414–6415, 1995; L. K. Johnson, S. Mecking, and M. Brookhart, Copolymerization of ethylene and propylene with functionalized vinyl monomers by palladium(II) catalysts, *J. Amer. Chem. Soc.*, 118:267–268, 1996.

Population dynamics

Populations of living organisms, whether animals, plants, or microbes, are in a perpetual state of change. This is true of a population's size, density, spatial distribution, age structure, and genetic structure. Changes occur on various time scales, from daily (for example, vertical movement in oceanic populations of plankton), through seasonal and annual, to long-term (for example, 11-year cycles of numbers in Canadian lynx populations). All such changes fall under the general heading of population dynamics.

Biotic versus abiotic factors. Throughout the history of research on population dynamics, an important distinction has been recognized between changes caused by biotic factors (predators, parasites, competitors, and the like) and those caused by abiotic factors (notably the weather). One important way in which these changes differ is that the effects of biotic factors are almost always density dependent. For example, predators tend to congregate near high-density populations of prey (the pantry effect), so these populations suffer from disproportionately high levels of predation. Also, while at low density there may be negligible competition for resources, such competition can become very fierce when densities are high. In contrast, the per capita effects of an unusually harsh winter are likely to be about equally severe in low-density and high-density populations.

Competition among species. Competitive interactions have provided a particular focus of attention in the study of population dynamics. Although it is now known that competition is not ubiquitous in nature (it should not be assumed that competition occurs without the necessary evidence), it is nevertheless a common phenomenon and has a major impact on many populations. Sometimes competition is largely or wholly intraspecific. However, many species belong to ecological guilds, groups of species utilizing a particular set of actually or potentially limiting resources; consequently, individuals may compete not only with others of their own species but also with individuals belonging to other species within their guild.

Seaweed flies. Much research in population dynamics generally, and on competition more specifically, is based on intensive case studies of particular species (or pairs or groups of species) living in particular habitats. The aim is to be able to draw out, from these case studies, aspects of general principle which have potentially wide applications. One especially fruitful case study is based on the seaweed flies (genus *Coelopa*) that live in and feed on large piles of decaying seaweed (wrack beds) that are thrown up on the shoreline by spring tides and storms.

Seaweed flies have many advantages as a study system. They are geographically widespread, locally abundant, and relatively easy to identify. Also, unlike many other animals, they are easy to work with in the laboratory, and have a short generation time (egg

to adult in approximately 2 weeks) which makes them good subject material for an experimental approach to population dynamics. In addition, a common British and European species, *C. frigida,* has a major chromosomal inversion polymorphism, which enables genetic effects of competition to be readily monitored alongside their ecological counterparts.

Several studies have been conducted since the mid-1970s on populations of the seaweed fly species *C. frigida* and *C. pilipes* inhabiting wrack beds on the coasts of the British Isles. Larvae of these two species are often found living together not just in the same wrack bed but even in the same microsites within it. Densities can get extremely high. For example, values in excess of five larvae per gram (143 larvae per ounce) of seaweed have been recorded. The wrack beds are typically of the order of 350–3500 ft^3 (10–100 m^3), and they generally persist for just a single tidal cycle, that is, about 1 month. Under these conditions, intra- and interspecific competition is likely to be periodically severe and to have a major impact on the populations concerned.

Evidence of competition. One of the classical types of evidence of competition is an inverse relationship between body size and population density. The usual interpretation of such a pattern is that it is ecophenotypic, that is, due to a direct effect of the environment on development; and more specifically, that the decrease in body size is caused by the limited amount of food available. However, correlations between two variables in the wild should be interpreted with care, and experimental demonstration of causality is highly desirable. Consequently, when a clearly inverse body size versus population density pattern was found in *Coelopa*, laboratory experiments were performed to test whether the pattern could be reproduced by varying only the number of larvae per unit of food resource. The results confirmed that patterns in the field are consistent with the occurrence of competition; that these can be reproduced in the laboratory; and that the overall competition in *Coelopa* has both intra- and interspecific components. Because female body size is closely correlated with fecundity, competition is a major determinant of birth rates in seaweed fly populations. Also, it seems likely that competition is a major contributor to population mortality rates, especially during periods when densities are high.

Competition and natural selection. While the patterns examined so far are ecophenotypic, it is also possible that competition causes genetic change in populations. Indeed, any major influence on birth and mortality (the two main components of fitness) is likely to be selective, given that virtually all populations exhibit considerable genetic and phenotypic variation. A selective effect of competition has been demonstrated to occur in *Coelopa* populations. However, the type of selection involved acts to maintain, rather than to alter, the population's genetic structure.

All populations of *C. frigida* are polymorphic for a large inversion on chromosome 1. Although the section of chromosome that is inverted has been estimated to contain up to 10% of the entire genome, in one sense it behaves as a single genetic unit, as inversions suppress crossovers. Consequently, a particular array of alleles at many loci characterizes one inversion sequence, while a quite different array characterizes the other. Indeed, it is through the association of alcohol dehydrogenase alleles with inversion types that the polymorphism may be most readily monitored, because running gels to distinguish *Adh* genotypes is a much more efficient way to operate, at the population level, than extensive microscopic scoring of chromosomal inversion types.

In general, this type of polymorphism exhibits an excess of inversion heterozygotes (sometimes called heterokaryotypes) compared with what would be expected under Hardy-Weinberg equilibrium. This finding, which is consistent with the results of similar studies on other organisms (for example, *Drosophila*), indicates a general superiority of the inversion heterozygote. However, in *Coelopa* the degree of heterozygote excess is strongly influenced by density. Laboratory data confirmed significant intra- and interspecific contributions to the density effect observed.

Natural selection can act in two rather different ways. It can distort a population's genetic structure in a directional manner, thereby causing microevolutionary change; or it can act in a balancing manner, maintaining genetic variation in the population concerned, thus preventing change. Heterozygote superiority is one of the main forms of balancing selection; and so in this case competition has a stabilizing effect on the genetic structure of populations of *C. frigida*. In some other species, competition appears to select in a directional manner, but the relative frequency of the two outcomes is not yet clear, and is an obvious subject for future investigation.

For background information *see* ALLELE; CHROMOSOME; ECOLOGICAL COMMUNITIES; ECOLOGY; POLYMORPHISM (GENETICS); POPULATION ECOLOGY in the McGraw-Hill Encyclopedia of Science & Technology.　　　　　　　　　　　　Wallace Arthur

Bibliography. R. K. Butlin, P. M. Collins, and T. H. Day, The effect of larval density on an inversion polymorphism in the seaweed fly *Coelopa frigida*, *Heredity*, 52:415–423, 1984; M. Leggett et al., The genetic effects of competition in seaweed flies, *Biol. J. Linn. Soc.*, 57:1–11, 1996; D. S. Phillips et al., Coexistence of competing species of seaweed flies: The role of temperature, *Ecol. Entomol.*, 20:65–74, 1995.

Printing

Recent advances in printing technology involve digital platemaking and digital color proofing.

Digital Platemaking

The typical method for preparing an offset printing plate is to place a film mask in intimate contact with a presensitized plate in a vacuum holddown frame and then to expose it to an intense source of ultraviolet light to change the chemistry of the photosensitive coating on the plate surface. Upon wet processing with a developer, the difference in solubility between the areas struck and not struck by light allows for dissolving away one area, revealing a substrate. This substrate becomes the nonprinting or white area; the undissolved area will take up ink for transfer to paper.

This method has been the practice for at least the last 50 years. However, this approach to platemaking is undergoing a significant change. The new technique involves writing onto the surface of the presensitized plate directly by means of a laser beam, bypassing the film step as well as several manual-mechanical operations. The terms digital platemaking, computer-to-plate, or simply direct-to-plate are all used to describe this technique.

First-wave systems. Early (first-wave) computer-to-plate systems (around 1970) involved scanning of a paste-up of the images to be printed and transferring the digitized information to a writing head in order to image the plate. Very powerful lasers were the norm because of the relative insensitivity of the plate imaging media available. These high-power lasers required heavy-duty water cooling, resulting in high costs and short laser life. Some of the first plate designs were unusual and included the creation of a mask by thermal ablation of a copper first layer and the thermal erosion of the revealed underlying polymer coating via a high-power carbon dioxide laser. This provided a letterpress relief printing plate. Later versions involved offset plate coatings that were sensitive to ultraviolet light and were imaged by means of high-power water-cooled ultraviolet lasers that had brief working lives and were expensive to replace. Nevertheless, a newspaper in Las Vegas, Nevada, utilized this system for several years. Another type of first-wave computer-to-plate system involved the use of electrophotographic materials based on either zinc oxide or organic photoconductors. This type was also employed in newspapers, and introduced the use of low-power long-life lasers using visible light. Such systems also served as the prototypes for the second-wave computer direct approaches; however, full electronic composition did not replace the scan-and-write approach until the late 1980s.

Second-wave systems. The first second-wave system used low-cost reliable laser diodes operating in the near-infrared, and not the visible-light, region; however, this approach, which utilized aluminum plates coated with organic photoconductors and liquid toner images, did not become widely accepted. The first commercial computer-to-plate system was based on a photopolymer-clad aluminum plate. After computer direct imaging, the plate was processed in an aqueous developer, similarly to conventional plates, and its performance on press was also equivalent to conventional plates. This combination of near-conventional plate characteristics and the use of reliable low-cost long-life lasers ushered in the age of commercial computer-to-plate processing. Aluminum-based plates are preferred for printing conditions that require long runs and dimensional stability to achieve critical registration of color. Additionally, many existing presses are designed with plate lockup systems that require the stretch resistance of metal plates, which possess high modules of elasticity. There are, however, many printing opportunities for short runs as well as black-and-white or spot color versus full-color printing. For these situations, computer-to-plate systems based on flexible base plates, either paper or polyester, have been very successful, particularly since the existing film imagesetters can be engaged as platesetters without modification. The plate coatings are based on silver halide; they exhibit high sensitivity and are processed in activator chemistry. This chemistry activates the silver development process by providing conditions complementary to the developer components contained in the silver gelatin coating. The alkalinity of the activator chemistry releases the developer components to effect silver-grain development. One advantage is that the plate carries the chemistry for development, and does not require a sophisticated replenishment system as is normally the case in development with silver coatings.

Silver-based coatings on aluminum plates, like their polyester counterparts, involve silver diffusion transfer as the imaging method. The aluminum-based silver plates provide the dimensional stability and the long run lengths required for several print markets, while exhibiting good compatibility with low-power, low-cost, long-life lasers.

In addition to the electrophotographic, photopolymer, and silver diffusion transfer-imaging schemes employed in the second-wave plates, a hybrid system has gained popularity. The hybrid plate is a two-layer plate in which a conventional rapid-access silver film emulsion is overcoated onto a normal presensitized offset plate. After imagewise laser exposure, in which radiant energy from the laser focused onto the photosensitive surface is turned on and off in a predetermined fashion, the plate passes through a processor that develops the silver emulsion image that serves as a mask. The next station of the processor provides overall flood exposure of the plate, and the laser-generated mask provides for imagewise exposure of the underlying normal plate coating. The final station of the processor provides for mask removal and normal plate development.

Each of these plate systems enjoyed some success, but none emerged as the dominant method. Some experts believe that the start-up difficulties of both the platesetters and the second-generation plates limited market acceptance, but an equally effective argument is that the success of the computer-to-plate system depends on multiple factors: electronic prepress readiness; a high-performance network, which can store and share information among various computer workstations; a centralized data server; a raster image process (RIP), which converts a data map into codified directions that allow the imaging device to record the information onto the sensitized surfaces; an imagesetter, which is a recorder whose function resembles that of a printer attached to a personal computer; a digital proofer, which provides a rendering of the expected final output based on the digital record from the originating computer record; a layout proofer, which provides a rendering of the geometric positioning of the job to be printed together with metrics needed to guide the operation prior to final printing; and a platesetter, which is an imagesetter dedicated to producing press-ready plates.

Many potential users of digital platemaking systems are early in their learning curves and in the accumulation of the skills and ancillary equipment necessary to ensure success. This situation is rapidly changing, and the digital plate manufacturers are introducing improved versions of second-wave products while preparing for the third wave.

Third-wave systems. These involve thermal imaging. Significant reductions in the cost of thermal laser diodes operating in the 1-W range opened the door for platesetter designs that could capitalize on some of the inherent advantages of thermal imaging. These advantages include handling under normal room illumination—a powerful incentive. Special darkroom hookups and the use of light-tight cassettes as well as automated plate-loading robotics added to the cost and burdens of visible-light laser platesetters; these are not necessary when thermal plates are used. Another potential advantage of thermal imaging is its simplified processing, or possibly no processing requirement at all. The first thermal plate/platesetter combination was the successor to an electroerosion plate approach in which a conductive layer under a silicone layer was spark imaged to produce a waterless plate. The spark imaging, in which an electrical spark discharge was used to record images, destroyed the adhesion of the silicone top coat, allowing for a simple rubbing development step. The successor thermal version simply replaced the low-resolution spark stylus with a much higher resolution thermal laser beam. This technology is available in both on-press and off-press configurations; it is used primarily for short-run color printing with waterless inks. Conventional offset printing is accommodated by alternative thermal computer-to-plate systems. These plates include designs that use thermal cross-linking, ablation, and thermal transfer imaging schemes. This third-wave technology is highly dependent on the laser type and imaging specifications of the platesetters available. The thermal transfer and cross-link systems are the most sensitive, while the plates based on ablation and adhesion destruction require higher energies.

The situation is a complex relationship of plate and platesetter, since with thermal imaging the power density and residence time of the laser imaging spot relates to the thermal time constant of the plate material employed; there is often a significant impact on the energy required for imaging. The most acceptable technology has not yet been determined, but developments are occurring rapidly.

Fourth-wave systems. These involve no-process methods. The advance of technology is most often accompanied by a simplification process that reduces steps. Computer-to-plate technology is expected to follow this paradigm. Two technologies have become operational that have the potential to produce plate products that can go directly to press or be produced on press without any processing. It is important to differentiate this approach from some current methods that do employ plate processing within the printing press, but involve design and operational complexities.

Thermal and ink-jet technologies will lead the way into the fourth wave. The thermal approach may result in rendering a water-soluble gum layer insoluble; or it may involve the removal of an ink-receptive layer by ablation, revealing an underlying hydrophilic substrate. The ink-jet approach may result in the imagewise deposition of an ink-jet image comprising cross-linkable components that would be activated to a durable state through heat or ultraviolet curing as part of the platesetter design.

The novelty of computer-to-plate designs will ultimately follow the advancements in computer-driven imaging devices. As these technologies advance, either through higher-power and lower-cost lasers or faster and higher-resolution ink-jet printers, so also will plate designs, until direct-to-paper systems can match the productivity and quality of commercial printing presses.   Robert Hallman

Digital Color Proofing

Until recently the term proof for color printing, especially a contact proof, meant a proof whose continuous tone (contone) image was rendered by halftone features. In the late 1980s, a shift occurred, as desktop computers with the capability of producing line art and halftones were becoming available. Laser semiconductor diodes and image-setting devices with open system applications were introduced. The crafts person began to give way to a computer specialist, and the standards of color printing changed. The new users recognized that the normal viewing distance of a proof was the blending distance that minimized the halftone pattern. At a specific viewing distance, a halftone appears to be a continuous tone. It was soon recognized that a desktop computer and color hard-copy device could be used to make ''easy color.''

Silver halide color proofs. In 1982, digital proofs were made by using color paper similar to that used for amateur color prints. The digital output of a large electronic graphic-arts laser exposure device and add-on processor both exposed and developed a color contone image. The wet chemistry used could injure the skin of the technician, and it was difficult to control. In 1983, a drier system known as instant process was introduced. A light-sensitive diffusion-transfer donor sheet containing transferable colors and developer, held in position by an anchor or ballast, was exposed to a continuous-tone image. Then the sheet was mated with a receptor sheet for development. An aqueous viscous layer containing alkali was positioned between the donor and receptor. After the sheets were peeled apart, the receptor contained the color image. Matching the dyes used in diffusion transfer to the colors of the ink pigments proved difficult. In 1995, a desktop device was announced whose output appeared dry, and whose color could be corrected by referring to a table. Diffusion-transfer technology was again employed. Development is accomplished by using only water, applied in a process with only the lightest of impressions, between the donor and receptor at the time of mating.

Ink-jet technology. In ink-jet printing, colorant droplets are forced through an orifice and propelled to a receiving surface, without intimate contact between the orifice and paper surface. There are a number of ink jet techniques, including continuous stream and impulse.

Continuous-stream technology. The first ink-jet technique available for digital proofing was continuous-stream ink jet. In this technology, water-borne colorant ink is directed through a nozzle. Irregularities in the nozzle or air currents associated with the ink jet tend to produce a chaotic irregularity in a continuous-stream device. Because of the lack of uniformity in the ink droplets, the use of such continuous-stream ink jet was abandoned.

The continuous-stream technique was improved by stimulation and drop steering. Applying vibration (stimulation) overcomes the chaotic nature of the stream. The extent of the vibration depends upon the size of this orifice, droplet speed, ink viscosity, and the surface tension of the ink. Once stimulated, the droplets are expelled from the nozzle and electrostatically charged, and then are steered to the paper. Uncharged droplets drop into the ink reservoir. In a binary deflection printer, the charged droplet is directed to its print position by means of electrodes and acquired charge.

Impulse technology. Other ink-jet systems involve impulse techniques. The phase-change impulse technique uses colored wax or hot-melt adhesives rather than dye in a binder or liquid. The imaging material is formed into ink pellets or sticks that can be placed in a well, head, or hollow heatable block with an exit orifice. Heat changes the colored binder from the solid to the liquid phase. The ink is heated to its boiling point and produces a high pressure. The droplet is forced out of the well as a series of spots onto a receptor. Generally, the formed image has less contrast and a visible irregular surface. A second impulse technique uses piezoelectric crystals, which change shape when they are subjected to an electrical voltage. Rapid on/off oscillations permit such crystals to be used as an electronic pump. When inks are positioned so that they flow through or past an activated piezoelectric transducer, the pumping action ejects ink droplets on demand.

Electrophotography. In color electrophotographic imaging, a photoconductive layer on a conductive support is electrostatically charged. Imagewise exposure to light dissipates the applied charge. Charged color-toner particles are attracted to the charged or uncharged areas, depending on the technique chosen; and a color image is produced. Four sequential exposures, each followed by toner development, produce cyan, magenta, yellow, and black images. Use of these "simple" systems proved difficult to control in a production environment. Both dry and liquid developer systems are used. Dry systems use very small toner particles. Liquid systems use toner particles suspended in a liquid-phase developer. Inadequate toner resolution limited the use of dry systems. The liquid hydrocarbon developer of the wet system introduced an environmentally unfriendly volatile organic compound.

Dye sublimation. Dye-sublimation color can be cleaner and more saturated than colors obtained from other transfer techniques. The dyes in a donor sheet, when heated, change phase (to a gas) and condense to a solid (sublime) on special receptor coatings. The backside of the sheets may be coated to reduce image misregister due to receptor slippage caused by the buildup of electrostatic charge. This technology lends itself to both contone and halftone imaging.

For contone images, the colorants yellow, magenta, cyan, and black are assembled as separate panels in a donor ribbon or single sheets. Some less expensive devices use only three colors. In that case, the three-color image file must be revised to provide the necessary four-color output; thus both the computer-output file and the image color itself differ from that used in printing the final image. Visual interpretation between simulated four-color proofs and actual four-color printing may prove difficult. During printing, the donor surface is placed in face-to-face contact with the receptor. The writing device is composed of a multitude of individual wires, each of which is capable of being heated (addressed individually). It is placed in contact with the backside of the dye-sublimation substrate to form an image. Computer-generated imagewise information heats the wires in an imagewise distribution, transferring color from donor to receptor.

For halftone images, the source of radiation is usually a semiconductor infrared laser diode array. Separate sheets of donor and receptor are supplied. The most popular dye subsystem uses infrared-absorbing compounds in the colorant layer in order to reduce the amount of heat required for imaging. The dye subprocess is a combination of sublimed and mass dye transfer. Beads are incorporated into the donor layer to minimize the mass-transfer effect. The halftone device uses a different raster image processor and dot generator than is used for film or plates. Both dot area and dot density may be manipulated to maximize appearance.

Color mass-transfer systems. In mass transfer, a colorant is completely transferred (go/no go process) from one surface to another. The color is in a binder (similar to a wax) and is coated on a donor ribbon. A multiwire addressable thermal head is also used. The backside of the ribbon substrate is heated imagewise and causes the transfer of binder plus ink from donor to receptor. The go/no go process produces high-contrast images. A recently developed color printer using a special thermal head with configured metal ribbon electrodes produces bilevel features that resemble a square halftone pattern; however, the dot pattern produced in the proof is not the same dot pattern as that which will be printed. It has been cosmetically altered to resemble the output dot pattern. Mass-transfer imaging can also be accomplished without using a donor ribbon.

Hybrid color-transfer systems. Most of these systems use color transfer from donor to receptor, with a donor sheet of colorant, alone or in a binder, and a heat sink. To thermally transfer the image, the heat necessary to transfer colorants can be supplied by any of several sources, including spark-gap discharge, xenon flash, infrared emission, and semiconductor laser radiation.

Prospects. Of all these color proofing technologies, none seems to completely satisfy printers and customers. The new concept of combining two calibrated proofers of the same technology (one at the printer site and the other at the customer site) into an interactive remote printing system seems to have promise of satisfying both. The technologies favored in these systems are ink-jet and dye sublimation.

For background information *see* PRINTING in the McGraw-Hill Encyclopedia of Science & Technology.   Richard Fisch

Bibliography. Graphic Arts Technical Foundation, *The Essentials of Computer-to-Plate Technology*, 1995; Society of Imaging Science and Technology, *Recent Advances in Ink Jet Printing*, 1996.

Production engineering

For a manufacturing system to achieve its maximum performance, the operations of its subsystem, including machines and personnel, as well as the flow of parts and products, should be well coordinated; this requires proper scheduling of the operations and flows. For example, if many parts are waiting to be processed by a certain machine, a decision must be made as to which part should be processed next when the machine finishes the current processing (job dispatching problem). When there are several machines capable of performing the same job, a machine must be selected for the task (machine selection problem). For a longer time scale, it is necessary to schedule when and how many of each product should be produced. The results of these decisions must satisfy certain production requirements. For example, the total time required to produce a certain number of products; the production delay from the due date (the time when each product should be completed); and the cost required for the personnel, machines, or inventory ideally should be minimized. In the recent trend for designing more flexible manufacturing systems, more and more decision-making points are introduced in order to deal with the demands for producing a large variety of products in small amounts and in frequently changing environments, where the proper scheduling plays an even more essential role.

Real-time production scheduling. Scheduling problems can be roughly divided into static problems and dynamic problems. In the static problems, all the information required for scheduling, such as the time when each part becomes available, the precise processing time of each job at each machine, or the number of available machines, is given beforehand. All the decisions can be made beforehand, and a timetable type of schedule can be set up; this is the typical case for operations such as railway scheduling. This type of scheduling is often called off-line scheduling. By contrast, in dynamic problems such information is given just partly or imprecisely beforehand, and the precise data can be obtained only in the course of actual production operations.

Scheduling problems of the typical manufacturing systems, especially those encountered in the flexible manufacturing systems, are dynamic problems; precise predictions cannot be made of the time required for each stage of the manufacturing process. The sequence of arriving parts, their arrival times and the processing time at each machine, and the production requirements may change. A certain job may be inserted or canceled abruptly, and machines may sometimes fail. In such a dynamic production environment, it is desirable for scheduling decisions to be made during the actual operations, based on the information obtained in real time from various parts of the system. This type of scheduling is called real-time scheduling.

Methodologies for real-time scheduling. In real-time scheduling, not only is the information given just partly but also the allowable time for decision making is very short, which necessitates a quite different approach from off-line scheduling.

The most popular method used for real-time scheduling is to make a decision only when it is needed, using a certain scheduling rule. A scheduling rule that is based on local real-time information of the manufacturing system dictates which operation should be performed next. For example, in the job-dispatching problem, one rule selects the job whose processing time at that machine is the shortest, while another rule selects the job whose due date is soonest, and so forth. A variety of such rules have been devised by using mathematical analysis or by human intuition or experience, and their performance in various manufacturing systems under various production requirements has been reported and compared. The strength of this method is its low computational load, ease in implementation, and a certain level of robustness to anomalies, such as machine failures. The shortcoming is that the best scheduling rule varies according to the characteristics of the

manufacturing system or current production requirements; therefore, it is essential to find an appropriate rule (or, where there are several decision points, a rule set) suitable for each application. In some cases, the best rule varies even according to the instantaneous situation of the manufacturing system, such as the content of the current buffer or machine status; it has been reported that the rule should be switched in real time, taking more global information of the system into account. This method is called dynamic scheduling. It requires some knowledge of which rule should be employed in which situation; this knowledge is provided by human experts (as in the case of the usual expert systems) or by machine learning. Simulations of the manufacturing system are often effectively utilized to find the appropriate scheduling rules or to acquire the necessary knowledge; therefore, the methods based on the scheduling rules have become known as simulation-based scheduling.

Another method employs an approach based on artificial intelligence, such as a rule-based system, case-based reasoning, or constraint-directed reasoning. In a rule-based system, the scheduling knowledge of a human expert is implemented in the form of if-then rules. During real-time operations, such rules whose if-parts match the current situation are selected and their then-parts are added to the current situation list or indicate the actions to be performed. By chaining if-then rules in this way, several actions are proposed, from which one is finally selected by a conflict-resolution algorithm. For this objective, various off-the-shelf expert-system development tools can be utilized as the inference engine, but the difficulty resides in how to obtain scheduling knowledge that is useful and sufficient. Recently, various machine-learning techniques have been utilized to compensate for this insufficiency of knowledge.

In case-based reasoning, past scheduling experience whose problem situation is most similar to the current one is retrieved from the memory, which provides some information as to how to solve the current problem. For example, past experience of successfully changing job assignments that reflects a failure of a certain machine can be used for selecting an action in similar failure situations. Difficult technological issues are how to define the similarity of the situations and how to derive the appropriate solution from the solution of a previous problem that is similar to, but not exactly the same as, the current problem.

In constraint-directed reasoning, a schedule is gradually modified so that the total constraint-satisfaction level may be increased. Here, constraints include not only the usual constraints, such as due dates, but also the problem definitions, such as the processing time or buffer capacity; and their satisfaction levels are defined in the form of penalties. The most important part is the algorithm that performs a sequence of modifications of the temporary schedule so that the sum of the penalties becomes sufficiently small. The difficulty resides in how to avoid being trapped by the local minimum of the penalty function and the reduction of computational load, so that real-time scheduling may be possible.

So far the discussion has treated the software aspect of the real-time scheduling problem. The hardware implementation of the scheduling system, such as the computer networking and the design of information flow, is also an important consideration. For example, even if the production scheduling using the global floor information has been found excellent, the hardware to enable the acquisition of the global information may sometimes be impossible or quite expensive. However, the production line that includes its information management system must be designed so that an appropriate scheduling can achieve the given production requirements. The hardware and software designs for the scheduling system must be made concurrently, so that they can be well balanced. For this purpose, recent attention was focused on a distributed scheduling architecture, an efficient, reliable, fault-tolerant, and flexible architecture that is easy to implement.

Similar scheduling problems. Various methodologies have been proposed to solve the problem of off-line scheduling. It differs from real-time scheduling in that all the required information for scheduling is provided; therefore, the constraint on the computational time is not so stringent, enabling an approach that is more mathematical and computation intensive. The most important contribution has been made from the field of operations research. It treats the scheduling problem as a combinatorial optimization problem and tries to determine the optimal solution in terms of a certain objective function. The strength of this approach is that once the scheduling problem is formalized in mathematical formulas, various invented algorithms can easily be applied. The required computational load, however, is known to increase exponentially as the problem size grows; in practice, truly optimal solutions can be obtained only for quite simple problems. In most scheduling problems in manufacturing systems, only such methods that attempt to find suboptimal solutions within practical computational loads can be utilized. Mathematical planning and dynamic programming fall into this category. Recently, various new technologies, such as genetic algorithms, Hopfield networks, or simulated annealing, have been applied to the scheduling problems in order to efficiently obtain suboptimal solutions. These approaches are sometimes called modern heuristics. Some real-time scheduling methods, such as those based on scheduling rules or artificial intelligence, are often combined with some of these new methods to improve the efficiency in determining suboptimal solutions.

Air-traffic control provides a scheduling problem that is quite similar to real-time production scheduling. Here, the movement of airplanes, which corresponds to the flows of parts or products in production scheduling, within a certain area (called a sector, such as the area surrounding a certain airport) must be scheduled. The decision items include which airplane should go, land, or depart first; the desired time each airplane passes through certain points (called way points); the altitude assignments to each airplane; and so forth. The objectives of scheduling are, in most cases, to reduce flight delays or fuel requirements without sacrificing flight safety. As is the case in most manufacturing systems, there are many uncertainties and anomalies; for example, an airplane in an emergency situation may sometimes fly into the sector, and the arrival time of each airplane is not so accurately predicted because of the delays at the departure airports, weather conditions, or other causes. Currently, most scheduling for air-traffic control is performed by a group of human controllers at control centers. It has been reported that they use certain scheduling rules, such as the first-in–first-out rule, which have been proved to be effective empirically, to make decisions in most cases, while they treat the special cases such as an emergency with their expertise. With the current trend toward the global information network and the need to deal with denser air traffic, the next generation of air-traffic control will require scheduling that utilizes more global information rather than that derived from only within one sector; in addition, it will need more computer-based or computer-aided scheduling architecture.

For background information *see* ARTIFICIAL INTELLIGENCE; DECISION THEORY; EXPERT SYSTEMS; FLEXIBLE MANUFACTURING SYSTEM; OPERATIONS RESEARCH; PRODUCTION ENGINEERING in the McGraw-Hill Encyclopedia of Science & Technology.　　Shinichi Nakasuka

Bibliography. M. S. Fox and S. F. Smith, ISIS: A knowledge based system for factory scheduling, *Expert Sys.*, 1:25–49, 1984; S. Nakasuka and T. Yoshida, Dynamic scheduling system utilizing machine learning as a knowledge acquisition tool, *Int. J. Prod. Res.*, 30 (2):411–431, 1992; F. A. Rodammer and K. P. White, A recent survey of production scheduling, *IEEE Trans. Sys. Man Cybern.*, 18:841–851, 1988.

Psychoacoustics

Spatial hearing is concerned with two aspects of sound perception: localization of emitted sound and echolocation, which is the sensing of the environment by means of reflected sound. This article discusses the first aspect, which has been the focus of most spatial hearing research involving humans.

The position of a sound source is specified in spherical coordinates: distance, azimuth (compass direction with respect to the facing direction of the head), and elevation (direction with respect to the ear-level plane, the horizontal plane passing through the ears). Most research on human spatial hearing has dealt with directional localization of stationary sources in anechoic (free of echoes) environments by a listener whose head is stationary. However, recent research is moving toward more natural lis-

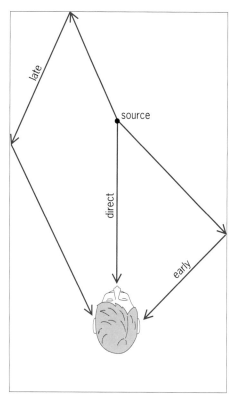

Fig. 1. Sound propagation between source and listener. Indirect paths may be of two types: early reflections and late reflections.

tening conditions that include multiple moving sources, echoic environments, and listeners who are free to undergo head rotation and translation. The changing focus of research is partly due to the emergence of virtual acoustics, the technology of producing realistic three-dimensional localization of sounds presented through earphones. Applications of virtual acoustics include conveying three-dimensional spatial information to blind people and informing aircraft pilots about potential hazards (for example, other aircraft and nearby terrain).

Virtual acoustics. The assumption underlying virtual acoustics is that a person who obtains the same binaural stimulation with earphones that he or she would obtain in the presence of real sound sources should have the same perceptual experience in both situations. Computer implementation of a virtual acoustic display involves (1) synthesizing binaural signals that mimic those from real sources after modification by the environment and by the observer's head and ears, (2) presenting these binaural signals to the listener using earphones, (3) tracking the rotations and translations of the listener's head, and (4) modifying the binaural signals in concert with these rotations and translations so that the signals are equivalent over time to what they would be if the person were in the presence of the real sources.

Causal chain. Sound arriving from a source positioned in front of the listener's head travels along direct and indirect paths (**Fig. 1**). The arriving sound

is modified by diffraction by the head, shoulders, and pinnae (the visible structures of the external ears). The sound entering each ear is then transmitted by the ear canal, eardrum, and middle-ear ossicles to the cochlea, which transduces it into neural activity that exits along the cochlear nerve. After transformation by various synaptic levels (nuclei) along the auditory pathway, the neural signals from the two ears arriving at the auditory cortex, along with signals from cortical centers associated with cognition, determine the perception of the sound (**Fig. 2**).

Directional localization. The stimulus cues for localizing sound in direction (azimuth and elevation) are well understood. The most important stimulus information can be described with reference to a spherical approximation of the head without pinnae. The aural axis is the extended imaginary line connecting the two ear canals. The angle between the aural axis and the source direction is referred to as the lateral angle.

Interaural intensity difference. Diffraction of the incoming sound by the approximately spherical head gives rise to the cue of interaural intensity difference (IID). If the source is off to one side of the midsagittal plane (the vertical plane that bisects the head through the nose), the ear on the opposite side receives a less intense signal than the ear on the same side. The IID cue is virtually nonexistent at low frequencies (below 300 Hz) but increases with frequency as diffraction by the head increases. For the spherical approximation, the IID cue is symmetric about the aural axis. A given lateral angle (for example, 30°) defines a conic surface about the aural axis; sources located on this surface produce approximately the same value of IID. A lateral angle of 90° corresponds to sounds in the midsagittal plane within which IID is zero at all frequencies.

Interaural time difference. The other primary cue for direction is interaural time difference (ITD). For a source off to one side of the listener's head, the path length that the sound must travel is greater for the ear on the opposite side than for the ear on the same side. When the sound lies on the aural axis off to one side (zero lateral angle), ITD is maximal, with a value of about 0.7 millisecond for the typical adult. As lateral angle increases toward 90°, the ITD decreases to zero. Under the spherical approximation, ITD, like IID, is constant for different locations having the same lateral angle.

The circular symmetry of IID and ITD about the aural axis means that sources equal in lateral angle should be impossible to discriminate by a listener whose head is stationary. While there is a tendency for a listener with a stationary head to confuse sources having equal lateral angles (for example, two sounds equally in front of and behind the aural axis, or two sounds equally above and below the aural axis), such a listener generally is able to discriminate between such sources much better than chance-level performance. Thus, the above description of IID and ITD using the spherical approximation is too simple.

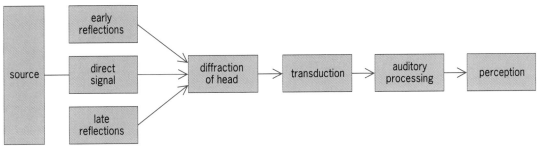

Fig. 2. Causal chain underlying localization of emitted sound by the listener.

Head-related transfer function. The complete mathematical specification of how sound reaching the head is modified by the head, shoulder, and pinnae is referred to as the head-related transfer function (HRTF). It represents the complex variations in IID and ITD that depend upon sound frequency and upon the azimuth and elevation of the source. The additional variations, beyond those due to the spherical model, are the result of diffraction by structures not encompassed in the model (for example, the pinnae and shoulder). A listener is most accurate in localizing the azimuth and elevation of a source when he or she listens with the normal HRTF. The basic IID and ITD cues associated with the spherical approximation of the head (without pinnae) are referred to as the simple HRTF.

Utilization of head rotation. Although confusion between left and right or between up and down is fairly common for a listener with a stationary head, unambiguous directional localization results when the stimulus is of sufficient duration to permit the listener to derive additional information by head rotations. The analysis of directional localization under head rotations reveals that perceived sound direction can be understood in terms of the sensed change in lateral angle relative to the sensed head rotation. For example, if a source is initially within the ear-level plane at 60° left azimuth, at the outset it has a lateral angle of 30°. If the listener rotates the head 10° clockwise about the vertical axis so that the lateral angle has diminished to 20°, then the 10° change in lateral angle for a 10° head rotation signifies to the listener that the source was initially in the ear-level plane and in front of the aural axis. This analysis implies that sounds should be localizable in azimuth and elevation by means of head rotations even without the supplementary directional cues provided by the pinnae, nonspherical head, and shoulders.

Distance localization. For an observer listening from a fixed position in space, localization of source distance is supported by at least three cues: (1) the inverse relationship between source distance and sound pressure at the ears; (2) the additional attenuation of higher frequencies associated with atmospheric absorption (for example, thunder crackles from a nearby lightning strike and rumbles from a distant one); and (3) the proportion of direct sound to reverberant sound in echoic environments; this

proportion is higher when the listener is closer to the source.

For an observer undergoing translational motion, there are two additional sources of distance information: (1) the motion parallax, which is the change in angular direction of a stationary source; that of nearby sources is greater than that of far sources, and (2) acoustic tau, which is instantaneous pressure at the ears divided by the time rate of change of pressure; it provides direct information about the distance of the source to a listener who knows his or her translational velocity.

Externalization of virtual sounds. Virtual acoustic displays using earphones are very effective in conveying source direction to a listener. Even those displays based on simple HRTFs (without pinnae cues) are quite effective, provided that the represented sounds are confined to a horizontal plane at ear level (requiring only azimuthal discrimination). In contrast, distance localization with virtual displays has proven much less effective. Just achieving externalization (external localization of the sound) has proven a challenge—computer-generated virtual sounds are often heard within the perceptual margin of the head. A prevalent view has been that externalization depends critically upon whether the full HRTF, including pinnae cues, is employed in the synthesis of the virtual sound. However, in view of the fact that real sounds are easily externalized even when the listener hears the sounds with a drastically altered HRTF (for example, listening with a helmet on, or listening with one ear blocked), employment of the full HRTF can be only a secondary determinant of externalization. Other compelling evidence for this conclusion is the fact that a person listening to environmental sounds with earphones still hears them as fully externalized. A simple demonstration can be observed by having a person listen to his or her surroundings through tight-fitting headphones that are fed by microphones worn on the headphone strap; even though the person is listening with significantly altered pinnae, externalization is complete. It is evident that the most important determinants of externalization, whether of real or virtual sound, involve using either the simple or full HRTF in creating the binaural signals, including early reflections rendered binaural by the HRTF, linking the binaural signals in real time to the rotations and translations of the listener, and maximizing for the listener the degree of concordance between the types of sounds being heard and the listener's surroundings.

For background information *see* AUDIOMETRY; BINAURAL SOUND SYSTEM; ECHOLOCATION; HEARING (HUMAN); PERCEPTION; PHONORECEPTION; PSYCHOACOUSTICS; SOUND in the McGraw-Hill Encyclopedia of Science & Technology.　　　　Jack M. Loomis

Bibliography. J. Blauert, *Spatial Hearing: The Psychophysics of Human Sound Localization*, 1983; N. I. Durlach et al., On the externalization of auditory images, *Presence: Teleoperators and Virtual Environments*, 1:251–257, 1992; H. Wallach, The role of head movements and vestibular and visual cues in sound localization, *J. Exper. Psychol.*, 27:339–368, 1940; F. L. Wightman and D. J. Kistler, Headphone simulation of free-field listening, II: Psychophysical validation, *J. Acous. Soc. Amer.*, 85:868–878, 1989.

Quality control

Quality is the collection of features and characteristics of a product or service that contribute to its ability to meet given requirements. Though quality concepts date back to ancient times, the early work and concentration in modern times was on methods and processes to create standards for producing acceptable products. By the middle 1950s, mature methods had evolved for controlling quality, including statistical quality control and statistical process control, utilizing sequential sampling techniques for tracking the mean and variance in process performance. During the 1960s, these methods and techniques were extended to the service industry. During 1960–1980, there was a major shift in world markets, with the position of the United States declining while Japan and Europe experienced substantial growth in international markets. Consumers became more conscious of the cost and quality of products and services. Firms began to focus on total production systems for achieving quality at minimum cost. This trend has continued, and today the goals of quality control are largely driven by consumer concerns and preferences.

Product quality characteristics. Though both products and services are subject to equivalent interpretations in terms of quality characteristics, it is convenient to restrict discussion to products. There are three views or bases for describing the overall quality of a product. First is the view of the manufacturer, who is primarily concerned with the design, engineering, and manufacturing processes involved in fabricating the product. Quality is measured by the degree of conformance to predetermined specifications and standards, and deviations from these standards can lead to poor quality and low reliability. Efforts for quality improvement are aimed at eliminating defects (components and subsystems that are out of conformance), the need for scrap and rework, and hence overall reductions in production costs.

Second is the view of the consumer or user, who can have different needs. To consumers, a high-quality product is one that well satisfies their preferences and expectations. This consideration can include a number of characteristics, some of which contribute little or nothing to the functionality of the product but are significant and influential in providing customer satisfaction. A third view relating to quality is to consider the product itself as a system and to incorporate those characteristics that pertain directly to the operation and functionality of the product. This approach should include overlap of the manufacturer and customer views.

An example of how the product quality of an automobile is examined is given in **Table 1**. From the point of view of the manufacturer, this product could be rated as high quality, since units leave the final assembly processes with only a 2% observed rate of defects. Assuming that few failures occur during the early stages of use and that the major failures are likely to occur after the warranty has expired, the vehicle seems to be manufactured within the engineering and production specifications for which it was designed. Moreover, the warranty expenses should be relatively low. The customer, however, might view this product differently, particularly in the absence of any highly attractive features through options that might offset his or her satisfaction level. While the early failures might be few in number, the failures that occur after warranty can be quite expensive for the customer. In this example, from the viewpoint of a product system, the product has more of a moderate level of quality based on the acceleration, braking distance, and fuel economy.

Product quality elements. Over the years a number of dimensions or factors have evolved that contribute to overall product quality as it is presently defined. Some dimensions are more dominant than others in describing a particular product. The following elements are most common in describing quality.

1. Operating performance comprises measures of actual operating performance of the product. Exam-

TABLE 1. Vehicle product quality data

Characteristic	Value
Warranty	3 years/36,000 mi (58,000 km)
Outgoing inspection	98%
Engine and power-train mean time to failure (MTTF)*	42,000 mi (67,000 km)
Acceleration 0–60 mi/h (0–100 km/h)	10 s
Braking distance 55–0 mi/h (88–0 km)	480 ft (140 m)
Fuel consumption	16 mi/gal (6.8 km/liter) town 20 mi/gal (8.5 km/liter) highway

* Usage (total distance traveled) is a common measure for vehicle failures, and is conventionally designated MTTF.

ples include automobile gasoline consumption in miles per gallon, brake horsepower for an engine, range of a weapon, and hours of life for a battery.

2. Durability refers to the amount of use before product deterioration, normally measured in the time or hours of usage before a product fails.

3. Reliability refers to the probability of a product failing within a specified period of time, having survived to that point. Primarily applied to durable goods, reliability is assessed by the average rate of failure, mean time between failures (MTBF), and related failure time measures.

4. Conformance is the degree to which design and operating characteristics meet preestablished standards, as measured through defect count, scrap rate, and amount of rework.

5. Serviceability refers to the degree or ease of restoring service, measured in time to restore service, frequency of calls, and mean time to repair (MTTR).

6. Usability refers to the ease of a consumer in developing the necessary preparation, skills, and proficiency for using a product as it was intended. The most common measure of usability is learning time.

7. Perceived quality concerns the overall image of a product among users and potential users. Though some objective measures are involved through surveys and opinions, perceived quality is generally quite subjective.

8. Esthetics concerns the way in which the product is sensed through looks, feel, sound, taste, and smell.

Many of the quality elements involve measurable product attributes that are easily quantified, but some are subjective and are influenced by selective group and individual preferences. This is particularly true of perceived quality and esthetics, which characterize the customer view of products.

Performance measures. It is convenient to think of product quality as a vector of attributes that span all important characteristics of the item relative to the overall performance according to the three bases—the manufacturer, the customer, and the overall product viewed as a system. Some of these elements are qualitative, but some are entirely quantitative. The challenge is to incorporate these elements into an overall measure of product performance suitable for monitoring progress in quality improvement and for making management decisions about the product quality and costs.

The overall performance of a product depends on each attribute separately. For example, if a product is considered with regard to two attributes (for example, quality and reliability), and if each of these attributes has three possible values (high, medium, and low), then there are $3 \times 3 = 9$ possible pairs of values of the two attributes, and each pair can give rise to a different value of overall performance. Moreover, the overall performance can vary with time as the values of the attributes vary, and it is often useful to graph this variation.

TABLE 2. Thresholds of overall quality and reliability		
Dimension	Threshold	Category
Quality (defects/1000)	<5	High
	5–20	Medium
	>20	Low
Reliability (mean time between failures, hours)	>200	High
	100–200	Medium
	<100	Low

For example, a new treadmill is brought to market, and the manufacturer has determined overall quality and reliability thresholds as shown in **Table 2**. The average time for customers to learn how to use all of the features on the unit ranges 2–5 h.

Here the most qualitative (subjective) element of interest is usability, because of the relative uncertainty in the training time. Therefore, the first step is to construct the set of combinations of low (L), medium (M), and high (H) attribute values for quality and reliability, giving {(L,L),(L,M),(L,H),(M,L),(M,M),(M,H),(H,L),(H,M),(H,H)}. To incorporate the usability dimension, the learning time is categorized into intervals, such as easy for times up to 2.5 h and normal for times between 2.5 and 5 h. This forms an expanded range of attribute values. By assigning variables to each of these combinations and employing a process of pairwise comparisons among these attributes, relative weights can be assigned to establish an overall measure of product performance $\Pi(t)$, as shown in the **illustration**.

Quality, reliability, and warranties. Two types of product failures can generally be distinguished: those that occur relatively soon after they are produced and sold to consumers, and those that occur later after the product has been in use, perhaps even without fault for many cycles or hours of usage. From a manufacturer's point of view, the failures that occur early in the product life are generally due to low quality and are attributed to a lack of conformance to manufacturing and production standards. Here quality of the product can be thought

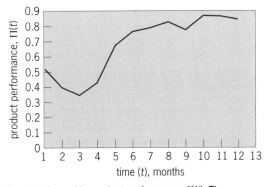

Sample of monthly product performance, $\Pi(t)$. The variable $\Pi(t)$ is an index of the level of performance, with 1 corresponding to perfect and 0 corresponding to worst level.

of as a state of acceptance that relates directly to the manufacturing processes, materials, and workmanship. Accordingly, quality is maintained and controlled through developed standards and specifications for characteristics such as dimensions and tolerances, composition and structure of materials, hardness and surface finish, and strength of linkages and mechanisms.

Those failures that occur later in the life or usage cycle of a product are commonly due to low reliability, caused by problems in the design and engineering of the item. These failures typically occur as stress fractures, fatigue, and shearing, which relate to problems in the materials, basic design, and production processes. Sound planning early in the design process is necessary to overcome these and other problems in order to achieve high product reliability.

Warranties require the producer to pay all or some of the expenses involved in the repair or replacement of products that fail within a prescribed period of time. Whether planned by the manufacturer as part of a market strategy or stipulated by liability legislation, the costs for warranties can represent a significant portion of the cost to manufacture a product. Low product quality caused by poor conformance, low reliability, or both will lead to high warranty costs.

Operational measures of product performance. Warranty costs are expenses that occur because of low product quality. Therefore, measures that relate to the warranty expenditures also relate to product quality. Common operational measures that are used to track quality-related performance in production include the percentage of failures or defects, the number of warranty claims per unit sales, and the average warranty costs per unit sold. These measures, which involve warranty claims, also reflect customer attitude toward the product. Quality improvement programs typically include goals for reducing warranty claims and associated costs because warranty data directly or indirectly impact most of the product quality dimensions.

For background information *see* MANUFACTURING ENGINEERING; PROCESS CONTROL; QUALITY CONTROL; RELIABILITY (ENGINEERING) in the McGraw-Hill Encyclopedia of Science & Technology.

Marlin U. Thomas

Bibliography. D. A. Garvin, Competing on the eight dimensions of quality, *Harvard Bus. Rev.*, 65(6):107–109, 1987; C. L. Karnes, S. V. Sridharan, and J. J. Kanet, Measuring quality from the consumer's perspective: A methodology and its application, *Int. J. Prod. Econ.*, 39:215–225, 1995; N. A. Morgan and N. F. Piercy, Market-led quality, *Ind. Market. Manag.*, 21:111–118, 1992.

Radio telescope

The Very Long Baseline Array (VLBA), a continentwide radio telescope system dedicated in 1993, has brought very high resolution imaging capability

Fig. 1. Very Long Baseline Array station at Pie Town, New Mexico. The antenna, identical to those at the other nine sites, weighs about 240 tons (220 metric tons) and is fully steerable. *(National Radio Astronomy Observatory/ Associated Universities, Inc.)*

to the astronomical community on a continuous, routine basis for the first time. This capability already has yielded important results and promises to provide new information on a very wide range of phenomena in the universe.

The Very Long Baseline Array, a facility of the U.S. National Science Foundation, consists of ten 25-m-diameter (82-ft) parabolic dish antennas (**Fig. 1**) distributed across United States territory from Mauna Kea, Hawaii, to St. Croix, in the Virgin Islands (**Fig. 2**). The Very Long Baseline Array is an aperture synthesis interferometric radio telescope, which gains the resolving power of a very large antenna by utilizing a number of smaller, separated antennas. Other aperture synthesis instruments include the Very Large Array (VLA) in New Mexico, the Australia Telescope, and the Westerbork Synthesis Radio Telescope in the Netherlands.

The technique of very long baseline interferometry (VLBI), first demonstrated in 1967, allows the use of antennas so widely separated that real-time communications among them are impracticable. Very long baseline interferometry requires extremely precise time standards at each antenna and very high density recording equipment, among other technologies. In the case of the Very Long Baseline Array, the separations of up to 8000 km (5000 mi) between antennas provide a maximum resolution of less than a thousandth of a second of arc.

Black holes and astrophysical jets. This high resolution, combined with great sensitivity, allowed astronomers using the Very Long Baseline Array to discover a dense disk of material orbiting what almost certainly is a supermassive black hole at the heart of a galaxy 2.1×10^7 light-years from Earth. The galaxy NGC 4258 (Messier 106) had long been

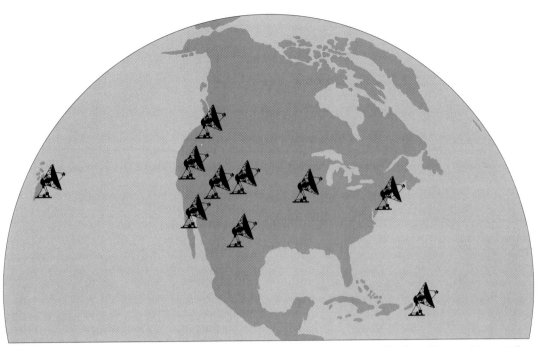

Fig. 2. Locations of the ten 25-m (82-ft) radio telescope antennas of the Very Long Baseline Array. From east to west, the locations are St. Croix, U.S. Virgin Islands; Hancock, New Hampshire; North Liberty, Iowa; Fort Davis, Texas; Los Alamos, New Mexico; Pie Town, New Mexico; Kitt Peak, Arizona; Owens Valley, California; Brewster, Washington; and Mauna Kea, Hawaii. *(National Radio Astronomy Observatory/Associated Universities, Inc.)*

known as a peculiar and interesting object, about the same size as the Milky Way but with unusual, helical jets of gas emerging from its nucleus. In 1982, radio astronomers discovered that the galaxy harbors powerful, naturally occurring microwave amplifiers consisting of heated water molecules. Such cosmic masers amplify microwave emissions in a manner similar to that in which a laser amplifies light. In 1992, the radio emission of some of these water masers was found to be strongly Doppler shifted in frequency, indicating that these masers have velocities up to 1000 km (650 mi) per second.

With the sharp radio vision of the Very Long Baseline Array, it was found that the water masers were part of a disk of material deep in the center of NGC 4258. The Very Long Baseline Array observations showed that the slightly warped disk rotates about the center of the galaxy according to Kepler's third law (the square of the rotation period equals the cube of the orbital radius), which allows a straightforward calculation of the mass of the central object. The surprising result was that the central object has a mass about 3.5×10^7 times greater than the Sun. Because the Very Long Baseline Array observations tightly constrain the volume within which this mass can exist, the resulting density, about 10^8 solar masses per cubic light-year, makes this object the best candidate yet for an extragalactic black hole.

Supermassive black holes such as that in NGC 4258 are believed to lie at the hearts of many galaxies and quasars. Many of these objects display powerful jets of material emerging from their cores, where an accretion disk of material is being pulled into the black hole. The material in the jets often is acceler-

ated to nearly the speed of light. The energy required to produce such acceleration probably comes from the gravitation of the black hole, but there is no generally accepted detailed model of this process.

An object fortuitously discovered in 1994 and extensively observed with the Very Long Baseline Array and other instruments may help improve the models. Located only 10^4 light-years distant within the Milky Way Galaxy, this binary-star system apparently has a black hole four to five times more massive than the Sun as one of its components. The Very Long Baseline Array was used to study jets of material emerging from this object, called GRO J1655-40. The resolving power of the Very Long Baseline Array showed the motion of condensations in the jets in as little as a few hours. The condensations moved at 92% of the speed of light. Most astronomers consider GRO J1655-40 to be a scaled-down version of the engines at the centers of galaxies.

Because of its nearby location, GRO J1655-40 provides an ideal target for scientists seeking to learn the details of such energy sources. It can be observed at wavelengths ranging from radio to x-ray, and frequently undergoes outbursts. The Very Long Baseline Array, along with numerous other instruments, both ground and space based, will aid in the use of this object to decipher the physics of black holes, accretion disks, and astrophysical jets.

Expanding supernova remnant. The Very Long Baseline Array is capable of working in conjunction with other radio telescopes around the world. When a supernova exploded in the galaxy Messier 81, 1.1×10^7 light-years away, in 1993, the Very Long Baseline Array, along with other instruments in Europe and the United States, began a long-term observational program. Understanding of the physics of supernova explosions and of the interaction of the explosion debris with the surrounding medium is best advanced by observing the early stages of these events.

Despite the distance to M81, the Very Long Baseline Array and its companion instruments were able to detect radio emission from the supernova and to clearly show the expansion of the explosion shell. A series of images, compiled over the course of a year's observations, showed a symmetrical expansion with no evidence for protrusions within the debris shell which are seen in older supernova remnants (**Fig. 3**). The measured expansion rate, nearly 16,000 km (10,000 mi) per second, also showed no indication that the shell has yet been slowed by interaction with the surrounding medium. The angular expansion rate, measured by the radio observations, combined with the actual speed of expansion, measured by optical observatories, provides an accurate value for the distance to M81. This is one of several ways the Very Long Baseline Array can contribute to improving the accuracy of the cosmic distance scale.

Measuring Earth and sky. By using extragalactic radio sources, typically quasars, as an inertial reference frame, practitioners of very long baseline interferometry can make extremely precise measurements of the relative positions of their antennas. With the Very Long Baseline Array, baselines between antennas routinely are measured with an accuracy of a few millimeters.

Geophysicists regularly use the Very Long Baseline Array to measure movement of tectonic plates, crustal deformation, and variations in the rotation of the Earth. Geodetic observing programs seek to improve the measurement techniques as well as to determine the Terrestrial Reference Frame. The Very Long Baseline Array has detected motion of 1–12 mm per year among the stations on the North American tectonic plate, a result that will help to characterize the degree of stability and rigidity of the plate and to improve knowledge of the probabilities and mechanisms of intraplate seismic activity. The Very Long Baseline Array also is used to make astrometric observations of compact extragalactic radio sources as part of a continuing program to establish an all-sky radio–optical celestial reference frame.

Cosmologists who seek to measure great distances in the universe have mounted continuing campaigns to discover new gravitational lens systems. With such a system, if an accurate model for the lensing mass can be determined, then measuring the time delay in variability among components of the lens can yield a distance estimate independent of other parameters, and thus help calibrate the Hubble constant, the primary yardstick for measuring very large distances in the universe. The Very Long Baseline Array is used to study candidate lens systems identified by instruments with lesser resolution, and thus distinguish the true lenses.

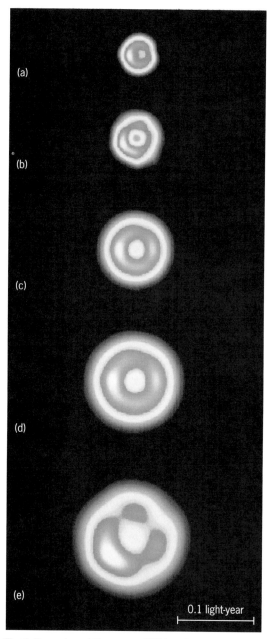

Fig. 3. Expansion of Supernova 1993J, as imaged by the Very Long Baseline Array and other radio telescopes. All images are to the same scale; the expansion is clearly seen despite the distance to the supernova of 1.1×10^7 light-years. (a) September 1993. (b) November 1993. (c) February 1994. (d) May 1994. (e) September 1994. *(From J. M. Marcaide et al., Expansion of SN 1993J, Science, 270:1475–1478, 1995)*

Very Long Baseline Array. The Very Long Baseline Array is controlled from the National Radio Astronomy Observatory's (NRAO) Array Operations Center in Socorro, New Mexico. A computer in Socorro communicates via the Internet with control computers at each of the 10 antenna sites. The array operator in Socorro controls the entire system and monitors the status of each station's equipment. Technicians at each station conduct maintenance and handle the magnetic tapes on which the observational information is recorded.

To provide the precise timing references required to allow the 10 stations of the Very Long Baseline Array to operate as a single instrument, each site has a hydrogen maser with a stability of one part in 2×10^{15} over time intervals ranging from 1000 to 10,000 s. The timing information is included in the data stored on the high-density magnetic tape.

Tapes are shipped from the stations to the Very Long Baseline Array correlator in Socorro. This correlator, a special-purpose digital machine designed and built by the NRAO, receives data from a bank of tape recorders. The tapes from all stations are played back simultaneously, and through the precise time tags placed on the tapes by the hydrogen masers at each site, the observation is essentially recreated for the correlator.

In the correlator, signals from all the sites are manipulated mathematically to make it appear that all observing stations are on the same plane surface perpendicular to the line of sight to the object observed, despite the great distances between the stations and the curvature and rotation of the Earth. Tiny differences in the arrival times of radio waves from different parts of the celestial body modify the combined signal detected by pairs of antennas. Data from all pairs of the array are combined to produce the information necessary to make an image of the observed object. The correlator's output can then be used by astronomers with workstations to perform postprocessing and image processing.

The Very Long Baseline Array correlator can accept data input from 10 additional observing stations, for a total of 20. In addition to other radio telescopes in the United States and abroad, the Very Long Baseline Array correlator was designed to be capable of reducing data from planned orbiting radio telescopes, such as the Japanese *VSOP* and the Russian *RadioAstron* satellites. The addition of baselines to such satellites will make the resolving power of the Very Long Baseline Array even greater in the future.

For background information *see* ASTROMETRY; ASTROPHYSICS, HIGH ENERGY; BLACK HOLE; COSMOLOGY; GEODESY; GRAVITATIONAL LENS; RADIO ASTRONOMY; RADIO TELESCOPE; SUPERNOVA in the McGraw-Hill Encyclopedia of Science & Technology.   David G. Finley

Bibliography. J. M. Marcaide et al., Expansion of SN 1993J, *Science*, 270:1475–1478, 1995; M. Miyoshi et al., Evidence for a black hole from high rotation velocities in a subparsec region of NGC 4258, *Nature*, 373:127–129, 1995; P. J. Napier et al., The Very Long Baseline Array, *Proc. IEEE*, 82:658–672, 1994; A. R. Thompson, J. M. Moran, and G. W. Swenson, *Interferometry and Synthesis in Radio Astronomy*, 1986.

Rangeland management

Vegetation burning is among the oldest of land management practices. Reduction in the natural intensity and frequency of burning is central among the complex of factors believed to be responsible for the replacement of grasslands with shrublands and woodlands. The science of fire ecology is relatively young, but the attitudes toward the systematic application of fire and its potential use have changed dramatically since the mid-1970s.

Early negative experiences with wildfires gave rise to a notion that fire suppression enhanced conservation. Yet grasslands and forests had evolved through centuries of burning. This contradiction stimulated research with controlled experiments that have demonstrated many benefits and safe applications of burning as a management tool. In fact, land managers now prescribe burning for improvement of the natural resources.

Burning as management practice. Prescribed burning is the systematically planned application of burning to meet specific management goals. The ultimate success of a prescribed burn hinges primarily on the skills of the person(s) responsible for planning and implementation. For improving grasslands, there are basically two types of prescription burns, defined by purpose: maintenance (perpetuation of the grassland) and reclamation (converting shrubland back to grassland).

Critical considerations for prescribed burning are weather, fuel, season, and time. Season, time of burning, and fuel preparation can be controlled. However, air temperature, wind (direction and speed), and moisture (relative humidity and precipitation) cannot be controlled, and their interactions determine fire behavior on any given day. Forest managers must learn to work with these environmental influences to apply prescribed burns successfully.

Fire plans. Prescribed burns must be applied according to a well-designed plan to accomplish specific goals and to minimize the probability of detrimental consequences. The planning process must ensure careful matching of the technique and timing of burns with specific management objectives and fuel conditions, weather, and topography of the targeted area.

The fire plan includes all preparations for the burns, selection of firing techniques and actual burn installation, and the necessary postburn activities which ensure a safe, effective burn. Primary considerations must include topography of the area to be burned, fuel, and weather. However, no two burns are exactly the same, so care must be taken to adapt any generalized fire plan to the specific situation. The simpler the fire plan, the greater the probability that it will be implemented with maximum effectiveness and minimum risk.

Fire fronts. There are two types of fire fronts: headfires and backfires. Headfires move with the wind, whereas backfires move directly against the wind. Variations include strip headfires and flank fires. The type of fire front chosen for the fire plan depends on the specific management objectives.

Backfires burn more slowly (use oxygen less rapidly), flames are less extended, and temperature maxima are generally lower, except perhaps at the soil surface, compared to headfires under the same environmental conditions. Since headfires move with the wind, they are tilted forward, maximizing the preheating of fine fuel ahead of the flame front. The stronger the wind, the greater the displacement of the flame front from a vertical position and the greater the preheating of the fuel.

Plant responses. Heat damages plant tissues by coagulating the protein in cells, followed by rupture of cell membranes. Therefore, morphological characteristics, especially those induced by increasing age, are important to the fire tolerance of perennial plants.

Fire tolerance varies among grasses. The aerial parts of perennial grasses serve as fuel for the fire, but their removal does not impart severe or lasting damage to the plant or diminish the prospect of the plant propagating itself. Perennial bunchgrasses are relatively tolerant of fire because their reproductive buds are protected just at or below the soil surface.

Bark is the primary insulator of woody plants and accounts for heat tolerance of many species of trees and shrubs. Fire resistance of trees and shrubs generally increases with age as lignification of the bark, trunk diameter, bark thickness, and elevation of the crown (canopy) increase. Sprouting ability in many woody species that occur in fire-type vegetation (plant communities in which fire is such a common occurrence that vegetation composition is shaped by fire) has likely been an evolutionary advantage. The ability to sprout from roots and other underground organs after fire breaks the inhibition of lateral bud growth (apical dominance) actually rejuvenates many shrubs and trees.

Influence on soils. Wildfires may bare the soil for prolonged periods, especially during dry weather, greatly reducing resistance to wind and water erosion. However, burning of grasslands under optimum conditions usually has minimal negative effects on the physical or chemical properties of soils. Because grassland fires move so rapidly, direct soil heating is much less extensive than that caused by burning of forests, especially where woody fuel loads are heavy. For example, temperature changes within the soil during grassland burning are short lived and are restricted to the immediate surface. Burning induces a general drying in the soil system, but does not necessarily cause drought stress. Evaporative losses may be increased, at least temporarily, from the bared surface, depending on the season of the burn; and water retention may be decreased in the short term because vegetative cover that traps water has been removed. Moreover, earlier warming of the soil of burned grasslands may stimulate earlier vegetation growth in the spring following burning in the winter, placing a greater water-use demand on the new, rapidly growing, postburn grassland system.

Influence on livestock. A primary objective of burning rangeland is to increase the amount of forage available to grazing animals (those animals that eat primarily herbs; grasses and herbaceous plants).

Given proper grazing management, burning can increase the proportion of highly productive plants for cattle grazing in the forage sward (grass-covered ground). Burning also may improve the acceptability of the available forage and increase its nutritional value to both grazing and browsing animals (those animals that eat the leaves and twigs of trees, shrubs, and woody plants).

Stages of burning. The forces associated with burning that influence any animal and its habitat may be partitioned into preburn, combustion (actual burning), shock (immediately after burnout), and recovery phases.

Preburn. The preburn phase is critical for accumulating a continuous fuel load to ensure effective prescribed burning. Changes in vegetation, composition, and structure of wildlife habitat during the preburn phase may be subtle, and likely have little effect on large animals.

Combustion. Public perception of fire effects on wild animals has been formed largely from a focus upon acute impacts, especially those during the combustion and shock (or aftershock) phases of wildfires. Direct effects of range fires on large animal populations are usually negligible with prescribed burns, since most animals react instinctively to escape harm. Most of the larger animal species can fly or run to safety.

Shock. The shock phase starts immediately following the combustion phase (burnout) and usually lasts for only a few weeks following winter burning of grasslands, depending on how rainfall and temperature conditions affect plant recovery.

Recovery. The recovery phase may last for several years or be essentially complete in a growing season, depending on the specific habitat burned, postburn weather, and burning efficacy (fire intensity, rate of spread, area covered).

Influence on animals. White-tailed deer, common to a large part of the United States, is an example of an especially resilient species favored by a mixture of successional vegetation stages in its habitat. Deer use woody plants for both food (browse) and cover. Burning can improve browse for such range animals by increasing the amount available (accessibility), its acceptability (palatability), and nutritional quality.

Upland game birds such as northern bobwhites (quail) use brush for cover, but prefer open grassy areas with scattered low shrubs for nesting. Prescribed burning of rangeland may become a useful tool for increasing food plants and insects for quail. Dirty (uneven or spotty) burns are best for meeting the habitat needs of quail. Such discontinuous burns may be applied to relatively large areas, several hundred acres in size, and result in positive impacts on the quail populations.

Insect management. Direct damage to insect populations during the combustion phase is minimal if the arthropods are soil dwellers or effective fliers. Greater damage is inflicted on surface dwellers which cannot escape by flight. The concentration and exposure of arthropods by burning may cause predators to congregate and reduce the insect populations immediately after burning. Conversely, burning may promote the recovery or return of some arthropod populations, such as desert termites.

Prescribed burning may effectively reduce populations of ticks. Postburn influences on the ticks' habitat, particularly increased temperatures and reduced relative humidities caused by removal of the standing vegetation and mulch, render the environment temporarily unfavorable for perpetuation of tick populations. Rate of vegetation replacement and size of burned area are important regulators of reinstatement of tick populations.

For background information *see* FOREST AND FORESTRY; FOREST FIRE; FOREST MANAGEMENT; FOREST REFORESTATION; FOREST SOIL; PLANT ANATOMY; PLANT TISSUE SYSTEMS in the McGraw-Hill Encyclopedia of Science & Technology.   C. J Scifres

Bibliography. A. A. Brown and K. P. Davis, *Forest Fire: Control and Uses*, 1973; S. J. Pyne and P. L. Andrews, *Introduction to Wildland Fire*, 2d ed., 1996; C. J. Scifres and W. T. Hamilton, *Prescribed Burning for Brushland Management*, 1983; H. A. Wright and A. W. Bailey, *Fire Ecology*, 1982.

Remote sensing

As the most Earth-like planet in the solar system, Mars is a subject of fascination and intense study. A primary objective of the *Viking* missions to Mars completed in the mid-1970s was to search for evidence of life. The landers detected extraordinary chemical reactivity in the Martian soil, but no organic material. Numerous theories, none universally accepted, explain the *Viking* results, usually invoking an inorganic oxidant in the Martian soil.

It is reasonable that the environment on Mars might be oxidizing; Mars's atmosphere, about 1% the density of Earth's, is composed mostly of carbon dioxide (CO_2), the most oxidized form of carbon. There is no ozone layer to block ultraviolet (UV) radiation, which can generate highly oxidizing species, such as ozone and peroxides. The characteristic red color of the Martian surface is probably due to one or more oxides of iron—relatives of rust—that could catalyze the oxidation of organic matter, particularly with the help of UV light.

The Mars Oxidant Experiment (MOx) was the United States' contribution to the Russian Mars '96 Mission, slated to be the first in 20 years to conduct experiments on the Martian surface. On November 16, 1996, the Russian spacecraft launched successfully but failed to leave Earth orbit, the larger part of it falling into the Pacific Ocean; the landers with their scientific payload are believed to have fallen into the jungles of South America.

Designed and built principally at the Jet Propulsion Laboratories using the micromirror technology invented at Sandia National Laboratories, the MOx

is shown on the Russian Lander in **Fig. 1**. The MOx was to investigate and characterize the chemical nature of the Martian surface and atmosphere, with particular emphasis on any oxidative properties.

Despite failure to reach Mars, the MOx instrument represents an exceptional state-of-the-art development of miniaturized chemical analysis instrumentation using optical sensor technology, and has a wide range of potential future applications. The small size and weight (about 2 lb or 1 kg) of this self-contained instrument, coupled with its capability to record general chemical reactivities of both solid materials and an ambient gas phase in a range of environments, suit it for various applications. Examples include the characterization of difficult-to-access hazardous waste sites to determine the general nature and extent of the pollution; recording the consequences of industrial emissions in a selected geographic region, particularly in terms of likely adverse effects on natural and human-made materials; describing the chemical environment of an underground cavity or other geological feature accessible only by a small well; and the study of both surface and atmospheric chemical characteristics of many of the solar system's planets and their satellites.

The 2-lb (1-kg) instrument includes the optical system and electronics to interrogate an array of 100 micromirror optical-chemical sensors, each comprising a tiny specimen of a reflective thin-film material selected for its potential to corrode, and hence change reflectivity, upon exposure to chemicals present in soil, dust, or atmosphere. For a given application, the pattern formed by the rate and extent of corrosion of the micromirror array can be thought of as the chemical flavor of the environment being sampled, whether it is the surface of another planet or a difficult-to-access waste site on Earth.

Chemically sensitive micromirrors. The MOx system was to rely upon surface chemical reactions to answer questions about the Martian soil and atmosphere, which have undergone eons of change. For a short-lived planetary mission or an assay of the chemistry of some terrestrial environment, the fact that surface and thin-film reactions can occur with high speed, even at low temperatures (the MOx was to make measurements at -70 to $0°C$ or -94 to $32°F$), enables measurements to be made quickly. A reactant requiring 1 million years to diffuse through a specimen 1 cm (0.4 in.) thick, for example, needs just 30 s to diffuse through a film of the same material 10 nanometers thick.

Though lacking the capabilities of full-size analytical instrumentation, chemical microsensors measure surface and interfacial reactions while consuming minimal power and occupying little space. To characterize a relatively unknown chemical environment, an array of microsensors, each bearing a chemically distinct coating, is coupled with mathematical pattern-recognition techniques. The goal is to "smell" or "taste" the environment in much the same way that the human olfactory and gustatory systems characterize the smell or taste of a complex

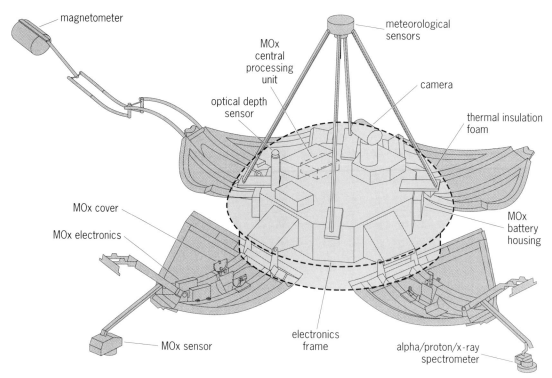

Fig. 1. Sensor head of the Mars Oxidant Experiment (lower left), associated components, and the other instruments carried by the lander of the Russian Mars '96 Mission. The central body of the lander is approximately 2 ft (0.6 m) in diameter.

mixture of chemicals, such as the aroma of coffee or the flavor of red wine. *See* MICROSENSOR.

The MOx instrument utilizes sensors based on chemically sensitive micromirror technology (**Fig. 2**). The end of a single optical fiber is coated with a thin chemically sensitive layer, often a semitransparent metal film 5–15 nm thick. Light passes down the fiber, some reflects from the film-coated tip, and the intensity of reflected light is measured. When a chemical reaction alters the front-surface reflectivity or the optical thickness of the film, the reflected intensity changes. In the laboratory, changes in metal thickness as small as 0.01 nm can be measured. When the reflective film is nonmetallic, for example, a polymer or metal-oxide layer, thickness changes of a few nanometers are necessary for detection; the

films are typically one to a few micrometers thick. An array of these micromirrors, each coated with a different chemically sensitive film, is the heart of the MOx instrument.

Selecting the array of sensing materials was a challenging part of the design of the MOx. Because the *Viking* results were ambiguous, the array of micromirrors was designed to respond to, and provide an identifying response pattern for, any of several postulated oxidants or mixtures. A further constraint was that all materials be deposited as high-quality optical thin films in small areas (less than 1×1 mm or 0.04×0.04 in. for each micromirror) without cross-contaminating one another. The materials selected to address these goals and constraints in the case of the Mars mission are listed in the **table**. The

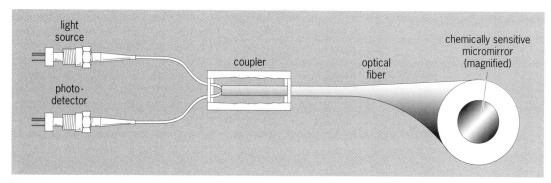

Fig. 2. Fiber-optic micromirror chemical sensor.

Micromirror materials for the MOx array

Coating	Purpose
Metals	
Magnesium	Very high reactivity to oxidants
Aluminum	High reactivity to oxidants; oxide is permeable, allowing intrafilm reaction
Titanium	Moderately high reactivity to oxidants; well studied in the laboratory
Vanadium	Moderate-to-high reactivity to oxidants; rich and variable oxide chemistry
Silver	Low reactivity, but extremely reactive to ozone, oxygen radicals, sulfur compounds
Palladium	Low reactivity, but sensitive to hydrogen, sulfides, unsaturated hydrocarbons
Thin gold	Frost indicator; reactive to sulfur compounds; organic adsorption indicator
Thick gold	Constant-reflectivity reference
Nonmetals	
Hydrocarbon-A	Analog of highly refractory kerogens (organics) found in meteoritic infall
Hydrocarbon-B	Analog of moderately refractory kerogens (organics) found in meteoritic infall
C_{60}	Carbonaceous material sensitive to combination of UV and oxidants
L-Cysteine	To detect enantiomeric preference in reactions with, or catalyzed by, Martian soil
D-Cysteine	To detect enantiomeric preference in reactions with, or catalyzed by, Martian soil
Thymol blue	pH indicator dye: $pK_1 = 2.0$, $pK_2 = 8.8$
Bromphenol blue	pH indicator dye: $pK = 4.0$
Bromcresol purple	pH indicator dye: $pK = 6.3$
Fluorescein	Fluoresces only at neutral or basic pH
Chlorophyllin	Ozone detection via ozonolysis of carbon-carbon double bonds
Iron porphyrin	May bind carbon monoxide (CO) with color change
Copper phthalocyanine	Well-characterized sensor material for oxidants
Lead sulfide	Reacts with hydrogen peroxide, giving large color change
Uncoated	Dust accumulation, surface film buildup, ambient-light-level reference

chemical diversity represented by the materials in the table also make this a good starting point for many anticipated non-Martian applications of MOx technology; nonetheless, replacing some of the films according to the specific application would often be desirable.

The films for the MOx are broadly categorized as metal or nonmetal. The seven metal films, ordered from most to least reactive in the table, offer wide variability in ease of oxidation (corrosion), and are therefore an effective probe of the corrosive potential of a monitored environment. Factors unique to individual species, such as the reactivity of silver to ozone, provide additional information.

The nonmetallic films can serve a number of purposes. Three carbon-based films probe the degradation of carbonaceous materials. (*Viking* failed to detect organic carbon and, at the very least, meteoritic infall on Mars is a source of carbonaceous compounds known as kerogens.) The rate of disappearance of the two kerogenlike films indicates how rapidly carbon-containing materials are oxidized. The form of carbon with the structure C_{60}, one of a class of materials known as fullerenes, is readily oxidized with the assistance of UV light. The role of photochemical oxidation on Mars, as well as some other planets and satellites, remains to be understood. *See* FULLERENE.

Both enantiomers (left- and right-handed versions of the same molecule) of a common amino acid, cysteine, are included to discern preference for the reaction of one over the other. While any enantiomeric preference could indicate metabolism by a living organism—much of life on Earth can metabolize only one of the two enantiomers of such mirror-image molecules—some transition-metal compounds (possibly present in Martian and other extra-

terrestrial soils) can catalyze the preferential reaction of one enantiomer over the other.

A series of three pH-indicating dyes and one fluorescent pH indicator gauge the acidic or basic character of an environment. A member of the chlorophyll family reacts with ozone. Two organometallic compounds may detect carbon monoxide (CO) as well as oxidizing species. Lead sulfide has exceptional sensitivity to hydrogen peroxide, one of the species theorized to account for some *Viking* results.

More than a year was to elapse from the deposition of the micromirror coatings in the laboratory until their deployment, so the films had to be kept clean and unreacted prior to arrival on Mars; keeping reactive materials fresh until deployment is important in many applications. A micromachined silicon nitride membrane about 0.5 μm thick served to seal the set of sensor films under an inert atmosphere, being broken upon deployment by a pinprick mechanism. To correct for the sensitivity of micromirror reflectivity to temperature variations, each sensing micromirror is complemented by an identical, reference micromirror that experiences the same thermal environment but is isolated from the soil and atmosphere by a permanent hermetic seal.

MOx instrument. The MOx instrument was to survive a landing shock some 250 times the Earth's gravity and diurnal temperature variations of nearly 100°C (180°F). Weighing just 850 g (30 oz), the instrument includes two light sources (590- and 870-nm light-emitting diodes), optical fibers, micromachined fixtures, an array of thin-film micromirrors, a linear diode-array photodetector, control-and-measurement electronics, a microprocessor, memory, a computer interface, batteries, and housing. The dual-wavelength feature is especially valuable with some organic coatings (such as the pH indicators) that

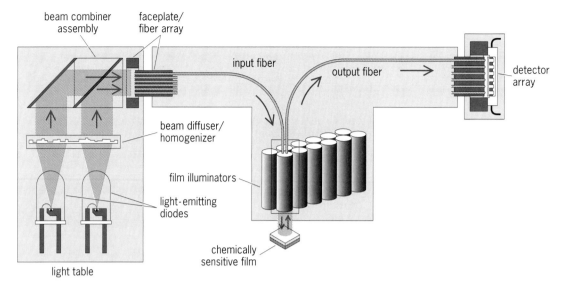

Fig. 3. MOx instrument optical system, housed inside the sensor head shown in Fig. 1.

change reflectivity at only a single wavelength; the unaffected wavelength can be used as a secondary reference.

The optical system (**Fig. 3**) is housed in the sensor head (at the end of the deployment arm in Fig. 1). Optical fibers carry light from the two light-emitting diodes to miniature graded index-of-refraction (GRIN) lenses focused on each of the 100 chemically sensitive micromirrors. Reflected light from the chemically sensitive films passes back through the GRIN lenses to the output optical fibers, which direct the light to the pixels of the linear diode-array photodetector. The system contains two identical sets of the micromirror coatings listed in the table. One set, located on the underside of the sensor head, is for soil measurements; these coatings were to directly contact the Martian soil. The second set of micromirrors, on top of the head, probes the gaseous ambient, for example, the Martian atmosphere. In situations where the instrument is exposed to sunlight, the micromirrors on top of the head respond to the combined stimulus of atmospheric gases and sunlight, while those immersed in the dust or soil receive little solar illumination.

Understanding the response. To fully understand the response pattern from the micromirror arrays, it is necessary to simulate the rates and extents of reaction of each of the coating materials by using a copy of the instrument and a simulated environment created in the laboratory. For the Mars mission, this was to be accomplished by using a unit called a Mars jar (a laboratory-scale Martian soil-plus-atmosphere mock-up). Such parameters as temperature, relative humidity, soil composition, and the mix of atmospheric gases must be tuned until the response pattern from the laboratory and field systems match as closely as possible. Thus, the price of the extraordinarily small size, weight, and power consumption of a sensor-based instrument such as the MOx is

often a more complex procedure to process and understand response data than is typical for full-sized laboratory instrumentation.

To obtain an accurate picture of interfacial chemistry on a remote planet by using a microsensor system weighing just 2 lb (1 kg) is unprecedented, and the success of the MOx was far from guaranteed. While the failure of the MOx to reach Mars is disappointing, the new technology that was developed has many promising future applications that reach from below the surface of the Earth far into outer space.

For background information *see* ACID-BASE INDICATOR; FIBER-OPTIC SENSOR; FULLERENE; KEROGEN; OPTICAL FIBERS in the McGraw-Hill Encyclopedia of Science & Technology.　　　A. J. Ricco; M. A. Butler; F. J. Grunthaner

Bibliography. M. A. Butler and A. J. Ricco, Chemisorption-induced reflectivity changes in optically thin silver films, *Appl. Phys. Lett.*, 53:1471–1473, 1988; F. J. Grunthaner et al., Investigating the surface chemistry of Mars, *Anal. Chem.*, 67:605A–610A, 1995; R. C. Hughes et al., Chemical microsensors, *Science*, 254:74–80, 1991; H. H. Kieffer et al. (eds.), *Mars*, 1992.

Ribonucleic acid (RNA)

The genes, or deoxyribonucleic acid (DNA), in bacterial and eukaryotic cells are transcribed into ribonucleic acid (RNA). Messenger RNA (mRNA) serves as a template for the translation of amino acid building blocks into proteins. This flow of genetic information from the gene through RNA to the protein has been called the central dogma of molecular biology. The discovery of processes that change the informational content of RNA (RNA editing) came as a considerable surprise to molecular biologists.

RNA editing can be mechanistically divided into

RNA editing

Organism	Genome	RNA	Change*
		Insertion/addition/deletion editing	
Kinetoplastids	Mitochondria	RNA	U addition or deletion
Physarum polycephalum	Mitochondria	mRNA	C, A, G, U insertion
		Substitution editing	
Physarum polycephalum	Mitochondria	mRNA	C to U
Acanthamoeba castellani	Mitochondria	tRNA	U to A, U to G, A to G
Spizellomyces punctatus	Mitochondria	tRNA	U to A, U to G, A to G
Vascular plants	Mitochondria	mRNA, rRNA,	C to U
	Chloroplasts	tRNA	U to C
Marsupials	Mitochondria	tRNA	C to U
Mammals	Nucleus	apoB mRNA	C to U
Mammals	Nucleus	Cation channel and serotonin receptor mRNAs	A to I
Mammals	Nucleus	tRNA	C to U, U to C
Human immunodeficiency virus (HIV)	Virus	Trans activation response RNA	A to I
Hepatitis delta virus (HDV)	Virus	mRNA	A to I

* A = adenine, C = cytidine, G = guanine, I = inosine, U = uridine.

processes in which the RNA backbone is cleaved and nucleic acid bases inserted or deleted, and substitution or modification editing in which the nucleic acid bases of the RNA are changed without cleavage of the RNA backbone (see **table**).

Insertion and deletion editing. The term RNA editing was first used in 1986 to describe the correction of defective mRNAs in trypanosomes. Trypanosomes are parasitic flagellate protozoa of the order Kinetoplastida. RNA editing has been found in all kinetoplastid species studied so far. Kinetoplastid protozoa parasitize mammals, birds, reptiles, and fish. RNA editing has not been seen in euglena, a divergent, free-living relative. These protozoa are of special biological interest because they are the most primitive extant eukaryotes (cells with nuclei) that contain mitochondria.

The mitochondrial genome of kinetoplastids is made of catenated maxicircles and minicircles of DNA (large and small circular pieces of DNA connected in a series of links forming a chain). Maxicircle DNA encodes genes, which have limited sequence identity to the mitochondrial genes from other organisms. Some genes contain coding sequence mistakes that would prevent them from being translated into protein, whereas others have been altered so much that they are almost unrecognizable.

Maxicircle transcripts are edited back to sequences found in the mitochondrial genes of other organisms, and produce a normal complement of mitochondrial proteins. RNAs are altered by the deletion of certain genomically encoded uridines (U) and the insertion of other noncoded uridines. The edited region contains up to 40 uridines. This process corrects localized anomalies in reading frames, or can create as much as half of the mRNA sequence, from those genes that had previously been hard to recognize.

An important insight into the mechanism of this RNA editing came from the discovery that some mitochondrial genes encode guide RNA (gRNA). Guide RNAs contain a 4–18-nucleotide anchor sequence that is the opposite of the sequence immediately downstream of the editing site on the unedited transcripts. Guide RNAs hybridize with the preedited RNA but are mismatched at the editing site. At the mismatch between the gRNA and the unedited pre-mRNA, the RNA backbone is cleaved by an endonuclease. Uridines are added (or deleted) directly as free uridine triphosphate (UTP). The corrected RNA is then re-ligated. Further clarification of the mechanistic components of this process can be anticipated from the recent development of RNA editing systems (extracts from cells that perform the editing reaction) in laboratory ware. *See* CANCER (MEDICINE).

The slime mold *Physarum polycephalum* modifies mitochondrial RNA by the insertion of nonencoded cytidine (C) bases and, at a lower frequency, guanine (G) and U residues at many precise sites. In addition, cytidine is substituted with uridine. Thus, *Physarum* displays mixed insertional and substitutional editing. The mechanisms of these editings are unknown.

Substitutional editing of nuclear transcripts. Apolipoprotein B (apoB) mRNA editing was discovered soon after the discovery in trypanosomes. Such editing is quite different from editing in trypanosomes. It is a discrete and highly specific process that occurs in the nucleus. A single cytidine is converted to uridine in a special form of apoB mRNA. This changes a codon for the amino acid glutamine (CAA) to a stop translation codon (UAA). The protein encoded by the edited mRNA is half the length of the protein encoded by the unedited mRNA. The two proteins have different functions. The short form is needed for dietary lipid absorption. The larger form transports fat and cholesterol synthesized in the liver. It is the sole protein in low-density lipoprotein (LDL), which is a major risk factor for coronary heart disease. *See* NUTRACEUTICALS.

apoB mRNA editing. The establishment of a laboratory system for apoB mRNA editing has allowed the catalytic subunit of the enzyme to be identified and the sequences associated with RNA editing site recognition to be characterized. The amino acid sequence and catalytic mechanism of the editing enzyme is similar to that of cytidine deaminase, an enzyme in the bacteria *Escherichia coli*. Both proteins have a similar structure, but significant gaps are present in the region of the active site of the editing enzyme compared to that of cytidine deaminase. These gaps may accommodate the much larger RNA substrate. Although no other targets for this editing deaminase have been identified, it is believed that they exist. The editing deaminase is produced in tissues that do not make apoB mRNA, indicating that other RNA targets are present in these tissues.

Wilms' tumor. The mRNA for the transcription factor (WT1) that confers susceptibility to Wilms' tumor (a malignant renal tumor) undergoes a U to C editing which changes the amino acid proline to leucine. This apparently alters the expression of the gene that is the target of the Wilms' tumor susceptibility protein. This could have a role in embryological development and tumorigenesis (the production of tumors).

RNA editing in plant mitochondria and chloroplasts. C to U editing (and occasionally U to C) of RNA is extensive in the mitochondria and chloroplasts of almost all land plants. This type of editing corrects multiple genomically encoded missense codons that deviate from the universal genetic code, thereby allowing the corrected RNAs to serve as templates for the synthesis of functional proteins.

No common denominator for editing site recognition has been identified, and preliminary reports of gRNAs have yet to be confirmed. However, in certain edited RNAs, transfer RNAs (tRNA), and unspliced introns, editing is found only in base-pairing stems, where it aligns base pairing between opposite strands. These editing sites are identified by mismatches with the other base-pairing strand. This is reminiscent of the gRNA-mediated editing site recognition in kinetoplastids. Interestingly, editing can occur in translation-defective organelles (organelles not able to produce functional proteins), so that the proteins required for editing must be imported from the nucleus. The catalytic mechanism of C to U conversion in plants is, like apoB mRNA editing, a cytidine deamination.

Other forms of RNA editing. Other forms of RNA editing occur in mitochondria. In *Acanthamoeba castellani* and the related fungus *Spizellomyces punctatus*, tRNA undergoes single-nucleotide conversions (U to A, U to G, A to G), which corrects mismatched base pairs to those found in normal tRNA. It is believed that this type of editing involves base-pair exchanges rather than modification.

Adenosine to inosine editing of RNA in nucleus. In the brain, the mRNAs encoding certain L-glutamate-activated cation channels, the receptors that mediate the majority of fast excitatory neurotransmission, have been found to be edited. These receptors are generated from four independently encoded subunits. The mRNA for one of these subunits, Glu RB, is edited from a glutamine codon (CAG) to an arginine codon (CGG). This editing markedly reduces the calcium permeability of the channel. It is a remarkably efficient process that causes most receptors to have low permeability. This same site is edited in the related family of kainate receptors.

A second type of mRNA editing affects the kinetic properties of the receptors of certain L-glutamate-activated cation channels. The mRNAs encoding the Glu RB, Glu RC, and Glu RD subunits also undergo an arginine (AGA) to glycine (GGA) substitution. The Glu RA mRNA is not edited. As a result, the edited channels recover much faster from desensitization and integrate incoming signals better. The mRNAs encoding certain potassium channels and serotonin receptors have also been found to undergo the same type of RNA editing.

This editing is biochemically a hydrolytic deamination (replacement of ammonia with water) of the base adenosine (A) to inosine (I). I is read as the normal coding base G by the ribosome (protein translation apparatus). The recognition sequence for this form of RNA editing consists of a duplex formed by an intronic sequence with exact complementarity to the exon centered on the unedited codon. The enzymes that perform these reactions are a family of double-stranded RNA adenosine deaminases. Specific editing enzymes apparently exist for each editing site. Edited transcripts have been found in other tissues, including the heart; and squid axon potassium channel mRNAs also undergo editing. Thus this activity is widespread in nature. Interestingly, the adenosine and cytidine mRNA editing deaminase have similar active site mechanisms and appear to have evolved from a common ancestor.

Evolution of RNA editing. It is evident that RNA editing is not a single process, but a series of distinct traits that originated separately and are distributed widely in the animal and plant kingdoms and in viruses. An important understanding about the biochemistry and evolution of different forms of RNA editing has been gained through the development of systems for editing RNA in the laboratory. From these systems, it has been established that all forms of RNA editing are mediated by protein enzymes and that RNA mediators, such as those that might have had their origins in an ancient RNA world (considered to have been the origin of life), are not involved.

The close similarity between the apoB mRNA editing enzyme and *E. coli* cytidine deaminase is of interest to molecular biologists as it suggests the possibility that C to U editing was established early in bacteria, and passed forward into eukaryotes. Plant organelles originated in bacteria. In the process of bacteria to organelle transformation, a massive transfer of bacterial genes to the nucleus occurred. The organelles now require the reciprocal transfer of proteins back to the organelles. Therefore it may be possible that the plant and apoB-type C to U editing had the same bacterial origin.

At first sight, the A to I editing deaminases would appear to be unrelated to C to U deaminases, but the double-stranded deaminases and the C to U editing enzyme of apoB share a conserved catalytic motif, similar to that in the *E. coli* cytidine deaminases. Evolution has engrafted on this motif the different substrate specificities and RNA binding characteristics of these two distinct forms of editing enzymes.

It is of interest to consider the origins of RNA editing, why the editing in organelles is more prevalent and why it is less so in the nucleus. The coincidence of RNA binding with a catalytic activity was presumably a necessary first step in the origin of editing. Fixation of this early editing by natural selection, followed by the creation of new editing sites by mutation and the conservation of these sites by natural selection, provided a route for the evolution of this type of editing. In the high-complexity nuclear genome, the spread of editing has presumably been limited; it possibly protects against too much deleterious change.

For background information *see* BACTERIA; BACTERIAL GENETICS; DEOXYRIBONUCLEIC ACID (DNA); EUKARYOTE; GENE; MOLECULAR BIOLOGY; PROTOZOA; RIBONUCLEIC ACID (RNA) in the McGraw-Hill Encyclopedia of Science & Technology.   James Scott

Bibliography. J. Scott, A place in the world for RNA editing, *Cell*, 81:833–836, 1995; L. Simpson and O. H. Thiemann, Sense from nonsense: RNA editing in mitochondria of kinetoplastid protozoa and slime molds, *Cell*, 81:837–840, 1995.

River

The Colorado River is considered to be one of the most regulated networks of natural, interacting water channels in the world. Beginning with the Bureau of Reclamation's high-dam-building period in the 1910s, 19 dams were erected along the Colorado River's main channel and tributary waters.

Glen Canyon Dam in Page, Arizona, has been the most controversial of the Colorado River high dams; it impounds Lake Powell and controls water flow through Grand Canyon National Park (**Fig. 1**). The closure and operation of Glen Canyon Dam drastically altered the hydrology, geomorphology, zoology, and botany along the 283-mi (471-km) reach of the Colorado River in Glen, Marble, and Grand canyons. All ecosystems downstream from the dam have been influenced by the release of cold (approximately 48°F or 9°C), sediment-free water throughout the year, with the water cooling slightly in its downstream trek in the winter and warming to approximately 64°F (18°C) during late summer.

Glen Canyon Dam. Glen Canyon Dam was completed in 1963 as a multipurpose facility. The Colorado River Storage Project Act of 1956 justified existence of the dam—for regulating flow of the Colorado River, water storage, reclamation of arid and semiarid lands, flood control, and hydroelectric power generation. No Environmental Impact Statement was filed regarding erection and operation of Glen Canyon Dam since construction took place before enactment of the National Environment Policy Act and emergence of Impact Statement requirements. Large daily water fluctuations from Glen Canyon Dam concerned federal, state, and tribal resource management agencies and additional agencies associated with fishing and rafting interests, and environmental groups attentive to detrimental effects on downstream culture, vegetation, wildlife, and other river-related resources.

Since completion of the dam, cold, sediment-free water has flowed through Grand Canyon, eliminating or endangering several native fish and wildlife species and allowing nonnative vegetation to spread. Beach erosion increased because of minimal sediment influx. Concern of Native Americans grew as numerous archeological ruins along the river's edge were exposed.

In response to these problems, the Bureau of Reclamation initiated the multiagency Glen Canyon Environmental Studies in 1982 to conduct field research and integrate the interests of agencies and individuals concerned with the condition of Grand Canyon National Park. Phase I of these studies was completed in 1988, and Phase II was begun in 1989 to collect more detailed data. Interim low, fluctuating flows were initiated in 1991 to reduce downstream impacts while the Environmental Impact Statement process was completed.

The Bureau of Reclamation was then instructed to prepare an Environmental Impact Statement to reevaluate operations at Glen Canyon Dam. The statement was necessary to determine options that could be implemented to minimize negative impacts—biological, cultural, and ecological—caused by dam releases, on the downstream canyon ecosystem. Cooperating agencies for preparation of the statement were the Arizona Game and Fish Department, Bureau of Indian Affairs, Bureau of Reclamation, Hopi Tribe, Hualapai Tribe, National Park Service, Navajo Nation, Pueblo of Zuni, San Juan Southern Paiute Tribe, Southern Utah Paiute Consortium, U.S. Fish and Wildlife Service, and Western Area Power Administration.

The final Environmental Impact Statement was issued in March 1995, resulting in selection of the preferred alternative—a modified low, fluctuating flow. This approach reduced daily flow fluctuations to protect downstream resources while allowing manageable power operations from Glen Canyon Dam.

Test flood. In addition to modified low flow levels, the preferred alternative of the Environmental Impact Statement included a periodic beach and habitat

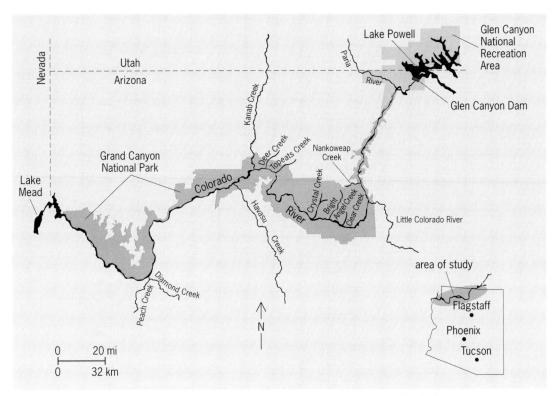

Fig. 1. Colorado River through Grand Canyon National Park. *(Glen Canyon Environmental Studies)*

maintenance flow to rejuvenate backwaters and rebuild beaches. Because of the presence of Glen Canyon Dam, flow regulation reduced annual flooding which decreased sand bar size, allowed vegetation to grow closer to the main channel, promoted expansion of debris fans, and filled in backwater areas which housed native fish species. With the hope of restoring the Grand Canyon ecosystem, the longest unimpeded riparian wildlife habitat in the West, the U.S. Department of the Interior initiated a test flood through Grand Canyon to focus on recreating predam flows, restoring natural beaches and wildlife habitat, and analyzing the effects of the controlled flood on flow, sediment transport, and water chemistry. The flood was approved, provided it would not impact three protected or endangered species: the humpback chub, the Kanab ambersnail, and the Southwestern willow flycatcher. Fish and Wildlife Service scientists implanted computer microchips into many humpback chub to trace their activity during the flood. Nearly 200 Kanab ambersnails were marked with a dot of nail polish and relocated on high ground before the test flow began. Rare and endangered birds' nests were located by orbiting satellites to monitor any impacts that changing flow conditions would impart.

The test flood was begun on March 22, 1996. A 4-day steady release of 8000 ft^3/s (226.5 m^3/s) was followed by an increase of 4000 ft^3/s (113 m^3/s) until a maximum flow of 45,000 ft^3/s (1274.3 m^3/s) was established, and this flow was held until April 2, 1996, when downramping (decreasing water releases) to 8000 ft^3/s (226.5 m^3/s) was initiated (**Fig. 2**). The flood was designed to contain enough water to lift sand that had accumulated along the river bed and to deposit it on the adjacent banks, while also depositing nutrients, restoring backwater areas, preventing return channels from becoming overgrown with vegetation, and restoring the natural dynamics of the Colorado River system. The flood began by opening peripheral river outlets (jet tubes) that channeled water around the dam's

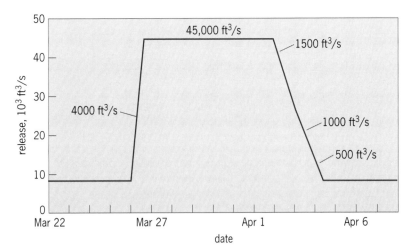

Fig. 2. Hydrograph for the March 22–April 8, 1996, beach- and habitat-building test flow. Increases and decreases in flow are measured in rates per hour. 1 ft^3/s = 0.0283 m^3/s. *(Glen Canyon Environmental Studies)*

power turbines and into the main channel of the Colorado River. The project incorporated studies performed by approximately 150 scientists who monitored the flood's actions, photographed the area, and surveyed the conditions that existed before the flood in order to compare them with changes created by the flood.

The U.S. Geological Survey monitored the controlled flood by using satellite telemetry to produce real-time streamflow data on the Internet. Geological Survey scientists injected 2200 lb (998 kg) of the nontoxic red dye Rhodamine WT into the Colorado River on March 27, 1996, to measure water velocity during the flood. Radio transmitters were implanted in holes drilled in rocks of the river bed to track rock movement during the flood.

Cross sections of the river were measured with a fathometer (depth-sounding device) to document the rate of change in sand storage across the river channel. Sidescan sonar was used to record the movement of sand waves and ripples on sandbar surfaces.

Results. Preliminary results indicate that the goals of the beach- and habitat-building flow were successful. Backwater habitats were scoured, allowing for increased breeding areas for many native fish species. Initially, approximately one-third of the additional sandbars and beaches were formed throughout the canyon. The flood deposited sediment from the river channel onto many eroded beaches along the river banks, and scoured vegetation and sediment from backwater habitats. According to the Glen Canyon Environmental Studies, 80% of the new beaches were created within the first 40 h of the artificial flooding. The indication is that future flooding events could possibly be shortened.

Pre- and postflood views of changing beach conditions at Kwagunt Marsh, located 55.5 river miles (89 km) downstream from Lee's Ferry, indicated that nearly 3.3 ft (1 m) of new sand was deposited on a sandbar portion while the mouth of the return channel grew larger because of scouring of an adjacent submerged sandbar by the flooding. At some downriver locations, 8–12 ft (2.4–3.6 m) of new sand were deposited along beaches, providing new living space for the 22,000 people who use the river annually.

Organic matter lifted from the river bed became incorporated into the reformed backwaters and marshes, providing food for the aquatic species. Trout, a nonnative fish of the cold river, appeared to be feeding on the nutrients incorporated into the main water flow. There was growth of the backwater habitats (particularly near the Little Colorado River), breeding grounds for the endangered humpback chub.

Further evaluation. Geographic Information Systems are being used by Bureau of Reclamation workers to map and overlay 40 different resource layers within the canyon, including topography, water quality, spawning areas and nesting sites, sediment deposits, climate variables, and archeological sites

for endangered fauna. Geographic Information Systems allow scientists to spatially analyze the influence of one canyon variable on another, and are the basis for a long-term monitoring program of 17 reaches of the canyon.

Ongoing monitoring of beaches and backwater habitats will disclose positive and negative effects of habitat maintenance test floods. Both water and wind have begun to influence the size and shape of many of the elevated beaches resulting from the test flow. However, more time will be required before the true impact of the flood can be quantified. The cooperation of diverse interest groups, such as environmentalists, Indian tribal units, and producers of hydroelectric power, has promoted scientific progress. Results of these studies will assist in providing future direction for other highly regulated river systems.

For background information *see* DAM; RIVER; RIVER ENGINEERING; WATER POWER in the McGraw-Hill Encyclopedia of Science & Technology.

Diane Stanitski-Martin

Bibliography. T. Adler, Healing waters: Flooding rivers to repent for the damage done by dams, *Sci. News*, 150(12):188–189, 1996; S. W. Carothers and B. T. Brown, *The Colorado River Through Grand Canyon: Natural History and Human Change*, 1991; Committee to Review the Glen Canyon Environmental Studies, Water Science and Techology Board, Commission on Geosciences, Environment, and Resources, *Colorado River Ecology and Dam Management*, 1991; D. Harper and A. Ferguson (eds.), *The Ecological Basis for River Management*, 1995; E. J. Hickin, *River Geomorphology*, 1995; U.S. Department of the Interior, Bureau of Reclamation, *Operation of Glen Canyon Dam Environmental Impact Statement*, 1995.

Saponins

Scientists are carefully scrutinizing saponins, which exert a host of biological effects on humans, plants, and animals. The diversity of structures; the challenges of isolation; the pharmacological, biological, and agricultural activities already found and those still to be discovered; and the promise of commercialization are driving the study.

Properties. Saponins are a class of natural products that are constructed of aglycones (triterpene or steroidal) and sugars (pentose, hexose, and uronic acid). An appropriate hydrolysis of saponins yields sugars and an aglycone—not necessarily the genuine aglycone. Saponins are biological detergents because of glycosylation of the hydrophobic aglycone and, when agitated in water, form a soapy lather. This unique ability to cause foaming has been used throughout the centuries for making cleaning solutions; extracts of soapberry and chinaberry serve as natural soaps. Indeed, saponins can act as an aid in identification of plant extracts. Triterpenoid, steroid, and steroid alkaloid glycosides are widely spread throughout the plant kingdom, and several have been found in marine animals. Some saponins have cardiac activity, hemolytic activity, activity as fish poisons, cholesterol-reducing ability, bitterness, ability to act as sweeteners, and cosmetics activity. They can also serve as allelochemicals (for plant defense and growth). Such behaviors are found in certain saponins rather than in all members of this chemical family. From the biological viewpoint, saponins have a diverse group of properties, some deleterious but many beneficial. Their use in plant drugs, folk medicines, and modern medicines has generated great interest in the chemical characterization of these molecules. In the East (particularly Japan), the literature on the isolation, purification, separation, structural elucidation, and biological activity of saponins attests to the skill of natural-product biochemists and chemists. It seems that the advantage of saponins to the producing plants is that they function as protecting agents, growth regulators, and allelochemicals.

Occurrences. Saponins are widely distributed in dicotyledonous plant species and also in monocots. They occur in foods (for example, beans, peanuts, oats, green peppers, garden peas, asparagus, garlic, onions, spinach, tomatoes, and potatoes) and animal feed (alfalfa and clover) as well as in some marine organisms. Saponins are also present in numerous herbal remedies, for example, ginseng, quilla, horse chestnut, and beans of various types. Saponin content depends upon many factors such as the cultivar, age, physiological state, and geographical location of the plant. When grown in different places, the same plant may vary in the composition and quantity of saponins.

Saponins are localized in organelles that have a high turnover rate, which implies that they not only are metabolically active but may be important regulatory substances in the development of an organism. The biosynthesis of saponins has been the subject of limited research activity, and mostly it is based on the classical biosynthetic pathways involving squalene-2,3-epoxide, which is cyclized to give cholesterol and the aglycones of saponins. Enzymes catalyzing the glycosylation of the saponins have been

isolated, but much remains to be done before an understanding of the biosynthesis and biodegradation (together they are called metabolism) are intimately known and understood.

Characterization. The nomenclature and the stereochemistry of saponins, which is difficult to determine, has led to the use of common (trivial) names, such as soyasaponin I [structure (**1**); 3-O-[α-L-rhamnopyranosyl-(1→2)-β-D-galactopyranosyl-(1→2)-β-D-glucuronopyranosyl]soyasapogenol B]. Detection, quantitative determination of saponins, and isolation through extraction and purification followed by chromatography of many types provide some of the methods; however, a combination of these steps is best for the very complicated saponin mixtures, such as ginseng, quillaja, gypsophila, and alfalfa. Structure determination for both the hydrolyzed and the nonhydrolyzed (naturally occurring according to the procedure used) saponins may include artifact formation, enzymatic hydrolysis, microbial hydrolysis, mild and selective cleavage methods, and nuclear magnetic resonance (NMR) and mass spectrometry as well as other techniques for structure elucidation (permethylation of sugars, infrared spectrometry, molecular optical rotation).

Applications. Of particular interest is use against the vector of the disease schistosmiasis, which is endemic in 76 countries and affects over 200 million people, mainly in tropical and subtropical regions. Other biological effects are cardiovascular, anti-inflammatory, antiexudative, immunomodulation, antiulcer, fungicidal, hemolytic, bitterness, sweetness, cholesterol-reducing ability in humans and animals, and effects on capillary fragility. Spiroconnazole possesses strong molluscicidal activity against all snail vectors. Toxicity is important in the use of saponins, because they are present in the diet (beans, peas, lentils, and so on); however, saponins are poorly absorbed by animals, and they are either excreted unchanged or metabolized in the gut. Consequently it is difficult to ascertain the pharmacokinetics of these molecules. The pharmacology and biological testing of the steroid saponins and steroid alkaloid saponins—including piscicidal activity (fish poison), insecticidal and antifeedant activity, plant growth inhibition (allelopathy), and cytotoxic and antitumor activity—has been reported. Some important commercial preparations are products of sarsaparilla root, licorice, aescine from horse chestnut (*Hederae folium*, which provides an expectorant), *Quillaja* bark, *Gypsophila* species, and ginseng. Some claims for the effectiveness of ginseng may be justified, particularly because the ginseng saponins act as adaptogens (stimulating the nonspecific resistance of an organism and building up general vitality).

A promising discovery comes from *Ornithogalum saundersiae*, a member of the lily family with no folkloric background of medicinal properties. From the bulb of this plant, Japanese workers have isolated a saponin tentatively named OSW-1 [structure (**2**)]. The compound has little toxicity to normal human

(1)

(2)

cells, but is remarkably toxic to malignant tumor cells. Laboratory assays show it to be up to 100 times as potent as anticancer agents now in clinical use, such as mitomycin C, adriamycin, cisplatin, camptothecin, and paclitaxel.

Certain saponins (for example, from alfalfa, soyabeans, mungbeans, and horse chestnut) can form a complex with cholesterol that inhibits absorption of cholesterol from the intestinal lumen. Other saponins may form large mixed micelles with bile acids, which also inhibit absorption from the small intestine leading to increased fecal excretion of the bile acids. Bile acids thus lost (and diverted from the enterohepatic cycle) are replaced by hepatic synthesis of cholesterol. Saponins may interact with cholesterol or bile acids or both, and the details depend largely upon their molecular structure. In either case, the net result is a decrease in the concentration of cholesterol in the blood plasma. Among the plant foods that contain significant levels of saponins, alfalfa saponins have been studied extensively, and their hypocholesterolemic activity is well documented. Alfalfa seeds have been shown to have as strong cholesterol-lowering activity in humans. The accumulated evidence shows that saponins from dietary sources confer protection from experimentally induced hypercholesterolemia (for example, with the inclusion of cholesterol and sodium cholate in animal diets). There are few estimates of the quantities of saponins in the diets of humans, although some estimates of saponins in the daily diet of 7193 households in London have been obtained. The mean daily intake of saponins was only 14.6 mg (0.515 oz); but in vegetarian European and Asian households it was 100–200 mg (3.5–7.0 oz). These latter levels are comparable with the daily intakes that were experimentally effective in reducing plasma cholesterol concentrations. Vegetarians are at least risk from cardiovascular disease.

Yucca schideraga, a plant native to North America, is known for its medicinal use by Native Americans and is recognized by the U.S. Food and Drug Administration as safe for human dietary use. In Japan, it is used as an additive to cosmetics, having the ability to cure infection by dermatophytic fungi and yeast. The extract of the rhizomes is utilized as a long-lasting foaming agent in carbonated beverages and as a flavor enhancer in foods. The saponins of

Quillaja bark and licorice root are utilized the world over. *Quillaja* saponin is used in large amount in photosensitized film as a surfactant; it is also used in shampoos, liquid detergents, toothpastes, and beverages as an emulsifier.

Abrusosides from leaves of *Abrus precatorius* (rosary pea) are saponins that are about 30–100 times as sweet as sucrose (sugar); a derivative, abrusoside E 6-methyl ester [structure (3)], is 150 times as sweet

(3)

as sugar. [In the natural product abrusoside E, the tinted group in (3) is a carboxyl group.]

Saponins may be used in commerce as pure compounds or mixtures. Some steroidal glycosides serve as starting materials for steroids used in oral contraceptives, and triterpene glycosides are used in fire extinguishers and as surfactants in soaps. Also, many products of major United States herbal manufacturers include saponin-containing plants.

For background information *see* ALLELOPATHY; BIOCHEMISTRY; PLANT GROWTH; STEROID; TERPENE in the McGraw-Hill Encyclopedia of Science & Technology.　　　　　　　　　　George R. Waller

Bibliography. K. Hostettman and A. Marston, *Saponins: Chemistry and Pharmacology of Natural Products*, 1995; D. Oakenfull and G. S. Sidhu, Could saponins be a useful treatment for hypercholesterolaemia?, *Eur. J. Nutrit.*, 44:79–88, 1990; G. R. Waller and K. Yamasaki, *Saponins Used in Food and Agriculture*, 1996; G. R. Waller and K. Yamasaki (eds.), *Saponins Used in Traditional and Modern Medicine*, 1996.

Satellite (spacecraft)

The *Infrared Space Observatory* (*ISO*) is an astronomical satellite launched in November 1995. Its infrared observations, which address virtually all fields of astronomy, represent a great advance over the *Infrared Astronomical Satellite* (*IRAS*).

Infrared astronomy. The wavelengths of infrared emission extend from 1 micrometer to several hundred micrometers (the far-infrared). All matter emits radiation, with a wavelength distribution that depends on the temperature: At 6000 K (10,400°F), the Sun emits mostly in the optical band. The particles of solid material that are mixed with tenuous gas in the space between the stars, the interstellar dust, are heated 10 to 50 K (−442 to −370°F), owing to absorption of stellar radiation; their emission peaks in the far-infrared.

Interstellar dust is made of condensed material processed in stellar nuclear furnaces. There is a continuous recycling of dust: New stars are born inside a cocoon of dust, recycle their matter through nuclear reactions, and expel it at the end of their lives. Dust is at the heart of the chemical evolution of the universe, including planets and very likely life itself.

The first infrared receivers were bolometers. Today, astronomers use far more sensitive solid-state detectors. However, there are two formidable obstacles to ground-based infrared astronomy: the Earth's atmosphere is opaque throughout much of the infrared band; and telescopes and the atmosphere are strong infrared emitters themselves, so that the faint signal from celestial sources is drowned in this background light. In space, the full infrared range is accessible. Furthermore, in the absence of radiation from the atmosphere it is possible to cool telescopes and detectors down to very low temperatures, of just a few degrees Kelvin, so that they are not themselves infrared emitters.

IRAS. The *Infrared Astronomical Satellite*, the first cryogenic space observatory, was a liquid-helium-cooled spacecraft. Launched in 1983, it mapped almost the whole sky in four bands, at 12, 25, 60, and 100 μm, during its lifetime of 10 months. *IRAS* discovered a new class of galaxies showing an excess of infrared emission, some 100 times more luminous in the infrared than in visible light. Subsequent observations at other wavelengths showed that these galaxies most often are involved in collisions, which trigger intense bursts of star formation.

Almost a third of the stars in the vicinity of the solar system measured by *IRAS* show an excess infrared luminosity as well. This radiation may be emitted by dust that is left over from the star's formation and that may be associated with planets around the star.

As expected, *IRAS* mapped interstellar dust clouds in the 60- and 100-μm bands; surprisingly, these cold clouds turned out to glow strongly in the 12-μm band as well. Although it was suspected that much of the 12-μm emission was due to spectral lines originating in hydrocarbons, *IRAS* lacked the spectral resolution to check this hypothesis.

Infrared Space Observatory. As the first results of *IRAS* were presented, the decision was made to fly a second-generation infrared cryogenic satellite, the *Infrared Space Observatory*. While *IRAS* had scanned the whole sky, *ISO* was designed to do detailed studies of selected regions, with better angular resolution, wider wavelength coverage, enhanced imaging and spectroscopic capabilities, and higher sensitivity.

Design. *ISO*'s optical design is a Ritchey-Chrétien telescope, of focal length 9 m (30 ft) and aperture 0.60 m (24 in.). Both the primary and secondary mirrors are made of fused silica, and their reflecting surfaces are gold plated. The primary mirror was ground to an accuracy of 0.1 μm. The mirrors and the scientific instruments are placed inside a large dewar bottle filled with liquid helium. Their temperature is kept at 1.8 to 3 K (−456 to −454°F) by the slow evaporation of liquid helium. The total amount of helium on board was 2140 liters (565 gallons) at launch; it was expected to last at least 24 months.

The service module, at a temperature of 290 K (62°F), contains the attitude-control system, data-handling systems, and radio equipment. It is attached to the payload module by glass-fiber struts to minimize heat transfer from the much colder telescope and detector environment.

Stable pointing capability of up to 10 h is achieved by the attitude system with data feedback from a star tracker. A guide star is acquired before the start of any *ISO* observation, and it remains in the field of view of the star tracker throughout the length of exposure. The pointing performance well exceeds specifications: the short-time jitter is better than 0.5 arc-second, the absolute pointing drift is better than 0.1 arc-second per hour, and the absolute pointing error is on the order of 4–5 arc-seconds.

The 600 watts of electrical power needed are provided by solar panels that always face the Sun. *ISO* itself never faces the Earth, the Sun, or the Moon, and spends little time looking at the giant planets, in order to spare its helium cargo and sensitive detectors. Its overall length is 5.3 m (17.4 ft) and its diameter is 2.5 m (8.2 ft); it weighed 2500 kg (5500 lb) at launch (**Fig. 1**).

Instruments. At the focal plane of *ISO*, a four-faced mirror feeds light to the four scientific instruments.

1. ISOCAM is a camera and polarimeter. Two 32 × 32-pixel array detectors together cover the spectral range 2.5–18 μm, allowing images to be obtained at four angular resolutions, ranging from 1.5 to 12 arc-seconds per pixel. Imaging spectrophotometry and spectroscopy are possible, using a set of discrete filters and a set of continuous variable filters.

2. ISOPHOT is a photopolarimeter that operates at various colors in the 2.5–240-μm range, and with a choice of apertures. It also contains a 3 × 3-pixel detector at 100 μm, a 2 × 2-pixel detector at 200 μm, and a low-resolution spectrometer.

3. The Short-Wavelength Spectrometer (SWS) operates in the wavelength range 2.5–45 μm, with a spectral resolution of 1000. Greater resolutions are possible with a Fabry-Perot interferometer.

4. The Long-Wavelength Spectrometer (LWS) operates in the wavelength range 45–180 μm, with a spectral resolution of 200. A Fabry-Perot subsystem is also available.

Launch and verification. *ISO* was launched by an Ariane L44 rocket from Kourou, French Guyana, on November 17, 1995. After several orbital maneuvers under its own power, *ISO* attained its final 24-h orbit, with

Fig. 1. *Infrared Space Observatory (ISO)* satellite at the test facility of the European Space Agency (ESA), before shipment to Kourou, French Guyana, for launch. Clearly visible are the solar panels, the payload module at the bottom, and its connecting rods to the service module. The two redundant star trackers are attached to the payload module, on the right side of the spacecraft. (*ESA/ISO*)

perigee at 1000 km (620 mi) and apogee at 70,000 km (43,000 mi), on November 24. *ISO*'s first observation was of the Whirlpool galaxy, seen by ISOCAM on November 28. Regions of star formation along the galaxy arms were immediately seen in the real-time monitor. As had been expected, ISOCAM is a thousandfold more sensitive and has 30 times sharper eyesight than *IRAS* at 12 μm. Satellite and verification phases lasted until February 3, 1996, when the science observations began.

Results. Virtually all fields of astronomy (planetary, stellar, interstellar, and extragalactic) can be addressed by *ISO*. The spectrometers have opened up previously unexplored spectral regions, very rich in line and band emission from a variety of ions and molecules in gas and solids, creating a small revolution in the field of interstellar chemistry of gas and dust.

Most interesting are the results on commonplace molecules that are ubiquitous in the interstellar medium but whose infrared emissions had not been detected previously. Hydrogen, the most abundant molecule, has such a faint emission at 28 μm, originating from its lowest rotational transition; the Short-Wavelength Spectrometer is the first instrument capable of detecting it. This and other transitions, seen in various galactic sources and in active star-forming regions in external galaxies, are an important tool toward determining the molecular hydrogen con-

tent of galaxies, the efficiency of star formation, and so forth.

Water is another molecule that can be detected by the *ISO* spectrometers. The thermal emission of water vapor from astronomical bodies is impossible to observe from the ground, or even balloons, because of its strong absorption in the Earth's atmosphere, but on *ISO* it is easily seen. *ISO* observes water in regions excited by shocks induced by stellar winds of evolved stars or in star-forming regions. The emission of water vapor plays an important role in the cooling of clouds which are undergoing gravitational collapse to form stars. For instance, the Long-Wavelength Spectrometer finds that a Herbig-Haro object, HH54, associated with the star-forming dark cloud Chameleon II, at a distance of 750 light-years, radiates in rotational lines of carbon monoxide (CO), the hydroxyl molecule (OH), and molecular hydrogen (H_2) essentially all the energy deposited by the weak shocks (with velocities of 10 km/s or 6 mi/s) in its surroundings.

An interesting Short-Wavelength Spectrometer detection is that of solid carbon dioxide (CO_2) in molecular clouds. At 4.27 μm, the observation is also impossible from the ground. Thus, *ISO* may at least make it possible to gauge the distributions of hydrogen, carbon, and oxygen in the gaseous and solid phases of the interstellar medium.

ISOCAM has obtained spectacular pictures of the interior of a nearby (distance of 500 light-years) dark cloud, opaque in visible light (**Fig. 2**). The ISOCAM image shows small-scale filaments near the rim of the cloud, illuminated by bright young stars in its vicinity. Dimmer stars just born inside the cloud, as

Fig. 2. ISOCAM image of the Earth's nearest star nursery, Ophiuchi, a dense cloud of gas and dust at a distance of 500 light-years. The image covers half a square degree and is a mosaic of 1500 pictures. Several stars recently born in the cloud are visible. The brightest blobs (middle right and lower right) are stars that have not yet shed their dust cocoons. (*CEA/IAS/ESA*)

Fig. 3. Whirlpool galaxy, at 2×10^7 light-years, imaged by ISOCAM at a wavelength of 15 μm. Bright spots in the spiral arms correspond to regions where star formation is proceeding on a large scale. A companion galaxy at the top of the image looks smaller than it does in visible light, because it experienced in the past a strong starburst very close to its nuclear regions. (*ESA/ISO, CEA Saclay, ISOCAM Consortium*)

well as dark patches of matter where stars are not yet formed, are also visible.

ISO has solved the mystery of the infrared glow of dark and diffuse clouds around 12 μm. ISOCAM and ISOPHOT, with their high sensitivity to diffuse objects, have obtained low-resolution spectra of these clouds and found the unmistakable features known to be the signature of hydrocarbons.

ISOCAM has produced beautiful maps of external galaxies, delineating regions of star formation better than at any other wavelength range (**Fig. 3**). These results are complemented by ISOPHOT, which has much lower angular resolution but reaches longer wavelengths, thus conveying information relevant to where the bulk of the energy is radiated.

Another example of the complementarity of the *ISO* payload is the study of the pair of colliding galaxies, NGC 4038 and 4039, known as the Antennae, 6×10^7 light-years away. In the region where the disks of the two galaxies crash into each other, the violence of the collision triggers a burst of star formation, whose compactness is revealed by ISO-CAM, and whose spectral characteristics and energetics emerge from the combined coverage of ISOCAM, ISOPHOT, and the Long-Wavelength Spectrometer.

For background information *see* HEAT RADIATION; HERBIG-HARO OBJECTS; INFRARED ASTRONOMY; SATELLITE ASTRONOMY; STARBURST GALAXY in the McGraw-Hill Encyclopedia of Science & Technology.

Diego A. Cesarsky; Catherine J. Cesarsky

Bibliography. Special issue on ISO results, *Astron. Astrophys. Lett.*, 315(2):L27–L400, 1996.

Satellite launch vehicle

Two major advances in satellite launch vehicles are the Ariane 5, the only new major rocket built in the 1990s, and the Sea Launch, the first major rocket to be launched from a semitransportable floating platform at sea.

Ariane 5

Ariane 5 is the newest of the Ariane family of expendable launch vehicles developed by the European Space Agency and operated commercially by Arianespace. It is the first new expendable launch vehicle to be designed since the mid-1980s.

Its design represents a radical departure from the more traditional Ariane 1–4 family of expendable launch vehicles, which have flown nearly 100 times since the first Ariane launch in 1979. Driven by the twin demands of higher reliability and lower cost, Ariane 5 is a vastly less complicated system than its relatives. Whereas the growth in payload lift capability by a factor of 2.5 from Ariane 1 to Ariane 4 was achieved by standard techniques such as lengthening the propellant tanks to enable the loading of more propellant and adding strap-on boosters, Ariane 5 began as a completely new design.

Payload lift capability. While reduced cost and complexity were major design goals of the Ariane 5 program, a major increase in payload lift capability was also a fundamental requirement. The steady upward trend in size and weight of powerful communications satellites was at the root of the decision to design Ariane 5 with a baseline capability of injecting two large satellites with a lift-off mass of 6600 lb (3000 kg) into the standard geostationary transfer orbit (GTO) used to deploy nearly all commercial communications satellites. This resulted in the configuration shown in **Fig. 1**, with a dual launch structure called SPELTRA, which houses both the lower satellite and the upper satellite, enclosed in the nose fairing. For applications demanding larger volume and mass, single payloads up to 15,000 lb (6800 kg) can be placed in geostationary transfer orbit. Low-Earth-orbit (LEO) missions can accommodate payloads up to more than 40,000 lb (18,000 kg), depending on orbital altitude. These performance increases are expected to be achieved at an operating cost per flight at or below that of the Ariane 4, after the learning process of the early flights and after higher production rates are in place. With a performance growth of 40–45% above Ariane 4, the cost per pound to orbit will be reduced by 30% or more.

Rocket engine reduction. To lower cost and reduce complexity, a major factor in improving reliability, the number of rocket engines was drastically reduced. The largest and most powerful 44L version of Ariane 4, with its four liquid-propellant strap-on boosters, has 10 turbopump-fed liquid engines, each of which must perform flawlessly, for as long as 12.5 min in the case of the third-stage cryogenic (liquid hydrogen/liquid oxygen) engine, for the mission to succeed. Ariane 5 uses only two liquid-propellant engines, augmented by two large solid-propellant boosters. The main core-stage cryogenic Vulcain engine is turbopump fed, while the upper-stage engine operates in the pressure-fed, or blow-down, mode, using high-pressure gas to force the hypergolic (self-igniting) propellants into the combustion chamber. This reduction in complexity, from as many as 10 turbopump-fed engines to only one, accounts for much of the reliability increase in the Ariane 5 design.

Inversion of stages. A second innovation is the use of cryogenic propulsion for the first stage rather than the more typical upper-stage applications, where the better performance of cryogenic propellants results in higher payload capability. This deliberate performance trade-off was made to allow for complete on-ground checkout of the cryogenic engine performance before ignition of the solid boosters and lift-off, since many past problems have involved cryogenic upper stages. Further, the performance is monitored for 8 s after ignition, far more than required for computers to determine proper operation but enough to ensure that no foreign objects obstruct the propellant feed lines.

The inversion of traditional stage configurations, which normally have the storable hypergolic stages below and the cryogenic stage on top, is carried further away with Ariane 5's upper stage. By using pressure-fed hypergolic propellants, the difficulties of cryogenic ignition are avoided, resulting in a further increase in reliability. The overall effect of these design changes is to increase the expected reliability from the current level of about 95% for Ariane 4, on a par with most of the world's mature launch vehicles, to 98.5% for Ariane 5. While this may appear to be only a small improvement, it represents a quantum increase in dependability to the users, namely, an expected failure rate of only 1 in 67 for Ariane 5 compared to 1 in 20 for Ariane 4.

Performance increases. Despite the fact that the lift capability of Ariane 5 will be among the best in the world, further performance increases are already being implemented. Under the Ariane 5 Evolution program, several design improvements are being made that will collectively increase performance by about 3300 lb (1500 kg) shortly after the year 2000. With a dual launch capacity of about 17,200 lb (7800 kg) to geostationary transfer orbit, Ariane 5 will be equipped to launch the largest commercial satellites presently envisioned, with masses in the 10,000–12,000 lb (4500–5500 kg) range, along with a second medium-sized satellite in the 5000–7000 lb (2300–3300 kg) class. Continued availability of the proven and flexible dual launch capability, to serve steadily growing commercial satellite demand, is a cornerstone of the sustained success of Ariane.

Launch preparation. Different approaches have not been confined to the launch vehicle itself. Detailed

Fig. 1. Ariane 5 vehicle with a cryogenic core stage, two large solid boosters, and a small upper stage to place one or more spacecraft into orbit. (*a*) Side view. (*b*) Cutaway view.

analyses of the entire operational concept have resulted in important changes from earlier practice. All launch vehicle and payload preparations are now conducted away from the launch pad, with the vehicle and its one or two spacecraft passengers fully integrated on a mobile launch platform in a final

Fig. 2. Ariane 5 vehicle and launch complex. Use of a mobile launch table incorporating an integral umbilical mast allows a so-called clean pad, requiring only propellant loading, flame deflectors, and water deluge provisions.

assembly building 1.8 mi (3 km) from the launch pad. The traditional mobile service gantry and fixed umbilical tower are gone, with the umbilical mast integrated onto the mobile launch table and the service gantry functions replaced by the final assembly building. Instead of the typical 4–6 weeks of vehicle assembly, payload integration, and on-pad activity, the total assembly will roll out of the final assembly building and move to the launch pad only 9 h prior to launch. The result of these innovative concepts is the first truly clean-pad launch facility. Ariane 5 on the mobile launch table, with its integral umbilical mast, is shown in **Fig. 2.** This overall view of the launch-pad complex also shows the water deluge tower, the flame deflectors, and the lightning protection towers.

First flight. Unfortunately, the first flight, on June 4, 1996, was a failure, attributed to a guidance program software error that was undetected by the extensive qualification and preflight test program. The unambiguous determination of the cause of the failure and its rapid correction provide high confidence that qualification flights will be successfully concluded. Ariane 5 is expected to be the mainstay of European space transportation capability well into the twenty-first century, continuing to provide dependable launch services at an even higher level of reliability and availability.   Douglas A. Heydon

Sea Launch Program

Sea Launch Company was formed in response to growing demand for a more affordable, reliable, and capable commercial satellite launching service. The partnership includes leading companies in Ukrainian, Russian, and American aerospace and in European shipbuilding. The use of proven launch systems developed by the former Soviet Union, combined with a novel marine-based operations concept, enables Sea Launch to lift off directly from the Equator to provide maximum performance to geostationary orbit, and to tailor the launch site for more inclined

orbital destinations. The first launch is scheduled for 1998.

The heart of the system is the Zenit rocket, built in Ukraine. Designed in the 1980s, the Zenit incorporates modern rocket technology with highly automated operations, enabling launch at sea. The upper stage is the Block DM, developed in Russia. The 4-m payload fairing, built in the United States, provides standard accommodations and environments for western communications satellites.

Launch sites and satellite orbits. Historically, commercial communications satellites have been destined for geostationary orbit, 22,300 mi (35,800 km) above the Earth's surface at the Equator. From this position, the spacecraft remains above the same spot on the Earth, and it is possible for three satellites properly spaced to provide worldwide coverage. To date, however, no launch site has been optimally located to take maximum advantage of the Earth's rotation without introducing inclination, which must be removed en route to the spacecraft's final destination, degrading the effective performance of the rocket. Sea Launch's ability to launch directly from the Equator eliminates this inclination, maximizing rocket performance.

With growing demand for mobile communications and increased bandwidth, a number of satellite systems are now planned for lower orbits. Although lower elevations reduce transmission time delay, it is necessary to have more satellites to provide the same coverage. These constellations tend to have inclined orbits, taking advantage of the Earth rotating beneath them to achieve coverage. Sea Launch's mobile launch pad also provides the optimum launch site for these inclined orbits. Finally, for constellations with both equatorial and inclined planes of satellites, Sea Launch has the unique capability to optimize performance for each orbit.

Prelaunch operations. Launch operations begin at the home port, in Long Beach, California. Here, spacecraft are delivered, tested, fueled, and encapsulated in the payload fairing. Then, the spacecraft is transferred to the Assembly and Command Ship for integration with the rocket.

The Assembly and Command Ship is a 660-ft-long (200-m), 106-ft-wide (32-m) vessel with facilities for launch vehicle processing, mission control, and communications; medical facilities; and accommodations for up to 250 customers and crew. Launch preparation activities include integration of the rocket segments in parallel with spacecraft encapsulation.

The integrated launch vehicle and spacecraft are then loaded onto the launch platform by a crane-and-truss mechanism. The launch platform is a floating launch pad, constructed on a modified oil drilling platform. Designed for precision oil drilling in the rigors of the stormy North Sea, this is among the largest semisubmersible platforms in the world. Pontoon length is 436 ft (133 m), deck width approximately 220 ft (67 m), and submerged displacement 50,700 tons (46,000 metric tons). Atop the deck are the hangar in which the launch vehicle resides, automated launch support equipment, and office and living accommodations for up to 20 essential crew. Once the ships are provisioned, launch team members board the Assembly and Command Ship, and both vessels head for the geostationary mission launch site, approximately 1000 mi (1600 km) south of Hawaii at the Equator.

Sea Launch vehicle. The Sea Launch vehicle is a three-stage liquid-propellant system capable of providing 1.6×10^6 lbf (7.1×10^6 newtons) of thrust. The first two stages are the Zenit, originally developed to reconstitute satellite constellations quickly with design emphasis on robustness, ease of operation, and fast turnaround. It uses horizontal assembly and roll-out, key factors in enabling launch at sea, in conjunction with high automation, requiring a very small launch crew. Zenit first flew in 1985 from the Baikonur Cosmodrome in Kazakhstan. As of September 1996, the Zenit had completed 24 of 27 missions, the last 10 consecutive missions being successful.

The Block DM upper stage has a long and successful history as the fourth stage of the Proton launch vehicle, having flown 181 successful missions out of 189 attempts. The Block DM-SL is a restartable upper stage, capable of being fired up to seven times during a mission.

Because the Zenit was designed to be flown with the Block DM, very few modifications are required. Those being implemented include adapting the fueling system of the Block DM to take advantage of the automated Zenit fueling capability, and updating the guidance computers on both the DM and the Zenit to take advantage of marine-based algorithms.

The Sea Launch payload fairing provides support and environmental protection for the spacecraft from encapsulation through flight. The composites-manufacturing technology developed for airplanes has been applied to produce the graphite epoxy/aluminum honeycomb fairing. The fairing has an internal diameter of 13 ft (3.9 m), with access to the spacecraft provided by hatches. The integrated launch vehicle rests horizontally in the launch platform hangar during the 11-day transit to the launch site.

Launch operations. At the launch site, the platform ballasts to its stable launch position, and an enclosed bridge is deployed to allow personnel to transfer between the two vessels. When all systems have been prepared, personnel board the Assembly and Command Ship, which moves to its control station 3 mi (5 km) away. Redundant communications lines provide continuous communications between the vessels. Dynamic positioning control systems ensure platform stability while the rocket is rolled out of its hangar and erected automatically, mating over 2000 electric connections and 25 fuel lines.

When all systems are ready, launch is controlled from the Assembly and Command Ship (**Fig. 3**). Upon ignition, the vehicle is held down until a thrust-to-weight ratio of 1.6-to-1 is attained, enabling a final

Fig. 3. Launch of a Sea Launch vehicle from a floating launch platform, controlled from the Assembly and Command Ship in the background.

checkout of the propulsion systems. The payload reaches low Earth orbit in approximately 9 min, and geostationary transfer orbit in less than an hour. Spacecraft and launch-vehicle telemetry are continuously collected and forwarded to customer and launch support sites. Following spacecraft insertion into the target orbit, the upper stage separates from the spacecraft and performs a contamination-and-collision-avoidance maneuver. Orbit determination is performed on board the Assembly and Command Ship and independently verified from the ground. After the launch, both vessels make the return voyage to the home port, while operations crews begin preparations for the next launch.

For background information *see* COMMUNICATIONS SATELLITE; COMPOSITE MATERIAL; LAUNCH COMPLEX; ROCKET PROPULSION; SATELLITE (SPACECRAFT); SPACE FLIGHT in the McGraw-Hill Encyclopedia of Science & Technology.   Amy L. Buhrig

Bibliography. Ariane: Europe's workhorse commercial launcher, *Aviat. Week Space Technol.*, 145(20):S1–S14, Nov. 11, 1996; International Astronautical Federation, *46th Congress Proceedings*, 1995; Sea Launch Limited Partnership, *Sea Launch User's Guide*, 1996.

Satellite navigation systems

The Global Navigation Satellite System (GLONASS) is a space-based radionavigation system being developed by the Russian Federation. Having much in common with the United States' Global Positioning System (GPS), GLONASS can provide accurate estimates of position, velocity, and time to users worldwide. Both systems were planned in the 1970s primarily for military use, and are operated by the respective defense departments. Each country has committed to make a subset of the signals available for civil use without any user fees. Detailed descriptions of these signals have been made available for each system so that receiver manufacturers can build and sell user equipment.

The basic idea behind position estimation with GLONASS, as with GPS, is both simple and ancient: Position can be determined if the navigator knows, or can measure, distances to objects whose position coordinates are known. It is easy to see that each such distance can be related to the three unknown position coordinates by an equation. Given distance measurements to three objects, there are three such equations which can be solved for the three unknowns. (Usually ranges to at least four satellites are needed to account for the unknown bias of the user receiver clock.)

The civil sector has been quick to recognize the extraordinary potential of satellite navigation technology in civil aviation, marine and surface navigation, surveying and geodesy, and recreational activities. The activity centered on GPS-based products and services has grown rapidly. Development of GLONASS would accelerate the pace of growth of applications, which range from precision landing of aircraft to guiding a motorist in an unfamiliar city.

GLONASS system elements. GLONASS, like GPS, can be thought of in terms of its three elements—the space segment, control segment, and user segment.

The space segment consists of 24 satellites in circular orbits around the Earth with a radius of 25,510 km (15,850 mi). The period of an orbit is about 12 h. The satellites are arranged in three orbital planes containing eight satellites each and inclined at 64.8° relative to the Equator. With this satellite constellation, more than 99% of the users worldwide have at least four satellites in view at any time. Each satellite transmits signals at two frequencies in the L band, one for civil use and the other intended exclusively for the military. Unlike GPS, each satellite transmits the same ranging code signal but at a different frequency in the range 1602.0–1615.5 MHz for civil use and 1246.0–1256.5 MHz for military use.

The control segment encompasses the functions and facilities required for operating GLONASS. This segment comprises satellite tracking stations to measure and predict satellite orbits, monitoring stations, data analysis centers, and a command and control center.

The user segment encompasses all aspects of design, development, production, and utilization of user equipment for the different applications.

GLONASS system status and plans. For the first time, on January 18, 1996, GLONASS had a full constellation of 24 satellites, all healthy and transmitting. This accomplishment was the culmination of an intensive effort during 1994–1995. The Russian government has announced plans for system development and

modification, and GLONASS is expected to be ready for operational use in the late 1990s.

The satellites launched so far have been prototypes with a design life of 3 years. The satellites launched recently appear to be meeting the specification on design life: seven of the satellites in the constellation on July 1, 1996, had been in service for over 3 years. The plans for system modification and enhancement leading up to a system to be called GLONASS-M include improved satellite service life (5 years), improved stability of the on-board clocks, improved ephemeris accuracy, expanded capabilities of the control segment, and provision for up to six on-orbit spares (two per orbital plane).

The frequency plan would be revised for GLONASS-M. Starting in the year 2005, the transmit frequencies are planned to be 1598.0625–1605.375 MHz and 1242.9375–1249.0625 MHz for civil and military signals, respectively.

Political and economic difficulties in the Russian Federation have continued to be a source of uncertainty among potential users of GLONASS. An immediate consequence has been the slow development of user equipment. GLONASS receivers were rare in 1995. The achievement of the full constellation and plans for system enhancements appear to have given receiver manufacturers the impetus to commit resources to produce GLONASS receivers. In mid-1996, two receiver manufacturers announced new products, and more were expected to follow.

GLONASS performance. The quality of position estimates obtained by a user of a satellite navigation system depends basically upon two factors. The first is the number of satellites in view and their spatial distribution, referred to as satellite geometry. The satellite geometry changes with time and user position as a result of the movement of the satellites. The number of satellites in the constellation and their arrangement determine how many satellites would be in view of the different users. The second factor is the quality of the measurements. The error in the range measurements, referred to as user range error, depends upon the stability of the clocks in the satellites, accuracy of the satellite orbital positions, atmospheric propagation anomalies, and receiver noise.

In an overall sense, the quality of the position estimates obtained from GLONASS is comparable to that achievable from GPS. The actual performance available currently to civil users from GPS is made considerably worse by the United States policy, to be discontinued by the year 2004, of degrading the signals available for civil use (selective availability). The two systems have similar satellite constellations, but they differ in the size of user range error. The root-mean-square value of the user range error for GLONASS is about 10 m (33 ft), and about 25 m (82 ft) for GPS with selective availability active. The positioning specifications for the two systems are given in the **table**. The GPS specifications are in the form of 95th percentile points of the horizontal and vertical position errors of users worldwide; GLO-

Positioning specifications for GPS and GLONASS

Type of position error	GPS		GLONASS	
	Error	Specification	Error	Specification
Horizontal	100 m (328 ft)	95%	60 m (197 ft)	99.7%
Vertical	156 m (512 ft)	95%	75 m (246 ft)	99.7%

NASS specifies 99.7th percentile points. According to GPS specifications, the horizontal error in a user's position estimate shall not exceed 100 m (328 ft) with a probability of 0.95.

The **illustration** gives a side-by-side comparison of positioning results obtained from GPS and GLONASS at one location. Position estimates were computed from each system with measurements taken 1 min apart over an entire day. The horizontal offsets in these estimates relative to the surveyed location are plotted. As noted previously, the quality of a position estimate depends upon the geometry of the satellites in view. Each position estimate is, therefore, represented with a symbol which characterizes the corresponding satellite geometry as good, fair, or poor to facilitate an apple-to-apple comparison of the positioning results from the two systems.

As can be seen in illus. *a* and *b*, when the satellite geometries were good or fair, the position estimates obtained from GLONASS were distinctly better than those from GPS. This is reflected in the significantly tighter shaded cluster of position estimates obtained from GLONASS versus that for GPS. However, there are more instances of poor geometries with GLONASS. The reason is that GPS had 24 working satellites on this day, while GLONASS had only 21. It is notable that the shaded cluster for GLONASS is off center by about 10 m (33 ft). This is a consequence of differences between the coordinate frames used

by GPS and GLONASS in which to express the positions of their users.

Combining measurements. Given that the signals are free, a user would prefer to utilize the combined set of signals from both GPS and GLONASS, and there are no basic problems in designing a receiver to do so. The main benefit is the increase in the number of satellites in view and improvement in the satellite geometry. This can be seen in illus. *c*, which gives position estimates based on measurements from the two systems combined. As expected, the satellite geometries are consistently good, and the position estimates in illus. *c* are superior to those obtained with either system alone, as shown in illus. *a* and *b*.

GPS and GLONASS are autonomous systems, each with its own system of reference for time and space. As time references, the two systems use their national standards, UTC(USNO) [an acronym for Universal Coordinated Time (U.S. Naval Observatory)] and UTC(SU), respectively. The clocks used to define these time references are not coordinated, and the two time scales have had an offset of several microseconds in recent years. This, however, creates no difficulty for the user as the time offset between the two systems at any instant can be estimated easily from the measurements themselves.

The coordinate frames adopted by the two systems to describe the locations of their satellites, and therefore of the users, are different: GPS provides

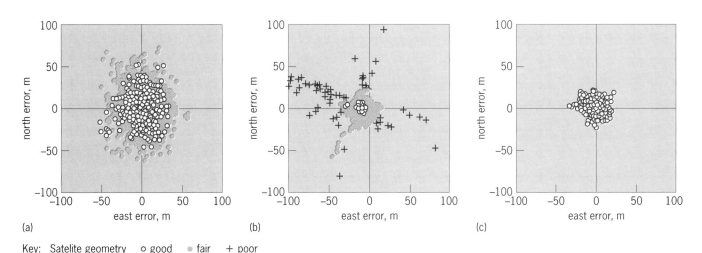

Key: Satelite geometry ○ good • fair + poor

GPS and GLONASS position estimates (1-min samples, July 1, 1996). 1 m = 3.3 ft. (*a*) GPS (24 satellites). (*b*) GLONASS (21 satellites). (*c*) GPS and GLONASS combined (45 satellites).

the satellite positions in the WGS 84 geocentric coordinate frame; GLONASS employs PZ-90. Clearly, the two reference systems must be reconciled before measurements from the two systems can be combined. Actually, there is no serious technical challenge associated with this problem. A transformation between the two coordinate frames can be estimated if coordinates of a set of points are known in both WGS 84 and PZ-90.

For background information *see* SATELLITE NAVIGATION SYSTEMS in the McGraw-Hill Encyclopedia of Science & Technology.   P. N. Misra

Bibliography. B. W. Parkinson et al. (eds.), *Global Positioning System: Theory and Applications*, vols. 1 and 2, 1996.

Seed germination

In recent years there has been a great deal of interest in allelopathic effects upon seed germination. Allelopathy is the direct or indirect harmful or beneficial effects of one plant on another through the production of chemical compounds that escape into the environment. Autotoxicity is an intraspecific form of allelopathy that occurs when a plant species releases chemical substances that inhibit or delay germination and growth of the same plant species.

Extracts. Aqueous extracts from various plant species have shown allelopathic effects, including autotoxicity, upon seed germination. These extracts have been found in many different plant parts, such as seeds, flowers, leaves, stems, and roots. In fact, when the entire plant is soaked with sterilized water, the aqueous extracts are usually more prevalent from plants in the reproduction stage of growth compared to the vegetative stage. This may be due to the higher concentration of allelochemicals in the flowering or reproductive plant parts than in other plant parts.

Species. Investigators have studied many plant species to determine allelopathic activity of one species upon itself or another species. Aqueous extracts of the herbage or roots of numerous species are inhibitory to the germination and seedling growth of a number of plant species. Grass species such as bermudagrass (*Cynodon dactylon*), johnsongrass (*Sorghum halepense*), orchardgrass (*Dactylis glomerata*), pangolagrass (*Digitaria decumbens*), redtop (*Agrostic gigantea*), rhodesgrass (*Chloris gayana*), ryegrass (*Lolium multiflorum*), smooth bromegrass (*Bromus inermis*), tall fescue (*Festuca arundinacea*), timothy (*Phleum pratense*), western wheatgrass (*Pascopyrum smithii*), rye (*Secale cereale*), sorghum (*Sorghum bicolor*), and wild oats (*Avena fatua*) produce from living plants or release substances from residue that are toxic to the growth of other species.

Some weed species that cause an allelopathic effect such as reduced germination of other species are velvetleaf (*Abutilon theophrastic*), ragweed (*Ambrosia psilostachya*), and lambsquarter (*Chenpodium album*). Several legume species have an allelopathic effect, such as alfalfa (*Medicago sativa*), hairy vetch (*Vicia villosa*), and crimson clover (*Trifolium incarnatum*). Walnut trees (*Juglans nigra*) and coffee (*Coffea arabica*) have shown the effect of reduced germination and seedling growth on a number of plant species.

Autotoxicity. Autotoxicity occurs when a plant species releases toxic chemical substances that inhibit germination and growth of the same species. Alfalfa is probably one of the most widely studied forage legumes that has autotoxicity effects. This is very important, especially when alfalfa is to be reestablished as a perennial forage crop. Many stands will remain productive for 6–8 years and then begin to die. If the producer wishes to maintain a highly productive, highly nutritious crop and finds it very difficult to reestablish alfalfa in an old stand, it is due to the release of auto-allelochemicals from the old plants.

The black walnut tree has been the most widely studied tree species that exhibits autotoxicity. The leaves contain an allelochemical called juglone. Juglone not only reduces germination and seedling growth of black walnut, an autotoxic effect, but also has a similar effect upon many other plant species, a heterotoxic effect; therefore, the dropping of black walnut leaves reduces not only the growth of walnut seedlings but also the growth of many other plant species under the tree canopy.

Effects. The effects of allelochemicals on germination and seedling growth are only the primary observed effects that are occurring at the molecular level. The mode is either a direct or an indirect action. A direct effect involves the various aspects of germination, seedling growth, or metabolism. An indirect effect includes the activity of harmful or beneficial organisms such as nematodes, insects, or microorganisms, or the nutritional status of the plant, or the alteration of soil properties.

An example of a direct allelopathic effect on germination and seedling growth is a drastic reduction of root hair formation and the number of root hairs in alfalfa (**Figs. 1** and **2**). In addition, within the root cortex cells there appears to be a deposit within the cells, thought to be a soluble starch. These deposits may reduce the uptake and transport of nutrients. Allelopathic effects are greater during cooler periods than in warmer weather.

Another effect is the release of allelochemicals from one plant species and its direct effect of inhibiting germination or early plant growth of competing species in the vicinity. There are many possible processes that may be altered or attacked by allelochemicals. The process may be a primary function, or it may be combined with another function to cause an allelopathic effect. Examples are a membrane and its permeability, mineral uptake, conducting tissue, water uptake or status within a plant, specific enzyme activity, protein synthesis, nitrogen fixation, respiration, stomatal movement, photosynthesis, germination of pollen, phytohormones and their activity, cytology or ultrastructure, anatomical struc-

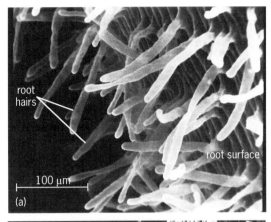

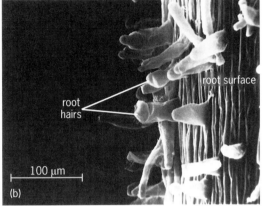

Fig. 1. Scanning electron micrographs of 5-day-old alfalfa primary roots. (*a*) Uninhibited root (control). (*b*) Root inhibited by alfalfa shoot aqueous extract.

alkaloids, amino acids, terpenoids, steroids, sulfides, glucosides, and unsaturated lactones. Other allelochemicals or secondary compounds are still to be discovered.

Allelochemicals may be important in minimum-tillage and no-till agriculture, where crop residues are left on the soil surface after harvest, or in intensive cropping systems such as intercropping and crop sequence. The allelochemicals that are released can be utilized as herbicides that reduce germination or seedling growth of a specific weed.

Allelochemicals may serve as multipurpose pesticides in an integrated pest management system. Increasing global concern about the indiscriminate use, overdependence, and hazards of synthetic pesticides has prompted exploration of natural plant products, since they are expected to be comparatively safe. Various allelochemicals have shown herbicidal and fungicidal activities in terms of inhibiting germination and growth of some weed species and inhibition of mycelial growth of some fungi.

Implications for crop practices. Allelopathy studies have encouraged crop rotation. Crop rotation allows the allelochemicals to be leached from the soil or decomposed over time. A possible future role will

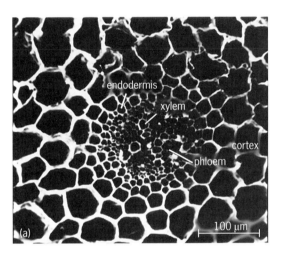

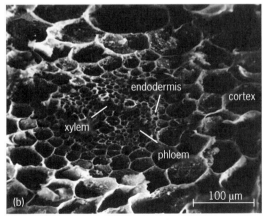

Fig. 2. Scanning electron micrographs of ethanolic cryofractured alfalfa root. (*a*) Uninhibited root (control). (*b*) Root inhibited by alfalfa shoot aqueous extract.

ture, or genetic material. Once the associated chemicals enter the environment, they may be retained, transformed, or transported within a susceptible species. There may be an interaction between an allelopathic substance and a nutrient.

Combining any of these various processes may affect the production of an allelochemical, its release into the environment, its absorption and translocation, or the site of action where it occurs; thus the study of allelopathic actions becomes very complex.

Allelochemicals. Several phytotoxic substances suspected of inhibiting germination and growth have been identified from plant tissues and soils; they are known as allelochemicals. Allelochemicals are usually considered to be secondary plant products or waste products of the main metabolic pathways in plants. These may be water-soluble substances that are released directly from living plants into the environment through leaching, root exudation, and volatilization; through decomposition of plant residues; or from seeds in the soil. Many large and complex molecules produced by microbial activity during decomposition of plant residue can also cause allelopathic effects. An incomplete list of various types of allelochemicals includes tannins, flavonoids, cinnamic acid, phenols, benzoic acid, quinones, organic acids, lactones, long-chain fatty acids, purines, be the utilization of certain species to provide weed control of the succeeding crop. Allelopathic effects of certain crops on the control of weeds, crops on crops, or weeds on crops may be possible. Some research indicates that there may be allelopathic control of certain insects. Use of crop rotation might eliminate the harmful allelopathic effects of crops on crops and provide beneficial reactions.

Germplasms of various crop species have been found to have autotoxic resistance and will be studied further. There are some instances where a certain crop must be continually grown. By determining what allelochemical is causing the allelopathic effect, it is possible to select or breed for resistance in that particular species. Therefore, genetics and biotechnology will play an important role in the future utilization of allelopathic research.

For background information *see* ALLELOPATHY; CHEMICAL ECOLOGY; PLANT GROWTH; ROOT (BOTANY); SEED in the McGraw-Hill Encyclopedia of Science & Technology.   Darrell A. Miller

Bibliography. A. Ozan, M. G. Nair, and G. R. Safir, *In vitro* effects of plant phenolics on root peroxidase activity of *Trifolium repens.*, *Int. J. Allelopathy*, 3:59–64, 1996; E. L. Rice, *Allelopathy*, 1984; S. J. N. Rizvi and V. Rizvi, *Allelopathy: Basic and Applied Aspects*, 1992.

Semiconductor

Electric fields are used to accelerate, redirect, or stop electrons in nearly all modern semiconductor devices. However, there are a number of other ways of altering the motion of electrons in semiconductors, such as with magnetic fields or temperature gradients. The use of local magnetic fields to control the motion of charge carriers will initiate new areas of fundamental research, and may have applications in microelectronics in the future.

External magnetic fields are already used in laboratory experiments to study and measure the properties of charge carriers in semiconductors, such as their mass, the Fermi surface, and their interaction with the semiconductor lattice. However, such experiments are done only with magnetic fields that are homogeneous throughout the sample. At the other extreme, the magnetic impurities in diluted magnetic semiconductors generate inhomogeneous magnetic fields on the angstrom scale (1 angstrom equals 0.1 nanometer, comparable to the size of an atom), which act as scattering centers and influence the resistance of the material. However, until recently, efforts at creating inhomogeneous magnetic fields on length scales between these two extremes (that is, on the micrometer and nanometer scales) had been unsuccessful.

Now, however, experiments have been carried out in which the motion of electrons in a two-dimensional electron gas is modified through the action of a spatially varying magnetic field. These experiments used devices based on a high-mobility two-dimensional electron gas, which was confined in a standard gallium arsenide/aluminum gallium arsenide (GaAs/AlGaAs) heterojunction. To realize the spatial modulation of the magnetic field, a periodic array of strips of ferromagnetic or superconducting material was deposited on the heterostructure surface.

Inhomogeneous magnetic fields. Figure 1 shows the magnetic field experienced by the electrons in the two-dimensional electron gas as the result of the presence nearby of a single ferromagnetic strip. The curves show the field when the electron gas is positioned close to the strip, as well as the fields at increasing separations. Thus, the electrons experience a spatially inhomogeneous magnetic field which can be controlled by changing the set-back distance or the width and height of the ferromagnetic strip.

Magnetic barriers. In the nonzero field region, the Lorenz force causes electrons to move in circular cyclotron orbits. The energy of an electron in such an orbit is quantized, and the energy states are called Landau levels. The kinetic energy of an electron in the lowest Landau level is larger than that of an electron outside the magnetic field by an amount equal to $\hbar\omega_c/2$. Here, $\hbar$ is Planck's constant divided by 2π, and ω_c, called the cyclotron frequency, is equal to eB/mc (e is the electron charge, B the strength of the magnetic field, m the electron mass,

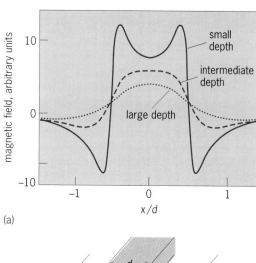

(a)

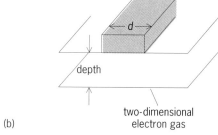

(b)

Fig. 1. Magnetic field produced by a single ferromagnetic strip, with perpendicular magnetization, in a two-dimensional electron gas. (*a*) Fields plotted at three values of the depth as functions of x/d, where x is the horizontal position and d is the width of the strip. (*b*) Diagram of the system. *(After F. Peeters, Microscopic magnetic manipulation of electron motion, Phys. World, pp. 24–25, October 1995)*

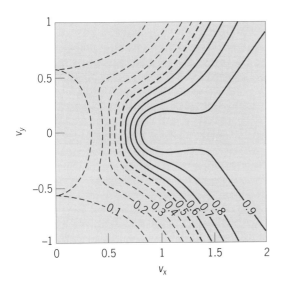

Fig. 2. Contour plot of the transmission probability of an electron through the magnetic barrier of Fig. 1 as a function of the electron velocity components, v_x and v_y, in the x direction, perpendicular to the barrier, and in the y direction, parallel to the barrier. The quantities v_x and v_y are in units of $B^{1/2}v_0$, where B is the magnetic field strength in tesla and v_0 is 4.5×10^4 m/s.

and c the speed of light). Because of the larger kinetic energy of electrons in nonzero field regions, such a region will act as a barrier to the electron motion. An electron moving toward such a magnetic barrier will be reflected, unless its energy exceeds a certain value and its angle of incidence at the barrier is less than a critical angle which depends on the energy.

There are two essential differences between these magnetic barriers and the potential barriers now used in tunneling devices. First, the magnetic barrier is a consequence of an increase in the kinetic energy of the electron rather than in its potential energy, as is the case for potential barriers. Second, the process of tunneling through a magnetic barrier is inherently two-dimensional, while tunneling in the absence of a magnetic field is a one-dimensional process. During the tunneling process the direction of the electron's motion is altered, but its energy is conserved. An important consequence is a strong dependence of the transmission probability on the direction of incidence of the electron on the magnetic barrier. This is illustrated in **Fig. 2**, showing a contour plot of the transmission probability as a function of the components of the incident electron velocity. Because of this strong angular dependence, such magnetic barriers can be used as electron wave-vector filters.

Demonstrations of magnetic modulation. In 1994, almost simultaneously, three different experimental groups succeeded in demonstrating the coupling between a two-dimensional electron gas and a magnetic field that was periodic in space. Their devices were based on a high-mobility two-dimensional electron gas confined in a standard GaAs/AlGaAs heterojunction, with a periodic array of strips on top of

the heterojunction to provide spatial modulation of the magnetic field. One group employed dysprosium metal gates, while a second group used nickel in order to produce a spatial modulation of the magnetic field. An external magnetic field could be used to increase the strength of these micromagnets. A third group took a different course, working with superconducting strips (lead or niobium). In this case the expulsion of the magnetic flux from the superconducting strips is due to the Meissner effect, and this expulsion leads to a nonuniform magnetic field underneath and near these strips.

All three experiments have demonstrated unambiguously the effect of the modulated magnetic field on the electron motion through the observation of commensurability oscillations in the resistivity. These are oscillations that result from resonant enhancement of the electron motion whenever the diameter of the cyclotron orbits of the electrons at the Fermi level matches the period of the modulation. Specifically, this matching occurs when the cyclotron diameter equals $n + 1/4$ times the modulation period, where n is an integer and the quantity $1/4$ plays the role of a phase factor. These oscillations were predicted in 1990, and are similar to oscillations associated with electric-field modulation, which were observed in 1989. Both phenomena are now called Weiss oscillations, but they are distinct from each other in their phase and amplitude.

Earlier efforts at observing these oscillations had failed because the effects of a strong modulation of the electric field overwhelmed those of magnetic modulation. During the cooling down of the device to the operating temperature, the periodic array of strips induces strain fields in the gallium arsenide which result in this electric field modulation. Improvements in reducing the residual modulation of the electric field, for example, by including a nickel-chromium layer between the superconducting strips and the two-dimensional electron gas, or depositing the strips on the planes of the gallium arsenide sample that are not piezoelectrically active, have made it possible to observe the oscillations in the resistivity.

Fields of random structures. Recently, there have been two further developments in reducing the size of nonuniform magnetic fields, which are now created by cluster and disklike structures. The group which had previously worked with dysprosium metal gates succeeded in microfabricating dysprosium disks of diameter 200 nm on top of a GaAs/AlGaAs heterojunction. Different routes were followed by two other groups, who inserted the magnetically active manganese atom in the semiconductor gallium arsenide (GaAs) and created gallium-manganese (GaMn) and manganese arsenide (MnAs) micromagnets. One of the groups bombarded gallium arsenide with manganese ions, while the other group used manganese implantation during molecular-beam-epitaxial (MBE) growth. An essential subsequent heat treatment resulted in the formation of gallium-manganese or manganese arsenide clusters, which became single-domain ferromagnets after

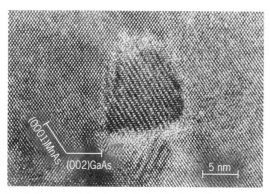

Fig. 3. High-resolution transmission electron microscope image of a single-crystalline manganese arsenide (MnAs) particle in a gallium arsenide (GaAs) crystal grown by molecular beam epitaxy. Crystallographic axes are indicated. *(From J. De Boeck et al., Nanometer-scale magnetic MnAs particles in GaAs grown by molecular beam epitaxy, Appl. Phys. Lett., 68:2744–2746, 1996)*

magnetization (**Fig. 3**). The size of these clusters was controlled by the temperature of the heat treatment and could be varied over the range 2–30 nm, while the group that bombarded gallium arsenide with manganese ions obtained clusters which were one order of magnitude larger.

An area of such randomly located single-domain magnets will make it possible to study the motion of electrons in a random magnetic field. Researchers are interested in learning how such randomness will influence the properties of the charge carriers and if it is possible to observe localization due to such magnetic field fluctuations. In applications, such micromagnets could be used as small memory elements in new computer architectures. The advantages are twofold: (1) because of the small size of such micromagnets, a large storage density can be achieved; (2) the active computing unit can now be put on the same chip as the memory elements, considerably reducing delay times for information retrieval from the memory elements.

For background information *see* CYCLOTRON RESONANCE EXPERIMENTS; FERMI SURFACE; ION IMPLANTATION; QUANTIZED ELECTRONIC STRUCTURE (QUEST); SEMICONDUCTOR; SEMICONDUCTOR HETEROSTRUCTURES in the McGraw-Hill Encyclopedia of Science & Technology.   François Peeters

Bibliography. H. A. Carmona et al., Two dimensional electrons in a lateral magnetic superlattice, *Phys. Rev. Lett.*, 74:3009–3012, 1995; J. M. Daughton, Magnetoresistive memory technology, *Thin Sol. Films*, 216:162–168, 1992; J. Heremans, Solid state magnetic field sensors and applications, *J. Phys. D: Appl. Phys.*, 26:1149–1168, 1993; A. Matulis et al., Wave-vector-dependent tunneling through magnetic barriers, *Phys. Rev. Lett.*, 72:1518–1521, 1994; J. Shi et al., Assembly of submicrometre ferromagnets in gallium arsenide semiconductors, *Nature*, 377:707–710, 1995; P. D. Ye et al., Electrons in a periodic magnetic field induced by a regular array of micromagnets, *Phys. Rev. Lett.*, 74:3013–3016, 1995.

Sexually transmitted disease

Sexually transmitted genital ulcerative diseases are a worldwide health problem and have been strongly associated with an increased risk of human immunodeficiency virus (HIV) transmission. Chancroid, caused by the gram-negative pathogen *Haemophilus ducreyi*, is the leading cause of genital ulcer disease in developing countries of Asia and Africa. In Europe and North America, *H. ducreyi* infections account for a smaller proportion of reported genital ulcers. However, recent studies in the United States suggest that many more cases of chancroid occur than are reported. *Haemophilus ducreyi* has been the focus of recent research aimed at understanding the mechanisms of pathogenesis of chancroid and its association with HIV transmission.

Pathogenesis of chancroid. *Haemophilus ducreyi* infection is initiated by bacterial entry through breaks in the skin that occur during sexual activity. As polymorphonuclear leukocytes respond to the presence of the bacteria, an inflammatory papule develops within the epidermis. Within several days, a pustule forms and ruptures, resulting in the characteristic soft, painful chancroid ulcer. Individual ulcers may fuse into larger sores, and organisms present in the ulcers may cause local cutaneous spreading through autoinoculation. Involvement of draining lymph nodes is often associated with *H. ducreyi* infection, but the organism does not penetrate into the blood or deeper tissues.

Observations of biopsies from naturally occurring chancroid lesions and recent studies using an experimental human model of *H. ducreyi* infection of nongenital skin have helped to explain the host responses in chancroid. *Haemophilus ducreyi* elicits an inflammatory response containing macrophages and CD4+ T cells in the skin. Resident skin cells and recruited T cells secrete a variety of messenger molecules known as cytokines in response to *H. ducreyi* infection. Despite this vigorous cellular response, infection may persist for months without antibiotic therapy.

HIV transmission. Genital ulcer diseases in general, and chancroid in particular, have been shown to be significant risk factors in the heterosexual transmission of HIV. Several mechanisms may explain this enhanced transmission. HIV has been detected in chancroid ulcers, and tissue destruction in the lesions may increase shedding of the virus. The disrupted skin barrier may also increase susceptibility to HIV infection. In addition, a higher proportion of HIV-susceptible cells present in chancroid lesions may facilitate HIV transmission. Both CD4+ T cells and macrophages recruited to the site of *H. ducreyi* infection can harbor HIV in infected individuals. Their location near the point of HIV entry or exit in a chancroid lesion may influence the spread of the virus.

Immunity. The patterns of cellular and cytokine responses that occur during chancroid contrast with those that commonly result in strong antibody responses to infection. Despite observations that individuals with chancroid have higher serum antibody levels to some *H. ducreyi* surface components than do uninfected controls, recurrent infections are a hallmark of this disease. In fact, Augusto Ducrey, for whom the organism was named, showed in the late 1800s that infection could recur at least 15 times in the same individual. Thus, little or no immunity develops in naturally acquired infection.

Potential virulence factors. Several bacterial products that may be important in the establishment and persistence of *H. ducreyi* infection have been identified. Like the cell envelopes of other gram-negative bacteria, the *H. ducreyi* cell envelope is composed of an inner cytoplasmic membrane and an outer membrane separated by a space known as the periplasm.

Lipooligosaccharide. The outer surface of the outer membrane of *H. ducreyi* contains a sugar-modified lipid molecule known as lipooligosaccharide. Although *H. ducreyi* lipooligosaccharide has been shown to cause considerable tissue damage when injected into animals, its role in pathogenesis of chancroid is not well understood. Recent studies have shown that lipooligosaccharide of *H. ducreyi* isolated from humans contains sialic acid, an abundant molecule in the human host. The addition of sialic acid (sialylation) to the terminal sugar residues of *H. ducreyi* lipooligosaccharide may serve to conceal the bacteria from the host immune response by preventing recognition of antigenic sites. Sialylated *H. ducreyi* are also more resistant to the bactericidal effects of normal human serum.

Pili. In addition to lipooligosaccharide, the *H. ducreyi* outer membrane contains several proteins that may be important in the pathogenesis of chancroid. Pili are filamentous appendages common on the surface of most gram-negative pathogens and are important for bacterial adherence to host cells and tissues. Pili are composed of many copies of a structural protein called pilin. The pilin protein synthesized by *H. ducreyi* is different from other bacterial pilins and forms unique fine, tangled pili on the surface of the organism. *Haemophilus ducreyi* attachment to host cells and extracellular matrix components and the involvement of pili in these processes are areas of active investigation.

Hemoglobin receptor. Iron is needed by all bacteria, and free iron inside the human host is scarce. When grown under iron-limiting conditions in the laboratory, *H. ducreyi* synthesizes several proteins that are not produced under iron-sufficient conditions. One of these new proteins is an outer membrane receptor that binds hemoglobin, which may serve as an iron source for growth during infection.

Hemolysins and cytotoxins. The characteristic tissue damage associated with chancroid ulcers suggests the production of one or more extracellular cytotoxins. A secreted hemolysin, which can lyse erythrocytes, has been identified. This molecule may also be termed a cytotoxin, since it has toxic effects on cultured keratinocytes and fibroblasts, two predominant cell types in human skin. Mutant *H. ducreyi*

cells that lack this hemolysin are unable to damage cultured fibroblasts, but retain some cytotoxic activity against keratinocytes. Thus, *H. ducreyi* may produce other cytotoxins that are important in ulcer formation.

Superoxide dismutase. Cytoplasmic superoxide dismutases are ubiquitous enzymes that protect cells from harmful oxygen radicals generated during normal metabolism. The most common forms of bacterial superoxide dismutases are manganese- and iron-containing cytoplasmic enzymes. *Haemophilus ducreyi* produces an unusual periplasmic copper- and zinc-dependent superoxide dismutase that may be important in protecting it against toxic oxygen radicals generated by polymorphonuclear leukocytes during the host response to chancroid.

Heat shock proteins. Heat shock proteins are a group of highly conserved proteins expressed by both prokaryotic and eukaryotic organisms in response to heat and other stressful stimuli. Studies have shown that bacterial heat shock proteins known as GroE proteins are major antigens produced by pathogens that may evoke damaging cross-reactive antibodies and cellular responses during infection. The immune response to infection is often a two-edged sword in that the same events that lead to bacterial killing and clearance may also damage host tissues and contribute to the disease process. Some evidence suggests that *H. ducreyi* GroE proteins are expressed and elicit an antibody response during infection. Thus, these may be important bacterial antigens contributing to the immune response to *H. ducreyi* and tissue damage in chancroid.

Chancroid models. The importance of potential *H. ducreyi* virulence factors in chancroid disease may best be understood by testing genetically defined mutant strains of *H. ducreyi* in appropriate models of infection. Comparison of the behavior of wild-type *H. ducreyi* and mutant strains lacking a particular postulated virulence determinant may provide insights into the roles of individual gene products in the pathogenesis of chancroid. Several recently developed models of *H. ducreyi* infection are described below.

Human models. Since *H. ducreyi* infects human skin, investigators have developed systems using human skin cells grown in the laboratory to study host–pathogen interactions. The two predominant cell types in human skin are keratinocytes, constituting the upper skin layers (the epidermis), and fibroblasts embedded in the dermis below the keratinocytes. These cell types have been cultured individually in monolayers and infected with *H. ducreyi* strains expressing or lacking individual virulence factors. The current understanding of the cytotoxic mechanisms of *H. ducreyi* has come from such studies.

An artificial skin system, in which keratinocytes and fibroblasts are cultured together in a physical arrangement that resembles intact human skin, is currently being evaluated as a model for *H. ducreyi* infection. Preliminary evidence suggests that *H. ducreyi* causes skin cells to secrete messenger cytokines that may be involved in the recruitment of immune cells to chancroid lesions.

Animal models. Although *H. ducreyi* is strictly a human pathogen in nature, several animal models of infection have been developed to study the pathogenesis of chancroid in a mammalian host. Rabbits housed at reduced temperatures, primates, and pigs can be infected with *H. ducreyi*, resulting in the development of lesions with similar characteristics to human chancroid ulcers. Studies using the rabbit model were instrumental in describing the toxic effects of *H. ducreyi* lipooligosaccharide. The swine model provides the opportunity to examine host immune cell responses to *H. ducreyi* infection, and the primate model may be useful to study interactions between chancroid and an HIV-like virus that infects monkeys. These newly developed animal models show great promise for a better understanding of the pathogenesis of chancroid.

Human challenge studies. The ultimate model for studying chancroid must employ the natural host for *H. ducreyi*: humans. Natural infection is often complicated by the presence of other bacteria, and obtaining sequential samples from individuals is often difficult or impossible. Thus, analysis of clinical samples has provided only limited information about *H. ducreyi* pathogenesis. An experimental human model of infection, in which subjects are infected on the upper arm, has proven safe and has provided considerable information regarding the human immune response to *H. ducreyi* infection.

Control of chancroid ultimately depends on understanding the basic mechanisms by which *H. ducreyi* causes disease and the host immune responses to infection. Taken together, the models described above and the growing number of potential *H. ducreyi* virulence factors that are being described and genetically characterized represent powerful tools with which researchers hope to unravel the complex relationships between *H. ducreyi* and its human host.

For background information *see* ACQUIRED IMMUNE DEFICIENCY SYNDROME (AIDS); BACTERIA; IMMUNITY; INFECTION; SEXUALLY TRANSMITTED DISEASES; VIRULENCE in the McGraw-Hill Encyclopedia of Science & Technology. Marcia M. Hobbs

Bibliography. Centers for Disease Control and Prevention, Chancroid detected by polymerase chain reaction—Jackson, Mississippi, 1994–1995, *Morbidity Mortality Week. Rep.*, 44:567, 573–574, 1995; V. T. DeVita et al. (eds.), *AIDS: Etiology, Diagnosis, Treatment and Prevention*, 4th ed., 1996; P. J. Lynch, *Genital Dermatology*, 1994; S. A. Morse, Chancroid and *Haemophilus ducreyi*, *Clin. Microbiol. Rev.*, 2:137–157, 1989; D. Plummer, G. Kovacs, and A. Westmore, *Sexually Transmitted Diseases*, 1995; S. M. Spinola et al., *Haemophilus ducreyi* elicits a cutaneous infiltrate of CD4 cells during experimental human infection, *J. Infect. Dis.*, 173:394–402, 1996; D. L. Trees and S. A. Morse, Chancroid and *Haemophilus ducreyi*: An update, *Clin. Microbiol. Rev.*, 8:357–375, 1995.

Smart structures

Smart structures is the name of an emerging technology with the potential to change the way that structures are used. Conventional structures are used primarily to carry mechanical loads and provide compartmental boundaries. Smart structures are highly integrated combinations of structures, sensors, actuators, and control systems, or information-processing systems.

Characteristics. The term smart structures has become a catchall to identify what are more accurately described as highly integrated, multifunctional structures. Multifunctional structures, in addition to carrying mechanical loads, may alleviate vibration, reduce acoustic noise, monitor their own health and environment, automatically perform precision alignments, or change their shape or mechanical properties on command.

Applications would include buildings that change their stiffness to resist earthquake loads and suppress transmission or generation of acoustic noise; bridges and airplane structures that monitor their own health to provide warning of incipient failure or to compensate for damage; and airplane wings that subtly change their shape to optimize fuel efficiency. Such applications have yet to be realized; however, more limited applications of smart structures have become operational. For example, there are smart-structure skis that have an embedded system to sense and suppress vibration (chatter) to help keep the skis from bouncing off the snow.

Smart structures make use of smart materials. These materials can significantly change their mechanical properties, such as shape, stiffness, and viscosity (fluids); or they can change their thermal, optical, or electromagnetic properties in a controllable manner. Construction of smart structures requires a wide combination of disciplines, including materials development; mechanical design, analysis, and manufacturing; structural dynamics; electronics; sensors and actuators; control systems; and information processing.

Smart materials. These materials are exceptionally good transducers, with a very strong coupling between their mechanical properties and one or more other properties such as optical, thermal, or electromagnetic. Although some of these materials are relatively new, many have been known and used for decades in the electronics industry. One of the most important classes of smart materials makes use of the piezoelectric effect, discovered over a century ago. The **table** summarizes characteristics of widely recognized smart materials. Currently, the most important smart materials are piezoelectric materials, particularly the ceramics composed of lead zirconate titanate, and the shape memory alloys.

Piezoceramics function very well as either sensors or actuators through the strong coupling of their electromechanical properties. This coupling is induced through the process of poling, in which a strong electric field is applied to the ceramic while

Characteristics of smart materials

Class	Primary transduction characteristic	Example	Primary uses in smart structures	Performance characteristics
Piezoelectric materials	Electrical field to mechanical strain or mechanical strain to electrical charge	Lead zirconate titanate (a ceramic)	As sensors, actuators, or both, primarily used for vibration suppression applications	Reasonably linear sensor/actuator capable of operating at high frequencies but low strain levels; bidirectional, but cannot operate statically
Electrostrictive materials	Electrical field to mechanical strain	Lead magnesium niobate (a ceramic)	As actuators for vibration suppression and static precision alignment	More strain capability than piezoelectric materials; can operate statically but is highly nonlinear, is unidirectional, and requires a mechanical preload
Magnetostrictive materials	Magnetic field to strain	Terbium-iron alloy (Terfenol)	Similar to electrostrictive materials	Similar to electrostrictive materials, requires a magnetic bias
Shape memory alloy	Temperature to mechanical strain and stiffness	Alloys of nickel and titanium (nitinol)	As actuators that provide large static shape change	Large strain capability limited to static or very low frequency
Electrorheological fluids	Electric field to fluid viscosity	Colloidal suspension of electrically charged particles	As adaptive shock absorbers for actively controlled viscous damping	Significant change in viscosity is provided, but usually requires relatively high-voltage electric fields
Magnetorheological fluids	Magnetic field to fluid viscosity	Colloidal suspension of magnetic particles	Similar to electrorheological fluids	High voltages are not required, but heavier systems are needed to generate magnetic fields

it is brought to a high temperature (above the Curie temperature) and then allowed to cool. When subsequently exposed to an electric field applied in the direction of the original poling field, the ceramic will expand or contract depending on the polarity of the field. Conversely, applying a strain to expand or compress the ceramic will generate an easily measured electric charge. The use of piezoceramics in sensors has been well established for over 50 years. They are attractive as actuators for smart structures because they can be fabricated as very small elements with a wide variety of shapes, and have the ability to generate large, precisely controllable forces in response to electrical voltage. Their principal shortcoming is their limited strain capability (about 0.02–0.04% for repeated and reliable operation). Also, like most ceramics, they fracture easily in tension.

Shape memory alloys make use of an effect invented and patented in 1965 at the Naval Ordnance Laboratory. The first, and still most widely used, is nitinol, an alloy of nickel and titanium. When heated above a critical transition temperature, nitinol undergoes a phase transformation in its crystalline state. In its low-temperature martensitic phase, nitinol is remarkably malleable. Upon heating above the transition temperature, nitinol changes to an austenitic phase and recovers plastic strains that were induced in the martensitic phase. The usual use is to train the alloy to a desired shape in the austenitic phase. The alloy can then be plastically deformed in the martensitic state. When nitinol is heated, the phase transition back to the austenitic state results in the recovery of up to 8% of the plastic deformation. A shape memory alloy appears to remember and return to its trained shape.

Actuators fabricated of shape memory alloys can generate both large forces and large displacements. Small changes in the alloy composition and addition of other metals allow the transition temperature to set in a range of -50 to $180°C$ (-58 to $356°F$). The phase change to austenitic is also accompanied by an increase up to a factor of 4 in Young's modulus and an increase up to a factor of 10 in yield strength. The primary shortcoming of actuators made of shape memory alloys is the need to change their temperature by up to $30°C$ ($54°F$) to induce the complete actuation cycle. The operational frequency of shape-memory-alloy actuators is thus limited by the rate at which heat can be removed to cool the alloy back to the martensitic phase. In most cases, this limits such actuators to quasistatic operation.

One of the most important areas of current development is the search for a smart material that can combine the characteristics of piezoceramics and shape memory alloys to provide a smart material actuator capable of both high strains and fast operation.

Synthesis. Smart materials, which on their own tend to be poor structural materials, are integrated into structures fabricated from more conventional structural materials to synthesize smart structures. Beyond the development of better actuator materials, a major developmental need is to provide a high degree of integration of the structures, sensors, and actuators, and ultimately of the control systems. The simple addition of smart materials to dumb structures has proven largely unsuccessful at producing smart structures. A fully integrated design approach to construction of smart structures is required and has been under development for a few years.

The widespread development and use of composite materials should enable the synthesis of smart structures. Composite materials are typically fabricated from layers of fibers that are bonded together by a matrix, usually a resin adhesive system. This fabrication technique permits facile embedding of actuators and components made of smart materials, thus enabling the high degree of integration needed for smart structures. The current research in synthesis of smart structures can be divided into four areas: design and modeling, sensors and actuators, manufacturing integration, and control systems.

Design and modeling developments are needed to accurately predict both the operation of smart structures and the reliability and long-term performance of the smart materials. Goals are to use smart materials for sensors and actuators. For the sensors, research is concentrated on development of low-cost sensor systems based on optical fibers and on development of new sensors to support monitoring of structural health. Current research on actuators is concentrated on new and improved smart materials and on innovative designs for devices that will increase the displacement of piezoelectric-based actuators. One such design is the inchworm actuator, which makes use of the small but very rapid displacements of piezoelectric actuators to provide much larger but slower displacements.

The integration of smart materials and structures has seen rapid development in the past few years with a primary emphasis on producing smart structures at an affordable cost. Methods for installing smart materials sensors and actuators during fabrication of composite structures and the interfacing of control systems to the embedded components represent a major technological breakthrough. The technology of feedback control systems is well developed for cases where the detailed response of the system can be accurately modeled. This is not currently the case for complex smart structures. In addition to the pursuit of better modeling methods, adaptive control systems that do not depend on highly accurate models are under development. Some of these adaptive control systems are based on the self-learning capabilities of artificial neural networks.

Active structures. Also called intelligent or adaptive structures, active structures are smart structures that utilize highly integrated actuators and advanced control systems to perform functions such as vibration control and noise control. One major application is suppression of vibration on orbiting spacecraft. Such vibrations, although very small, frequently disturb sensitive instruments and reduce their performance. Suppression of these vibrations (termed jitter in optical systems) and the precision alignment possible

with smart structures are enabling technologies for some space missions such as large orbiting astronomical observatories based on optical interferometry. Aircraft performance will also reap large benefits from the development of active structures. Small adjustments in the shape of airfoils can have major benefits in reducing drag and increasing fuel efficiency. Current aircraft are controlled by movement of discrete control surfaces such as ailerons, elevators, flaps, and rudders. Smart structures will eventually do away with these discrete surfaces by controlling aircraft by warping entire wings (reminiscent of the methods used by the Wright brothers). The benefits will include reduced drag, higher speeds, and improved stealth capability.

Structural health monitoring. Aging infrastructure (buildings, bridges, and the commercial airline fleet) coupled with developments in smart structures that include integrated sensors has prompted a strong development effort in the field of structural health monitoring. The sensor systems of smart structures will be used to monitor and record loading conditions and to detect and provide advance warning of incipient structural failure. The technological challenge is that while current sensors are capable of directly measuring parameters such as load conditions, the measured parameters provide only an indirect indication of structural integrity and cannot detect incipient structural failures. In metallic structures the primary detection problems involve corrosion and fatigue-induced cracking. In composites the primary detection problems involve delaminations and microcracking, usually caused by low-velocity impact damage. Detection of these conditions is based both on new sensors and on development of information-processing methods that correlate information provided by a large number of sensors.

For background information *see* COMPOSITE MATERIAL; CONTROL SYSTEMS; EMBEDDED SYSTEMS; PIEZOELECTRICITY; SHAPE MEMORY ALLOYS; STRESS AND STRAIN; TRANSDUCER in the McGraw-Hill Encyclopedia of Science & Technology.   Edward V. White

Bibliography. B. Culshaw, *Smart Structures and Materials*, 1996.

Soil

Recent advances in time domain reflectometry (TDR) soil-water measurement technology have produced instrument systems which operate reliably in a variety of difficult soils, including saline soils, heavy clay soils, soils with high organic content, and soils having extremely nonuniform vertical profiles. Some of these systems can measure water content profiles to a depth of 47 in. (120 cm) with a single probe, or to greater depths with two or more profile probes. A single instrument can be used to automatically interrogate multiple probes at distances up to 984 ft (300 m) by using inexpensive technology originally developed for the cable television industry. Time domain reflectometry systems can provide high accuracy at all water contents, yet require no soil calibration for nonclay applications. Measurement accuracy has been documented, and simple methods have been outlined that allow users to calculate accuracy for specific applications. The simplest measurement probes consist of two or three parallel stainless steel rods that are inserted into the soil (**Fig. 1**a). More complex probes are made from epoxy and stainless steel. Since about 1980, laboratory-based time domain reflectometry systems have evolved into robust systems appropriate for field use. These employ sophisticated electronic techniques that allow automatic and reliable collection of data.

Applications. Reliable and accurate measurements of soil-water content are required for a wide range of commercial and scientific applications. For example, accurate determination of water use by crops or trees allows more effective irrigation planning and scheduling. Recent trials on agricultural crops have demonstrated that irrigation scheduling, based on the measurement of vertical water-content profiles, can lead to a 20% reduction in water use and a concomitant increase in yield. Automatic multipoint water-content profile measuring systems are being used at landfill sites to monitor the movement of dissolved leachates. In other applications, time domain reflectometry systems are being evaluated for transpiration-type biological tertiary water-treatment facilities for sewage systems and nuclear power plant cooling water. In these applications, the soil-water content of a soil-plant basin is held at the optimum level to maximize the total transpiration. Reliable measurements of soil water are also critical in the forest industry, whether to determine if soils are dry enough to resist compaction from the use of heavy machinery, or to identify planting windows for reforestation.

Water-content measurement. A time domain reflectometry system employs circuits and techniques capable of measuring the speed of light. Its operation is similar to radar, except that pulses are transmitted along wires or cables rather than through free space. In conventional use, time domain reflectometry is employed to measure the distance from a transmitter to a discontinuity (such as a break) in a radio-frequency transmission line cable. Any discontinuity in the cable results in a distinct reflection of the transmitted pulse from the location of the discontinuity. This application is sometimes known as wire radar. For soil-water-content use, time domain reflectometry is employed to measure the time for the transmitted pulse to travel between two artificial discontinuities that are a known distance apart in a transmission line surrounded by soil.

A basic time domain reflectometry water-content measurement system consists of parallel metal rods, typically about 0.2 in. (6 mm) in diameter, held in place by an insulated header, and attached at one end to a time domain reflectometry instrument by a coaxial cable. The rods form a radio-frequency transmission line and, in a typical installation, might

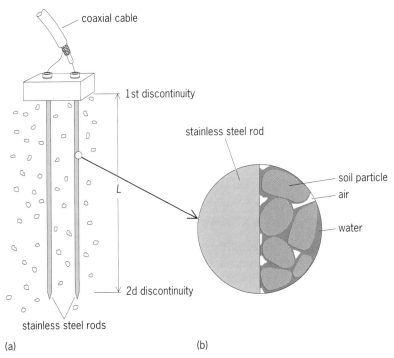

Fig. 1. Basic time domain reflectometry (TDR) probe. (a) Macroscopic view. Water content is proportional to the time for the transmitted pulse to travel between the two discontinuities. The first discontinuity is at the air–soil interface, and the second is at the rod ends. (b) Enlargement of portion of a. Microscopic view of water distribution within the soil particle matrix.

be 12 in. (30 cm) long and 1.2 in. (3 cm) apart. The instrument applies a very fast (0.2-nanosecond rise time) voltage pulse to the rods. If the rods are surrounded by air, the pulse travels down the pair of rods at the speed of light (c), is reflected from the end of the line, and returns up the rods at the same speed and back to the instrument. The total time taken by a pulse traveling along the rods in both directions through air (T_{AIR}) is $2L/c$, where L is the rod length. If the rods are surrounded by a medium other than air, the total travel time (T) is longer than T_{AIR}. The ratio T/T_{AIR} is 1.55 for a typical oven-dried agricultural soil and 9 for water. The fact that water has a much higher ratio than soil allows accurate measurement of water content, even in unknown soil types. Time domain reflectometry measures the average water content over the entire length of the rods, and out from the rods for a distance of about one rod diameter.

Accuracy and reliability considerations. A soil time domain reflectometry system provides a measure of volumetric water content, that is, the fraction of water in a given volume of soil. It is a dimensionless quantity but is sometimes expressed as cubic meter of water per cubic meter of soil, percent of water, or even centimeter of depth of water per centimeter of depth of soil. For many soils there is a linear and predictable relation between water content and T/T_{AIR}. If the transmission line assembly (the probe) is placed in an empty container, T/T_{AIR} will be 1.0. If the container is then filled with water, T/T_{AIR} will

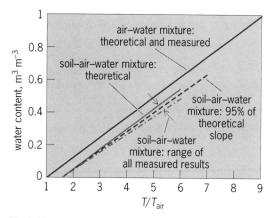

Fig. 2. Measured and theoretical curves to convert measured time intervals into water content for simple air–water mixtures and soil–air–water mixtures; each of these curves has the same slope; a good slope estimate for all soils is 95% of theory.

be 9.0 (the square of this number gives the dielectric constant, which for water is 81 at 68°F or 20°C). If the container is half full, T/T_{AIR} will be 5.0, and so forth, as shown by the air–water curve in **Fig. 2.** The slope of this curve is determined by the dielectric constant of water, which is only slightly temperature dependent. Such a container makes a very useful standard to verify the accuracy of a time domain reflectometry system.

Time domain reflectometry measurements in soil are not complicated. In its simplest form, wet soil can be considered to be a mixture of air, water, and mineral matter. For dry soil, the fraction of the total volume occupied by air is known as the porosity, a dimensionless quantity which is typically about 0.40–0.45 (for peat soils this fraction can be as high as 0.8–0.9). A useful way to think about the mixture is to assume that water added to the mixture displaces the air in the interstices between the mineral matter (Fig. 1b), and when all the air has been replaced with water the soil has reached saturation. Thus the water content of a saturated soil is typically 0.40–0.45 $m^3 \, m^{-3}$.

It may be assumed that the water is uniformly distributed throughout the volume of soil. The theoretical relationship between water content and T/T_{AIR}, derived only from these simple assumptions, is shown by the soil–air–water curve. This curve has a zero intercept which can easily be measured by using oven-dried soil, and a theoretical slope identical to the slope of the air–water curve. The measured intercept for all agricultural soils falls within a narrow range around 1.55 (1.4–1.77). The measured slope for an ideal soil such as sand is identical to the theoretical slope, and the measured slopes for all soils published since about 1980, with the exception of a few clay soils, fall within the range of 100–90% of the theoretical slope. The latter data include soils with high organic, high salt, and sig-

Comparison of time domain reflectometry (TDR), oven-drying, and neutron probe water-content measurement methods			
Consideration	Time domain reflectometry	Oven drying	Neutron probe
Requires soil calibration?	No	No	Yes (otherwise hydrogen atoms from nonwater media such as organic material are counted as water)
Requires special adjustment upon installation?	May require parameter adjustment in heavy clay, very saline soils	No	No
Linear response to water content?	Yes	Yes	No for modern, high-sensitivity instruments (effect of detection chamber)
Time required to obtain a reading?	~1 min	~48 h	~1 min
Soil disturbed with each measurement?	No	Yes	No
Nearby soil compacted by trampling of operator?	No if cable is used	Yes	Yes
Easily measured close to the surface?	Yes	Yes	No
High spatial resolution?	Yes	Yes	No
Volume of soil interaction?	Relatively small, adjacent to rods	Small to large, depending on volume sampled	Large, but decreases with increasing water content
Remote monitoring of rapidly change water content?	Yes	No	No
Profile measurements?	Yes	Yes with difficulty	Yes (manual)
Continuous and automatic monitoring of a water-content profile?	Yes	No	No
Automatic multiplexing of many probes to one instrument?	Yes	No	No
Maximum distance from probe to instrument?	300 m	Not applicable	Not applicable
Easy to verify accuracy in the laboratory?	Yes, using water and sand	Yes	No

nificant clay content. Thus a reliable method to convert from the measured time interval T to water content is to assume an intercept of 1.55 and a slope equal to 0.95 of the theoretical slope. For certain clay soils, where there is a significant physical or chemical interaction between the water and the soil particles, the relationship between water content and T/T_{AIR} is yet to be fully established.

The principal historical problem with time domain reflectometry systems has been that many soils having even moderate clay, salt, or organic content attenuate and distort the natural pulse reflections necessary for making the time-interval measurements. Modern techniques use active electronic elements in the probe to create artificial reflections, or special insulating sheaths around the probes to isolate the probe from the effects of ions in the soil. These techniques increase reliability in ordinary soils, greatly reduce the problems with difficult soils, and allow the fabrication of single-piece profile probes. Time domain reflectometry does require care in difficult soils to ensure that the detection parameters are correctly set to match the pulse reflections. Measurement systems are now able to achieve accuracies of 0.025–0.04 m^3 m^{-3} over the full measurement range and resolutions of better than 0.01 m^3 m^{-3}. Accuracy can be conveniently verified by using containers filled with dry sand, saturated sand, and water.

Oven-drying and neutron probe techniques. A standard technique for determining soil-water content in the field involves the collection of a soil sample which is then sealed and transported to a laboratory, where it is weighed, oven-dried for 2 days, and then reweighed. This provides a measure of the gravimetric water content, that is, the weight of water per unit weight of soil. The bulk density of the soil must be determined to convert the gravimetric water content to volumetric water content. This requires that the volume of the soil sample be known. A comparison of oven-drying (gravimetric) with time domain reflectometry techniques is given in the **table**. Critical disadvantages of the oven-drying technique, however, are its destructive nature and the long time required to process samples.

An alternative technique, involving the measurement of the density of hydrogen atoms, has been widely used in the field since the early 1960s. Fast neutrons are generated in a small radioactive source and radiated in all directions. When a neutron strikes a hydrogen atom, it is slowed (moderated) to a thermal velocity and backscattered. A detection chamber is used to count the thermal neutrons over a fixed period. For field measurements, an aluminum access tube is inserted into the ground, and the source/ detection assembly is then lowered into the tube to a given position. A count is made, and then the detection assembly is moved to a different depth. The relation between the thermal neutron count and soil-water content is soil and site specific. Therefore, extensive calibration is required to obtain acceptable measurement accuracy. Another disadvantage of the neutron probe technique is that, like oven drying, it cannot be automated and does not allow the continuous monitoring of many points.

Advantages of both oven-drying and neutron probe methods include a larger sampling volume than that used in time domain reflectometry, although this volume is poorly defined for the neutron probe.

For background information *see* AGRICULTURAL SOIL AND CROP PRACTICES; RADAR; SOIL; SOIL CONSERVATION; SOIL ECOLOGY; TRANSMISSION LINES in the McGraw-Hill Encyclopedia of Science & Technology.   W. R. Hook; N. J. Livingston

Bibliography. W. R. Hook et al., Remote diode shorting improves measurement of soil water by time domain reflectometry, *Soil Sci. Soc. Amer. J.*, 56:1384–1391, 1992; W. R. Hook and N. J. Livingston, Errors in converting time domain reflectometry measurements of propagation velocity to estimates of soil water content, *Soil Sci. Soc. Amer. J.*, 60:35–41, 1996; G. C. Topp, J. L. Davis, and A. P. Annan, Electromagnetic determination of soil water content: Measurements in coaxial transmission lines, *Water Resour. Res.*, 16:574–582, 1980; N. Wilson, *Soil Water and Ground Water Sampling*, 1995; S. J. Zegelin, I. White, and D. R. Jenkins, Improved field probes for soil water content and electrical conductivity measurements using time domain reflectometry, *Water Resour. Res.*, 25:2367–2376, 1989.

Soil conservation

Most studies of soil erosion have focused on cultivated lands. From such studies, much is known about the erosional characteristics of cultivated soils and about the erosional effects of different crops and cultivation practices, especially in countries such as the United States where soil erosion research has been extensive. The research has also been beneficial because a better understanding of the processes of soil erosion has aided the development of more effective soil conservation measures. However, serious problems involving soil erosion continue to threaten such key activities as food and hydroelectric power production throughout the world. As reservoirs fill with sediment, it becomes apparent that patterns of soil erosion in entire watersheds, not just the cultivated portions, need to be examined.

Mountain-land erosion. Mountain lands are especially vulnerable to soil erosion because the steep slopes accelerate the downward movement of water and soil. Mountain regions may be sites of hydroelectric power production as well as agriculture and other economically important activities, so soil loss on mountain slopes, and the consequent input of sediment into mountain rivers and reservoirs, threatens local and national economies. One characteristic that makes erosion research particularly challenging is that components of the mountain landscape— slopes, soils, vegetation, and human uses—vary dra-

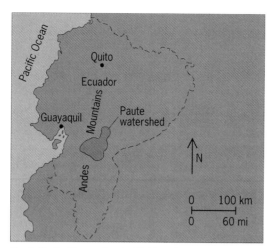

Fig. 1. Paute River watershed in Ecuador.

matically over short distances. Thus, results obtained from a study done in one erosion plot may not apply to other locations, even on the same mountainside.

Portable rainfall simulators. One way to better study soil erosion conditions in mountain regions is with portable rainfall simulators. Recent experiments in the Andes Mountains of South America using a portable rainfall simulator to simulate small storms at many different sites provide important new insights into the erosional effects of different mountain-land uses and challenge existing watershed management strategies.

Using a portable instrument enabled researchers to study soil erosion at over 100 sites in the Paute watershed in the Andean country of Ecuador (**Fig. 1**). The portable rainfall simulator (**Fig. 2**) is made of Plexiglass, is mounted on a standard surveyor's

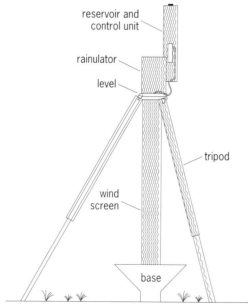

Fig. 2. Portable rainfall simulator.

tripod, requires only a few quarts (liters) of water per rainstorm, and can be carried to remote areas. To conduct a rainfall experiment, the simulator is leveled, a metal base piece is inserted in the soil directly below the tripod, and water is poured into the Plexiglass reservoir. Water flows from the reservoir into the open-topped rainulator cylinder, where a constant depth of water is maintained so the pressure to form raindrops remains constant throughout the experiment. Drops form on 91 surgical steel wires, each wire fitting through a carefully machined orifice in the base of the rainulator. When a drop reaches a critical size, it falls to the ground, protected by a cylindrical Plexiglass wind screen.

Example from Ecuadorian Andes. The Paute River watershed in Ecuador occupies 2002 mi² (5186 km²) of a populated mountain region containing a variety of land uses, including agriculture (predominantly corn), hydroelectric power production for the national power grid, local industry, and an increasingly urban population. People have lived, farmed, and built roads on Ecuadorian mountainsides for thousands of years. The mountain locations attracted settlement because of their favorable climate, fertile soil, and access to water and firewood. Settlement in the mountains also reflected positive cultural factors, such as maintaining the integrity of distinct community or tribal groups, and negative factors, such as coercion by or escape from dominant political forces. Where population density was low and soils were deep, soil erosion was not an important problem in Ecuadorian mountain settlements, but today soil erosion and reservoir sedimentation have reached critical proportions.

Land uses tested in the Paute watershed soil erosion study were fields of mature corn plants, recently plowed agricultural fields, pastures, forests, and abandoned or fallowed agricultural lands. Each experiment involved a 30-min simulated rain; the median rainfall intensity was a relatively modest 1 in./h (25 mm/h). At 5-min intervals, the rainfall intensity was checked, and the ground surface inside the metal base piece was observed to see whether all of the simulated rain had infiltrated into the soil or whether some of the water was pooled on the surface. In the latter case, the water was collected by using gentle suction, and its volume recorded. During a natural rainstorm, with no base piece, this water would have become rainfall runoff. The simulated runoff was then filtered to capture the sediment entrained in it, and the sediment was later dried and weighed. At the end of each 30-min experiment, sediment that had splashed onto the flared rim of the base piece was also collected. Combining sediment from the simulated runoff and the base piece gave a total (dry) weight of sediment detached by each 30-min simulated rainfall. This procedure allowed the researchers to compare the relative erosion potential of sites throughout the watershed.

As had been expected, no soil erosion occurred in forested sites, and very little on pasture lands with vigorous grass cover (see **table**). The most dramatic

Results of rainfall simulation experiments in Paute watershed, Ecuador

| | Eroded sediment, oz/in. (g/mm) of simulated rain | | Runoff, % of simulated rain | |
Land use	Median	Maximum	Median	Maximum
Abandoned or fallow	0.4 (0.4)	3.3 (3.7)	0.35	0.89
Cultivated	0.4 (0.4)	1.5 (1.7)	0.02	1.00
Recently plowed	0.3 (0.3)	1.0 (1.1)	0.08	0.97
Pasture	0.1 (0.1)	0.4 (0.4)	0.38	1.00
Forest	0.0 (0.0)	0.0 (0.0)	0.00	0.00

revelation of this study was that abandoned lands are important contributors of eroded sediment and rainfall runoff. The greatest amount of sediment detachment in a single experiment occurred on the bare soil of an abandoned field. Although sediment detachment on abandoned lands was not significantly greater than that from cultivated lands, abandoned lands absorbed significantly less of the rainfall, yielding more runoff than any land use except well-trampled pasture.

Soil erosion and abandoned fields. Abandoned lands in this Andean region do not quickly regrow a vegetative cover that protects them from the erosional forces of rain, flowing water, and wind. To understand why, it is important to know more about the climate, the ongoing activities in rural areas, and the reasons that lands are abandoned.

If a vegetative cover is established soon after land abandonment, the soil is protected from erosion. In Andean Ecuador, the combination of a marginally dry climate and the effective moisture loss as rainfall becomes surface runoff creates difficult conditions for plant growth, especially on lands that are already degraded by soil erosion and have little of their initial moisture-holding capacity. Whether or not land was badly degraded at the time of abandonment, its ability to support plants is also compromised by the unregulated grazing of domestic livestock, most often sheep and cattle. The animals eat the few plants able to sprout, and compact the soil by trampling so that rainfall runoff is high, accelerating soil erosion.

One reason why people abandon agricultural lands is impoverished soil. As the soil becomes nutrient depleted or lost to erosion, the productivity of the land declines, and its inhabitants are forced to purchase expensive fertilizers or seek work elsewhere. Another reason is the reduction of the farming population. Throughout the history of the world, famine, war, and plagues have dramatically reduced local populations, causing land abandonment. Today land is sometimes left behind by the younger generation who choose the amenities of urban life over the isolation, hard work, and risk of a rural life-style in relatively remote areas.

Geographers have come to recognize that depopulation of mountain lands is driven more by socioeconomic forces outside the mountain region than by changes in the region itself. Better transportation systems in mountain regions enable the inhabitants more economically to purchase than to grow food and fodder. The mechanization of farming leads to abandonment of fields too steep or inaccessible to tractors and other farm machinery. Furthermore, radio and television, now found throughout the countryside, profoundly shape the desires and priorities of rural people, often motivating them to leave the land.

Watershed management policy. As rural to urban migration throughout the world leads to further land abandonment, whether temporary or permanent, there will tend to be an increase in erosion-caused problems of decreasing productivity, increasing runoff, desertification, and downstream sedimentation. Although soil erosion may lead to land abandonment in some cases, it is accelerated by land abandonment in marginally dry landscapes with unregulated grazing, leading to a rapid and long-lasting decline of productivity of abandoned or even fallow lands. Sustainable watershed management decisions require understanding the hydrologic and erosional behavior of soils under different land uses. Although accelerated soil erosion in mountain regions is often linked to cultivation practices on steep hillsides, the Ecuadorian study demonstrates that abandonment of cultivated fields can pose an even greater risk of rapid runoff and soil erosion. It also shows the ineffectiveness of allowing previously cultivated lands to remain idle without both planting a cover crop and protecting the crop from grazing animals. The rapid decline in soil productivity and regional effects of erosion associated with abandoned mountain croplands challenge traditional approaches to watershed management which, in democratic countries such as Ecuador, have been based on the cooperation of on-site landowners.

For background information *see* AGRICULTURAL SOIL AND CROP PRACTICES; EROSION; FOREST SOIL; LAND-USE CLASSES; SOIL; SOIL CONSERVATION in the McGraw-Hill Encyclopedia of Science & Technology.

Carol P. Harden

Bibliography. P. Blaikie and H. Brookfield, *Land Degradation and Society*, 1987; C. Harden, Interrelationships between land abandonment and land degradation: A case from the Ecuadorian Andes, *Mount. Res. Dev.*, 16(3):274–280, 1996; N. Hudson, *Soil Conservation*, 3d ed., 1995; I. McQueen, *Devel-

opment of a Hand-portable Rainfall Simulator In-filtrometer, USGS Circ. 482, 1963; G. O. Schwab et al., *Soil and Water Conservation Engineering*, 4th ed., 1992; J. Thompson and J. Pretty, Sustainability indicators and soil conservation, *J. Soil Water Conserv.*, 51(4):265–273, 1996.

Soil microbiology

Measurement of the quantity of adenosine triphosphate (ATP) in a soil sample is a useful method for estimating soil microbial biomass, because ATP is closely correlated with other estimates of soil microbial biomass (see **illus.**). Adenosine triphosphate is an important energy compound in the metabolism of all living organisms. It is present only in living cells, and it occurs in the soil microbial biomass in a remarkably constant concentration. It has only a transient exocellular existence in soil; essentially, all measured soil ATP comes from living organisms. Thus, microbial biomass carbon (Bc) can be calculated from soil ATP content by Eq. (1), when Bc is

$$Bc = 86 \cdot ATP \qquad (1)$$

expressed as micrograms per gram of soil oven dry basis (dry weight of soil) and ATP as nanomoles per gram of soil oven dry basis.

Soil microbial biomass. The soil microbial biomass is the mass of all soil organisms less than 200 micrometers in diameter. In the biomass concept, all the microorganisms (bacteria, fungi, actinomycetes, protozoa, unicellular algae, yeast, and so forth) are considered collectively as a single living unit. This is called the holistic approach to soil microbiology.

Determination of the size of the soil microbial biomass and the activity of microorganisms in soil is of fundamental importance in soil ecology. Moreover, measuring the microbial biomass and its activity in soil makes it possible to estimate the total amount of energy stored in the biota and to monitor the fluxes of nutrients through it [for example, carbon (C), nitrogen (N), phosphorus (P), and sulfur (S)].

Measurement of the ATP content of soil entails preparation of the soil for microbiological analyses,

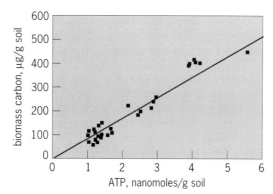

Relationship of ATP and microbial biomass carbon in English arable soils; Bc = 86 · ATP.

extraction of ATP from the soil, and determination of ATP in soil extracts.

Soil preparation. Preparation comprises sampling, sieving, storing, and incubation. Mishandling of soil at this crucial stage is one of the largest sources of errors in these microbiological analyses.

Sampling. Soil can be sampled with an auger or a soil corer. Soils must be adjusted to 40–50% water-holding capacity for analysis; this requires drying down if the soil is too wet or rewetting if too dry.

Sieving. To homogenize the soil and to remove small stones and gravel, the soil is passed through a 2-mm (0.08-in.) sieve. This operation does not affect the amount of biomass. During sieving, roots and small organisms (such as worms and insects) are removed by hand picking.

Storing. It is advisable to use the soil as soon as possible after sampling and sieving; storage should not exceed 3 months. If the soil has to be stored prior to the analyses, it must be kept moist in loosely sealed plastic bags at 5°C (41°F). Air-drying the soil invalidates the results, as this causes some biomass to be killed and the ATP content of the surviving biomass to fall.

Incubation. Incubation of soil permits the effects of sampling to subside before measurements of microbial biomass are performed. Incubation, under standard conditions of moisture and temperature, is also essential if ATP measurements are to be used as an index of microbial biomass. The bags of moist soils (about 40% water-holding capacity) are placed in an airtight drum together with a jar of soda-lime to keep the concentration of carbon dioxide (CO_2) low, and with some open jars of water to maintain humidity. The soil is incubated at 25°C (77°F) for 7–10 days prior to extraction of ATP.

Extraction of ATP. During ATP extraction from soil, it is necessary to disrupt the microbial cells in order to release their ATP, and to quickly inactivate the enzymes involved in its degradation; moreover, it is very important to minimize the amount of ATP adsorbed on soil surfaces (mineral colloids and organic matter). Effective extractants have to meet three requirements: complete cell lysis, effective and fast enzyme inactivation, and protection from adsorption on colloids. In addition, the extractant must not interfere with the subsequent quantitative analysis of ATP in soil extracts.

Extractants. Several different types of extractants have been proposed and tested for extraction of ATP from soil. Neutral or alkaline extractants such as chloroform–sodium bicarbonate and tris hydroxymethyl aminomethane (TRIS) + ethylenediamine tetraacetic acid (EDTA); organic solvents such as butanol, acetone, and dimethyl sulfoxide (DMSO); acidic extractants such as sulfuric acid (H_2SO_4), perchloric acid ($HClO_4$), and trichloroacetic acid (TCA); and commercial detergent–based extractants (nucleotide release agents) have been used.

There is no agreement as to which is the most effective extractant. Probably no single extractant is universally applicable to all soils. It is now generally accepted that a soil extractant for ATP has to be strongly acidic in order to inactivate endocellular and exocellular soil enzymes. Some extractants contain detergents or chloroform to improve lysis of microbial cells.

Ultrasonics is used to disrupt soil aggregates, homogenize the soil, and lyse microbial cells. Strong acids are included to inactivate soil enzymes. The most suitable are those with protein denaturation properties. Neutral or alkaline extractants are not effective in enzyme denaturation. Since the ATP molecule is amphoteric in acidic solution (that is, contains both positively and negatively charged functional groups), it is important to protect its sorption on both negative and positive charges on soil.

To protect ATP from adsorption onto positively charged soil surfaces, orthophosphate (PO_4^{3-}) is frequently added to the extractant either as the salt or as phosphoric acid. To compete with negative charges in soil, paraquat is added; this strongly cationic compound has two permanent positive charges on the molecule at pH 2, and it binds irreversibly to soil constituents. It has been proposed to use adenosine to compete for the same sites of adsorption of ATP.

Internal standard. During extraction from soil, ATP is lost partly by hydrolysis (enzymatic or chemical) and partly by adsorption onto soil colloids. To assess these losses and to make an appropriate correction for them, an internal standard should always be used. The function of the internal standard is to evaluate the apparent losses of native ATP resulting from the different sources of analytical interference. Three different approaches with the internal standard have been proposed: addition of ATP as the pure chemical compound; addition of ATP as microbial cells; and addition of radiolabeled ATP, for example, carbon-14 (^{14}C), hydrogen-3 (tritium; ^{3}H), and phosphorus-32 (^{32}P).

It has been found that the most accurate method for introducing an internal standard is to add ATP to the extractant and then measure its percentage of recovery. This datum is then used to correct the native ATP for incomplete extraction. Temporal and uncontrolled fluctuations in cellular ATP levels of microorganisms as measured in the laboratory make the use of live microbial cells unsatisfactory as an internal standard. The use of radiolabeled ATP does not permit consideration of all possible analytical errors in estimating ATP, because the procedure and the instrumentation to measure labeled and unlabeled ATP are different. Radiolabeled ATP can provide an estimate of the relative amounts of added ATP lost by dephosphorylation (removal of a phosphate group) and by adsorption during extraction, when the recovery of ^{14}C-ATP is compared to the recovery of unlabeled internal standard ATP.

Soil extractant ratio. The usual soil-to-solution ratio is 5 g of soil to 25 ml of extractant solution (that is, 1:5), but wider ratios are often used (such as 1:10 and 1:20). Particular attention should be paid to calcareous soils (that is, containing more than 10% calcium carbonate). This amount of calcium carbonate could be sufficient to neutralize up to 40% of TCA. This causes more counts per unit of ATP than in the acid reagent. In using the TCA extractant, the soil solution pH must never rise over 2.0; otherwise, enzyme inactivation would be insufficient, and the standard curve for ATP measurement would be invalid.

Extraction procedure. The extraction procedure has four steps: the actual extraction, cooling, filtration, and measurement. Two different sets of test tubes containing a precisely weighed amount of soil are prepared. The extractant is then added just before sonication, which is applied at full power for 2 min. Prior to extraction, all test tubes are kept on ice to prevent the warming of the sample. After sonication, the tubes are kept on ice for at least 5 min, and the suspension is then filtered. The filtrate is transferred to a scintillation vial along with an arsenate buffer and a firefly luciferin–luciferase system. A series of blanks (test tubes for each replicate for each soil sample) are sonicated for each new batch of extractant. Soil extracts can be stored at $-18°C$ ($-0.4°F$) for 1 month until ATP measurements.

In order to use the internal standard and evaluate its recovery after extraction, two batches of extractant are used: one containing only the extractant reagents (extractant A); the second containing 25 picomoles of ATP per 50 microliters of extractant, a concentration similar to that expected to be in the soil, usually 0.5–1.0 micromole (extractant B).

Analysis of ATP. Various methods—chromatographic, electrophoretic, and enzymatic—are available for the quantitative determination of ATP in soil extracts. Currently, the firefly bioluminescence method is the most rapid, sensitive, and reproducible assay for ATP in soil extracts. It is based on the light-emission reaction catalyzed by luciferase and its substrate, luciferin, in presence of ATP, as shown in reaction (2), where O_2 = dioxygen, AMP = adeno-

$$ATP + luciferin + O_2 \xrightarrow{\text{luciferase}}$$
$$AMP + PP_i + CO_2 + oxyluciferin + light \quad (2)$$

sine monophosphate, and PP_i = inorganic pyrophosphate. Light (as photon counts) can be accurately measured by luminometers or scintillation counters. The limit of detection of this method is about 1×10^{-15} mole.

Recently, a chromatographic method, high-performance liquid chromatography, has been proposed to measure ATP in soil extracts. Since ATP is present in soil extracts in very low concentrations (0.01–1.0 micromole)—about 100 times less than the detection limit of an ultraviolet-visible spectrophotometer—the ATP has to be chemically derivatized to a fluorescent compound. Fluorescent detectors have a limit of detection of around 10 picomoles. The derivatization has a poor reproducibility and introduces another source of error in ATP measurement, so this technique is not commonly applied.

For background information *see* ADENOSINETRI-PHOSPHATE; BIOLUMINESCENCE; BIOMASS; SOIL ECOLOGY; SOIL MICROBIOLOGY in the McGraw-Hill Encyclopedia of Science & Technology.   Marco Contin

Bibliography. J. M. Bollag and G. Strotzky (eds.), *Soil Biochemistry*, vol. 6, 1990; D. M. Karl, Cellular nucleotide measurements and applications in microbial ecology, *Microbiol. Rev.*, 44:739–796, 1980; P. Nannipieri and K. Alef, Estimation of adenosine triphosphate in soil, *Methods in Applied Soil Microbiology and Biochemistry*, pp. 194–203, 1995; J. R. Wilson (ed.), *Advances in Nitrogen Cycling in Agricultural Ecosystems*, 1988.

Solar energy

At the close of World War II, as high-power microwave technology became available, engineers and scientists reexamined the idea, originally proposed by N. Tesla, of transmitting electric power to distant locations by means of radio waves. In the 1960s, these efforts resulted in the idea of a solar power satellite, a concept that was extensively developed in the late 1970s. Field experiments concerning microwave energy transmission were carried out in the late 1980s and the 1990s.

Solar power station. An important milestone in the history of microwave power transmission was the 3-year study called the DOE/NASA Satellite Power System Concept Development and Evaluation Program, started in 1977. This program studied the concept of a solar power satellite, which would beam down an electric power of 5–10 GW from each such satellite to a rectenna site on the ground. (A rectenna is an antenna with a built-in rectifying system.) This concept was first proposed in 1968 to meet both space-based and Earth-based power needs (**Fig. 1**).

Fig. 2. MILAX microwave-driven airplane experiment. A microwave power of 1.25 kW is transmitted toward a model airplane from the transmitter car.

The solar power satellite would generate electric power of the order of several hundreds to thousands of megawatts by using photovoltaic cells of sizable area, and would transmit the generated power via a microwave beam to the receiving rectenna site. The realization of such a satellite requires not only the technological development of microwave power transmission with high efficiency and high safety but also scientific analysis of the impact of microwaves on the space plasma environment.

Field and space experiments. A program to develop a long-endurance, high-altitude platform called the Stationary High-Altitude Relay Platform (SHARP) was proposed in Canada in the 1980s. The goal was a crewless, lightweight airplane to circle for a long period at an altitude of about 21 km (13 mi) for the purpose of relaying radio communications signals over a wide area. On September 17, 1987, a 1/8-scale prototype SHARP flew on beamed microwave power for 20 min at an altitude of about 150 m (500 ft). The microwave beam was transmitted by a 4.5-m-diameter (15-ft) parabolic antenna transmitting 10 kW of microwave power at a frequency of 2.45 GHz. Two water-cooled magnetrons, each with 5-kW output power, were used. The parabolic antenna mechanically tracked the airplane, which flew inside a 50° cone.

A similar project, called Stratospheric Radio Relay Systems (SRRS), was proposed in Japan in 1990. Its objectives are similar to those of SHARP. In the SRRS project, five crewless airplanes are to be launched over Japan, so that these five platforms can cover most of the areas where communication demands are heavy. In parallel with the SRRS working group, a microwave-driven airplane experiment called MI-

Fig. 1. Concept of solar power satellite, designed to meet both Earth-based and space-based power needs.

Fig. 3. Payload section of ISY-METS rocket experiment.

LAX (Microwave Lifted Airplane Experiment) was conducted on August 29, 1992. The MILAX flew successfully at an altitude of about 15 m (50 ft) for 40 s, covering a distance of 400 m (1300 ft) over a straight course along a car-driving test range (**Fig. 2**). The microwave power beam was radiated and electrically controlled to point toward the fuel-free MILAX airplane by an active phased-array antenna. The total radiation capability was 1.25 kW, and the frequency used was 2.411 GHz in the ISM (industry, science, and medical use) frequency band.

On February 18, 1993, a rocket experiment to investigate microwave power transmission between mother and daughter units was carried out by using a Japanese S-520-16 sounding rocket. The experiment was named ISY-METS (Microwave Energy Transmission in Space during the International Space Year). One objective was to investigate nonlinear effects of high-power microwave radiation on the ionospheric plasma, in a manner somewhat like the MINIX (Microwave Ionosphere Nonlinear Interaction Experiment) rocket experiment conducted in 1983. This subject of research concerns one of the important problems that need to be overcome before the realization of a full-fledged solar power satellite project in the twenty-first century. In the ISY-METS, a newly developed active phased-array microwave transmitter, which had been used in the MILAX, was used to transmit the microwave power toward the daughter section. **Figure 3** shows the transmitting section with four paddles of the active phased-array antennas.

Spin-off technologies from research on microwave power transmission for the solar power satellite have recently attracted interest with a view to practical application. One technology is small-scale ground-to-ground power transmission without wires toward distant locations where wired power distribution networks either are unavailable or provide poor service. A collaborative field experiment was conducted in Japan to collect fundamental data on microwave power transmission. The transmitter was connected to a parabolic antenna 3 m (10 ft) in diameter, driven by a 5-kW magnetron. The rectenna array of size of 3.5 × 3.2 m (11.5 × 10.5 ft) was placed 42 m (138 ft) away from the parabolic transmitting antenna. The microwave frequency used was 2.45 GHz. The experiment began in October 1994 and ended successfully in March 1995 (**Fig. 4**).

Microwave power transmission toward an airship was demonstrated in October 1995 at the Wireless Power Transmission Conference in Kobe, Japan. The purpose of this experiment was almost the same as that of the MILAX as a prototype for the SRRS. Microwave power transmission toward a helicopter called SABER (Semiautonomous Beam Rider) was also demonstrated.

Technological development. The success of these field experiments is mainly due to the advance of the technology of microwave energy transmission. In the MILAX and ISY-METS experiments, an active phased-array system was adopted for the microwave transmitter. Two charge-coupled-device (CCD) cameras were used to trace the airplane or the daughter rocket. After the image of the target was recognized by using pattern recognition technology implemented on a microcomputer, the microwave phases fed to each antenna element were adjusted by a microcomputer to keep the microwave beam well focused on the target. The MILAX active phased-

Fig. 4. Transmitting parabolic antenna and receiving rectenna array in microwave power transmission experiment.

array transmitter was composed of five-stage gallium arsenide (GaAs) semiconductor amplifiers, 4-bit digital phase shifters, and circular microstrip antenna arrays. The transmitter was divided into 96 subarrays, each consisting of three antennas, one phase shifter, and one gallium arsenide amplifier. Each subarray could supply a 13-W microwave output.

In addition to computer-controlled beam steering, two systems called retrodirective transmitters were developed, in 1987 and 1996. The retrodirective system uses a phase-conjugate circuit which reverses the phase of an incoming pilot signal. In the 1996 system, one-third of the microwave power beam was used for the frequency of the pilot signal.

The receiving antennas used for the MILAX rectenna resembled neither the dipole-type antennas used in the JPL/Goldstone Ground-to-Ground Power Transmission Experiment (1975) nor those used in the MINIX and SHARP experiments. They were examples of a new type of microstrip circular patch antenna. For the airship experiment in Kobe, a dual-polarization patch rectenna was developed. The maximum radio-frequency-to-direct-current conversion efficiency of this rectenna was reported to be 81%. The weight–power ratio was 3.8 g/W.

A new element for rectifying the microwave energy beam has been proposed, called the cyclotron wave converter. It is a high-power rectifier that uses an electron beam, and is expected to rectify 100-kW microwave power to direct current with 85–90% efficiency. The cyclotron wave converter needs a direct-current power source to drive the system, unlike the conventional passive rectenna.

In the past, studies have almost always been conducted with 2.45-GHz microwaves. However, a few technologies for microwave energy transmission at other frequencies have been developed. A 10-GHz rectenna was developed with a radio-frequency-to-direct-current conversion of about 60%. Theoretical analysis of a 35-GHz rectenna has been carried out with a mathematical model.

A study of microwave energy transmission directed toward a micromachine in a small pipe has been initiated. The micromachine is a small robot with a small CCD camera, which moves in the pipe. The frequency is 14–14.5 GHz, and the diameter of the pipe is 15 mm (0.6 in.). The micromachine successfully moved in the pipe, powered only by the microwave energy.

Advanced projects. There are other advanced projects concerning solar power satellites and microwave energy transmission. In Japan, a small project is under study to launch the SPS2000, a 10-MW-class solar power satellite, into low Earth orbit. In 1995, a functional model of this satellite was developed. It is composed of solar cells, semiconductor amplifiers, phase shifters controlled by a microcomputer, and transmitting antennas. A 2.45-GHz microwave beam was generated based on the direct-current power generated by the solar cells, and was transmitted from the SPS2000 model to rectennas separated from it by about 2 m (6.5 ft).

In France, there are plans for ground-to-ground microwave energy transmission over a distance of 3 km (2 mi). The transmitting power is to be about 10–100 kW. The planned site for the experiment is Réunion Island, a department of France located in the Indian Ocean.

There are some United States projects which are relevant to solar power satellite microwave energy transmission technology. Besides the SABER project, the WISPER (Wireless Space Power Experiment) project would use both laser and 35-GHz microwave beams to transmit power from the ground to a satellite in low Earth orbit.

The most recent efforts toward solar power satellite microwave energy transmission are related to the International Space Station. Feasibility studies are under way for an experiment on the exposed platform of the Japanese Experimental Module (JEM) on the space station.

For background information *see* ANTENNA (ELECTROMAGNETISM); MICROWAVE; SOLAR CELL; SOLAR ENERGY in the McGraw-Hill Encyclopedia of Science & Technology.   Hiroshi Matsumoto

Bibliography. W. C. Brown, The history of power transmission by radio waves, *IEEE Trans. Microw. Theory Techniques*, MTT-32:1230–1242, 1984; X. Claverie (ed.), *Proceedings of SPS'91*, 1991; N. Kaya (ed.), *Proceedings of WPT'95*, 1995; Z. Van Daele and P. Delogne (eds.), *Space and Radio Science: 75th Anniversary of URSI*, 1995.

Soliton

In 1973 it was suggested that optical fibers could propagate solitons. An optical fiber made of silicon dioxide glass with a germanium-doped core is dispersive in that the group velocity of optical waves (that is, electromagnetic waves, as predicted by Maxwell's equations) depends on wavelength. The fiber is nonlinear because of the Kerr effect: The optical (refractive) index (and hence phase velocity) is a function of the intensity of the optical wave. Optical solitons can form in glass if the dispersion and the Kerr effect balance. The dispersion is called positive if the group velocity increases with increasing wavelength, and negative if it changes in the opposite way. The Kerr effect in a fiber is positive, meaning that the optical index increases with intensity. A positive Kerr effect can balance negative dispersion, making the existence of solitons possible. Optical fibers have negative dispersion for wavelengths longer than 1.3 micrometers, but dispersion can be controlled somewhat by the proper control of the index profile of the fiber core and the surrounding cladding. The core has a slightly higher index than the cladding, thus providing guidance of the optical beam by total internal reflection. A single-mode fiber propagates one characteristic pattern of a mode with one single intensity maximum of the intensity profile transverse to the direction of propagation. Solitons in such single-mode fibers were predicted at wavelengths longer than 1.3 μm. *See* NONLINEAR OPTICS.

Communication with solitons. Optical fibers have another remarkable property: They have very low loss at a wavelength of 1.5 μm. Light propagating at this wavelength loses only a few percent of its power within a fiber 1 km (0.6 mi) in length. In 1984 the use of solitons for long-distance repeaterless transoceanic fiber-optic digital signal transmission was proposed. A soliton pulse stands for a one, an empty time interval for a zero. The transmission would be repeaterless in that the signal would not be detected and retransmitted every 100 km (60 mi) or so, as was done exclusively until 1995 along transoceanic fiber cables. The loss in the fiber would be compensated by amplification, but no detection and regeneration would be employed. The pulse stream would be maintained over transatlantic distances (5000 km or 3000 mi) or transpacific distances (9000 km or 5500 mi). This new method of communication would allow changes of the bit rate of transmission by modification of the transmitter and receiver, without modification of the cable—changes impossible in a repeatered fiber communication system.

Solitons have another remarkable property. If a soliton collides with another soliton with a different carrier wavelength (center frequency), traveling at a different speed because of the group velocity dispersion, the two pulses recover completely their original shape after the collision. **Figure 1** shows a computer simulation of such a collision in terms of the intensity profile of two solitons as a function of time, at successive positions along the fiber. This property of solitons makes wavelength division multiplexing (WDM) of soliton communications possible. If the wavelength-division multiplexing system is properly designed, solitons in different wavelength channels can pass through each other without affecting each other, and therefore without causing crosstalk, an effect that arises generally when the transmission medium is nonlinear.

Fiber amplifiers in the wavelength range of minimum fiber loss were developed in the 1980s by using erbium as the active medium, the dopant. They are pumped by compact semiconductor laser diodes, the size of a grain of salt, operating at a (shorter) wavelength (1.48 μm or 0.98 μm) in one of the higher-lying (in energy) absorption bands of the 1.54-μm amplifying wavelength. The diode pump lasers are highly efficient, emitting powers of the order of 100 mW, sufficient to pump a fiber amplifier. The pumping diodes of 200 amplifiers, roughly the number required for a transoceanic one-way transmission, consume only about 500 W total electrical power. The amplifiers compensate for the fiber loss by the process of stimulated emission. Signal photons stimulate a transition of erbium atoms from an excited energy level to a lower-lying level, thereby amplifying the signal coherently, but not without a cost. In addition to the stimulated emission, there is spontaneous emission, as was realized by A. Ein-

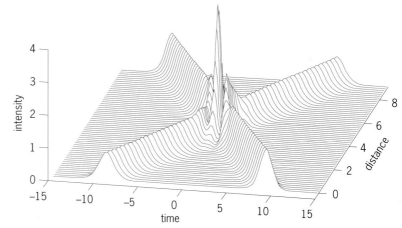

Fig. 1. Two solitons before, during, and after collision. The scales of distance, time, and intensity are in arbitrary units.

stein. An excited atom can decay spontaneously, and incoherently with respect to the signal, thereby adding noise. *See* LASER.

The implementation of repeaterless transmission took some time. In a repeaterless system, in which the signal is not regenerated, the amplifier noise is not periodically removed, but propagates along with the signal, causing errors in detection. It was shown that the amplifier noise causes small changes in pulse carrier wavelength that generate a so-called random walk of the pulse carrier frequencies and pulse velocities. This random walk is transformed into timing jitter, which is called the Gordon-Haus effect. Pioneering experiments verified many predicted properties of soliton propagation, including this effect. Since a fiber 5000 km (3000 mi) in length with 200 amplifiers would make an experiment excessively costly, a loop was used of the order of 100 km (60 mi) in length, with three amplifiers. A digital signal composed of a pseudorandom sequence of solitons and empty intervals was launched in this loop, and was then recirculated as many times as necessary to simulate propagation at multiples of 100 km (60 mi).

The Gordon-Haus effect stood in the way of high-bit-rate (10 gigabit-per-second or greater) long-distance propagation until it was shown that this effect could be controlled and the propagation distances extended. This was demonstrated in experiments with amplitude modulators that retimed the pulses at intervals, a scheme that is termed active since it requires electronically driven modulators. These experiments inspired research for simpler implementations. Soon it was discovered that passive filters inserted at intervals in the amplifier stages of the cable prevented the full buildup of the random walk of soliton carrier frequency, thus reducing the Gordon-Haus effect. The cascade of filters, even though each has a bandwidth much larger than the soliton bandwidth, results in a net transmission bandwidth much narrower than the inverse bit rate. Yet, soliton communications is possible at the nominal bit rate because the soliton spectrum, even though narrowed by each of the filters, is rebroadened by the nonlinearity of the Kerr effect of the fiber. Purely linear transmission through the filter cascade, at the same bit rate, would be impossible. However, narrow-band amplifier noise could still pass the filters at the filter center frequency, causing errors in detection. The ingenious idea of skewing the center frequencies of the filters from one end of the cable to the other allows the linear additive noise to be attenuated by the filters at all frequencies. With this so-called sliding guiding filter scheme, unprecedentedly high bit rates and long transmission distances have been achieved experimentally.

Existing repeaterless cables. However, there is no commercial soliton communication system currently in existence. As often happens, successes in one technology encouraged developments in another. The bold idea of repeaterless long-distance soliton communications has found partial realization in the laying of repeaterless transoceanic cables between the United States, Europe, and Japan without the use of solitons. The scheme is linear in the sense that it does not utilize the Kerr effect in the fiber. Indeed, in this case the Kerr effect is a nuisance, causing signal deterioration that has to be combatted by changing the fiber dispersion along the cable from slightly positive to slightly negative and back. This dispersion management destroys the coherence among the phase delays induced by the fiber nonlinearity (that is, by the Kerr effect).

The bit rates of optical communications will undoubtedly increase in the future. For fixed signal-to-noise ratio, the intensity of the transmitted optical radiation must increase proportionally to the bit rate, aggravating the Kerr effect. Thus, at higher bit rates, solitons that balance the nonlinearity of the fiber by dispersion may find practical applications. It is also likely that some switching, clock-recovery, multiplexing, and demultiplexing functions will be performed all-optically, since electronic switching is limited in speed.

All-optical switching with solitons. A generic all-optical switch using solitons is illustrated in **Fig. 2a**. It consists of a coupler and a fiber loop, containing a section of erbium fiber for amplification. An incoming pulse is split by the coupler into two pulses countertraveling around the loop. In the absence of amplification, the two pulses have equal intensities. They acquire identical Kerr-induced phase shifts as they travel around the loop. As they meet upon return to the coupler, they interfere constructively to exit into port A, from which they entered. In the presence of gain, the clockwise-traveling pulse amplified at its entry experiences a larger effective index and hence a larger optical phase shift due to the Kerr effect than the counterclockwise pulse, which is amplified after passing the passive fiber. The two pulses are dephased with respect to each other. If the amplification is properly adjusted, the pulse exits in fiber B. This kind of intensity-dependent limiter

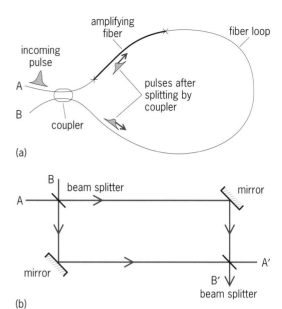

(a)

(b)

Fig. 2. Nonlinear amplifying loop mirror. (*a*) Diagram of device. (*b*) Analogous Mach-Zehnder interferometer.

can be used to remove noise background from a pulse stream: The low-intensity noise radiation is reflected into port A, whereas the pulses exit at port B. The device, which is called a nonlinear amplifying loop mirror (NALM), is really a Mach-Zehnder interferometer (Fig. 2b) folded back on itself. The Sagnac interferometer ports A and B perform the functions of both the input ports A and B and the output ports A′ and B′ of the Mach-Zehnder interferometer. A regular Mach-Zehnder interferometer with nonlinear propagation media in each of its arms and an amplifying medium in one arm performs the same function. However, it would be impossible to maintain the balance of the Mach-Zehnder interferometer, where the two interfering radiations travel through different arms, against environmental fluctuations. The virtue of the loop mirror is that both pulses traverse the same fiber segments. Changes of index induced by fluctuations in the environment are suppressed if these fluctuations are slow compared with the transit time.

Modifications of the nonlinear loop mirror can be used to perform other functions. The loop can operate as a pulse-activated switch. **Figure 3** shows one example. A control pulse is fed into the loop at one end and coupled out at the other end. If the signal pulse has a different carrier frequency or a different polarization than the control pulse, and the couplers are properly designed, the signal pulse is unaffected by the couplers. The clockwise-propagating signal pulse travels along with the control pulse and experiences a greater phase shift than the counterclockwise propagating pulse. If the intensity of the control pulse is chosen properly, the signal pulse can be made to exit through port B.

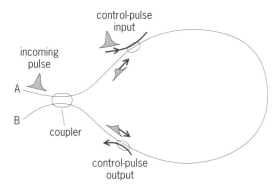

Fig. 3. Pulse-activated all-optical switch.

Self-guiding waves. Only the optical soliton phenomenon in which group velocity dispersion is held in balance by the Kerr nonlinearity has been discussed. There is current research in spatial solitons in which diffraction is held in balance by the Kerr nonlinearity, so that optical radiation can form its own waveguide in an initially uniform optical medium. There are, as yet, no practical applications of this phenomenon.

For background information *see* CROSSTALK; DISPERSION (RADIATION); INTERFEROMETRY; KERR EFFECT; OPTICAL COMMUNICATIONS; OPTICAL FIBERS; SOLITON in the McGraw-Hill Encyclopedia of Science & Technology.　　　　　　　　Herman A. Haus

Bibliography. N. J. Doran and D. Wood, Nonlinear-optical loop mirror, *Opt. Lett.*, 13:56–58, 1988; M. E. Fermann et al., Nonlinear amplifying loop mirror, *Opt. Lett.*, 15:753–754, 1990; H. A. Haus, Optical fiber solitons, their properties and uses, *Proc. IEEE*, 81:970–983, 1993; L. F. Mollenauer, Soliton transmission speeds, *Opt. Photonic News*, pp. 15–19, April 1994.

Space flight

While the United States space budget remained stable in 1996, international space flight activities were characterized by trends, recognizable in the two prior years, of incisive reductions in public spending for space programs, with increasing pressure on governments to shift such responsibilities to the private sector.

Despite shifts in economic priorities, 1996 was one of the most active years for sustaining operations and advanced developments in space, mainly in the United States with the progressive development of the International Space Station, as well as in Russia with the continued utilization of the space station *Mir*. The year also brought new discoveries by the Hubble Space Telescope, the European Infrared Space Observatory, and the deep-space probe *Galileo* at Jupiter, as well as the beginning of the National Aeronautics and Space Administration's (NASA) in-tensive robotic exploration program of Mars after a 20-year hiatus.

Commercial companies that were engaged in providing space services with a variety of expendable carrier rockets continued to prosper. However, some countries, particularly European nations, Russia, and China, encountered setbacks in commercial space activities.

The United States' space shuttle and Russia's Earth-orbiting *Mir* added two rendezvous-and-docking flights to the two joint missions of 1995, bringing the total flights of phase 1 of the development program of the International Space Station to four, with five more scheduled for 1997 and 1998. These missions are conducting joint onboard research preparatory to space station operations, gathering practical experience in joint on-orbit activities in the interest of risk mitigation, and testing assembly procedures for the space station construction phase.

A total of nine crewed missions from the two major space-faring nations carried 48 humans into space, including 4 women. Thus, the number of people (counting repeaters) launched into space since 1958 totaled 715, including 63 women (or 362 different individuals, including 30 women).

Significant launches for 1996 are listed in **Table 1**. The total number of successful launchings by various countries, amounting to 69 (out of 77 attempts), is broken down in **Table 2**.

United States Activity

The four reusable shuttle vehicles of the United States Space Transportation System continued carrying people and payloads to and from Earth orbit for science, technology, and operational research. The International Space Station program moved forward briskly toward inception of orbital assembly.

Space shuttle. During 1996, NASA successfully completed seven space shuttle missions, the same number as in the three preceding years.

STS 72. On *Endeavour*'s tenth sortie, January 11–20, the crew retrieved Japan's *Space Flyer Unit* (*SFU*) satellite launched by an H-2 rocket from Tanegashima on March 18, 1995. On January 14, they released the Office of Aeronautics and Space Technology (OAST)-Flyer, a Spartan-type free-flying platform carrying four autonomous experiments. The satellite was retrieved 2 days later. Also on January 14, two of the flight crew performed a 6 h 9 min spacewalk to evaluate a new portable work platform and to build spacewalking experience for assembly of the International Space Station. A second spacewalk by two astronauts took place on January 17.

STS 75. *Columbia*'s nineteenth flight was launched on February 22. On February 25, the crew began to deploy the Tethered Satellite System on the end of a thin 12.5-mi-long (20-km) tether line—a reflight from STS 46 in July 1992. However, subsequent experiments, including the generation of electricity by induction as the satellite and tether passed through the Earth's magnetic field, as well as research into the

TABLE 1. Some significant space launches in 1996

Mission designation	Launch date	Country	Main payload or mission
STS 72 (*Endeavour*)	Jan. 11	United States	Tenth flight of *Endeavour*; retrieval of *Space Flyer Unit* (*SFU*) satellite launched by Japan on March 18, 1995
NEAR	Feb. 17	United States	*Near Earth Asteroid Rendezvous*: first probe to study an asteroid (Eros) at close range, and first of NASA's small, low-cost Discovery-class spacecraft
Soyuz TM 23	Feb. 21	Russia	Mir 21 crew of two, trained with Shannon Lucid for joint work on *Mir*
STS 75 (*Columbia*)	Feb. 22	United States	Deployed 12.5-mi (20-km) Tethered Satellite System (TSS) to 12 mi (19 km), when an electric arc broke the tether; TSS was lost and reentered atmosphere
Polar	Feb. 24	United States	Last United States element of International Solar-Terrestrial Program research probes
IRS-P3	Mar. 21	India	Second successful launch of India's four-stage Polar Satellite Launch Vehicle (PSLV-D3); remote-sensing payload *IRS-P3*
STS 76 (*Atlantis*)	Mar. 22	United States	Third *Mir* link-up of International Space Station (ISS) phase 1; delivered Shannon Lucid and freight
Priroda	Apr. 3	Russia	Proton launch of last science module of multielement *Mir* complex, planned over a decade earlier
AMOS	May 16	Israel	First communications satellite of Israel; launched on Ariane 4
STS 77 (*Endeavour*)	May 19	United States	New record of four rendezvous events with two satellites: *Spartan* and *PAMS(STU)*
Cluster/Ariane 5	June 4	Europe	First test flight of new European heavy launcher; exploded on launch
STS 78 (*Columbia*)	June 20	United States	New duration record of almost 17 days; Life and Microgravity Sciences (LMS) Spacelab
Soyuz TM 24	Aug. 17	Russia	Mir 22 crew of three, with French researcher Claudie André-Deshays
STS 79 (*Atlantis*)	Sept. 16	United States	Fourth *Mir* link-up; delivered John Blaha and cargo, returned Shannon Lucid who set new world women's and United States crews' space record of 188 days 5 h
MGS	Nov. 7	United States	*Mars Global Surveyor*; first of new class of robotic Mars orbiters, launched on Delta 2 for Mars arrival in September 1997
Mars '96	Nov. 16	Russia	Mars Lander-Penetrator mission; lost in Pacific after failure of Proton rocket
STS 80 (*Columbia*)	Nov. 19	United States	Crew of five; unable to conduct two planned spacewalks because of jammed airlock hatch; deployed and retrieved two satellites: *Orfeus-Spas* and *Wake Shield Facility*
Mars Pathfinder	Dec. 4	United States	Second United States Mars probe; lander and six-wheeled minirover Sojourner; landing targeted for July 4, 1997 at Ares Valley
Bion 11	Dec. 24	Russia	Life sciences research capsule; launched on Soyuz-U rocket from Plesetsk with two monkeys aboard; one died after capsule return and recovery

dynamics of long tethers, could not be completed because the cable snapped near the shuttle's cargo bay about 5 h after deployment, with 12 mi (19 km) of tether reeled out, and the satellite was lost in an initially higher orbit. Later evaluations showed that the probable cause was arcing from the electrically charged tether (3500 V) to the deployment mechanism at an unsuspected small defect in the tether's insulation sheath. Activities then focused on the operation of the third Microgravity Payload (USMP 3), continuing research aimed at improving basic knowledge on materials under conditions of weightlessness. *Columbia* returned to Kennedy Space Center on March 9.

STS 76. For the third link-up with *Mir*, *Atlantis* lifted off on March 22, carrying a crew of six, including two women, and the private-industry-owned Spacehab module on its first commercial flight. After docking and hatch opening on March 23, the combined crew of eight conducted joint operations and transferred 5000 lb (2268 kg) of supplies and water to *Mir*. On March 27, a 6 h 2 min spacewalk, the first by United States astronauts with a shuttle attached to the station, mounted four experiments to the station's exterior for assessing its environment. After concluding science and technology experiments in Spacehab, *Atlantis* undocked on March 28. Astronaut Shannon Lucid, the first United States crew member to be ferried to *Mir* by shuttle transport, was left aboard the station to join Yuri Onufrienko and Yuri Usachev as an official crew member for the next 188 days. *Atlantis* returned on March 31 (see **illus**.).

STS 77. On May 19, *Endeavour* lifted off with six persons and, as primary payload, the commercial Spacehab module on its second flight, with nearly 3000 lb (1370 kg) of experiments and support equip-

TABLE 2. Space launches and attempts in 1996

Country	Number of launches	Number of attempts
United States (NASA, Department of Defense, commercial)	32	33
Russia	23	27
Europe (European Space Agency, Arianespace)	10	11
People's Republic of China	2	4
Japan	1	1
India	1	1
Total	69	77

Astronaut Shannon Lucid reading a book in the *Spektr* module aboard the *Mir* space station. *(NASA; Russian Space Agency)*

ment, an Inflatable Antenna Experiment on the free-flying *Spartan 207* platform, and a suite of four experiments called Technology Experiments for Advanced Missions (TEAMS). The *Spartan* was released on May 20, and the 50-ft-diameter (15-m) Inflatable Antenna Experiment inflated on its three 92-ft-long (28-m) struts—the most complex and precise inflatable space structure ever and the largest since the *Echo 2* satellite in 1964. Later, the structure was separated from *Spartan* and burned up in the atmosphere. The *Spartan* was retrieved on May 21, followed by further experiments, including observations of the behavior of and rendezvous with the passive aerodynamically stabilized, magnetically damped satellite *PAMS/Satellite Test Unit* (*STU*). Having set a new record of four rendezvous maneuvers with two satellites, *Endeavour* returned on May 29.

STS 78. A new duration record of almost 17 days was flown by *Columbia* from June 20 to July 7. Its main payload, the Life and Microgravity Sciences (LMS) Spacelab, was equipped to study both living organisms in the low-gravity environment, highlighting musculoskeletal physiology, and the subtle influences at work during the processing of various substances, such as alloy materials, when gravity effects are greatly reduced.

STS 79. *Atlantis* was launched on September 16, for the fourth docking with Russia's *Mir*, after several weeks of delays. It was the third flight to carry a commercial Spacehab in the cargo bay and the first featuring its double (two-segment) version. Docking occurred on September 18. The main objectives of the mission were the exchange of astronaut Lucid for astronaut John Blaha for a 4-month stay on *Mir*; hauling 4600 lb (2100 kg) of logistics, including food, clothing, experiment supplies, and spare equipment, to the space station; and returning 2200 lb (1000 kg) of Russian, European Space Agency (ESA), and United States science samples and hard-

ware to Earth. Five different experiments were turned off on the shuttle and rapidly transferred to *Mir*—the first such transfer of powered scientific apparatus. Hatch closure and separation of the two vehicles took place on September 23. When *Atlantis* landed on September 26, Lucid held the new world record for longest space flight for women and also for United States astronauts (188 days 5 h; see illus.).

STS 80. On its November 19–December 7 mission, *Columbia* carried the 7900-lb (3600-kg) reusable German *Orfeus-Spas II* ultraviolet observatory and the 4600-lb (2100-kg) *Wake Shield Facility*, on its third flight, with its 4750-lb (2160-kg) carrier system. The *Orfeus-Spas* satellite was released on November 19 and recovered on December 3. The free-flying *Wake Shield Facility* was deployed on November 22 and retrieved on November 25. Both satellite missions were successful. The crew also conducted biomedical and genetic experiments in weightlessness. The mission established a new duration record of 17 days 15 h 53 min for the space shuttle fleet.

International Space Station. In 1996, the International Space Station program was still targeting the first-element launch in November 1997. The development team of government and industry dealt with a myriad of details: major qualification and development testing, completion of flight components, delivery of software releases, and bringing all this activity together to ensure the timely availability of components and modules at the Kennedy Space Center for processing and integration with the space shuttle.

Phase 1. Phase 1 of the program, the joint United States–Russian effort to expand cooperation in human space flight, made significant progress with the two additional shuttle-*Mir* space missions. These link-ups ferried supplies required on board the space station as well as United States astronauts Lucid and Blaha. Lucid's arrival aboard *Mir* in March 1996 inaugurated a new period of permanent presence of United States astronauts in space. Embodied within phase 1 are a range of activities which provide the framework for United States–Russian cooperation in space, and the mechanism for facilitating the integration of Russia into full partnership in the program.

Phases 2 and 3. By the end of 1996, design and fabrication of flight elements and software for the first six United States assembly flights were essentially complete, with qualification testing under way. The first such assembly flight was scheduled for December 1997. Consisting of Node 1, with two pressurized mating adapters and two stowage racks, its hardware had been structurally completed and was being outfitted and tested.

Development programs continued in other countries of the 15-nation partnership as well. Significant changes to the United States–Russian cooperative plans were agreed to at the Joint Commission on Economic and Technological Cooperation meeting between Russian Minister Chernomyrdin and Vice President Gore in July. Russia reaffirmed its willingness to meet its obligations with regard to critical early elements, but will defer some subsequent elements until the year 2000 to relieve its launch burden.

Advanced launch vehicle activities. In its ongoing research toward an advanced space launch vehicle to eventually replace the space shuttle, reduce the high cost of space transportation, and reduce turnaround time, NASA in 1996 went from phase I to phase II. Phase I is a concept exploration effort to define both an operational single-stage-to-orbit (SSTO) reusable launch vehicle and an SSTO demonstrator designated X-33. Phase II (1996–1999) is a technology demonstration program involving design, fabrication, and flight test of the subscale (53% scale) X-33 and ground tests of full-scale components. The remotely piloted, suborbital X-33 is expected to fly at altitudes of 50 mi (80 km) and speeds of Mach 15. On July 2, NASA selected Lockheed Martin and their *VentureStar* concept.

Space sciences and astronomy. Several automated and remotely directed research missions added a very broad range of important discoveries to knowledge about the universe.

Hubble Space Telescope. The performance of the Hubble Space Telescope in 1996 continued to meet, and perhaps exceed, the original expectations. A wide range of discoveries at distances reaching from billions of light-years to those of nearby planets surprised cosmologists and astronomers. Images from the telescope included the first direct image of a star other than the Sun, the red supergiant Betelgeuse (Alpha Orionis), in ultraviolet light. An enormous hot spot, at least 2000 K (3600°F) hotter than the surface of the star itself, was revealed.

Other highlights of the continuing exploration with the telescope included revelation of hundreds of previously undiscovered galaxies in various stages of evolution that probably date back to near the big bang; assembly of a mosaic of images covering an area one-thirtieth the diameter of the Moon, the deepest and most detailed optical view ever taken of the universe; an image of the planetary nebula NGC 7027, about 3000 light-years from Earth, showing a star in its death throes; and the first observations of the surface of Pluto, revealing a complex object with more large-scale contrast than any planet except Earth.

Hubble discovered thousands of gaseous tadpole-shaped fragments in the Helix Nebula, a ring of glowing gases blown off the surface of a sunlike star late in its life. Termed cometary knots, they are probably the result of the dying star's final outbursts.

When Comet Hyakutake passed by Earth just 9.3×10^6 mi (15×10^6 km) away late in March, the telescope probed its inner region and the small icy nucleus, estimated to measure only 1–2 mi (1–3 km) across. The telescope also produced numerous images of the heart of the Crab Nebula over several months, showing dramatic changes in its dynamic interior.

Hubble measured the diameters of the pulsating Mira variables R Leonis and W Hydrae. The results suggest that these gigantic old stars are egg shaped. It revealed galaxies under construction in the early universe, part of a long-sought ancient population of galactic building blocks. The telescope showed dust storms on Mars in September and October, near the edge of the northern polar cap. *See* SPACE TELESCOPE.

Galileo. On December 7, 1995, NASA's deep-space probe *Galileo* inserted itself in an orbit around Jupiter. In early 1996, it completed transmitting the relayed data from its Probe, which had entered Jupiter's atmosphere on the same day. The Probe found the entry region of Jupiter to be much drier than anticipated, and failed to detect the three-tiered cloud structure that most researchers had postulated. Only about one-half of the expected amount of helium was observed. Extremely strong winds and very intense turbulence were detected.

On March 14, the last firing of *Galileo*'s main engine sent the spacecraft to flybys with Jupiter's four largest satellites: Ganymede, Io, Callisto, and Europa. The first such flyby, on June 27, revealed that Ganymede's surface has been extensively bombarded by comets and asteroids and dramatically wrinkled and torn by tectonic forces, and that Ganymede, unlike the Moon, possesses its own magnetic field. Also in June, the automated explorer started taking pictures of Io, showing new layers of sulfur and sulfur dioxide frost deposited from volcanoes since the *Voyager* flyby in 1979.

Remarkable images from Europa, in August, showed clear evidence of "warm ice" or even liquid water that may have existed, or perhaps still exists, beneath the cracked icy crust. In a second pass by Ganymede on September 6, the spacecraft obtained three-dimensional pictures of giant icy fissues, and on November 4 the spacecraft flew close to crater-studded Callisto. *See* JUPITER.

NEAR. With the launch of the *Near Earth Asteroid Rendezvous (NEAR)* spacecraft on February 17, NASA began the first attempt to study an asteroid at close range. The 1800-lb (816-kg, half of it propellant) probe was also the first in an anticipated series of inexpensive, quickly built interplanetary excursion craft named Discovery. *NEAR* is targeted for the 25-mi-long (40-km) asteroid 433 Eros, to spend a year in an orbit around it.

Polar. Launched on February 24 from Vandenberg Air Force Base into a near-polar orbit of the Earth, *Polar* is the last United States element of the International Solar-Terrestrial Physics program. The mission is designed to measure the complete plasma, energetic particles, and fields in the high-latitude polar region, the solar energy input, and the characteristics of the auroral plasma outflow, to better understand the Sun's myriad effects on the Earth and its space environment. *Polar*'s sister probe, *Wind*, was launched November 1, 1994, toward the lagrangian equilibrium point L1 between the Sun and Earth.

Compton. The *Compton Gamma-Ray Observatory*, dedicated to the observation of gamma-ray emissions across a broad spectrum of energies, has made many significant discoveries since its launch in 1991. Among them was the sudden appearance, in December 1995, of a type of object never seen before, which bursts and pulses at the same time and is currently the brightest source of gamma rays and x-rays in the sky. At the end of 1996, all four instruments of the spacecraft were still performing near design specifications.

Mars exploration. In 1996, Mars became the object of increased public interest following the discovery of possible lifeforms in its remote past. Indications pointing to the possible existence of primitive biota on Mars more than 3.6×10^9 years ago were discovered deep inside a meteorite from Mars, designated ALH84001, which fell to Earth 13,000 years ago. These indications consisted of organic molecules, traces of several mineral compounds associated with biological activity, and possible microscopic fossils of primitive organisms—some ovoid, others tubular or wormlike and jointed, similar to but tiny in comparison to certain terrestrial microbes. A similar discovery was reported in October in a considerably younger Mars meteorite, EETA79001. *See* MARS.

These discoveries came too late to enable appropriate modifications to the research machines launched to Mars in 1996, and those of the next synodic period 2 years thence. Modifications of later missions, however, for 2001 and thereafter, have begun.

On November 7, NASA launched *Mars Global Surveyor* (*MGS*) as the first United States spacecraft bound for Mars since the ill-fated *Mars Observer* in 1992. It is also the first in a planned series of surveyor-type Mars explorer probes, and with over a ton of mass it is also the heaviest of them. *Mars Global Surveyor* was scheduled to arrive at Mars on September 12, 1997, where it is to establish itself eventually in a polar, Sun-synchronous orbit of 236 mi (378 km) altitude. Its objectives include the complete mapping of the Martian surface with a sophisticated camera system.

Planned to be the first Mars lander since the *Viking* probes in 1976, NASA's *Mars Pathfinder* was launched on December 4. Consisting of a lander and a remote-controlled six-wheeled minirover named Sojourner, the 1900-lb (870-kg) spacecraft was due to touch down in the Ares Valley region of Mars on July 4, 1997, using braking rockets and inflated airbags to soften the impact. After its deployment, the 24-lb (11-kg) rover was to explore the vicinity of the landing site, equipped with two black-and-white television cameras and an alpha-proton–x-ray spectrometer for soil investigations.

Ulysses. The NASA-ESA solar-polar explorer *Ulysses* continued its return journey from its overflight of the Sun's north pole in June 1995, heading out to the orbit of Jupiter. Its measurements over the solar south pole in November 1994, reported in 1996, found a surprisingly small increase in the amount of the isotope helium-3 since the formation of the solar system; this allows a more precise estimate of the amount of dark matter in the universe.

Department of Defense activities. Efforts continued to make space a routine part of military operations across all service lines. Military launches included nine intelligence and communications satellites on four Titan 4 launchers, three *Navstar GPS-IIA* satellites on Delta II rockets, the Navy's *UHF Follow-on F7* communications satellite on an Atlas II, and two technology development payloads, *MSX* and *MSTI3*, on a Delta and a Pegasus air-launched carrier, respectively. *MSX* (*Midcourse Space Experiment*) tested infrared sensors for space- and ground-based missile defense systems, while *MSTI3* conducted similar tests for infrared, visible, and ultraviolet signatures of ballistic missile launches. In its program to reduce the operational cost of space transportation by one-half, the U.S. Air Force selected Lockheed Martin and McDonnell Douglas as finalists to build a new family of Evolved Expendable Launch Vehicles, intended to replace the medium-lift Delta and Atlas boosters and heavy-lift Titans now in use.

On December 3, the Department of Defense announced the potential discovery of frozen water deep inside the Aitken Basin at the south pole of the Moon. The ice formation was inferred by the radar signal returns of the Ballistic Missile Defense Organization's *Clementine* spacecraft launched to the Moon in 1994. However, there was general agreement that more definitive measurements are needed to confirm the assertion.

Commercial space activities. More than a dozen companies have plans to develop worldwide or regional communication networks using satellites. With them, developers hope to realize the goal of unhindered mobile telephone and facsimile traffic as well as data transfer from any place on Earth to any other. A gigantic gap exists between telephone connections in the industrial countries (400 million sets for 800 million people) and the Third World (200 million sets for 5 billion people), and mobile satellite systems would enable developing countries to expand telecommunications quickly and economically into areas where extensive infrastructures are lacking.

In the lead of mobile satellite network projects are Big LEO (low Earth orbit) systems such as Globalstar, Iridium, Odyssey, and Aries/ECCO, with from 12 to 66 satellites planned. Particularly ambitious are plans for the Teledesic system, comprising up to 840 satellites. *See* COMMUNICATIONS SATELLITE.

In 1996, commercial launch attempts of expendable space carriers totaled 22. There was only one failure: the sixth flight of Pegasus XL, a finned rocket launched from a Lockheed L-1011 aircraft. There were ten successful launches of Delta vehicles, six Atlas IIA launches, and one Atlas I. Including the failed flight on November 4, Pegasus XL took to the air five times in 1996.

Russian Activity

Despite continuing economic problems and a severe cash shortage for its space program, Russia in 1996 continued its space operations at a brisk pace, even if at a considerably reduced level. Payments from other countries for use of the space station *Mir* continued to help support Russia's space operations. However, without increased funding, Russia will be unable to maintain even its 1996 level in 1997.

Mir. By the end of 1996, Russia's seventh space station, *Mir,* in operation since February 20, 1986, had circled Earth approximately 62,094 times at altitudes of 238–245 mi (380–393 km) in an orbit inclined 51.65° to the Equator. Since its inception, it has been visited 28 times, including four times by United States space shuttles. To resupply the occupants, the space station was visited in 1996 by three automated Progress M cargo ships, bringing the total of Progress and Progress M ships launched to *Mir* and the two preceding space stations, *Salyut 7* and *Salyut 6,* to 76, with no failure. In 1996, demands for repair and maintenance of *Mir* became increasingly burdensome, and the station became increasingly dependent on the space shuttle for its logistics, because of setbacks in Russia's capability to finance Progress resupply missions.

In November, United States astronaut Blaha, aboard *Mir,* completed the first Bioreactor experiment run to produce a three-dimensional cartilage, and harvested wheat grown in space in a greenhouselike structure built by Slovakia. This represented the first time that an important agricultural crop and primary candidate for a future plant-based life-support system successfully completed an entire life cycle in the space environment.

Priroda. The 43,000-lb (20-metric-ton) crewless *Priroda* module was launched on a Proton rocket on April 3. Docking to *Mir* occurred on the first attempt on April 26, despite the failure of two of the module's batteries the previous day. The module carried a set of remote-sensing instruments and some microgravity experiments from Russia and the United States. It also served as a cargo carrier for various equipment from these countries. The addition of the module increased the station's total mass (excluding transport vehicles) to about 242,500 lb (110 metric tons) and its useful volume to nearly 14,000 ft^3 (390 m^3), finally exceeding the 12,500 ft^3 (350 m^3) available in NASA's 176,000-lb (80-metric-ton) Skylab in 1973 and 1974.

Soyuz TM 23. The Mir 21 crew, Onufrienko and Usachev, launched on February 21 and docked to the *Mir* station on February 23, joining three crew members of Mir 20, who returned in *TM 22* on February 29. In the first month, onboard activities focused on national research, followed by the United States science program NASA-2 when astronaut Lucid joined the crew on March 23 (see illus.). Onufrienko and Usachev performed their second and third spacewalks on May 21 and 24, to install and unfurl the Cooperative Solar Array brought up by *Atlantis* (STS 74), and a fourth on May 30 to install a European earth resources camera on the *Priroda* module. On June 13, they installed a truss structure called Rapana on *Kvant 1,* taking the place of the similar Strela. Later, they worked on the French research program

Cassiopeia when Mir 22 arrived with French researcher-cosmonaut Claudie André-Deshays. Onufrienko and Usachev returned with her on September 2.

Soyuz TM 24. The Mir 22 crew, including André-Deshays, was launched on August 17, after the primary crew, Gennadiy Manakov and Pavel Vinogradov, were grounded because of a cardiac irregularity of Manakov, and replaced by the back-up crew, Valeriy Korzun and Aleksandr Kaleri. The three cosmonauts joined the Mir 21 crew on board *Mir* on August 19. Prior to their arrival, on August 1, the crewless cargo ship *Progress M-31* was undocked and deorbited over the Pacific; *Progress M-32* arrived the next day.

Satellite and space-probe launches. Russia launched about 29 military, scientific, and telecommunication satellites. In a setback involving more than 20 participating nations, Russia's attempt at launching the large, complex Mars-exploration probe *Mars '96* failed catastrophically shortly after its launch on November 16, when the fourth stage of its Proton launch vehicle malfunctioned.

Russian commercial activities. Still in its early stages of entering commercial space markets, Russia opened a new chapter in its cooperation with the United States when it launched, on April 8, the United States–built European communication satellite *Astra 1F* on a Proton rocket. This was the first of 20 satellite launches planned under an agreement between Russia's Proton builder, Khrunichev, and the joint venture International Launch Services, with Lockheed Martin and the Russian company RKK Energia. On the second commercial launch for International Launch Services, a Proton carried the communications satellite *Inmarsat III F-2* into space.

European Activity

With France's reliable Ariane 4 family of expendable launch vehicles (92 flights with 7 failures), the commercial operator Arianespace carried out 10 launches from Kourou, French Guyana, in 1996. Europe's space activities, however, suffered a setback when the new heavy lifter Ariane 5 was lost shortly after its first test launch on June 4. Also destroyed was the *Cluster* payload, a set of four complex German-built satellites to study plasma streams in the solar wind and their interaction with the Earth's magnetosphere. *See* SATELLITE LAUNCH VEHICLE.

On February 29, the German-ESA cosmonaut Thomas Reiter returned from EUROMIR 95, an ESA-Russian mission aboard *Mir,* with *Soyuz TM 22.*

The *Infrared Space Observatory* (ISO), launched in 1995 by Ariane, in February reported the first infrared detection of cold molecular hydrogen in deep space. Other discoveries by *ISO* included the first galactic water vapor found in deep space; hundreds of previously invisible galaxies; a galactic collision between the twin Antennae Galaxies, 6×10^7 light-years from Earth; the birth of stars in the spiral arms and nucleus of the Whirlpool Galaxy, 2×10^7 light-years away; and the death of a star, by ejection of layers of material, in NGC 6543, about 3000 light-years from Earth.

Asian Activity

In 1996, space activities continued in Japan, the People's Republic of China, and India.

Japan. On August 17, Japan's National Space Development Agency (NASDA) launched its fourth H-2 vehicle carrying the powerful *Advanced Earth Observing Satellite (ADEOS)*, the first in a series of new Japanese environmental platforms. Japan's largest spacecraft, at 7700 lb (3500 kg), went into a circular Sun-synchronous polar orbit after requiring several days to overcome off-nominal orbit conditions due to a thruster malfunction. The U.S. National Oceanic and Atmospheric Administration (NOAA) distributes *ADEOS*-generated data worldwide for use in weather forecasting as well as environmental monitoring, but by year's end *ADEOS* had developed serious problems with some instrumentation.

Having achieved an independent launch capability with its H-2 heavy-lift launcher, Japan is well on its way to becoming a major space power. The country is also developing a major part of the International Space Station, the Japanese Experiment Module, consisting of a pressurized laboratory module, a space-exposed platform facility (main porch), a remote manipulator system, a pressurized logistics module, and an exposed-experiment logistics module. Its development accelerated in 1996, moving into flight hardware development. According to current plans, five space shuttle missions will carry the Japanese hardware to the station in 2000 and 2001. Two other major Japanese station-support programs were initiated: studies of a new data relay satellite system and the development of an automated, crewless station resupply vehicle, the H-2 heavy-booster Transfer Vehicle (HTV).

Efforts are under way to launch an automated winged shuttle vehicle called *Hope* on top of the H-2. To this end, Japan launched a suborbital hypersonic flight experiment (HYFLEX) on February 12. Although the test model sank in the ocean, the desired flight measurements were obtained via telemetry. Since *Hope* will have a maximum landing weight of 33,000 lb (15,000 kg), the payload capability of the H-2 would have to be increased considerably. Plans are to augment the current version with two liquid-fuel strap-on boosters.

China. The space program of the People's Republic was again set back in 1996 when two of its three commercial missions failed. The program seemingly had recovered from the loss of a Long March (Chang Zheng) 2E early in 1995 with the subsequent successful launch of two of these rockets. However, China's efforts to establish a major commercial space-launch program suffered another setback when a Long March 3B, on its first flight, exploded 20 s after takeoff from the Xichang Satellite Launch Center on February 14. The accident destroyed the *INTELSAT 708* satellite which was to be used primarily for direct broadcast satellite services to Latin America. At least 6 people were killed and 57 injured. A second Long March 3 failed on August 18, when its payload, the *ChinaSat 7* (*Zhongxing 7*) satellite, was left in an unusable orbit due to an incomplete burn of the rocket's third stage.

A Long March 3 was successfully launched on July 3, carrying China's communications satellite *Apstar 1A*. On October 20, the remote-sensing satellite *FSW 2* was carried into space by a smaller Long March 2D.

India. On March 21, India successfully launched its new four-stage Polar Satellite Launch Vehicle (PSLV), carrying the 2033-lb (922-kg) Indian earth resources satellite *IRS-P3*. The flight was the second success in a row for the 145-ft (44-m) booster. The Indian Space Research Organization (ISRO) satellite was placed into a 500-mi (800-km) Sun-synchronous polar orbit, carrying a wide-field scanner and a German electrooptical scanner for ocean data, as well as a payload for x-ray astronomy. Earlier, India announced that the three cameras of its *IRS 1C* satellite, launched on December 28, 1995, were successfully switched on and were functioning as expected. This brings the number of Indian-built Earth-observing satellites in the last 9 years to five.

For background information *see* ASTEROID; COMMUNICATIONS SATELLITE; GAMMA-RAY ASTRONOMY; INFRARED ASTRONOMY; JUPITER; MARS; MILITARY SATELLITES; SATELLITE ASTRONOMY; SPACE BIOLOGY; SPACE FLIGHT; SPACE PROBE; SPACE PROCESSING; SPACE STATION; SPACE TECHNOLOGY in the McGraw-Hill Encyclopedia of Science & Technology. Jesco von Puttkamer

Bibliography. Euroconsult 1995 world space markets survey, *Space Commun.*, 12(5):19–22, September–October 1996; *Jane's Space Directory, 1995–1996*; NASA Public Affairs Office, *News Releases '96*, 1996.

Space telescope

Launched in 1990, the Hubble Space Telescope has invigorated astronomy with major discoveries. Its contribution to the understanding of the universe has taken place at a rate nearly unprecedented for any single new astronomical instrument.

In a low Earth orbit, above the blurring effects of the atmosphere, the Hubble Space Telescope provides images of celestial objects that are ten times sharper than those typically seen from ground-based telescopes. The telescope can observe objects across a broad swath of the electromagnetic spectrum, from ultraviolet light, through the visible, and down to near-infrared wavelengths. It can detect faint objects down to the 30th magnitude, which is comparable to the limit for the largest ground-based telescopes.

Flaw and fix. The Hubble Space Telescope's long-anticipated, exquisite optical performance was initially crippled by a flaw in its 8-ft (2.4-m) primary mirror. After the nature of the flaw was precisely determined, a team of engineers and astronomers devised an optical correction. In December 1993, astronauts on the space shuttle *Endeavour* retrofitted this correction to the existing instruments and installed a replacement camera containing corrective optics. This fix will also be employed in future instruments planned for the telescope.

Science program. The Hubble Space Telescope has viewed nearly 10,000 celestial objects and made more than 100,000 observations, resulting in over 2.5×10^{12} bytes of data as of October 1996, which have been stored on 375 optical disks. Every year, over 100 observing proposals are accepted from astronomers who need to use the telescope's unique capabilities.

Galaxy evolution. By resolving the shapes of distant galaxies, the Hubble Space Telescope has provided the first direct visual evidence for galaxy evolution. This accomplishment is possible because distant galaxies existed when the universe was only a fraction of its present age, and their light is just now reaching the Earth. Understanding how galaxies developed into their current elliptical and spiral shapes has challenged astronomers ever since the discovery, early in the twentieth century, of galaxies beyond the Milky Way. Hubble observations show that elliptical galaxies formed early in the universe, less than 10^9 years after the big bang. By contrast, spiral galaxies may have taken longer to form. The Hubble Space Telescope also helped identify an ancient population of faint blue star–forming galaxies, which have since disappeared either through mergers or fading.

Hubble Deep Field. To study the evolution of galaxies, the Hubble Space Telescope was used to construct the deepest pictures of the universe ever made. The picture is a result of 10 consecutive days of observing a small patch of sky just above the handle of the Big Dipper.

This observation yielded an unprecedented view of nearly 3000 galaxies in a patch of sky no larger than a grain of sand held at arm's length. The image, called the Hubble Deep Field, revealed remote galaxies as faint as the 30th magnitude (**Fig. 1**). At this faint threshold, the image shows a veritable zoo of galaxies that have a bewildering variety of shapes and sizes, besides the classical spiral and elliptical-shaped galaxies. Many appear to be interacting, either through near-encounters or collisions. Others appear fragmentary, and may be intermediate steps in galaxy formation.

Before determining how these galaxies fit into an evolutionary sequence, astronomers will need to do complementary observations with ground-based telescopes to measure how the light of these galaxies has been redshifted by the expansion of space. The greater the redshift, the farther the galaxy is from the Earth. Once distances are known, it will be possible to study how galaxy shapes, brightness, and other evolutionary parameters changed between epochs.

Images from the Hubble Deep Field have shown that star birth was not instantaneous in the early universe, but accelerated toward a peak. The results support the emerging view, first gleaned from

Fig. 1. Hubble Deep Field, the result of Hubble Space Telescope observations of a small patch of sky for 10 consecutive days. *(R. Williams; Hubble Deep Field Team; NASA)*

ground-based data, that the universe had an active, dynamic youth, when stars formed out of dust and gas at a great rate, ten times greater than is seen today.

Galaxy building blocks. The Hubble Space Telescope uncovered the long-sought primordial building blocks of galaxies. Spectral redshifts of 18 objects in a medium-deep Hubble view of the heavens resolving galaxies down to the 29th magnitude found them to be at the same distance of 1.1×10^{10} light-years. These objects, clusters of 10^9 stars each, are close enough to each other to eventually merge to become one or more galaxies. This idea is reinforced by the realization that such objects are not seen in the local universe, and so could represent a missing link in the stages of galaxy formation.

Expansion rate of the universe. As a result of the big bang, the universe is expanding uniformly in all directions. The galaxies appear to be moving away from the Earth at a rate proportional to their distance. This relationship, called the Hubble constant, establishes the expansion rate of the universe. This value is a prerequisite for estimating the age of the universe. The faster the expansion, the younger the universe.

The Hubble constant cannot be precisely calculated without first knowing accurately the distances to galaxies. For decades, astronomers tried to extend the range of distance measurements by looking for members of a class of stars called Cepheid variables. The rhythmic pulsation of these aged, bright stars yields their intrinsic brightnesses, which in turn are compared with their observed brightnesses to calculate their distances.

Because the Hubble Space Telescope can detect Cepheid variables ten times farther away than ground-based telescopes can, it can survey a volume of space 1000 times larger, and potentially probe thousands of galaxies in clusters adjacent to the Local Group, whereas ground-based telescopes were restricted to a few dozen galaxies in the immediate neighborhood of the Milky Way Galaxy. In these clusters of galaxies, Cepheid variables are being used to calibrate even more distant markers, such as supernovae, which are so bright they can be seen out to much greater distances.

Because of this capability, the Hubble Space Telescope has moved the decades-old debate over the value of the Hubble constant toward a resolution. The first findings have suggested that the universe is younger than some earlier estimates. To keep these values in agreement with the 1.3×10^{10}-year age of the oldest stars, which provides an independent estimate of the age of the universe, it may be necessary to rework those models of the expanding universe that require a critical density of matter. This density results in space expanding at an ever-slowing rate but never quite coming to a halt. *See* COSMOLOGY.

Quasars revealed. The universe's most energetic objects, quasars, are among the most baffling because of their small size but prodigious energy output. Theoreticians have proposed monstrous black holes, gobbling up stars, dust, and gas, as the energy source (or engine) powering quasars. Galaxy collisions offer an effective process for fueling the black hole. However, definitive evidence has eluded astronomers for three decades because quasars are so bright that they mask any details of the environments where they exist.

The Hubble Space Telescope has helped solve the mystery of quasars, discovered in 1963, by clearly revealing these environments. However, its results have also compounded the mystery by showing that quasars occur in a variety of galaxies, both elliptical and spiral. Although many of these galaxies are interacting, which explains how the quasar is fueled, others appear normal. This complicated picture suggests there may be a variety of mechanisms, some quite subtle, for activating a quasar.

Black holes. A black hole is a celestial object that squeezes a tremendous amount of material into a very small space. The resulting gravitational pull is so intense that anything passing nearby, even light, is trapped forever.

A black hole's presence is inferred by the effects of its powerful gravity on the motion of neighboring stars, or infalling dust and gas. Once the speed of the entrapped material is measured, the mass of the black hole can be calculated by using simple laws of gravity. If it turns out that there is far more mass present than there are stars, the matter must be tucked away in something that is invisible and compact—a black hole.

Previous ground-based telescopic observations of black-hole candidates gave ambiguous lower limits to the central mass, while the Hubble observations are decisive in accurately measuring the mass and ruling out all other possible explanations.

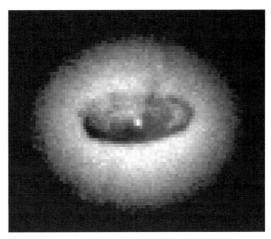

Fig. 2. Core of the galaxy NGC 4261, an 800-light-year-wide spiral-shaped disk of dust fueling a massive black hole in the center of the galaxy, located 10^8 light-years away. *(L. Ferrarese; NASA)*

These observations have established the presence of black holes in several active and quiescent galaxies.

The first such black-hole confirmation followed the discovery of a spiral disk of gas swirling around the hub of the giant elliptical galaxy M87. The gas velocity indicates the presence of a black hole of 2.2×10^9 solar masses. Similar observations of trapped dust in the galaxy NGC 4261 (**Fig. 2**) yield a black hole of 1.2×10^9 solar masses. Hubble observations of the motions of stars in the galaxies NGC 3115 and M32 yield black holes with 2×10^9 and 4×10^6 solar masses, respectively.

Embryonic planetary systems. The Hubble Space Telescope provided the first visual evidence that the first steps in planet formation may be common in the Milky Way Galaxy. Dozens of stars in the Orion region, which is just one of thousands of star-birth regions in the Milky Way Galaxy, have disks of dust (**Fig. 3**). Such disks were proposed around 1755 by Immanuel Kant to explain the simple fact that all the planets of solar system lie in nearly the same plane, and so were born from a primordial disk that provided the raw material for planet growth. Although it is not clear if the Orion disks will go on to condense into planetary systems, their abundance alone is solid visual evidence that the first small steps toward planet formation are very common. *See* PLANET.

Solar system. Although the solar system has been largely reconnoitered by crewless space probes since the 1960s, the Hubble Space Telescope has made important additions to the list of discoveries, including the following:

1. The Hubble Space Telescope provided the first images showing details on Pluto, revealing a remarkably varied surface of bright and dark regions with even more contrast than is observed on the Moon.

2. Taking advantage of a rare alignment that allows Saturn's rings to be seen edge-on, the Hubble Space

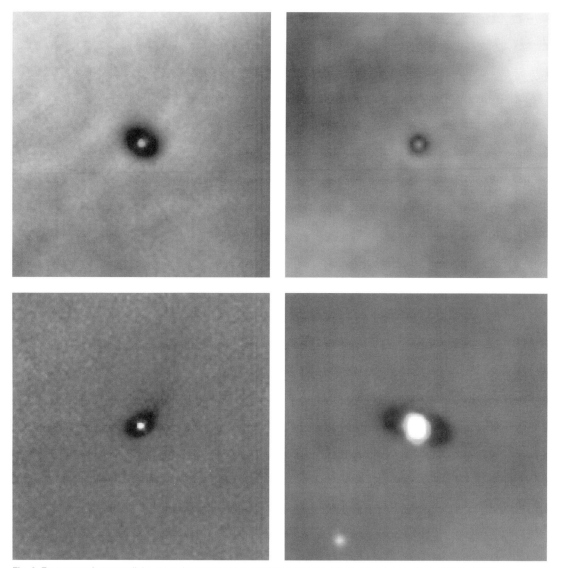

Fig. 3. Four protoplanetary disks around young stars in the Orion Nebula. *(M. McCaughrean; C. R. O'Dell; NASA)*

Telescope uncovered evidence for dynamic events in which satellites near the visible outer edge of the ring system are shattered into clouds of orbiting debris, presumably by comet collisions. The telescope also found that the rings are no more than 50 ft (15 m) thick.

3. Hubble observations revealed that the dark spot in Neptune's atmosphere is transient. It disappeared from one hemisphere, and a new spot emerged in the opposite hemisphere.

4. Hubble discovered that Jupiter's icy moon, Europa, has a thin oxygen atmosphere.

5. Hubble showed that a belt of hundreds of millions of comets encircles the solar system just beyond the orbit of Pluto.

For background information *see* BLACK HOLE; CEPHEIDS; COSMOLOGY; GALAXY, EXTERNAL; HUBBLE CONSTANT; QUASAR; SATELLITE ASTRONOMY in the McGraw-Hill Encyclopedia of Science & Technology.

Ray Villard

Bibliography. D. Fischer and H. Duerbeck, *Hubble: A New Window on the Universe*, 1996; J. Mitton and S. P. Maran, *The Gems of Hubble*, 1996; C. C. Petersen and J. C. Brandt, *Hubble Vision*, 1995; R. Smith, *The Space Telescope*, 1993.

Stainless steel

For economic production of stainless steel, high-carbon (5–8%) ferrochrome and scrap are charged into the melting furnace. After melting, the carbon must be removed to prevent intergranular free corrosion (due to a fall in the free chrome content below the critical content, needed for corrosion resistance). In the conventional production process this removal is achieved by blowing oxygen into the furnace. The steel must be raised to a very high temperature in order to achieve the required low-carbon (<0.03%) content.

A recently developed vacuum oxygen decarburization (VOD) process has several advantages compared to other stainless steel production processes. They include low chromium oxidation; very low final carbon and nitrogen contents, as required for superferritic stainless steel grades; and high metallurgical flexibility of the equipment, which is especially needed for a mixed production schedule.

In the vacuum oxygen decarburization process, carbon is removed by blowing oxygen onto the surface of the melt held at a reduced pressure. This method avoids the extremely high temperature required by the conventional method and simultaneously minimizes the amount of oxidized chromium (Cr). The carbon (C) and oxygen (O) form carbon monoxide (CO), as in reaction (1).

$$C + O \rightarrow CO_{gas} \tag{1}$$

Theoretical background. The benefits of oxygen blowing under vacuum become evident by examining the equilibrium curves of the carbon and oxygen at different pressures (P_{CO}). Taking into account the influence of the alloying elements (mainly chromium) by the introduction of the activity coefficients (a_C, a_O) and the interaction coefficients (f_C, f_O) for carbon and oxygen, the equilibrium curves are defined by the law of mass action: the Vacher-Hamilton equation (2), where k is a constant whose value is only slightly dependent on the melt temperature.

$$a_C \cdot a_O = k \cdot P_{CO} \tag{2a}$$

$$f_C \cdot \%C \cdot f_O \cdot \%O = k \cdot P_{CO} \tag{2b}$$

In conventional practice, for example, the decarburization of a heat (the metal in the furnace) containing 18% Cr starts in the electric arc furnace at the equilibrium point: 0.30% C, 0.065% O, and 3100°F (1700°C). By blowing oxygen at atmospheric pressure, carbon is removed as gaseous carbon monoxide. Simultaneously chromium is oxidized while going into the slag, which is mainly calcium oxide (CaO). Stepwise the content of chromium in the melt fails to 14%. As the Cr-O reaction is exothermic, the temperature of the melt rises to approximately 3600°F (1980°C). Only at that temperature is the goal of 0.03% C obtainable. As this temperature obviously causes refractory problems in the electric arc furnace, the heat must be cooled down very quickly after the oxygen blow. The chromium oxide (Cr_2O_3) is reduced by addition of ferrosilicon (FeSi) to the slag, as in reaction (3), where SiO_2 is silicon dioxide and Fe is metallic iron.

$$2Cr_2O_3 + 3FeSi \rightarrow 4Cr + 3SiO_2 + 3Fe \tag{3}$$

In the VOD process, oxygen is blown under vacuum onto the melt. The decarburization is usually started with 0.6% C at a pressure of 250 torr (3.3 × 10⁴ pascals) and a temperature of 2900°F (1590°C), which is well below the chromium oxidation line. Toward the end of decarburization, the pressure is lowered incrementally. To achieve the same final carbon of 0.03%, oxygen blowing is stopped at a level of around 0.05% C. The temperature of the heat is around 3100°F (1704°C), which is considerably less than that prevailing during conventional processing. The oxygen content of the melt is about 0.05%, which is sufficient for further natural decarburization by a vacuum treatment at a pressure of 1 torr (133 Pa). If the oxygen blow is halted at around 0.03% C, a final carbon value of 0.01% or lower is achieved. The lower carbon content further improves corrosion resistance of the final product.

VOD configuration. A VOD unit typically consists of a stationary tank, which can be situated on the plant floor or in a pit, and a tank cover (see **illus.**). The cover is mounted on a car that travels on rails. It is moved over the tank and lowered to close the tank vacuumtight. An off-take duct is welded to the tank to connect it to the vacuum-pump system for evacuation. The system has several powerful steam-ejector stages to create the vacuum. In situations where a sufficient amount of steam is unavailable, a combination of steam-ejector stages and water-ring pumps is used.

Mounted on the cover are several pieces of processing equipment, including the oxygen lance, the alloy additions lock, and the sampling measurement mechanism. Oxygen is blown onto the melt through a water-cooled lance. Attached to the lance port on the cover is a seal box, inside of which the lance is sealed vacuumtight during the treatment. Ferroalloys and slag formers can be added to the bath under vacuum by means of the additions lock, a vacuum lock. Normally temperature readings and chemical samples are taken at atmospheric pressure by means of a sample lance mechanism. Also, wire materials can be fed into the bath through a cover access port at atmospheric pressure. Inside the tank cover, a special water-cooled heat shield is suspended. As the cover is lowered, the heat shield is mounted on the ladle top and serves to protect the cover during processing.

VOD process. The VOD process comprises three periods: preparation of the melt, including the actual melting; the vacuum treatment period; and final alloying and temperature adjustment, which is done at atmospheric pressure.

Preparation of melt. The preparation of the melt begins when the melting furnace (usually an electric arc furnace) is charged. High-carbon ferrochrome (50–70% Cr, 5–8% C) and stainless steel scrap are used for ferritic grades. For austenitic grades, nickel is added. The quantity of the charged metal is limited in order to achieve, after tapping in the ladle, a freeboard of 50–60 in. (130–150 cm). Freeboard is defined by the distance from the liquid metal to the top of the ladle. Melting down will be done with electrodes only. When the charged metal becomes liquid, the gaseous oxygen burners are used. Utilizing gaseous oxygen with a natural gas (for example, methane, CH_4) or another gas provides a mixture to support the melting capacity of the arc of the electrodes. Oxygen is blown for a short time to get a carbon content between 0.5 and 1%. Since some chromium is oxidized during oxygen blowing, the

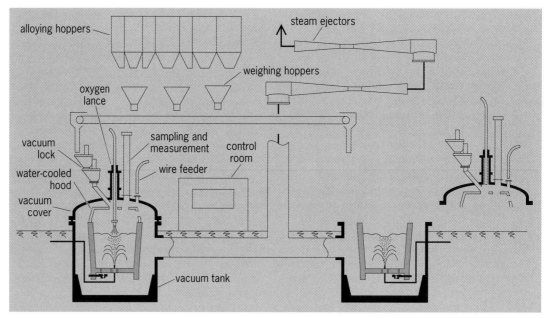

Vacuum oxygen decarburization (VOD) twin-tank system, used for large-scale production.

slag is treated with FeSi to reduce the chromium oxide. The heat is then tapped, together with slag, into a preheated basic lined ladle (lined with bricks produced from dolomite or from magnesia). The ladle is then transferred to a deslagging station, and from there to the VOD tank. Argon gas is added through the porous plug in the ladle and is visually adjusted. The argon gas added into the melt improves mixing; it does not react with the melt but escapes at the surface. The flow depends on ladle size; for example, for a 100-metric-ton melt, about 5 standard cubic feet per minute (SCFM) of argon is required. Usually the melt contains a 0.10% Si. Between 10 and 20 lb (4.5 and 9 kg) of lime per ton are added to cover the melt and to bind silicon as $2 \cdot CaO \cdot SiO_2$, which will be formed during oxygen blowing.

Vacuum period. After reducing the pressure to 250 torr (3.3×10^4 Pa), the oxygen lance is lowered to 40–70 in. (100–180 cm) above the melt surface. The oxygen flow for a 100-ton melt will be between 800 and 1000 SCFM. Vacuum pressure, argon flow, oxygen flow, and waste-gas analysis are recorded for the operator in the control room. Based on consumption of oxygen and waste-gas analysis, a computer calculates the speed of carbon removal. A typical vacuum treatment procedure is divided into six steps:

1. Pump-down to operation pressure is achieved.

2. Oxygen blowing and oxidation of silicon lasts a couple of minutes, during which the pressure drops further. Oxygen flow is increased continuously until the desired flow is reached. After oxidation of silicon, the oxidation of carbon starts.

3. The rise in pressure in the tank is caused by the generation of CO and carbon dioxide (CO_2).

Since these combustion reactions are exothermic, the temperature of the melt increases.

4. Decarburization is almost complete when the vacuum pressure starts to decrease. The oxygen blow is discontinued according to the specified final carbon content desired. The total oxygen blow time depends on the initial carbon content and the oxygen flow rate. For a 100-ton heat size and a starting carbon level of 0.60%, oxygen must be blown for about 45 min.

5. Further decarburization is initiated by pumping down to a vacuum pressure of about 1 torr (133 Pa). The dissolved oxygen in the bath/slag system reacts with the remaining carbon. The speed of carbon removal increases again for a short time to a new maximum. The duration of this deep-vacuum step is 10 min. When the oxygen blow is halted at 0.04% C, carbon can be reduced during this deep-vacuum step to less than 50 parts per million (ppm). Simultaneously, nitrogen is removed. This effect can be improved by increasing the argon flow.

6. Still under vacuum, the reduction of the oxidized chromium and desulfurization are accomplished. For this purpose, FeSi, aluminum (Al), CaO, and calcium fluoride (CaF_2) are added according to the output data of a computer model. For a good slag-metal reaction, the argon flow rate should be increased for the 100-ton ladle to 32 SCFM.

Final alloying and temperature adjustment. After breaking vacuum, a sample is taken, and the bath temperature is measured. Based on the actual chemistry and in accordance with the computer calculation, ferroalloys are added. For some grades, such as nitrogen-alloyed austenitics, a second sample is taken. Final alloying is performed, and cooling scrap is added to achieve the desired temperature of the melt.

Before subjecting the heat to teeming (casting), calcium silicide (CaSi) wire can be fed into the bath. This is often required for ferritic grades to improve the castability of the steel by modification of the shape of the oxide inclusions in the melt; it also results in a better cleanness rating (lower total oxygen content) in the final product. Then the ladle is removed from the tank and transferred to the casting bay.

Metallurgical results. These involve removal of carbon, oxygen, sulfur, and nitrogen, as follows.

After the oxygen-blowing period, enough oxygen is dissolved in the melt to reduce carbon from 400 ppm to less than 50 ppm. A final carbon of 20 ppm is obtainable.

Oxygen samples taken at the caster from the tundish show a total oxygen level of 40 ppm for stainless steel grades with 0.040% Si and only 0.005% Al. Lower total oxygen content is achievable with higher aluminum content; for example, with 0.02% Al, 20–30 ppm total oxygen is achievable.

At the end of the oxygen blowing period, the basicity of the slag is 1.8–2.5. Combined with strong argon stirring, this slag provides effective sulfur removal.

During the production of stainless steel, nitrogen is removed mainly (1) during the decarburization step, by the high amount of carbon monoxide gas, which creates a low nitrogen partial pressure in the bubbles of argon–carbon monoxide; and (2) during the slag reduction step, in which oxygen and sulfur are removed to very low levels. These surface-active elements interfere with the diffusion of nitrogen. Low-nitrogen specifications are mostly required for ferritic and superferritic stainless steel grades, which should have a very high corrosion resistance. The obtainable final nitrogen content depends on the chromium content in the melt.

For background information *see* IRON ALLOYS; STAINLESS STEEL; STEEL; STEEL MANUFACTURE; VACUUM METALLURGY in the McGraw-Hill Encyclopedia of Science & Technology.   Rainer Teworte; Gerd Stolte

Bibliography. W. Pulvermacher, G. Stolte, and J. Rushe, Production of high-chrome steel using vacuum metallurgical technology, *Steel Times*, pp. 269–274, June 1987.

Steel manufacture

Superplasticity refers to the ability of a material to develop unusually high tensile elongations at temperatures between 0.4 and 0.6 of its melting point in kelvins. Strains in excess of 1000% are frequently obtainable. By contrast, elongations often considerably less than 100% strain are typical for materials in a nonsuperplastic state when deformed at the same strain rates and temperatures where superplasticity is obtained in these materials.

Use of superplasticity. Having a material capable of developing superplastic properties provides the opportunity economically to produce complex components to their final shape by either press forming or gas-pressure forming while using only one or two operations. Savings result from reductions in machining, material wastage (scrap), and energy consumption as well as from lower costs in the handling and transportation of the material from the initial billet to the final product.

Commercial applications of superplastic forming of complex net-shape components (which are produced in their final shapes by forming or casting, without the need for additional operations such as machining or welding) or nearly net-shape components (which need some minor additional operation) are well established for nickel, titanium, and aluminum alloys. Superplasticity also has been demonstrated in a large number of other metallic materials as well as in ceramics.

Classes of superplastic steels. Considerable work has focused on developing superplasticity in steels. These alloys can be divided into three classes: duplex stainless steels (high-chromium-nickel-molybdenum-iron base); hypoeutectoid steels (contain less than 0.77% carbon); and hypereutectoid steels (contain more than 0.77% carbon). The development of superplasticity has also been demonstrated for cast irons and for Hadfield manganese steels. Recently, superplastic duplex stainless steels have been made commercially available. Superplasticity in hypoeutectoid steels is still primarily of academic interest. Extensive studies have been performed on the hypereutectoid steels, including collaboration with industrial concerns, and these steels show considerable promise for commercialization. The requirement of having a fine microstructure that is stable during superplastic forming is common to developing superplastic properties in most of these materials. Such microstructure is developed through thermomechanical processing, phase transformations, or both. The hypereutectoid (ultrahigh-carbon) steels present an excellent example of processing to achieve a microstructure capable of producing superplasticity.

Evidence of superplasticity. In the tensile testing of a nonbrittle, nonsuperplastic metal, failure is preceded by localized reduction (necking) of the cross section of the test sample. With the differential increase in the stress level in the necked region, the rate of deformation is accelerated here relative to the rest of the sample; further deformation is then confined mainly to the neck, leading to failure. Under this behavior, deformation is controlled by crystallographic slip mechanisms where the rate of deformation has a relatively minor influence on the flow stress (resistance to deformation). By contrast, superplastic behavior is obtained when an increase in the deformation rate results in an appreciable increase in the flow stress. If large enough, this increase in the resistance to deformation arrests the growth of any incipient necking and allows the material to reach high levels of strain prior to failure.

The relationship between flow stress and deformation (strain) rate can be expressed by $\sigma = k\dot{\varepsilon}^m$,

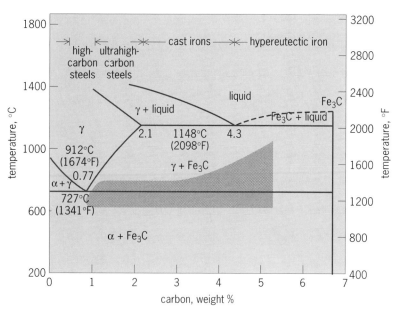

Fig. 1. Iron-carbon phase diagram. Shaded area is the range of temperature and composition over which superplasticity has been demonstrated in high-carbon steels and white cast irons.

where σ is the flow stress, k is a constant, $\dot{\varepsilon}$ is the strain rate, and m is the strain-rate sensitivity exponent. For nonsuperplastic behavior, where deformation occurs by crystallographic slip, the exponent (m) will have a low value, such as 0.05. For superplastic behavior, where deformation occurs predominantly by grain-boundary sliding, m would have a value of at least 0.4; for newtonian flow, n is equal to 1. As an example of the effect of the magnitude of m on ductility, the strain to failure can increase from about 5 to 1000% as m increases from about 0.05 to 0.4.

Of the number of microstructural conditions that must be met in order to achieve superplasticity, four are most important. (1) The grain size must be sufficiently fine that grain-boundary sliding is the dominant deformation mode. (2) Grain growth during the superplastic forming process must be minimized so as to maintain the grain-boundary sliding mechanism as the principal deformation mode. (3) The grains must be reasonably equiaxed; that is, they must not have any directional shape, such as being elongated or distorted; for example, an elongated grain structure will hinder the boundary-sliding process. (4) Brittle microstructural networks must be absent. Stabilizing against grain growth is made possible by the presence of contiguous grains of two or more phases or by the presence of a dispersion of particles that can pin the grain boundaries so as to prevent them from expanding into and adsorbing neighboring grains of the same phase. The development of an ultrafine structure generally involves a sequence of deformation and recrystallization steps that may include one or more phase-transformation steps, the latter being important for controlling the mixture of phases for both grains and particles.

Ultrahigh-carbon steels. In the mid-1970s a research group at Stanford University first showed that steels having a sufficiently high carbon content could be made to behave superplastically by using thermomechanical processing of the steels. These ultrahigh-carbon steels are low-alloy plain-carbon steels containing 1–2.1% carbon (15–32 vol% iron carbide, Fe_3C). A typical composition may be 1.6% carbon, 1.6% aluminum (or 3% silicon), 1.5% chromium, 0.5% manganese, with impurities that are common to most steels, with the balance being iron. The steel microstructure that would be amenable to superplastic behavior would consist of submicrometer-size particles of iron carbide dispersed in a matrix of ferrite grains (smaller than 5 micrometers). In addition to providing superplastic properties at the temperatures required for superplastic forming, the ultrafine microstructure leads to significant improvements in room-temperature mechanical properties over those normally exhibited by high-carbon steels processed by conventional methods.

The composition and processing for obtaining superplastic properties in ultrahigh-carbon steels is directed toward developing an ultrafine microstructure consisting of ferrite (iron phase) grains having a dispersion of iron carbide (Fe_3C phase) particles and free of any carbide network. The composition of the ultrahigh-carbon steels is selected to maximize the proportion of carbides and to help minimize the formation of carbide networks.

Iron-carbon phase diagram. The composition and temperature regions where superplasticity has been observed in dilute-alloyed ultrahigh-carbon steels, as well as in cast irons, are shown on the iron-carbon phase diagram of **Fig. 1.** The lines indicate the boundaries between the one-phase γ (austenite), α (ferrite), and Fe_3C regions and two-phase $\gamma + Fe_3C$, $\gamma + \alpha$, and $\alpha + Fe_3C$ regions. The addition of small amounts of alloying elements, as present in the ultrahigh-carbon steels, will shift the boundaries somewhat, as well as introducing small three-phase regions. In the binary Fe-C system of Fig. 1, at the eutectoid temperature of 727°C (1340°F), the three phases ferrite, austenite, and carbide coexist at 0.77 wt % carbon. On cooling from above to below 727°C (1340°F) the austenite phase normally transforms to a structure consisting of alternate layers of ferrite and iron carbide referred to as pearlite. The boundary line at 727°C (1340°F), above which austenite is present, is referred to as A_1.

At carbon contents below the eutectoid composition (0.77 wt % carbon), the transformed structure will consist of pearlite with a ferrite network, ferrite grains, or both. At carbon contents above the eutectoid composition, on cooling through the austenite/austenite + carbide boundary (referred to as A_{cm}), the carbides will normally precipitate as a network along the austenite grain boundaries; and then on cooling below A_1, the structure that is developed will consist of pearlite with carbide networks (**Fig. 2**). With the carbide being brittle and tending to be continuous, the material in this

Fig. 2. Microstructure of ultrahigh-carbon steel slowly cooled from the austenite range, showing precipitation of proeutectoid carbides as a continuous grain-boundary network. (*a*) Extensive network formation. (*b*) Some areas of the lamellar pearlitic colonies at higher magnification.

condition is likely to be extremely brittle. By contrast, the pearlite-ferrite structure is relatively ductile; however, it will not exhibit superplasticity and must be converted to a structure that can become superplastic.

Processing. A sequence of steps was developed for the processing of ultrahigh-carbon steels. Depending on the alloy composition, the cast ingot or billet is soaked in the temperature range 1125–1225°C (2057–2237°F) in the austenite field for about 6 h and then continuously hot-and-warm worked as the ingot cools to below the A_1 temperature. The soaking and hot-working step is performed to homogenize and break down the as-cast structure. Normally, in conventional practice the deformation is discontinued at about 950°C (1740°F). The continuation of working down into the warm-working range results in preventing the formation of the carbide network as the carbides precipitate out of the austenite. The final structure consists of a dispersion of spherical carbide particles and pearlite; the former formed during the warm-working stage, while the latter formed from the transformation of austenite on cooling below A_1 (**Fig. 3***a*).

The next step is to convert the pearlite into fine ferrite grains and fine carbide particles. The steel is heated into the austenite region, about 30–40°C (50–70°F) above the A_1 temperature for a period of time just sufficient to develop through-thickness temperature (that is, uniform temperature throughout the thickness of the part) and to minimize homogenization of the austenite. The proeutectoid carbide particles are partially dissolved in the austenite during this period. These particles restrict the growth of the austenite grains. On subsequent controlled cooling, instead of forming the lamellar pearlitic structure, the fine austenite transforms to produce fine ferrite grains with a dispersion of carbide particles and a complete absence of pearlite (Fig. 3*b*). This transformation is referred to as a divorced-eutectoid transformation. The nonuniform carbon distribution in the austenite and the residual undis-

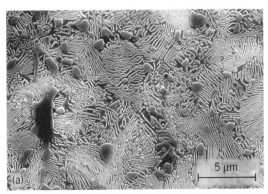 

Fig. 3. Stages in formation of the final structure of ultrahigh-carbon steel. (*a*) Microstructure after hot- and warm-working, showing spheroidal proeutectoid carbides dispersed in a lamellar pearlitic matrix. (*b*) Microstructure of ferrite grains and carbide particles after hot- and warm-working followed by a divorced-eutectoid transformation.

solved proeutectoid particles are sites for the precipitation and growth of any carbides formed during the cooling process. Too high an austenitizing temperature, too long a holding period, or too rapid a cooling rate may adversely affect the efficiency of the process; these factors give an incomplete divorced-eutectoid transformation in that some pearlite is formed. The presence of pearlite will reduce the temperature and strain-rate range for superplasticity, with the range shrinking rapidly with small increases of pearlite.

Extensive studies showed that the temperature range over which the transformation of austenite occurs influences the degree to which the divorced-eutectoid transformation takes place. For example, depending on the exact composition and the part thickness, air cooling can lead to too rapid a cooling rate to prevent the formation of at least some pearlite. Alternative processing steps have been developed to obtain a complete divorced-eutectoid transformation as well as to further refine the structure.

Addition of a ferrite stabilizer. An important consideration in the commercialization of a superplastic material is the strain-rate and temperature regions over which superplasticity exists. An increase in the fineness of the microstructure and an increase in temperature lead to an increase in the maximum achievable superplastic strain rate. Addition of ferrite stabilizers such as silicon and aluminum raises the A_1 temperature. With silicon being more effective than aluminum, the early studies with the ultrahigh-carbon steels were performed with additions of silicon. It was subsequently discovered that the presence of aluminum (greater than about 1 wt %) largely prevented the formation of carbide networks on cooling through the A_{cm} boundary following the hot-working step. The use of aluminum eliminated the need for continuing into the warm-working step; therefore, many of the more recent developments were made with the aluminum addition.

Although additions of aluminum facilitate the elimination of carbide networks, some difficulties have been encountered in continuous casting of ultrahigh-carbon steels containing aluminum. Casting experience with other silicon-bearing steels suggests that these difficulties would not be encountered with additions of silicon. Chromium (1.5 wt %) was added to prevent the possibility of graphitization of the carbide, which could be induced by additions of either silicon or aluminum.

For background information *see* STEEL; STEEL MANUFACTURE; STRESS AND STRAIN; SUPERPLASTICITY in the McGraw-Hill Encyclopedia of Science & Technology.   Alfred Goldberg

Bibliography. D. R. Lesuer et al., The case for ultrahigh-carbon steels as structural materials, *JOM*, 45:40–46, 1993; Y. Maehara and T. G. Langdon, *J. Mater. Sci. Eng.*, A128:1–13, 1990; O. D. Sherby et al., Superplastic ultrahigh-carbon steels, *Scripta Meta.*, 9:569–574, 1975; M. J. Strum et al., Transformation Process for Production of Ultrahigh Carbon Steels and New Alloys, U.S. Pat. 5,445,685, August 20, 1995.

Stratospheric ozone

Stratospheric ozone (O_3) has decreased significantly in the polar regions and perhaps has decreased slightly in the tropics since the late 1970s (**Fig. 1**), as documented by both ground-based and satellite instruments. In the northern middle latitudes where most people live, the total ozone column (the sum of ozone molecules per unit area between Earth and the Sun) has decreased 4% per decade since 1979. This decrease in the total ozone column comes mostly from ozone decreases in the lower stratosphere at heights between 12 and 25 km (7.4 and 16 mi). Ozone has not decreased smoothly, but had steeper rates of decrease in 1982 and 1992, with recovery within a few years to a downward trend line (**Fig. 2**). This ozone decrease may be natural; it may be linked to human activity, just as the Antarctic ozone hole has been.

Ozone in middle latitudes. The amount of stratospheric ozone depends on the local production and loss plus the amount of ozone that is transported from elsewhere. Ozone is produced when solar ultraviolet light decomposes dioxygen molecules (O_2) into oxygen (O) atoms; these then combine with other O_2 molecules to form O_3 molecules, as in reactions (1) and (2). This sequence occurs mainly at

$$O_2 \rightarrow 2O \qquad (1)$$

$$2(O + O_2 \rightarrow O_3) \qquad (2)$$

tropical higher altitudes where the solar ultraviolet light is strongest. Ozone is lost by certain chemical reactions wherever sunlight can energize these reactions. At the same time, stratospheric transport tends to carry tropical air poleward and downward toward the high latitudes and into the troposphere. During this transport, ozone is continually being created and destroyed, although destruction dominates production toward high latitudes.

Air entering the tropical stratosphere contains a little water vapor (H_2O), methane (CH_4), nitrous oxide (N_2O), the chlorofluorocarbons (CFCs), carbon tetrachloride (CCl_4), methyl chloroform (CH_3CCl_3), methyl chloride (CH_3Cl), methyl bromide (CH_3Br), and the bromine-containing fluorocarbons known as halons. All the chlorine compounds except methyl chloride, 80% of the total, and about 50% of the bromine compounds are human made. The others are mostly natural. These chemicals are decomposed in the stratosphere by solar ultraviolet light and the subsequent photochemical reactions.

The decomposition products include both reactive compounds that destroy ozone and less reactive compounds, called reservoir compounds. These compounds are usually grouped into families according to their origin: hydrogen compounds from water vapor and methane; nitrogen compounds from nitrous oxide; and halogen compounds. The chemical compounds both within and among chemical families are kept in balance with each other by photochemical reactions. The relative importance of different reactions depends on the altitude, latitude, and season, as well as on the abundance of each chemical. Of these factors, the chemical abundance has shown the most distinct long-term changes. Chlorine has increased by far the most since 1950; its stratospheric abundance has gone from 0.6 part per billion of the total number of air molecules in 1950 to 3.5 parts per billion in 1996 because of the widespread use of chlorofluorocarbons.

The reactive compounds of each chemical family destroy ozone by chemical catalytic cycles, sequences of chemical reactions in which reactive chemicals are regenerated. In the lower stratosphere outside the polar regions, catalytic cycles involving the reactive hydrogen compounds (HO_x), the hydroxyl radical ($OH\cdot$), and the hydroperoxyl radical ($HO_2\cdot$) destroy the most ozone. Catalytic cycles involving the reactive nitrogen compounds (NO_x), nitrogen dioxide (NO_2), and nitric oxide (NO) are less important than catalytic cases involving hydrogen compounds.

The catalytic cycles involving the halogen compounds, chlorine monoxide (ClO), and bromine monoxide (BrO) destroy about as much ozone as HO_x. Chlorine is important, even though most chlorine exists as less reactive reservoir compounds, HCl and $ClONO_2$, with only a few percent as ClO. However, about 50% of the bromine is in the form of BrO. This partitioning determines how much ozone can be destroyed by chlorine and bromine.

The nitrogen oxides, NO and NO_2, react with the reactive halogen and hydrogen compounds to produce reservoir compounds, particularly HNO_3, $ClONO_2$, and $BrONO_2$. Two other reservoir compounds, HCl and HBr, come from reactions of chlorine and bromine with CH_4. Thus, while reactive nitrogen destroys ozone, it also holds down the levels of reactive halogen and reactive hydrogen. The nitrogen oxides play this dual role, because their stratospheric abundance is greater than that of the hydrogen and halogen compounds, but their ozone-destroying capability per molecule is less. Thus, any process that partitions the nitrogen family into reservoir and reactive compounds will affect the partitioning of the other chemical families and, as a result, the ozone loss.

The partitioning of both nitrogen and halogen compounds is influenced not only by gas-phase chemistry but also by heterogeneous chemistry on stratospheric aqueous particles. These reactions convert nitrogen compounds from reactive to reservoir forms, and halogens from reservoir to reactive forms. In the Antarctic ozone hole, heterogeneous chemistry on polar stratospheric clouds rapidly and efficiently produces reactive halogen compounds

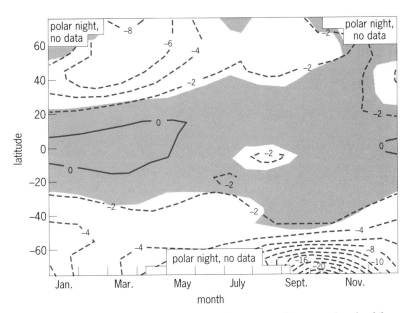

Fig. 1. Total ozone trends, 1978–1993. The trends (percentage change per decade of the total ozone column) are deduced from measurements of the satellite-borne Total Ozone Mapping spectrometer. Shaded areas have no statistically significant trend. *(After S. M. Hollandsworth et al., Ozone trends deduced from combined Nimbus 7 SBUV and NOAA SBUV/2 data, Geophys. Res. Lett., 22:905–908, 1996)*

and removes reactive nitrogen compounds. The result is the dominance of halogen chemistry unfettered by reactive nitrogen chemistry and thus a rapid ozone collapse. Particles exist in the lower stratosphere in the middle latitudes and tropics as well, but these particles, composed of sulfuric acid and water, are smaller and have less reactive surface than polar stratospheric clouds. Two heterogeneous reactions, (3) and (4), occur that are efficient and

$$N_2O_5 + H_2O^* \rightarrow 2HNO_3 \tag{3}$$

$$BrONO_2 + H_2O^* \rightarrow HOBr + HNO_3 \tag{4}$$

practically temperature independent (the asterisk indicates liquid water). Nitrogen pentoxide (N_2O_5)

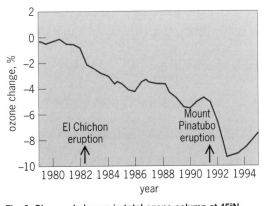

Fig. 2. Observed change in total ozone column at 45°N, 1979–1994. The effects of the annual cycle and the Quasibiennial Oscillation have been removed. *(After S. Solomon et al., The role of aerosol variations in anthropogenic ozone depletion at northern midlatitudes, J. Geophys. Res., 101(D3):6713–6727, 1996)*

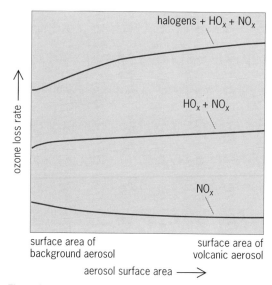

Fig. 3. Ozone loss rates caused by the different chemical families, as a function of aerosol surface area. Background aerosol is the typical aerosol long after a volcano has erupted. Volcanic aerosol has about 10 times more surface area than background aerosol. *(After S. Solomon et al., The role of aerosol variations on anthropogenic ozone depletion at northern midlatitudes, J. Geophys. Res., 101(D3):6713–6727, 1996)*

is a nighttime reservoir for NO and NO_2 and is quickly decomposed into NO_2 by sunlight. In this way, even the sulfate (SO_4^{2-}) aerosol at middle and tropical latitudes converts reactive nitrogen compounds to reservoir compounds, and reservoir halogen compounds to reactive halogen compounds.

Volcanic injections of sulfur dioxide (SO_2) into the stratosphere increase the reactive surface area of the sulfate aerosol by a factor of 20 or more. The heterogeneous reactions go faster, and even greater amounts of nitrogen and halogen compounds are converted. At a high enough surface area, reaction (5) becomes important.

$$ClONO_2 + H_2O^* \rightarrow HOCl + HNO_3 \qquad (5)$$

These heterogeneous reactions reduce the reactive nitrogen by converting it into reservoir forms. Reactive nitrogen both destroys ozone and regulates the amounts of the halogen and hydrogen compounds that destroy ozone. The net effect of decreasing the reactive nitrogen is that more ozone is destroyed, primarily because more chlorine is in the reactive compound ClO. Thus, at large aerosol surface areas the heterogeneous reaction rate is greater, resulting in less reactive nitrogen. As a result, the ozone loss rate is larger (**Fig. 3**).

This basic picture of stratospheric chemistry fits the observations made by instruments on the ground, high-altitude aircraft, satellites, and helium-filled scientific balloons. These observations encompass much of the globe in all seasons; several chemical species, particle characteristics, and local meteorology are measured simultaneously.

Causes of middle latitude ozone decrease. Any proposed mechanism for the ozone decrease at middle latitudes must be consistent not only with the observed ozone loss but also with the other available observations. The mechanism that best describes all observations is halogen chemistry. First, the downward ozone trend resulted from the upward trend in stratospheric chlorine. Second, the ozone loss occurs in the lower stratosphere where the ubiquitous stratospheric aerosol reduces the control of the nitrogen oxides on halogens and hydrogen compounds. Third, the loss is greatest in winter and at higher latitudes because more N_2O_5 is converted to nitric acid, lessening the control on halogens and hydrogen even more. Fourth, in the tropics, ozone production exceeds destruction, and little tropical ozone change occurs. Finally, ozone decreased the most in 1982 and 1992 because major volcanic eruptions from El Chichon in Mexico in 1982 and Mount Pinatubo in 1991 injected large amounts of aerosol-forming sulfur into the stratosphere, temporarily reducing the control of the nitrogen oxides on halogen and hydrogen compounds until the volcanic aerosol settled out of the stratosphere.

Photochemical-transport models are used to test ozone trends resulting from the halogen chemistry. The observed ozone decrease in the summertime is simulated accurately by these models, but the observed wintertime decrease is almost twice the modeled trends. Thus, the local halogen chemistry cannot explain all the observed ozone decrease at middle latitudes.

A possible additional contribution to wintertime ozone loss is the export of air from inside the polar vortex. Polar vortex air is affected by the polar stratospheric clouds; it has high levels of reactive chlorine and bromine, low levels of reactive nitrogen, and possibly decreased levels of ozone. Polar air is observed to peel away from the vortex and mix into the middle latitudes. However, only about 5–25% of polar air in middle latitudes came from the polar vortex during the course of recent winters, so that this process can account for only a portion of the unexplained wintertime ozone decrease at middle latitudes.

A second possibility is that air from the middle latitudes on the edge of the polar vortex gets cold enough that polar stratospheric clouds form, enhancing halogen chemistry. This air then travels to lower latitudes, spreading ozone destruction. However, studies show that the air on the vortex edge only occasionally gets cold enough to form polar stratospheric clouds. Thus, this process makes almost no contribution to the ozone decrease at middle latitudes.

A third possibility is that atmospheric dynamics have changed. Ozone variations result from meteorological variations such as the El Niño–Southern Oscillation and the Quasibiennial Oscillation in the tropical stratosphere. These meteorological changes are quasiperiodic, like the solar sunspot cycle, and each causes a quasiperiodic ozone variation of a few percent. However, some measures of dynamical activity at middle latitudes have changed during the last decade, possibly forced by changes in atmospheric circulation due to long-term climate variability. These dynamical effects on the ozone trend remain unquantified but may be important.

A fourth possibility is that some unknown chemistry remains undetected. Ozone-destroying iodine chemistry was recently proposed, but measurements of the reactive iodine-oxygen compound IO and its catalytic reaction rates show that iodine is unlikely to contribute to the ozone trend. At this time, no other potential missing chemistry has been identified.

With the possible exception of dynamics, every likely or proposed contributor to the downward ozone trend involves halogen chemistry, particularly chlorine chemistry. Because most stratospheric chlorine is derived from chlorofluorocarbons, the downward ozone trend is most likely linked to human activity. The Montreal Protocol, an international treaty for protecting stratospheric ozone, has led to a peaking in the atmospheric levels of chlorofluorocarbons. A downward trend in stratospheric chlorine should soon follow, which should result in a reversal to ozone's downward trend. *See* CHLOROFLUOROCARBON.

For background information *see* AEROSOL; ATMOSPHERIC CHEMISTRY; ATMOSPHERIC OZONE; HALOGEN ELEMENTS; HALOGENATED HYDROCARBON in the McGraw-Hill Encyclopedia of Science & Technology.

W. H. Brune

Bibliography. C. A. Ennis (coord. ed.), *Scientific Assessment of Ozone Depletion: 1994*, World Meteorological Organization, Global Ozone Research and Monitoring Project, Rep. 37, 1995; S. Solomon et al., The role of aerosol variations in anthropogenic ozone depletion at northern midlatitudes, *J. Geophys. Res.*, 101(D3):6713–6727, 1996.

Superlattices

In solid-state materials, the term superlattice usually means a synthetically prepared set of lattice planes, of variable thickness and composition, lying parallel to those of the host (substrate). (Such a structure is also known as an artificially layered structure or artificial crystal.) Similarly, in the case of extended molecular adlayers (adsorbed layers), interactions between the terminal atoms of the adsorbed molecule and the solid surface, acting in concert with strong intramolecular interactions, can lead to formation of a self-assembled monolayer. The monolayer structure is, in a sense, a superlattice with only one artificial lattice plane extending from the substrate, with a composition and thickness that can be controlled by choice of adsorbate.

The term superlattice arises because there is an artificial periodicity, caused by the preparative methodology, that determines the structural and electronic properties of the superlattice material. In general, a superlattice consists of a series of artificially

prepared layers, parallel to one of the substrate surface planes. Ordinarily, these superlattice planes differ from one another both in thickness and in composition.

Types. The various types of superlattices include periodically layered structures, doping superlattices, and structures in which the superconducting and insulating layers are alternated.

Periodically layered structures. The idea of an artificial superlattice was first put forward in 1969 by L. Esaki and T. Tsu, who discussed two sorts of superlattices. In the first, now often called a periodically layered structure or compositional superlattice, the artificial planes alternate in thickness and composition. For example, a plane of the first material of thickness D1 is followed by a plane of the second material of thickness D2, which in turn is followed by another plane of the first material of thickness D1. This results in a material whose lattice spacing, in the direction perpendicular to the substrate, is fixed by preparative methodology.

Ordinarily, these lattice planes are prepared by molecular beam epitaxy, by metal organic chemical vapor deposition, or by sputtering. Molecular beam epitaxy produces very high quality films, whose slow growth rate (usually of the order of 0.1 nanometer per second) results in very few defects. With these methods, layers spanning two or three atomic layers to layers of micrometer thickness can be prepared.

These periodically repeated structures possess unusual electronic properties that can be understood in terms of the simple Kronig-Penney model of a series of periodic square wells. The periodic repetition introduces gaps among allowed energy bands, with the gap width and spacing due to the modulated potential caused by the periodic superlattice structure. By control of this structure, it is possible to control the positions and widths of the allowed bands.

This unusual band structure results in unusual properties, including optical properties and magnetic susceptibilities. Perhaps most striking, because of gaps within the band structure, differential negative resistance is anticipated (and indeed can be observed); that is, the current tunneling through the superlattice may actually decrease with an increase in applied potential, as the tunneling passes from a band region into a gap region. This behavior is the basis for the resonant tunneling diode, a structure of substantial promise in very small scale electronics.

For ordinary compositional superlattices, close lattice compatibility between the two structures is required. For example, there has been extensive study of situations in which the first material is gallium arsenide (GaAs) and the second is a gallium-aluminum arsenide ($Ga_xAl_{1-x}As$); here the lattice spacings are nearly the same.

Doping superlattices. A second kind of structural lattice discussed by Esaki and Tsu consists of doping superlattices, in which the same host material is used, but differential doping can result in regions that are of *n*-type, *p*-type, or undoped (*i*-type) character. For example, it is possible to prepare a superlattice such as *n-i-p-n-i-p-n-i-p* and so forth, within a homogeneous structure. Then, again, the band gap and the electronic and conductive properties can be varied by varying the choice of concentrations and thicknesses of the doped regions.

Alternating layers and strained layers. A third kind of superlattice alternates superconductor and insulator. It has been studied extensively in connection with flux quantization in high-temperature superconductivity. Finally, strained-layer superlattices, in which the lattice constants of the two components are quite dissimilar, have been studied for particular applications, such as control of the bandgap.

Intercalated and polymeric materials. In passing from atomic and ionic materials to molecular materials, there has been extensive investigation of intercalated materials. For example, the layered structure of titanium disulfide (TiS_2) has alternating layers of Ti-S-S-Ti-S-S-Ti. Intercalated materials are soft and fracture easily along the directions perpendicular to the lattice planes because of the very weak S . . . S interactions. These interactions are of largely van der Waals type, and other species can be intercalated between the sulfur layers. This van der Waals' intercalation gap can accommodate very large organic molecules, resulting in an artificial structure of superlattice type, where D1 is now fixed by the length of the organic molecule, and D2 by the S . . . Ti . . . S distance.

Fully artificially structured polymeric materials, corresponding to polymeric superlattices, have been prepared and are of interest for a number of applications involving optics, sensors, and separation of mixtures of chemicals.

Self-assembled structures. Self-assembled monolayers (SAMs) are films formed by the chemisorption of a single layer of adsorbate molecules from solution or the gas phase onto a surface of interest. They have been extensively studied because of their utility in tailoring surface properties. Through choice of adsorbate molecule, the reactivity, wetting, adhesive, electrical, and structural properties of a surface can be controlled. Self-assembled monolayers have been used to develop novel surface-patterning methodologies, fabricate new types of chemically sensitive devices, study interfacial electron-transfer processes, and prepare a variety of unusual and potentially useful electronic and photonic materials, as well as redox-active materials (from which electrons can be reversibly removed or added). In addition, self-assembled monolayers have been used as passivating layers to stabilize numerous materials, including the very reactive class of cuprate-based high-temperature superconductors. Thus far, self-assembled monolayers have been prepared by using a wide range of adsorbate functional groups and substrates, including thiols, disulfides, sulfides, and phosphines on gold (Au); carboxylic acids on metal oxides; trichloro- and trialkoxysilanes on oxide surfaces; phosphonates on metal phosphonate surfaces;

Fig. 1. The $(\sqrt{3} \times \sqrt{3})$R30° structure for a linear alkanethiol self-assembled monolayer on Au(111). Open circles denote gold atoms. Filled circles denote R-S groups (alkyl chains with attached sulfur atoms).

thiols on GaAs, cadmium selenide (CdSe), cadmium sulfide (CdS), copper (Cu), and silver (Ag); thiols, olefins, and isonitriles on platinum (Pt); and amines and thiols on yttrium barium copper oxide (YBa_2Cu_3O-).

Some adsorbate molecules not only form strong chemical bonds with surfaces but also self-organize into highly ordered, even crystalline, structures. Although somewhat unusual, these superlattice structures have been demonstrated for a variety of adsorbate molecules, including linear alkanes, fluorinated linear alkanes, extended rigid-rod aromatic molecules, hybrid linear alkaneazobenzenes, and fullerenes. *See* FULLERENE.

The superlattice formed from the adsorption of linear alkanethiols onto Au(111) substrates has been the most extensively studied system (**Fig. 1**). This structure has been studied by in-plane x-ray diffraction (XRD), helium diffraction, atomic force microscopy (AFM), scanning tunneling microscopy (STM), low-energy electron diffraction (LEED), and reflection-absorption infrared spectroscopy. Most data thus far are consistent with a structure, denoted $(\sqrt{3} \times \sqrt{3})$R30°, that is commensurate with the underlying Au(111) substrate (Fig. 1). In such a structure, each alkyl group may rest over a threefold

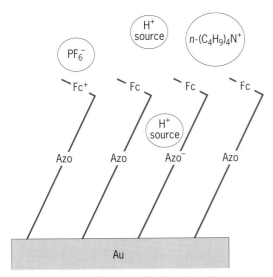

Fig. 2. Schematic of the ion-gating behavior of a self-assembled monolayer of *p*-ferrocenylazobenzene-butanethiol (PF₆⁻). H⁺ = hydrogen cation, Azo = azobenzene group, fc = ferrocenyl group, *n*-(C₄H₉)₄N⁺ = tetra-*n*-butylammonium cation, Au = gold substrate.

ticated adsorbate molecules, supramolecular properties are becoming evident and, in some cases, are quite useful. A key question is whether an organized monolayer (or multilayer constructed from monolayers) can have a set of properties distinct from the adsorbate molecules that compose it and, if so, whether these properties can be correlated with film structure.

Ion gating. Ion gating is one example of supramolecular behavior in self-assembled monolayers. Certain redox-active self-assembled monolayers, such as

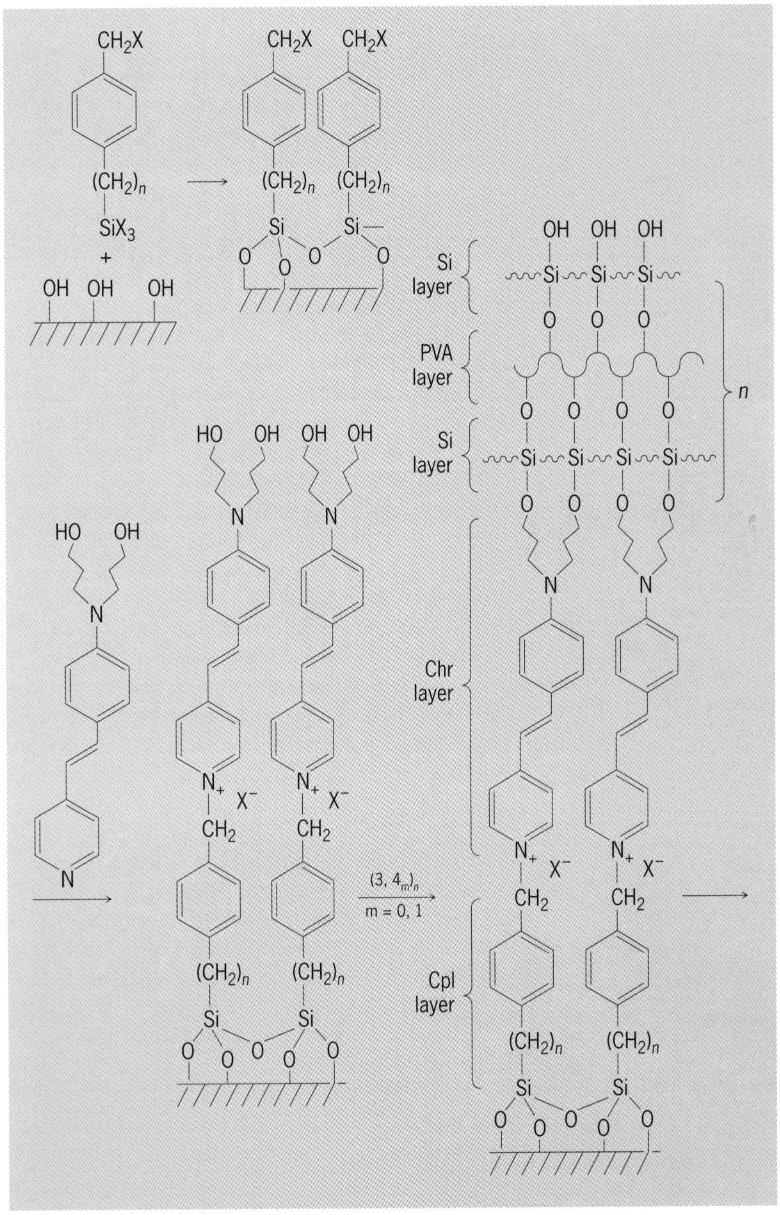

hollow site on the underlying Au(111) face. The nature of the surface attachment may involve an Au . . . thiolate interaction; however, it should be noted that some workers recently have raised the possibility of the R-S groups (where R denotes an alkyl chain and S is sulfur) dimerizing on the surface to form disulfides. Shorter-chain linear alkanethiols are believed to have a similar structure, but the extent of order decreases as chain length decreases. Indeed, the weak van der Waals interactions between the adjacent alkyl chains in these structures, becoming increasingly more important with increased chain length, are partly responsible for overcoming entropy and driving the self-assembly or self-organization process. With several adsorbate molecules and substrates, incommensurate structures are sometimes formed. This again emphasizes the importance of intermolecular interactions in the formation of these monolayer superlattice structures.

A key set of issues regarding self-assembled monolayers pertains to the chemical and physical consequences of adsorbate surface attachment, dense packing, and self-organization. In the molecular biology literature, the term self-assembly implies not only the formation of an ordered, definable structural unit but also an element of supramolecular function. With proteins and lipid bilayers, this can be understood through the complex functions that these structures provide in spite of the relatively simple molecules used to construct them. In contrast, there are few recognized examples of supramolecular behavior in self-assembled monolayers. This, in part, could be due to the relatively simple molecules used thus far to construct such films. With the design and synthesis of more sophis-

Fig. 3. Schematic of the synthesis of a self-assembled multilayer that shows noncentrosymmetric structure, and therefore nonlinear optical response. The lower structure is a glass surface. The multilayer contains polyvinyl alcohol (PVA), chromophore (Chr) layers, and coupling (Cpl) layers. X = halide counterions, Si = silicon. (*After T. J. Marks and M. A. Ratner, Design, synthesis, and properties of molecule-based assemblies with large second-order optical nonlinearities, Angew. Chem. Int. Ed. Engl., 34:155–173, 1995*)

those formed from alkanethiols with ferrocene-substituted azobenzene on the end of the alkyl thiol, with the formula $S-(CH_2)_4-C_6H_5NNC_6H_4C_5H_4FeC_5H_5$, exhibit electron-transfer properties that can be triggered through the choice of charge-compensating ions (**Fig. 2**). The structure of the self-assembled monolayer positions the redox-active azobenzene centers several tenths of a nanometer within the film and orients them perpendicular to the Au(111) electrode surface. As a result, electrochemical reduction of the azobenzene groups is prohibited with large cations such as tetra-*n*-butylammonium $[(C_4H_9)_4N^+]$ that cannot penetrate the film and compensate the charge created in the electrochemical process. However, the reduction process can be facilitated with small cations such as H^+, because they can penetrate the film. Groups such as ferrocenyl are always electrochemically active, since they reside at the solution–monolayer interface and are accessible to charge-compensating anions. Therefore, film structure can be used to control ion transport from solution to the azobenzene redox centers within the monolayer; in so doing, it can effectively turn on or off electron-transfer processes between the electrode surface and the azobenzene groups within the film. Furthermore, electron-transfer processes between the gold and the azobenzene redox centers within these novel films may be gated by controlling the size and concentration of the charge-compensating cations or by manipulating the structure of the film.

Structural control. Structural control of the self-assembled monolayer can also be important in other areas—indeed, the self-assembled monolayer itself may include subcomponents. For example, a very strongly ordered film can be made by covalent incorporation of a noncentrosymmetric chromophore species, denoted D-A, in a self-assembled-monolayer host. The ordering of the self-assembled monolayer then orders the D-A species to be strongly parallel to one another. In this case, the self-assembled monolayer may be extended to a multilayer structure (**Fig. 3**). This sort of adlayer architecture is of great interest in the manufacture of nonlinear optical films, electrochemical protection layers, and optoelectronic devices.

For background information *see* ARTIFICIALLY LAYERED STRUCTURES; BAND THEORY OF SOLIDS; CRYSTAL STRUCTURE; INTERCALATION COMPOUNDS; KRONIG-PENNEY MODEL; SEMICONDUCTOR HETEROSTRUCTURES; SUPRAMOLECULAR CHEMISTRY; VAPOR DEPOSITION in the McGraw-Hill Encyclopedia of Science & Technology.   Mark Ratner; Chad Mirkin

Bibliography. D. J. Campbell et al., Ion gated electron transfer in self-assembled monolayer films, *J. Amer. Chem. Soc.*, 118:10211–10219, 1996; G. Einspruch and W. R. Frensley (eds.), *Heterostructures and Quantum Devices*, 1994; W. P. Kirk and M. A. Reed (eds.), *Nanostructures and Mesoscopic Systems*, 1992; A. Ulman, Formation and structure of self-assembled monolayers, *Chem. Rev.*, 96:1533–1554, 1996.

Superstring theory

To find the theory that unifies the electromagnetic, strong nuclear, weak nuclear, and gravitational forces within the framework of quantum mechanics is a central goal of physics. A promising idea in this direction is superstring theory, where the basic objects are vibrating loops, and the basic dynamical process is the splitting of one loop into two or the joining of two into one. In principle, the whole spectrum of elementary particles can be obtained as different states of vibration of the string, and all the forces can be understood as consequences of the same splitting–joining interaction.

The theory is incomplete, however, in that it is well understood only when the splitting–joining interaction (the coupling) is weak, so that quantum-mechanical perturbation theory can be applied. When the coupling is strong, approximation methods break down, new phenomena are likely to occur, and even a precise definition of the theory is not known. The strong-coupling behavior of strings must be understood in detail before any detailed tests of the theory are possible. Since early 1995 there has been substantial progress in this direction through the discovery of a new principle, string duality.

Duality. The strong-coupling problem arises in many parts of physics. In some cases the physics of a strongly coupled system can be understood because of a special property known as weak/strong duality. Duality refers to an equivalence between two seemingly different physical systems. In weak/strong duality, one system at strong coupling is equivalent to the other system at weak coupling. Roughly speaking, when the particles of the first system become strongly interacting, there are certain bound states (solitons) whose interactions become weak and which behave precisely like the particles of the dual system.

Weak/strong duality is known to occur in some relatively simple systems, specifically low-dimensional quantum systems. In 1978 it was conjectured to be a property of certain supersymmetric gauge theories. Gauge theories are a generalization of the theory of electromagnetism and are now known to describe the strong and weak nuclear forces as well. Supersymmetry is an additional mathematical structure that has not been seen in nature but is widely believed to be a property of the unified theory at very short distances. For some time the evidence for these duality conjectures was limited and circumstantial. In 1994 it was discovered that supersymmetry could be exploited in a systematic way to determine much of the strong-coupling behavior of gauge theories. Much new evidence for weak/strong duality and many new examples were found. This phenomenon is now well established. An interesting feature is that weak/strong duality interchanges electric and magnetic fields. In gauge theories the basic particles carry electric charge (the electron, for example, in electromagnetism). In the dual theory the basic particles are magnetic monopoles.

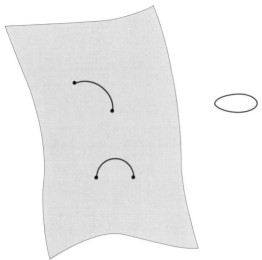

Two strings trapped on a D-brane, and one moving freely.

In 1995 the same methods were applied to superstring theories. There are five basic superstring theories, which differ in the nature of their supersymmetry and the geometry of the strings. However, each theory has many thousands of stable ground states, each physically distinct, effectively multiplying the number of string theories by a large factor. Weak/strong duality—in this context called string duality—was found to be a universal property of these theories. That is, in every case the strongly coupled behavior appears to be equivalent to that of some other weakly coupled system. The weakly coupled system is in most cases some other string theory. In fact, all the different string theories are now understood to be connected by string duality combined with variation of other parameters. That is, there is only one theory, and all the seemingly different string theories just describe different weakly coupled limits of this theory. Notably, some weakly coupled limits of the theory are not described by string theories but by something new that has been provisionally named M-theory. It is not understood in detail, but its high degree of symmetry allows many of its properties to be established. It is a generalization of an earlier theory known as supergravity.

D-branes. String duality has many implications. One of these is the existence in string theory of a new kind of object known as a D-brane. D-branes have the property that strings, which normally move freely, can become trapped on them. D-branes can be points, curves, or (as shown in the **illustration**) surfaces; in higher-dimensional space-times they can be higher-dimensional surfaces. At weak coupling D-branes are heavier than strings, but at strong coupling they are lighter and become the particles or strings of the weakly coupled dual theory. Their properties are calculable by using the standard string theory techniques, and give detailed information about the strongly coupled theory.

The fundamental constants of quantum mechanics, relativity, and gravity (Planck's constant $\hbar$, the speed of light c, and the gravitational constant G) can be combined to form a fundamental length scale known as the Planck length, $L_P = \sqrt{\hbar G/c^3}$. This length, roughly 10^{-35} m, is where the effects of gravity and quantum mechanics are simultaneously strong; at longer distances, gravity behaves classically (non-quantum-mechanically). An important issue in any theory that unifies quantum mechanics and gravity is the nature of space-time at this very short distance. In string theory, there is an effective minimum distance. This is set by the size of a typical string loop and is roughly 10 times the Planck length. However, studies of the dynamics of D-branes show that they appear to be associated with a somewhat shorter distance scale, closer to the Planck length. In other words, they are more effective than strings at probing the shortest and most fundamental distances. The implications of string duality and D-branes for the nature of space-time and the physics at very short distances is a subject under active development.

Black holes. In the mid-1970s it was discovered that the classical and quantum behavior of black holes satisfies laws that are directly analogous to the laws of thermodynamics. Ordinary thermodynamics is a macroscopic theory, meaning that it describes the average behavior of large collections of atoms. The corresponding microscopic theory, dealing with the behavior of the individual atoms, is statistical mechanics. For black-hole thermodynamics, no such microscopic theory was known; finding it has been an outstanding challenge. Following the discovery of string duality, it has been understood that at least in some cases black holes can be regarded as collections of large numbers of D-branes. The thermodynamic properties of black holes can then be derived from the statistical mechanics of their constituent D-branes. Roughly speaking, D-branes seem to be the underlying "atomic theory" of black holes.

Closely associated with this is a paradox known as the black-hole information problem. When a collection of matter collapses under the gravitational force to form a black hole, all details of the matter become hidden in the black hole except for its total mass, electric charge, and spin. The final black hole looks the same independent of what sort of matter collapsed to form it. Further, as shown by S. Hawking, the black hole will eventually decay into ordinary radiation, which again does not depend on the details of the original matter. The net effect is that many different initial states end up in the same final state. This is not consistent with the laws of quantum mechanics. That is, attempting to combine the theories of quantum mechanics and gravity leads to a paradox, where one of the existing laws must be modified.

In the past, such paradoxes have been very fruitful in finding new physical theories, and played a role in the discoveries of special relativity, general relativity, and quantum mechanics. The black-hole information paradox has therefore been a subject of great interest. It requires a significant modification in the understanding either of quantum mechanics or of space-time, but there is not yet a consensus as to which. The theory of D-branes, providing a microscopic theory of black holes and new information about the nature of space-time, has led to new approaches to this problem. *See* BLACK HOLE.

Outstanding problems. Individually, the electromagnetic, strong nuclear, weak nuclear, and gravitational forces are well understood experimentally. Experimental information about the unified theory, however, is very limited. In its absence, developments are largely theoretical. A consistent framework is sought that incorporates all the known laws. Fortunately, this requirement is very restrictive and has led to a unique candidate, superstring theory, that does incorporate these laws in a framework of great mathematical elegance.

Development has been slowed by the absence of experimental information, but even so, there is rapid progress at present. String duality has unified disparate ideas, solved long-standing problems, and produced many surprises (including the unification of all string theories, the existence of D-branes and M-theory, and new understanding of the nature of space-time). A key problem, now that all string theories are known to be limits of a single theory, is to find the central principle that defines that theory. Undoubtedly, further surprises lie ahead, so that it is impossible to predict whether the current developments will lead to a complete theory in the near future or whether major new ideas will be needed. At the same time, experiments that might probe the unified theory proceed; the search for supersymmetry at particle accelerators is the most prominent of these.

For background information *see* BLACK HOLE; GAUGE THEORY; MAGNETIC MONOPOLES; PERTURBATION (QUANTUM MECHANICS); QUANTUM MECHANICS; SOLITON; STATISTICAL MECHANICS; SUPERGRAVITY; SUPERSTRING THEORY; SUPERSYMMETRY; THERMODYNAMIC PRINCIPLES in the McGraw-Hill Encyclopedia of Science & Technology.   Joseph Polchinski

Bibliography. M. Mukerjee, Explaining everything, *Sci. Amer.*, 274(1):88–94, January 1996; J. Polchinski, String duality, *Rev. Mod. Phys.*, 68:1245–1258, 1996; G. Taubes, A theory of everything takes shape, *Science*, 269:1511–1513, 1995; E. Witten, Reflections on the fate of spacetime, *Phys. Today*, 49(4):24–29, April 1996.

Supersymmetry

Supersymmetry is a conjectured enhanced symmetry of the laws of nature which would relate the two fundamental observed classes of particles, bosons and fermions. All the known elementary bosons are mediators of the fundamental forces (the photon for electromagnetism, the graviton for gravity, the $W^\pm$ and Z^0 bosons for the weak nuclear force, and the gluons for the strong nuclear force), while all known elementary fermions (such as electrons, quarks, and neutrinos) constitute the matter in the universe. Thus, this symmetry would be a remarkable step forward in understanding the interrelationships between various aspects of the physical world. Recent theoretical and experimental developments indicate that an experimental confirmation of the existence of supersymmetry may be within reach.

Supersymmetry is, basically, conjectured to be a symmetry of the laws of physics. There should exist elementary particles related by supersymmetry transformations, just as the points of an equilateral triangle drawn on a sheet of paper are related by rotations through $120°$. If supersymmetry were realized as an exact symmetry, the particles so related should have almost all their characteristics, such as mass and charge, preserved. However, a unique property of supersymmetry among the known or conjectured symmetries of the laws of nature is that the spin of the related particles can differ by $\hbar/2$. (Particles with odd-half-integral intrinsic angular momentum or spin, $J = \hbar/2, 3\hbar/2, \ldots$, are fermions and obey the Pauli exclusion principle, while all particles with integer spin, $J = 0, \hbar, 2\hbar, \ldots$, are bosons. Here, $\hbar$ is the fundamental unit of angular momentum, Planck's constant divided by 2π.)

The end result of supersymmetry is that any fermion of spin $\hbar/2$ has a boson superpartner of spin 0, while any gauge boson of spin $\hbar$ has a fermion superpartner of spin $\hbar/2$. This is apparently a disaster for the idea of supersymmetry since it predicts, for instance, that there should exist a spin-0 boson partner of the electron, the selectron, with electric charge, and mass, equal to that of the electron. Such a particle is certainly ruled out by very many experiments.

Spontaneously broken symmetry. The crucial caveat to this negative result is the condition that supersymmetry be realized as an exact symmetry. One fundamental concept of modern physics is spontaneously broken symmetry. Physics displays many examples of symmetries that are exact symmetries of the fundamental equations describing a system, but not of their solutions. An everyday analogy involves a roulette wheel which, if it is fair, is rotationally invariant (the equations describing the motion of the ball on the spinning wheel do not depend on angle), but nevertheless has the property that on each spin the ball ends up in only one of the pockets, thereby in effect breaking the rotational symmetry. In particle physics the spontaneous breaking of a symmetry usually results in a difference in the masses of the particles related by the symmetry. The amount of breaking can be quantified by this mass difference.

The discovery of a spontaneously broken symmetry of nature is of fundamental importance, since it still gives information about the underlying symmetry properties of the laws of nature. If supersymmetry is broken by a very large amount, then all the superpartners have masses very much greater than the particles currently observed, and there is little hope of seeing evidence for supersymmetry. However, evidence for only moderate breaking of supersymmetry comes from examination of the properties of the fundamental forces at high energy.

Strengths of fundamental forces. Of the four forces, the three excluding gravity are very similar in their basic formulation; they are all gauge theories, generalizations of the quantum theory of electromagnetism, quantum electrodynamics. In quantum electrodynamics, the forces between electrically charged particles are mediated by the exchange of quanta of light, or photons, massless particles with $J = \hbar$. The strength of electrical interaction between two electrons, both of electric charge $-e$, can be quantified in terms of a dimensionless coupling constant, the fine-structure constant, defined in Eq. (1). (Here,

$$\alpha_{em} \equiv \frac{e^2}{4\pi\varepsilon_0\hbar c} \approx \frac{1}{137.036} \quad (1)$$

c is the speed of light, and ε_0 is the permittivity of free space.)

However, the quantity α_{em} is actually not a constant. It depends on the energies at which the interaction strength is measured. The reason is that in gauge theories there exists an analog of the phenomenon of dielectric screening. Dielectric screening is the name given to the property of materials whereby, if a charge is introduced into their bulk, the charged constituents of the material rearrange themselves so as to reduce the net electric field as the distance from the original charge is increased. Now, in the quantum theory of gauge interactions the vacuum itself responds to the introduction of a charge so as to reduce the field at long distances. Equivalently, if the charge of the electron is measured at shorter distances, it is found to increase in magnitude (as has been verified by many experiments). Physicists most often speak of the charge increasing with increasing energy or momentum of the probe being used to measure the charge. That this is equivalent to the above statement in terms of distance can be seen by use of the Heisenberg uncertainty relation (which states that the minimal distance probed in an experiment is inversely proportional to the momentum transferred) and relativistic mechanics.

The weak nuclear force and the strong nuclear force (described by quantum chromodynamics) are very similar in nature to quantum electrodynamics. Again, $J = \hbar$ gauge bosons, analogs of the photon quantum of light, mediate the respective forces by interacting with charged particles. (The charge is not electric charge, but some new, generalized form of charge.) However, these two forces differ from electromagnetism in one significant respect: The strengths of the interactions can decrease with increasing energy, rather than increase, as in the case of quantum electrodynamics. Whether this occurs depends on the amount of charged matter in the theory. This phenomenon, called asymptotic freedom, is a result of the self-interactions among the gauge bosons (which do not occur in quantum electrodynamics).

The strengths of the three gauge interactions (electromagnetic, weak nuclear, and strong nuclear), labeled α_1, α_2, and α_3, respectively, have been accurately measured at the Large Electron-Positron Collider (LEP) facility at CERN in Geneva, Switzerland, at an energy scale of 91 GeV (the mass, M_Z, of the Z^0 gauge boson, roughly 100 times the mass of the proton, expressed in terms of energy). However, rather than express these fine-structure constants directly, physicists choose to express α_1 and α_2 in terms of α_{em} and θ_w, the so-called weak angle, defined by Eq. (2). The reason for this apparently strange

$$\sin^2\theta_w = \frac{3\alpha_1/5}{(3\alpha_1/5)+\alpha_2} = \frac{\alpha_{em}}{\alpha_2} \quad (2)$$

choice is that the electromagnetic and weak forces actually unify into an electroweak theory, in which quantum electrodynamics is a mixture of interactions with strengths α_2 and α_1. (This mixing explains why α_{em} is different from α_1.) In any case, the values measured at LEP are given in Eq. (3).

$$\alpha_{em}(M_Z) = \frac{1}{127.90 \pm 0.09}$$
$$\sin^2\theta_w(M_Z) = 0.2315 \pm 0.0004 \quad (3)$$
$$\alpha_3(M_Z) = 0.118 \pm 0.003$$

The value of α_{em} given in Eq. (3) is greater than the value given in Eq. (1). This difference is due precisely to the energy dependence of the electric charge mentioned above, since the value quoted in Eq. (1) is that measured at zero energy (that is, at very large distances) rather than at a high energy as at LEP.

Unification of couplings. It is possible to calculate the way in which the interaction strengths, α_1, α_2, and α_3, depend on energy, μ. To leading order they satisfy some differential equations, called renormalization group equations, whose coefficients depend on the amount of fundamental matter that exists (with mass at or below the energy scale, μ) and that is charged with respect to each of the three interactions. Now, if the fundamental particles include not only the observed particles but also their superpartners, taken to have masses not greater than about 1000 GeV, then from the renormalization group equations, the couplings are predicted to depend on energy, μ, as shown in **Fig. 1**. The remarkable feature of the dependence in Fig. 1 is that the couplings meet (unify) at a single point, at an energy scale around 2×10^{16} GeV. (For the purposes of Fig. 1, the difference in mass between the observed particles and their hypothesized superpartners is taken to be $M_Z = 91$ GeV. However, the three couplings similarly unify at a single point, within experimental errors, for superpartner masses below 1000 GeV.) In contrast, **Fig. 2** illustrates the predicted energy dependence of the couplings if the superpartner contributions are not included—the appropriate situation if either supersymmetry is not an underlying symmetry of the world, or if it is very badly broken so that the superpartners are very massive

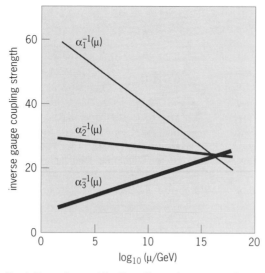

Fig. 1. Dependence of the three (inverse) gauge coupling strengths, α^{-1}, α^{-2}, and α^{-3}, upon energy scale, μ, within the supersymmetric extension of the standard model. The difference in mass between the observed particles and their hypothesized superpartners is taken to be 91 GeV. The thickness of the lines corresponds to the experimental uncertainties in the measured initial values of the (inverse) couplings.

indeed. In this case the couplings fail to unify at a single point.

It is possible to turn this chain of logic around. If it is assumed that the couplings do unify at some unknown single energy scale, then since there are two unknowns (this unification scale, as well as the value of the unified coupling) but three measurable quantities, it is possible to make a prediction for one of the measured quantities. A prediction of $\sin^2\theta_w(M_Z)$ in the case with relatively light superpart-

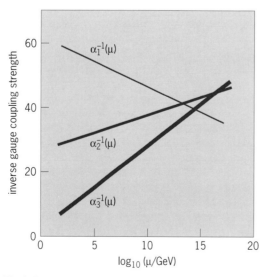

Fig. 2. Dependence of the three (inverse) gauge coupling strengths upon energy scale, μ, within the unextended, nonsupersymmetric standard model.

ners yields the value given in Eq. (4), which is in

$$\sin^2\theta_w(M_Z) = 0.231 \pm 0.003 \qquad (4)$$

truly remarkable agreement with the measured value, given in Eq. (3). (The nonsupersymmetric prediction is very far from the observed value.)

The result in Eq. (4) led to tremendous excitement for two reasons: It was the first successful prediction of a fundamental parameter of the standard model of particle physics, and it strongly suggested that the masses of the additional superpartner particles implied by supersymmetry might be within the experimental reach of the upgraded Tevatron at Fermilab in Illinois, and certainly the planned Large Hadron Collider facility at CERN.

Hierarchy and dark-matter problems. Although the unification of couplings is the single most significant indication that supersymmetry is relevant to the laws of nature, there are a number of other hints in this same direction. Two of the most interesting are the hierarchy and dark-matter problems.

The hierarchy problem arises in the electroweak theory. The predictions of this theory, such as the existence of the massive Z^0 and $W^\pm$ gauge bosons as well as their interactions, have been strikingly confirmed by experiments since the early 1980s. However, one aspect of this theory has not been directly tested, the mechanism by which particles get their mass. It is widely believed that the Higgs mechanism is responsible for the masses of the observed particles. In the simplest formulation of this mechanism, there exist (some number of) additional spin-0 particles called Higgs bosons, whose interactions with the observed particles lead to the generation of their masses. However, if the electroweak theory, together with these Higgs bosons, is embedded in a larger theory (such as a unified theory of all gauge interactions), then there exist enormously large corrections to the mass of the Higgs bosons, and necessarily to the masses of all the known particles—a disaster for the theory. Although it is possible to imagine other solutions to this problem, one very attractive feature of the supersymmetric extension of the standard model is that if the superpartner masses are not very much above 1000 GeV, these enormously large corrections to all the masses are almost entirely (and automatically) canceled, leaving a benign correction. Thus, most particle physicists believe that this cancellation also indicates that supersymmetry should be experimentally accessible.

Finally, the dark-matter problem involves the missing mass of the universe. By observing the large-scale motions of the galaxies, the average density of large volumes of the universe can be deduced, resulting in a value that is substantially greater than that directly observed in luminous matter (such as stars and hot gas). Therefore, a substantial fraction of the mass of the universe must be composed of some form of dark matter. Remarkably, many attractive models of supersymmetry predict that the lightest of all the superpartners is a weakly interacting massive particle (WIMP) with just the proper characteristics to be this dark matter (again, as long as the superpartner masses are not too large). It would certainly be one of the greatest triumphs of theoretical physics if theories developed to explain the interactions of particles at the smallest distances also turned out to imply the existence of the dominant constituent of the universe.

For background information *see* COSMOLOGY; ELECTROWEAK INTERACTION; GRAND UNIFICATION THEORIES; HIGGS BOSON; QUANTUM CHROMODYNAMICS; QUANTUM ELECTRODYNAMICS; QUANTUM FIELD THEORY; SPIN (QUANTUM MECHANICS); STANDARD MODEL; SUPERSYMMETRY; SYMMETRY BREAKING in the McGraw-Hill Encyclopedia of Science & Technology.

John March-Russell

Bibliography. S. Dimopoulos, S. Raby, and F. Wilczek, The unification of couplings, *Phys. Today*, 44(10):25–33, October 1991; P. G. O. Freund, *Introduction to Supersymmetry*, 1986; G. Ross, *Grand Unified Theories*, 1985.

Synchrotron radiation

In 1995 the Advanced Photon Source at Argonne National Laboratory near Chicago, Illinois, ended a 5-year construction program and began providing insertion-device and bending-magnet radiation in the hard x-ray spectral region for research on the structure and composition of matter. Users of the Advanced Photon Source include scientists from universities, industry, and federal laboratories. They apply advanced synchrotron x-ray technology to experiments in materials science; biological science; chemical science; agricultural science; environmental science; geoscience; atomic, molecular, and optical physics; and the development of novel synchrotron x-ray instrumentation. The Advanced Photon Source belongs to the class of third-generation synchrotron light sources, which also includes the Advanced Light Source in California, the European Synchrotron Radiation Facility in France, and the 8-GeV Super Photon Ring in Japan. All of these facilities afford new opportunities for research by extending those characteristics that make synchrotron-based x-ray beams highly useful to a broad spectrum of disciplines.

Advanced Photon Source facility. The Advanced Photon Source (**Fig. 1**) occupies an 80-acre (32-hectare) site at Argonne National Laboratory. The facility itself (**Fig. 2**) comprises a particle-beam injection and storage system, an annular-shaped experiment hall with a circumference of approximately ⅔ mi (1 km), laboratory–office modules for researchers adjacent to the experiment hall, a central office and laboratory building for staff, utility buildings, and a nearby residence for researchers staying at the Advanced Photon Source on a short-term basis.

Generation of x-ray beams. X-ray beams are generated at the Advanced Photon Source by a multistep process that utilizes synchrotron accelerator technology. A beam of positrons (positive-charge electrons)

is accelerated inside a system of aluminum alloy chambers containing a vacuum of ~10^{-9} torr (10^{-7} pascal), a state where all but one atom out of every 10^{12} present in the normal atmosphere is removed. Positrons were chosen over electrons because the positive charge of the former repels positive ions from residual gases that might otherwise detune the beam.

A beam of electrons is accelerated to 200 MeV (million electronvolts) in a first-stage (electron) linear accelerator (linac). The beam is focused on a 0.3-in-thick (7-mm), water-cooled tungsten disk, producing electron–positron pairs. The resulting 8-MeV positrons are accelerated to 450 MeV by the second-stage (positron) linac. The total length of both linacs is 190 ft (58 m).

Accelerated positrons are collected in a 100-ft-circumference (30.7-m) positron accumulator ring. This dc ring collects 10–12 linac pulses at 60 Hz in 0.5 s. These pulses are combined into a single, high-emittance bunch and compressed longitudinally. Positron emittance in the positron accumulator ring is damped substantially. Positron bunches are transferred to the booster, a racetrack-shaped, 1207-ft-circumference (368-m) ring of electromagnets. Four 5-cell, 352-MHz radio-frequency cavities accelerate the beam from 450 MeV to 7 GeV (billion electronvolts) in 0.25 s.

Twice per second, positron bunches are injected into a 3622-ft-circumference (1104-m) ring of electromagnets (the storage ring) located within the experiment hall. This ring is divided into 40 sectors. A typical sector includes two dipole magnets, ten quadrupole magnets, seven sextupole magnets, eight corrector magnets, nine radio-frequency beam-position monitors, and six extruded aluminum vac-

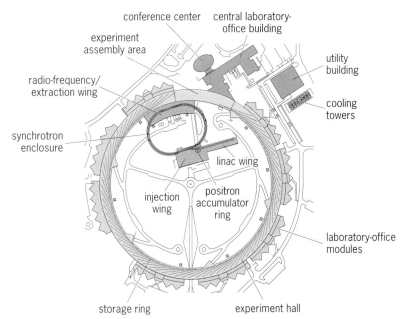

Fig. 1. Plan view (not to scale) of the Advanced Photon Source facility. *(Argonne National Laboratory)*

Fig. 2. Advanced Photon Source experiment hall, central laboratory–office building, and conference center. *(Argonne National Laboratory)*

uum chambers. The storage ring is designed to run at a nominal energy of 7 GeV, with all systems capable of achieving 7.5 GeV. Beam energy is maintained by four groups of four single-cell radio-frequency cavities in the storage ring. As the positron current drops off over time, the storage ring is resupplied from the injector.

Synchrotron radiation is produced by bending magnets and insertion devices located in the storage ring, which is configured according to the Chasman-Green arrangement, or lattice. The lattice was designed by R. Chasman and G. K. Green. In the early 1970s, their ideas on the periodic arrangement of all the bending and focusing electromagnets in a storage ring produced a particle beam of very small size and low angular divergence, beam qualities that are highly prized by users of synchrotron light sources. In the Advanced Photon Source storage ring, this sequencing yields 40 periods and 40 corresponding sectors. Five of these sectors are used for radio-frequency and beam injection and transfer equipment. Each of the remaining 35 sectors contains a bending-magnet and insertion-device pair, for a total of 70 research beam lines. The bending-magnet and insertion-device beam lines emerge onto the experiment hall floor through exit ports in the ratchet-shaped storage ring shield wall.

Insertion devices are the source of the high-brilliance x-ray beams generated by the Advanced Photon Source. Two linear arrays of permanent north–south magnets are arranged so that the poles alternate in direction along the path of a positron beam passing between them. This strong, alternating field causes the beam to oscillate rapidly, fixing beam energy and increasing beam brilliance. Insertion devices can be configured either as wigglers or undulators, depending upon the effect they are to have on the movement of the positron beam. Wigglers produce very intense radiation over a wide range of energies, while undulators yield radiation of selected (or tuned) energy at high brilliance. One of the two bending magnets in each sector (in addition to steering the positron beam) serves as a source of synchrotron radiation that exhibits a broad distribution of energies.

Radiation characteristics. Synchrotron x-rays are very intense, permitting atomic-level studies of extremely small samples at submicrometer resolution. The wavelengths of light from synchrotron radiation cover a broad range of energies in the invisible portion of the spectrum, including ultraviolet radiation and x-rays. These wavelengths are perfect for probing the structure of matter and studying various physical, chemical, and biological processes, because their lengths are comparable to the dimensions of the corresponding features of atoms, molecules, crystals, and cells. The pulsed-time structure of synchrotron radiation allows rapid imaging of reactions and processes that occur on a time scale measured in nanoseconds.

At the Advanced Photon Source, these characteristics are maximized by increasing the photon-beam qualities of flux and brilliance. Both are based on a measure of the number of photons per second in a narrow-energy bandwidth and in a unit of solid angle in the horizontal and vertical directions. Flux (the number of photons per second passing through a defined area) is the measure for experiments utilizing the entire, unfocused x-ray beam. Brilliance (a measure of the intensity and directionality of a given x-ray beam) determines the smallest spot onto which an x-ray beam can be focused. In addition, these beams are available from a broad, continuously tunable energy range of 1–100 keV (kiloelectronvolts), and they travel in a pulsed-time structure measured in billionths of a second.

Research applications. The net effect of these gains is the ability to carry out investigations on smaller samples in shorter periods of time, and to image physical, chemical, or biological processes that occur too rapidly for capture when light from other sources is used.

Materials science. Materials scientists can determine the precise positions of atoms and how the atomic structures influence the properties of materials. This information can then be used to improve existing materials, and to engineer new materials (such as semiconductors, polymers, ceramics, superconductors, composites, metallic glasses, and artificially layered structures) for specific applications in computing, microminiaturization, robotics, space exploration, communications, manufacturing, and energy technologies.

Molecular biology. Molecular biology has long provided extensive opportunities for x-ray research. Crystallographers have used x-rays diffracted from single crystals of atoms, nucleic acids, proteins, and viruses to determine both their three-dimensional atomic structure and their function. Within the complex atomic topography of molecules are the sites that control reactive biological functions. In order for these reactions to occur, the peaks and valleys of each molecule must fit together precisely. This form-and-function relationship is at the heart of structural biology, and plays an important role in medicine, the pharmaceutical industry, and biotechnology. Biological studies carried out at the Advanced Photon Source map the atomic structures of biomolecules in order to determine how they fit together and react. This information can then be applied to the creation of new pharmaceuticals.

Soil materials studies. The study of soil materials plays an important role in understanding and controlling the environment, the food and water supply, and even the earth that supports buildings, roads, and bridges. Environmental and agricultural scientists coupling Advanced Photon Source x-ray beams to a variety of research techniques can gain a better understanding of the chemical processes and atomic-level crystalline structures in soils that govern the availability of plant nutrients, the movement of water through substrata, the migration of environmental pollutants, and the characterization of soil and water responses to fertilizers and waste disposal. Studies

of soil materials also yield information on geological processes such as earthquakes, volcanic eruptions, and continental formation and drift. These changes often occur under conditions of very high temperature and extreme pressure, which can be replicated at the Advanced Photon Source on very small scales by using present research techniques and new experiments not feasible with previous x-ray and synchrotron-radiation sources.

Chemical science. Users engaged in chemical science at the Advanced Photon Source anticipate obtaining the first high-resolution structural information on samples smaller than a 1-micrometer cube. Critical catalytic materials such as zeolites exist only in microcrystalline form. Images of atomic-level structures in catalytic materials, obtained with third-generation light source x-ray beams, will show the precise mechanics of these materials and give clues to methods for engineering new, application-specific catalysts.

For background information *see* PARTICLE ACCELERATOR; SOIL CHEMISTRY; SYNCHROTRON RADIATION; X-RAY CRYSTALLOGRAPHY; ZEOLITE in the McGraw-Hill Encyclopedia of Science & Technology.

David E. Moncton

Bibliography. Argonne National Laboratory, *Chemical Applications of Synchrotron Radiation*: *Workshop Report*, ANL/APS-TM-4, 1989; Argonne National Laboratory, *7-GeV Advanced Photon Source*: *Conceptual Design Report*, ANL-87-15, 1987; Argonne National Laboratory, *Synchrotron X-ray Sources and New Opportunities in the Soil and Environmental Sciences*: *Workshop Report*, ANL/APS/TM-7, 1990; D. E. Moncton, Status of the Advanced Photon Source at Argonne National Laboratory (invited), *Rev. Sci. Instrum.*, vol. 67, no. 9, 1995; H. Winick (ed.), *Synchrotron Radiation Sources: A Primer,* Series on Synchrotron Radiation Techniques and Applications, vol. 1.

Thunderstorm

Transient luminous phenomena over thunderstorms have been reported occasionally for many years, but solid quantitative documentation has been lacking. In the 1980s, a research group at the University of Minnesota recorded broad luminous shapes in the dark sky over a thunderstorm 250 km (150 mi) distant. In the early 1990s, laterally extensive short-lived illumination over thunderstorms was documented in video images recorded on the NASA space shuttle. These two sets of pioneering observations quickly motivated two independent field campaigns to characterize the phenomena: a ground-based experiment from an observatory in Colorado and an aircraft-based program over the central United States led by researchers associated with the University of Alaska. The detailed results of these studies inspired two names for transient luminous events over thunderstorms, each with a distinct physical origin: sprites and elves.

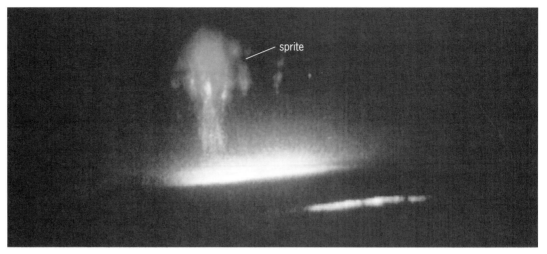

Fig. 1. Video image of a sprite taken from an airplane in the vicinity of a mesoscale convective system. *(D. Sentman and E. Wescott, University of Alaska)*

Sprites and elves. Figure 1 is a video image of a sprite over a laterally extensive thunderstorm illuminated internally by energetic lightning. The top of this sprite is 85–90 km (51–54 mi) above the Earth's surface. The maximum diameter of the sprite body is 20 km (12 mi). In the original color image, the body of the sprite is seen to be red, the result of the excitation of molecular nitrogen by electron collision. A similar color is commonly seen in the aurora in the upper atmosphere at high latitudes. The lower end of the sprite shows blue "tendrils" that extend to lower altitudes (about 45 km or 27 mi), usually far short of the thunderstorm cloud top (20–25 km or 12–15 mi) beneath. This blue color is likely a spectral emission from ionized nitrogen. The evidence that sprites scatter human-made electromagnetic waves at very low frequencies of 20–30 kHz indicates that the sprite is a plasma with an electron density enhanced relative to its surroundings. Sprites occur in a variety of shapes and sizes, and have been named carrots, columns, angels, jellyfish, and so forth. The reasons for these shapes are poorly understood at present.

Figure 2 shows an example of an elve—a broad disk of illumination at an altitude of 85–90 km (51–54 mi) with a thickness of about 6 km (4 mi). The elve shape is considerably simpler than that of the sprite, a difference probably attributable to the difference in its physical origin by lightning. A theory for this shape has been proposed. Elves exhibit both the blue emission associated with ionized nitrogen and the same red emission as the upper regions of sprites. Sprites and elves have been observed separately, but they often occur together in systematic sequence: an elve followed by a sprite.

The large-scale images (Figs. 1–2) in the visible portion of the electromagnetic spectrum immediately raise the question of why these events were not reported and documented extensively until the 1990s. Their brevity, dimness, obscuration by cloud and aerosol, relative infrequency, and typical time of day have probably contributed to their elusiveness. In comparisons with ordinary lightning having typical durations of hundreds of milliseconds, elves illuminate for less than a millisecond and the most persistent sprites, a few tens of milliseconds. In peak brightness, sprites and elves are three to five orders of magnitude dimmer than a typical lightning stroke to ground. Sprites and elves are also far less frequent than ordinary lightning. Finally, the meteorological conditions favored for sprites often place them very late in the evening and into the early morning hours when fewer observers are awake.

Characteristics. Sprites are clearly associated with lightning and thunderstorms, but not with ordinary lightning and ordinary isolated thunderstorms. As in Fig. 1, the parent thunderstorm below the sprite is laterally extensive—more like 200 km (120 mi) in diameter instead of the 10–20-km (6–12 mi) diameter typical for isolated thunderstorms. These larger thunderstorms are called mesoscale convective systems, occur later in the day, and likely contain greater quantities of electric charge than the more

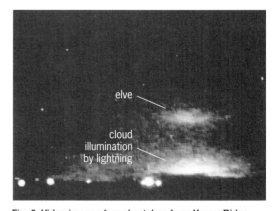

Fig. 2. Video image of an elve taken from Yucca Ridge Observatory in Colorado. The illumination of the thunderstorms by lightning beneath can also be seen. *(W. Lyons, FMA Research)*

common isolated thunderstorms. Sprites (and very likely elves) appear to be associated exclusively with cloud-to-ground lightning of positive polarity. This polarity is far less common than negative lightning and is generally characterized by greater charge transfer. Positive ground flashes associated with sprites have also been shown to launch electromagnetic waves called slow tails and Q-bursts (large-amplitude electromagnetic transients at extremely low frequencies of 3–500 Hz) that propagate around the world in the natural waveguide between the conductive Earth and the ionosphere. The electromagnetic energy in the sprite events can dominate all other lightning on the planet for several hundred milliseconds, thereby emphasizing the exceptional nature of these flashes.

In 1956, C. T. R. Wilson, a Scottish physicist, published ideas that largely account for the production of sprites by lightning. Electric fields are necessary to accelerate charged particles (that is, electrons) to ionize atoms and molecules and thereby stimulate light emission. Wilson's idea, illustrated in **Fig. 3**, was that cloud-to-ground lightning flashes serve to intensify the electric field above the storm by transporting negative charge to Earth, leaving behind the upper positive charge. Wilson also suggested the possible production of x-rays by thunderstorms. Recent theoretical work on electric field–driven runaway electrons supports this idea. Recent observations of gamma-ray bursts from space over a thunderstorm making positive lightning extends this picture.

Positive-polarity ground flashes and mesoscale convective systems had not been identified in Wilson's time. When the present-day observations on sprites and elves and their associated meteorological conditions are considered, Wilson's basic picture requires slight modification. The observed sequence of events in **Fig. 4** includes (*a*) the development of lightning in the cloud ; (*b*) the descent of a positively charged leader toward ground; (*c*) the interception of the descending leader by an upward-moving negative streamer to form the electromagnetic-pulse–producing positive return stroke with very high peak current (50–200 kA); (*d*) the appearance of the elve at 90 km (54 mi) altitude 300 microseconds (speed-of-light time) after the return stroke ; (*e*) the continued intrusion of the return-stroke current upward and, outward within the cloud, serving to transfer electric charge between ground and cloud in a "continuing current"; (*f*) and the appearance of the sprite several milliseconds after the return stroke.

Propagation. This sequence of events, with the elve preceding the sprite and appearing at a time consistent with speed-of-light propagation from the return-stroke channel to elve altitude, is strong evidence that the electric field of importance to elve illumination is driven by the electromagnetic pulse associated with the explosive increase of current in the vertical channel segment to ground. The electric field necessary for the subsequent sprite is probably of electrostatic origin, the charge transferred by the

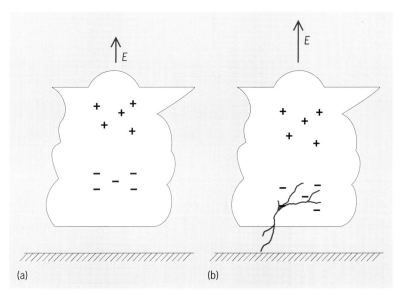

Fig. 3. Picture proposed by C. T. R. Wilson for creating a cloud-to-ionosphere discharge by cloud-to-ground lightning beneath an ordinary thundercloud. (*a*) Typical thunderstorm with dominant upper positive charge and lower negative charge, with upward-pointing electric field (*E*) above. (*b*) Cloud-to-ground lightning lowering negative charge and thereby increasing the electric field above the cloud.

continuing current immediately following the return stroke. This continuing current, with duration ranging from milliseconds to hundreds of milliseconds, is probably responsible for the slow-tail characteristics at a few hundred hertz and the Q-burst characteristics at extremely low frequency (8 Hz). The association of sprites and elves with distinct features of the parent lightning provides an explanation for the occurrence of each phenomenon individually, but sometimes together.

Both sprites and elves occur in the mesosphere at considerable elevation (50–90 km or 30–54 mi) above the parent thunderstorm. Both the electrostatic and electromagnetic fields of lightning are larger closer to the storm; thus there is a question as to why sprites and elves occur at such great altitude. The explanation probably lies in the trade-off between the ease of ionization and the electrical conductivity of the upper atmosphere.

Electrons must be accelerated by electric fields to threshold energies to produce light (and additional electrons) when the electrons collide with atoms and molecules. For a given electric field, arising either electrostatically or from the electromagnetic pulse, these thresholds are more readily exceeded when the paths available for particle acceleration are large. Long paths are favored in an atmosphere with lower density. The density of air diminishes exponentially with altitude, and therefore the tendency for air to support ionization and the production of light by electric fields increases with altitude. Wilson first noted that the dielectric strength of the atmosphere, the ability of air to prevent an electrical discharge, diminished with altitude more rapidly than the decline of electric field with distance from its thunderstorm source.

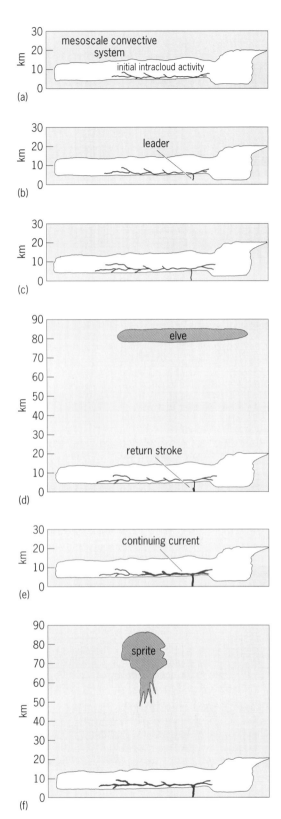

Fig. 4. Sequence of events in the troposphere and mesosphere associated with the formation of elves and sprites by positive cloud-to-ground lightning. 1 km = 0.6 mi. (a) Initial intracloud lightning. (b) Descent of positive leader toward ground. (c) Connection with ground and initiation of high-current return stroke. (d) Appearance of elve, centered on return stroke. (e) Continuing current, probably responsible for slow-tail and Q-burst excitation. (f) Appearance of sprite, centered on maximum charge transfer by continuing current.

Sprites and elves have not been detected at altitudes greater than about 95 km (57 mi). This apparent upper altitude limit is probably constrained by the electrical conductivity of the nighttime atmosphere that increases exponentially with altitude, owing to ionizing radiation from the Sun and from space. The electric fields necessary for light emission are naturally excluded from conductive regimes unless the electric field is imposed (by the action of lightning) in a time short compared to the local electrostatic relaxation time. The relaxation time at 80 km (48 mi) altitude is of the order of milliseconds, but at 100 km (60 mi) altitude it has diminished to microseconds.

Given these well-known dielectric properties of the upper atmosphere, sprites and elves are giant equivalents of the glow discharge tubes used to test high-voltage equipment or the low-pressure neon tubes used extensively for advertising. The mesosphere is the low-density tube, and the thunderstorm is the high-voltage source. The increasing tendency for a filamentary structure of the sprite at lower altitude in air of higher density is consistent with the change in form of laboratory discharges over a similar range of gas density. The presence of downwardly branched filaments beneath the sprite body suggests that the sprite is initiated in air of lower density and subsequently propagates downward.

Rough estimates of the global rate of sprite production are presently based on the rate of occurrence of the larger-amplitude Q-bursts that have been associated with sprites and elves. These rates are in the range 1–10 per minute globally. (Ordinary lightning occurs at a rate of 10–100 flashes per second globally.) The volume of atmosphere affected by individual sprite events may easily exceed 10^3–10^4 km^3 (10^2–10^3 mi^3), and current work is focused on the implications for global atmospheric chemistry. The evidence for ionization in sprites and elves opens up the possibility for modifications in the mesospheric budget of a form of nitrogen oxide (NO_x), a greenhouse gas substance and a major ingredient in ozone chemistry. Though sprites and elves are centered at altitudes above the ozone layer at 20–30 km (12–18 mi), further studies are needed to quantify their global impact.

For background information *see* ATMOSPHERIC ELECTRICITY; DYNAMIC METEOROLOGY; LIGHTNING; UPPER-ATMOSPHERE DYNAMICS in the McGraw-Hill Encyclopedia of Science & Technology.   Earle Williams

Bibliography. D. J. Boccippio et al., Sprites, ELF transients, and positive ground strokes, *Science*, 269:1088–1091, 1995; W. R. Corliss, *Handbook of Unusual Natural Phenomena*, 1983; G. J. Fishman et al., Discovery of intense gamma ray flashes of atmospheric origin, *Science*, 264:1313–1316, 1994; R. C. Franz, R. J. Nemzek, and J. R. Winkler, Television image of a large upward electrical discharge above a thunderstorm system, *Science*, 249:48–51, 1990; C. T. R. Wilson, A theory of thundercloud electricity, *Proc. Roy. Meteorol. Soc. A*, 236:297–317, 1956.

Tillage systems

Recent research involving tillage systems has included studies of responses of crops and soils to conservation tillage practices, and of the changes in the rooting patterns of plants that result from different types of tillage.

Conservation Tillage

Tillage is the mechanical manipulation of soil to prepare seedbeds, to control weeds and brush, to bury residues or diseased plants, to loosen and mix the soil for better aeration and water infiltration, and to cause fast breakdown of organic matter and release of minerals for plant nutrition. There are several types of tillage practices because no single one is suitable for all soils and climates. One practice that became important in the mid-1990s is conservation tillage. It is commonly defined as any tillage and planting system that maintains at least 30% of the soil covered by residue after planting to reduce water erosion; or, where erosion by wind is of primary concern, maintains at least 1000 lb/acre (1120 kg/hectare) of flat, small-grain residue equivalent on the surface during the critical wind erosion period. This definition is useful in the United States when determining if a field complies with the 1985 and 1990 Farm Bills.

Variants. Conventional tillage consists of primary and secondary tillage. Primary tillage is often done with tools that invert the soil (that is, moldboard plows or disk plows); secondary tillage is generally performed with disks, subsurface sweeps, or harrows. Weed control is accomplished with herbicides incorporated into the soil, cultivation, or postemergence spray herbicides. Variants of conservation tillage practices include ridge tillage, stubble mulch tillage, and no-tillage.

In ridge tillage, seedbeds are prepared on ridges that are produced with tillage implements. Crop residue is left on the soil surface between ridges. Generally, only one seed row is planted on each ridge. Weed control is accomplished with herbicides or cultivation.

In stubble mulch tillage, or simply mulch tillage, plant residues are minimally disturbed during tillage and seedbed preparation so as to leave a protective cover on the soil surface. Subsurface sweeps are the general tillage implements used. Weed control can be chemical or mechanical.

In the no-tillage variation, the soil is left undisturbed, except for the opening of the soil with the seed drill to place the seed at the intended depth. Previous crop residue remains undisturbed by tillage, and weed control is accomplished with herbicides.

Conservation tillage practices are gaining some acceptance among farmers but are far from being the most popular practices. For example, of approximately 1.6 million hectares (4 million acres) of wheat grown in Montana, nearly 30% receives some form of conservation tillage, but only 7% is no-tillage.

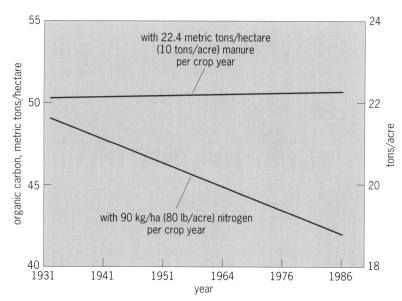

Soil organic carbon at 0–30-cm (0–1-ft) depth in a fallow-winter-wheat cropping system in the Pacific Northwest. *(After P. E. Rasmussen and W. J. Parton, Long-term effects of residue management in wheat-fallow: I. Inputs, yield, and soil organic matter, Soil Sci. Soc. Amer. J., 58:523–530, 1994)*

Various tillage practices have profoundly different influences on the physical and chemical properties of soils because of variations in soils and climate. A given tillage practice employed in different locales may cause different soil responses. Certain soil and crop response factors are particularly important in the semiarid agriculture of North America.

Organic matter. Organic matter is an important constituent of mineral soils. Its influence on physical and chemical properties of soils far outweighs its proportionate presence in soils. Long-term tillage causes large decreases in organic matter content of soils. As compared with virgin equivalent grassland soils, cultivated soils lose 30–50% of their original organic matter during the first 40–50 years of cultivation. To maintain in cultivated soils the levels of the original virgin organic matter, or to revert to them, is impossible and unnecessary; but a certain level of organic matter must be maintained for productive agriculture.

Maintaining productive soils is a function of soil and crop management. The fallow-wheat production system in semiarid agriculture is particularly prone to organic matter loss. Few crop residues are returned to the soil, soil erosion is common, and microbes readily oxidize organic matter. Organic matter can be maintained in semiarid agriculture by heavy application of manures where available (see **illus.**), or by soil and crop management practices that allow annual cropping, which returns proportionally more crop residue to the soil as compared with alternate fallow-crop production.

Bulk density. Soil bulk density, the dry weight of a unit volume of soil, includes both solids and pores, so loose and fluffy soils have lower bulk densities than more compact soils. Depending on soil types and soil conditions, bulk densities range from about 1.0 to 1.8 g/cm^3 (0.6 to 1.0 oz/in.^{3}) in surface soils. The higher bulk densities are found in sandy loams and sands. Crop and soil management practices influence bulk density. Additions of large amounts of manure or organic residues lower bulk densities, whereas intensive cultivation, which includes heavy-machinery traffic, increases bulk densities.

Results from recent comparative tillage studies in the United States have been inconsistent with respect to bulk-density effects. Some studies found that bulk densities increased under no-tillage as compared to conventional (plow) tillage. Results from other studies show a decrease in bulk densities under no-tillage. Data from a 10-year study on the northern semiarid Great Plains showed a decrease under no-tillage. On the southern Great Plains, results were mixed; some studies showed a decrease in bulk densities following a period of no-tillage, others an increase. When tillage practices are changed, it takes several years for significant permanent changes to take place in soil properties such as bulk density.

Soil penetration resistance. Resistance of a soil to penetration can be measured with a cylindrical rod that has a blunt or tapered probe tip. Penetrometer resistance is an integrated index of soil density, soil water content, soil texture, and type of clay minerals present in the soil. The most important factor influencing penetrometer readings appears to be soil-water content. Therefore it is important when comparing penetration resistance among tillage treatments that soil-water content and bulk density also be determined. Soil penetrometer measurements are more sensitive to small changes in soil strength and density than bulk density. Soil bulk density is generally measured by taking relatively large soil volume samples.

Penetration resistance generally followed the same trends as bulk density in studies conducted on the northern semiarid Great Plains. Resolution of differences among tillage treatments was generally higher with penetrometer measurements. Annually cropped no-tillage wheat lowered penetration resistance as compared with conventional fallow-wheat crop sequence. Results from the southern Great Plains were again mixed, with no difference between no-tillage and tilled treatments in some cases, and increased penetration resistance under no-tillage in other cases.

Infiltration. Water infiltration refers to the downward movement or entry of water into the soil. Infiltration can be estimated in various ways, from ponding water on the soil surface to using rainfall simulators. Data from a study of rainfall simulator infiltration on the northern Plains showed that there was little difference in infiltration between no-tillage and tilled treatments. On the southern Plains, rainfall simulator infiltration was highest on tilled treatments. Surface crusting is a serious problem on many soils, hindering infiltration. The low infiltration on no-tillage treatments was ascribed to this soil crust formation. Tillage loosens the soil surface and creates small basins for water to accumulate, thereby increasing infiltration. In no-tillage, the residue is left

undisturbed and, along with other factors associated with no-tillage and tillage treatments, reduces soil-water evaporation from the soil surfaces. Thus, overall soil-water conservation was greater for no-tillage treatments.

Crop response. Crop yields across the United States are generally favorable for conservation tillage practices, including no-tillage. On the southern Plains, no-tillage practices are viable for conserving soil water, they have no adverse effects on soil properties, and wheat yields have been similar on conventional and no-tillage fields. On the northern Plains, a 10-year study shows that no-tillage management of annually cropped wheat was the most profitable practice when compared to other management practices, including traditional fallow-crop rotation. The study included five low-yielding years (as low as 2 bushels/acre or 135 kg/ha), and five high-yielding years (as high as 51.5 bushels/acre or 3465 kg/ha).

Fallow practices may have their place, particularly during drought years; and infrequently, it is necessary to incorporate into the soil fertilizers that do not readily move in the soil solution, for example, phosphorus. Large applications of phosphorus will satisfy plant requirements for many years. Nitrogen, however, must be applied every year; and depending on the type of nitrogen fertilizer, it must be either broadcast or injected into the soil. Injection causes soil disturbance. Weed control can usually be accomplished with herbicides, regardless of the tillage method used. There may be occasions when a particularly noxious weed infestation must be mechanically controlled. With modern, improved farming techniques, including appropriate use of herbicides, conservation of soil water, and protection of soil from erosion, the general conclusion is that no-tillage and reduced-tillage farming are viable practices and that successful farming on the semiarid Great Plains need not include a season of summer fallow.

J. Kristian Aase

Tillage and Root Development

A soil–root association is extremely complex and delicate, and can be altered in countless ways by tillage-induced disturbance. In considering the vastly different soil types, climates, and crop plant species that are found in modern agriculture, it is obvious that the type and magnitude of crop root responses to tillage systems are extremely variable. Differences among soil tillage systems influence root growth and development in a variety of ways that may also impact plant performance and, ultimately, crop yield.

As farming becomes more specialized and sophisticated, and as costs continue to rise, a greater understanding of plant–soil interactions within modern cropping systems is essential. The urgent need to reduce soil erosion and runoff of agricultural chemicals while reducing costly inputs, such as fuel and labor, has led to the recent development of a variety of alternative agricultural tillage systems. The response of particular crop plants specifically in terms of their root systems has not often been considered when alternate tillage systems have been developed and improved. Tilling and disturbing a soil can have profound effects on its composition, structure, tilth, air and water content, temperature, flora, fauna, and other properties. Changes in any of these soil properties may directly or indirectly impact the growth and development of the plant root system. Scientific studies are beginning to confirm that different tillage systems do indeed affect root systems and thus overall performance of specific crops in a variety of ways. Understanding and manipulating many of these factors may help optimize the beneficial aspects of these systems and maximize crop yields.

Roots. Roots are the plant organs directly affected by any changes in a soil ecosystem, but the consequences of these effects may be manifested throughout the plant. The growth, volume, depth, distribution, health, and efficacy of root systems, even within the same species, may be significantly altered under different tillage and soil conditions. In considering the properties and rooting patterns of different crop plant species, and the various ways in which they may interact with different soils and tillage systems, the relationships become even more complex. Tillage is performed primarily to prepare seedbeds for good germination, to kill weeds, and to incorporate materials such as lime, fertilizers, and plant residues into the soil. In compacted, crusted, or anaerobic soils, proper soil tillage can greatly assist root development and plant growth of many crop species. Tillage, however, significantly disrupts the soil ecosystem. It must be performed and managed properly in order to enhance root growth and development while minimizing erosion, compaction, and long-term degradation of the soil.

Soils. Soils comprise an extremely variable and complex association of living and dead organisms, minerals, water, gases, and other components. An undisturbed soil ecosystem has generally developed a very organized and balanced system of layered materials that conserve an optimum amount of water, nutrients, and air, given the parent rock materials, climate, and indigenous species. When a soil is tilled, these naturally developed structures and functions are disrupted, and often virtually turned upside down. Some of the most direct and important changes when soils are tilled result from direct exposure to harsh elements, such as wind, rain, cold, and sunshine. The question arises as to what is specifically different about methods of tilling (or not tilling) the soil that causes crop roots to develop and respond differently.

Moisture and temperature. Numerous scientific studies have revealed important differences in crop root and plant performance among various tillage systems. Why such differences occur is only beginning to be understood. Perhaps the two most important tillage-induced factors that impact root system development are changes in soil moisture and soil temperature. Tillage often reduces soil water, because the water-holding layers are disrupted and exposed directly to drying winds and sunlight; an untilled soil will likely retain more soil water within the pro-

tected network of channels and pores below the surface. Tillage that removes or buries protective plant residues and exposes bare dark soil directly to sunlight promotes warming earlier each morning and earlier each spring compared with an untilled soil. Soil temperature or water conditions that encourage a seed to begin germinating earlier than an identical seed under different soil conditions may create important differences in the growth and development of the two root systems.

Other factors. Other components within the complex soil ecosystem may be affected by any tillage activity and may directly or indirectly impact root growth and development. After a given soil has been repeatedly tilled over a number of years, an impermeable layer of compacted subsoil, called a tillage pan, often develops just below where the tiller blades reach. A tillage pan often becomes impenetrable to plant roots, including strong tap roots, not only because of the very hard layer but also because of poor drainage and a lack of oxygen that results from compaction. In addition, certain soils that are frequently tilled, especially those that are finely pulverized in the process, may develop a hard crust at the soil surface that is difficult for germinating seedlings and certain roots, such as brace roots in corn, to penetrate. Such a crusted, sealed surface can also cause rains, fertilizers, and pesticides to run off a field rather than soak into the soil, further affecting root growth and development. Studies have also shown that root growth and root numbers can be substantially reduced in field areas that become compacted under wheel traffic from heavy agricultural equipment used for tillage and other operations.

Plant nutrients, fertilizers, and other materials such as lime and pesticides, whether added to the soil or occurring naturally, move and react within the soil substrate in manners unique to each substance. Tillage may promote or inhibit movement, availability, and breakdown of such materials in the soil; the effect may be either beneficial or detrimental to a root system, depending on the substance as well as the plant species. Tillage systems may also impact root growth and development because of differences in how and where fertilizers are applied within a specific tillage system, as roots tend to seek out and proliferate in areas of high fertility. The effects of tillage on many soil microbiological processes that impact soil fertility and plant nutrition, including nodulation of certain legume roots by *Rhizobium* bacteria, are largely unknown, but they are almost certainly influenced by various methods of tilling (or not tilling) the soil.

Beneficial and harmful bacteria, fungi, nematodes, insects, mammals, weeds, and other soil-dwelling organisms, as well as a crop plant's response to such organisms, may be affected by tillage. In turn, there may be important consequences in terms of root system health and overall plant development. For example, studies have revealed that a variety of plant diseases, especially root rots, may become prevalent under some tillage management systems and not under others, and that populations of beneficial soil mycorrhizal fungi may be reduced under certain tillage systems. In addition, root-damaging insects, such as grubs and rootworms, may be able to adapt to and actually thrive under certain tillage systems, while other methods of tilling may assist in their eradication.

For background information *see* AGRICULTURAL SOIL AND CROP PRACTICES; PLANT GROWTH; PLANT-WATER RELATIONS; ROOT (BOTANY); SOIL; SOIL ECOLOGY in the McGraw-Hill Encyclopedia of Science & Technology.   Andrew L. Thomas

Bibliography. J. K. Aase and J. L. Pikul, Jr., Crop and soil response to long-term tillage practices in the northern Great Plains, *Agron. J.*, 87:652–656, 1995; J. W. Bauder, G. W. Randall, and R. T. Schuler, Effects of tillage with controlled wheel traffic on soil properties and root growth of corn, *J. Soil Water Conserv.*, 40(4):382–385, 1985; K. Y. Chan and J. A. Mead, Tillage-induced differences in the growth and distribution of wheat roots, *Aust. J. Agr. Res.*, 43(1):19–28, 1992; M. D. Johnson and B. Lowery, Effect of three conservation tillage practices on soil temperature and thermal properties, *Soil Sci. Soc. Amer. J.*, 49:1547–1552, 1985; T. C. Kaspar, H. J. Brown, and E. M. Kassmeyer, Corn root distribution as affected by tillage, wheel traffic, and fertilizer placement, *Soil Sci. Soc. Amer. J.*, 55:1390–1394, 1991; P. E. Rasmussen and W. J. Parton, Long-term effects of residue management in wheat-fallow: I. Inputs, yield, and soil organic matter, *Soil Sci. Soc. Amer. J.*, 58:523–530, 1994; D. L. Sparks (ed.), *Advances in Agronomy*, vol. 51, 1993; P. W. Unger, Soil physical properties after 36 years of cropping to winter wheat, *Soil Sci. Soc. Amer. J.*, 46:796–801, 1982.

Transportation engineering

The concept of intelligent transportation systems (ITS; previously termed intelligent vehicle highway systems) focuses on the application of computers, communications, and other advanced technologies to surface-transportation problems. Twenty-nine user services, grouped in seven bundles, are listed in the **table**. These define a series of transportation-related products and services that may be provided for travelers and for managers of transportation systems. Since the passage of the Intermodal Surface Transportation Efficiency Act (ISTEA) of 1991, the deployment of intelligent transportation systems has become widespread, and there has been significant development in many areas.

User services. The seven bundles into which user services are grouped in the table are travel and transportation management, travel demand management, public transportation operations, electronic payment, commercial vehicle operations, emergency services, and advanced vehicle control and safety systems.

Travel and transportation management. En-route driver information, route guidance, traveler information, traffic control, incident detection and management, and emissions testing and mitigation form the core of a group of services intended to improve the efficiency and convenience of surface transportation. En-route driver information is being provided to drivers in a number of ways. Roadside changeable message signs are used to provide timely information on a range of topics including congestion, alternative routes, parking, and weather and other road hazards. In-vehicle navigation systems, which are widely used in Japan, are emerging in the United States. Systems that provide turn-by-turn guidance for drivers to their destinations are now available in rental cars in major metropolitan areas. A recent test of these systems in Orlando, Florida, called TravTek, demonstrated the utility and popularity of systems that provide detailed information to drivers in unfamiliar areas.

Traffic control and incident detection systems are combined with driver information and advanced roadway surveillance capabilities into advanced transportation management centers. Approximately 80 centers exist at various levels of sophistication in the United States. They are generally staffed by operators and representatives from agencies responsible for traffic control, law and traffic enforcement, and emergency response; some incorporate repre-

Intelligent transportation user services*

Bundle	User services
Travel and transportation management	En-route driver information
	Route guidance
	Traveler services information
	Traffic control
	Incident management
	Emissions testing and mitigation
Travel demand management	Demand management and operations
	Pretrip travel information
	Ride matching and reservation
Public transportation operations	Public transportation management
	En-route transit information
	Personalized public transit
	Public travel security
Electronic payment	Electronic payment services
Commercial vehicle operations	Commercial vehicle electronic clearance
	Automated roadside safety inspection
	On-board safety monitoring
	Commercial vehicle administrative processes
	Hazardous materials incident response
	Freight mobility
Emergency management	Emergency notification and personal security
	Emergency vehicle management
Advanced vehicle control and safety systems	Longitudinal collision avoidance
	Lateral collision avoidance
	Intersection collision avoidance
	Vision enhancement for crash avoidance
	Safety readiness
	Precrash restraint deployment
	Automated highway system

* After National ITS Program Plan, ITS America, March 1995.

sentatives from public transit agencies. These centers allow operators to monitor traffic conditions and respond to detected incidents quickly and efficiently. The effect is smoother traffic flow with reduced delay to travelers, and a reduction in the number of secondary crashes following incidents and accidents. The 1996 Olympics in Atlanta, Georgia, were used to demonstrate a variety of traveler information delivery systems including personal digital assistants (small handheld computers that provide individual users with calendars, phone lists, and other databases), cable television, kiosks, and in-vehicle navigation systems.

Travel demand management. Travel demand management incorporates a variety of tools and techniques intended to influence the travel choices of users and to carry out transportation policy. Systems which provide users with advance information on travel options and travel conditions can influence their travel choices. A growing number of agencies are using a range of information technologies, including the World Wide Web and cable television, to disseminate information on transit routes and schedules, traffic congestion, and alternative routes. The World Wide Web contains sites which show real-time congestion and travel-time maps, transit routes and schedules, and transit trip planning software.

Public transportation operations. Transit agencies and other transportation service providers are using a range of systems to improve their operations. Public transportation management systems incorporate technologies such as advanced vehicle location systems to schedule buses more efficiently. Operators can track the locations of buses and direct bus drivers to wait or to proceed on a specific route. This keeps buses evenly spaced on heavily traveled routes and avoids the problems associated with the bunching that occurs on bus routes. Similar systems provide operators of demand-responsive systems (such as taxi fleets) with the ability to react more quickly to requests for service and to more efficiently use their fleet of vehicles.

Electronic payment. Electronic payment systems include both automated toll collection on roadways and the development of single payment medium that can be used for a variety of transportation, and perhaps nontransportation, purposes. Currently, more than 20 toll agencies in the United States use electronic toll collection. Tags are mounted on the windshields of vehicles and are automatically read as the driver passes through toll plazas. The amount of the toll is deducted from the driver's account. Electronic toll systems are well accepted by commuters and provide considerable cost savings to toll agencies. A growing number of transit agencies are using smart-card technologies to allow travelers to use a single payment medium for a range of transportation purposes. With smart cards, travelers can, for example, pay for parking, bus fare, and light-rail or subway fare with a single card. Reduced reliance on cash payments increases efficiency, convenience, and security for passengers and operators.

Commercial vehicle operations. These systems are designed to increase the safety and efficiency of commercial vehicles. Commercial vehicle electronic clearance systems are designed to reduce the amount of time spent by drivers in obtaining clearance at state borders. Projects such as Advantage I-75 and Help-Crescent integrate weigh-in-motion systems, which allow trucks and other vehicles to be weighed as they travel at normal speeds, and electronic credentialing, which allows truckers to bypass state-line inspection and weigh stations.

Emergency management. Intelligent transportation systems can help provide more timely responses to accidents and other incidents. Mayday systems provide a means for rapidly summoning assistance to the site of an accident or incident. Most systems incorporate both automated vehicle location technology and two-way communications to notify an emergency response center, to pinpoint the vehicle's location, and to allow the emergency center to speak with motorists to tailor needs to the circumstances. One commercially available system is patterned after home security systems in that calls are routed through a privately operated security center. Several operational tests are under way which provide manual or automated distress signals to a response center. Mayday systems are available as an option on two vehicle models.

Advanced vehicle control and safety. Advanced vehicle control and safety systems incorporate advanced sensing, control, and warning technologies into systems designed to avoid crashes or to enhance driver capabilities. A number of collision-avoidance systems are in the prototype stage. Intelligent cruise-control systems enhance the capabilities of current cruise control by helping a driver maintain a specified minimum spacing between his or her vehicle and the preceding vehicle. Lateral collision-avoidance systems, such as blind-spot detection systems, are emerging for commercial vehicles (trucks) and school buses. Those intended for private vehicles are in the prototype stages. The Automated Highway Systems Project is developing a prototype system that supports hands-off–feet-off driving on dedicated highway lanes.

Intelligent Transportation Infrastructure. The U.S. Department of Transportation has developed the concept of an Intelligent Transportation Infrastructure (ITI), consisting of traffic detection and monitoring, communications, and control systems required to support a variety of products and services of intelligent transportation systems in metropolitan and rural areas. This infrastructure provides the basic building blocks needed to effectively deploy and operate intelligent transportation systems, as locally appropriate.

The infrastructure is defined in terms of nine basic elements, all of which are possible with existing technology: traffic signal control, transit management, freeway management, electronic toll collection, regional multimodal travel information, electronic fare payment, railroad grade crossings, emergency management services, and incident management.

National systems architecture. The concept of a system architecture for developing and supporting complex electronic systems has evolved over many years. The intelligent transportation systems architecture began with the 29 user services listed in the table as user requirements for a system design. It identifies the subsystems of an intelligent transportation system, defines the basic function performed by each subsystem, and identifies the data that must be transferred between them. The objective is to support the development of interoperable, compatible systems. The architecture purposely does not provide the actual design; rather, it provides a framework by which to build and integrate the intelligent transportation system and other systems.

Goal for ITS. The National Surface Transportation Goal for Intelligent Transportation Systems was developed in order to focus and accelerate the deployment of intelligent transportation systems in the United States. Since its development in 1996, the goal has been adopted or is under consideration by a range of associations, organizations, and user groups with an interest in the development of improved transportation systems. This goal calls for the deployment of the basic services of intelligent transportation systems for consumers of passenger and freight transportation across the United States by 2005 in the areas of travel information, transportation management, intermodal freight, commercial vehicle operations, and in-vehicle and personal information products.

For background information *see* TRAFFIC-CONTROL SYSTEMS; TRANSPORTATION ENGINEERING in the McGraw-Hill Encyclopedia of Science & Technology.   James Costantino; Donna C. Nelson

Bibliography. Intelligent Transportation Society of America, *National Surface Transportation Goal for Intelligent Transportation Systems* (ITS), 1996; U.S. Department of Transportation, *Building the ITI: Putting the National Architecture into Action*, 1996; U.S. Department of Transportation, *Intelligent Transportation Infrastructure Benefits: Expected and Experienced*, 1996.

Transuranium elements

Elements 110 and 111 were discovered in late 1994, and element 112 in early 1996. They are still unnamed.

Element 110. Element 110 should be a heavy homolog of the elements platinum, palladium, and nickel. It is expected to be the eighth element in the 6d shell.

Searches for this element began in 1985. Experiments performed at Dubna, Russia; at GSI (Gesellschaft für Schwerionenforschung), Darmstadt, Germany; and at Lawrence Berkeley Laboratory, Berkeley, California, failed to provide reliable evidence for a successful synthesis. However, at GSI-Darmstadt on November 9, 1994, a decay chain was observed which proved the existence of the isotope $^{269}110$ (the isotope of element 110 with mass number 269). The isotopes were produced in a fusion reaction of a nickel-62 projectile with a lead-208 target nucleus. The fused system, with an excitation energy of 13 MeV, cooled down by emitting one neutron and forming $^{269}110$, which by sequential alpha decays transformed to $^{265}108$, $^{261}106$, $^{257}104$, and ^{253}No (nobelium-253). All these daughter isotopes were already known, and four decay chains observed in the following 12 days corroborated without any doubt the discovery of the element. **Illustration** *a* shows the first decay chain observed, which ended in $^{257}104$. The isotope $^{269}110$ has a half-life of 0.2 ms, and it is produced with a cross section of about $3 \cdot 10^{-36}$ cm^2.

A second isotope, $^{271}110$, was produced in a subsequent 12-day experiment by fusion of nickel-64 and lead-208. Nine atoms, with an excitation energy of 12 MeV, were produced, and by sequential alpha decay they transformed to the known isotopes $^{267}108$, $^{263}106$, $^{259}104$, and ^{255}No (nobelium-255). The half-life of $^{271}110$ is 1.1 ms, and its production cross section amounts to $1.5 \cdot 10^{-35}$ cm^2.

The methods used to produce element 110 were the same as those already used to synthesize the three preceding elements, 107, 108, and 109. Improved beam intensity and quality, improvement of the detection efficiency, and a new detector system allowing nearly complete chain reconstruction made possible the discovery after an extensive search for the optimum bombarding energy. The total sensitivity for finding a new species was increased by a factor of 20.

Element 111. Element 111 should be a homolog of the elements gold, silver, and copper. It is expected to be the ninth element in the 6d shell.

The element was discovered on December 17, 1994, at GSI-Darmstadt by detection of the isotope $^{272}111$, which was produced by fusion of a nickel-64 projectile and a bismuth-209 target nucleus after the fused system was cooled by emission of one neutron. The optimum bombarding energy for producing $^{272}111$ corresponds to an excitation energy of 15 MeV for the fused system. Sequential alpha decays to $^{268}109$, $^{264}107$, $^{260}105$, and ^{256}Lr (lawrencium-256) allowed identification from the known decay properties of $^{260}105$ and ^{256}Lr. In the decay chain in illus. *b*, the first three members are new isotopes. The isotope $^{272}111$ has a half-life of 1.5 ms, and is produced with a cross section of $3.5 \cdot 10^{-36}$ cm^2. Altogether, three chains were observed during the 17 days of irradiation. The new isotopes $^{268}109$ and $^{264}107$ are the heaviest isotopes of these elements currently known. Their half-lives of 70 ms and 0.4 s, respectively, are longer than those of the previously known isotopes of these elements. The methods used to produce element 111 were the same as those used in the discovery of element 110.

Element 112. Element 112 should be a heavy homolog of the elements mercury, cadmium, and

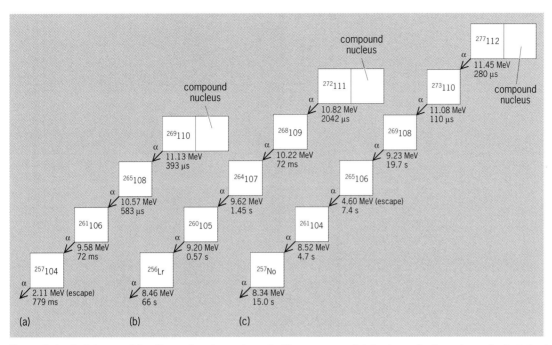

Decay chains that document the discoveries of new elements. The sequence of alpha decays is shown for each element. Numbers below boxes are alpha energies and correlation times. (a) Element 110, produced in the reaction ^{62}Ni + ^{208}Pb → ^{269}110 + 1n. (b) Element 111, produced in the reaction ^{64}Ni + ^{209}Bi → ^{272}111 + 1n. (c) Element 112, produced in the reaction ^{70}Zn + ^{208}Pb → ^{277}112 + 1n.

zinc. It is expected to be the last element in the $6d$ shell.

The element was discovered on February 9, 1996, at GSI-Darmstadt by detection of the isotope ^{277}112, which was produced by fusion of a zinc-70 projectile and a lead-208 target nucleus following the cooling down of the fused system by emission of a single neutron. The fused system was observed at an excitation energy of 12 MeV. Sequential alpha decays to ^{273}110, ^{269}108, ^{265}106, ^{261}104, and ^{257}No (nobelium-257) allowed unambiguous identification by using the known decay properties of the last three members of the chain. In the decay chain in illus. c, the first three members are new isotopes. The isotope ^{277}112 has a half-life of 0.24 ms, and it is produced with a cross section of $1.0 \cdot 10^{-36}$ cm^2. The new isotopes of elements 110 and 108 are of special interest. Their half-lives and alpha energies are very different, as is characteristic of a closed-shell crossing. At the neutron number $N = 162$, a closed shell was theoretically predicted, and this closed shell is verified in the decay chain observed. The isotope ^{269}108 has a half-life of 9 s, which is long enough to allow studies on the chemistry of this element. The methods used to produce element 112 were the same as those used for the two preceding elements, 110 and 111. Two decay chains of the new element were observed in an irradiation time of about 3 weeks. The cross section measured is the smallest observed in the production of heavy elements.

The crossing of the neutron shell at $N = 162$ is an important achievement in the field of research on superheavy elements. The stabilization of su-

perheavy elements is based on high fission barriers, which are due to corrections in the binding energies found near closed shells. The shell at $N = 162$ is the first such shell predicted, and is now verified. The predicted shells at proton number $Z = 114$ and neutron number $N = 184$ are still elusive, and until now no increase in half-lives approaching the predicted shell at $Z = 114$ has been observed. Moreover, the trend toward smaller cross sections has continued.

For background information *see* ELEMENT 107; ELEMENT 108; ELEMENT 109; NUCLEAR REACTION; NUCLEAR STRUCTURE; TRANSURANIUM ELEMENTS in the McGraw-Hill Encyclopedia of Science & Technology.

Peter J. Armbruster

Bibliography. S. Hofmann et al., The new element 111, *Z. Phys. A*, 350:281–282, 1995; S. Hofmann et al., The new element 112, *Z. Phys. A*, 354:229–230, 1996; S. Hofmann et al., Production and decay of ^{269}110, *Z. Phys. A*, 350:277–280, 1995.

Tuberculosis

Tuberculosis of cattle and deer is caused by the bacillus *Mycobacterium bovis,* which infects all warm-blooded animals, including humans. The most common route of transmission among cattle and deer is aerosol inhalation. Transmission may also occur in utero or by ingestion of contaminated water or feed and by the nursing of infected dams by calves. Tuberculosis recently has been diagnosed with increasing frequency in humans, cattle, farmed deer and elk, bison, llamas, and other animals.

Emerging health problems. The reemergence of this disease, combined with the current fiscal constraints on most national regulatory policies and with the limitations of knowledge and technology for dealing with tuberculosis, poses food-safety, public-health, and economic problems for livestock industries. Tuberculosis has reemerged as a major animal health problem due to multiple causes, including (1) endemic tuberculosis in some very large dairy herds; (2) isolated, severe outbreaks of tuberculosis in dairy and beef cattle; (3) deficiencies of current diagnostic test technologies; (4) low prevalence of bovine tuberculosis, which limits the effectiveness of routine skin testing as a surveillance tool; (5) expanding trade among lower- and higher-prevalence nations, resulting in the importation of increased numbers of animals infected with *M. bovis*; (6) inadequate definition of the molecular basis for the processes, and of mechanisms by which *M. bovis* infects and causes disease in livestock; (7) rapid expansion of the farmed deer and elk industries; and (8) lack of realistic alternatives (options or supplements) or novel approaches (such as vaccination programs or breeding of resistant livestock) to the traditional test-and-slaughter approach for control and eradication of the disease.

Cattle and deer tuberculosis. Most national animal tuberculosis regulatory programs were begun during the early decades of this century. These programs consisted of systematic testing of all cattle within a defined area. Cattle which tested positive for the tuberculin skin test were removed and slaughtered. In 1918 about 1 of every 20 animals tested positive for *M. bovis* in the United States, but in 1990 only 1 of every 6800 cattle tested positive. During the 1960s the lower number of *M. bovis*–infected animals led to a restructured eradication program, with primary emphasis on nationwide slaughter surveillance, coupled with epidemiologic investigation or trace-back to the origin of infected animals and trace-outs of movements of potentially infected animals into other herds.

Since initiation of the slaughter surveillance program in the United States, more intensive animal livestock production has increased animal-to-animal contact. Currently, 75% of all cattle slaughtered spend time in a feedlot. Large feedlots account for 54% of total feedlot capacity. Slaughter establishments are specialized and geographically concentrated. In 1991, 87% of all federally inspected cattle slaughter occurred in only 6% of all slaughter plants. Dairy operations are also intensive. Thus, while the national tuberculosis control program in the United States has greatly reduced the prevalence of bovine tuberculosis over the last several decades, the livestock industry may have become overconfident, and federal funds for research and disease surveillance may have been reduced excessively.

Nontraditional animal production, such as deer and elk farming and ranching, is undergoing tremen-

dous growth. Imports, breeding, specialized auctions, and trading have led to the introduction of other nontraditional farmed species, including llamas, alpacas, and other exotic ruminants. In addition, the wild population of elk has grown by 62% over the last 30 years. Such growth has resulted in captive, wild, and domestic animals residing in closer geographical proximity to each other.

Since 1991, 31 captive deer and elk herds in the United States have tested positive for *M. bovis*. These animals may rapidly develop extensive tuberculosis lesions. The first known cattle herd to be infected by association with a tuberculous elk was detected in Nebraska during 1991. In 1992, similar scenarios occurred in New York and Pennsylvania, causing those states to lose their accredited-free status for bovine tuberculosis. Investigations and awareness increased in the United States after Canada reported in late 1990 that cattle tuberculosis had been discovered in one or more of their captive elk herds. By the end of 1991, bovine tuberculosis had been confirmed in ten cervid herds: three in Montana and one each in Idaho, Colorado, Oklahoma, Nebraska, Texas, Wisconsin, and New York. Deer were infected at seven of the locations, elk at five, and both deer and elk were infected at two locations. Recently, wild white-tailed deer have been confirmed to be infected with *M. bovis* in the northern United States, where 26 infected white-tailed deer from a 182-mi^2 (471-km^2) area in Michigan resulted in massive tuberculin testing of deer, cattle, goats, sheep, pigs, and llamas in the 500-mi^2 (1295-km^2) surrounding area without disclosing transmission to domesticated livestock. These infections suggest that *M. bovis* is being transmitted from deer to deer and maintained in the deer population without apparent involvement of infected domesticated livestock. Until this occurrence, there was no known documented instance in the United States where *M. bovis* was maintained in wild populations of cervids once the source of the infection in either cattle or free-ranging bison was eliminated.

Impacts. The potential tuberculosis crisis in cattle and deer as perceived by the public may greatly jeopardize the sale of meat, milk, and live animals in domestic and international markets. Failure to eliminate tuberculosis in large herds has dramatized the inadequacies of the tuberculin skin test as a diagnostic tool, and the other deficiencies in the understanding of the tuberculosis disease process. For example, bovine tuberculosis has been endemic in the El Paso, Texas, milkshed since 1985, with several large herds containing animals that test positive for the bacterium. Over 200,000 cattle have received tuberculin skin tests, but until recently only about 50 (0.025%) of these animals tested positive. At the same time, some infected animals have not been identified correctly by the skin test, allowing the disease to persist and spread in these herds. This experience has clearly demonstrated a low level of correlation between individual animal skin-test results with confirmed cases of *M. bovis* infection, reflecting deficiencies in the tuberculin skin test, in terms of the specificity (failure to correctly identify uninfected animals) and the sensitivity (failure to correctly identify infected animals).

The natural history of this Texas area highlights the difficulty of ridding infected herds of tuberculosis by tuberculin skin-testing procedures alone. The long-term history of the program has shown that about one-third of all herds not depopulated can be expected to remain infected when released from quarantine. The advent of large dairy herds that have not been depopulated has increased the overall recurrence rate of tuberculosis in herds to about 50%. Until tests are developed that improve the detection rate of infected individual animals, it is not likely that the success rate will be improved.

Government regulations. State and federal animal health regulations require that a thorough epidemiological investigation be conducted in all cases of bovine tuberculosis to determine the source of introduction of disease, and to identify and trace all potentially exposed contact animals and herds. Investigations resulting from recent cases include mandatory tests of all herds identified as potential sources, test or slaughter of all exposed animals which have previously been sold from affected herds, testing of adjacent cattle herds, and areawide tests of all dairies.

Molecular epidemiology. Molecular epidemiology using restriction-fragment-length polymorphisms to type each of the isolates is providing a new approach to tracking the transmission of the organism. The dramatic differences in observed reactor rates and lesion rates between herds may be accounted for by differences in herd susceptibility, considering management factors, age structure, nutrition, herd genetics, and culling rates, and the duration of disease in the herd. However, empirical evidence suggests that consideration of these rates, along with noted differences in pathological observations, may indicate differences in the virulence of the organism as well. If strain differences of *M. bovis* do occur, identification of specific types by restriction-fragment-length polymorphism typing would likely provide information relating to source of exposure, prognosis for the herd, risks associated with decisions to elect test-and-slaughter rather than depopulation, and risks associated with traced-exposed animals lost in trade channels. Additional risk factors which have been identified but not adequately investigated include the practice of back-grounding, that is, conditioning dairy heifers in plots of land on which livestock are fattened for market (feedlots); geographical proximity of the heavily infected milksheds; and the potential of human, wild, and feral animals as vectors or reservoirs of *M. bovis*.

Because livestock populations that are free of tuberculosis are essential in order to maintain food safety, public health, and safe intrastate, interstate, and international trade, control leading to the eradication of the disease from livestock remains the paramount goal for most nations. This goal remains attainable, but its timely achievement will require increased commitment from its major stakeholders: the affected industries, state and federal animal health agencies, veterinarians, educators, and research scientists.

Control. Educating, training, and informing the public are very important aspects of developing an effective disease control program. In the past, the progress of the tuberculosis eradication program has reduced the prevalence of the disease, and has placed most associated people at least two generations removed from acquiring an understanding of the disease and gaining experience with its impact in both humans and animals. Currently, it is unlikely that beef or dairy producers have seen clinical cases of the disease in cattle or know people who have contracted the disease. The same applies to the associated industries: most veterinarians have not had direct experience with the disease, and slaughterhouse workers and inspectors are unfamiliar with actual cases.

The primary method for detecting infection in traditional national control programs has involved tuberculin skin testing of all cattle herds within defined geographic areas with slaughter of all positive animals, but this approach often is not feasible because of the high cost of replacement animals. Previously identified issues in the current eradication program which require consideration include improved methods and delivery of disease surveillance, indemnity for destruction of infected and exposed animals, regulatory modifications to minimize risks associated with infected imported animals, and educational efforts to improve awareness and appreciation of bovine tuberculosis as an integral part of routine herd health problems.

For background information *see* BEEF CATTLE PRODUCTION; EPIDEMIOLOGY; PUBLIC HEALTH; TUBERCULOSIS in the McGraw-Hill Encyclopedia of Science & Technology.   L. Garry Adams

Bibliography. W. M. Dankner et al., *Mycobacterium bovis* infections in San Diego: A clinicoepidemiologic study of 73 patients and a historical review of a forgotten pathogen, *Medicine*, 72:11–37, 1993; M. A. Essey and M. A. Koller, Status of bovine tuberculosis in North America, *Vet. Microbiol.*, 40:15–22, 1994; S. W. Martin et al., *Livestock Disease Eradication: Evaluation of the Cooperative State-Federal Bovine Tuberculosis Eradication Program*, 1994; V. S. Perumaalla et al., Molecular epidemiology of *Mycobacterium bovis* in Texas and Mexico, *J. Clin. Microbiol.*, 34:2066–2071, 1996.

Turbulent flow

Recent advances in the study of turbulent flow include the development of techniques of active control for suppressing turbulence, and the development of the technique of large-eddy simulation for the prediction of complex turbulent flows.

Control of Turbulence

Preventing the transition of a laminar flow to turbulence and suppressing the fluctuations of a turbulent flow, perhaps with the ultimate goal of returning the turbulent flow to a laminar state, are problems of great theoretical as well as practical importance. From the viewpoint of practical engineering, suppression of turbulence through appropriate control actions would provide a dramatic reduction of viscous drag and consequently of transportation costs. From the theoretical viewpoint, control of turbulence is a test of the completeness of physical theory, and understanding the physical mechanism of turbulence should lead either to methods of control or to an explanation of why control fails in some cases.

The turbulence control problem is made more difficult by the fact that theory does not specify the magnitude and structure of the perturbations that necessarily lead to a transition between the laminar and turbulent states. However, there is ample experimental evidence that sufficiently small perturbations fail to induce a transition, at least for Reynolds numbers less than 10^5, in simple pressure-driven shear flows in pipes. (The Reynolds number, Re, is a single dimensionless number characterizing the flow. It measures the ratio of the typical force required to produce observed accelerations in the fluid flow to the viscous force, and is defined for a fluid with characteristic velocity U, length scale L, density ρ, and viscosity μ by $Re = \rho UL/\mu$.) This observation suggests that a control strategy that adequately suppresses the fluctuations may succeed in maintaining the laminar flow in some flow geometries.

Perturbation suppression. Perturbation energy in turbulent flows is concentrated in erratically recurring structures called coherent structures. Very near the bounding surface, these coherent structures are counterrotating streamwise vortices accompanied by regions of high and low velocity elongated in the direction of the flow, referred to as streamwise streaks. Farther from the boundary the streaks give way to vortices distorted into the form of horseshoes. Current attempts at active and passive control of turbulence aim to suppress or to cancel these developing coherent structures.

Because the exact degree of perturbation suppression necessary to maintain laminar flow is not known theoretically, turbulence control strategies can be verified only by experiment or by direct numerical simulation of the turbulence (which currently provides accurate turbulence simulations only at moderate Reynolds numbers of the order of 10^3, while turbulent-flow Reynolds numbers may be of the order of 10^6). A physical theory accounting for the emergence of coherent structures in the turbulent fluid is particularly valuable because it may lead to new control strategies.

Active versus passive control. Flow manipulation can proceed by passive or active means. In passive control the flow is modified by actions which do not require unsteady external input. Probably the most effective currently available means for passive suppression of turbulent fluctuations is introduction of polymers into the flow, which can lead to reduction of viscous drag by as much as 80%. Surface modifications, with grooves (riblets), large-eddy breakup devices, or convex bounding surfaces, have been shown to lead to passive reductions of viscous drag as large as 20%. The use of flexible boundaries, which deform in response to the overlying flow, has at times been claimed to lead to drag reduction, but the reductions have not been verified.

In active control, there is a dynamic input to the flow which either can be determined in advance (in open-loop control) or can be the coordinated reaction to real-time measurements of the actual flow state (in closed-loop or feedback control). An example of open-loop control is oscillation of the bounding surface in the spanwise direction (across the flow), which has been recently shown to lead to drag reduction of as much as 40%.

In feedback control the dynamic adjustment is accomplished by a combination of an observation produced by a device that detects some aspect of the flow state and a subsequent response produced by a mechanical activator. For example, real-time observation of the normal velocity near the wall followed by opposing wall transpiration (by applying blowing or suction) recently has been shown to lead to drag reductions of the order of 20%.

It is expected that with recent advances in technology, in particular the design of microelectronic devices that can measure fluid properties and be programmed to provide feedback in a designed manner, it will be possible to implement a wide variety of active control strategies in the near future. *See* MICROSENSOR.

Optimal active control. It has been proposed recently that the present capabilities of numerical simulation of turbulence suffice for the identification of the optimal active control. Optimal control is the flow manipulation that minimizes a cost functional. In turbulence research, this functional is often taken to be the drag. The well-developed optimal control theory commonly applied in the past to low-dimensional dynamical systems provides the technical underpinnings of this strategy. The optimal control is approached iteratively by estimating the sensitivity of the cost functional to various active control strategies. This sensitivity can be found by integrating the fluid equations forward, after which adjoint analysis is used to identify the directions of greatest sensitivity which identify the direction of optimal flow manipulation. This technique has reduced drag in computer simulations by an order of 20% using wall transpiration (blowing or suction) in antiphase with the observed vertical velocity of the flow above the boundary.

Theory of transition to turbulence. A widely accepted theory of transition to turbulence envisions unstable two-dimensional wave structures called Tollmien-Schlichting waves growing exponentially until they fall victim to secondary three-dimensional instabilities. These instabilities, in turn, give way to a cascade of further instabilities, and the ensemble of these exponential instabilities supports the turbulent state. Control strategies proceeding from this paradigm include passive means to lower the growth rate of the primary unstable Tollmien-Schlichting wave by using suction to change the velocity profile and by altering the viscosity of the flow through heating or cooling the surface. While attempts to depress growth rates using compliant boundaries have met little success, direct active cancellation of the Tollmien-Schlichting waves by introduction of antiphase perturbations has produced a delay in transition in carefully controlled experiments.

However, observations demonstrated that during transition the energetic nonmodal (perturbations which are not a single mode of the system) three-dimensional disturbances dominate the perturbation variance. This led researchers to propose a bypass theory of transition to turbulence according to which transition can occur through transient perturbation growth without the intercession of exponentially unstable Tollmien-Schlichting waves. Moreover, recent advances in a generalization of stability theory have identified the disturbances (called optimal perturbations) most likely to cause transition by growing the most over the time span during which transition occurs. These optimal initial perturbations were shown to evolve into the coherent structures which are typically observed during the transition to turbulence. Active control strategies following from this paradigm attempt to cancel or reduce the growth of these optimal initial perturbations rather than the growth of unstable Tollmien-Schlichting waves.

Control of existing turbulence. Turning from prevention of transition from the laminar to the turbulent state to suppression of already existing turbulence, the recent theoretical description of the coherent structures in the turbulent state enables the researcher to identify actions that can impede the growth of these structures. This development suggests that it may be possible to actively suppress turbulence by boundary forcing appropriately configured so as to cancel the emerging coherent structures which maintain the turbulent flow.

Petros J. Ioannou

Large-Eddy Simulation

Accurate prediction of the turbulent flows encountered in engineering practice remains the principal challenge of computational fluid dynamics. One primary difficulty with simulation and modeling is that turbulence comprises a wide range of length and time scales. The largest scales of motion are responsible for most of the momentum and energy transport and are strongly dependent on the flow configuration, while the smallest eddy motions tend to depend only on viscosity and are more universal (that is, similar, even in different flows) than the large eddies.

Large-eddy simulation is rapidly emerging as a viable technique for prediction of complex turbulent flows. In large-eddy simulation the contribution of

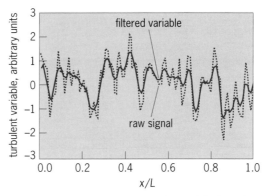

Fig. 1. Filtered variable computed in large-eddy simulation by spatially filtering the raw signal (such as density, velocities, or pressure). Here, x represents position, and **L** is the length scale of the system under study.

the large, energy-containing scales of motion is computed directly, and only the effect of the smallest scales of turbulence is modeled. Since the small scales are more homogeneous and less affected by boundary conditions than the large eddies, it is possible to model them by using simpler models than those required in other techniques.

Motivation. Large-eddy simulation was originally developed because a direct solution of the equations describing the transport of mass, momentum, and energy—the Navier-Stokes equations—is beyond the capacity of even the largest supercomputers for flows of engineering interest. It is possible to characterize the computational cost of a direct simulation in terms of the Reynolds number, Re, which is the ratio of inertial to viscous forces and characterizes the state of fluid motion. Since the ratio of the largest to smallest turbulence length scale is proportional to $Re^{3/4}$, the number of grid points for a three-dimensional computation is proportional to $Re^{9/4}$. Because the number of time steps scales like $Re^{4/3}$, the cost of directly simulating the Navier-Stokes equations is at least proportional to Re^3. Even with the continued development of supercomputers, the rapid increase in cost with increases in Reynolds number limits application of direct numerical simulation to simple flows at low Reynolds number.

Filtering. In large-eddy simulation, transport equations are derived for the large eddies by spatially filtering the Navier-Stokes equations. Application of a filter to a representative turbulent variable is shown in **Fig. 1**. Variations of the fluctuating variable occurring on longer wavelengths (that is, larger scales) are preserved, while the shorter wavelength variation is removed by the filter. The mean value of the raw signal is retained following the filtering operation, but other statistics, such as cross correlations, will differ between the filtered and raw signals because of the absence of the small-scale contribution to the filtered signal. However, in turbulent flows many correlations of interest for engineering prediction are mostly influenced by the large scales of

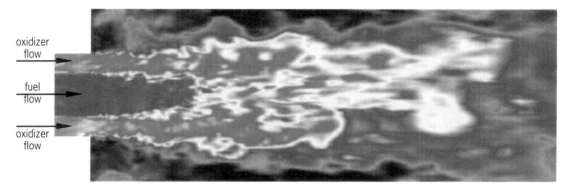

oxidizer
flow

fuel
flow

oxidizer
flow

Fig. 2. Contours from one plane for the fuel mass fraction of a coannular axial jet, as computed by large-eddy simulation. Fuel is supplied in the central pipe; oxidizer is supplied in the annular pipe. The Reynolds number, based on bulk velocity and diameter of the chamber, is 38,000. *(From K. Akselvoll and P. Moin, Large-eddy simulation of turbulent confined jets, J. Fluid Mech., 351:387–441, 1996; visualization by C. K. Pierce)*

motion, and the loss of information arising from the filtering operation is typically not large.

The filtering procedure leads to turbulent stresses which must be modeled. In large-eddy simulation the model need account only for the effect of those motions not resolved by the computational mesh, known as the subgrid scales. Since the small eddies tend to be more homogeneous and respond to external perturbations more rapidly than the large scales, it is possible to model their effect on the resolved motions with relatively simple approaches. In fact, the vast majority of models used in large-eddy simulation are simple algebraic expressions in which an eddy viscosity is introduced to relate the turbulent stresses to the strain rate of the large eddies.

Dynamic modeling. The main limitation of large-eddy simulation has traditionally been the inability of models to account for changes in turbulence structure without arbitrary adjustment. Dynamic modeling of the turbulent stresses, introduced in 1991, represents a new approach since, rather than being provided as input in advance, model coefficients are computed during the course of the calculation. Dynamic models are fundamentally different from other approaches since the model reflects local properties of the flow and is sensitive to changes in external conditions. Large-eddy simulation and dynamic models have now been applied to the prediction of a wide array of geometrically simple turbulent flows, yielding agreement with experimental measurements as good as or better than other approaches requiring arbitrary adjustment.

Applications. The focus of current applications of large-eddy simulation is to flows subject to the complicating effects encountered in practice. These include strong pressure gradients, as in variable-area ducts, streamline curvature, flow separation from a surface, and flows in which the mean velocity profile is three dimensional. These effects severely distort the turbulence structure and are a stringent test for models. In addition to complications arising from these physical effects, recent applications have addressed numerical issues important for advancing large-eddy simulation as a predictive technique for complex flows. Turbulent flows of engineering interest exhibit spatial growth, and it is therefore necessary to prescribe at the inflow boundary of the computational domain a realistic, time-dependent description of the dependent variables. Recent work has shown that it is optimal to prescribe turbulent inflow conditions which are themselves solutions of the Navier-Stokes equations. This can be accomplished by performing a separate calculation whose results are subsequently fed into the inflow boundary of the main simulation.

Coannular jets. Recent applications of large-eddy simulation to prediction and scientific study of complex turbulent flows include the calculation of turbulent confined coannular jets to study mixing of jets discharging into a sudden expansion (**Fig. 2**). This flow is important to areas such as gas-turbine manufacturing since the combustion which occurs in similar configurations is strongly affected by the fluid-dynamical processes governing mixing. In combustion applications, fuel is normally supplied through the central pipe, and oxidizer (air) is supplied through the surrounding annular pipe. A region of recirculating flow is established by the sudden expansion and is important for assisting in stabilizing the flame. Results of large-eddy simulation show that pockets of fuel-rich fluid from the central jet are capable of crossing the annular jet into the recirculation zone and that large-scale structures are responsible for a zone of intense mixing downstream of the expansion. Similar information is very difficult to extract from measurements and virtually impossible to obtain with other simulation techniques. *See* GAS TURBINE.

Flow over a bump. Another second representative application of large-eddy simulation is the prediction of the turbulent boundary layer in flow over a bump (**Fig. 3**). Boundary-layer turbulence in the flow over a bump is subject to combinations of perturbations in pressure gradient and curvature typical of that encountered in many engineering applications. Changes in pressure gradient and curvature can en-flow

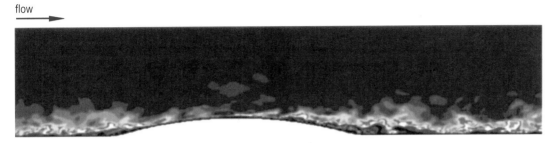

Fig. 3. Contours from one plane of the streamwise velocity of the turbulent boundary layer in flow over a bump as computed by large-eddy simulation. The Reynolds number, based on momentum thickness at the inflow boundary, is 1500. *(From X. Wu and K. D. Squires, Large eddy simulation of the turbulent flow over a bump, ASME FED, 237:583–588, 1996)*

hance or suppress turbulence levels. While the separate effect of pressure gradients and streamline curvature on turbulent boundary layers is reasonably well understood, boundary-layer response under their combined influence is difficult to predict. The flow thus poses a significant challenge to large-eddy simulation and subgrid models. In addition to its being difficult to model, an important issue for strongly distorted boundary-layer turbulence is the process of recovery following removal of perturbations in external conditions. The flow over the bump is unique since experimental measurements show that the downstream recovery of the flow is rapid. An important contribution of the large-eddy simulation was in demonstrating that a key feature of the rapid recovery is due to initiation of an internal layer, a thin region close to the wall, which grows rapidly downstream.

Validation. The visualizations shown in Figs. 2 and 3 are of the large scales of motion; that is, the smallest scales of motion are absent from the figures. In spite of this, both figures provide, on a qualitative level, an appreciation of the detail which is available from large-eddy simulations of complex flows and which is very difficult to extract with other approaches. These examples are representative of several now being obtained. A key feature of the simulations is validation through comparison with experimental measurements. Once the accuracy of the simulations has been established, the calculations may be used to provide detailed understanding of the complex dynamical processes governing turbulent flows. These studies also raise new questions and issues requiring research to further advance large-eddy simulation as a predictive tool for scientists and engineers. One example of an issue currently being addressed in the large-eddy simulation of turbulent boundary layers is the development of models of the near-wall region which may be used in conjunction with large-eddy simulation of the outer part of the boundary layer. Since approximately 30–40% of the grid points in a calculation are often required to resolve less than 10% of the boundary layer, development of near-wall models will substantially improve the viability of large-eddy simulation for the simulation of complex flows.

For background information *see* BOUNDARY-LAYER FLOW; COMPUTATIONAL FLUID DYNAMICS; FLUID FLOW; NAVIER-STOKES EQUATIONS; OPTIMAL CONTROL THEORY; REYNOLDS NUMBER; TURBULENT FLOW in the McGraw-Hill Encyclopedia of Science & Technology.

Kyle D. Squires

Bibliography. D. M. Bushnell and C. M. McGinley, Turbulence control in wall flows, *Annu. Rev. Fluid Mech.*, 21:1–20, 1989; D. M. Bushnell and K. J. Moore, Drag reduction in nature, *Annu. Rev. Fluid Mech.*, 23:65–79, 1991; B. F. Farrell and P. J. Ioannou, Turbulence suppression by active control, *Phys. Fluids*, 8:1257–1268, 1996; M. Lesieur and O. Métais, New trends in large-eddy simulations of turbulence, *Annu. Rev. Fluid Mech.*, 28:45–82, 1996; P. Moin and T. Bewly, Feedback control of turbulence, *Appl. Mech. Rev.*, 47(no. 6, part 2):S3–S13, 1994; W. C. Reynolds, Computation of turbulent flows, *Annu. Rev. Fluid Mech.*, 8:183–208, 1976; R. S. Rogallo and P. Moin, Numerical simulation of turbulent flows, *Annu. Rev. Fluid Mech.*, 16:99–137, 1984; H. Tennekes and J. L. Lumley, *A First Course in Turbulence*, 1972.

Underground mining

Mining engineers have long had the goal of controlling underground mining machines from the surface and thus removing personnel from hazards of the underground environment. This is becoming more important as mining operations increasingly need to go deeper, for example, beyond 2000 m (6600 ft). As a result of the geothermal gradient, deep mining is confronted by hot and humid working conditions, where virgin-rock temperatures may exceed 40°C (100°F). In such environments, underground tunnels are unstable because of the high rock stresses. The stresses may generate rockbursts, which are seismic events resembling small earthquakes, generated by the violent failure of brittle rock around the mine openings. Other underground hazards include dust and gases, use of explosives, and radioactivity if the mine is extracting uranium minerals. A further incentive to automate mining is to increase productivity and reduce operating costs, for mining companies need to maintain competitiveness in the global economy.

Communication. The primary obstacle to underground automation has been the inability to communicate effectively through the rock. Early underground communications systems were telephones and fixed pagers. In order to communicate with mobile machines and personnel, however, efforts in the 1970s focused on radio systems. One popular technique used conventional two-way radio technology (walkie-talkies) as used on the surface. A radio repeater, used to send signals from one portable radio to another, was connected to a coaxial cable with a loosely braided outer shield. This "leaky feeder" or "radiating cable" became the antenna that was run wherever communications to mobile machines was required underground. The demand for increased capacity led to attempts to develop communications systems having up to 16 channels, used primarily for the remote monitoring and control of stationary equipment such as pumps and ventilation fans. Automated mines, however, require more utility in the form of local-area networks (LANs), closed-circuit television, high-speed programmable-logic-controller communication, and automation and teleoperation (operation from a remote location) of mobile machines.

An advanced communications system developed by Inco Ltd. in Canadian nickel mines comprises a backbone utility, installed down the mine shaft and out into the mine workings; it is a broadband, multiple-channel coaxial cable system, similar to that used to provide cable television services to homes. This system can provide a variety of services on different channels: telephone and paging; two-way radio; closed-circuit television; supervisory control and data acquisition (SCADA); networking with a programmable logic controller; and communications by mainframe and personal computers and local area networks. It is connected to a distributed antenna translator (a minicell, similar to the cell transmitters used in cellular phone systems) from which the leaky-feeder cable is run out into the mine workings to provide services to the mobile machines and personnel. This system makes the underground mine equivalent to a large, dispersed manufacturing plant, with communications being sufficiently reliable and economical to support the automation of routine mine production, using teleoperated or autonomous machines.

Mine automation. Early attempts to automate underground mining related to line-of-sight remote control. In coal mining in the 1960s, attempts to remotely control the excavation of coal within longwall mining systems had only limited success. The coal sector then focused research and development activities on mechanical-electrical design of face machinery itself, mine communications, and technology for monitoring machines and the underground environment. The aim was to develop environmental and production monitoring to exploit the capital-intensive, high-productivity programs of face mechanization as they evolved. More recently, a strategy for coal mine automation was developed by the U.S. Bureau of Mines, which sees continuous-mining, roof-bolting, and continuous-haulage machines, together with ventilation systems, being transformed into computer-assisted, remotely supervised machines operating in well-understood situations.

Advanced automation recently achieved in Saskatchewan potash mines in Canada is based on systems for continuous-mining machines, using computers, programmable logic controllers, and sensors for navigation-guidance-monitoring to allow for remote monitoring and control. The electric mining machines typically weigh 250 metric tons (275 tons), and are propelled by individual hydraulic-powered track pads. Each machine bores an opening in the potash ore, 8 m (26 ft) wide by 2.4 m (8 ft) high, with a four-rotor rock-cutting head over distances around 2 km (1.2 mi). The mining machine can be fully automated by using sensors that include laser spot tracking to keep the machine on line; gyroscopic and ultrasonic positioning for navigation; tracking of clay seams, using digital analysis of video images of the tunnel as it is bored; and an ore-grade analyzer, based on gamma-radiation sensors, to keep the miner in the richest part of the orebody. Control systems are also used for other mining machinery (including continuous-haulage, loadout, and ventilation systems); a broadband radio communications system; and a central control and dispatch system (providing a human–machine interface and control of other systems). In a uranium mine in Saskatchewan, a leaky-feeder communications system was installed, motivated by the need to monitor alpha radiation and airflow for mine safety, and to control underground pumps and fans.

In 1993, the Inco Ltd. communications system demonstrated the first production application of multiple-machine teleoperation from the surface. At a mine in Sudbury, Ontario, two mobile mining machines at 600 and 900 m (2000 and 3000 ft) below the Earth's surface were operated simultaneously from Toronto, 450 km (280 mi) away, via a satellite link. A miner teleoperated the two machines as they loaded and discharged broken rock. Otherwise, the machines independently traversed the mine tunnels under automatic optical guidance. A pair of onboard video cameras (one looking front, the other aft), a microphone, and 18 sensors for machine monitoring provided the operator with the necessary dashboard information for machine control. A continuous-rope lighting system, fixed to the tunnel roof, was tracked by using the video system, enabling an onboard programmable logic controller to control the machine's steering cylinders for automatic guidance. Another form of automatic guidance for mobile underground machines has been developed using a scanning laser guidance system. It employs front- and rear-mounted lasers that continuously scan for an overhead reflective guide tape and instructions in the form of bar codes at strategic locations. Collision avoidance is

Automatic electric truck in the Little Stobie Mine in Canada. *(Inco, Ltd.)*

achieved through a combination of a vehicle-based transmitter and passive transponders. In a recent breakthrough, the Noranda Technology Centre in Montreal, Quebec, was able to automate the process of loading fragmented rock into such a vehicle, so the complete automation of loading, dumping, and movement of such vehicles is now feasible. These autonomous machine systems will be used at first as "islands of automation." Inco Ltd., for example, started development in 1983 of its 64-metric-ton-capacity (70-ton) automatic haulage truck, using an electric trolley system and computer control (see **illus.**). It has operated as a very reliable autonomous island, solely responsible for the entire ore haulage of 1.8 million metric tons (2 million tons) at one mine for over 2 years (as of early 1997).

Prospects. There are a number of visions of the underground automated mine of the future. In 1992, the Finnish mining industry announced its Intelligent Mine Technology Programme, whose goal is real-time processlike control of an automated mine. Projects included real-time resources management and production control; a minewide information network; new machines and automation; and automation of production and maintenance.

South African investigators are focusing on a combination of several innovative technologies in future deep mining: concentrated mining; impact fragmentation; hydropower; continuous scraping; ground-penetrating radar; radiowave tomography; gold analysis; grade prediction depositional simulation software (software which aims to simulate the manner in which an ore deposit was originally formed, as a basis for predicting the location and grade or value of new, undiscovered ore in the vicinity of the mine); a geographic information system for real-time three-dimensional planning and monitoring systems; rock-burst control and a portable seismic system; and backfilling and hoisting technology.

Australian recommendations for research and development on innovative underground mining technology identified two important objectives: minimization of ore loss and dilution; and continuous

mining systems for selective mining and narrow ore-bodies.

Another goal is that by the year 2050 a Canadian hardrock mine in a remote, northern location will have a mining operation in which explosives are replaced with mechanical rock fragmentation, and ore sorting and mineral processing are carried out underground. The underground machines will be operated via a satellite link from Sudbury, Ontario, with no miners working underground.

Studies have considered the extent of the benefits and costs of removing workers from underground. In the economics of underground machine teleoperation, significant cost reduction is thought to arise from productivity, work quality, and throughput time improvement. Comparisons have been drawn with factory automation, comparing underground machine control with computer numerically controlled manufacturing. Costs of rock support may be reduced if the machines require less protection or induce less damage to the surrounding rock. Significant savings in power costs have already been demonstrated through automation of underground mine ventilation and development of the concept of ventilation by demand. Many underground mines are challenged to maintain competitiveness in increasingly difficult mining conditions associated with deeper, though not necessarily richer, reserves. Where the environment is too hazardous for human entry—for example, in severe ground conditions or high-grade uranium mining—the only alternative may be a nonentry automated system, provided it is adequately robust and economic.

For background information *see* AUTOMATION; CLOSED-CIRCUIT TELEVISION; COMPUTER NUMERICAL CONTROL; DISTRIBUTED SYSTEMS (COMPUTERS); PROGRAMMABLE CONTROLLERS; ROCK BURST; UNDERGROUND MINING in the McGraw-Hill Encyclopedia of Science & Technology.   Malcolm Scoble

Bibliography. L. Ozdemir and K. Hanna (eds.), *Proceedings of the 3d International Symposium on Mine Mechanization and Automation*, 1995; L. Ozdemir, R. King, and K. Hanna (eds.), *Proceedings of the 1st International Symposium on Mine Mechanization and Automation*, 1991; M. J. Scoble, Canadian underground mine automation: Progress and issues, *Bull. Can. Inst. Min. Metal.*, 89(966):29–32, 1996; N. Vagenas, M. Scoble, and G. Baiden, A review of the first 25 years of mobile machine automation in underground hard rock mines, *Bull. Inst. Min. Metal. Petrol.*, 90(1006):57–62, 1997.

Underwater noise

In the early 1990s there was a surge of public concern that human activities producing high levels of underwater noise could seriously affect marine animals, especially whales. Although this concern had been identified in the early 1970s and some research had been started in the early 1980s, negative publicity on acoustic programs in oceanography and the military focused attention on the issue.

Sound level is currently used as the primary measure for estimating noise impact. For example, sound intensity level is measured in decibels (dB), calculated as 10 times the logarithm of the ratio of the intensity of the sound of interest relative to a reference intensity. Confusion arises since the sound-level reference for measuring sound intensity in water (1 micropascal) is different from that in air (20 μPa). For this reason, underwater sound intensity levels are properly given with an intensity reference, and source levels include a reference to the range (for example, 175 dB re 1 μPa at 1 m). Interpretation of comparisons between animal hearing responses in air and water are confounded by existing uncertainties as to whether marine mammal ears are most sensitive to pressure, intensity, or energy.

The basis for the concern over noise impact stems from the possible conflict between marine animals that depend on sound for communicating, navigating, and finding food and the dramatic increase in human production and use of low-frequency (less than 400 Hz) underwater sound. The greatest perceived threat from human-made, low-frequency underwater sound is to the baleen whales (mysticetes). Whales depend upon sound production and perception for survival. Their ears and central nervous system are remarkably well adapted for detecting and encoding low-frequency sounds and for dealing with the high sound levels from naturally occurring underwater noises. Recent data from U.S. Navy hydrophones in the deep ocean reveal that whales are acoustically prolific throughout the oceans during all months of the year and can be detected at hundreds of miles with a single hydrophone.

Commercial shipping traffic generates enormous amounts of noise, particularly in the 10–40-Hz frequency band, raising the background noise level by as much as 10–40 times that in a preshipping ocean. Since the 1960s, humans have learned to probe the ocean with sounds for oil exploration, three-dimensional oceanographic feature mapping (acoustic tomography and acoustic thermometry), and military surveillance (sonar and low-frequency active acoustics). From an evolutionary perspective, the highly adapted bioacoustic systems of whales, dolphins, and porpoises that have evolved over the last 30 million years to detect natural underwater sounds are now in jeopardy because these sounds are competing with human-made underwater noises introduced in the last 100 years. Changes in ambient noise conditions due to shipping have the potential to mask biologically important sounds and reduce the ability of animals to hear each other. The increased exposure to a suite of human-made, extremely loud, low-frequency sounds has the potential to affect animal behavior or even cause hearing damage.

Sounds of whales. The 11 species of whales in the mysticete group produce a wide variety of vocalizations that include frequency-modulated and ampli-

tude-modulated sounds, and complex mixtures of frequency-modulated, amplitude-modulated, and broadband sounds. How whales specifically use these sounds remains mostly unknown. Only a few cases of communication have been documented, and there is circumstantial evidence to suggest that some whales use low-frequency sounds for navigation and orientation.

The best-known sounds are from coastal species: the bowhead (*Balaena mysticetus*), humpback (*Megaptera novaeangliae*), and right (*Eubalaena* sp.) whales. Signal documentation has been more difficult for the pelagic group that includes blue (*Balaenoptera musculus*) and fin (*B. physalus*) whales, which produce the loudest sounds of all animals at 190 dB re 1 μPa at 1 m. **Figure 1** gives examples of low-frequency sounds from four different species.

Sounds from deep-water (depths greater than 3300 ft or 1000 m) whales are relatively simple compared to shallow-water (depths less than 1000 ft or 300 m) species and consist of highly patterned sequences of loud, low-frequency, frequency-modulated signals. Many sounds are infrasonic (below the threshold of human hearing) and in the frequency band with the greatest amount of shipping noise.

What is emerging from comparative analysis of sounds from deep-water and shallow-water species is that deep-water whales produce sounds that are lower in frequency, are louder, have greater stereotypy, and are more redundant than the more variable sounds of the coastal species. Such acoustic characteristics of the deep-water species greatly increase sound detectability at great range and reduce chances of masking from naturally occurring ambient noise (such as that from surf, storms, earthquakes, and air bubbles from breaking waves). This observation is one of the major pieces of evidence supporting the conclusion that deep-water species use sound for long-range communication and are susceptible to the effects of human-made underwater noise.

Sounds from human-made activities. Since the 1970s, major technical advances have led to the increased production of loud, low-frequency sounds at various ocean depths for prospecting for oil and gas, for studying oceanographic processes such as temperature variability and thermal flux, and for submarine detection and tracking. All these methods take advantage of the physical properties of the ocean that allow for the efficient propagation of low-frequency sound—the very same properties for which whale sounds are so well adapted. Worldwide geophysical prospecting in the ocean routinely involves the use of arrays of airguns to produce extremely loud (greater than 250 dB re 1 μPa at 1 m peak pressure level), controlled explosions repeated every 9–40 s. Airgun array ships operate for many months at a time, and in some regions the intensity of the acoustic prospecting is so great that the ocean is never quiet. In the Acoustic Thermometry of Ocean Climate (ATOC) project, a loud (195 dB re 1 μPa at

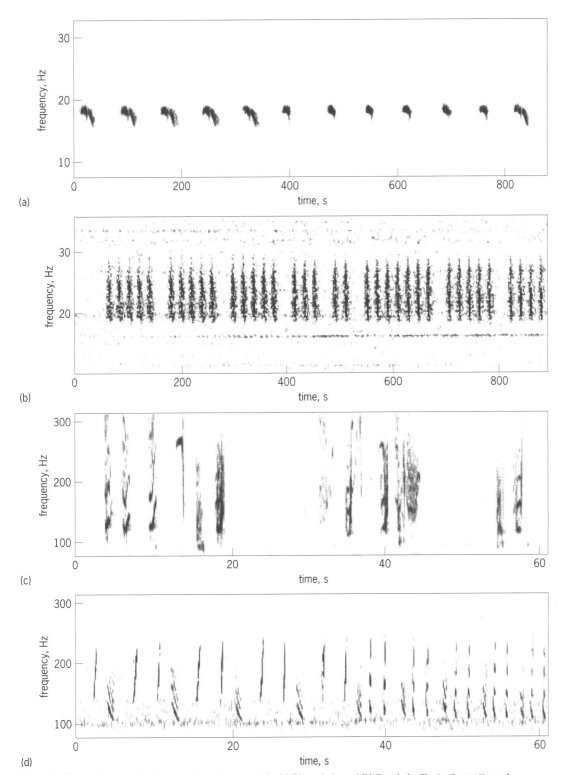

Fig. 1. Audio spectrograms for four species of great whale. (*a*) Blue whale and (*b*) fin whale, illustrating patterned, infrasonic sounds of deep-water species. (*c*) Right whale and (*d*) humpback whale, illustrating higher-frequency, more variable sounds of coastal species.

1 m), specially coded 75-Hz sound is being used to measure changes in average ocean temperatures throughout the northern Pacific Ocean. Many fear that this loud signal could seriously affect whales, dolphins, seals, and sea turtles. Sonar systems rely on the production of a very loud chirp sound and the reflection of the sound off a target. Most sonar systems operate at frequencies above several thousand hertz. Since the 1980s, with the dramatic quieting of submarines, navies have begun to develop long-range submarine detection systems. Such systems produce very loud, low-frequency sounds

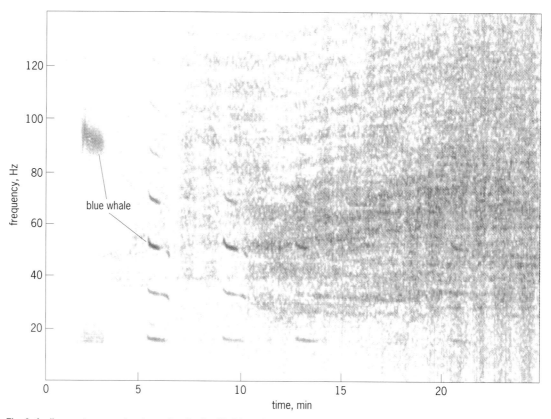

Fig. 2. Audio spectrogram showing calls of a Pacific blue whale being masked by typical shipping noise. The shipping noise is the broadband smearing starting at about 7 min, while the blue whale voice starts with the distinct blob at 90 Hz and 2 min, and continues with very structured (ladderlike) moans at 6, 9, 13, 17, and 21 min.

that have features remarkably similar to those of whales.

Potential impact. Research on the subject of impact by human-made sounds on whales is minimal, and the evaluation and prediction of potential impact of human-made noises on marine mammals is difficult. Estimating acoustic environmental impact involves interpretation and integration of results from many disciplines, including the study of how sound waves interact with the environment (physical acoustics), how animals hear sounds with their ears (anatomy and physiology), and how animals use sounds for communicating, navigating, and finding food (bioacoustics, psychoacoustics, and behavioral ecology).

Response to noise. There is evidence that certain human-made sources of underwater noise, including the noises from industrial activities, can change whale behavior or cause whales to abandon their habitat. Although minor behavioral reactions may appear to have no impact over the short term, there is no way of knowing whether these could have long-term impacts and affect reproductive success and survival. For example, loss of breeding, calving, or feeding habitat could have a serious impact on survival.

The present approach to evaluating and predicting the possible impact of human-made noises on animals considers that there are four stages of response related to received sound level or an ani-

mal's distance from a source. At great distance (low sound level) impact is assumed to be least, while at very close range (high sound level) impact is assumed to be greatest.

The first stage of response occurs when an animal can hear the noise but does not respond to it. The distance at which an animal first hears the sound is limited by either the ambient noise level or its hearing sensitivity.

The second stage of response occurs when an animal changes its behavior. There is no standardized method for estimating or predicting when this occurs, and this effect is best determined from behavioral studies. The most widely used value of received sound level for estimating response has been ~115–120 dB re 1 μPa, derived by observing whale responses to playback of industrial sounds. Bowhead whales have been observed to detour around industrial noise sources, including seismic survey and drilling operations at ranges of up to 20 mi (32 km). Beluga whales have been observed to react to noises from icebreakers at distances of 50 mi (80 km). Tests with sounds of the ATOC project are still ongoing, but so far whales show no overt response to this sound.

The third stage occurs when the human-made sound raises the total ambient level so that an animal cannot hear biologically important sounds. **Figure 2** shows the effect of masking from shipping noise on the detection of blue-whale sounds. Masking also

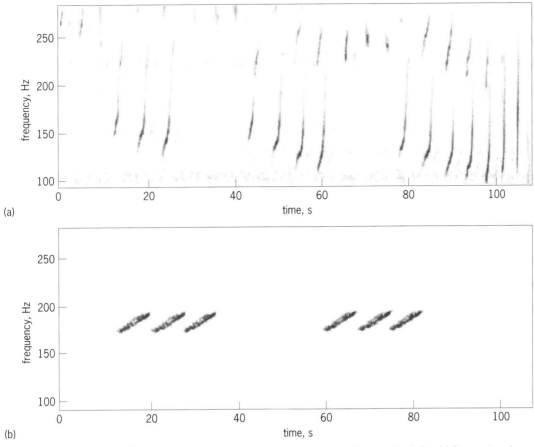

(a)

(b)

Fig. 3. Audio spectrograms illustrating potential for interference from human-made acoustic clutter. (a) Song notes of a humpback whale. (b) Human-made sound probe.

includes cases when the human-made sound has features similar to those of biological importance. This type of sound is referred to as clutter. **Figure 3** is an example of human-made acoustic clutter and shows the similarity between a human-made sound probe and humpback song notes.

The fourth stage occurs when an animal is exposed to very high levels of noise that can cause hearing loss, pain, or injury. Exposure to high levels of noise can cause either a temporary threshold shift or a permanent threshold shift. The extent of the hearing loss, discomfort, or injury is determined by the sound intensity level, the duration of the exposure, the number of occurrences of the exposure, and the sound's frequency. There are still no published data on these types of effects as they relate to whales, although some recent studies are beginning to provide preliminary information on dolphins and seals.

Efforts to provide guidelines for behavioral-response, temporary-threshold-shift, and permanent-threshold-shift levels for different species are ongoing. Working criteria are that behavioral response is 70 dB above the hearing level of greatest sensitivity, or best threshold; temporary threshold shift is 80–100 dB above the best threshold; and permanent threshold shift is 155 dB above the best-threshold level. However, for most species the best-threshold levels are not known.

For background information *see* ACOUSTIC NOISE; BEHAVIORAL ECOLOGY; CETACEA; ECHOLOCATION; LOUDNESS; MASKING OF SOUND; PHONORECEPTION; PHYSIOLOGICAL ACOUSTICS; PSYCHOACOUSTICS; SOUND; ULTRASONICS in the McGraw-Hill Encyclopedia of Science & Technology.   Christopher W. Clark

Bibliography. R. Payne (ed.), *Communication and Behavior of Whales*, 1983; W. J. Richardson et al., *Marine Mammals and Noise,* 1995; J. A. Thomas and R. A. Kastelein (eds.), *Sensory Abilities of Cetaceans: Laboratory and Field Evidence*, 1990; R. J. Urick, *Ambient Noise in the Sea*, 1986.

Virtual manufacturing

The term virtual manufacturing became prominent in the early 1990s, in part as a result of the U.S. Department of Defense Virtual Manufacturing Initiative. The concept has now gained international acceptance and has broadened in scope. For the first half of the 1990s, pioneering work in this field was done by a few major organizations, mainly in the

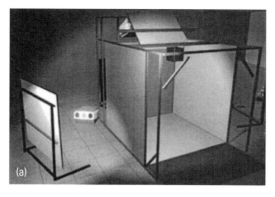
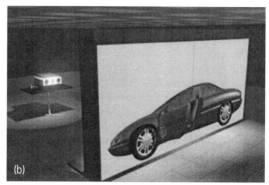

Use of virtual manufacturing technology by General Motors. The technology used is the Cave Automatic Virtual Environment (CAVE), developed and trademarked by the University of Illinois at Chicago. It features three room-sized rear-projected walls and a top-projected floor, each driven by Silicon Graphics Onyx workstations. (*a*) CAVE virtual reality display. (*b*) View of rear-projected, full-size car. (*c*) Virtual car interior, with photographic textures. (*d*) Virtual view of simulated car crash. (*From R. C. Smith et al., Really getting into your work: The use of immersive simulations, in P. Banerjee, L. Freitag, and S. Mehrotra, eds., Virtual Reality in Manufacturing Research and Education, Proceedings of a symposium sponsored by the National Science Foundation, 1996*)

aerospace, earthmoving equipment, and automobile industries, plus a few specialized academic research groups. In 1996, accelerating worldwide commercial interest became evident, fueled by improvements in the required hardware and software technologies and by increased awareness of the huge potential of virtual manufacturing.

Virtual manufacturing broadly refers to the modeling of manufacturing systems with effective use of audiovisual or other sensory features to simulate or design alternatives for an actual manufacturing environment, or the prototyping and manufacture of a proposed product mainly through effective use of computers. The motivation is to enhance the ability to predict potential problems and inefficiencies in product functionality and manufacturability before real manufacturing occurs.

Framework. The concepts underlying virtual manufacturing include virtual reality, high-speed networking and software interfaces, agile manufacturing, and rapid prototyping.

Virtual reality. This is the ability to create and interact in cyberspace, a space that represents an environment very similar to the actual environment. Virtual reality is closely associated with a so-called virtual environment.

Virtual environment systems differ from previously developed computer-centered systems in the extent to which real-time interaction is facilitated. The perceived visual space is three dimensional rather than two dimensional, the human–machine interface is multimodal, and the user is immersed in the computer-generated environment. The screen separating the user and the computer appears to become nonexistent to the user.

The presently available means of simulating a virtual environment for virtual manufacturing is through immersion in computer graphics coupled with acoustic interface and domain-independent interacting devices such as wands, and domain-specific devices such as the steering and brakes for cars or earthmovers or instrument clusters for airplanes. There are two main techniques for creating immersion in current virtual reality systems: head-mounted displays and stereoscopic projectors. A head-mounted display is a tracking device incorporating liquid-crystal displays or miniature cathode-ray tubes mounted on the head of the user to provide information on head movements for updating the visual images. An example is the Binocular Omni-Oriented Monitor (BOOM) developed commercially. The stereoscopic projectors provides a three-dimensional sensation of depth by utilizing screen projections of two distinct views of an object separated by a small angle in which the object is viewed with the left eye only and with the right eye only. The users normally wear liquid-crystal-display shutter glasses. An example is the Cave Automatic Virtual Environment (CAVE) developed by the University of Illinois at Chicago (**illus.** *a*).

High-speed networking and software interfaces. The issues here are concerned with portability among systems of models produced with computer-aided design (CAD), trade-offs of high-detail models versus real-time interaction and display, rapid prototyping, collaborative design using virtual reality over distance, use of the Web for small- or medium-business virtual manufacturing, use of qualitative information (illumination, sound levels, ease of supervision, handicap accessibility) to design manufacturing systems, use of intelligent and autonomous agents in virtual environments, and determining the validity of virtual reality versus actual reality (quantitative testing of virtual versus real assemblies and equipment). With regard to hardware, the two near extremes (in terms of cost and performance) are personal computers using Virtual Reality Modeling Language (VRML) and the Internet, and specialized workstations and the CAVE library and asynchronous-transfer-mode networks with data transfer speeds up to 155 megabits per second (Mbps).

Agile manufacturing. Sometimes mentioned in the context of virtual manufacturing, agile manufacturing is a structure within which agility is achieved through the integration of three primary resources: organization, people, and technologies. One approach is through innovative management structures and organization, a skill base of knowledgable and empowered people, and flexible and intelligent technologies. While agility focuses on the ability to make rapid changes in products and processes based on customer demand, virtual manufacturing provides a means for doing so.

Rapid prototyping. Another area in which virtual manufacturing has made an impact is rapid prototyping or layered manufacturing. In processes such as stereolithography, selective laser sintering, and fused deposition modeling, a CAD drawing of a part is processed to create a file of the part in slices. The slice file then directs a source of energy to build the part one slice (or layer) at a time, thus precisely depositing layer upon layer of material.

Virtual collaborative environments. Global virtual manufacturing extends the definition of virtual manufacturing to emphasize the use of Internet/Intranet global communications networks for virtual component sourcing, and multisite multiorganization virtual collaborative design and testing environments. Companies that commit to global virtual manufacturing may develop the potential for dramatically shortening the time to market for new products, cutting the cost of prototyping and preproduction engineering, enabling many more variations to be tried out before committing to manufacture, and increasing the range and effectiveness of quality assurance testing. Virtual prototypes can be virtually assembled, tested, and inspected as part of production planning and operative training procedures. They can be demonstrated, market tested, used to brief and train sales and customer staff, transmitted instantly from site to site via communication links, and modified and recycled rapidly in response to feedback.

Manufacturers and their worldwide subcontractors and main suppliers can establish agile manufacturing teams who will work together on the design, virtual prototyping, and simulated assembly of a particular product, at the same time establishing confidence in the virtual supply chain. Using the most advanced virtual reality systems, geographically remote members of the team can meet in the same virtual design environment to discuss and implement changes to virtual prototypes.

Examples of recent developments in virtual collaborative environments include projection of gestures and movements of multiple remote designers as voice-activated models to help explain the intention of the designer to others in real time using high-speed asynchronous-transfer-mode networks.

Complex manufacturing systems. For monitoring and control of complex manufacturing systems, four dimensions can be conceived to express complexity: space, time, process, and network. The first dimension, space, permits examination of the physical location, layout, and flow issues critical in all manufacturing operations. The second dimension, time, permits consideration of facility life cycle and operational dynamic issues, beginning with concurrent engineering of the production process and testing facilities during product design, extending through production and decline of the initial generation products, and cycling through the same process for future generation products. The third dimension, process, allows study of the coherent integration of engineering, management, and manufacturing processes. It permits examination of the important yet intricate interplay of relationships between classically isolated functions. Examples are relationships between production planning and purchasing, production control and marketing, quality and maintenance, and design and manufacturing. Processes involve decisions ranging from long-range operational planning to the short-term level of planning and control of machines and devices. The integration between various levels of aggregation is essential. The fourth dimension, network, deals with organization and integration of the infrastructure. While the third dimension focuses on the actions, the fourth dimension concentrates on the actors and their needs and responsibilities. Clearly including personnel, the set of actors also includes all devices, equipment, and workstations; all organizational units, be they cells, teams, departments, or factories; and all external interactors, such as customers, vendors, subcontractors, and partners. Issues such as contrasting hierarchically controlled networks with non-hierarchical, autonomous agent networks must be addressed.

The techniques of virtual manufacturing facilitate understanding of the four dimensions by addressing issues such as designing products that can be evaluated and tested for structural properties, ergonomic functionality, and reliability without having to build actual scale models; designing products for esthetic value that will meet individual customer preference; ensuring compliance of the facility and equipment with various federally mandated standards, permitting remote operation and control of equipment (telemanufacturing and telerobotics); developing processes to ensure manufacturability without actually having to manufacture the product (for example, avoiding destructive testing); developing production plans and schedules and simulating their correctness; and educating employees on advanced manufacturing techniques worldwide, with emphasis on safety.

Virtual manufacturing already has some commercial applications. An aircraft manufacturer has used the BOOM technology to design new models. The CAVE technology developed at the University of Illinois has been used in a variety of manufacturing settings, including virtual designing of the interior of automobiles and virtual crash testing (illus. c and d). Other applications of this technology include building electronic prototypes of earthmoving equipment instead of physical prototypes, and exploration of virtual collaborative environments using multiple CAVEs to coordinate designs with suppliers and showrooms. A manufacturer is using this technology to visualize boiler designs in order to reduce the likelihood of expensive litigation stemming from boiler failures. Another is exploring use of this technology for drug manufacturing to rapidly conform to the requirements of the U.S. Food and Drug Administration (FDA). A major manufacturer of computer components is comparing the impact of employees trained with head-mounted devices in a virtual reality environment with those trained in the real environment. Other applications of virtual manufacturing are under active consideration.

For background information *see* CONCURRENT PROCESSING; MANUFACTURING ENGINEERING; MODEL THEORY; VIRTUAL REALITY in the McGraw-Hill Encyclopedia of Science & Technology.   Pat Banerjee

Bibliography. P. Banerjee, L. Freitag, and S. Mehrotra (eds.), *Virtual Reality in Manufacturing Research and Education*, Proceedings of a symposium sponsored by the National Science Foundation, 1996; C. Cruz-Niera, D. J. Sandin, and T. A. DeFanti, The CAVE: Audio visual experience automatic virtual environment, *Commun. ACM*, 35(6):65–72, 1992.

Virus

Hundreds and possibly thousands of different types of viruses are capable of infecting humans. Viruses are obligate parasites that must enter host cells to replicate. Because of their relatively simple structures, they replicate rapidly (generally in less than a day, sometimes in 2 to 3 h) and prolifically (each infected cell produces 100–10,000 new viruses).

The combination of a low fidelity of replicating their genetic information and, with some viruses, the ability to incorporate host genes into their own genome results in an extraordinary ability to mutate.

The formidable threat posed by viruses probably contributed to the evolution of the vertebrate immune system. The importance of protecting the host against lethal virus infections becomes obvious in diseases such as acquired immunodeficiency syndrome (AIDS), where depression of just one arm of the immune system (CD4+ thymus-derived lymphocytes or T cells) can result in death. The battle between the host's immune system and viruses is best regarded as a dynamic standoff, with each thrust of an opponent parried by a countermeasure from the other. In recent years it has become evident that viruses have developed at least two strategies to reduce the ability of the immune system to limit their spread: avoiding recognition by immune cells, and producing proteins that deactivate the immune response. Additionally, it is possible that the ability of viruses to infect and disable immune cells confers evolutionary selective advantages for the virus.

Avoiding recognition. Viruses employ two basic strategies to avoid recognition by the immune system. First, they can mutate the proteins that are recognized by the immune system. This is most effectively used to avoid the action of antibodies. Antibodies are proteins, secreted by B lymphocytes into body fluids, that bind to specific areas on viral proteins. When the antibody interferes with the ability of the virus to infect host cells, it is said to be neutralizing. Antibodies can also block viral replication by enabling the virus to be recognized by cells specialized in the destruction of foreign material. Often, mutation of a single residue in a target viral protein is sufficient to prevent the binding of a given antibody. For example, the rapid mutation of influenza virus in this manner (termed antigenic drift) has precluded the production of an effective, broad-acting vaccine. Changes in human immunodeficiency virus (HIV) have similarly frustrated attempts to produce vaccines.

A variation of this strategy is to block antibody access to vulnerable regions of the viral surface by altering the sequence of the viral protein in a way that causes the host cell to add an oligosaccharide group. Since the added oligosaccharides are identical to those on host proteins, they cannot be recognized by the immune system. This may explain why the surface protein of HIV (gp160) has an abnormally large number of oligosaccharides.

Second, viruses can produce proteins that interfere with the presentation of viral proteins to T lymphocytes. Unlike antibodies, T lymphocytes recognize fragments of viral proteins in a complex with a cellular protein, termed a major-histocompatibility-complex class I or class II molecule. Class II molecules bearing viral peptides activate CD4+ T lymphocytes, which play an important role in regulating immune responses. Viral peptides complexed with class I molecules activate CD8+ T lymphocytes,

Viral immunomodulatory proteins			
Protein	Virus family	Virus species	Function
E3-gp19K	Adenoviridae	Adenovirus 2	Binds and retains class I molecule in endoplasmic reticulum
ICP47	Herpesviridae	Herpes simplex virus	Binds transporter associated with antigen processing (TAP) and prevents peptide transport
US11	Herpesviridae	Cytomegalovirus	Targets major-histocompatability-complex (MHC) class I molecule for rapid degradation
EBNA1	Herpesviridae	Epstein-Barr virus	Blocks generation of antigenic peptides
VCP	Poxviridae	Vaccinia virus	Inhibits complement C4 enhancing effect
gC	Herpesviridae	Herpes simplex virus	Inhibits complement C3 receptor activity
HVS-CCPH	Herpesviridae	Herpesvirus saimiri	Inhibits complement C3 convertase activity
HVS-CD59	Herpesviridae	Herpesvirus saimiri	Inhibits terminal complement inhibitor CD59
gE	Herpesviridae	Herpes simplex virus	Binds IgG Fc receptor and inhibits antibody activity

which release molecules that either kill virus-infected cells before the virus can complete its replication cycle or alter the cell in a manner that reduces viral replication.

Viruses are known to interfere with the presentation of viral peptides to CD8+ T lymphocytes by multiple strategies. Adenoviruses produce a protein that retains class I molecules at their sites of synthesis inside the cell and prevents them from informing the immune system of the presence of viral proteins (see **table**). The herpes simplex virus produces a protein that blocks the ability of cells to deliver viral peptides to class I molecules. Another herpesvirus (cytomegalovirus) produces a protein that diverts newly synthesized class I molecules from the endoplasmic reticulum to the cytosol, resulting in its rapid degradation. A protein from Epstein-Barr virus (another herpesvirus) that would be an excellent source of antigenic peptides is protected by the presence of a glycine-alanine repeat sequence that blocks the generation of antigenic peptides.

Less is known about viral interference with presentation of viral peptides to CD4+ lymphocytes. One herpesvirus is known to produce a protein released from infected cells that binds class II molecules and activates T cells. How this affects immune responses to the virus is uncertain.

Deactivating immune responses. Viruses have evolved numerous gene products that act to decrease the effectiveness of immune responses. As with gene products that interfere with antigen presentation to T cells, most are produced by deoxyribonucleic acid (DNA) viruses with large genomes (herpesviruses, poxviruses, adenoviruses). The increased genome size enables the virus to acquire genes from host cells, and indeed, viral genes that modify immune responses usually are quite obviously removed from the host genome based on genetic homologies of viral and host genes.

DNA viruses are clearly interested in cytokines and chemokines. These are small proteins released from cells that affect the activity of other cells, usually by binding to a specific cell surface receptor that transmits a signal to the interior of the cell. Cytokines and chemokines are used to regulate immune responses and to alter the metabolism of nonimmune cells in a manner that inhibits viral replica-

tion. Viruses interfere with cytokines in a wide variety of ways. For example, infected cells may secrete cytokine homologs that exhibit either agonist or antagonist activity by binding to a cytokine receptor on the surface of a cell intended to receive the cytokine.

Alternatively, homologs of cytokine receptors may be secreted that bind to cytokines and neutralize their activity by compromising their interaction with the intended receptor. Together, these soluble factors are called virokines and are of special interest to the pharmaceutical industry, since they provide evolution-tested, off-the-shelf reagents for modulating the effects of cytokines in autoimmune or inflammatory diseases. Viruses may also produce cytokine receptor homologs displayed on the cell surface called viroceptors. The precise function of many viroceptors remains to be determined. Some may act simply to sequester cytokines. Others may maintain the capacity to transmit signals to the cell interior upon cytokine binding, and probably subvert the normal signaling process.

Viruses also express proteins that act in the cell interior to modify the cytokine signaling pathway. Similarly, viruses express proteins that inhibit apoptosis, which is the programmed self-destruction of cells, often triggered by specific cytokines. Apoptosis entails the fragmentation of all DNA in cells (including viral DNA) and cessation of cellular processes required for viral replication. By blocking apoptosis, viruses can prevent fragmentation of their DNA and also can buy time for the production of progeny.

A complementary strategy utilized by viruses is the production of proteins that block the function of immune effector molecules. Herpesviruses and poxviruses express proteins that interfere with the complement system. The complement system is composed of a number of proteins in body fluids that have a myriad of functions in immune responses, including enhancing the activities of antibodies, destroying membranes, and regulating inflammation and B-cell responses. The poxvirus protein inhibits the enhancing effect of complement on antibody-mediated neutralization of viral infectivity. Herpesviruses are known to produce three complement interacting proteins, each acting in a different manner.

One blocks complement-mediated destruction of viral membranes; the others block the enhancing effects of complement on antibody-dependent destruction of viruses. Herpesviruses also produce a protein that directly interferes with antibody activity by binding the antibody.

Disabling immune cells. Many viruses are known to infect and disable cells of the immune system. The most notorious is HIV, which infects CD4+ T cells and macrophages. Infection leads to cell death directly by viral cytopathogenicity or indirectly through the action of other elements of the immune system acting to destroy virus-infected cells. As HIV infection progresses, CD4+ T cells are severely depleted, leading to infection by organisms that are normally controlled easily by the intact immune system. This advanced condition is recognized clinically as AIDS and is nearly always lethal. While the depleting effect of HIV on T cells is obviously of great clinical importance, it is unclear whether this contributes to the spread of the virus, and hence is of evolutionary significance to the virus. Similarly, while many other viruses are known to infect and disable immune cells, it is uncertain whether this results in evolutionarily significant modification of the immune response or whether the immune cell is utilized simply as an efficient means of viral replication.

For background information *see* ACQUIRED IMMUNE DEFICIENCY SYNDROME (AIDS); CYTOMEGALOVIRUS INFECTION; EPSTEIN-BARR VIRUS; HERPES; IMMUNITY; VIRUS; VIRUS INTERFERENCE in the McGraw-Hill Encyclopedia of Science & Technology.   Jack R. Bennink; Jonathan W. Yewdell

Bibliography. G. McFadden (ed.), *Molecular Biology Intelligence Unit: Viroceptors, Virokines and Related Immune Modulators Encoded by DNA Viruses*, 1995; D. D. Richman, R. J. Whitely, and F. G. Hayden (eds.), *Clinical Virology*, 1997; M. K. Spriggs, One step ahead of the game: Viral immunomodulatory molecules, *Ann. Rev. Immunol.*, 14:101–130, 1996.

Voice analysis

Traditional voice analysis is based on extraction of temporal and spectral features from a microphone signal. This observation-at-a-distance via an airborne signal has limitations, especially if sound production at the larynx and sound transmission through the airways are to be studied separately. The microphone signal is an unfortunate mixture of the combined properties of the source of sound and the propagation of sound through the airways (known as the filter). Thus, speech scientists have been looking for additional information to augment acoustic recordings of human voices. The source–filter theory of voice production could then be studied in greater detail. In particular, the use of fiber-optic viewing of the vocal folds has provided important information about the source, while magnetic reso-

nance imaging (MRI) and electron beam computed tomography (EBCT) have produced three-dimensional shapes of the vocal-tract airways (the filter). Both of these techniques still have limitations with temporal and spectral resolution, but results are promising.

Fiber-optic imaging of vocal folds. Fiber-optic imaging of the larynx, and more specifically of the vocal folds that vibrate to produce the sound, falls into two categories: stroboscopic imaging at normal video rates of 30 frames per second (**Fig. 1**), and high-speed video imaging at 1000–10,000 frames per second. Since the vocal folds typically vibrate at 100–1000 Hz in speech and singing (male and female ranges included), it is clear that a 30-Hz frame rate will at best capture one view of the folds every 3 to 4 cycles, and at worst one view every 30 to 40 cycles. This is acceptable if the vibration pattern is periodic. A stroboscopic flash can then be synchronized with the frequency of the vocal folds (as detected on the skin of the neck), and the phase of the flash can be moved around to observe specific points within the vibration cycle. The frequency of the strobe flash can also be detuned slightly from the frequency of the vocal folds to observe the vibration in slow motion. This type of observation has given clinicians useful information about the amplitude of vibration, the degree of contact (collision) between the folds, and the modes of vibration. But when the modal pattern is complex and variable from cycle to cycle, the stroboscopic technique fails, and high-speed imaging must be invoked. With this technique, 10–100 images can be obtained within each cycle of vibration, depending on the frequency of vibration and the frame rate of the high-speed video system. This temporal resolution is often adequate to capture transient responses (such as in a cough or a sudden pitch jump) and complex vibratory modes (such as in a hoarse voice or a high whistle register of a singer).

Imaging of vocal tract. Magnetic resonance imaging and electron beam computed tomography images

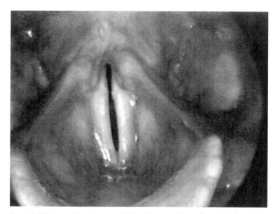

Fig. 1. Fiber-optic image of the larynx (top view) during phonation. The dark vertical slit is the glottis, an airspace between the vocal folds, which vibrate at frequencies of about 100–1000 Hz.

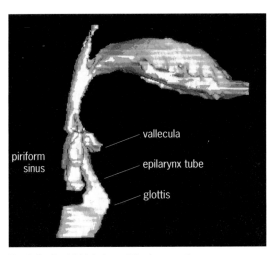

Fig. 2. Sagittal (side) view of the human airway as obtained from electron beam computed tomography. The vowel shape is an |a| as in father. The larynx is at the bottom and the mouth is on the top right.

of the vocal tract (**Fig. 2**) are basically static, affording no temporal resolution of dynamically changing airway structures in speech or singing. But the spatial resolution of these images is so good that even these static shapes help to uncouple the propagation characteristics from the source characteristics. A subject typically spends many hours in the magnetic resonance imaging machine to map out his or her articulatory space by mimicking the vowels and consonants in speech and holding each steady for several minutes. A complete volumetric scan, with millimeter accuracy, is obtained for each shape, including the oral cavity, the nasal cavity with the sinuses, the pharynx, the larynx, and the trachea. In and around the larynx are small pockets of air that make interesting resonators for sound (the piriform sinus, vallecula, epilarynx tube, and glottis; Fig. 2).

Special features of singing voice. Recent research has shown that trained singers are able to optimize their source–filter interactions to obtain easier and more efficient voice production. Singers open their mouths wide, particularly at high notes. This has the acoustic benefit of producing a better impedance match from the airway to free space, much like the flare of a trumpet. Impedance (the ratio of acoustic pressure to airflow) is high at the larynx and low in free space, requiring a megaphonelike transformation. A wide mouth also has the benefit of getting the jaw out of the way of the larynx. Often a tense jaw restricts the larynx in finding a position that can be held over a wide range of pitches.

A related point about trained singers is that the frequent practice of vocalises (little nonsense songs) on a variety of chosen vowels and consonants and over wide pitch ranges builds a certain source–articulator independence that untrained vocalists do not have. Voice quality, pitch, and loudness can then be changed or held constant (at will) when the words of a song impose their own vowel and consonant structure. Speech articulation thus be-

comes a mere modulation of an instrumentlike carrier sound that is little affected by this modulation.

Another benefit of training is the use of vocal-tract resonances to enhance the acoustic output power of selected partials (overtones). Males and females use somewhat different strategies. In males, there is generally a paucity of acoustic energy in the 3000-Hz region because the fundamental frequency is low (100–500 Hz). At 100 Hz, for example, the thirtieth partial is at 3000 Hz, whereas at 500 Hz the sixth partial is at 3000 Hz. These partials, as produced by the larynx, have considerably less energy than the fundamental, thus tilting the overall frequency spectrum toward the low end. In terms of a woofer-tweeter analogy, the woofer carries excessive amounts of power. By creating a small resonance tube above the vocal folds (in the region of the ventricular folds), the singer can boost the high-frequency energy selectively in the 3000-Hz region. This is perceived as the operatic ring in the voice, one of the most valued qualities in male voices.

Female singers have less of a need to boost the high-frequency portion of their acoustic spectrum because their fundamental frequencies are generally twice as high as males in song. Nevertheless, they sometimes use vocal-tract resonance to boost selective portions of the spectrum. In particular, when the soprano uses a high lifting or floating quality, there is a desire to make the sound pure, almost free of harmonic structure. The natural resonance of the entire length of the vocal tract (from larynx to the lips) is then used to enhance the fundamental rather than an overtone. The result is a flutelike sound rather than a brasslike sound.

Makeup of premier singing voices. The singing voice characteristics mentioned so far, as well as others such as the use and control of vibrato, are achievable by most people with voice training. However, the makeup of a premier voice, one that appears only once in a decade and is universally recognized as special, is still speculative; but some hints are beginning to emerge. Naturally great voices tend to have a nonencumbrance of skeletal structures in and around the larynx. This involves, at a minimum, the shoulders, the neck, and the jaw. As a wide range of pitches and loudnesses are accessed, soft and hard tissues do not interfere with each other. There is enough room for expansion, contraction, and linear displacement of the larynx, rib cage, abdomen, diaphragm, and airways as necessary. These organs do not move a lot—on the contrary, a firm equilibrium position (posture) is desirable—but the critical movements must be unencumbered. This applies particularly to the use of opposing muscles (agonist-antagonist pairs). They must not fight each other, but be able to turn on and off gradually (like a dimmer switch) to move structures and change tensions precisely and differentially. Jerky on–off movements are seldom seen in a premier singer. Rather, there is a deathlike calmness on the surface, underneath which huge muscular efforts are expended.

Within the larynx, there are likely to be some

morphological differences between ordinary and premier singers, although direct verification by inspection of the organs of deceased singers has not been possible. Scientists have relied on simulation, therefore, to test the optimal structures for sound production. Symmetry between the left and right vocal folds seems to play an important role. In principle, the two vocal folds have their own characteristic modes of vibration (like drums, bells, or strings). These modes depend on the viscoelastic properties of the vocal-fold tissues and the boundaries that surround the tissues (the cartilages). If either the boundary structures or the internal tissue properties of the vocal folds are asymmetric, different modes (with different natural frequencies) can be excited. These modes can fight each other. A common airflow between the vocal folds does help to entrain the modes, but there is a limit to this entrainment. If large ranges of pitch and loudness are to be achieved, a highly symmetric pair of vocal folds has a much better chance of avoiding chaotic oscillation.

Computer simulation and physical construction of self-oscillating models of the vocal folds have also shown that a large benefit is obtained by having a thick, pliable mucosa as a covering of the vocal folds. This mucosa propagates a surface wave while the vocal folds are vibrating. In fact, it is the surface wave that facilitates the energy transfer from the airstream between the vocal folds to the tissue itself, thereby producing self-oscillation. Highly gifted singers probably have the genetic construct of a thick and pliable vocal-fold mucosa, although direct histological verification is yet pending.

Underneath the loose, pliable mucosa must be a tough ligament that can support large tensions, much like a piano or violin string. For high pitches, this ligament absorbs most of the tension in the vocal folds. The amount and the type of collagen and elastin fibers that make up this vocal ligament may again be genetically determined. Thus, some people may be born with better material properties than others, much like certain woods or metals are more desirable for musical instrument design.

For background information *see* ACOUSTIC IMPEDANCE; FIBER-OPTICS IMAGING; MEDICAL IMAGING; SPEECH; STROBOSCOPE in the McGraw-Hill Encyclopedia of Science & Technology.   Ingo R. Titze

Bibliography. B. Story, I. Titze, and E. Hoffman, Vocal tract area functions from magnetic resonance imaging, *J. Acous. Soc. Amer.*, 100:537–554, 1996; J. Sundberg, *Science of the Singing Voice*, 1987; I. R. Titze, *Principles of Voice Production*, 1994; I. Titze, S. Mapes, and B. Story, Acoustics of the tenor high voice, *J. Acous. Soc. Amer.*, 94:1133–1142, 1994.

Volcanic islands

Some of the largest landslides involve mass failures on the flanks of oceanic volcanoes, including both seamounts and those volcanoes that have grown sufficiently to form islands. The widespread occurrence of mass failures as a major process in determining the morphology of oceanic volcanoes was not recognized until acoustic swath-mapping techniques, such as multibeam-sounding and long-range side-looking sonar, could provide comprehensive views of the submerged parts of the volcanoes. The term mass failure is commonly applied to gravitationally driven movement of sediment and rock in the form of slides, slumps, and debris flows. Even for the largest oceanic volcanoes, such as Mauna Loa on the island of Hawaii, which stands 13,000 ft (about 4000 m) above sea level, continuing volcanic activity above sea level rapidly buries most landslide features, leaving the impression that the products of volcanic eruptions are almost solely responsible for shaping the volcanic edifice. Mauna Loa, however, stands 30,000 ft (about 9000 m) above the surrounding sea floor, and its submerged flanks cover an area much larger than is exposed on the island. The dominant submarine morphologic features have resulted from a series of very large mass failures of the volcano flanks. The importance of landslides in shaping Mauna Loa is typical not just of the Hawaiian Islands but also of oceanic volcanoes in general.

Landslide characteristics and distribution. Large-scale mass failure of the flank of a Hawaiian volcano was first suggested more than a century ago, but no unequivocal evidence from the deep-sea floor was available until surveys of the Hawaiian Exclusive Economic Zone (EEZ) documented the enormous scale of submarine slope failures along the entire Hawaiian Ridge. The surveys used long-range side-looking sonar and were known as the Geological Long-Range Inclined ASDIC (GLORIA) surveys. Slumps and debris avalanches were observed to cover nearly 30,000 square nautical miles (100,000 km^2) of the Ridge flanks and adjacent sea floor just within the segment from Kauai to Hawaii, which encompasses an area more than five times the combined land area of the State of Hawaii. Along the 1500-nmi (2800-km) length of the Hawaiian Ridge, which extends from the island of Hawaii (154°W longitude) to Midway Island (near 180°W longitude), at least 68 major landslides of more than 11 nmi (20 km) in length have been imaged (**Fig. 1**).

The Hawaiian Ridge formed as a result of volcanic activity above a mantle hot spot as the Pacific lithospheric plate moved west-northwest over the hot spot at a rate of 3.5 in. (9 cm) per year. Midway Island formed about 28 million years ago; and the volcanoes of the Ridge become progressively younger toward the island of Hawaii, which is still actively forming. This means that a major landslide occurred approximately every 350,000 years along the Hawaiian Ridge. Thus, the Hawaiian EEZ surveys provide a comprehensive overview of the characteristics of oceanic volcanic landslides.

Some of the individual landslides are more than 110 nmi (200 km) long and about 800 nmi^3 (5000 km^3) in volume, ranking them among the largest in

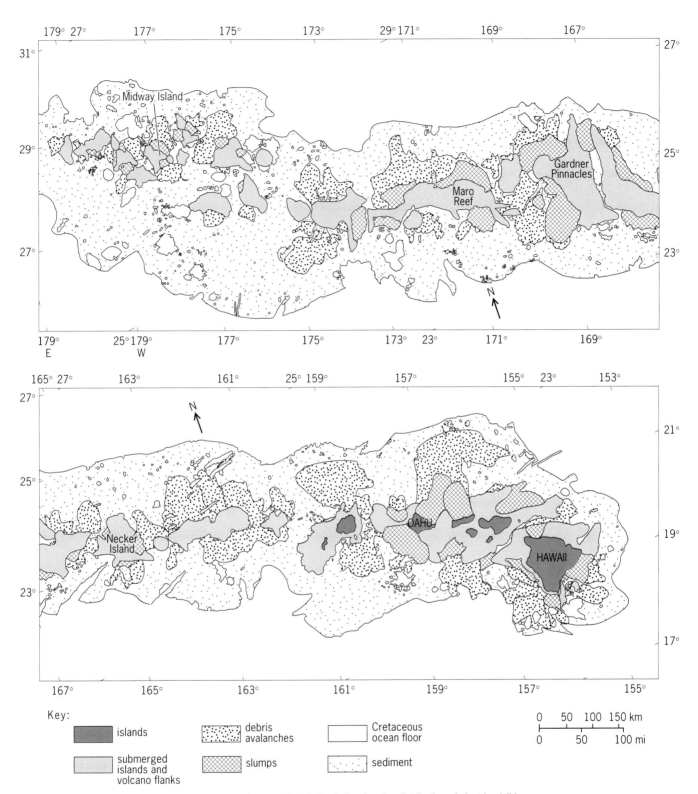

Fig. 1. Generalized geologic map of Hawaiian Ridge (left and right halves) showing the distribution of giant landslides, both slumps and debris avalanches, generated by mass failure of the flanks of volcanoes. *(From J. G. Moore, W. R. Normark, and R. T. Holcomb, Giant Hawaiian landslides, Annu. Rev. Earth Planet. Sci., 22:119–144, 1994)*

the world. These slope failures begin early in the history of individual volcanoes, when they are small submarine seamounts; culminate near the end of volcanic shield-building activity above sea level; and may continue long after dormancy. Most of the land-slides moved into, and locally across, the axis of the Hawaiian Deep, which is a moat feature flanking the Hawaiian Ridge. The Hawaiian Deep forms as the weight of the growing volcanoes depresses the oceanic crust. Therefore, mass failure redistributes vol-

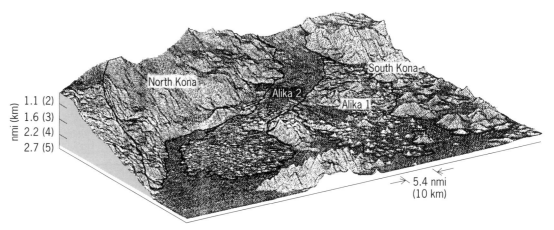

Fig. 2. Physiographic view, based on NOAA multibeam bathymetry, of the southwest submarine flank of Mauna Loa volcano, Hawaii, showing four major landslides. The mass failure that formed the Alika 2 debris avalanche is the youngest (about 105,000 years ago) and is thought to have generated both a giant tsunami and a turbidity current. *(From J. G. Moore, W. R. Normark, and R. T. Holcomb, Giant Hawaiian landslides, Annu. Rev. Earth Planet. Sci., 22:119–144, 1994)*

canic material into the Deep and widens the base of the Ridge. The eruptive processes of the volcanoes are affected by landsliding; and the geometry of the landslides is related to the structural features of the volcanoes, including rift zones and fault systems. The landslides are of two general types: slumps and debris avalanches.

Slumps. Large slumps are deeply rooted in the volcanic edifice and extend seaward from the volcanic rift zones; they can include nearly the entire volcanic pile. Slumps, which are as much as 60 nmi (110 km) wide and 6 nmi (10 km) thick, are slow moving. Apparently they may creep continuously over extended periods as they move in response to the increasing load of volcanic material erupted on their upper part. Although slumps have an overall gradient of little more than 3°, the upper tensional parts of the slumps exhibit substantial relief along transverse normal faults, which are marked by scarps that commonly bound a few large tilted blocks up to tens of miles in length and several miles in width. The lower part of a slump is a compressional regime marked by broad seaward-protruding bulges in bathymetric contours, closed depressions, and a steep toe area at the termination of the slump. The slumps generally maintain an overall coherent lobate shape by which they are identified in GLORIA images. Parts of the slumps may collapse and produce a debris avalanche extending downslope beyond the lower end of the slump.

The Hilina slump on the south flank of Kilauea volcano is actively moving and may be indicative of oceanic volcano slump behavior. During a magnitude 7.2 earthquake in 1975, a length of coastline of more than 30 nmi (about 60 km) along the south coast of Kilauea subsided as much as 11 ft (3.5 m) and moved seaward as much as 26 ft (8 m). On the flank of Kilauea above sea level, the arcuate fault scarps of the Hilina fault system were reactivated over a length of 15 nmi (nearly 30 km), with normal faults downthrown toward the sea. Similar displace-

ments along Kilauea's south flank were qualitatively described by eyewitnesses during an even larger earthquake in 1868. In addition to sudden displacements during earthquakes, which are accompanied by small, locally observed tsunamis, the south flank of the Kilauea volcano also is creeping seaward continuously, as indicated by frequently repeated geodetic measurements. The rift zone itself seems to define the pull-apart zone (landslide headwall area) wherein the region to the south (and extending to the coast and beyond) is moving south at about 2.4 in. (6 cm) per year.

Most of the movement on slumps occurs along a major slip surface at or near the base of the volcanic edifice. Close to the rift zone, which is under the thickest portion of the volcanic pile, slump movement is probably facilitated by a hot plastic zone related to magma within the rift zone. Additionally, movement of the main body and toe of the slump is aided, because the volcanic pile rests on a layer of poorly consolidated, mostly pelagic sediment deposited on the oceanic crust by the background rain of particles before the volcanic pile began to build. This thin, weak layer provides a lubricated slip surface.

Debris avalanches. Debris avalanches are faster moving than the giant slumps, and are formed by more surficial processes. They have a thickness of 100–7000 ft (30–2100 m) or more and a length of as much as 125 nmi (230 km), with an overall gradient commonly less than 3°. The distal fringe of the debris avalanches on GLORIA images is recognized by the occurrence of discrete blocks resting on a generally smooth sea floor (**Fig. 2**). Thus, debris avalanches are longer (relative to their width), thinner, and less steep in longitudinal profile than the slumps. Photographs from deep-towed survey systems show that some debris avalanches appear to have extensive marginal aprons of smaller blocks that are below the resolution of the GLORIA side-looking sonar technique. In some cases, the fringe areas of the avalanche deposits can consist of landslide-associated gravel and fine-grained sediment. These marginal areas of small blocks and sediment are easily masked by later pelagic deposits. Hence, the mapped areas of all but the youngest debris avalanches are probably underestimated when based only on acoustic (sonar) images.

Rapid, single-event movement of debris avalanches is indicated by their thinness and great length; the uphill movement of individual blocks in their outer reaches, suggestive of tremendous inertia generated during the collapse of the volcano flank; and the nature of their hummocky, fragmented distal regions that closely resemble that of known catastrophic volcanic landslides above sea level. Catastrophic movement of the debris avalanches is also indicated by the sedimentary deposits left by giant tsunamis and turbidity currents generated during these rapid mass-failure events. A series of gravels that include coral debris have been mapped locally as much as 1070 ft (326 m) above sea level on the islands of Oahu, Lanai, Molokai, and Maui, and Kohala volcano on Hawaii. These gravels are believed to have been deposited by giant tsunamis generated by the mass failures that formed the debris avalanches. Such deposits on Lanai have been dated at 105,000 years, and both their age and location suggest that these deposits were emplaced as a result of the movement of the Alika 2 debris avalanche on the west side of Mauna Loa. This is the youngest failure known for one of the Hawaiian Islands. Coastal erosion on the southeast coast of Australia has been attributed to the effects of the tsunami generated by the Alika 2 avalanche. Cores from Leg 136 of the Ocean Drilling Program recovered volcanic sediment from the island of Hawaii in turbidite deposits 160 nmi (300 km) from the debris avalanche that generated the turbidity current. These turbidite deposits lie 1200 ft (370 m) above the floor of the Hawaiian Deep, where the avalanche came to rest. In order to deposit sediment at this elevation and great distance, the turbidity current must have been very large.

Future study needs. In addition to the giant landslides, many medium-size landslides that occurred in shallower water near the coast have not been adequately imaged with existing multibeam-sonar or side-looking-sonar imagery. Young volcanoes undergo small- to medium-scale mass failures, the evidence for which is lost as the edifice continues to grow. For example, there are three landslides on the flanks of Loihi Seamount, which is the newest Hawaiian volcano but is still nearly 3300 ft (1000 m) from reaching sea level. The landslides are of unknown age, but Loihi itself is probably only 100,000 years old. Even major landslides on the hot spot (southeast) end of the Hawaiian Ridge can go undetected, because continued growth of new volcanoes over the hot spot will bury the evidence on older volcanoes.

Thus, the potential hazard from submarine landsliding and resulting tsunamis on the Hawaiian Ridge

(and other active oceanic volcanoes) is incompletely known, because the frequency of destructive landsliding cannot be estimated from the limited information available. Only the largest events are easily recognized on GLORIA acoustic images or from detailed multibeam sounding data. There have been at least three major failures of the west flank of Mauna Loa since it grew above sea level, suggesting a recurrence interval of about 100,000 years. Smaller failures probably occur on time scales of about 1000 years. The recurrence interval of landslides on oceanic volcanoes is, in part, limited by the large ratio of landslide volume to eruption rate, especially for the largest failures. Because large landslides have been identified at numerous oceanic volcanoes in addition to the Hawaiian Islands, and because their effects can be catastrophic, it seems advisable to undertake further surveys of potential landslides on oceanic volcanoes in order to understand the processes and the initiating causes.

For background information *see* AVALANCHE; HOT SPOTS (GEOLOGY); LANDSLIDE; MARINE GEOLOGY; PLATE TECTONICS; SEAMOUNT AND GUYOT; TSUNAMI; TURBIDITY CURRENT in the McGraw-Hill Encyclopedia of Science & Technology.   William R. Normark

Bibliography. A. Borgia and B. Treves, Volcanic plates overriding the oceanic crust: Structure and dynamics of Hawaiian volcanoes, *Geol. Soc. London Spec. Pub.*, no. 60, pp. 277–299, 1992; R. T. Holcomb and R. C. Searle, Large landslides from oceanic volcanoes, *Mar. Geotechnol.*, 10:19–32, 1991; J. G. Moore, W. R. Normark, and R. T. Holcomb, Giant Hawaiian landslides, *Annu. Rev. Earth Planet. Sci.*, 22:119–144, 1994; J. G. Moore, W. R. Normark, and R. T. Holcomb, Giant Hawaiian underwater landslides, *Science*, 264:46–47, 1994.

Wave mechanics

Recent advances in laser and electrooptical technology have made it possible to directly observe the motion of electrons within atoms. Specifically, because of the development of ultrashort laser pulses with durations of less than 100 femtoseconds (1 fs $= 10^{-15}$ s), instruments can be created that act as cameras with extremely fast shutters to view electrons as they move about the nucleus. In addition, these same laser pulses can be used to alter the distribution of electronic charge about the atom to take on a particular configuration at a specified time. This type of control over charge distributions may eventually make it possible to alter the outcome of chemical reactions between atoms and molecules.

Classical and quantum physics. During the early twentieth century, it was discovered that atoms exhibited many interesting features which could not be explained with classical physics. In a classical atom, negatively charged, pointlike electrons orbit the nucleus just as the planets in the solar system orbit the Sun. However, the development of quantum mechanics in the 1920s made it possible to understand

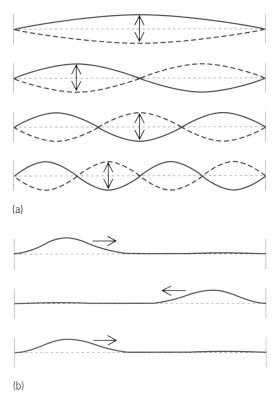

(a)

(b)

Fig. 1. Waves on a string with two fixed end points. (*a*) The first four standing waves which can exist on the string, analogous to the stationary-state wave functions of an atom. These waves have different vibrational frequencies just as the states of an atom have different energies. The positions of the maxima and minima do not move along the length of the string. (*b*) Analogs of wave packets produced by taking the square of a superposition of the four standing waves from part *a* at three different times, shown in succession from top to bottom. A clear maximum in the wave-packet amplitude can be seen to move back and forth along the string.

the atom in terms of waves instead of classical particles. In a quantum atom, the electrons are described in terms of discrete and continuous energy levels and their corresponding wave functions. Each wave function maps out the probability for finding an electron in a particular region of space with a given energy, angular momentum, and so forth. These wave functions are said to describe stationary states because the probability distribution does not change in time.

There is little resemblance between the classical atom with its electrons whizzing about the nucleus and the stationary-state quantum description. However, the stationary states of an atom are only a small subset of the possible wave functions. The stationary-state wave functions are analogous to standing waves in a string. The positions of the nodes and antinodes of the wave do not move in time (**Fig. 1***a*). However, a traveling wave or pulse which moves along the length of the string can also be produced (Fig. 1*b*). Similarly, nonstationary electron wave-packet states that move can be produced in atoms. As in a string, such a wave packet or pulse is created by superimposing stationary waves that

alternately interfere constructively and destructively as time goes by. Wave packets, then, are the connection between quantum and classical physics. In fact, atoms and molecules can be studied by monitoring the motion of wave packets as a function of time, a technique that can be more physically appealing than measuring electronic structure as a function of energy.

Wave packets in the laboratory. Wave packets have been of interest in atomic physics for many years. Until recently, no way had been devised of viewing the extremely rapid motion within atoms, which typically occurs over picosecond (10^{-12} s) and femtosecond time scales. Now laser pulses with durations as short as 10 fs can provide a fast shutter for observing the internal motion. These ultrashort laser pulses can be used to create wave packets as well. Because of the uncertainty principle, a pulse that has a short duration must have a large frequency spread. Since the structure of the wave packet depends on the number of superimposed waves, the use of short laser pulses makes it possible to excite coherently many atomic levels to produce wave packets that move about the atom in specific ways.

Creation of localized wave packets. For instance, localized electronic wave packets in atoms have been produced in which there is a high probability that the electron resides at a particular distance from the nucleus and there is negligible probability of finding it elsewhere. Typically, localized electron wave packets are created by exposing a low-energy stationary state to a short laser pulse. The laser pulse excites part of the stationary electronic wave function to a band of more energetic levels. Because the initial state has such low energy, it is very unlikely that the electron will be found far from the nucleus during the laser excitation. Therefore, the wave packet which is produced initially has a high probability of being located near the nucleus. However, as time evolves, the wave packet uses the extra kinetic energy it gained from the laser pulse to move away from the nucleus, in analogy to a rocket leaving the surface of the Earth. The wave packet travels far away, but will eventually be pulled back by the positively charged nucleus. A short laser pulse is required for the excitation to ensure that the part of the wave packet which is produced by the leading edge of the laser pulse does not complete an orbit before the end of the pulse. Under these conditions, the wave packet's localized bundle of electron probability moves about the nucleus in close analogy to the motion of a classical electron. However, the energy spread of the stationary states that make up the wave packet causes the wave packet to spread, producing quantum-mechanical interference which cannot be described with classical physics. Therefore, studying the evolution of localized wave packets provides new insight as to the depth of correspondence between classical and quantum physics.

Observation of wave packets. Of course, it is not enough to create the wave packet; it is also necessary to observe its evolution to learn from it. Several tech-

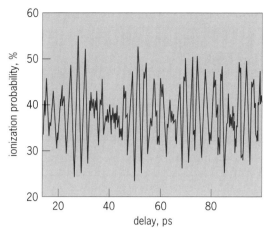

Fig. 2. Probability for ionizing a wave packet as a function of time using an ultrashort unipolar field pulse. The oscillations in the ionization probability are due to the changing position and velocity of the wave packet as it moves about the atom. By using an advanced technique, curves similar to this one make it possible to experimentally determine the velocity of the wave packet as a function of time.

niques can be implemented to observe wave-packet evolution. Each method requires a fast shutter so that the rapid periodic motion of the wave packet does not lead to a blurred observation. Typically, the fast shutter is another short laser pulse. Photoionization of the wave packet is a particularly simple observation method. In general, the photoionization probability is proportional to the probability that the electron passes near the nucleus at any time during the laser pulse. Therefore, by monitoring the ionization probability as a function of time between the pulse which excites the wave packet and the pulse which ionizes it, the probability for finding the wave packet near the nucleus can be measured as a function of time. The periodic motion of wave packets was observed by this method for the first time several years ago. Subsequently, other observation techniques were developed which give additional information on wave-packet evolution. One such method is bound-state interferometry, which measures the similarity between two wave packets by using quantum-mechanical interference. The technique is called interferometry because of its similarity to standard optical interferometers, where the similarities between two or more light beams are measured by using interference. In addition, more sophisticated techniques are being developed which will make it possible to directly monitor the evolution of the entire wave packet, not just the part near the nucleus. For example, by ionizing a wave packet using a very short, unipolar electric field pulse, the full velocity distribution of the wave packet can be measured (**Fig. 2**).

Other wave-packet types. Other types of wave packets are produced in many situations of interest in atomic physics, and studying the motion of these wave packets can aid in the understanding of complex phenom-

ena. For instance, anti or dark wave packets, where a small hole or notch is placed in a nonlocalized, stationary wave function, have been produced in atoms. A short laser pulse ionizes or deexcites a small part of an extended electronic wave function, leaving a hole near the nucleus. The motion of this localized notch or antiwave packet about the nucleus can be identical to that of an ordinary localized wave packet. Instead of residing in a small region of space where the electron probably is located, the antiwave packet defines a small volume where the electron probably is not located. Studying the evolution of these antiwave packets can lead to better understanding of the process of photoionization by extremely intense laser pulses. Experiments have also examined the creation and subsequent evolution of wave packets in strong static and time-varying electromagnetic fields, which can significantly alter the classical dynamic motion within atoms. By launching localized electron wave packets and monitoring their return experimentally, it is possible to study the transition from regular to chaotic motion in these complicated quantum systems. Wave packets can also be used to examine collisions between electrons in multielectron atoms.

Alteration of atomic behavior. Because the distribution of charge about an atom is determined by the electronic wave function, the production of appropriate wave packets can be used to alter the response of atomic systems to subsequent transients. The creation of designer wave packets that exhibit specific dynamic behavior can be achieved, in principle, by controlling the laser light that excites the wave packet. Several experiments have explored the use of tailored light pulses to produce wave packets with interesting properties. For example, by creating the appropriate wave packet it is possible to enhance or inhibit photoionization at specific times. Properly designed wave packets might be used to alter the course of chemical reactions between atoms and molecules, making a particular set of desirable byproducts significantly more abundant than all the others.

In sum, the study of wave packets has yielded new insights into the relationship of these quantum-mechanical entities to their classical analogs as well as the complicated dynamic evolution of atoms in the presence of strong external fields. Ultimately, it is the possibility of controlling atomic and molecular processes which makes this field of research highly promising.

For background information *see* ATOMIC STRUCTURE AND SPECTRA; INTERFEROMETRY; NONRELATIVISTIC QUANTUM THEORY; OPTICAL PULSES; PHOTOIONIZATION; QUANTUM MECHANICS; STANDING WAVE; ULTRAFAST MOLECULAR PROCESSES; WAVE MOTION in the McGraw-Hill Encyclopedia of Science & Technology.

Robert R. Jones

Bibliography. R. R. Jones, Creating and probing electronic wavepackets using half-cycle pulses, *Phys. Rev. Lett.*, 76:3927–3930, 1996; G. M. Lankhuijzen and L. D. Noordam, Streak camera probing of rubidium wavepacket decay in an electric field, *Phys. Rev. Lett.*, 76:1784–1787, 1996; M. Nauenberg, C. Stroud, and J. Yeazell, The classical limit of an atom, *Sci. Amer.*, 250(6):44–49, June 1994; W. S. Warren, H. Rabitz, and M. Dahleh, Coherent control of quantum dynamics: The dream is alive, *Science*, 259:1581–1589, 1993.

Yeast

Early Egyptians and Mesopotamians discovered that if barley grains were wetted, allowed to germinate, and subsequently dried, the barley would be sweeter and less perishable. When water was added, a mildly alcoholic beverage was produced through the action of naturally occurring yeast. The first recorded evidence of brewing was found in Egypt in 6000 B.C. From there, the technology passed to Greece, to Rome, and to the rest of the Roman Empire. By the thirteenth and fourteenth centuries, brewing was well established in the monasteries of northern Europe. Later in Europe and in early America, most households brewed their own beer. As towns and cities developed, home brewing was replaced by the small brewery, which eventually disappeared during Prohibition in the United States. Thereafter, the commercial brewing industry emerged. A good economy, improved technology, and better ingredients have allowed home brewing to flourish again as a leisure-time hobby.

Brewing. The basic ingredients of beer are water, fermentable sugars, hops, and living yeast, which is classified as a fungus. Yeast cells are round to ellipsoidal and 2.5–10 micrometers in length. Although they resemble bacteria, yeast cells have the fundamental characteristics of higher (eukaryotic) cells, including an organized nucleus with a nuclear membrane and cytoplasmic organelles.

There are approximately 86 yeast genera, about 40 of which are found in the subdivision Ascomycetes. The species of yeast used in brewing are in the genus *Saccharomyces*. All members of this genus reproduce by budding at many sites on the cell wall; all produce one to four resistant ascospores in saclike structures called asci under poor nutritional conditions; and all are capable of converting sugars to ethanol.

Scientific names. The scientific name for beer yeast originated in the late 1830s, when living organisms were observed in beer. *Saccharomyces cerevisiae* means "sugar fungus of beer." By the beginning of the twentieth century, investigators interested in alcoholic fermentation had isolated and named new species of yeast. More than 80 different names have been given to species of *Saccharomyces*. By modern molecular taxonomy methods, only four species are recognized; two are associated with brewing.

Growth and reproduction. Increase in yeast cell mass and cell number is accomplished through budding. The emergence of a bud formed by a localized outgrowth of the yeast cell wall indicates that the cell has entered a new division cycle. After formation of a dividing wall between the bud and the mother cell, separation gives rise to a daughter cell with all the cellular components and functions necessary for vegetative growth.

Saccharomyces species have a well-characterized sexual cycle with a haploid and a diploid phase, as found in higher plant and animal cells. However, yeast used for brewing is usually polyploid (having extra sets of chromosomes) or aneuploid (having an abnormal number of chromosomes) and does not undergo sexual reproduction.

Yeast metabolism. The metabolism of yeast is not designed for making beer, but for making more yeast. During the brewing process, yeast cells go through their entire life cycle, gaining and storing energy, expending stored energy, and then preparing for dormancy when the food and energy sources have been consumed. A yeast population will multiply three or four times during active fermentation.

Yeast cells are facultative anaerobes, capable of growing in the presence or absence of oxygen. As an energy source, brewer's yeast utilizes a variety of simple sugars (monosaccharides), including glucose, fructose, mannose, and galactose. Wort contains primarily the disaccharide maltose, a sugar derived from the enzymatic breakdown of the starch in germinated barley grains. Yeast can transform maltose into glucose. In addition to a carbon source, yeast requires a nitrogen source (for making structural proteins and enzymes), sulfur, phosphorus, metals (such as magnesium, calcium, zinc, iron, and copper), and growth factors.

When yeast is first added to the wort, there is a lag time during which it adapts to the new environment. Aerobic respiration follows, whereby yeast metabolizes low levels of glucose to carbon dioxide and water in a manner similar to most plant and animal tissues. No alcohol is formed during this period. As oxygen is rapidly taken up and the nutrients (especially phosphate) are utilized, the pH drops.

Under anaerobic conditions (absence of oxygen), fermentation is initiated. Glucose is rapidly metabolized to ethanol and carbon dioxide. During this phase, which exhibits exponential growth, the yeast in suspension continues to reproduce until an optimum population of about 2×10^6 cells/oz (5×10^7 cells/ml) is attained. The ability of yeast to ferment the sugars in the wort is termed attenuation. Yeast with a high attenuation percentage produces slightly lower final specific gravities and often a drier-flavored beer than yeast with a lower attenuation percentage. Fermentation generally lasts 3–7 days, sometimes longer. Although fermentation is a much less efficient process for producing energy than aerobic respiration, it forms the basis of the brewing industry.

As nutrients are depleted and retarding concentrations of fermentation products accumulate, the yeast population enters a stationary phase. During this period, flocculation and sedimentation occur. Certain strains of brewer's yeast spontaneously aggregate to form clumps (flocs) that settle to the bottom of the container. This characteristic is genetically determined but is also affected by environmental factors. Yeast with medium or high flocculation settles out more readily than yeast with low flocculation. The flocculent nature of a strain is important for the removal of yeast at the end of fermentation and may be a major factor in the selection of a strain for the production of a given beer.

Ales and lagers. The first fermented barley beverages were probably prototypes of ales. The strains of *S. cerevisiae* involved in ale production are less temperature sensitive than lager strains. Ale yeast grows well at temperatures up to 100°F (38°C), although the typical fermentation temperature for ales ranges 55–75°F (13–24°C). Ale yeast tends to flocculate at the surface of the fermenting beer in a dense head during the first few days of fermentation and is referred to as top-fermenting yeast. Eventually, ale yeast settles on the bottom of the fermenter. Lager beers are fermented at lower temperatures, 55°F (13°C) down to 32°F (0°C) by strains of *S. pastorianus*. Lager yeast, also known as bottom-fermenting yeast, grows less rapidly and with less surface foam. It tends to flocculate to the bottom of the container as fermentation nears completion. The distinction between top and bottom yeast is a reflection of the yeast's ability to associate with the bubbles of carbon dioxide, which rise to the surface under certain conditions.

Good brewing yeast. A species of yeast may include a number of strains, all of which are taxonomically identical but display differences that are important to the particular kind of beer produced. Strain variations may include growth rate, temperature tolerance, flocculation character, and production of many flavor compounds. A brewer selects beer yeast based on these criteria. A genetically stable yeast strain ensures consistent product quality. Yeast transforms some 90% of fermented wort sugar into alcohol and carbon dioxide; the remaining 10% is divided about equally between yeast biomass and compounds that produce the flavor and aroma.

Compounds that produce the fermentation bouquet depend upon the nature and concentration of a large number of minor metabolites formed by yeast in small quantities at different points during fermentation. Different yeast strains impart flavors ranging from dry to sweet, spicy to mild, as determined by levels of alcohols, acids, esters, aldehydes and ketones, and sulfur compounds. Esters give beer a fruity aroma. A mixture of isobutyraldehyde and methylglyoxal produces an earthy, musty aroma. Undesirable butter-smelling diacetyl is especially problematic in lager beers. Other compounds with undesirable flavors include hydrogen sulfide with a potent, vegetablelike taste and dimethyl sulfoxide with a sweet corn or garlic smell.

Home brewing. The home brewer can choose yeast commercially prepared in several packaging styles: dry yeast in foil pack, liquid in foil pack, and liquid in plastic tubes. There is an optimum storage temperature and reactivation temperature for each packaging style. A great deal of variation in the number and types of organisms is found in commercial preparations; dry yeast preparations are more likely to be contaminated than the others. Liquid culture yeast sealed in a foil pouch is theoretically the best starter. However, propagating one's own yeast and making a starter liquid culture that is inoculated from a pure culture stored on agar slants (test tubes which contain a solid agar medium surface that is slanted to provide a greater surface area) or petri dishes is the most desirable, although it is the most difficult method for the home brewer.

The only difference between the potential quality of home-brewed beer and commercially brewed beer is the vast amounts of money spent on consistency and quality control. Commercial breweries want their beer to turn out exactly the same every time. Superb beer can be made by the home brewer, but it will vary slightly from batch to batch.

For background information *see* ASCOMYCOTINA; ETHYL ALCOHOL; FERMENTATION; FUNGAL BIOTECHNOLOGY; FUNGAL GENETICS; FUNGI; MALT BEVERAGE; MYCOLOGY; YEAST in the McGraw-Hill Encyclopedia of Science & Technology.   Jeannette M. Birmingham

Bibliography. C. P. Kurtzman and J. W. Fell, *The Yeasts: A Taxonomic Study*, 4th ed., 1997; A. H. Rose and J. S. Harrison (eds.), *The Yeasts*, 2d ed., vols. 1 and 2, 1987; Special beer and yeast issue, *Zymurgy*, vol. 2, no. 4, 1989.

Contributors

The affiliation of each Yearbook contributor is given, followed by the title of his or her article. An article title with the notation "in part" indicates that the author independently prepared a section of an article; "coauthored" indicates that two or more authors jointly prepared an article or section.

A

Aase, Dr. J. Kristian. *U.S. Department of Agriculture, Agricultural Research Service, Kimberly, Idaho.* TILLAGE SYSTEMS—in part.

Adams, Dr. L. Garry. *Department of Veterinary Pathobiology, College of Veterinary Medicine, Texas A&M University, College Station.* TUBERCULOSIS.

Agris, Dr. Paul F. *Department of Biochemistry, North Carolina State University, Raleigh.* DEOXYRIBONUCLEIC ACID (DNA)—coauthored.

Antranikian, Prof. G. *Institute for Biotechnology/Technical Microbiology, Technical University of Hamburg-Harburg, Hamburg, Germany.* MICROBIOLOGY.

Archibald, Prof. J. David. *Department of Biology, College of Sciences, San Diego State University, San Diego, California.* EXTINCTION (BIOLOGY).

Armbrust, Dr. Dean V. *U.S. Department of Agriculture, Agricultural Research Service, Kansas State University, Manhattan.* EROSION.

Armbruster, Prof. Peter J. *Gesellschaft für Schwerionenforschung mbH, Darmstadt, Germany.* TRANSURANIUM ELEMENTS.

Aronson, Dr. Richard B. *Senior Marine Scientist, Dauphin Island Sea Lab, Dauphin Island, Alabama.* MACROEVOLUTIONARY DYNAMICS.

Arthur, Dr. Samuel. *Research Associate, DuPont Central Research and Development, Experimental Station, Wilmington, Delaware.* POLYMER.

Arthur, Prof. Wallace. *Director, Northumbrian Water Ecology Centre, University of Sunderland, United Kingdom.* POPULATION DYNAMICS.

B

Bacon, Dr. Michael P. *Department of Marine Chemistry and Geochemistry, Woods Hole Oceanographic Institution, Massachusetts.* OCEANOGRAPHY—coauthored.

Ball, Dr. J. B. *Senior Forestry Officer, Forest Resources Development Service, Forest Resources Division, Food and Agriculture Organization of the United Nations, Rome, Italy.* FORESTRY—in part.

Banerjee, Dr. Pat. *Associate Professor, Department of Mechanical Engineering, University of Illinois, Chicago.* VIRTUAL MANUFACTURING.

Beckman, Dr. Eric J. *Associate Professor, Department of Chemical and Petroleum Engineering, University of Pittsburgh, Pennsylvania.* CARBON DIOXIDE—coauthored.

Bennink, Dr. Jack R. *Department of Health and Human Services, National Institutes of Health, Bethesda, Maryland.* VIRUS—coauthored.

Birmingham, Dr. Jeannette M. *Biologist/Information Scientist, Mycology and Protistology Program, American Type Culture Collection, Rockville, Maryland.* YEAST.

Bly, Dr. Benjamin Martin. *Department of Psychology, Harvard University, Cambridge, Massachusetts.* BRAIN—coauthored.

Bolkan, Dr. Steven A. *Church & Dwight Co., Inc., Princeton, New Jersey.* CHLOROFLUOROCARBON—in part.

Brazel, Dr. Anthony J. *Chair, Department of Geography, State Climatologist for Arizona, Arizona State University, Tempe.* AIR POLLUTION—coauthored.

Brown, Dr. D. Ann. *Research Associate, Department of Geological Sciences, University of Manitoba, Winnipeg, Manitoba, Canada.* MINERALOGY.

Brune, Prof. William H. *Department of Meteorology, Pennsylvania State University, University Park.* STRATOSPHERIC OZONE.

Buhrig, Dr. Amy L. *Boeing Commercial Space Company, Seattle, Washington.* SATELLITE LAUNCH VEHICLE—in part.

Bunnett, Prof. J. F. *Department of Chemistry and Biochemistry, University of California, Santa Cruz.* HAZARDOUS WASTE.

Butler, Dr. Michael A. *Technical Staff, Microsensor R&D Department, Sandia National Laboratories, Albuquerque, New Mexico.* REMOTE SENSING—coauthored.

Butler, Dr. R. Paul. *Department of Physics and Astronomy,*

San Francisco State University, San Francisco, California. PLANET.

C

Calderone, Dr. Richard. *Department of Microbiology and Immunology, Georgetown University School of Medicine, Washington, D.C.* MEDICAL MYCOLOGY.

Caramazza, Prof. Alfonso. *Department of Psychology, Harvard University, Cambridge, Massachusetts.* BRAIN—coauthored.

Cassens, Prof. Robert G. *Department of Animal Sciences, Muscle Biology Laboratory, College of Agricultural and Life Sciences, University of Wisconsin, Madison.* FOOD MANUFACTURING.

Castro, Dr. Bernardino G. *Assistant Professor of Ecology, Department of Natural Resources and Environment, Faculty of Sciences, University of Vigo, Pontevedra, Spain.* ECOLOGICAL INTERACTIONS.

Cavanaugh, Prof. Colleen M. *Department of Organismic and Evolutionary Biology, Harvard University, Cambridge, Massachusetts.* HYDROTHERMAL VENT—coauthored.

Cesarsky, Dr. Cathrine J. *DSM/CEA-Saclay, Giff CEDEX, France.* SATELLITE (SPACECRAFT)—coauthored.

Cesarsky, Dr. Diego A. *IAS/CNRS-Orsay, Institut d'Astrophysique Spatiale, Université de Paris Sud, Orsay CEDEX, France.* SATELLITE (SPACECRAFT)—coauthored.

Chamberlain, Prof. Steven C. *Department of Bioengineering and Neuroscience, Institute for Sensory Research, Syracuse University, Syracuse, New York.* HYDROTHERMAL VENT—in part.

Chambers, Dr. Frederick B. *Assistant Professor, Department of Geography/Geology and Center for Environmental Sciences, University of Colorado, Denver.* GLACIOLOGY.

Chan, Dr. Winston. *Department of Electrical and Computer Engineering, University of Iowa, Iowa City.* OPTOELECTRONIC INTEGRATION.

Chang, Dr. C. T. Philip. *Assistant Research Scientist, Department of Chemical Engineering, Texas A&M University, College Station.* FLUID FLOW—coauthored.

Clark, Dr. Christopher W. *Director, Bioacoustics Research Program, Cornell Laboratory of Ornithology, Ithaca, New York.* UNDERWATER NOISE.

Cohill, Dr. Andrew Michael. *Blacksburg, Virginia.* INTERNET—in part.

Contin, Dr. Marco. *Azienda Agraria Sperimentale "A. Servadei," Udine, Italy.* SOIL MICROBIOLOGY.

Costantino, Dr. James. *President, Intelligent Transportation Society of America, Washington, D.C.* TRANSPORTATION ENGINEERING—coauthored.

Crabtree, Dr. Robert H. *Department of Chemistry, Yale University, New Haven, Connecticut.* CHLOROFLUOROCARBON—in part.

Cummins, Dr. Christopher C. *Department of Chemistry, Massachusetts Institute of Technology, Cambridge.* MOLYBDENUM.

Czarnik, Dr. Anthony W. *Senior Director, Chemistry, IRORI Quantum Microchemistry, La Jolla, California.* COMBINATORIAL CHEMISTRY—in part.

D

Deleplanque, Dr. M. A. *Lawrence Berkeley Laboratory, Nuclear Science Division, University of California, Berkeley.* GAMMA-RAY DETECTORS—coauthored.

Detera-Wadleigh, Dr. Sevilla D. *Clinical Neurogenetics Branch, National Institute of Mental Health, Bethesda, Maryland.* BEHAVIORAL GENETICS—coauthored.

Deutsch, Dr. Clayton V. *Associate Professor, Department of Petroleum Engineering, School of Earth Sciences, Stanford University, Stanford, California.* PETROLEUM EXPLORATION.

Dodson, Dr. Peter. *Department of Animal Biology, School of Veterinary Medicine, University of Pennsylvania, Philadelphia.* DINOSAUR.

Dougherty, Dr. Brian A. *The Institute for Genomic Research (TIGR), Rockville, Maryland.* GENE—coauthored.

Dresselhaus, Prof. M. S. *Department of Electrical Engineering and Computer Science and Department of Physics, Massachusetts Institute of Technology, Cambridge.* NANOSTRUCTURE—coauthored.

Drueckhammer, Dr. Dale G. *Assistant Professor, Department of Chemistry, Stanford University, Stanford, California.* HYDROGEN BOND.

Dugan, Dr. Frank. *Mycology and Protistology Program, American Type Culture Collection, Rockville, Maryland.* AGRICULTURE.

Dunlap, Dr. Jay C. *Department of Biochemistry, Dartmouth Medical School, Hanover, New Hampshire.* BIOLOGICAL CLOCKS.

E

Ebel, Dr. Denton S. *Department of the Geophysical Sciences, University of Chicago, Illinois.* METEORITE.

Eble, Dr. Gunther J. *Department of Paleobiology, National Museum of Natural History, Smithsonian Institution, Washington, D.C.* ANIMAL EVOLUTION.

Ebstein, Dr. Richard P. *Director of Research, S. Herzog Memorial Hospital, Jerusalem, Israel.* BEHAVIORAL GENETICS—in part.

Edmonds, Dr. James S. *MCM Enterprises Ltd., Bellevue, Washington.* ELECTRIC POWER GENERATION.

Elbestawi, Dr. M. A. *Professor and Chair, Department of Mechanical Engineering, McMaster University, Hamilton, Ontario, Canada.* MACHINABILITY OF METALS.

Elias, Dr. Nicholas. *U.S. Naval Observatory, Astrometry Department, Navy Prototype Optical Interferometer, Washington, D.C.* INTERFEROMETRY.

Ellers, Prof. Olaf. *Division of Biological Sciences, Section of*

Evolution and Ecology, University of California, Davis. BIVALVIA—in part.

Elsayed, Prof. Elsayed A. *Professor and Chairman, Department of Industrial Engineering, College of Engineering, Rutgers, State University of New Jersey, Piscataway.* COMPUTER-INTEGRATED MANUFACTURING—in part.

Epstein, Dr. Arthur J. *Director, Center for Materials Research, and Professor of Physics and Chemistry, Ohio State University, Columbus.* MAGNET—coauthored.

Eskin, Prof. N. A. Michael. *Department of Foods and Nutrition, University of Manitoba, Winnipeg, Manitoba, Canada.* NUTRACEUTICALS.

F

Filip, Dr. Gregory M. *Associate Professor, Department of Forest Science, Oregon State University, Corvallis.* FOREST PESTS.

Finley, David G. *Public Information Officer, National Radio Astronomy Observatory, Socorro, New Mexico.* RADIO TELESCOPE.

Fisch, Dr. Richard S. *3M Printing Division, Saint Paul, Minnesota.* PRINTING—in part.

Fitch, Dr. Mark J. *Research Associate, Office of Climatology, Department of Geography, Arizona State University, Tempe.* AIR POLLUTION—coauthored.

Fleischmann, Dr. Robert D. *The Institute for Genomic Research (TIGR), Rockville, Maryland.* GENE—coauthored.

Fox, Prof. Richard H. *Department of Agronomy, College of Agricultural Sciences, Pennsylvania State University, University Park.* CHLOROPHYLL—coauthored.

François, Dr. Roger. *Department of Marine Chemistry and Geochemistry, Woods Hole Oceanographic Institution, Massachusetts.* OCEANOGRAPHY—coauthored.

Frankham, Dr. Richard. *Associate Professor, Key Centre for Biological Diversity and Bioresources, School of Biological Sciences, Macquarie University, Sydney, New South Wales, Australia.* ENDANGERED SPECIES.

Fraser, Dr. Claire M. *The Institute for Genomic Research (TIGR), Rockville, Maryland.* GENE—coauthored.

Fraser, Dr. Nicholas C. *Curator of Vertebrate Paleontology, Virginia Museum of Natural History, Martinsville.* INSECTA—coauthored.

Friedman, Arnold. *Executive Director, DataComm Satellites, Space Systems/Loral, Palo Alto, California.* COMMUNICATIONS SATELLITE—coauthored.

Fry, Prof. Edward S. *Department of Physics, Texas A&M University, College Station.* LASER—in part.

G

Gabrielse, Prof. Gerald. *Department of Physics, Lyman Laboratory, Harvard University, Cambridge, Massachusetts.* ANTIMATTER.

Galli, Prof. Stephen J., M.D. *Director, Division of Experimental Pathology, Harvard Medical School, Boston, Massachusetts.* INFECTION.

Gardner, Dr. Julian W. *Director of Nanotechnology Centre, Department of Engineering, University of Warwick, Coventry, United Kingdom.* MICROSENSOR—coauthored.

Gardner, Dr. Scott Lyell. *Associate Professor and Curator of Parasitology, Harold W. Manter Laboratory of Parasitology, University of Nebraska State Museum, Lincoln.* EPIDEMIOLOGY.

Gershon, Dr. Elliot S., M.D. *Clinical Neurogenetics Branch, National Institute of Mental Health, Bethesda, Maryland.* BEHAVIORAL GENETICS—coauthored.

Ghenciu, Dr. E. G. *Department of Chemical and Petroleum Engineering, University of Pittsburgh, Pennsylvania.* CARBON DIOXIDE—coauthored.

Goldberg, Dr. Alfred. *Lawrence Livermore National Laboratory, Livermore, California.* STEEL MANUFACTURE.

Goldburg, Prof. Walter I. *Department of Physics and Astronomy, University of Pittsburgh, Pennsylvania.* FLUID FLOW—in part.

Granstrom, Prof. David E. *Associate Professor, College of Agriculture, Veterinary Science, University of Kentucky, Lexington.* EQUINE PROTOZOAL MYELOENCEPHALITIS.

Grilli, Prof. Stephan T. *Department of Ocean Engineering, Naragansett Bay Campus, University of Rhode Island, Naragansett.* FLUID FLOW—in part.

Grimaldi, Dr. David A. *Department of Entomology, American Museum of Natural History, New York, New York.* INSECTA—coauthored.

Grunthaner, Dr. Frank J. *Center for Space Microelectronics Technology, Jet Propulsion Laboratory, Pasadena, California.* REMOTE SENSING—coauthored.

Guenther, Dr. Richard. *Department of Biochemistry, North Carolina State University, Raleigh.* DEOXYRIBONUCLEIC ACID (DNA)—in part.

Gunasekera, Prof. Jay S. *Moss Professor and Chair, Department of Mechanical Engineering, College of Engineering and Technology, Ohio University, Athens.* COMPUTER-INTEGRATED MANUFACTURING—coauthored.

Gusek, James J. *Senior Engineer, Knight Piesold LLC, Denver, Colorado.* MINING.

H

Halliday, Dr. Alex N. *Department of Geological Sciences, University of Michigan, Ann Arbor.* EARTH, AGE OF.

Hallman, Dr. Robert W. *Corporate Research and Development, Polychrome Corporation, Carlstadt, New Jersey.* PRINTING—in part.

Hanstorp, Dr. Dag. *Department of Physics, Gothenburg University and Chalmers University of Technology, Gothenburg, Sweden.* NEGATIVE ION.

Happer, Prof. William, Jr. *Department of Physics, Joseph Henry Laboratories, Princeton University, Princeton, New Jersey.* MEDICAL IMAGING.

Harden, Dr. Carol P. *Associate Professor and Head, Department of Geography, University of Tennessee, Knoxville.* SOIL CONSERVATION.

Hassibi, Dr. Babak. *Information Systems Laboratory, Stanford University, Stanford, California.* ADAPTIVE SIGNAL PROCESSING — coauthored.

Haus, Prof. Hermann A. *Department of Electrical Engineering and Computer Science, Massachusetts Institute of Technology, Cambridge.* SOLITON.

Heaney, Dr. Peter J. *Assistant Professor of Mineralogy, Department of Geosciences, Princeton University, Princeton, New Jersey.* MINERAL.

Heinrich, Dr. Rudol. *Chief, Forest Harvesting, Trade and Marketing Branch, Food and Agricultural Organization of the United Nations, Rome, Italy.* FORESTRY — in part.

Heller, Dr. Michael J. *Nanogen Inc., San Diego, California.* BIOTECHNOLOGY — in part.

Hemming, Dr. Sidney. *Geochemistry, Lamont-Doherty Earth Observatory, Palisades, New York.* MARINE SEDIMENTS — in part.

Heydon, Douglas A. *President, Arianespace, Inc., Washington, D.C.* SATELLITE LAUNCH VEHICLE — in part.

Hobbs, Dr. Marcia M. *Department of Microbiology and Immunology, University of North Carolina School of Medicine, Chapel Hill.* SEXUALLY TRANSMITTED DISEASE.

Holland, Dr. Murray J. *JILA and University of Colorado, Boulder.* BOSE-EINSTEIN STATISTICS.

Hook, Dr. W. R. *Department of Biology, University of Victoria, British Columbia, Canada.* SOIL — coauthored.

Hsu, Dr. Yong. *Associate Scientist, Assistant Research Professor, Michigan Molecular Institute, Midland.* DENDRITIC MACROMOLECULES — coauthored.

Hung, Dr. Chung-Chih. *Analog VLSI Laboratory, Department of Electrical Engineering, Ohio State University, Columbus.* ELECTRONICS — coauthored.

Hurley, Dr. Neil F. *Department of Geology and Geological Engineering, Colorado School of Mines, Golden.* PETROPHYSICAL LOGGING.

Hwang, Dr. Changku. *Micrys, Inc., Columbus, Ohio.* ELECTRONICS — coauthored.

I

Imperiali, Dr. Barbara. *Division of Chemistry and Chemical Engineering, California Institute of Technology, Pasadena.* PEPTIDE.

Ioannou, Dr. Petros J. *Associate Professor, Department of Applied Sciences, Harvard University, Cambridge, Massachusetts.* TURBULENT FLOW — in part.

Isgur, Dr. Nathan. *Thomas Jefferson National Accelerator Facility, Newport News, Virginia.* NUCLEAR PHYSICS.

Ismail, Prof. Mohammed. *Analog VLSI Laboratory, Department of Electrical Engineering, Ohio State University, Columbus.* ELECTRONICS — coauthored.

J

Joel, Amos E., Jr. *Retired Consultant, AT&T Bell Laboratories, South Orange, New Jersey.* INTEGRATED SERVICES DIGITAL NETWORK (ISDN).

Johnson, Dr. Torrence V. *Jet Propulsion Laboratory, California Institute of Technology, Pasadena.* JUPITER — in part.

Jones, Dr. Robert R. *Jesse W. Beams Laboratory of Physics, University of Virginia, Charlottesville.* WAVE MECHANICS.

Jonsson, Dr. Jan Ake. *Department of Analytical Chemistry, University of Lund, Sweden.* LIQUID MEMBRANES — coauthored.

K

Kailath, Prof. Thomas. *Information Systems Laboratory, Stanford University, Stanford, California.* ADAPTIVE SIGNAL PROCESSING — coauthored.

Kamarthi, Dr. Sagar V. *Assistant Professor of Mechanical, Industrial and Manufacturing Engineering, Northeastern University, Boston, Massachusetts.* MANUFACTURING.

Kelly, Dr. James P. *Division of Otolaryngology, Health Sciences Center, University of New Mexico, Albuquerque.* ENVIRONMENTAL NOISE — coauthored.

Kemper, Prof. Susan. *Department of Psychology, University of Kansas, Lawrence.* ALZHEIMER'S DISEASE.

Khalaila, Dr. Isam. *Department of Life Sciences, Ben-Gurion University of the Negev, Beer-Sheva, Israel.* CRAYFISH — coauthored.

Kimball, Dr. Bruce A. *U.S. Department of Agriculture, Agricultural Research Service, Water Conservation Laboratory, Phoenix, Arizona.* GREENHOUSE EFFECT.

Kluth, Dr. Edward. *Sensor Dynamics Ltd., Hants, United Kingdom.* PETROLEUM RESERVOIR PRODUCTION.

Klutke, Dr. Georgia-Ann. *Associate Professor, Department of Industrial Engineering, Texas A&M University, College Station.* MAINTAINABILITY OF SYSTEMS — coauthored.

Knowles, Dr. Martyn. *Applications Manager, Materials Processing Group, Oxford Lasers Ltd., Abingdon, Oxon, United Kingdom.* LASER — in part.

Kuchner, Dr. Olga. *Department of Chemical Engineering, Calfornia Institute of Technology, Pasadena.* ENZYME.

Kulkarni, Dr. Shrinivas R. *Department of Astronomy, California Institute of Technology, Pasadena.* BROWN DWARF — coauthored.

Kupitz, Dr. Juergen. *Head, Nuclear Power Technology Development Section, Division of Nuclear Power and the Fuel Cycle, International Atomic Energy Agency, Vienna, Austria.* DESALINATION.

L

Lancaster, Dr. Nicholas. *Research Professor, Desert Research Institute, Quaternary Sciences Center, University and*

Community College System of Nevada, Reno. CLIMATE HISTORY.

Lang, Prof. Kenneth R. *Department of Physics and Astronomy, Tufts University, Medford, Massachusetts.* ASTRONOMICAL OBSERVATORY.

Larsen, Dr. Finn. *Department of Physics and Astronomy, University of Pennsylvania, Philadelphia.* BLACK HOLE.

Lee, Dr. I. Y. *Lawrence Berkeley Laboratory, Nuclear Science Division, University of California, Berkeley.* GAMMA-RAY DETECTORS—coauthored.

Libson, Dr. Joseph. *Beckman Institute, University of Chicago at Urbana-Champaign, Illinois.* INTERNET—in part.

Livingston, Dr. Nigel J. *Department of Biology, University of Victoria, British Columbia, Canada.* SOIL—coauthored.

Loomis, Prof. Jack M. *Department of Psychology, University of California, Santa Barbara.* PSYCHOACOUSTICS.

M

McCarthy, Dr. J. Howard. *U.S. Geological Survey, Mackay School of Mines, University of Nevada, Reno.* GEOCHEMICAL PROSPECTING.

Macek, Robert M., Jr. *Manager, Product Sourcing/Quality Assurance, Mitsubishi Materials U.S.A. Corporation, Irvine, California.* MACHINING.

McKay, Dr. David S. *Johnson Space Center, Houston, Texas.* MARS.

McWilliams, Prof. Michael O. *Department of Geophysics, Stanford University, Stanford, California.* GEOMAGNETISM.

Malone, Dr. Kathleen E. *Division of Public Health Sciences, Fred Hutchinson Cancer Research Center, Seattle, Washington.* CANCER (MEDICINE)—coauthored.

March-Russell, Dr. John. *School of Natural Sciences, Institute for Advanced Study, Princeton, New Jersey.* SUPERSYMMETRY.

Mathiasson, Dr. Lennart. *Department of Analytical Chemistry, University of Lund, Sweden.* LIQUID MEMBRANES—coauthored.

Matsumoto, Prof. Hiroshi. *Director, Radio Atmospheric Science Center, Kyoto University, Uji, Kyoto, Japan.* SOLAR ENERGY.

Menasce, Prof. Daniel A. *Department of Computer Science, George Mason University, Fairfax, Virginia.* COMPUTER SYSTEM.

Miller, Prof. Darrell A. *Department of Crop Sciences, University of Illinois at Urbana-Champaign.* SEED GERMINATION.

Miller, Harold E. *Gas Turbine Programs, General Electric Company, Schenectady, New York.* GAS TURBINE.

Miller, Prof. Joel S. *Department of Chemistry, University of Utah, Salt Lake City.* MAGNET—coauthored.

Mirkin, Dr. Chad. *Associate Professor, Department of Chemistry, Northwestern University, Evanston, Illinois.* SUPERLATTICES—coauthored.

Misra, Dr. P. N. *Lincoln Laboratory, Massachusetts Institute of Technology, Lexington.* SATELLITE NAVIGATION SYSTEMS.

Moncton, Dr. David E. *Argonne National Laboratory, Argonne, Illinois.* SYNCHROTRON RADIATION.

Moore, Dr. Chester G. *Arbovirus Disease Branch, Department of Health and Human Services, National Center for Infectious Diseases, Division of Vector-Borne Infectious Diseases, Centers for Disease Control, Fort Collins, Colorado.* ENCEPHALITIS.

Mou, Dr. Duen-Gang. *Development Center for Biochemistry, Taipei, Taiwan.* PHARMACOLOGY.

Mueller, Dr. Daniel L., M.D. *Assistant Professor of Medicine, Department of Medicine, University of Minnesota Medical School, Minneapolis.* ANERGY.

Mullins, Dr. Joe H. *Department of Mechanical Engineering, University of New Mexico, Albuquerque.* ENVIRONMENTAL NOISE—coauthored.

N

Nakasuka, Dr. Shinichi. *Research Center for Advanced Science and Technology, University of Tokyo, Komaba, Meguro-ku, Tokyo, Japan.* PRODUCTION ENGINEERING.

Nelson, Dr. Donna C. *Director of Operations, Intelligent Transportation Society of America, Washington, D.C.* TRANSPORTATION ENGINEERING—coauthored.

Normark, Dr. William R. *Geologic Division, U.S. Department of the Interior, Menlo Park, California.* VOLCANIC ISLANDS.

Novac, Dr. B. M. *Institute of Atomic Physics, IFTAR Lab 22, Bucharest, Romania.* IMPULSE GENERATOR—coauthored.

O

O'Brien, Dr. Stephen. *SDL Incorporated, San Jose, California.* LASER—in part.

Oppenheimer, Dr. Ben R. *Department of Astronomy, California Institute of Technology, Pasadena.* BROWN DWARF—coauthored.

Osterberg, Prof. Ulf L. *Associate Professor of Engineering, Thayer School of Engineering, Dartmouth College, Hanover, New Hampshire.* NONLINEAR OPTICS.

Ostrander, Dr. Elaine A. *Clinical Research Division, Fred Hutchinson Cancer Research Center, Seattle, Washington.* CANCER (MEDICINE)—coauthored.

Ott, Prof. Edward. *Department of Electrical Engineering, Institute for Plasma Research, University of Maryland at College Park.* CHAOS.

P

Peck, Dr. Lloyd S. *Head, Nearshore Marine Biology, British Antarctic Survey, High Cross, Cambridge, United Kingdom.* BRACHIOPODA.

Peeters, Dr. François. *Department of Physics, University of Antwerp, Belgium.* SEMICONDUCTOR.

Piekielek, William P. *Research Support Associate, Department of Agronomy, College of Agricultural Sciences, Pennsylvania State University, University Park.* CHLOROPHYLL—coauthored.

Pierce, Prof. Allan D. *Department of Aerospace and Mechanical Engineering, Boston University, Boston, Massachusetts.* FUZZY-STRUCTURE ACOUSTICS.

Pike, Dr. Andrew C. *Neotronics Scientific Ltd., United Kingdom.* MICROSENSOR—coauthored.

Polchinski, Prof. Joseph. *Institute of Theoretical Physics, University of California, Santa Barbara.* SUPERSTRING THEORY.

Potvin, Dr. Jean. *Department of Science and Mathematics, Parks College of Engineering and Aviation, Saint Louis University, Saint Louis, Missouri.* PARACHUTE.

R

Ram, Dr. Jeffrey L. *Department of Physiology, Wayne State University, Detroit, Michigan.* BIVALVIA—in part.

Ratner, Prof. Mark A. *Department of Chemistry, Northwestern University, Evanston, Illinois.* SUPERLATTICES—coauthored.

Reardon, Dr. Joyce T. *School of Medicine, Department of Biochemistry and Biophysics, University of North Carolina at Chapel Hill.* DEOXYRIBONUCLEIC ACID (DNA)—in part.

Reigosa, Dr. Manuel J. *Department de Recurses Naturais Med. Ambiente, Area de Bioloxia Vegetal, Facultade de Ciencias, Universidade de Vigo, Spain.* FOREST ECOSYSTEM.

Ricco, Dr. Antonio J. *Senior Member, Technical Staff, Microsensor R&D Department, Sandia National Laboratories, Albuquerque, New Mexico.* REMOTE SENSING—coauthored.

Robinson, Jonathan J. *Department of Organismic and Evolutionary Biology, Harvard University, Cambridge, Massachusetts.* HYDROTHERMAL VENT—coauthored.

S

Sage, Prof. Andrew P. *Founding Dean Emeritus and First American Bank Professor, University Professor, School of Information Technology and Engineering, George Mason University, Fairfax, Virginia.* CLIENT-SERVER SYSTEMS.

Sagi, Dr. Amir. *Department of Life Sciences, Ben-Gurion University of the Negev, Beer-Sheva, Israel.* CRAYFISH—coauthored.

Saito, Dr. R. *Department of Electronics Engineering, University of Electro-Communications, Chofugaoka, Chofu, Tokyo, Japan.* NANOSTRUCTURE—coauthored.

Sanchez, Dr. Anthony. *Research Officer, Department of Health and Human Services, Centers for Disease Control and Prevention, Atlanta, Georgia.* EBOLA VIRUS.

Sanders, Dr. Alan R. *Clinical Neurogenetics Branch, National Institute of Mental Health, Bethesda, Maryland.* BEHAVIORAL GENETICS—coauthored.

Sayed, Prof. Ali H. *Department of Electrical Engineering, University of California, Los Angeles.* ADAPTIVE SIGNAL PROCESSING—coauthored.

Schaffner, Dr. William, M.D. *Chairman, Department of Preventive Medicine, Vanderbilt University School of Medicine, Nashville, Tennessee.* EHRLICHIOSIS—coauthored.

Schopf, Dr. Kenneth M. *Museum of Comparative Zoology, Harvard University, Cambridge, Massachusetts.* PALEOECOLOGY.

Schreyer, Dr. W. *Institute for Mineralogy, Ruhr University, Bochum, Germany.* HIGH-PRESSURE MINERAL SYNTHESIS.

Schuster, Prof. David I. *Department of Chemistry, New York University, New York, New York.* FULLERENE.

Scifres, Dr. Charles J. *Dean and Associate Vice President for Agriculture, Dale Bumpers College of Agricultural, Food and Life Sciences, University of Arkansas, Fayetteville.* RANGELAND MANAGEMENT.

Scoble, Prof. Malcolm. *Department of Mining and Metallurgical Engineering, McGill University, Montreal, Quebec, Canada.* UNDERGROUND MINING.

Scott, Prof. James. *Chairman, Department of Medicine, Royal Postgraduate Medical School, Hammersmith Hospital, London, United Kindom.* RIBONUCLEIC ACID (RNA).

Shamaly, Dr. John J. *Vice President, Technology and Marketing, SVG Lithography Systems, Inc., Wilton, Connecticut.* INTEGRATED CIRCUITS.

Sharma, Rohini P. *Senior Mine Engineer, Cyprus Miami Mining Corporation, Claypool, Arizona.* OPEN PIT MINING.

Shockey, Kathy A. *DataComm Satellites, Space Systems/Loral, Palo Alto, California.* COMMUNICATIONS SATELLITE—coauthored.

Shoshani, Dr. Jeheskel. *Department of Biological Sciences, Elephant Research Foundation, Wayne State University, Detroit, Michigan.* ELEPHANT.

Sigman, Dr. Daniel M. *Department of Geology and Geophysics, Woods Hole Oceanographic Institution, Massachusetts.* OCEANOGRAPHY—coauthored.

Silk, Prof. Joseph. *Professor of Astronomy and Physics, Department of Astronomy, University of California, Berkeley.* COSMOLOGY.

Simonotto, Dr. Enrico. *Department of Physics, University of Genoa, Italy.* PERCEPTION.

Smith, Dr. Andrew B. *Curator of Paleobiology, Natural History Museum, London, United Kingdom.* PHYLOGENY.

Smith, Prof. Ivor R. *Department of Electronic and Electrical Engineering, Loughborough University, Leicestershire, United Kingdom.* IMPULSE GENERATOR—coauthored.

Song, Prof. Bang-Sup. *Department of Electrical Engineering, Coordinated Science Laboratory, University of Illinois at Urbana-Champaign.* CIRCUIT (ELECTRONICS).

Squires, Prof. Kyle D. *Department of Mechanical Engineering, University of Vermont, Burlington.* TURBULENT FLOW—in part.

Standaert, Dr. Steven M., M.D. *Division of Infectious Diseases, Department of Medicine, Vanderbilt University, Nashville, Tennessee.* EHRLICHIOSIS—coauthored.

Stanitski-Martin, Dr. Diane. *Department of Geography, Arizona State University, Tempe.* RIVER.

Stanley, Dr. George D., Jr. *Department of Geology, University of Montana, Missoula.* CORAL REEF.

Stephens, Dr. Frank S. *Lawrence Berkeley Laboratory, Nuclear Science Division, University of California, Berkeley.* GAMMA-RAY DETECTORS—coauthored.

Stewardson, H. R. *Department of Electronic and Electrical Engineering, Loughborough University, Leicestershire, United Kingdom.* IMPULSE GENERATOR—coauthored.

Stipp, Dr. Susan L. S. *Geological Institute, Copenhagen University, Copenhagen, Denmark.* CALCITE.

Stolte, Dr. Gerd. *Vacmetal GmbH, Dortmund, Germany.* STAINLESS STEEL—coauthored.

Streilein, Prof. Wayne J., M.D. *Professor and Vice-Chair for Research, Schepens Eye Research Institute, Department of Ophthalmology, Harvard Medical School, Boston, Massachusetts.* IMMUNITY.

Swallow, Dr. Jacinda. *Department of Biochemistry, North Carolina State University, Rayleigh.* DEOXYRIBONUCLEIC ACID (DNA)—coauthored.

T

Tarlinton, Dr. David. *Walter and Eliza Hall Institute of Medical Research, Victoria, Australia.* IMMUNE SYSTEM.

Tesh, Dr. Vernon L. *Department of Medical Microbiology and Immunology, College of Medicine, Texas A&M University, College Station.* FOOD-BORNE DISEASE.

Teworte, Dr. Rainer. *Vacmetal GmbH, Dortmund, Germany.* STAINLESS STEEL—coauthored.

Thomas, Andrew L. *Research Associate, Horticulture, Southwest Missouri Center, College of Agriculture, Food, and Natural Resources, University of Missouri-Columbia, Mount Vernon.* TILLAGE SYSTEMS—in part.

Thomas, Prof. Marlin U. *Head, School of Industrial Engineering, Purdue University, West Lafayette, Indiana.* QUALITY CONTROL.

Thorp, Dr. H. Holden. *Associate Professor, Department of Chemistry, University of North Carolina at Chapel Hill.* DEOXYRIBONUCLEIC ACID (DNA)—in part.

Titze, Dr. Ingo R. *Director, National Center for Voice and Speech, Department of Speech Pathology and Audiology, University of Iowa, Iowa City.* VOICE ANALYSIS.

Tomalia, Dr. Donald A. *Research Professor, Director of Nanoscopic Chemistry and Architecture, Michigan Molecular Institute, Midland.* DENDRITIC MACROMOLECULES—coauthored.

Tramo, Dr. Mark Jude, M.D. *Department of Neurology and Department of Neurobiology, Harvard Medical School, Boston, Massachusetts.* PHYSIOLOGICAL ACOUSTICS.

Tuck, Edward F. *Kinship Venture Management, West Covina, California.* COMMUNICATIONS SATELLITE—in part.

V

Venugopal, Dr. S. *Indira Gandhi Centre for Atomic Research, Department of Atomic Energy, Tamil Kadu, India.* COMPUTER-INTEGRATED MANUFACTURING—coauthored.

Villard, Dr. Ray. *Space Telescope Science Institute, Baltimore, Maryland.* SPACE TELESCOPE.

von Puttkamer, Dr. Jesco. *Office of Space Flight, NASA Headquarters, Washington, D.C.* SPACE FLIGHT.

W

Wallace, Dr. Paul J. *Ocean Drilling Program, Science Operations, Texas A&M University Research Park, College Station.* MARINE SEDIMENTS—in part.

Waller, Prof. George R. *Department of Biochemistry and Molecular Biology, Oklahoma State University, Stillwater.* SAPONINS.

Watson, Prof. A. Ted. *Department of Chemical Engineering, Texas A&M University, College Station.* FLUID FLOW—coauthored.

Welham, Dr. Chris J. *Druck Ltd., United Kingdom.* MICROSENSOR—coauthored.

Wen, Dr. Xiao-Gang. *Department of Physics, Massachusetts Institute of Technology, Cambridge.* HALL EFFECT.

White, Edward V. *Manager, Smart Structures and Systems, Advanced Systems and Technology, McDonnell Douglas Aerospace, Saint Louis, Missouri.* SMART STRUCTURES.

Wiik, Dr. Bjorn H. *Deutsches Elektronen-Synchrotron, Hamburg, Germany.* PARTICLE ACCELERATOR.

Wildt, Dr. David E. *National Zoological Park, Smithsonian Institution, Conservation and Research Center, Front Royal, Virginia.* BIOTECHNOLOGY—in part.

Williams, Dr. Earle. *Ralph M. Parsons Laboratory, Department of Civil Engineering, Massachusetts Institute of Technology, Cambridge.* THUNDERSTORM.

Winograd, Prof. Nicholas. *Department of Chemistry, Pennsylvania State University, University Park.* COMBINATORIAL CHEMISTRY—in part.

Winter, Prof. Hannspeter. *Institut für Allgemeine Physik, Technical University of Vienna, Austria.* ION-SOLID INTERACTIONS.

Wortman, Dr. Martin A. *Associate Professor, Department of Industrial Engineering, Texas A&M University, College Station.* MAINTAINABILITY OF SYSTEMS—coauthored.

Y

Yewdell, Dr. Jonathan W. *Department of Health and Human Services, National Institutes of Health, Bethesda, Maryland.* VIRUS—coauthored.

Young, Dr. Richard E. *National Aeronautics and Space Administration, Ames Research Center, Moffett Field, California.* JUPITER—in part.

Yu, Ein-Fen. *Department of Earth Sciences, National Taiwan Normal University, Taipei, Taiwan.* OCEANOGRAPHY—coauthored.

Index

Index